PRIMATE ADAPTATION AND EVOLUTION

FOURTH EDITION

PRIMATE ADAPTATION AND EVOLUTION

FOURTH EDITION

JOHN G. FLEAGLE
Stony Brook University, New York, USA

ANDREA L. BADEN
Hunter College of the City University of New York, New York, USA
The Graduate Center of the City University of New York, New York, USA

CHRISTOPHER C. GILBERT
Hunter College of the City University of New York, New York, USA
The Graduate Center of the City University of New York, New York, USA

AMSTERDAM • BOSTON • HEIDELBERG • LONDON • NEW YORK • OXFORD • PARIS
SAN DIEGO • SAN FRANCISCO • SINGAPORE • SYDNEY • TOKYO
Academic Press is an imprint of Elsevier

Academic Press is an imprint of Elsevier
125 London Wall, London EC2Y 5AS, United Kingdom
525 B Street, Suite 1650, San Diego, CA 92101, United States
50 Hampshire Street, 5th Floor, Cambridge, MA 02139, United States

ISBN: 978-0-12-815809-8

For Information on all Academic Press publications
visit our website at https://www.elsevier.com/books-and-journals

Publisher: Peter B. Linsley
Acquisitions Editor: Emma Hayes
Editorial Project Manager: Lindsay Lawrence
Publishing Services Manager: Deepthi Unni
Production Project Manager: Nandhini Thanga Alagu
Cover Designer: Matthew Limbert

Typeset by MPS Limited, Chennai, India

Printed in India

Last digit is the print number: 9 8 7 6 5 4 3 2 1

Contents

Preface

Primatology and primate evolution have changed considerably in the three and a half decades since the first edition of *Primate Adaptation and Evolution* was written. Like all other areas of knowledge, the data informing these subjects have increased dramatically, and the published literature has increased manyfold. There are more species, more sites, more studies, more techniques, more analyses, more articles, more journals, and, hopefully, more understanding. But there is also more evidence of just how limited our current knowledge actually is, and how much it is likely to change in the future. This, like every other book, is perhaps best viewed as a progress report rather than a synthesis.

In this edition every chapter has been revised and rewritten, some much more than others. All of the tables have been redone; there are many new figures; and many of the references are new. Some of these changes deserve further explanation.

The most obvious change between previous editions and the fourth edition is the change in authorship. Given the exponential growth in our knowledge of extant and fossil primates through the increasing number of practicing primatologists and paleontologists, it has become nearly impossible for a single author to continue authoring a book as broad as *Primate Adaptation and Evolution*. To solve this problem and provide a book that remains useful to the widest audience, Andrea Baden and Chris Gilbert have added their knowledge and expertise, with Andrea providing significant updates to Chapters 3 through 7, resulting in a more modern view of the natural history, behavior, and ecology of the living primates, and Chris providing significant updates to the chapters covering primate anatomy, primate adaptations, and the primate fossil record.

Among primatologists, the number of recognized primate species continues to grow dramatically. There are many reasons for this. Partly, it reflects an extensive increase in fieldwork in remote parts of the world that has generated a greater appreciation of the details of primate biogeography and diversity. In addition, the continuing influence of molecular systematics has generated new insights into the genetic diversity among primate populations. Finally, the widespread use of the Phylogenetic Species Concept has had a major effect on the abilities and willingness of systematists to describe and diagnose new or forgotten taxa. In general, we have used the IUCN Red List website to create the tables of extant species in this volume. However, this increase in recognized primate species has created problems in the use of older literature for retrieving information about the behavior, ecology, body mass, or limb proportions of individual taxa. For example, data that in previous decades, or previous editions of this book, were attributed to the single species of woolly lemur, *Avahi laniger*, may well have been derived from one of several other species now recognized as distinct in that genus. Readers should thus view some of the data in the tables as rough estimates for the purpose of broad comparisons, not detailed analyses. Where possible, we have updated these data in light of more current taxonomy. However, it is still the case that some of the values listed in the tables for a given taxon may represent data from populations that are now recognized as multiple species. Data for intermembral indices, for instance, were taken from a variety of sources, including previous editions of *Primate Adaptation and Evolution*, the compilations provided by the *All the World's Primates* website (https://www.alltheworldsprimates.org), Godfrey et al. (2010) for subfossil lemurs, and other recent published literature. Body mass estimates for extant taxa are largely derived from Smith and Jungers (1997), the *All the World's Primates* website, Jungers et al. (2008) for subfossil lemurs, Delson et al. (2000) for living cercopithecoid monkeys, and the various entries in *The Mammals of Africa, Vol. II: Primates* edited by Butynski et al. (2013).

In previous editions this book tried to provide body mass estimates for most species of fossil primates derived from a single regression formula based on molar tooth dimensions. The third edition relied more on estimates from a wide variety of sources in the literature, based on many different parameters. In this edition we have tried to use both approaches, relying on targeted equations for major primate fossil groups where available and on a variety of sources in the literature for other groups where a single equation seemed inappropriate. Many estimates are thus methodologically comparable within broad clades or grades but not across all primates. Body mass estimates for fossil prosimians (i.e., plesiadapiforms, adapoids, and omomyoids) are based on the lower M_1 area prosimian regression by Conroy (1987), and early anthropoids are largely based on the lower M_1 area anthropoid regression by Conroy (1987).

Fossil platyrrhines were generally estimated using the M_1 area platyrrhine equation in Kay and Meldrum (1997), and body mass estimates for fossil cercopithecoids are based on the estimates and equations found in Delson et al. (2000). Estimates for the fossil ape chapter cover a large range of taxa and fossil primate grades, ranging from stem catarrhines to great apes. Thus estimates for this chapter come from a variety of sources, including Sankhyan et al. (2017) for many pliopithecoids, Harrison (2010) and Grabowski and Jungers (2017) for many hominoids, and original published descriptions or the transitive property for a number of other stem catarrhine and/or hominoid taxa. Hominin estimates are largely derived from Grabowski et al. (2015, 2018) and Ruff et al. (2018). In this way, estimates within each chapter should generally be consistent, and any interested reader can look up the equations provided in the references listed above to derive the original tooth area and apply them to another equation if they wish. While we recognize that body size estimates derived from the dentition can sometimes be less accurate than those derived from other anatomical regions, teeth are the most common elements found in the fossil record and provide mass estimates across the widest array of taxa. In this way, the body mass values provided in this book are meant to give the reader an appreciation for the sizes of a wide range of extinct taxa in a general sense, but they are perhaps not suitable for detailed analyses, depending on the question being asked. If a reader has a question about a given body mass estimate (or the data behind it) for any extant or fossil primate, please contact Chris Gilbert: cgilbert@hunter.cuny.edu.

As in previous editions, we have included two types of references for each chapter. There are general references that provide broad reviews of the topics covered in that chapter. These are designed to provide a more detailed documentation and discussion and, in some cases, alternative views on the material discussed in that chapter. In addition, there are numerous citations within the text of each chapter that are listed as cited references at the end of the chapter. These are not meant to provide a comprehensive or even representative documentation for the contents of the chapter. Rather, they are meant to provide the readers with an entry into the literature regarding particular facts and ideas that we found interesting and/or significant. In particular, we have cited relatively recent publications that may not appear in the broader General References section. However, we fully expect any reader will be able to find numerous additional references to any topic in this book through an online search.

This edition of *Primate Adaptation and Evolution* contains a number of additional illustrations. As with previous editions, we have limited these to line drawings and black-and-white photos with an emphasis on comparisons rather than documentation and description. Nevertheless, we appreciate that these do not capture the remarkable beauty and diversity of living primates or the details of morphology that are available in various other media, including videos and three-dimensional figures. Readers are urged to look more widely for additional illustrative materials, and we especially recommend *All the World's Primates* (www.alltheworldsprimates.org).

This edition has benefited from the generous advice, assistance, and expertise of many people. The efforts and contributions of those listed in previous editions are still greatly appreciated. For help with this edition, we thank the following people, in no particular order: Alfred Rosenberger, Todd Disotell, Callum Ross, Colin Groves, Richard Kay, James Rossie, Tim Smith, Chris Kirk, Mark Coleman, Stephanie Maiolino, Doug Boyer, Steve Leigh, Andreas Koenig, Carola Borries, Charles Janson, Tim Clutton-Brock, Katie Hinde, Erin Vogel, Peter Lucas, Nate Dominy, Vivek Venkataraman, Diane Doran-Sheehy, Scott Suarez, Herman Pontzer, Patricia Wright, Chia Tan, Mireya Mayor, Shawn Lehman, Rachel Jacobs, Laurie Godfrey, Tim Ryan, Bill Jungers, Brigitte Demes, Betsy Dumont, Suzanne Strait, Sara Martin, Anja Deppe, Ian Tattersall, Myron Shekelle, Dan Gebo, Marion Dagosto, Anna Nekaris, Anne Yoder, Christian Roos, Russ Mittermeier, Sharon Gursky, Peter Kappeler, Tony DiFiore, Marilyn Norconk, Paul Garber, Anthony Rylands, Leila Porter, Mark Van Roosmalen, Barth Wright, Karen Wright, Scott McGraw, Joan Silk, Eric Sargis, Alice Elder, Wendy Erb, David Fernandez, Jessica Rothman, Jessica Lodwick, Michael Steiper, Richard Wrangham, John Mitani, Dan Lieberman, Sarah Hrdy, Kristen Hawkes, Kim Hill, Kaye Reed, Jason Kamilar, Sandy Harcourt, Oliver Schulke, Julia Oster, Jon Bloch, Philip Gingerich, Frank Brown, Thure Cerling, Craig Feibel, Ian McDougall, Mary Silcox, Stephen Chester, Gregg Gunnell, Xijun Ni, Matt Cartmill, Ken Rose, Lawrence Flynn, Chris Heesy, Elwyn Simons, Nancy Stevens, Jorn Hurum, Blythe Williams, Walter Hartwig, Jonathan Perry, Marc Godinot, Chris Beard, Mark Klinger, Lauren Halenar, Siobhan Cooke, Alexa Krupp, Castor Cartelle, Ross MacPhee, Terry Harrison, Bill Sanders, Iyad Zalmout, Jay Kelley, John Kappelman, David Alba, Sergio Almecija, Salvador Moya-Sola, Isaac Casanovas-Vilar, David Pilbeam, Ellen Miller, Ari Grossman, Nina Jablonski, Rajeev Patnaik, Russ Ciochon, Brenda Benefit, Eric Delson, Martin Pickford, Mauricio Anton, Meave Leakey, The Turkana Basin Institute, Richard Leakey, Carol Ward, Michael Plavcan, Peter Ungar, The Kenya National Museum, Michel Brunet, Franck Guy, Bill Kimbel, Adam Gordon, Bernard Wood, Brian Richmond, Chris Stringer, Randall Susman, Fred Grine, Karen Baab, Philip Rightmire, David Strait, Ian Wallace, Gunter Brauer, Susan Larson, Zeray Alemseged, Tim White, John

Shea, Lee Berger, Steve Frost, Kelsey Pugh, Julia Arenson, Erik Seiffert, Biren Patel, Chris Campisano, and many others we may have overlooked.

As with previous editions, the heart of this book is the illustrations. Most of these are due to the long-term efforts and unfailing patience of Stephen Nash and Luci Betti-Nash. In their talented hands, even the most muddled ideas are somehow transformed into illustrations that are crisp and understandable.

References

Butynski, T.B. Kingdon, J. Kalina, J. (Eds.), 2013. In: The Mammals of Africa, vol. II. Primates, Bloomsbury, London, pp. 545.

Conroy, G.C., 1987. Problems of body-weight estimation in fossil primates. Int. J. Primatol. 8, 115–137.

Delson, E., Terranova, C.J., Jungers, W.L., Sargis, E.J., Jablonski, N.G., Dechow, P.C., 2000. Body mass in Cercopithecidae (Primates, Mammalia): estimation and scaling in extinct and extant taxa. Anthropol. Pap. Am. Mus. Nat. Hist. 8, 1–159.

Godfrey, L.R., Jungers, W.L., Burney, D.A., 2010. Subfossil lemurs of Madagascar. In: Werdelin, L., Sanders, W.J. (Eds.), Cenozoic Mammals of Africa. University of California Press, Berkeley, pp. 351–368.

Grabowski, M., Jungers, W.L., 2017. Evidence of a chimpanzee-sized ancestor of humans but a gibbon-sized ancestor of apes. Nature Communications 8, 880. https://doi.org/10.1038/s41467-017-00997-4

Grabowski, M., Hatala, K.G., Jungers, W.L., Richmond, B.G., 2015. Body mass estimates of hominin fossils and the evolution of human body size. J. Hum. Evol. 85, 75–93.

Grabowski, M., Hatala, K.G., Jungers, W.L., 2018. Body mass estimates of the earliest possible hominins and implications for the last common ancestor. J. Hum. Evol. 122, 84–92.

Harrison, T., 2010. Dendropithecoidea, Proconsuloidea, and Hominoidea. In: Werdelin, L., Sanders, W.J. (Eds.), Cenozoic Mammals of Africa. University of California Press, Berkeley, pp. 429–470.

Jungers, W.L., Demes, B., Godfrey, L.R., 2008. How big were the "giant" extinct lemurs of Madagascar? In: Fleagle, J.G., Gilbert, C.C. (Eds.), Elwyn Simons: A Search for Origins. Springer, New York, pp. 343–360.

Kay, R.F., Meldrum, D.J., 1997. A new small platyrrhine from the Miocene of Columbia and the phyletic position of Callitrichinae. In: Kay, R.F., Madden, R.H., Cifelli, R.L., Flynn, J.J. (Eds.), Vertebrate Paleontology in the Neotropics. Smithsonian Institution Press, Washington, D.C., pp. 435–458.

Ruff, C.B., Burgess, M.L., Squyres, N., Junno, J.-A., Trinkaus, E., 2018. Lower limb articular scaling and body mass estimation in Pliocene and Pleistocene hominins. J. Hum. Evol. 115, 85–111.

Sankhyan, A.R., Kelley, J., Harrison, T., 2017. A highly derived pliopithecoid from the Late Miocene of Haritalyangar, India. J. Hum. Evol. 105, 1–12.

Smith, R., Jungers, W.L., 1997. Body mass in comparative primatology. J. Hum. Evol. 32, 523–559.

1

Adaptation, Evolution, and Systematics

Adaptation

Adaptation is a concept central to our understanding of evolution, but the term has proved very difficult to define in a simple phrase. One of the most succinct definitions has been offered by Vermeij (1978, p. 3): "An adaptation is a characteristic that allows an organism to live and reproduce in an environment where it probably could not otherwise exist." In the following chapters, we examine extant (living) and extinct (fossil) primates as a series of **adaptive radiations** – groups of closely related organisms that have evolved morphological and behavioral features, enabling them to exploit different ecological niches. Adaptive radiations provide especially clear examples of evolutionary processes. The adaptive radiation of finches on the Galapagos Islands of Ecuador played an important role in guiding Darwin's views on the origin of species.

Adaptation also refers to the process through which organisms obtain their adaptive characteristics. The primary mechanism of adaptation is natural selection. **Natural selection** is the process whereby any heritable features, anatomical or behavioral, that enhance the fitness of an organism relative to its peers increase in frequency in the population in succeeding generations. **Fitness**, in an evolutionary sense, is reproductive success. It is important to remember that natural selection acts primarily through differential reproductive success of individuals within a population (Williams, 1966). However, there is considerable debate regarding the extent to which selection can also act at higher levels, including groups **(group selection)** and species **(species selection)**.

Evolution

Evolution is modification by descent, or genetic change in a population through time. Although biologists consider most evolution to be the result of natural selection, there are other, non-Darwinian mechanisms that can and do lead to genetic change within a population. **Genetic drift** is change in the genetic composition of a population from generation to generation due to chance sampling events independent of selection. **Founder effect** is a more extreme change in the genetic makeup of a population that occurs when a new population is established by only a few individuals. This new population may sample only a small part of the variation found in the ancestral population. Thus, recessive alleles that are not expressed in the larger population may become more common or even fixed in the new population. In this way, the chance characteristics of a founder population can have dramatic effects on the subsequent evolution and adaptive diversity of a group of organisms.

The fact that the diversity of life is the result of evolution means that all organisms are related by virtue of sharing a common genetic ancestry in the distant past. However, it is also clear that living organisms are not a continuous spread of variation. The living world is composed of distinct kinds of organisms that we recognize as species. Although virtually all biologists recognize species as the natural units of life, defining exactly what a species is or how species form are more difficult problems. These are the problems that Darwin (1859) set out to explain, and they are still the subject of intense study and debate (e.g., Kimbel and Martin, 1993; Groves, 2001a,b, 2012; de Queiroz, 1998, 2007; Rosenberger, 2012; Tattersall, 1992).

Until recently, most biologists and anthropologists generally accepted the **Biological Species Concept (BSC)**, in which species are defined as "groups of actually or potentially interbreeding natural populations that are reproductively isolated from other such groups" (Mayr, 1942). Although appealing, in that it emphasizes the genetic and phyletic distinctiveness of species through reproductive isolation, the Biological Species Concept is obviously impossible to apply to fossils or allopatric populations of living animals, and even difficult to apply to sympatric populations without detailed data on mating behavior and fertility (Tattersall, 1989; Groves, 2001a,b, 2012). Moreover, as more and more "species" have been sampled genetically, it has become clear that hybridization between presumed species has been very common in primate evolution (e.g., Detwiler, 2002; Zinner et al., 2011).

Many students of living organisms are more comfortable with a **Mate Recognition Concept** (Paterson, 1978, 1985; Masters, 1993). In this concept, species are defined as "the group of individuals sharing a common fertilization system" (Paterson, 1985). A particularly important aspect of this common fertilization system is a specific mate-recognition system. Members of a species recognize one another as potential mates through such behavior or morphological features as vocalizations, mating displays, or ornamentation. Like the Biological Species Concept, the Mate Recognition Concept is virtually impossible to apply to extinct organisms.

In contrast with the Biological Species Concept and the Mate Recognition Concept, which are based on information about reproductive behavior, many paleontologists prefer a species concept based on morphological differences. The **Phylogenetic Species Concept** (Cracraft, 1983) is commonly adopted by students of phylogenetic systematics. In this approach, a species is "the smallest diagnosable cluster of individuals within which there is a parental pattern of ancestry and descent." In this concept, a species is defined on the basis of morphological or genetic distinctions from other taxa (Cracraft, 1983). In principle, this could be based on a single feature. At present, most phylogenetic analyses, and most systematic revisions, generally follow a Phylogenetic Species Concept as species are identified by detailed morphological features or aspects of their DNA. However, the question of how many genetic differences are needed to identify a separate species is a critical, but largely unresolved, issue in primate phylogeny (Groeneveld et al., 2009).

Other researchers, using what may be called a **Phenetic Fossil Species Concept**, would argue that species defined morphologically should have approximately the same amount of metrical variation as extant populations (e.g., Gingerich and Schoeninger, 1979; Cope, 1993). This is often a very useful criterion when one has a continuously changing time-successive lineage in the fossil record, in which the endpoints may be very different but individual samples overlap.

Most biologists agree that a species is a distinct segment of an evolutionary lineage, and many of the differences among species concepts reflect attempts to find criteria that can be used to identify species based on different types of information, such as behavioral observations of living populations, genetic sequences, or morphological information from teeth, skull, or pelage. Some of the diversity of species concepts may be more useful in distinguishing species at different phases of their formation (de Quieroz, 1998, 2007). Paradoxically, the greatest challenge to species identification often

comes not from incomplete information, but from those rare paleontological instances in which there is a continuous temporal sequence of populations undergoing directional selection (e.g., Rose and Bown, 1993). As noted above, the endpoints are clearly differentiable, but any species boundary is necessarily arbitrary (see Chapter 18).

Phylogeny

Evolutionary change within a population can take place at different rates and can yield different results. There are several terms available to describe how and why different patterns of evolutionary change occur. What is more, numerous theories exist about how common these different patterns of evolutionary change are in the history of life. The pattern in which a lineage undergoes gradual change over time is called **anagenesis.** By contrast, **cladogenesis** is the division of a single lineage into two lineages. Gradual change in the morphology of a population of organisms through time, either anagenetic or cladogenetic, is often called **phyletic gradualism**. This type of evolutionary pattern is very common in the fossil record, and many biologists believe that most evolutionary change has been of this type.

The rates of evolutionary change that take place in populations through time may vary considerably, theoretically over several orders of magnitude. In addition, directional selection may shift over the course of relatively short time periods due to climatic fluctuations. Some paleontologists argue that the most common type of evolutionary change is **punctuated equilibrium**, in which the morphology of most species is essentially stable (in equilibrium) for long periods of time, and that speciation events are the result of rapid morphological and genetic shifts (or punctuations) over brief periods of evolutionary time so that intermediate forms are rarely recovered (e.g., Gould and Eldridge, 1993). Unfortunately, the punctuated equilibrium model is very difficult to distinguish from a discontinuous fossil record and is generally defined in a way that virtually precludes the possibility of determining how often it actually takes place. Overall, it is now well documented that evolutionary change has taken place at many different rates.

The branching pattern of successive species that results from numerous cladogenetic events is a **phylogeny**. Because this book deals with the adaptive radiations of primates, we are interested in reconstructing the evolutionary branching sequence, or phylogeny, of various primate groups to see how they are related to one another. Although some of us can trace our own genealogies (or those of our pets) through several generations, tracing the genealogical relationships among all primates is a much more daunting undertaking. The evolutionary radiation of primates has taken place over tens of millions of years of geological time and has involved thousands of species, millions of generations, and billions of individuals. Moreover, the records available for reconstructing primate phylogeny are meager, consisting of individuals of fewer than 500 living species and occasional bony remains of several hundred extinct species drawn from various parts of the world at various times during the past 66 million years.

The methods we use to reconstruct phylogenies are based primarily on identifying groups of related species through similarities in their morphology and in the molecular sequences of their genetic material. However, rather than just looking at overall similarity, most biologists agree that organisms should be grouped together on the basis of shared specializations (or shared derived features) that distinguish them from their ancestors. For example, body hair is a shared specialization that unites humans, apes, monkeys, and cats as mammals and distinguishes them from other types of vertebrates, whereas the common possession of a tail by many monkeys, lizards, and crocodiles is an ancestral feature that is of no particular value in assessing the evolutionary relationships among these organisms, since their common ancestor had a tail. On the other hand, the absence of a tail in apes and humans represents a derived specialization that sets us apart from our ancestors (Fig. 1.1). The common possession of a group of specializations by a cluster of species or genera is interpreted as indicating that this cluster shares a unique heritage relative to other related species. Most current studies of primate phylogeny are based on extensive analysis of morphological features and/or genetic sequences from parts of a species' genome, using sophisticated computer programs to determine the pattern of shared derived similarities

(Kay et al., 1997; Disotell, 2008; Perelman et al., 2011; Kuderna et al., 2023).

Unfortunately, not all derived similarities among organisms are indicative of a unique heritage. Animals have frequently evolved morphological similarities independently. In addition to apes and humans, for example, a few monkey species and a few strepsirrhine primates (as well as frogs) have lost all or most of their tail. The biologist's task in reconstructing phylogeny is to distinguish those specializations that are the result of a unique heritage from those that are not. The sameness of features that results from common ancestry is called **homology**, and identical features that are the result of a common ancestry are **homologous**. In contrast, the presence of similar features in different species that is not the result of a common inheritance is described as **homoplasy**.

Homoplasy is a very common phenomenon in evolution (Sanderson and Hufford, 1996). Most analyses of morphological evolution find that nearly half of all similarities are the result of homoplasy rather than common ancestry. There are several types of homoplasy and many factors that cause it to be such a common phenomenon. The evolution of superficially similar features in different lineages, such as the wings of bats and the wings of birds, is often called **convergence**. In contrast, **parallelism** refers to the independent evolution of identical features in closely related organisms, such as the independent evolution of white fur in many lineages of monkey. **Reversal** refers to an evolutionary change in an organism that resembles the condition found in an earlier ancestor, such as the presence of hair on the fingers of humans descended from apes with hairless digits. In many cases, homoplasy is the result of natural selection for similar functional adaptations. For example, larger

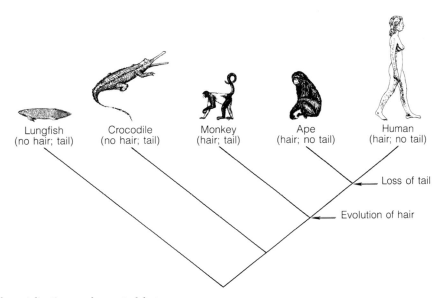

Lungfish
(no hair; tail)

Crocodile
(no hair; tail)

Monkey
(hair; tail)

Ape
(hair; no tail)

Human
(hair; no tail)

Loss of tail

Evolution of hair

FIGURE 1.1 Shared specializations and ancestral features.

animals may face similar mechanical problems. In addition, homoplasy may also reflect the fact that evolutionary pathways are constrained by development and available genetic potential. In molecular evolution, the potential for homoplasy is dictated by the limited number of amino acids making up the genetic code. Although many people feel that some morphological regions, such as teeth, skulls, or postcranial elements, are more prone to homoplasy than others, most comparative assessments show that this is not the case. Moreover, features that show considerable homoplasy in one group of animals may show no homoplasy in other groups. Most significantly, homoplasy is a phenomenon that can only be identified in a retrospective analysis after a phylogeny has been constructed (Begun, 2007). Although frequently treated as an undesirable distraction in attempts to reconstruct phylogeny, homoplasy is an important evolutionary phenomenon that deserves greater study and consideration (Lockwood and Fleagle, 1999; Kay and Fleagle, 2010).

TABLE 1.1 Classification of the white-fronted capuchin monkey.

Kingdom	Animalia
Phylum	Chordata
Class	Mammalia (mammals)
Order	Primates (primates)
Semiorder	Haplorhini
Suborder	Anthropoidea (higher primates)
Infraorder	Platyrrhini (New World monkeys)
Superfamily	Ceboidea (New World monkeys)
Family	Cebidae (capuchins, squirrel monkeys, owl monkeys, and marmosets)*
Subfamily	Cebinae (capuchins and squirrel monkeys)*
Genus	*Cebus* (capuchins)
Species	*Cebus albifrons* (white-fronted capuchin)

*Indicates only one of several common classifications (see Chapter 5).

Taxonomy and Systematics

Taxonomy is a means of ordering our knowledge of biological diversity through a series of commonly accepted names for organisms. If scientists wish to communicate about animals and plants and to discuss their similarities and differences, they need a standard system of names both for individual types of organisms and for related groups of organisms. For example, the white-fronted capuchin monkey of South America, known to many people as the organ grinder monkey, goes by over a dozen different names among the different tribal and ethnic groups of South America. To scientists around the world, however, this species is known by a single name, *Cebus albifrons*. The practice of assigning every biological species, living or fossil, a unique name composed of two Latin words was initiated by Carolus Linnaeus, a Swedish scientist of the eighteenth century whose system of biological nomenclature is universally followed today.

Under the Linnean system, *Cebus* is the name of a **genus** (pl. **genera**), or group of closely related species, in this case several kinds of capuchin monkey. (The name of a genus is always capitalized). The word *albifrons*, the **species** name, refers to a particular type of capuchin monkey, the white-fronted capuchin monkey. (A species name always begins with a lowercase letter.) Each genus name must be unique, but species names need to be unique only within a particular genus so that the combination of genus and species names is unique and refers to only one kind of organism. (The name is always written in italics, or else underlined.) Somewhere in a museum there is a preserved skeleton (or skull, or

skin) that has been designated as the **type specimen** for this species. The type specimen provides an objective reference for this species so that any scientist who thinks he or she may have discovered a different kind of monkey can examine the individual on which *Cebus albifrons* is based.

The Linnean system contains a hierarchy of levels for grouping organisms into larger and larger units (Table 1.1). Within the genus *Cebus*, for example, there are several species: *Cebus albifrons*, the white-fronted capuchin; *Cebus capucinus*, the capped capuchin; *Cebus olivaceous*, the weeper capuchin; and others. Genera are grouped into **families**, families into **orders**, orders into **classes**, and classes into phyla. For particular lineages, these basic levels are often further subdivided or clustered into **semiorders, suborders, infraorders, superfamilies, subfamilies, tribes, subgenera**, or **subspecies**. For convenience, names at different levels of the hierarchy are often given distinctive endings. Family names usually end in *idae*, superfamily names in *oidea*, and subfamily names in *inae*.

In the science of classifying organisms, **systematics**, we attempt to apply the tidy Linnean system to the untidy, unlabeled world of animals. Fig. 1.2, the higher-level classification used in this book, is the result of one such attempt. Although biologists agree to use the Linnean framework for naming organisms, they frequently disagree about the proper classification of particular creatures. They may disagree as to whether each of the gibbon types on different islands in Southeast Asia is a distinct species or only a subspecies of a single species. Some authorities may feel that gibbons and great apes should be placed in a single family, others that they

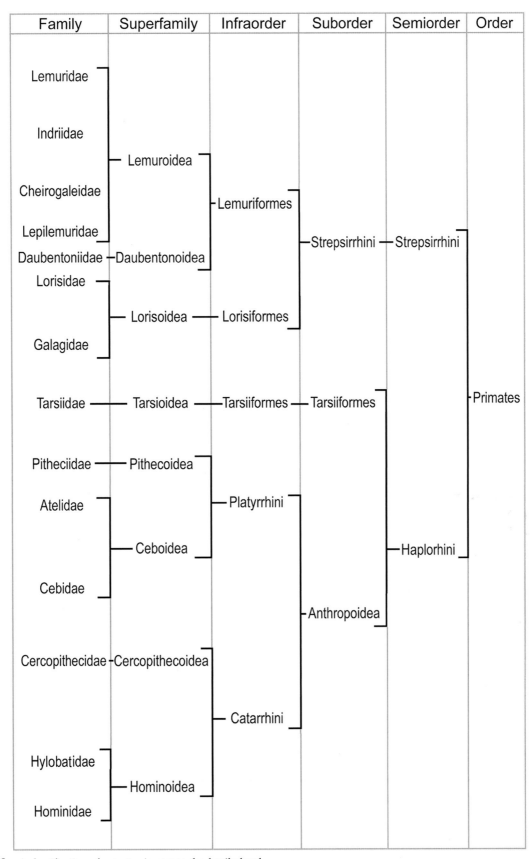

FIGURE 1.2   A classification of extant primates to the family level.

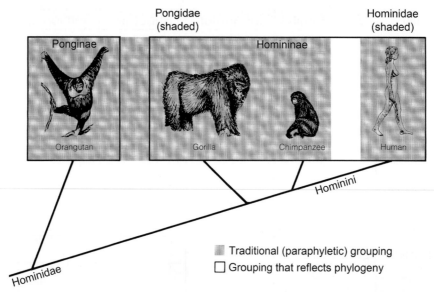

FIGURE 1.3 A strictly phyletic classification recognizes that humans, chimpanzees, and gorillas are more closely related to each other than any of them are to orangutans; the latter are therefore grouped separately as the only pongids. A more traditional classification recognizes adaptive differences; in this case, chimpanzees and gorillas are classified with orangutans (pongids), and humans are grouped separately (hominids) because of the great degree of adaptation that distinguishes humans from even their closest primate relatives.

should be placed in separate families. Once they have learned the Linnean hierarchy, many students are understandably frustrated and annoyed to find that textbooks often do not agree on the classification of different species. There are, however, usually good reasons for the disagreements about primate classification, as we see in the following chapters.

One reason for disagreements about primate classification is that the rules for distinguishing a genus, a family, or a superfamily are somewhat arbitrary. Scientists usually set their own standards. The only generally accepted rules are for species. However, as already discussed, many different ways have been proposed for distinguishing what a species is. Living species of mammals are remarkably consistent in their metric variability (Gingerich and Schoeninger, 1979; but see Tattersall, 1993), and we can use this standard to identify species in the fossil record. The limits for genera and families are, however, much more arbitrary.

It is generally agreed that classification should reflect phylogeny, and that taxonomic groups such as families, superfamilies, and suborders should be **monophyletic** groups: that is, they should have a single common ancestor that gave rise to all members of the group. Many also feel that taxonomic groups should be **holophyletic** groups as well: they should contain all the descendants of their common ancestor, not just some of them. But it is often not practical or possible to achieve this unambiguously, and classifications are often compromises compatible with several possible phylogenies. In addition, some biologists feel that classification should reflect not only phylogeny but also major

adaptive differences, even among closely related species. For example, most biologists now agree that humans are much more closely related to chimpanzees and gorillas than to orangutans. Thus, a true phyletic taxonomy would group humans with the African apes and the orangutan in a separate grouping. In spite of this, some experts still place humans in a separate family, the Hominidae, and all living great apes in a common family, the Pongidae, because they believe that humans have departed further from the common ancestor of humans and great apes than have chimpanzees and gorillas. In this arrangement, shown by the shading in Fig. 1.3, the family Pongidae is called a **paraphyletic** grouping because some of its members (chimpanzees and gorillas) are more closely related to a species (humans) placed in another family than they are to other members (orangutans) of their own family (Fig. 1.3). The taxonomy used in this book (Fig. 1.2) uses a more natural grouping in which all great apes and humans are placed in a single family, Hominidae, with orangutans in one subfamily, the Ponginae, and the African apes and human in a separate subfamily, Homininae. However, paraphyletic groups are commonly used when taxonomic schemes include both extant forms and fossils that may be broadly ancestral to many later groups (e.g., Cartmill, 2012).

References

Begun, D.R., 2007. How to identify (as opposed to define) a homoplasy: Examples from fossil and living great apes. J Hum Evol 52, 559–572.

Cartmill, M., 2012. Primate origins, human origins, and the end of higher taxa. Evol Anthropol 21, 208–220.

Cope, D., 1993. Measures of dental variation and indicators of multiple taxa in samples of sympatric *Cercopithecus* species. In: Kimbel, W.H., Martin, L.B. (Eds.), Species, Species Concepts, and Primate Evolution. Plenum Press, pp. 211–238.

Cracraft, J., 1983. Species concepts and speciation analysis. Curr Ornithol 1, 159–187.

Darwin, C., 1859. On the Origin of Species by Means of Natural Selection, or the Preservation of Favoured Races in the Struggle for Life. John Murray.

de Queiroz, K., 1998. The general lineage concept of species, species criteria, and the process of speciation: A conceptual unification and terminological recommendations. In: Howard, D.J., Berlocher, S.H. (Eds.), Endless Forms: Species and Speciation. Oxford University Press, pp. 57–75.

de Queiroz, K., 2007. Species concepts and species delimitation. Syst Biol 56, 879–886.

Detwiler, K.M., 2002. Hybridization between red-tailed monkeys (*Cercopithecus ascanius*) and blue monkeys (*C. mitis*) in East African forests. In: Glenn, M.E., Cords, M. (Eds.), The Guenons: Diversity and Adaptation in African Monkeys. Kluwer Academic/Plenum Publishers, pp. 79–97.

Disotell, T.R., 2008. Primate Phylogenetics. Encyclopedia of Life Sciences. John Wiley & Sons.

Gingerich, P.D., Schoeninger, M.J., 1979. Patterns of tooth size variability in the dentition of Primates. Am. J. Phys. Anthropol 51, 457–566.

Gould, S.J., Eldridge, N., 1993. Punctuated equilibrium comes of age. Nature 366, 223–227.

Groeneveld, L.F., Weisrock, D.W., Rasoloarison, R.M., Yoder, A.D., Kappeler, P.M., 2009. Species delimitation in lemurs: Multiple genetic loci reveal low levels of species diversity in the genus *Cheirogaleus*. BMC Evol. Biol. 9, 30.

Groves, C., 2001a. Primate Taxonomy. Smithsonian Institution Press.

Groves, C., 2001b. Why taxonomic stability is a bad idea, or why are there so few species of primates (or are there?). Evol. Anthropol 10, 192–198.

Groves, C., 2012. Species concept in primates. Am. J. Primatol 74, 687–691.

Kay, R.F., Fleagle, J.G., 2010. Stem taxa, homoplasy, long lineages and the phylogenetic position of *Dolichocebus*. J. Hum. Evol 59, 218–222.

Kay, R.F., Ross, C., Williams, B.A., 1997. Anthropoid origins. Science 275, 797–804.

Kimbel, W.H., Martin, L.B. (Eds.), 1993. Species, Species Concepts, and Primate Evolution. Plenum Press.

Kuderna, L.F.K., Gao, H., Janiak, M.C., et al., 2023. A global catalog of whole-genome diversity from 233 species. Science 380, 906–913.

Lockwood, C.A., Fleagle, J.G., 1999. The recognition and evaluation of homoplasy in primate and human evolution. Yrbk Phys. Anthropol 42, 189–232.

Masters, J.C., 1993. Primates and paradigms: Problems with the identification of genetic species. In: Kimbel, W.H., Martin, L.B. (Eds.), Species, Species Concepts, and Primate Evolution. Plenum Press, pp. 43–64.

Mayr, E., 1942. Systematics and the Origin of Species. Columbia University Press, New York.

Paterson, H.E.H., 1978. More evidence against speciation by reinforcement. S. Afr. J. Sci 74, 369–371.

Paterson, H.E.H., 1985. The recognition concept of species. In: Vrba, E.S. (Ed.), Species and Speciation. Transvaal Museum, Pretoria, pp. 21–29.

Perelman, P., Johnson, W.E., Roos, C., et al., 2011. A molecular phylogeny of living primates. PLoS Genetics 7, e1001342.

Rose, K.D., Bown, T.M., 1993. Species concepts and species recognition on Eocene primates. In: Kimbel, W.H., Martin, L.B. (Eds.), Species, Species Concepts, and Primate Evolution. Plenum Press, pp. 299–330.

Rosenberger, A.L., 2012. New World monkey nightmares: Science, art, use, and abuse (?) in platyrrhine taxonomic nomenclature. Am J Primatol 74, 692–695.

Sanderson, M.J., Hufford, L., 1996. Homoplasy: The Recurrence of Similarity in Evolution. Academic Press.

Tattersall, I., 1989. The roles of ecological and behavioral observation in species recognition among primates. Hum Evol 4, 117–124.

Tattersall, I., 1992. Species concepts and species identification in human evolution. J Hum Evol 22, 341–349.

Tattersall, I., 1993. Speciation and morphological differentiation on the genus *Lemur*. In: Kimbel, W.H., Martin, L.B. (Eds.), Species, Species Concepts, and Primate Evolution. Plenum Press, pp. 163–176.

Vermeij, G.J., 1978. Biogeography and Adaptation: Patterns of Marine Life. Harvard University Press.

Williams, C.C., 1966. Adaptation and Natural Selection: A Critique of Some Current Evolutionary Thought. Princeton University Press.

Zinner, D., Arnold, M.L., Roos, C., 2011. The strange blood: Natural hybridization in primates. Evol. Anthropol. 20, 96–103.

Further Reading

Groves, C.P., 2001. Primate Taxonomy. Smithsonian Institution.

Rowe, N., Myers, M. (Eds.), 2016. All the World's Primates. Pogonias Press. http://www.alltheworldsprimates.org/

2

The Primate Body

Primate Anatomy

Fossil and living primates are an extraordinarily diverse array of species. Some are among the most generalized and primitive of all mammals; others show morphological and behavioral specializations unmatched in any other mammalian order. This diversity in structure, behavior, and ecology and its evolutionary history are the subjects of this book. The purpose of this chapter is to establish an anatomical frame of reference – a survey of features common to all (or almost all) primates. This chapter, then, provides pictures and descriptions of primate anatomy and preliminary indications of those anatomical features that have undergone the greatest changes in primate evolution.

Compared to most other mammals, we primates have retained relatively primitive bodies. Some of us are specialized in that we have lost our tails, and many have a relatively large brain. But no primates have departed so dramatically from the common mammalian body plan as bats, whose hands have become wings; horses, whose fingers and toes have reduced to a single digit; or baleen whales, who have lost their hindlimbs altogether, adapted their tails into flippers, and replaced their teeth with great hairlike combs. The anatomical features that distinguish the bones and teeth of primates from those of many other mammals are the result of subtle changes in the shape and proportion of **homologous** elements, rather than major rearrangements, losses, or additions of body parts. We generally find the same bones and teeth in all species of primates, with only minor differences reflecting different diets or locomotor habits. The fact that humans are constructed of the same bony elements as other primates (and generally other mammals) is a major piece of evidence demonstrating our evolutionary origin.

Size

Size is a basic aspect of an organism's anatomy and plays a major role in its ecological adaptations (Jungers,

1985). It is a feature that can be readily compared, both among living species and between living and fossil primates. Adult living primates range in size from mouse lemurs and pygmy marmosets, with masses of less than 100 g, to male gorillas with a body masses of over 200 kg (Fig. 2.1). The fossil record provides evidence of a few extinct primates from early in the Age of Mammals that were much smaller (probably as small as 20 g) and at least one, *Gigantopithecus blacki* from the Pleistocene of China (see Chapter 15), that was much larger (probably over 300 kg). In their range of body sizes, primates are one of the more diverse orders of living mammals. As a group, however, primates are rather medium-sized mammals (Fig. 2.2) – larger than most insectivores and rodents and smaller than most ungulates, elephants, and marine mammals.

FIGURE 2.1   A mouse lemur (*Microcebus*) and a gorilla (*Gorilla*), the smallest and largest living primates.

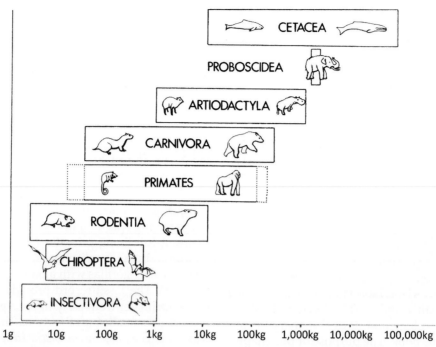

FIGURE 2.2   Size ranges for various orders of mammals, including primates. The solid lines include all living species; the dotted lines include all known living and fossil primates.

Cranial Anatomy

The anatomy of the head, or **cranial** region, plays a particularly important role in studies of primate adaptation and evolution. Many of the anatomical features that have traditionally been used to delineate the systematic relationships among primates are cranial features, and most of our knowledge of fossil primates is based on this region.

Broadly speaking, living **strepsirrhines** (lemurs, lorises, and galagos; Chapter 4) have skulls that resemble other generalized mammals in terms of shape and show relatively little diversity compared to **haplorhines** (Fig. 2.3; Fleagle et al., 2010, 2015). Haplorhine taxa (tarsiers, Old and New World monkeys, and apes), on the other hand, have skulls that are quite diverse and distinctive, probably as a result of the reorganization of the olfactory and visual systems that took place early in their evolution (Fig. 2.3; Fleagle et al., 2010, 2016; see also Chapter 12). Relative to the lemurs, lorises, and galagos, the skulls of higher primates are generally characterized by smaller eye sockets that tend to be closer together, greater cranial flexion, larger brains, and increased overall size.

Bones of the Skull

The adult primate skull (Fig. 2.4) consists of many different bones that together form a hollow, bony shell that houses the brain and special sense organs and also provides a base for the teeth and chewing muscles. Only the lower jaw, or **mandible**, and the three bones of the middle ear are separate, movable elements; the others are fused into a single unit, the **cranium**. This unit can be roughly divided into two regions: a more posterior braincase, or **neurocranium**, and a more anterior facial region, or **splanchnocranium**.

The braincase serves as a protective bony case for the brain, a housing for the auditory region, and an area of muscle attachment for the larger chewing muscles and the muscles that move the head on the neck. Three paired flat bones – the **frontal**, **parietal**, and **temporal** bones – make up the top and sides of the braincase. (The temporal bone is a relatively complicated bone with several distinct parts.) The posterior and inferior surfaces of the braincase are formed by a single bone, the **occipital**, which also has a number of distinct parts. A complex, butterfly-shaped bone, the **sphenoid**, forms the anterior surface of the braincase and joins it with the facial region.

The facial region is formed by the **maxillary** and **premaxillary bones**, which contain the upper teeth; the **zygomatic bone**, which forms the lateral wall of the **orbit**, or eye socket; the **nasal bones**, which form the bridge of the nose; and numerous small bones that make up the orbit and the internal nasal region. The lower jaw, or **mandible**, contains the lower teeth. In many mammals, and in most strepsirrhines and tarsiers, the two halves of the mandible are loosely connected anteriorly in such a way that they can move somewhat independently of one another. This joint is called the **mandibular symphysis**. In anthropoid primates, including humans, the two sides of the lower jaw are fused to form a single bony unit.

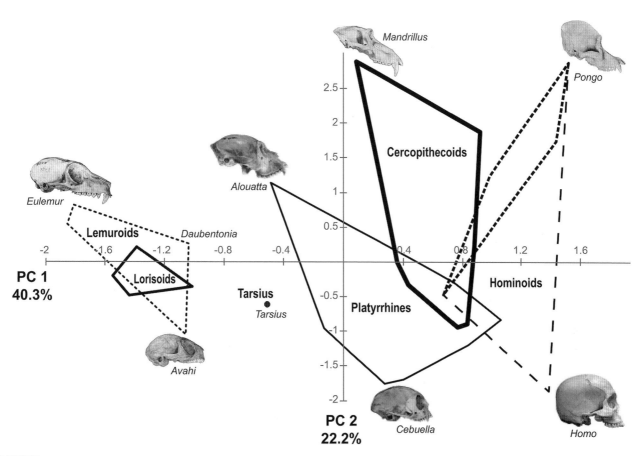

FIGURE 2.3 Principal components analysis (PCA) of 18 cranial landmarks characterizing male primate crania and demonstrating the extremes of extant primate cranial morphology. Wireframe configurations were used to help visualize the changes taking place along the principal component (PC) axes with the percentage of variation explained by each axis. PC 1 is associated with increased orbital convergence, increased cranial flexion, increased brain size, decreased eye size, and increased overall size as PC scores increase, and PC 2 tracks increased relative snout length, increased relative palate length, decreased neurocranial height, and increased overall size as scores increase. Polygons outline the distribution of major groups among strepsirrhines and anthropoids. Note the increased diversity of haplorhines and major anthropoid groups compared to strepsirrhines. Human crania with their large, tall neurocranium, flexed cranial base, and flat face represent one notable outlier compared to other primate and hominoid crania. *(Modified from Fleagle, J.G., Gilbert, C.C., Baden, A.L., 2010. Primate cranial diversity. Am. J. Phys. Anthropol. 142, 565–578.)*

Although all primate skulls are made up of these same components, they can have very different appearances, depending on the relative size and shape of individual bones (Fig. 2.4). The skull functions as a base and structural framework for the first part of the digestive system and as a housing for the brain and special sense organs of sight, smell, and hearing. Much of the diversity in primate skull shape reflects the need for this single bony structure to serve numerous, often conflicting functions. For example, although the size of the orbits is most directly related to the size of the eyeball and to whether a species is active during the day or the night, it influences the shape and position of the nasal cavity and the space available for chewing muscles.

Teeth and Chewing

Many parts of the head and face are important in the acquisition and initial preparation of food. The lips,

cheeks, teeth, mandible, tongue, **hyoid bone** (a small bone suspended in the throat beneath the mandible), and muscles of the throat all participate in this complex activity; and many of these same parts also play a role in communication and sound production. The parts of the skull that can be linked most clearly to dietary habits are the teeth, the mandible, and the chewing muscles that move the lower jaw.

Teeth, more than any other single part of the body, provide the basic information underlying much of our understanding of primate evolution. Because of their extreme hardness and compact shape, teeth are the most commonly preserved identifiable remains of most fossil mammals. But teeth are more than just plentiful: they are also complex organs that provide considerable information about both the phyletic relationships and the dietary habits of their owners. Because of the importance of teeth in evolutionary studies, there is an extensive but fairly simple terminology for dental anatomy.

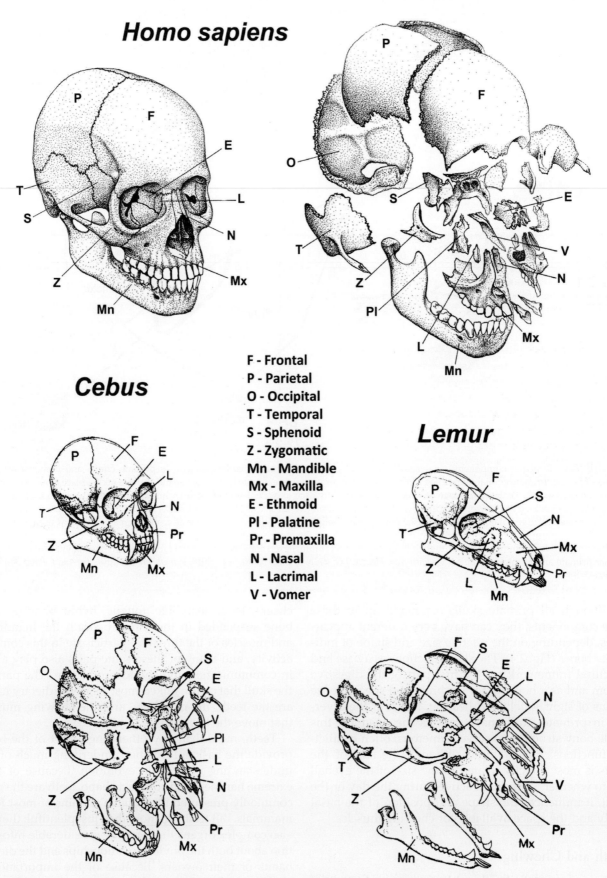

FIGURE 2.4   Top: The skull of a human (*Homo sapiens*). Bottom: Skulls of a capuchin monkey (*Cebus*) and a lemur (*Lemur*) showing how differences in the size and shape of individual bones contribute to overall differences in skull form.

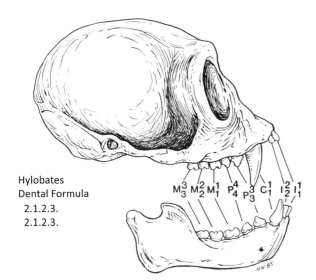

Hylobates
Dental Formula
2.1.2.3.
2.1.2.3.

$M^3_3 M^2_2 M^1_1 P^4_4 P^3_3 C^1_1 I^2_2 I^1_1$

FIGURE 2.5 The dentition of a gibbon (*Hylobates*) showing two incisors (I), one canine (C), two premolars (P), and three molars (M) in each dental quadrant for a dental formula of 2.1.2.3/2.1.2.3.

All primates have teeth in both the upper jaw (**maxilla** and **premaxilla**) and the lower jaw (**mandible**), and, like most features of the primate skeleton, primate teeth are **bilaterally symmetrical** – the teeth on one side are mirror images of those on the other. Each primate jaw normally contains four types of teeth (Fig. 2.5). These are, from front to back, **incisors**, **canines**, **premolars**, and **molars**. The number of teeth a particular species possesses is usually expressed in a **dental formula**. The human dental formula is 2.1.2.3./2.1.2.3., indicating that we normally have two incisors, one canine, two premolars, and three molars on each side of both the upper and the lower jaw for a total of 32 adult teeth. In most primate species, formulae for the upper and lower dentition are the same. In addition to **adult** (or **permanent**) **teeth**, primates have an earlier set of teeth, the **milk** (or **deciduous**) **dentition**, which precedes the adult incisors, canines, and premolars and occupies the same positions in the jaws. The human milk dentition, for example, contains two deciduous incisors, one deciduous canine, and two deciduous **premolars** (often called "milk molars") in each quadrant, for a total of 20 deciduous teeth.

The three main cusps of an upper molar (Fig. 2.6) are the **paracone**, the **metacone**, and the **protocone**. The triangle formed by these cusps is called the **trigon**. Many primates have evolved a fourth cusp distal to the protocone, called the **hypocone**. Small cusps adjacent and lingual to these major cusps are called **conules** (the **paraconule** and the **metaconule**). Accessory folds of enamel on the **buccal** (cheek-side) surface of the tooth are called **styles**, and an enamel belt around the tooth is referred to as a **cingulum**. Shallow areas between crests are called **basins**.

The basic structure of a lower molar in a generalized mammal (Fig. 2.6) is another triangle, this one pointing toward the cheek side. The cusps have the same names as those of the upper molars but with the suffix *-id* added

(**protoconid**, **paraconid**, and **metaconid**). This basic triangle in the front of a lower molar is called the **trigonid**. In primates and all but the most primitive mammals there is an additional area added to the distal end of this primitive trigonid. This extra part, the **talonid**, is formed by two or three additional cusps: the **hypoconid** on the buccal side, the **entoconid** on the **lingual** (tongue) side, and, in many species, a small, distal-most cusp between these two, the **hypoconulid**.

Primate dentitions are involved in two different aspects of feeding. The anterior part of the dentition, the incisors and often the canines (together with the lips and often the hands), is primarily concerned with **ingestion** – the transfer of food from the outside world into the oral cavity in manageable pieces that can then be further prepared by the cheek teeth (the molars and premolars). The subsequent breakdown of food items by chewing is called **mastication**.

The molars and premolars of primates break down food mechanically in three ways: (a) by puncture-crushing or piercing the food with sharp cusps; (b) by shearing the food into small pieces, that is, by trapping particles between the blades of enamel that are formed by the crests that link cusps; and (c) by crushing or grinding food in mortar-and-pestle fashion between rounded cusps and flat basins. Different types of food require different types of dental preparation before swallowing, and it is possible to relate the various characteristics of both the anterior teeth (for obtaining and ingesting objects) and the cheek teeth (for puncturing, shearing, or grinding) to diets with different consistencies (as discussed in Chapter 9).

The movement of the lower jaw relative to the skull in both ingestion and chewing (mastication) is brought about by four major chewing muscles that originate on the skull and insert on different parts of the lower jaw (Fig. 2.7). The largest is the **temporalis**, which has a fan-shaped origin on the side of the skull and inserts onto the **coronoid process** of the mandible. The second large muscle is the **masseter**, which originates from the **zygomatic arch** and inserts on the lateral surface of the ascending ramus of the mandible. Both of these muscles close the jaw when they contract. There are two smaller muscles on the inside of the jaw: the **medial** and **lateral pterygoids**. Much of the bony development of the primate skull seems to be related to the size and shape of these muscles and to the magnitude and direction of the forces generated in the skull during chewing. These muscular differences have in turn evolved to meet mechanical demands associated with dietary differences.

Tongue and Taste

In primates, as in all mammals, the tongue forms the floor of the oral cavity. Primate tongues vary considerably in shape (Fig. 2.8). Part of this variation reflects differences

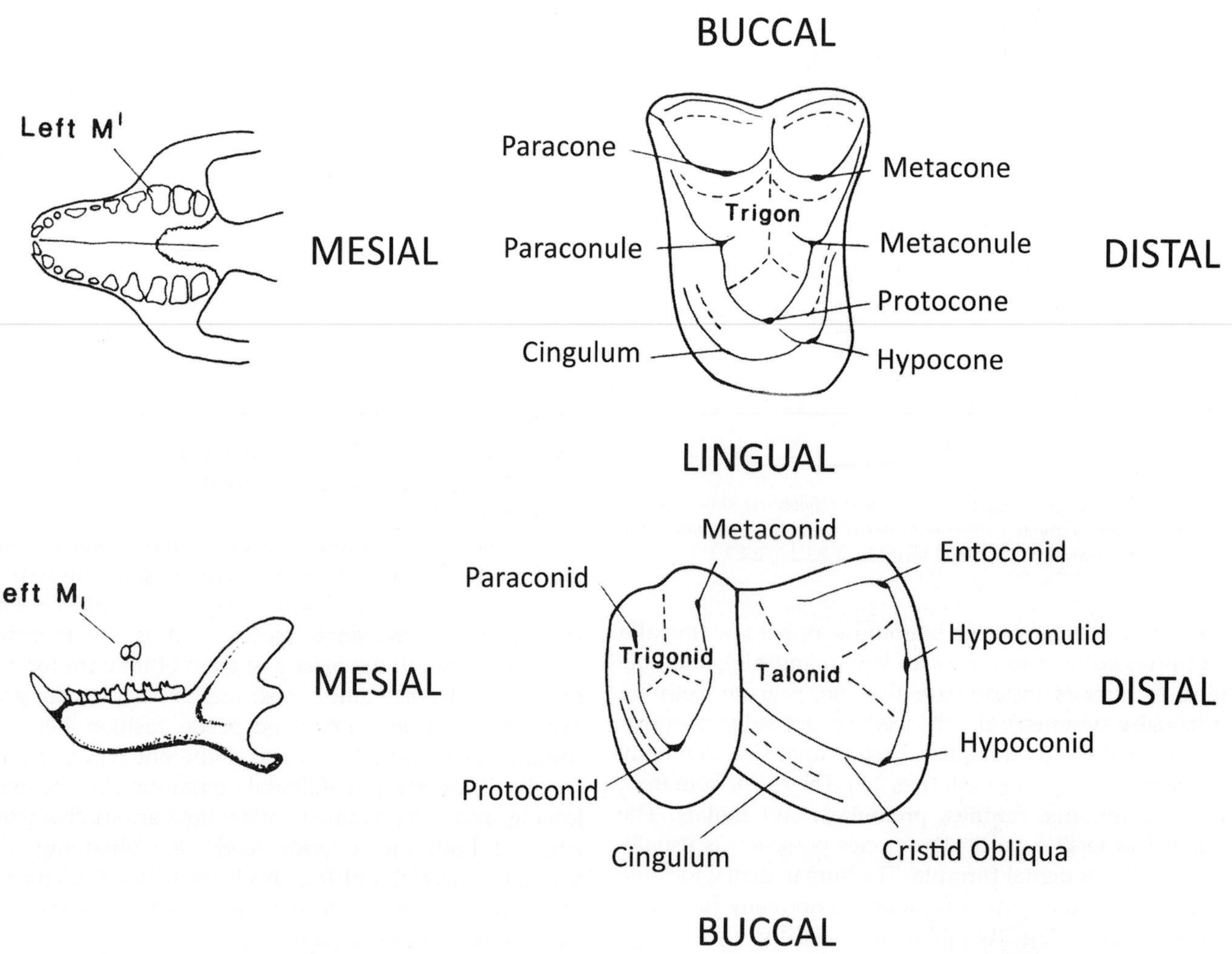

BUCCAL

Left M¹

MESIAL

Paracone — — Metacone

Trigon

Paraconule — — Metaconule   DISTAL

— Protocone

Cingulum — — Hypocone

LINGUAL

Metaconid

Paraconid — — Entoconid

Trigonid   Talonid   — Hypoconulid   DISTAL

MESIAL

Protoconid — — Hypoconid

Cingulum — Cristid Obliqua

BUCCAL

FIGURE 2.6   Major parts of the upper and lower teeth of a primitive primate.

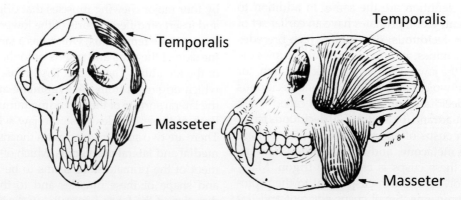

FIGURE 2.7   Anterior and lateral views of a primate skull showing the major chewing muscles.

in the shape of the oral cavity itself and is determined by the relative breadth of the anterior dentition. However, part of this variation is related to differences in the function of the tongue as an organ of taste, touch, and vocalization. For example, in all strepsirrhines there is a distinctive projection from the underside of the tongue, the **sublingua**, which serves to clean the tooth comb (Fig. 2.9). The brown lemur, *Eulemur fulvus*, has a series of highly sensitive

conical structures at the tip of its tongue; feathery projections have been described on the tip of the tongue of *Eulemur rubriventer* that seem to be associated with its habit of licking nectar from flowers (Hofer, 1981; Overdorff, 1992); and the fork-marked lemur, *Phaner*, has a long, muscular tongue that it uses while feeding on exudates (gums and resins). Among higher primates, humans have a very unusual tongue, in that it extends very far posteriorly and

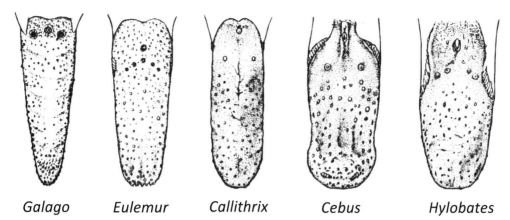

Galago   Eulemur   Callithrix   Cebus   Hylobates

FIGURE 2.8   Superior view of the tongues of five primates showing differences in overall shape and in the distribution of taste papillae.

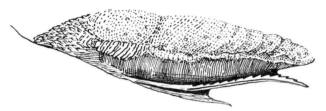

FIGURE 2.9   The tongue of a ring-tailed lemur (*Lemur catta*) showing the serrated sublingua that is used to clean the toothcomb.

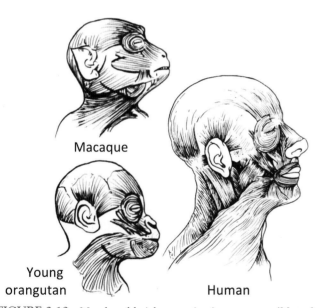

Macaque

Young orangutan   Human

FIGURE 2.10   Muscles of facial expression in a macaque (*Macaca*), a young orangutan (*Pongo*), and a human (*Homo sapiens*). Note the increasing differentiation of individual muscles, which enables finer control of expressions.

plays a major role in vocalization by changing the shape of the throat. Although all primate tongues contain the taste buds (**papillae**) that are responsible for the sense of taste, the size, shape, and number of different taste sensors vary considerably from species to species (Fig. 2.8; see also Hladik and Simmen, 1997; Alport, 2009; Muchlinski et al., 2011).

Muscles of Facial Expression

Another aspect of cranial anatomy in primates that deserves special consideration is facial musculature (Fig. 2.10). Among primates, and especially in humans, the muscles of facial expression are more highly developed and differentiated into separate units than among any other groups of mammals (Huber, 1931). It is these muscles that make possible the range of visual expressions that characterizes and facilitates the complex social behavior of primates (Dobson 2009, 2012).

The Brain and Senses

The structural shape of the skull – the development of bony buttresses and crests, as well as the relative positioning of the face and the neurocranium – seems to be greatly influenced by the size and functional requirements of the masticatory system. However, the relative sizes of many parts of the skull, such as the neurocranium and the orbits, as well as the size and position of various openings in the skull, seem more directly related to the skull's role in housing the brain and the sense organs responsible for smell, vision, and hearing. The ways in which different primates use these senses in their daily activities are critical for understanding the variation among species in their anatomy (Dominy et al., 2001).

The Brain

The brain is the largest organ in the head, and its relative size is an important determinant of skull shape among primates. Relative to body weight, primates have the largest brains of any terrestrial mammals; only marine mammals are comparably brainy. There are,

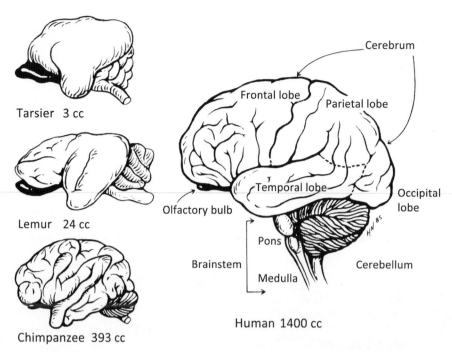

FIGURE 2.11   Lateral view of the brains of a lemur (*Lemur*), a tarsier (*Tarsius*), a chimpanzee (*Pan*), and a human (*Homo sapiens*), showing differences in relative size of the parts of the brain. Note especially the differences in size of the olfactory bulb and the size and development of convolutions on the cerebral hemispheres.

however, differences in relative brain size among primates. Lemurs, lorises, galagos, and tarsiers all have relatively small brains compared to monkeys and apes, and human brains are relatively enormous. Still, the brain is a complex organ with many parts, and although some parts of primate brains are relatively large by mammalian standards, others are relatively small. In gross morphology, a primate brain can be divided into three parts (Fig. 2.11): the **brainstem**, the **cerebellum**, and the **cerebrum**. Each part has very different functions, and each is made up of many different functionally distinct sections.

The brainstem forms the lower surface and base of the brain. It is an enlarged and modified continuation of the upper part of the spinal cord and is the part of the primate brain that differs least from that found among other mammals and lower vertebrates. The brainstem is concerned with basic physiological functions such as reflexes, control of heartbeat and respiration, and temperature regulation, as well as the integration of sensory input before it is relayed to "higher centers" in the cerebrum. Many of the cranial nerves, which are responsible for innervation of such things as the organs of sight and hearing and the muscles of the orbit and face, arise from the brainstem. Very little of the primate brainstem is visible in either a lateral or a superior view; it is covered by two areas that have become large and specialized: the cerebellum and the cerebral hemispheres.

The **cerebellum**, which lies between the brainstem and the posterior part of the cerebrum, is a developmental outgrowth of the caudal region of the brainstem. It is concerned primarily with control of movement and with motor coordination. Compared to other mammals, primates have a relatively large cerebellum.

The paired **cerebral hemispheres** are the part of the brain that has undergone the greatest change during primate evolution. It is in this part that we find the greatest differences between primates and other mammals and the greatest differences among living primates (Fig. 2.11). Gigantic cerebral hemispheres are one of the hallmarks of human evolution. Anatomically, this part of the brain is divided into **lobes** named for the bones immediately overlying them – **frontal**, **parietal**, **temporal**, and **occipital**. In most primates, the surface of the cerebral hemispheres is covered with convolutions made up of characteristic folds, or **gyri**, which are separated by grooves, or **sulci**. The development of these convolutions is most apparent in larger species and reflects the fact that the most functionally significant part of the cerebrum, the **gray matter**, lies at the surface. The convolutions or foldings of the brain surface provide a greater surface area for the cerebral hemispheres, relative to brain or body volume, than would be provided by a smooth surface.

Overall, the cerebral hemispheres are involved with recognition of sensations, with voluntary movements, and with mental functions such as memory, thought, and interpretation. Different regions of the cerebrum (i.e., specific gyri) can be related to particular functions (Fig. 2.12). The central sulcus, for example, separates an anterior gyrus related to voluntary movement from a more posterior

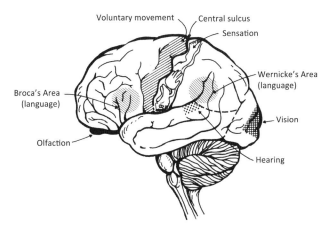

FIGURE 2.12 Important functional areas of the human brain.

gyrus concerned with sensation. Within each of these areas it is possible to identify more specific regions concerned with voluntary movement or sensation of particular parts of the body. In addition, there are other parts of the cerebral hemispheres, called **association areas**, which are related to the integration of input from several different senses (such as hearing and vision) and to specific tasks, such as language and speech. Two particularly well-developed association areas in the human brain are those related to language: **Broca's area** in the frontal lobe and **Wernike's area** in the parietal and temporal lobes.

Although the brain is a soft structure, primate brains often leave their mark on the bony morphology of the skull. Size (in particular, volume) is an obvious feature of a primate brain that can be determined from a skull. Furthermore, in many species, sulci and gyri also leave impressions on the internal surface of the cranium. Such impressions on fossil skulls can provide limited information about the development of different functional regions on the cerebral hemispheres of extinct primates.

The nerves that take signals to and from the brain enter and leave the cranial cavity through various holes, called **foramina**, in the skull bones. The largest of these holes is the **foramen magnum**, through which the spinal cord passes. The many smaller foramina vary considerably in size and position among living primates and are widely used in systematics. In a few cases, it seems possible to correlate the size of a foramen carrying a specific nerve to the development of a particular function or anatomical region.

Foramina also serve as passages for the arteries that supply blood to the brain and other cranial structures, and for the veins that drain those same structures. The pathway of the blood supply to the brain shows a number of distinctly different patterns among living primates (Fig. 2.13). Although we know little about the functional significance of these differences, they have proved useful in sorting the phyletic relationships among many living and fossil primate species. The major blood supply to the head in primates comes from two branches of the common carotid artery at the base of the neck. The **external carotid** is responsible primarily for supplying structures in the neck and face; the **internal carotid** (along with the vertebral arteries) supplies the brain and the eye. The internal carotid artery enters the cranial cavity as two distinct arteries, a **stapedial artery** passing through the stapes bone and a **promontory artery** that generally lies medial to the stapedial artery and crosses the **promontorium**, a raised surface in the middle ear, to enter the cranial cavity further anteriorly. In most lemurs, for example, the stapedial is the larger artery; in tarsiers, New World monkeys, Old World monkeys, apes, and humans, the stapedial is generally absent in adults and the promontory provides most of the blood supply to the brain. The size of the promontory artery (when dominant) appears closely correlated with brain size, suggesting this blood supply route may be associated with the evolution of large brains in haplorhine primates (Boyer et al., 2016). In lorises, galagos, and cheirogaleids, a branch of the external carotid artery, the **ascending pharyngeal**, provides the major blood supply to the front part of the brain (Fig. 2.13).

Nasal Region and Olfaction

In many mammals, smell is the dominant sensory mode. It provides much of the information on which animals rely to find their way around, locate their food, locate potential predators, communicate with their kin and neighbors, and determine the sexual status of potential mates. Among more **diurnal** (active during the day) higher primates, smell seems to be less important for some of these functions than other senses such as vision, but even for these species, this most basic of senses has not been abandoned. In most primate species, it still plays an important – albeit relatively poorly understood – role in reproduction, communication, and food evaluation.

The sensation of smell is carried by the **olfactory nerves**, which end in paired swellings, the **olfactory bulbs**, which lie under the large frontal lobes of most primates (Fig. 2.11). The olfactory nerves receive their input from the special sensory membranes lining the scroll-like **turbinates** of the internal nasal cavity. The development of the nasal part of the olfactory system and its position with respect to the orbits show two distinctly different arrangements among primates (Fig. 2.14). In lemurs and lorises, as well as in most other mammals, the nerves responsible for olfaction pass between the orbits from the internal cavity to the brain. Within the nasal cavity, large numbers of turbinates are attached to several different bones, including several derived from the ethmoid bone that lies in the **sphenoid recess**,

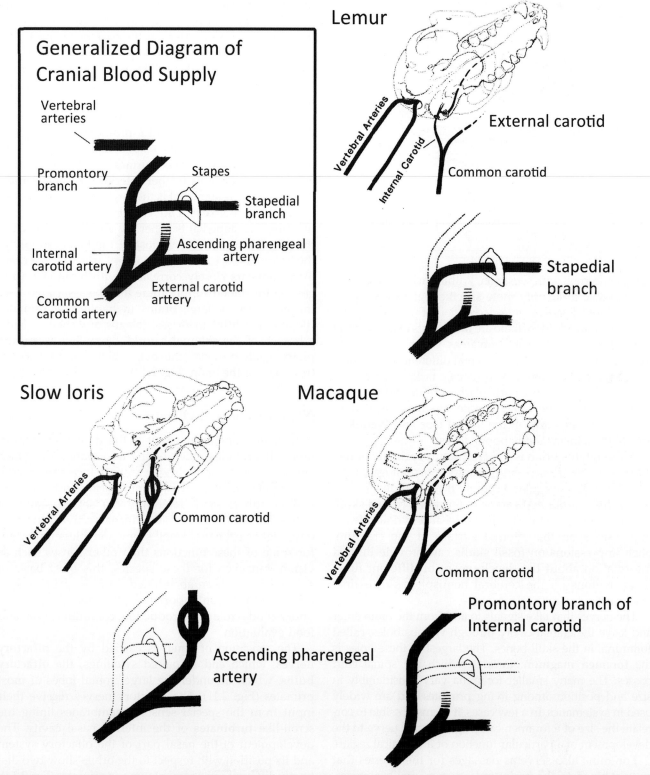

FIGURE 2.13 Cranial external blood supply in several types of living primates. In all living primates, the vertebral arteries supply blood to the brain; however, species differ considerably in the relative contributions of the stapedial and promontory branches of the internal carotid artery, and of the ascending pharyngeal branch of the eternal carotid artery. In the lemur (*Lemur*), the stapedial branch provides the major arterial supply to the anterior part of the brain; in a slow loris (*Nycticebus*), the blood to the front of the brain comes from a large ascending pharyngeal artery; in macaques (*Macaca*) and all higher primates, the promontory branch of the internal carotid provides the major arterial blood supply to the anterior part of the brain.

The Brain and Senses   19

Lemur

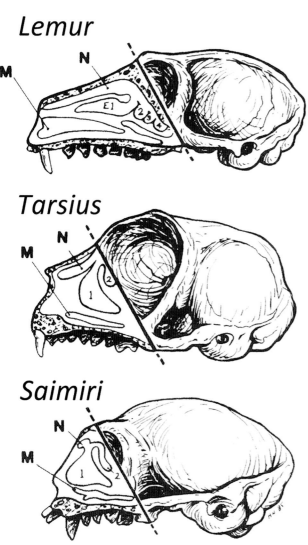

Tarsius

Saimiri

FIGURE 2.14   Structure of the interior nasal region of a lemur (*Lemur*), a tarsier (*Tarsius*), and a squirrel monkey (*Saimiri*). Note the reduction in number and relative size of the turbinates in *Tarsius* and *Saimiri*. M, maxilloturbinate; N, nasoturbinate; E, ethmoturbinates (numbered).

a special cul-de-sac. In tarsiers, monkeys, apes, and humans, the structure of this region is greatly simplified. The olfactory nerves pass over the interorbital septum, rather than between the orbits, and the sphenoid recess and posteriormost two turbinates are missing or greatly reduced. In apes and humans, this region is even further reduced.

Although primate noses and the tissue-lined passages that make up their internal structure are associated primarily with olfaction, they also play important roles in respiration and temperature regulation by warming and humidifying the air that passes over them.

In addition to their sense of smell, lemurs, lorises, tarsiers, and many New World monkeys (but apparently not Old World monkeys, apes, or humans) have an additional sense that seems to be particularly important in sexual communication. The **vomeronasal organ** (or **Jacobson's organ**) is a chemical-sensing organ that lies in the anterior part of the roof of the mouth in many mammals and leaves a bony depression on the maxilla known as the vomeronasal groove (Garrett et al., 2013). It is stimulated by substances found in the urine and scent marks of primates, and permits individuals to chemically determine information – including identity, relatedness, and reproductive status – about others (see Chapter 3).

In addition to these "internal" differences in nasal structure, there is a major dichotomy among primates in the structure of the external nasal region. The **strepsirrhine** primates (lemurs, lorises, and galagos) have a nose with a median cleft and moist region that extends from the base of the nasal opening to the inside of the upper lip (Fig. 2.15), as in many mammals, such as dogs and cats. It has generally been argued that the strepsirrhine condition is related to the function of the Jacobson's organ. In contrast, tarsiers and anthropoids, the **haplorhines**, have a dry nose, continuous upper lip and often a hairy region between the lip and the base of the nasal opening (Fig. 2.15). Associated

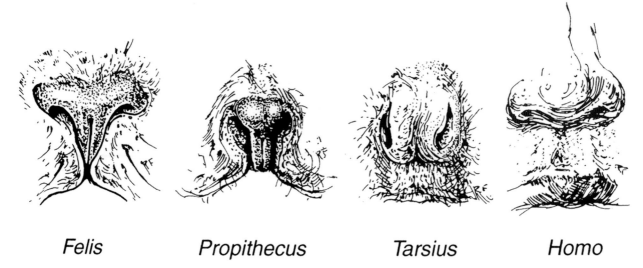

Felis   Propithecus   Tarsius   Homo

FIGURE 2.15   The nostrils and upper lip of two strepsirrhine mammals – a cat and a lemur – and two haplorhine primates – a tarsier and a human. Note the midline cleft in the lip and nose of the cat and the lemur.

with the presence of a moist **rhinarium** of strepsirrhines is a recently discovered bony feature of the nasal cavity. The **nasolacrimal duct** is a small tube that transmits tears from the orbit to the nasal cavity in all primates. However, the length and orientation of the duct differ in the two suborders of primates (Rossie and Smith, 2007; Fig. 2.16). In strepsirrhines the lower end of the duct extends anteriorly to direct moisture toward the external nose, but in haplorhines, this anterior portion is absent, so the duct is vertical.

Eyes and Vision

Primates rely extensively on vision to understand the world around them. Nevertheless, there are considerable differences among primate species in many aspects of their visual systems, both in the bony structure of the orbit and in the soft anatomy of the eye and the parts of the brain related to sight. As a group, primates have larger eyes than other mammals, but primate eyes vary strikingly in relative size (Kirk, 2006; Ross and Kirk, 2007). **Nocturnal** (active during the night) species have relatively larger eyes and bony orbits than do **diurnal** (active during the day) species.

In addition to their differences in size, the bony orbits of primate skulls show important differences in construction (Fig. 2.17). In most mammals, each eye lies nestled in a pocket of tough but flexible connective tissue

on the side of the skull, medial to the zygomatic bone. The lateral side of the orbit is formed by a fibrous ligament, rather than bone. In all living primates, however, the zygomatic bone and the frontal bones join to form a lateral strut, or **postorbital bar**, so that the eye is surrounded by a complete bony ring. In anthropoid primates, and to a lesser extent in tarsiers, the orbit is

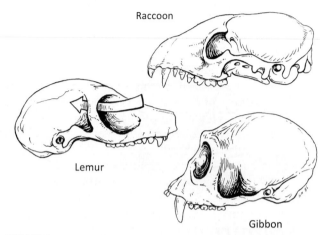

FIGURE 2.17   The bony structure of the orbit in a raccoon, a lemur, and a gibbon. In the raccoon, the orbit is open laterally. In the lemur, the orbit is surrounded by a bony ring, but is open posteriorly. In the gibbon, the posterior opening of the orbit is closed off so that the eye is surrounded by a bony cup.

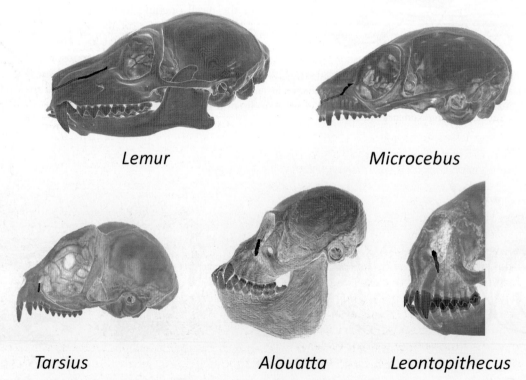

FIGURE 2.16   The nasolacrimal duct shown as a black line, is strepsirrhine and haplorhine primates. In strepsirrhines, the duct extends anteriorly toward the rhinarium; in haplorhines, it is short and vertical. *(Courtesy of J. Rossie.)*

further walled off behind by a bony partition, the **postorbital plate** (or **postorbital septum**); thus, the eyeball lies within a bony cup. This is described as **postorbital closure**. The functional significance of the postorbital bar and postorbital plate is regularly debated, with no clear solution. The most important function seems to be to isolate the orbital contents from the effects of chewing muscles (Heesy et al., 2005). In addition to these major differences in the mechanical structure of primate orbits, there is considerable variation among primate species in the arrangement of the mosaic of small bones forming the medial wall of the orbit, and in the size of the eyeball relative to the size of the bony orbit.

The overall structure of most primate eyeballs is similar; the main differences lie in the structure of the **retina**, the film-like sheet of light-sensitive cells that lines the back of the eye. Two types of cells make up the retina in most primates: **rods**, which are very sensitive to light but do not distinguish color; and **cones**, which are sensitive to color. In many nocturnal primates, the retina is composed predominantly of rods. Furthermore, in lemurs and lorises we find an additional feature characteristic of many nocturnal mammals: the retina contains an extra layer that reflects light. This layer, the **tapetum lucidum**, reduces visual acuity but enhances an animal's ability to see at night by "recycling" incoming light. In tarsiers, monkeys, apes, and humans (all of which lack a tapetum lucidum), we find a different modification of the retina – a specialized area called the **fovea**, in which the light-sensitive cells are packed extremely close together, allowing very good visual acuity. Aspects of primate visual acuity are correlated with the relative size of the optic foramen, which transmits the optic nerve from the eye to the brain, and this relationship can be used to reconstruct visual acuity in extinct taxa (Kay and Kirk, 2000).

Although most diurnal primates have color vision, there is diversity in the color vision abilities found among different species and among individuals within species. All Old World monkeys and apes that have been tested have color vision similar to that found in most humans, with three types of cones, each sensitive to a different part of the visual spectrum. However, Malagasy lemurs and New World monkeys are much more diverse: within many species, only some individuals have three types of cones, whereas others have only two, so they have color vision similar to that of "color-blind" human males. Most, but not all, nocturnal primates have a single type of cone. There are also indications that some primates have vision in the ultraviolet range (Perry et al., 2007; Melin et al., 2012). The adaptive significance and evolutionary history of this diversity in color vision is a hot topic of current research, but remains largely unexplained.

Ears and Hearing

Hearing plays an important role in many aspects of primate life. Many species, especially those active at night, use hearing to locate insect prey, and most use their ears to listen for approaching predators and to receive the vocal signals emitted by their family and neighbors. Although we know much about the anatomy of the auditory system, the physiological significance of many anatomical differences among primate ear regions is poorly understood (see Coleman and Colbert, 2010).

Anatomically, the primate ear can be divided into three parts: the **outer ear**, the **middle ear**, and the **inner ear** (Fig. 2.18). The outer ear is composed of the **external ear**, or **pinna**, and a tube leading from that structure to the **eardrum**, or **tympanic membrane**. Primate pinnae are extremely variable in size, shape, and mobility (Fig. 2.19). In many nocturnal primates that rely extensively on hearing to locate prey, the outer ear is often a large, membranous structure that can be moved in many directions by a distinct set of muscles. In other species, it is smaller, often only slightly movable (as in humans), and may even be totally hidden under fur. The outer ear collects sounds, localizes them with respect to direction, and funnels them into the auditory canal, where they set the tympanic membrane in motion.

The tympanic membrane, or eardrum, a sheet of connective tissue spread over a bony ring formed by the tympanic bone, forms the boundary between the outer ear and the middle ear. It changes the moving air that makes up sound into mechanical movements that are passed along the three ossicles of the middle ear (the **malleus, incus**, and **stapes**). The last of these, the stapes, transfers this motion to the fluid-filled inner ear.

The inner ear contains three functionally different parts within the petrous portion of the temporal bone. The part concerned with hearing is the **cochlea**, a coiled, snail-shaped bony tube. Within the fluid-filled cochlea is a pressure-sensitive organ that registers the movement of the fluid and sends impulses to the brain through the acoustic nerve. The other two parts of the inner ear, three **semi-circular canals** and two other fluid-filled chambers (the **utricle** and **saccule**), are responsible for sensing movement and for orientation with respect to gravity (see Chapter 9).

Apart from differences in relative sensitivity to particular frequencies, all primate ears seem to function in much the same way. There are, however, considerable architectural differences in the way the bony housing of the middle ear is constructed (Fig. 2.20). In all living primates, the inferior surface of the middle ear is covered by a thin sheet of bone, the **auditory bulla**, derived from the petrous part of the temporal bone. In some primates, this bulla is inflated

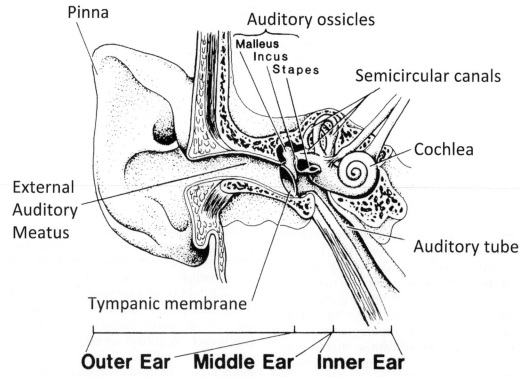

FIGURE 2.18   A primate ear showing the three major parts and the individual elements in each part.

FIGURE 2.19   External ears (pinnae) of several primates.

or balloon-like, and is often divided into many compartments; in others, it is flatter. The physiological significance of the different types of ear architecture is poorly understood. The inflated auditory bullae of many small nocturnal primates seem to increase perception of low-frequency sounds and may be associated with nocturnal predation on flying insects.

The spatial relationship between the tympanic ring and the auditory bulla differs considerably among major groups of living primates (Fig. 2.20). In lemurs, the ring lies within the cavity formed by the bulla; in lorises, the ring is attached to the inside wall of the bulla; and in New World monkeys, it is attached to the outside wall of the bulla. In tarsiers and **catarrhines**, the ring is also attached to the wall of the bulla but extends laterally to form a bony tube, the **external auditory meatus**.

The Trunk and Limbs

Whereas the skull is concerned primarily with sensing the environment, with communication, and with the ingestion and preparation of food, the part of the skeleton behind the skull, the **postcranial skeleton**, as it is often called, serves quite different functions. Obviously, it provides support and protection for the organs of the trunk, but its primary functions, and those that seem to account best for the major differences in skeletal shape, are related to locomotion. In this capacity, the postcranial skeleton provides both a structural support and a series of attachments and levers to aid in movement. The primate postcranial skeleton (Figs. 2.21 and 2.22) is relatively generalized by mammalian standards. Primates have retained many bones from their early mammalian

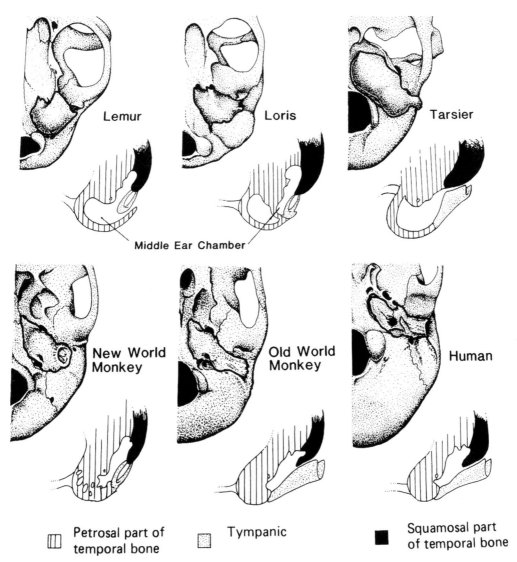

Middle Ear Chamber

Petrosal part of temporal bone

Tympanic

Squamosal part of temporal bone

FIGURE 2.20   The structure of the tympanic (ectotympanic) bone surrounding the eardrum and its position in relation to the bones surrounding the middle ear cavity vary considerably among living primate species (inferior view above, cross-sectional view of the middle ear below). In a lemur (*Lemur*), the tympanic bone is ring shaped and suspended within the bony bullar cavity. In lorises, the tympanic bone lies at the edge of the middle ear cavity and is connected to the wall of the bulla; note also that the bulla cavity is divided. In tarsiers (*Tarsius*), the tympanic bone is elongated to form a bony tube at the lateral edge of the bullar cavity. In New World monkeys, the tympanic bone is a ring-like structure fused against the lateral wall of the bulla. In catarrhines, represented by an Old World monkey and a human, the tympanic bone is extended laterally to form a bony tube.

ancestors that other mammals have lost. For example, most primates have a primitive limb structure, with one bone in the upper (or **proximal**) part of each limb (the humerus or femur), a pair of bones in the lower (**distal**) part (the radius and ulna or tibia and fibula), and five digits on the hands and feet. Primate skeletons can be divided into three parts: the **axial skeleton** (the backbone, tail, and ribs), the **forelimbs**, and the **hindlimbs**. To facilitate descriptions of anatomical features, we use a standard terminology for directions with respect to an animal's body (Fig. 2.23).

Axial Skeleton

The backbone, which is made up of individual bones called **vertebrae**, is divided into four regions (Fig. 2.21). In almost all mammals, the **cervical**, or neck, region contains seven vertebrae. The first two vertebrae, the **atlas** and the **axis**, are specialized in shape and serve as a support and pivot for the skull. The second region of the backbone is the **thorax**. Most of the rotational movements of the trunk involve movements between thoracic vertebrae. Primates have between 9 and 13 thoracic vertebrae, each of which is attached to a rib. The ribs are connected anteriorly to the

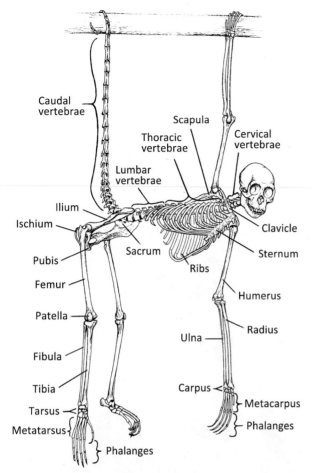

FIGURE 2.21 The skeleton of a spider monkey (*Ateles*). This species is unusual in having a very small thumb and a prehensile tail.

sternum to enclose the thoracic cage, within which lie the heart and lungs. The outside of the thorax is covered by the muscles of the upper limbs. The thoracic vertebrae are followed by the lumbar vertebrae. There are no ribs attached to the lumbar vertebrae, but there are very large **transverse processes** for the attachment of the intrinsic back muscles that extend the back. In most primates, these transverse processes arise from the body of the vertebrae; however, in apes, humans, and some fossil lemurs they arise from parts of the vertebral arch. Most of the flexion and extension of the back takes place in the lumbar region, and the length of this part of the back varies considerably in association with differences in locomotion and posture, as discussed in later chapters.

The next lower region of the backbone is the **sacrum**, a single bone composed of three to five fused vertebrae. The **pelvis**, or hipbone, is attached to the sacrum on its two sides, and the tail joins it distally. The last region of the spine, the **caudal region**, or tail, varies from a few tiny bones fused together (the **coccyx**, in humans) to a long, grasping organ of as many as 30 bones in some species (Fig. 2.21).

Upper Limb

The primate upper limb, or **forelimb**, is divided into four regions, most of which contain several bones. The most proximal part, nearest the trunk, is the **shoulder girdle**, which is composed of two bones: the **clavicle** anteriorly and the **scapula** posteriorly. All primates have a clavicle, in contrast to many other mammals – particularly fast, terrestrial runners such as dogs, cats, horses, and antelopes, which have lost this bone. The clavicle is one of the primitive skeletal characteristics of primates. This small S-shaped bone, attached to the sternum anteriorly and to the scapula posteriorly, provides the only bony connection between the upper limb and the trunk.

The flat, triangular scapula is attached to the thoracic wall only by several broad muscles. It articulates with the single bone of the upper arm, the **humerus**, by a very mobile ball-and-socket joint. Most of the large propulsive muscles of the upper limb originate on the chest wall or the scapula and insert on the humerus. The muscles responsible for flexing and extending the elbow originate on the humerus (or just above, on the scapula) and insert on the forearm bones.

There are two forearm bones that articulate with the humerus: the **radius**, on the lateral or thumb side, and the **ulna** on the medial side. The elbow joint is a complex region where these three bones meet. The articulation between the ulna and the humerus is a hinge joint that functions as a simple lever. The radius forms a more complex joint; this rodlike bone not only flexes and extends but also rotates about the end of the humerus. There are two articulations between the radius and the ulna, one at the elbow and one at the wrist. Because of its rotational movement, the radius can roll over the ulna. The movement of the radius and ulna is called **pronation** when the hand faces down and **supination** when the hand faces up. The muscles responsible for movements at the wrist and for flexion and extension of the fingers originate on the distal end of the humerus and insert on the two forearm bones. Distally, the radius and the ulna articulate with the bones of the wrist. The radius forms the larger joint between the forearm and the wrist, and in some primates (lorises, humans, and apes) the ulna does not even contact the wrist bones.

Primate hands (Figs. 2.24 and 2.25) are divided into three regions: the **carpus**, or wrist; the **metacarpus**; and the **phalanges** (singular, **phalanx**). The wrist is a complicated region consisting of eight or nine separate bones aligned in two rows. The proximal row articulates with the radius, and the distal row articulates with the metacarpals of the hand. Between the two rows of bones is another composite joint, the **midcarpal joint**, which has considerable mobility in flexion, extension, and rotation.

The five rodlike metacarpals form the skeleton of the palm and articulate distally with the phalanges, or finger

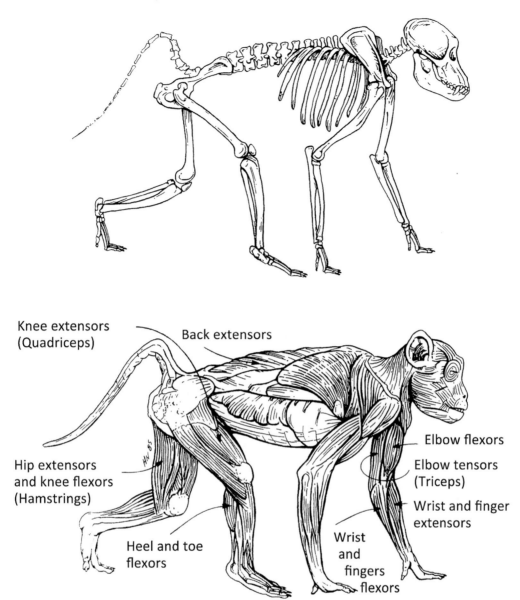

FIGURE 2.22 The skeleton of a baboon (*Papio*) and the superficial limb musculature of the same species showing the major muscle groups responsible for locomotion.

bones, of each digit. The joints at the base of most of the metacarpals are formed by two flat surfaces, offering little mobility, but the joint at the base of the first digit, the **pollex**, or thumb, is more elaborate and allows the more complex movements associated with grasping. The joints between the metacarpal and the proximal phalanx of each finger allow mainly flexion and extension and a small amount of side-to-side movement (**abduction** and **adduction**) for spreading the fingers apart. There are three phalanges (proximal, middle, and distal) for each finger except the thumb, which has only two (proximal and distal). The interphalangeal joints are purely flexion and extension joints.

As just noted, the muscles mainly responsible for flexing and extending the fingers and thumb lie within the forearm and send long tendons into the hand that insert on the middle and distal phalanges. The only muscles that lie completely within the hand are those forming the ball of the thumb, which are responsible for fine movements of that digit, a smaller group forming the other side of the palm, and a series of small muscles within the palm that aid in complex movements of the digits. The **palmar** (relating to the palm) surfaces of primate hands and feet are covered with **friction pads**, a special type of skin covered with **dermatoglyphics** (fingerprints), and sweat glands. In most living primates the

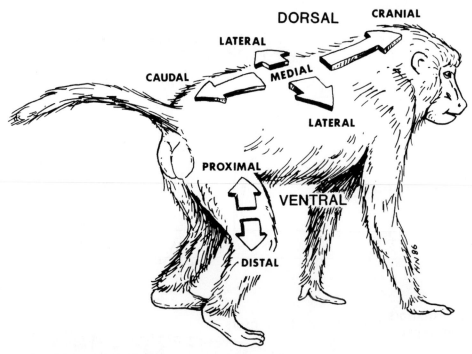

FIGURE 2.23   Terminology for anatomical orientation.

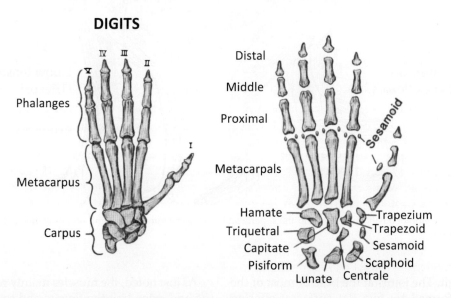

FIGURE 2.24   Dorsal view of the bony skeleton of a baboon's left hand.

tips of the distal phalanges have flattened nails, in contrast to the claws on the digits of most primitive mammals or the hooves of ungulates. A few primates have specialized claws on some of their digits.

Although primate hands usually have approximately the same number of bones, the relative sizes of the hand elements can vary greatly in conjunction with particular needs for locomotion or manipulation (Fig. 2.25). The

slow-climbing loris, for example, has a robust thumb and long lateral digits for grasping branches; the more suspensory, hanging primates, such as the gibbon and spider monkey, have very long, slender fingers. Primates that use their hands for manipulating food, such as mountain gorillas, capuchins, and macaques; or tools, such as humans, have well-developed thumbs that can be opposed to the fingers.

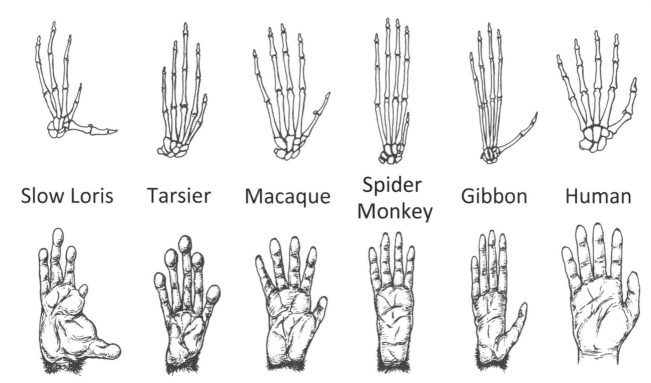

Slow Loris   Tarsier   Macaque   Spider Monkey   Gibbon   Human

FIGURE 2.25   Dorsal views of the left-hand skeleton and palmar views of the right hand of six primate species.

Lower Limb

The primate lower limb, or **hindlimb**, can be divided into four major regions: pelvic girdle, thigh, leg, and foot. These regions are comparable to the shoulder girdle, arm, forearm, and hand of the forelimb.

The primate **pelvic girdle** is composed of three separate bones on each side (the **ilium**, **ischium**, and **pubis**) that fuse to form a single rigid structure, the **bony pelvis**. In contrast to the pectoral girdle, which is quite mobile and loosely connected to the trunk, the pelvic girdle is firmly attached to the backbone through a nearly immobile joint between the sacrum and the paired ilia. The primate pelvis, like that of all mammals, serves many roles. Forming the bottom of the abdominopelvic cavity, the internal part supports and protects the pelvic viscera, including the female reproductive organs, the bladder, and the lower part of the digestive tract. The bony pelvis also forms the birth canal through which the newborn must pass. In conjunction with this requirement, many female primates (including women) have a bony pelvis that is relatively wider than that of males. Finally, the pelvis plays a major role in locomotion: it is the bony link between the trunk and the hindlimb bones, and it is the origin for many large hindlimb muscles that move the lower limb.

The ilium is the largest of the three bones forming the bony pelvis. A long, relatively flat bone in most primates, it lies along the vertebral column and is completely covered with large hip muscles, primarily those responsible for flexing, abducting, and rotating the hip joint. The rodlike ischium lies posterior to the ilium; the **hamstring** muscles responsible for extending the hip joint and flexing the knee arise from its most posterior surface, the **ischial tuberosity**. This tuberosity is the primate sitting bone. In Old World monkeys it is expanded and covered by a tough fatpad, the **ischial callosity**. The pubis lies anterior to the other two bones and gives rise to many of the muscles that adduct the hip joint. The ischium and pubis join inferiorly and surround a large opening, the **obturator foramen**. The relative sizes and shapes of the ilium, ischium, and pubis vary considerably among different primate species in conjunction with different locomotor habits.

The hip joint is a ball-and-socket joint that allows mobility in many directions. The socket part of the bony pelvis, the **acetabulum**, which articulates with the head of the femur, lies at the junction of the three bones.

The single bone of the thigh is the **femur**. The prominent features of this long bone are a round **head** that articulates with the pelvis, the **greater trochanter** where many hip extensors and abductors insert, the **shaft**, and the **distal condyles**, which articulate with the tibia to form the knee joint. Most of the surface of the femur is covered anteriorly by the quadriceps muscles, which are responsible for extension of the knee. Attached to the tendon of this set of muscles is the third bone of the knee, the small **patella**.

Two bones make up the lower leg: the **tibia** medially and the **fibula** laterally. The tibia is larger and participates in the knee joint; distally, it forms the main articulation with the ankle. The fibula is a slender, splintlike bone that articulates with the tibia both above and below, and also forms the lateral side of the ankle joint. In tarsiers, the tibia and fibula are fused to form a single element in the distal half of the leg, and in some small prosimians and New World monkeys, much of the distal part of the fibula is pressed against the shaft of the tibia. Arising from the surfaces of the tibia and fibula (and also from the distalmost part of the femur) are the large muscles responsible for movements at the ankle, and those that flex and extend the toes during grasping.

Like the hand, the primate foot (Figs. 2.26 and 2.27) is made up of three parts: **tarsus**, **metatarsus**, and **phalanges**. The most proximal two tarsal bones are part of the ankle – the **talus** above and the **calcaneus** below. The head of the talus articulates with the **navicular bone**. This boat-shaped bone articulates with three small **cuneiform bones**, which in turn articulate with the first three metatarsals. The body of the talus sits roughly on the center of the calcaneus, the largest of the tarsal bones. The tuberosity of the calcaneus extends well posterior of the rest of the ankle and forms the heel process, to which the Achilles tendon from the calf muscles attaches. This process acts as a lever for the entire foot. Anteriorly, the calcaneus articulates with the **cuboid**, which in turn articulates with the metatarsals of the fourth and fifth digits.

In nonhuman primates, the digits of the foot resemble those of the hand (Figs. 2.24 and 2.25). Each of the four lateral digits has a long metatarsal with a flat base and a rounded head, followed by three phalanges. The shorter first digit, the **hallux**, is opposable, like the thumb, and has a mobile joint at its base for grasping. Primate feet, like primate hands, show considerable differences from species to species in the relative proportions of different elements in association with different locomotor abilities. The climbing loris has a grasping foot, the tarsier has a very long ankle region for rapid leaping, and the suspensory gibbon and spider monkey have long, slender digits for hanging. With their short phalanges and lack of an opposable hallux, human feet are propulsive levers most suitable for walking and running on flat surfaces. The feet of primates, like our hands, have friction pads on the sole and digits and nails (or claws) on the distal phalanges.

Claws and Nails

As a group, primates differ from other mammals in that most have flat nails on the distal phalanges rather than claws or hooves. However, some primates also show modifications of the nails for different purposes. The type of nail is reflected in the shape of the distal phalanx (Maiolino et al., 2011, 2012). The distal phalanx underlying grooming "claws" (which are not really claws) differs from that underlying the claws of

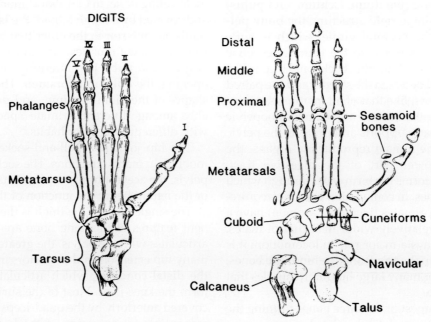

FIGURE 2.26   Dorsal view of the skeleton of a baboon's left foot.

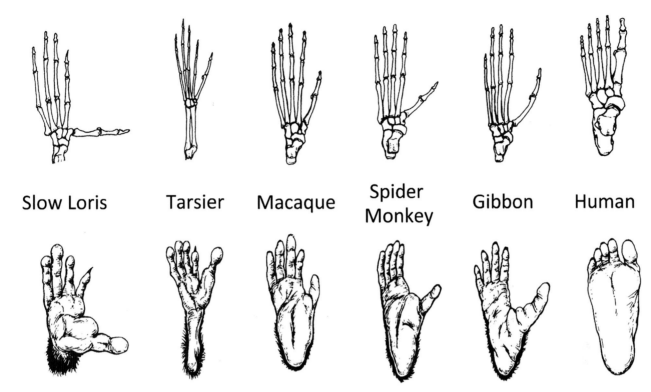

FIGURE 2.27   Dorsal views of the left foot skeleton and plantar views of the right foot of six primate species.

nonprimates in being relatively flatter and, most significantly, in being angled dorsally compared to the other digits. The clawlike structures (**tegulae**) found in marmosets and tamarins resemble true claws in being narrow, but the distal phalanx is much shallower (Fig. 2.28).

Limb Proportions

Primates vary dramatically in their overall body proportions. Some species have forelimbs longer than hindlimbs; others have hindlimbs longer than forelimbs. Some have limbs relatively long for the length of their trunk; others have relatively short limbs. These proportional differences are often described by a **limb index**, a ratio of the length of one part to the length of another part of the same animal. Table 2.1 gives the formula for some of the most commonly used indices. Of these, the **intermembral index**, a ratio of forelimb length to hindlimb length, is especially useful for describing the body proportions of a species, and also seems to be correlated with locomotor differences in many primates (see Chapter 9). In general, leapers have a low intermembral index (longer hindlimbs), suspensory species have a high intermembral index (longer forelimbs), and quadrupedal species have intermediate indices (forelimbs and hindlimbs similar in size).

Soft Tissues

Primates are composed of more than just bones and teeth, but these are the parts usually preserved in the fossil record and in most museum collections of extant primate species. For extinct species, our knowledge of other aspects of anatomy must be based on inferences derived from our knowledge of the relationships between bony anatomy and the softer structures associated with that bony anatomy. For example, we can often reconstruct details of muscular attachments in extinct species from scars on bones. However, for understanding the adaptations and phylogenetic relationships of living primates, details of "soft" anatomy are often very important (Gibbs et al., 2002; Diogo and Wood, 2011). In addition to primate musculature, several other primate organ systems have been well studied and provide insight into the evolution and adaptations of living primates.

Digestive System

Earlier in this chapter, we discussed the first part of the digestive system: the dentition and structures of the oral cavity. These cranial parts are involved in ingestion and the initial mechanical and chemical preparation of food items. The remainder of the digestive system

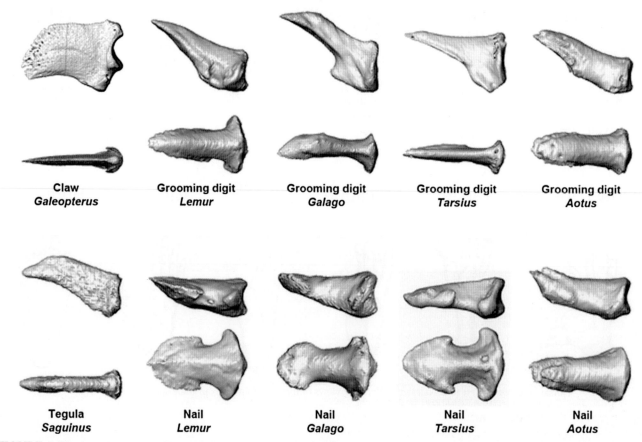

FIGURE 2.28 Morphological differences in the distal phalanx of digits bearing claws, nails, tegulae, and grooming digits, usually called "claws." Grooming "claws" are distinctive in dorsal elevation of the distal part of the digit and are not true claws. *(From Maiolino, S., Boyer, D.M., Rosenberger, A., 2011. Morphological correlates of the grooming claw in distal phalanges of platyrrhines and other primates: A preliminary study. Anat. Rec. 294, 1975–1990.)*

TABLE 2.1   Skeletal proportions.

Intermembral index	$\dfrac{\text{Humerus length} + \text{radius length}}{\text{Femur length} + \text{tibia length}} \times 100$
Humerofemoral index	$\dfrac{\text{Humerus length}}{\text{Femur length}} \times 100$
Brachial index	$\dfrac{\text{Radius length}}{\text{Humerus length}} \times 100$
Crural index	$\dfrac{\text{Tibia length}}{\text{Femur length}} \times 100$

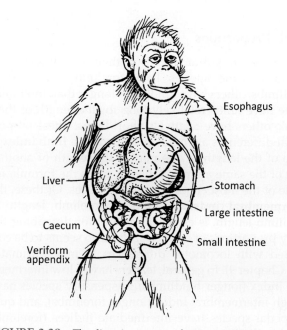

FIGURE 2.29   The digestive system of an orangutan.

(Fig. 2.29) lies primarily in the abdominal cavity and is concerned with further chemical preparation of food, absorption of nutrients, and excretion of wastes. The primate digestive system, like that of all vertebrates, is basically a long tube with some enlarged areas (stomach and large intestine), some coiled loops (small intestine), one cul-de-sac (cecum), and two developmental outgrowths of the digestive tract: the liver and pancreas, which produce various digestive enzymes. Although there is considerable variation among primate species in the relative size, shape, and number of separate parts of individual organs in this system, which are largely associated with their different diets (see Chapter 9), the organs themselves and their functions are relatively similar throughout the order.

After food is prepared in the oral cavity, it is passed through the **esophagus**, a narrow muscular tube that traverses the thoracic cavity, into the abdominal cavity, where it empties into the **stomach**. Here the food undergoes chemical preparation by digestive juices. The most specialized primate stomachs are those of the colobine monkeys; in these primates this organ is divided into several sections that function as fermenting chambers in which colonies of microorganisms break down cellulose (see Chapter 9).

From the stomach, the food passes to the **small intestine**, where further chemical preparation takes place. Here, digestive juices from the liver and pancreas are mixed with the food, and much of the nutrient absorption takes place in this part of the gut. The small intestine is normally the longest part of the digestive tract. In most primates, it is several times as long as the animal's body and is usually folded into a series of loops within the abdominal cavity. At the end of the small intestine, the unabsorbed food and wastes are passed to the large intestine.

The **large intestine** is larger in diameter than the small intestine but usually shorter. It is involved primarily with further absorption of nutrients and water and, in its final parts, with excretion of solid wastes. At the beginning of the large intestine is the **cecum**, a cul-de-sac that varies considerably in size among different primate species and serves several special digestive functions. Like the colobine stomach, this out-of-the-way segment of the digestive tract is an ideal place for harboring the microorganisms that break down food items that primates cannot normally digest, such as leaves or gums. The remainder of the large intestine, the **colon**, is usually divided into several parts on the basis of position within the abdominal cavity. From this last part of the large intestine, solid wastes leave the body through the **rectum** and **anus**.

Many of the adaptive differences in the digestive systems of living primates are discussed further in Chapter 9. It is worth noting here, however, that different groups of primates have evolved quite different visceral adaptations for similar digestive functions. Leaf-eating colobines, for example, have evolved an enlarged stomach for digesting leaves, whereas leaf-eating primates of Madagascar have evolved an enlarged colon. Like all parts of primate anatomy, the digestive system often reflects the interaction of evolutionary history and adaptation.

Reproductive System

All primates have a characteristically mammalian reproductive system, in which the egg is fertilized internally and the embryo develops within the female's uterus for many months before it is born. This basic mammalian pattern of extensive investment by the mother during development, and of infant nourishment for months or years after birth, has important implications for the evolution of primate social behavior (discussed in Chapter 3). In this chapter we briefly review the anatomy underlying primate reproduction.

The anatomical structures associated with primate reproduction are similar to those found in other mammals (Fig. 2.30). Like other mammals, male primates have paired **testicles** that normally lie suspended in a pouch, the **scrotum**, at the lower end of the anterior abdominal wall. Male primates differ from species to species in the position of the scrotum, which is usually behind the penis but may be in front, and in the timing of the descent of the testes from their fetal position within the abdomen to the reproductive position in the scrotum. There are also considerable differences in the relative size of the testes, which are related to mating systems (see Chapter 9), and in the size and external appearance of the penis. In most nonhuman primates there is a bone in the penis, the **baculum**.

Like other mammals, female primates have paired **ovaries** and paired **fallopian** – or **uterine** – **tubes** extending laterally toward the ovaries from the midline uterus (Fig. 2.30). There is considerable variation among primate species in the relative size of the fallopian tubes and the body of the uterus. Among lemurs and lorises, the fallopian tubes are large relative to the body of the uterus, a condition normally found among mammals that have multiple births. Among tarsiers, monkeys, apes, and humans, the fallopian tubes are relatively slender and the body of the uterus is much larger, the condition normally found among mammals that give birth to single offspring.

The **vagina** lies below the uterus and opens onto the **perineum**, where the external genitalia are found. The external genitals of female primates generally consist of two sets of **labia**, on either side of the vaginal opening, and the **clitoris**, anterior to the vagina. The clitoris of female primates varies in size and shape: in some species it is small and hidden beneath a hood; in others it is large and pendulous, in some cases larger than the male's penis. In addition, many female primates have areas of sexual skin surrounding the external genitalia that change color and size during the sexual cycle. In some species, such as baboons and chimpanzees, these sexual swellings are extremely large and provide a rather spectacular advertisement of an individual's reproductive condition.

Primates vary considerably in the periodicity of their reproductive physiology. At the extremes are Malagasy lemurs, in which reproductive activity in both males and

FIGURE 2.30   The male and female reproductive organs in gorillas.

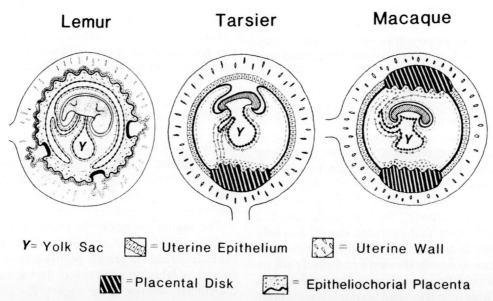

FIGURE 2.31   Fetal membranes in three primates. In the lemur, several layers of tissue separate the uterus of the mother from the diffuse epitheliochorial membrane of the fetus. In the tarsier and the macaque, the developing embryo forms one or two placental disks that invade the lining of the uterus to become embedded in the uterine wall, providing a more intimate interchange between fetal and maternal circulation.

females is limited to one to a few days per year, and most higher primates, in which male sperm production seems to be relatively constant throughout the year and female ovulation occurs regularly at approximately monthly intervals. There are also numerous intermediate species in which both male and female reproductive activity (sperm production and ovulation) is limited to one or two seasons each year, often in response to environmental cues such as food availability and day length.

There are considerable differences among living primate species in the form of the placenta and other structures associated with the developing fetus within the mother's womb (Fig. 2.31). In most lemurs and lorises, the placental membranes are spread diffusely throughout the uterine cavity, and fetal circulation is separated from maternal circulation by several tissue layers: this is an **epitheliochorial placenta**. In tarsiers and in all higher primates the placenta is localized into one or two discrete disks, and there is a much closer approximation between fetal and maternal blood supplies: this is a **hemochorial placenta**. In great apes and humans, the intimacy of fetal and maternal circulation reaches its greatest degree and provides the most efficient transfer of nutrients to the fetus.

Growth and Development

In this chapter we have discussed primate bodies as if they were fixed entities. Obviously, our bodies are constantly changing. During the early part of an individual's lifespan, it undergoes both an increase in size (**growth**) and changes in shape and function of many organs (**development**). Together, these changes are referred to as **ontogeny**. Compared to most other mammals of a similar size, primates are characterized by slow metabolic rates and long periods of growth and development (Pontzer et al., 2014). The extremes are the great apes and humans, in which individuals reach adult size and become sexually mature only after 10 or more years of growth (Fig. 2.32). Primates

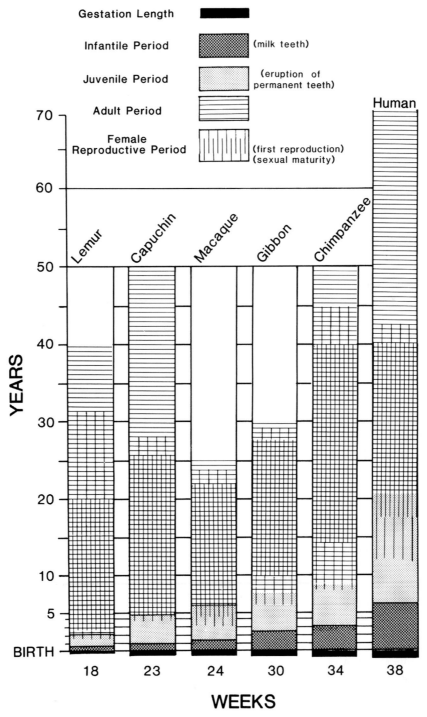

FIGURE 2.32 The timing of life history events in several primates. (*Modified from Schultz, A.H., 1956. Postembryonic age changes. Primatologia 1, 887–964.*)

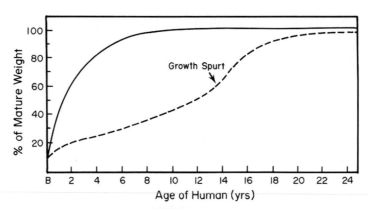

FIGURE 2.33   A human growth curve (*dashed line*) and a generalized growth curve of a nonprimate mammal (*solid line*). In humans and other primates, there is a long period of slow childhood growth followed by the adolescent growth spurt. In contrast, most animals have a growth curve that decreases in rate from birth onward. (*Modified from Watts, E.S., 1986. Skeletal development. In: Dukelow, W.R., Erwin, J. (Eds.), Comparative Primate Biology. vol. 3. Reproduction and Development, Alan R. Liss, New York, pp. 415–439*).

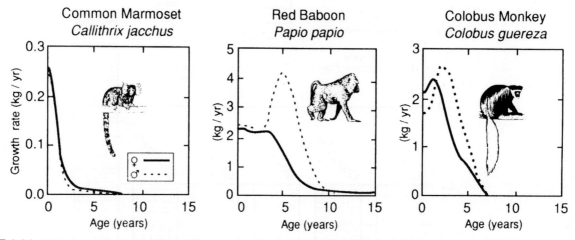

FIGURE 2.34   Primate species show striking differences in patterns of somatic growth, both between species and between sexes of a single species. Marmosets show no adolescent growth spurt; baboons show considerable dimorphism in the growth spurt; and colobus monkeys show a spurt in both sexes. (*Courtesy of S. Leigh.*)

differ from many other mammals in that the rapid growth during infancy is followed by a long childhood in which growth is relatively slow; just prior to sexual maturity there is a phase of rapid growth called the **adolescent growth spurt** (Fig. 2.33). Although many primates have an adolescent growth spurt, the timing and magnitude vary considerably both among different species and between males and females. Indeed, differences in the timing and rates of growth are often responsible for different patterns of sexual dimorphism (Fig. 2.34).

Systematically collected data on primate growth and development are available for only a few species, and most of our knowledge of primate growth and development comes from isolated, anecdotal observations and the many works of Adolph Schultz (1956). Although information on growth in body weight is available for many species,

skeletal maturation has been studied in detail in many fewer taxa.

In contrast, data on the development and eruption of the dentition has been studied in many primates (Fig. 2.35). In general, larger, longer-lived species are characterized by later eruption of the last two molar teeth, and humans are unusual in the early eruption of their canine teeth. These differences have been important in providing clues to the timing of major events such as gestation, weaning, and first reproduction in the lives of extinct taxa, including human ancestors.

Aspects of growth, development, and aging are collectively referred to as the **life history** of a species. As we discuss in the next chapter, these maturational and reproductive characteristics vary considerably among primates and can be related to ecological differences among species.

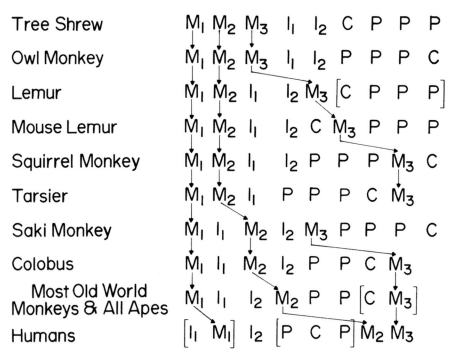

FIGURE 2.35 Dental eruption sequences in a variety of primates. Arrows highlight eruption of the molar teeth to illustrate the sequence differences between general. C, Canine; I, incisor; M, molar; P, premolar. *(Redrawn and modified from Schultz, A.H., 1960. Age changes in primates and their modification in man. In: Tanner, J. M. (Ed.), Human Growth. Pergamon Press, New York, pp. 1–20.)*

References

Alport, L.J., 2009. Lingual fungiform papillae and the evolution of the primate gustatory system. PhD dissertation. University of Texas, Austin.

Boyer, D.M., Kirk, E.C., Silcox, M.T., et al., 2016. Internal carotid arterial canal size and scaling in Euarchonta: Re-assessing implications for arterial patency and phylogenetic relationships in early primates. J. Hum. Evol. 97, 123–144.

Coleman, M.N., Colbert, M.W., 2010. Correlations between auditory structures and hearing sensitivity in non-human primates. J. Morphol. 271 (5), 511–532.

Diogo, R., Wood, B., 2011. Soft-tissue anatomy of the primates: Phylogenetic analyses based on the muscles of the head, neck, pectoral region and upper limb, with notes on the evolution of these muscles. J. Anat. 219, 273–359.

Dobson, S.D., 2009. Socioecological correlates of facial mobility in non-human anthropoids. Am. J. Phys. Anthropol. 139, 413–420.

Dobson, S.D., 2012. Coevolution of facial expression and social tolerance in macaques. Am. J. Primatol. 74, 229–235.

Dominy, N.J., 2001. The sensory ecology of primate food perception. Evol. Anthropol. 10, 171–186.

Fleagle, J.G., Gilbert, C.C., Baden, A.L., 2010. Primate cranial diversity. Am. J. Phys. Anthropol. 142, 565–578.

Fleagle, J.G., Gilbert, C.C., Baden, A.L., 2015. Cranial diversity of mammals: past and present. J. Vert. Paleontol. 35 (Suppl), 124.

Fleagle, J.G., Gilbert, C.C., Baden, A.L., 2016. Comparing primate crania: The importance of fossils. Am. J. Phys. Anthropol. 161, 259–275.

Garrett, E.C., Dennis, J.C., Bhatnagar, K.P., et al., 2013. The vomeronasal complex of nocturnal strepsirhines and implications for the ancestral condition in primates. Anat. Rec. 296, 1881–1894.

Gibbs, S., Collard, M., Wood, B.A., 2002. Soft-tissue anatomy of the extant hominoids: A review and phylogenetic analysis. J. Anat. 200, 3–49.

Heesy, C.P., Ross, C.F., Demes, B., 2005. Oculomotor stability and the functions of the postorbital bar and septum. In: Ravosa, M., Dagosto, M. (Eds.), Primate Origins: Adaptations and Evolution. Kluwer Academic/Plenum Publishers, New York.

Hladik, C.M., Simmen, B., 1997. Taste perception and feeding behavior in nonhuman primates and human populations. Evol. Anthropol. 5, 58–71.

Hofer, H., 1981. Microscopic anatomy of the apical part of the tongue of *Lemur fulvus* (primates, Lemuriformes). Gegenbaurs morph. Jahrb. 127, 343–363.

Huber, E., 1931. Evolution of facial musculature and cutaneous field of trigeminus. Pt. II. Q. Rev. Biol. 5, 389–437.

Jungers, W.L., 1985. Size and Scaling in Primate Biology. Plenum Press, New York.

Kay, R.F., Kirk, E.C., 2000. Osteological evidence for the evolution of activity pattern and visual acuity in primates. Am. J. Phys. Anthropol. 113, 235–262.

Kirk, E.C., 2006. Effects of activity pattern on eye size and orbital aperture size in primates. J. Hum. Evol. 51, 159–170.

Maiolino, S., Boyer, D.M., Rosenberger, A., 2011. Morphological correlates of the grooming claw in distal phalanges of platyrrhines and other primates: A preliminary study. Anat. Rec. 294, 1975–1990.

Maiolino, S., Boyer, D.M., Bloch, J.I., et al., 2012. Evidence for a grooming claw in a North American adapiform primate: Implications for anthropoid origins. PLoS One 7, e29135.

Melin, A.D., Moritz, G.L., Fosbury, R.E., et al., 2012. Why aye-ayes see blue. Am. J. Primatol. 74, 185–192.

Muchlinski, M.N., Docherty, B.A., Alport, L.J., Burrows, A.M., Smith, T.D., Paesani, S.M., 2011. Behavioral and ecological consequences of sex-based differences in gustatory anatomy in *Cebus apella*. Anat. Rec. 294, 2179–2192.

Overdorff, D., 1992. Differential patterns in flower feeding by *Eulemur fulvus rufus* and *Eulemur rubriventer* in Madagascar. Am. J. Primatol. 28, 191–204.

Perry, G.H., Martin, R.D., Verrelli, B.C., 2007. Signatures of functional constraint at aye-aye opsin genes: The potential advantage of adaptive color vision in a nocturnal primate. Mol. Biol. Evol. 24, 1963–1970.

Pontzer, H., Raichlen, D.A., Gordon, A.D., et al., 2014. Primate energy expenditure and life history. Proc. Natl. Acad. Sci. USA 111, 1433–1437.

Ross, C.F., Kirk, E.C., 2007. Evolution of eye size and shape in primates. J. Hum. Evol. 52, 294–313.

Rossie, J.B., Smith, T.D., 2007. Ontogeny of the nasolacrimal duct in Haplorhini (Mammalia: Primates): Functional and phylogenetic significance. J. Anat. 210, 195–208.

Schultz, A.H., 1956. Postembryonic age changes. Primatologia 1, 887–964.

Schultz, A.H., 1960. Age changes in primates and their modification in man. In: Tanner, J. M. (Ed.), Human growth. Pergamon Press, New York, pp. 1–20.

Watts, E.S., 1986. Skeletal development. In: Dukelow, W.R., Erwin, J. (Eds.), Comparative Primate Biology. vol. 3. Reproduction and Development. Alan R. Liss, New York, pp. 415–439.

Further Reading

Aiello, L., Dean, C., 1992. Human Evolutionary Anatomy. Academic Press, London.

Ankel-Simons, F., 2007. Primate Anatomy, third ed. Academic Press, San Diego.

Coleman, M.N., Colbert, M.W., 2010. Correlations between auditory structures and hearing sensitivity in non-human primates. J. Morphol. 271 (5), 511–532.

Diogo, R., Wood, B.A., 2012. Comparative Anatomy and Phylogeny of Primate Muscles and Human Evolution. CRC Press, Jersey, UK.

Gebo, D.L. (Ed.), 1993. Postcranial Adaptation in Nonhuman Primates. Northern Illinois Press, DeKalb.

Gebo, D.L., 2014. Baltimore Primate Comparative Anatomy. Johns Hopkins University Press.

Gregory, W.K., 1922. Origin and Evolution of the Human Dentition. Williams and Wilkins, Baltimore.

Hartman, C.G., Straus Jr., W.L., 1933. The Anatomy of the Rhesus Monkey. Williams and Wilkins, Baltimore.

Hill, W.C.O., 1953–1970. Primates, vol. 1–8. Edinburgh University Press, Edinburgh.

Hofer, H., Schultz, A.H., Stark, D., 1956–1973. Primatologia, vols. 1–4. Karger, Basel, Switzerland.

Martin, R.D., 1990. Primate Origins and Evolution. Princeton University Press, Princeton, N.J.

Plavcan, J.M., Kay, R.F., Jungers, W.L., van Schaik, C.P., 2002. Reconstructing Behavior in the Primate Fossil Record. Kluwer Academic/Plenum Publishers, New York.

Swindler, D.R., 2002. Primate Dentition: An Introduction to the Teeth of Non-human Primates. Cambridge University Press, Cambridge.

Ungar, P.S., 2010. Mammal Teeth: Origin, Evolution and Diversity. Johns Hopkins University Press, Baltimore.

3

Primate Lives

In the previous chapter, we discussed the physical characteristics of primates. The purpose of this chapter is to enliven the primate body by introducing the basic features of primate behavior, ecology, and life history – where primates live, what they eat, how they move, and how they organize their social life and lives in general. In later chapters we see how these parameters vary from species to species; in this one we introduce terminology and general principles.

Primate Habitats

Nonhuman primates today are found naturally on five of the seven continents (Fig. 3.1). There are no living primates other than humans on either Antarctica or Australia, and no evidence that primates ever inhabited either continent before the relatively recent arrival of humans. Although nonhuman primates occupy only marginal areas of Europe (Gibraltar) and North America (Central America and southern Mexico), they were formerly much more widespread on both continents. For the present, however, Africa, Asia, South America, and their nearby islands are the home of most living nonhuman primates.

A few hardy primate species live in temperate areas where the winters are cold, such as Nepal and Japan, but these are exceptional. The vast majority of primate species and individuals are found in tropical climates, where daily fluctuations in temperature between day and night far exceed the changes in average temperature from season to season. In these climates, seasonal changes in

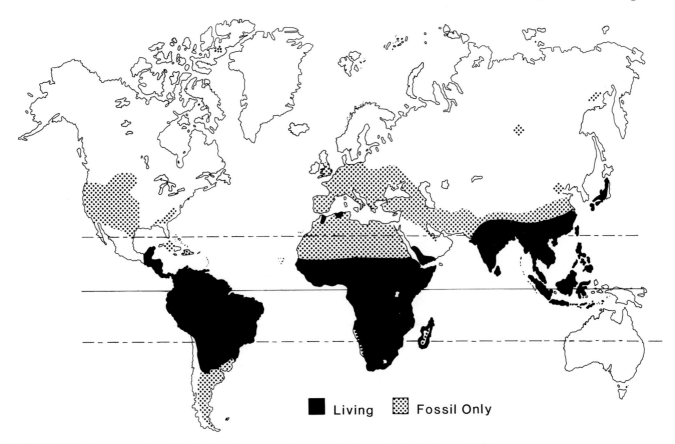

FIGURE 3.1 The geographical distribution of extant nonhuman primates and extinct primate species.

rainfall have a much greater effect on the vegetation and on the primates than do any seasonal differences in temperature or day length.

Forest Habitats

Within their geographic range, living primates are found in a variety of habitats, ranging from deserts to tropical rainforests. Only a few hardy types such as chimpanzees, baboons, and Senegal bushbabies manage to successfully ply their primate trade year after year in drier, more poorly vegetated areas. The majority of primate species and individuals live in tropical forests of one sort or another. The forests come in many shapes, with variations in climate, altitude, topography, and soil type as well as in the characteristic flora and fauna. A few of the more distinctive forest types are illustrated in Fig. 3.2.

Primary rainforests are usually characterized by the height of the trees (up to 80 m) and the relatively continuous canopy that results from intense competition between many tree species for access to light. The dark understories of primary rainforests, which are usually quite open, are made up primarily of trunks and vines. The canopies of these rainforests are punctuated by occasional emergent trees, which stand above the rest, and by gaps resulting from tree falls. It is through these gaps that light reaches the forest floor, enabling the forests to renew themselves.

Secondary forests, like the areas around tree falls, are characterized by denser, more continuous vegetation because of increased availability of light. The canopy

structure is less distinct and is often characterized by an abundance of vines and short trees. Because of the high light levels in secondary forests, leaves and fruit can be very abundant.

African **woodlands** are made up of relatively shorter, often deciduous trees. Between individual trees are continuous growths of grasses and low bushes. As the trees become sparser, woodland gives way to **bushlands**, **scrub forests**, and **savannah**.

In many relatively dry tropical regions, forests are concentrated around rivers. These **gallery forests** can contrast strongly with surrounding areas in the types of animals they support.

There are other ways of categorizing forests. We find highland rainforests and lowland rainforests, as well as swamp forests, montane forests, and bamboo forests. Each of these environments presents a primate with a different array of substrates on which to move, different places to sleep, and different things to eat from season to season. The primates that inhabit these forests must meet these different demands. Many of the behavioral differences among living primate populations reflect adaptations to this diversity of habitats.

Habitats Within the Forest

Equally as diverse as the types of forests primates inhabit are the different niches primate species may occupy within a single forest (Fig. 3.3). In a tropical forest, which may reach heights of 80 m or more, the temperature and

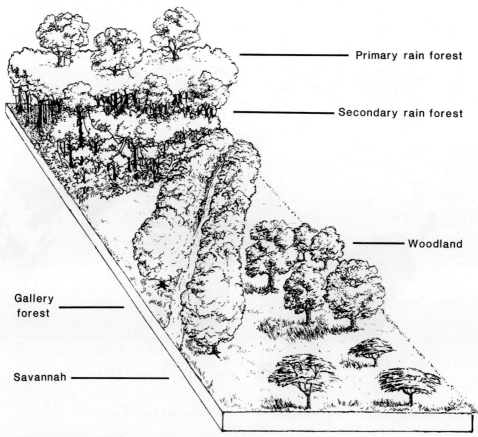

FIGURE 3.2 The diversity of habitats occupied by extant primates.

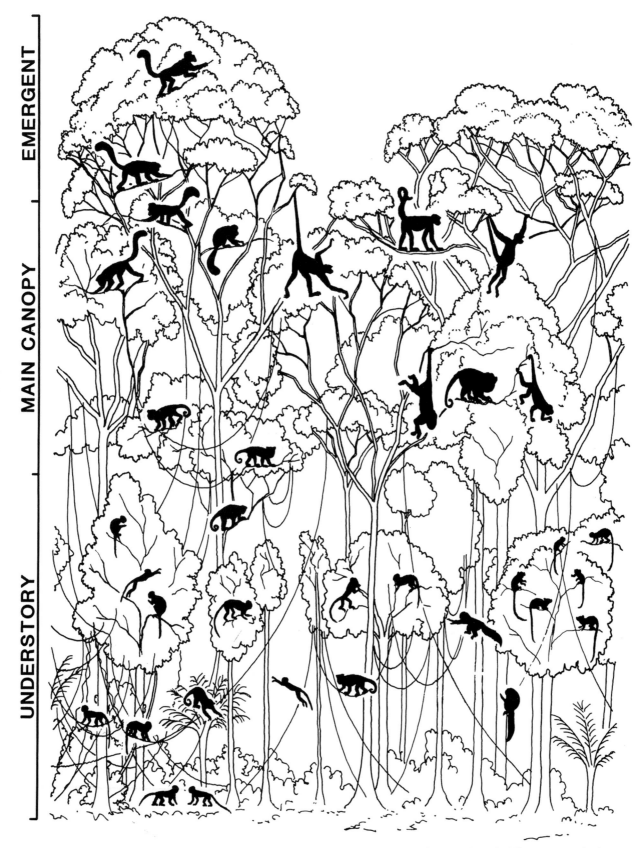

FIGURE 3.3 A rainforest scene from Suriname showing the different levels of a tropical forest, each with different types of substrate and each occupied by different primate species.

EMERGENT

MAIN CANOPY

UNDERSTORY

humidity, the shapes of the branches, the kinds of plant foods, and the types of other animals a species encounters are usually quite different on the ground, 20m above the ground, or 40m up in the canopy. Near the ground, there is little light, there are many vertical supports (such as small lianas and young trees), and there are terrestrial predators. Higher in the canopy, there are more horizontally continuous supports that provide convenient highways for arboreal travel, and a greater abundance of leaves and fruits. Still higher, in the emergent layer, the canopy again becomes discontinuous, heat from the sun may be quite intense, and individuals are exposed to aerial predators. Primates and many other arboreal animals, often move and feed in specific forest levels and show adaptations to these different demands and opportunities.

Primates also specialize on different types of trees within a forest – trees that may have distinctive structures or produce foods with unique characteristics. Some species rely on bamboo patches within the forest, others on palms, and still others on vines. Primates often seem to specialize on trees with characteristic sizes and productivity; some species seem to feed primarily on small trees that produce small quantities of fruit, whereas others concentrate on the forest giants that produce huge bonanzas of fruit during seasonal punctuations of productivity. In sum, there are many different **niches** within any single forest habitat, each of which offers a slightly different way for primates to make a living.

Primates in Tropical Ecosystems

The tropical forests inhabited by most primates are the most complex ecosystems on earth, often containing thousands of plant species, hundreds of vertebrates, and innumerable insects and other invertebrates. It is important to remember that the evolution of living primates has taken place in conjunction with the evolution of other members of these complex environments. Plants, for example, are not the passive structural elements of the forest they might appear to be. Natural selection in plants has led to the evolution of elaborate and complex mechanisms for obtaining the necessary resources of light and nutrients, including defending leaves from herbivores, attracting pollinators, and ensuring that seeds are adequately dispersed and prepared for germination. The brightly colored, sweet, juicy fruits that form the diet of many primates have probably evolved those attributes for the purpose of attracting primates and other bird and mammal dispersers: once the fruits have been eaten, the seeds they contain will be scattered about the forest. At the same time that many plants have evolved ways of enticing animals to help them disperse their seeds, they have also evolved mechanisms to protect their leaves and immature fruits and seeds from predators: they may cover them with spines, for example, fill them with indigestible materials or toxic substances, or encase them in

hard shells. Primates, in turn, have evolved ways to overcome many of these plant defenses. Thus, the dietary and foraging adaptations of living primates have evolved hand in hand with features in the tropical plants that affect their dietary choices.

In their roles as competitors, predators, and prey, the other animals of the forest have also had an important influence on the evolutionary history of primates. In Manu National Park, a pristine rainforest environment in the upper Amazon basin of Peru, **frugivorous** (fruit-eating) monkeys account for only about one-third of the biomass of frugivorous vertebrates (Terborgh, 1983). Birds, bats, various carnivores, and numerous rodents eat many of the same fruits as the primates and are often found in the same trees at the same time. There has certainly been competition among these different animals for access to the various food items in the forest.

Many of the animals that inhabit the same forests as primates are interested not in the monkeys' food but in the monkeys as food. Large felids (including lions, tigers, leopards, jaguars, and pumas) prey on primates, as do many large birds and snakes (Isbell, 2009). The presence of predators has exerted an important influence on the evolution of many aspects of primate ecology and behavior, including activity patterns, social organization, choice of sleeping sites, vocalizations, and coloration patterns.

Only a few primates – the slender lorises and tarsiers of Asia – seem to rely exclusively on other animals for their food; but many primates, especially the smaller species, include various invertebrates and small vertebrates, such as lizards or birds, as a regular part of their diet. Larger species such as chimpanzees prey on larger vertebrates, including other primates. As we shall see, primate species have evolved a number of unique predation strategies to exploit different types of prey in distinct parts of the forest structure. Capturing flying insects requires keen eyes and quick hands, and locating cryptic insects that live beneath the bark of trees or in leaf litter requires a keen sense of smell or hearing. Often, such cryptic prey can be reached only by gnawing through the bark with specialized teeth, by ripping it open with strong hands, or by probing in crevices with slender fingers. Again, the evolution of primate adaptations reflects an interaction with the evolutionary history of other organisms in the forest.

Habitat Use

Primates live in a complex environment with many constantly changing variables. One way in which groups of primates deal with this complexity is to restrict their activities to a limited area of forest that they know well. Thus, we find that primates are very conscious of real estate. In contrast to many birds or other mammals that have seasonal migrations, most primates spend their days, years, and often most of their entire lives in a single, relatively restricted area of forest. To exploit this area effectively, they must know many things about it – the different food trees and their seasonal cycles; the best pathways for moving; the best water sources; and the safest places to sleep. Many researchers have suggested that it is this need for knowledge of their environment that is responsible for the evolution of primate mental abilities. Nevertheless, although most researchers are convinced that primate groups have a very detailed knowledge of the distribution of resources in their area of forest, we still know very little about what type of **mental map** they may have of the geographical and temporal patterning of food resources, or how this information is maintained or communicated (Di Fiore and Suarez, 2007; Janson and Byrne, 2007; Erhart, 2008).

There is a standard terminology that is used to describe the normal patterns of land use by primates and other animals (Fig. 3.4). The distance an individual or

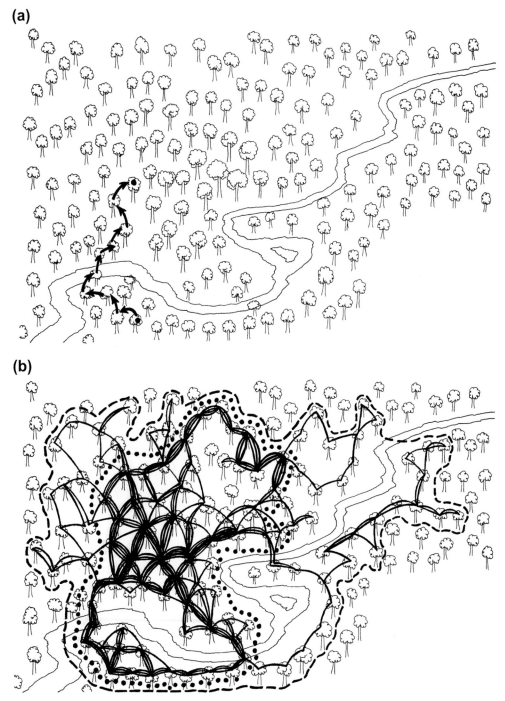

FIGURE 3.4   Primate land use: (a) The path an individual or group travels in a day is called a **day range** or **daily path length**. (b) If all day ranges are combined, the total area utilized by the group is its **home range** (*dashed line*). The part of the home range that is most heavily used is called the **core area** (*dotted lines*).

group moves in a single day (or night) is called a **day range** or **daily path length** (Fig. 3.4a, *arrows*). If we map all the day ranges for a primate group, we can see the total area of land used over a longer period of time, for example, a year. This area is called the **home range** (Fig. 3.4b, *dashed line*). Often, a group uses one part of its home range intensively, with only occasional – usually seasonal – forays into other parts. This heavily used area is called the **core area** (Fig. 3.4b, *dotted line*). Frequently, the home ranges of neighboring groups of the same species overlap. In other instances, there is almost no overlap, and adjacent groups actively defend the boundaries of their home ranges with actual fighting or vocal battles. Such defended areas are called **territories**.

Activity Patterns

Most primates limit their activities to one particular segment of each 24-hour day. Most mammals are **nocturnal**, i.e., they are active primarily at night and sleep during the day. In contrast, most birds are **diurnal**, meaning they are active during daylight hours and sleep when it is dark. Some mammals are **crepuscular**: they are most active in the hours around dawn and dusk, when light levels are low. Nearly three-quarters of living strepsirrhines (lemurs, lorises, and galagos) are nocturnal, as are tarsiers and *Aotus*, the owl monkey. All other haplorhines and many lemurs are diurnal. Many primate species show peaks of activity at dawn and dusk

and have a rest period either at midday (for diurnal species) or at midnight in nocturnal species; however, none seem to be crepuscular. There are also primates with quite variable activity patterns. Rather than being strictly diurnal or nocturnal, they seem to be active at intervals throughout a 24-hour day, an activity pattern that is called **cathemeral** (Tattersall, 1988, 2006). Several lemurs, including several species of brown lemurs and even ring-tailed lemurs, show this activity pattern (Parga, 2011; Donati et al., 2013; LaFleur et al., 2014).

Each of these ways of life has its advantages and disadvantages (Fig. 3.5). Diurnal species presumably have a better view of where they are going, of available food, and of potential mates, friends, competitors, and predators. At the same time, they have a greater risk of being seen by predators. Nocturnal species are better concealed from many predators, except owls, and have fewer direct primate or avian competitors. They avoid heat stress due to sunlight, and may even avoid diurnal parasites. However, they have difficulties in feeding and social communication associated with restricted visual abilities, though their vocal communication may be better during the hours of darkness, and olfactory communication seems to be enhanced by the humid night air (Wright, 1996). Thus, it is not surprising that nocturnal primates tend to live in small groups or alone, and to communicate primarily through smells and sounds. A cathemeral activity pattern enables a species to exploit the advantages of both diurnality and nocturnality in conjunction with changes in temperature or food availability. The

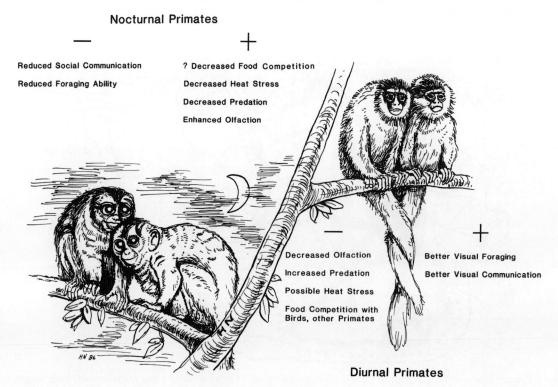

FIGURE 3.5   Potential benefits (+) and costs (–) of diurnality and nocturnality for two New World monkeys: the dusky titi monkey (*Callicebus*) and the owl monkey (*Aotus*).

mongoose lemur, for example, is most active during daylight hours for the part of the year in which it feeds on fruits and new leaves; in the dry season, however, when these food items are scarce, it becomes more active at night and feeds on nectar (see Chapter 4).

A Primate Day

In addition to such drastic differences in lifestyle as diurnality and nocturnality, primates show differences in the way they spend each day or night. For most primates, the day is generally divided among three main activities: feeding, moving, and resting. Activities such as sex, grooming, and territorial displays usually occupy a relatively small part of each day. There are exceptions, of course. During their breeding season, males of some lemur species may spend half of their waking hours engaged in fights with other males for the opportunity to mate with females during their annual period of sexual activity. For most primates, however, these activities are just occasional punctuations in long sequences of resting, feeding, and travel.

The distribution of activities throughout the day, or an animal's **activity budget**, is usually not random (Fig. 3.6). Many diurnal primates generally travel early and late in the day and rest in the middle of the day when temperatures are highest. Most begin and end each day with a long feeding period. Gibbons, and many other species, show a temporal pattern in food preference: they eat fruit in the morning and leaves in the evening. This preference for fruit early in the day reflects

a need for the quick energy available in fruits because of their high sugar and low fiber composition. Their choice of leaves in the evening perhaps reflects an attempt to maximize available digestion time (overnight) and to maximize the various nutrient contents of leaves. Because plants cannot photosynthesize in the dark, the protein level in leaves increases during the morning and the sugar levels increase throughout the day (Ganzhorn and Wright, 1994).

Primate Diets

Variation in the choice of foods on a daily, seasonal, and yearly basis is one of the greatest differences among living primates and one that has far-reaching effects on virtually all aspects of their life and morphology. Primate diets have generally been divided into three main food categories: fruits, leaves, and fauna (usually insects and arachnids). Species that specialize on one of these dietary types are sometimes referred to as **frugivores**, **folivores**, and **insectivores** (or **faunivores**), respectively. These dietary categories accord well with the structural and nutritional characteristics of primate foods, and thus frugivores, folivores, and faunivores have characteristic features of teeth and guts that enable them to process their different diets (see Chapter 9).

In addition to particular nutritional and mechanical features, primate foods may vary considerably in their distribution and availability in both space and time (Fig. 3.7). Foods may be found in small patches that can accommodate one or two small animals, or in large

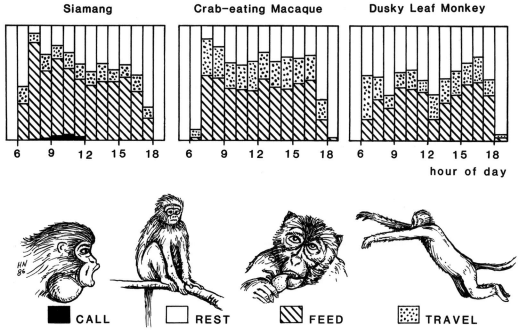

FIGURE 3.6 Primate activity budgets showing the proportion of each hour of the day spent calling, resting, feeding, and traveling by three Asian primates.

Small Patches   Large Patches

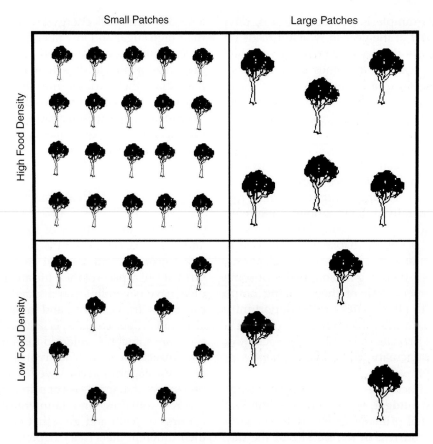

High Food Density

Low Food Density

FIGURE 3.7   Different patterns of spatial distribution of potential food resources.

patches that contain enough food to satiate many individuals of a large species. Similarly, in each habitat, some foods may be distributed either densely or relatively sparsely. In addition to differences in spatial distribution, foods may show many types of temporal availability. Some foods may be available all year in some environments, others may be seasonal, and still others may be available only every 2 or 3 years. The distribution of food items is often correlated with primate grouping and ranging patterns. Folivores, for example, tend to have smaller home ranges and day ranges than frugivores, because leaves are more uniformly distributed in both time and space than are fruits (but see Snaith and Chapman, 2007).

Like any categorization, these gross descriptions gloss over many subtle differences in the types of foods primates eat and the different problems they must overcome to obtain a balanced diet from day to day (Lambert and Rothman, 2015). For example, new leaves and mature leaves often have very different chemical, textural, and nutritional compositions and may be available during different seasons of the year. Some fruits appear in large clumps; others are more evenly scattered in small numbers over a large area. As already noted, flying insects must be hunted differently from burrowing insects. In addition, feeding on foods such as gums (**gummivory**), seeds (**gramnivory**), or nectar (**nectivory**) is an important

aspect of the dietary behavior of many primates and often requires unique adaptations, but such behaviors do not fit easily into these three categories. Finally, there are many different ways of evaluating the importance of different food items in a primate's diet that may yield very different views of its dietary specializations. For example, the food that is eaten most often throughout most of the year may seem to be the most important from an energetic perspective, but other foods – like soil or tree bark (e.g., Rothman et al., 2006) – may contain rare nutritional elements not otherwise available in a primate's diet. In many habitats, food availability may be highly seasonal, and the food that a species eats during the worst part of the year may be the most critical item. These limiting foods are called **fallback foods** (Marshall et al., 2009).

The many, intricate ways in which primates obtain their food are usually referred to as **foraging strategies** – so called because many factors are involved, and the behavior of any species is probably the result of compromises and decisions among an array of potential behaviors, each with unique costs and benefits. Thus, within any one dietary category, such as frugivory, different species may have quite different foraging strategies. One species may specialize on fruits that are regularly available in small amounts throughout the forest, whereas another species may specialize on fruits that are found in more irregularly spaced, but larger, clumps. We would expect two such species to be similar in their dentition and digestive system but to have very different ranging patterns and social dynamics. Many of the descriptions of individual primate species in later parts of this book emphasize the subtle differences in foraging strategies that have been found among primate species within the same general habitat. These subtle differences in feeding habits demonstrate the richness of primate adaptations that have evolved over the past 60 million years.

Locomotion

A major aspect of the foraging strategy of any species, and an aspect of behavior that shows considerable variation among primates, is **locomotion**, or the way animals move. No other order of mammals displays the diversity of locomotor habits seen among primates. Like diet, primate locomotor habits can be crudely divided into several major categories (Fig. 3.8), each characterized by

FIGURE 3.8   Examples of primate locomotor behavior.

different patterns of limb use: leaping, arboreal and terrestrial quadrupedalism, suspensory behavior, and bipedalism. Each of these ways of moving may provide a primate with better access to a particular type of forest structure, or may be more efficient for traveling on a particular type of substrate.

Leaping (**saltation**) allows arboreal species to move between discontinuous supports; for example, between separate trees or between tree trunks in the understory. **Arboreal quadrupedalism** is more suitable for movement on a continuous network of branches and is probably less hazardous than leaping, especially for larger species. **Terrestrial quadrupedalism** enables a primate to move rapidly on the ground. **Suspensory behavior** allows larger species to spread their weight among small supports and also to avoid the problem of balancing their body above a support. Finally, **bipedalism** allows a species to progress on a continuous, level substrate while freeing the hands for other tasks.

As with dietary categories, these locomotor categories do considerable injustice to the actual diversity of primate movements. Some species leap from a vertical clinging position (so-called "**vertical clingers and leapers**"), others from more horizontal supports. Quadrupedal walking and running may involve different gaits in the trees and on the ground. Suspensory behavior includes many different activities, including **brachiation** (swinging by two arms), climbing, and bridging. As with dietary groups, this lumping of behaviors is mainly for the purpose of examining general patterns in either morphology or ecology (see Hunt et al., 1996 for a more detailed classification).

In addition to locomotion, primatologists also pay careful attention to differences in primate postures – the way primates sit, hang, cling, or stand while they obtain their food, rest, or sleep (Fig. 3.9). In many instances, feeding postures may be as important in the evolution of the species as locomotion. Primates that feed on gums or other tree exudates, for example, often must cling to the side of a large trunk. This clinging ability may be more important than the method by which the tree is reached. Likewise, the suspensory locomotion of many primates may be just a byproduct of their need to hang below supports to feed on food sources at the end of small branches. The combination of locomotion and posture is often called **positional behavior**.

Communication

Given the retention of an enhanced olfactory apparatus (Chapter 2), most lemurs, lorises, and galagos communicate heavily through **olfactory cues**. Most strepsirrhines have at least one scent gland, typically in their anogenital region, though some have as many as three or more. Scent marks have been shown to change

seasonally (Scordato et al., 2007), vary individually (Palagi and Dapporto, 2006; Scordato et al., 2007), and advertise genetic quality and relatedness (Charpentier et al., 2008). Scent-marking by female lorises may also help to select males with high competitive ability (Fisher et al., 2003).

By contrast, monkeys and apes typically rely less on olfactory signals, and instead rely on diverse **visual signals** in many aspects of their daily lives. This includes eyelid flashes or yawns to assert dominance; grimaces to concede submission; facial expressions to initiate play; or hand gestures to request food. Moreover, color can be an honest signal of rank, as in the brightly colored faces of male mandrills (Setchell and Dixson, 2001), or fertility, as in the sexual skin coloration of macaques (Dubuc et al., 2009; Higham et al., 2010). Coloration can also be an important mechanism for species mate recognition (e.g., in guenons; Allen et al., 2014).

In most, if not all primates, visual signals are accompanied by **vocal signals** as a means of responding to and relaying information about the world around them. Many pair-living primates, like gibbons and indri, rely on vocal duets to advertise their territories; contact calls help group-living primates to coordinate movement and group cohesion; screams can elicit help from conspecifics during conflicts; and **alarm calls** can be used, both to alert conspecifics to the presence of predators (Seyfarth et al., 1980), and in some cases, to deceive them (e.g., capuchins, Wheeler, 2009). Some primates, like tarsiers, even communicate in the ultrasonic domain (Ramsier et al., 2012).

It is rare that primates rely solely on one mode of communication. Instead, most communicate via multiple signal elements simultaneously, something known as **multimodal signaling** (Higham and Hebets, 2013).

Social Life

The size and composition of the groups in which primates carry on their daily activities and how they interact within their physical and social environments are the most extensively studied aspects of primate behavior and ecology. All primates are social animals: they interact regularly with other members of their species in various ways. However, primate species vary considerably in the size, composition, and cohesion of the groups that they feed, travel, and sleep with on a daily basis.

Social System

A primate society, or **social system**, is a set of conspecifics whose members interact with each other more than with members of other such societies (Struhsaker, 1969), and is an emergent property resulting from three main

FIGURE 3.9 Examples of primate feeding postures.

types of behavioral interactions and strategies: social organization, mating system, and social structure (Hinde, 1976; Kappeler and van Schaik, 2002).

A primate's **social organization** refers to the demographic composition of primate groups – the numbers of males and females – and their spatiotemporal cohesion. Primate social organization differs considerably from species to species. Several distinct types of grouping patterns are particularly common (Fig. 3.10).

The simplest dichotomy in primate social organization is between **solitary** species that normally feed and travel alone and **gregarious** species that normally feed and travel in groups. Many nocturnal species are solitary and most diurnal species and some nocturnal ones are gregarious. However, there are many subtleties and complications in this seemingly simple distinction, because solitary species are still social and maintain a variety of interactions with other "solitary" individuals in the same general area. They may share sleeping places,

either regularly or occasionally; they may occasionally travel together for part of a night; and in almost all cases, they have consistent patterns of home range overlap. For example, among the nocturnal and "solitary" cheirogaleids, there is an extraordinary diversity in patterns of individual ranging and interactions (Fig. 4.6).

The simplest, and certainly most primitive, social grouping among **solitary** primates is the **noyau**, which seems to characterize most primitive nocturnal mammals (Charles-Dominique, 1983). The basic unit of this arrangement is the individual female and her offspring. In the noyau, adult males and females do not form permanent mixed-sex groups: rather, individual males have ranges that overlap several different female ranges. However, even though the two sexes do not travel together regularly, they interact often enough for males to monitor the reproductive status of the females and for females to have a choice of potential mates. This type of spatial arrangement may be associated with several

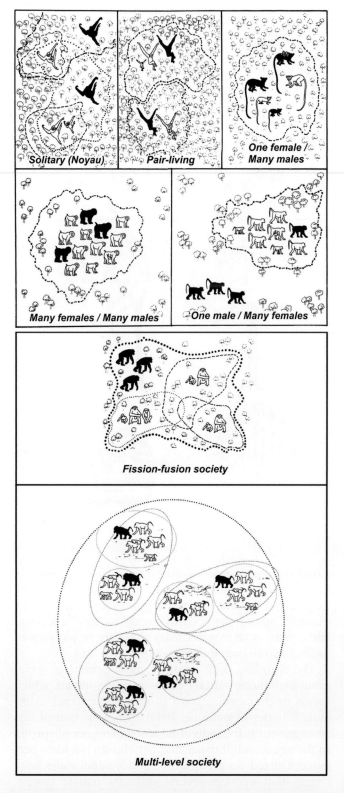

FIGURE 3.10 Common types of primate social grouping patterns. Light figures are females; dark figures are males.

The next simplest grouping, at least in terms of numbers, is **pair-living**, wherein groups consist of one adult female, one male, and their offspring. Nonhuman primates that live in these families often show intense territorial competition between adjacent groups. Most of this competition appears to be intrasexual – males compete to exclude other males and females compete to exclude other females. The role of males in pair-living primate groups may vary considerably. In some species, males play a prominent role in carrying infants (owl monkeys, titi monkeys, siamangs); in other pair-living species, males seem to contribute very little to infant care. However, resident males may help deter infanticide by other males seeking to take over access to the female (van Schaik and Dunbar, 1990). Alternatively, the male's presence may be largely an effort at mate-guarding rather than directly related to infant care (van Schaik and Paul, 1997).

Most other primates live in groups that include three or more adult males and/or females in varying combinations. **Group-living** primates can be further categorized based on their levels of spatiotemporal cohesion (Fig. 3.10). Under naturalistic conditions, marmosets and tamarins (small New World monkeys) and some gibbons live in cohesive groups consisting of a single adult female and several adult males. In these groups, several males, as well as many other group members, may participate in the care of offspring, and some authors have suggested that these groups are better viewed as a "cooperative breeding system" (Garber, 1997; Erb and Porter, 2017). Most gregarious primates, however, live in groups consisting of one or more males and many females.

Many primate species live in cohesive groups consisting of a single adult male along with several females and their offspring. These **one-male groups**, often referred to as one-male units or "OMUs," are often based on **matrilineal** social systems. In some species, adult males not living with females may band together to form separate **all-male "bachelor" groups**, and in other species, they live alone. Although the resident male in the one-male groups of Asian langurs seems to be defending reproductive access to the females in the troop, in some cercopithecine species the mating season is characterized by a large influx of other males. Clearly, there can be considerable diversity in social relationships and reproductive activity within similar demographic patterns.

In contrast to these one-male groups, many species live in large **multimale–multifemale groups** that include several reproductively active adult males, females, and offspring, which can be further distinguished by their degree of spatiotemporal cohesion. Among cohesive multimale–multifemale groups, activities are coordinated and members forage together as a group. Such groups are characterized by complex intratroop politics and competition. The distinction between one-male groups and such multimale–multifemale groups is very

different types of mating systems and patterns of intrasexual competition, depending on the synchrony of female receptivity.

difficult to make for many species of living primates, and can vary among species populations. As the young males mature, many one-male groups seem to become multimale groups. Also, the composition of primate groups within a species may depend on factors such as troop size and population density. Because of this blurred distinction, some authorities have introduced an intermediate type of social group, the **age-graded group**.

Among several primate taxa, multimale–multifemale groups can be more fluid, and the size and composition of groups may vary hourly, daily, or sometimes seasonally. In some species, group or **community** membership is stable, but its members form subgroups that vary in size, composition, and cohesion, depending on the availability of resources and on individual relationships. These groups, traditionally called **fission–fusion societies**, are better described as exhibiting high levels of **fission–fusion social dynamics** (Aureli et al., 2008). Although best recognized in chimpanzees and spider monkeys, experts now recognize fission–fusion dynamics as falling along a continuum, such that even groups that are typically considered cohesive can and do exhibit fission and fusion events. For example, we now know that group-living taxa as diverse as ruffed lemurs, uakaris, howler monkeys, and snub-nosed monkeys all incorporate fission–fusion dynamics, albeit to varying degrees (reviewed in Baden et al., 2016).

Multimale–multifemale groups may also form **multilevel or modular** primate societies, where social organization is spatiotemporally complex and layered (Greuter et al., 2012). In these cases, subunits are stable through time, and vary instead in their patterns of higher-level organization. This social organization is common among three major primate clades: the papionins, Asian colobines, and humans. Hamadryas baboons, for example, forage all day in small one-male units consisting of one adult male, one or a few females, and their offspring. Several one-male units sometimes join to form **clans** led by male-bonded and probably related adult males. Each evening, however, dozens of these small groups congregate on a single sleeping cliff into higher-level units known as **bands**, and sometimes into even larger aggregations of several bands, referred to as **troops** or **herds**. Social interactions occur both within and between the social levels, but relationships are strongest within the first tier, which also typically functions as the reproductive unit (e.g., Swedell and Saunders, 2006).

In still other species, including some colobus monkeys, golden monkeys, and proboscis monkeys, separate groups seem to come together amicably on a regular basis to form **supertroops**, either during foraging or during rest. The organization of these supertroops is not altogether clear. They may represent "family reunions"

of related individuals, or they may represent some type of foraging strategy to maximize individual knowledge of food resources, to preempt competition, or to exploit superabundant resources.

There is often a close, but imperfect relationship between a primate's social organization and its **mating system**, which describes the actual patterns of reproduction within a social group. These distinctions are important because many studies have shown that there is not always an obvious relationship between group membership (social organization) and patterns of reproduction among member of a social group (mating system).

Solitary primates typically use a form of **scramble** or **sperm competition**, where males compete indirectly over access to mates. This form of mating system is typical of orangutans, as well as the small, nocturnal strepsirrhines. In these species, males find and mate with as many females as possible. In these species, most notably the small, nocturnal cheirogaleids, it is common to see a more than seven-fold increase in testicular volume in the months preceding mating (Wrogemann and Zimmermann, 2001).

Other mating systems include **monogamy**, where males and females typically mate with only one member of the opposite sex and have roughly equal variances in reproductive success. Strictly monogamous species include the pair-living owl monkeys (*Aotus*), indri (*Indri*), and red-bellied lemurs (*Eulemur rubriventer*). Monogamy is often accompanied by extensive paternal investment. **Polyandry**, which is characteristic only of the marmosets and tamarins, occurs when one reproductive female mates with several sexually mature males. In these cases, it is common for the dominant female to prevent other females in the group from reproducing, probably by emitted pheromones that keep subordinate females from cycling, something known as **reproductive suppression**. **Polygyny**, a mating system where a male mates with several reproductive females, is common among one-male units, including baboons and gorillas, and leads to high **reproductive skew**. Lastly, **polygynandry** is probably the most common primate mating system, wherein mating occurs among multiple males and multiple females within a social group.

Finally, **social structure** describes the patterning and nature of social interactions among members of a social group. Because males and females differ in behavioral strategies, social structure is typically considered from three main perspectives: female–female, male–male, and female–male social relationships. Variation in the nature, frequency and intensity of affinitive, affiliative, and agonistic interactions related to acquiring food or mates ultimately leads to differences in social relationships amongst group members.

Of course, categorizing the vast diversity of primate grouping patterns can be challenging and problematic.

Increasing data from long-term field studies continue to expand our understanding of primate behavioral diversity, blurring lines between these different classifications. Moreover, these categories, like all such classifications, provide us mainly with a convenient framework for comparing different species. The ultimate goal of such classification is to facilitate investigation of the factors that have given rise to this diversity in primate societies, which are undoubtedly the result of many selective factors, each of which influences in a different way the size, composition, and dynamics of the social group. It is the dynamics of interindividual interaction, and the genealogical relationships among individuals and groups of individuals, rather than just the numbers of males and females, that provide the real clues to understanding primate social systems.

Dispersal

Although most primates are regularly found to live in species-specific types of social units, most individuals do not spend their entire lives in the same social group. In many primate species, females stay in their natal group, and males emigrate to other groups (Kappeler, 1997). In chimpanzees and bonobos, however, most males stay in their natal group, and females migrate. In many other species (howling monkeys, tamarins, and many lemurs), both sexes migrate from group to group and dispersal is said to be unbiased. Interestingly, this last pattern can give rise to a situation in which a single primate group continues to occupy the same home range from year to year even though there is no continuity in the group membership. Thus, from the perspective of the individuals that comprise them, primate social groups are not stable, permanent units: they are complex and dynamic networks that continually change as individuals are born, mature, emigrate, immigrate, mate, reproduce, and die. Factors such as inbreeding avoidance, resource distribution, and protection from infanticide all seem to be important in determining the patterns of immigration and emigration in different species. Selective factors affect individual group members in different ways. Factors that are of critical importance to one individual may be less significant to another of a different sex, age, or kinship, or to the same individual from month to month or year to year as its reproductive status or its relationships with other individuals change.

Initially, most of our early knowledge of primate social behavior came from studies that lasted only one or two years. There are now numerous long-term behavioral field studies (Kappeler and Watts, 2012), as well as many new and innovative (genetic and physiological) approaches to studying primate behavior (Higham, 2016; Thompson, 2017; Anderson et al., 2020; Orkin et al., 2021) that provide us with a more biologically realistic perspective. Rather than yielding simple answers, this increase in information is demonstrating the many factors involved and their potential interactions in determining the social interactions among primates. This dynamic nature of primate groups reinforces the view that primate social groups are the result of many selective factors acting on each individual, and that to understand the diversity of behavior we find within our order, we have to consider all of the potential costs and benefits of group living to an individual as well as the phylogenetic history of each species.

Why Primates Live in Groups

Compared to most other types of mammals, primates are extremely social animals. This behavior is evident not only in the diverse types of social groups just described but also in the elaborate systems of scents, postures, facial expressions, and vocalizations that primates have evolved for communicating with their conspecifics.

Primate social behavior has evolved through natural selection. Like all other primate adaptations, social behavior can be viewed as the result of a complex and dynamic balance of selective advantages and disadvantages. From an evolutionary perspective, the fitness of an individual animal, or its evolutionary success, is equivalent to its **reproductive success** – the number of reproductively successful offspring it contributes to the next generation. Thus, in evaluating the importance of different factors that might select for group living in primates, we must try to determine how group living would affect the fitness of an individual. The predominant theoretical approach to understanding primate social organization is **socioecology**, which relates primate grouping patterns and interactions to the abundance and dispersal of food resources (Sterck et al., 1997). However, it has become increasingly clear that food dispersal is one of many factors that must be considered when understanding the complexity of primate social behavior (Clutton-Brock and Janson, 2012; Janson, 2000; Thierry, 2008; Koenig and Borries, 2009).

From the point of view of the individuals that make up primate groups, there are four potential advantages to group living: greater protection from predators, improved access to food, better access to mates, and assistance in caring for offspring. Each of these is likely to have greater selective value for some individuals than for others, depending on the individual's age and sex, the reproductive physiology of the species, and the ecological environment. Each potential advantage also must be balanced against the potential disadvantages of group living: greater visibility to predators, and increased competition with other individuals for these same resources of food, mates, and assistance in rearing offspring. The behavioral and physiological adaptations individual primates have evolved for maximizing their survival and that of their offspring in this maze of advantages and disadvantages are referred to as reproductive strategies.

Protection From Predators

Predator avoidance seems to be the major factor that selects for group living in diurnal primates, although actually measuring the influence of predation on living populations is extremely difficult (Janson, 2003). Individuals living in groups gain increased protection in several ways (Janson, 1992). One way is through dilution and geometry of the selfish herd; as group size increases, an individual's risk of being caught decreases by pure numbers, and is further reduced by placing itself toward the center of the group (Hamilton, 1971). Groups of individuals can also gang up on a potential predator. Should the group have to flee, any individual in the group is less conspicuous and more difficult for a predator to isolate and attack. Most significantly, though, each individual benefits from the eyes, ears, and warning calls of other individuals; thus, individuals in larger groups are safer, and each can spend less time in vigilance behavior (the so-called **"vigilance effect"**; Caraco, 1979). Finally, it has been argued that members of groups may have a better knowledge of their home range and the likely whereabouts of predators than nongroup-living individuals.

In contrast, it has also been suggested that group living may, in some cases, make an individual more susceptible to predation. Because a group makes more noise and occupies a larger space, it may be easier for a predator to locate than an isolated individual, and some predators may have large appetites that lead them to prefer "clumped prey." Thus, some species may adopt a **cryptic strategy** to avoid predators.

Improved Access to Food

The reproductive success of any individual depends ultimately on its ability to obtain enough food for itself and its offspring. As noted earlier, the ways animals obtain and select their diets from the array of potential food sources are often referred to as **foraging strategies**. Foraging strategies are, in a larger sense, just one part of an individual's reproductive strategy. Overall, it would seem most likely that group living would be detrimental to the foraging success of an individual, because each individual would have increased competition for food resources from other group members. Thus, individuals living in groups should need to find larger food sources and/or travel farther and visit more food sources each day.

The most important factor in determining the size of the groups in which primates live seems to be the distribution of food resources in time and space (Fig. 3.7; Sterck et al., 1997). Primate species relying on foods that are found in small, evenly scattered patches, such as gums or many small forest fruits, usually live in small groups. Those that specialize on foods such as figs, which are usually found in gigantic but erratically spaced patches, tend to live in large groups. It is easy to see how the distribution of food resources can limit the size of groups that are able to feed on any single resource patch.

However, there are also several ways in which group living may provide individuals with better access to food than they might be able to obtain by foraging alone. Many primate species actively defend food sources – in some cases individual food trees, in others the troop's entire range. In general, disputes over food resources are often resolved by group size: larger groups can displace smaller groups in preferred food trees or in preferred areas. By joining a group, individuals gain access to its resources. There is obviously a fine balance between a group size that is small enough for the group to subsist on a particular resource or set of resources and one that is still large enough for the group to defend those resources from other groups. It should not be surprising that many primate groups that defend their food resources are composed of closely related individuals, usually females and their offspring. Living in groups may also help primates locate food. Individuals may benefit in several ways from communal knowledge about the location of food sources, either through the memory of other individuals or through food calls given by other group members who are foraging semi-independently. There are also suggestions that primates feeding on insects may benefit from the disturbance caused by other troop members who inadvertently flush out insects as they move.

As already discussed, primate foods may vary considerably in both their patchiness and their density. Accordingly, foods with different patterns of distribution should select for different types of grouping and ranging patterns. Furthermore, the distribution of food in the environment also affects dominance patterns within groups. Where food is found in dense patches and can be monopolized by one or two individuals, hierarchies and dominance relationships are expected to form within groups in order to better contest access to these food resources. Where food is distributed in larger, more evenly spaced areas that cannot be monopolized, more tolerant relationships and a lack of strict hierarchies might be expected (Sterck et al., 1997).

Access to Mates

Sexual reproduction requires that each reproductively successful male and female should find a mate of the opposite sex. The reproductive strategies of males and females are, however, quite different for virtually all sexually reproducing animals. A critical aspect of primate reproduction that influences individual reproductive strategies is the marked asymmetry in the roles played by males and females during the early development of offspring. Female primates, like all female mammals, nourish and carry developing young for many months before birth, and also provide milk for the infant for months or years after birth. In contrast, the investment by a male primate in its offspring during this part of development is much less – and theoretically could be as little as a single sperm cell.

There are several consequences of this dramatic difference in the time and energy required of male and female primates, or **reproductive asymmetries**. First, because of the time required by gestation, the maximum number of potential offspring a female primate can have in a lifetime is far less than the number that can be sired by a male, and the female's offspring must necessarily be more evenly spaced in time. With unlimited food resources, a single female with a 20-year period of reproductive fertility, a litter size of one, and a 6-month gestation period could theoretically (but not actually) produce 40 offspring in her lifetime. To achieve this reproductive success, she would (again theoretically) need to associate with a male for mating purposes only briefly every 6 months. In many primate species, a male theoretically could father the same number of offspring in a week (or even a day) if that number of receptive, fertile females are available. Thus, in the number of offspring they can physiologically produce, females are limited primarily by time and food resources, whereas males are limited by their access to females.

There are other consequences of this asymmetry in required investment in early reproductive investment. One is that females are always sure that the offspring they bear are their own. An individual male, on the other hand, can never be sure that he is the father of a newborn. Only by limiting the access of his mates to other males can he increase the likelihood that the offspring they produce are his own.

From these physiological differences in the relative minimal investment required to produce offspring, and the relative certainty of parentage, we can predict that the theoretically optimal strategies of males and females for maximizing their reproductive success will be very different. The most successful male is the one that mates with the greatest number of females and excludes other males from mating with these same females in order to ensure that all offspring are his own progeny. Females, on the other hand, have fewer obvious strategies for producing greater numbers of offspring. Female reproductive strategies seem to emphasize the quality rather than the quantity of offspring. Because every offspring involves such a large investment in time and energy,

female strategies are concerned with ensuring that the male that sires the offspring is likely to engender healthy, strong progeny through paternal investment in such forms as protection from predators and access to food resources.

From these considerations, we would expect male reproductive behavior to involve more intensive competition with other males for access to reproductively active females. The relatively greater intensity of male–male competition for access to mates over that expected in females is generally regarded as a major cause of sexual dimorphism in body size and in the size of canine teeth: features that are important in fighting and in dominance displays (Chapter 9). Thus, it seems that access to mates plays an important role in the reproductive strategies of males, and is a major selective factor in males joining groups. In general, it is the distribution of available females that determines male ranging behavior (e.g., Wrangham, 1979).

Access to many potential mates seems to be a less important factor favoring group living for females of many primate species. Indeed, in most primate groups, adult females outnumber males. However, there is certainly competition among females for access to mates, and in species that live in groups with numerous males, females frequently mate with multiple males. For females, a major important factor is the contribution that one or more of these males can make toward the survival of the offspring.

Assistance in Protecting and Rearing Offspring

Mating is only the first step in successful reproduction. An individual's reproductive success is determined by the number of offspring that live to reproduce themselves, not by the number of conceptions. Offspring that do not survive to successfully reproduce are, from an evolutionary perspective, a wasted effort. For primates that give birth to relatively helpless young that require a relatively long time to reach adulthood, parental investment in the growing offspring is a particularly important aspect of reproductive behavior.

Because of their greater initial investment in offspring and the certainty of maternity, females always make a substantial contribution toward the upbringing of infants in a primate group. Milk is expensive to produce, and females may eat twice as much food when they are lactating, although there is considerable variation in both milk quality and nursing strategies (Hinde and Milligan, 2011). Thus, it is not surprising that most female primates usually solicit and receive help in raising offspring from other troop members (Isler and van Schaik, 2012). There is, however, considerable variability among primates in the contributions of males, females, and other, less closely related troop members to the care and rearing of immature animals. Investment in infants and dependent young seems to be correlated with the degree to which individuals are likely to be related to the offspring. In some monogamous species, such as titi monkeys, owl monkeys, and especially callitrichines, males often contribute as much or more to the care of infants as do females. In larger, more complex social groups, adults of both sexes often assist the mother in caring for infants. Indeed, this is a major characteristic of human societies (Hrdy, 2009). In many primate societies, the adult females in the group are probably related, so infants are the "nieces" and "nephews" of other troop members. In addition, female primates have evolved many behavioral strategies to ensure assistance in rearing infants. By mating with several males, for example, females in multimale troops can confuse the issue of paternity and perhaps elicit some investment from all of the males, since none can exclude the possibility that an infant is his offspring.

Perhaps more significantly, a male's willingness or ability to care for offspring may be a prerequisite for future matings. Baboon females are more likely to mate with males who have helped care for their offspring in the previous year (Smuts, 1985; but see Palombit, 2009), and some researchers have suggested that because paternity is so uncertain for most male primates, male care is probably best seen as a mating strategy, even in monogamous species. Access to help in rearing offspring from other individuals of all age-sex classes is probably a more important factor favoring group living by females than is access to mates.

In addition to direct care through infant carrying or babysitting, males of many species may provide critical assistance to female consorts through deterring **infanticide** by other males. It has been argued that, in many species, this infanticide protection is actually more important than any infant care (van Schaik and Janson, 2000).

Polyspecific Groups

In addition to living regularly in social groups composed of members of their own species, some primates are commonly found in groups composed of several species. These **polyspecific groups**, or mixed-species groups, are particularly common among species of *Cercopithecus* (Fig. 6.13) and *Saguinus* (Fig. 5.22). In many cases, these mixed groups may forage and sleep together for several days; in other instances, the interactions may be for only a few hours. It seems likely that two of the factors that select for intraspecific grouping behavior – predator avoidance and shared information about food resources – are responsible for these polyspecific associations as well.

Phylogeny

Although the details of grouping behavior and social interactions among primates are undoubtedly the result of natural selection for maximum individual fitness, evolution by natural selection has not operated on a blank slate for each species. Rather, each species has evolved with a particular phylogenetic background. There may be more than one satisfactory solution to particular ecological problems, and it is quite evident that closely related taxa are often similar to one another in their social behavior and that different primate radiations often show distinctive patterns of social organization (Di Fiore and Rendall, 1994; Rendall and Di Fore, 2007). Thus, many of the most revealing studies of the factors determining social adaptations involve comparisons of closely related species, or even of the same species in different environments. In these studies it is possible to control for phylogenetic differences and tease out the ecological parameters responsible for the social differences.

Primate Life Histories

In the previous sections we have discussed primate lives and activities from the relatively static perspective of an adult individual at a single point in time. What does it do in the course of a 24-hour day? What does it eat? How does it move? What type of social group does it live in? How does it communicate? In addition, primate lives, like those of all organisms, have a temporal dimension over which the features of their life take place. Many aspects of the lives of individual primates, and the way in which they expend their energies, are constantly changing from conception to birth to sexual maturation to death. Moreover, the timing of developmental and reproductive events varies dramatically from species to species (Fig. 2.29). Consider, for example, the timing of reproduction in the smallest living primate, the mouse lemur, and in the largest, the gorilla. In the 10 years it takes a female gorilla to reach sexual maturity, a female mouse lemur born on the same day could theoretically have left 10 million descendants. Why do gorillas take so long to grow up and produce offspring so slowly? Why do some species produce twins twice a year and others a single offspring every 4 or 5 years?

Analyses of the timing of events in the lives of individuals of different species are called **life history studies**. The temporal aspects of primate biology discussed in this and the previous chapter fall under the category of life history variables, including growth and development, and patterns of immigration and emigration of individuals over the course of their lifespan. However, most studies of life history tend to focus specifically on the allocation of energy and time towards reproduction. The patterns that characterize individual species or different sexes within a species are called life history strategies, because they involve a balance of diverse selective factors. Life history studies are one of the most exciting areas in evolutionary biology (Kappeler and Pereira, 2003; Leigh and Blomquist, 2011). Primates show considerable diversity in many aspects of life history. Gestation length may vary from about 60 days to over 250 days, and birth mass varies from less than 10 g in the smallest species to more than 2000 g in apes and humans. Most primate species, like humans, give birth to one offspring at a time, but for some species twins are the norm, and a few regularly produce triplets. Once they are born, primates show very diverse ways of caring for their infants. In most species, the offspring are carried around after birth by their parents or other relatives. However, in some species, newborns are kept in nests, and in others they are parked on nearby branches while the mother feeds (Tecot et al., 2013). The time primate mothers spend nursing their offspring varies from less than two months in some species to more than four years in others (Lee, 1997). In some species an individual may reach sexual maturity in less than a year, whereas in others it may take over 10 years, with males and females following very different schedules. Finally, the expected lifespan of an individual is less than 10 years in some species, but over 50 in humans.

There are numerous problems in trying to make sense of all this diversity. Many aspects of life history, such as lifespan or weaning age, are poorly known for most species, or known only for captive individuals in others. There are considerable analytical problems involved in comparing features among members of closely related species. Perhaps most importantly, there are simply a great many potential variables to consider at the same time. Nevertheless, studies of primate life history have been tremendously productive, and there are a number of general patterns in the way primate lives are organized. In some aspects of our life histories, primates are very similar to other mammals. For example, in primates, as among virtually all mammals, weaning weight seems to be fixed at four times birth weight, regardless of the timing of growth or the size of the species (Lee et al., 1991). However, there are also ways in which primates are unusual as an order, and likewise, different families or species may have their own unique life history features.

General Patterns

The most obvious correlate of life history variation is body size. In general, larger species have longer gestations, fewer and larger infants, longer weaning ages, and longer time to sexual maturity. Their lives are lived at a much slower reproductive pace than those of smaller animals. Primates follow these trends found in mammals as a class. However, the factors related to reproduction seem to be more highly correlated with one another than with size alone. Thus, even when body size is held constant, analyses of life history differences still show a fast–slow continuum between and among primate species, as well as among mammals as a group, suggesting that overall variation in life history features is determined by environmental factors, such as age-specific mortality rates.

Primate Patterns

In a broad comparison of life history variables, Charnov and Berrigan (1993) found that the relative amount of energy an individual of a species devotes to either growth or the production of offspring seems to be a constant function of body size for both growing juveniles and reproducing adults (Fig. 3.11). Thus, the amount of energy that a mother can expend on the production of offspring is determined by her size. However, the relative contribution allocated to growth and reproduction varies considerably among groups of vertebrates.

Compared to other mammals, primates invest remarkably little of their daily energy budgets in growth and reproduction throughout their life (Charnov and Berrigan, 1993). For their size, primates have smaller litters, take longer to reach sexual maturity, and live longer than members of other orders (Fig. 3.12). These differences are affected by their lower rates of prenatal and postnatal growth. Correlates of this slow rate of postnatal growth are the relatively low nutritional content of primate milk compared with that of many other mammalian mothers (e.g., Oftedal, 1991; Hinde and Milligan, 2011) and the long juvenile growth period (e.g., Janson and van Schaik, 1993). The ultimate reasons for the slow rate of primate lives are less clear. One possibility is the high energetic cost of their large brains; another is that primates suffer relatively low rates of juvenile and adult mortality compared to other mammals. More broadly, however, our primate pattern of low investment in growth and reproduction may be part of a general strategy of low energetic expenditure relative to body size across the order (e.g., Pontzer et al., 2010).

Within this general trend of low-level reproductive investment for our order, there are a number of patterns of life history variation associated with particular phylogenetic groups, and a few broad associations between life history variation and aspects of ecology. For example, among primates, strepsirrhines have a much lower neonatal mass and larger litters than haplorhines; and

Model Mammal Life History

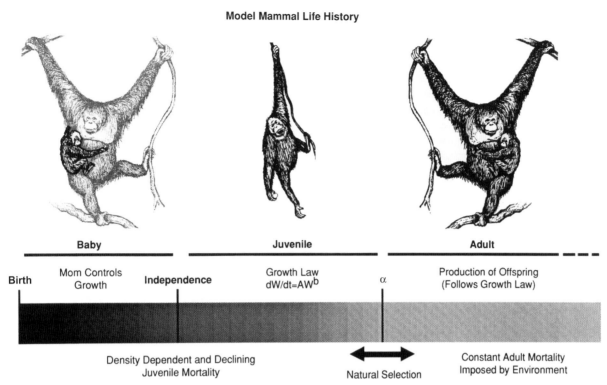

FIGURE 3.11 Mammalian life histories can be represented as a production model in which individuals allocate a portion of their resources to growth or reproduction as a function of their body weight. The growth rate of an infant is determined by the resource allocation of the mother; after weaning, the energy devoted to growth is proportional to body size; at adulthood, the growth energy is allocated to reproduction. According to this model, primates contribute a relatively small amount of the energy budgets to either growth or reproduction. *(From Charnov, E.L., Berrigan, D., 1993. Why do female primates have such long lifespans and so few babies? Or life in the slow lane. Evol. Anthropol. 6, 191–193.)*

within strepsirrhines, lorises have long gestation times and slow growth rates, whereas cheirogaleids have short gestation times and fast growth rates compared to lemurs. Among haplorhines, callitrichids have very high postnatal growth rates.

Because demographic and mortality data are available for very few natural populations of primates, correlations between life history patterns and the ecology of wild populations are few. Nevertheless, there are some broad associations emerging. Folivorous anthropoids grow more rapidly and have shorter postnatal growth periods than nonfolivorous species (Leigh, 1994). There are also associations between growth rates and social interactions: some species, such as colobines and callitrichids, in which infants are cared for by individuals other than their mother, show higher growth rates than species with only maternal care (Fig. 3.13). This pattern does not seem to hold for lemurs (Tecot et al., 2012). Also, within radiations of Old World monkey groups, forest species have slower growth rates and later ages of first reproduction than species that live in secondary forests or more open habitats (Ross, 1992, 1998). As life history variables become increasingly well documented for primates, the next step is to obtain good demographic information on wild populations and to relate life history variation to details of ecology and behavior.

Primate Communities

The ability of primate species to specialize on different canopy levels or on different types of food within a single forest habitat – and to do so during different parts of each day – often permits several species to thrive in the same habitat. Two species that are found in the same area are said to be **sympatric**; species whose ranges do not overlap are **allopatric**. Studies of groups of sympatric species are particularly important for our understanding of primate adaptations because they allow direct comparison of ecological variables (locomotion, diet, social organization) within one environment. Such studies of primate communities are essentially natural experiments in which the climate, the forest, and the competing species (such as predators, parasites, or other arboreal mammals or birds) are held constant. Thus, the observer can see how changes in one ecological variable are correlated with changes in other variables.

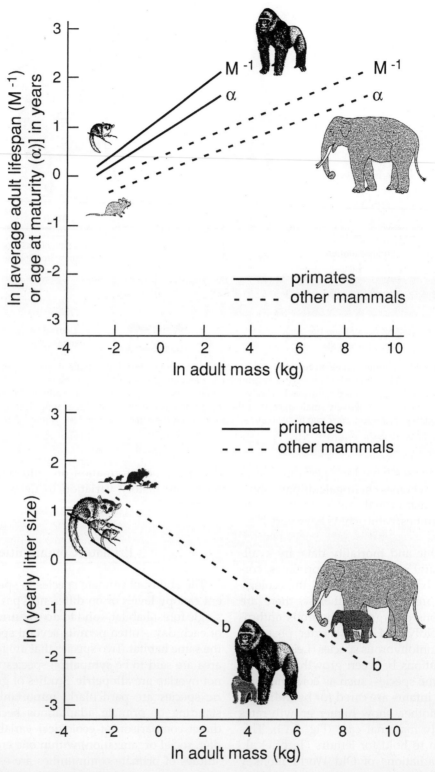

FIGURE 3.12 Compared with other mammals, primates have a longer adult lifespan and higher age of maturity (*top*); primates also produce smaller litters than other mammals (*bottom*). *(From Charnov, E.L., Berrigan, D., 1993. Why do female primates have such long lifespans and so few babies? Or life in the slow lane. Evol. Anthropol. 6, 191–193.)*

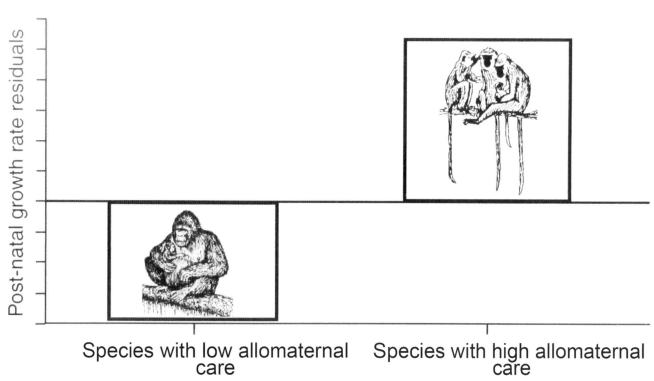

FIGURE 3.13 Primates with high amounts of allomaternal care generally show relatively higher rates of infant growth than species with low amounts of allomaternal care. *(Courtesy of C. Ross.)*

In the following chapters we generally discuss primate species individually, but we often also compare related species that are sympatric. In this way, we not only highlight the diversity of related animals but also see how differences in one parameter, such as diet, change in relation to other parameters, such as locomotion or social organization. Many of this book's illustrations show the ecological features of several sympatric species, enabling comparisons both within selected communities and across various communities. In each taxonomic chapter we discuss the diversity of adaptations within a particular evolutionary radiation, and in Chapter 8 we compare the primate communities within and between major continental areas. Only by looking at primates from several perspectives can we appreciate how adaptation and evolution have produced the diversity of species we see today.

References

Allen, W.L., Stevens, M., Higham, J.P., 2014. Character displacement of Cercopithecini primate visual signals. Nat Commun. 5 (1), 4266.

Anderson, J.A., Vilgalys, T.P., Tung, J., 2020. Broadening primate genomics: new insights into the ecology and evolution of primate gene regulation. Curr. Opin. Genet. Dev. 62, 16–22.

Aureli, F., Schaffner, C.M., Boesch, C., et al., 2008. Fission–fusion dynamics: New research frameworks. Curr. Anthropol. 49, 627–654.

Baden, A.L., Webster, T.H., Kamilar, J.M., 2016. Resource seasonality and reproduction predict fission–fusion dynamics in black-and-white ruffed lemurs *(Varecia variegata)*. Am. J. Primatol. 78 (2), 256–279.

Caraco, T., 1979. Time budgeting and group size: a test of theory. Ecology 60 (3), 618–627.

Charles-Dominique, P., 1983. Ecology and social adaptations in didelphid marsupials: Comparison with eutherians of similar ecology. In: Eisenberg, J.E., Weiman, D. (Eds.), Advances in the Study of Mammalian Behavior. American Society of Mammalogists, CITY, pp. 395–422.

Charnov, E.L., Berrigan, D., 1993. Why do female primates have such long lifespans and so few babies? Or life in the slow lane. Evol. Anthropol. 6, 191–193.

Charpentier, M.J.E., Boulet, M., Drea, C.M., 2008. Smelling right: the scent of male lemurs advertises genetic quality and relatedness. Mol. Ecology 17, 3225–3233.

Clutton-Brock, T., Janson, C., 2012. Primate socioecology at the crossroads: Past, present and future. Evol. Anthropol. 21, 136–150.

Di Fiore, A., Rendall, D., 1994. Evolution of social organization: A reappraisal by using phylogenetic methods. P. Natl. Acad. Sci. 91, 994–995.

Di Fiore, A., Suarez, S., 2007. Route-based travel and shared routes in sympatric spider and woolly monkeys: cognitive and evolutionary implications. Anim. Cogn. 10, 317–329.

Donati, G., Santini, L., Razafindramanana, J., et al., 2013. Un expected nocturnal activity in "diurnal" *Lemur catta* supports cathemerality as one of the key adaptations of the lemurid radiation. Am. J. Phys. Anthropol 150, 99–106.

Dubuc, C., Brent, L.J.N., Accamando, A.K., et al., 2009. Sexual skin color contains information about the timing of the fertile phase in free-ranging rhesus macaques. Int. J. Primatol. 30, 777–789.

Erb, W.M., Porter, L.M., 2017. Mother's little helpers: What we know (and don't know) about cooperative infant care in callitrichines. Evol. Anthropol.: Issues, News, and Reviews 26 (1), 25–37.

Erhart, E.M., 2008. Spatial memory during foraging in prosimian primates: *Propithecus edwardsi* and *Eulemur fulvus rufus*. Folia. Primatol. 79, 185–196.

Fisher, H. S., Swaisgood, R. R., Fitch-Snyder, H., 2003. Odor familiarity and female preferences for males in a threatened primate, the pygmy loris Nycticebus pygmaeus: applications for genetic management of small populations. Naturwissenschaften, 90, 509–512.

Ganzhorn, J.U., Wright, P.C., 1994. Temporal patterns in primate leaf eating: The possible role of leaf chemistry. Folia. Primatol. 63, 203–208.

Garber, P.A., 1997. One for all and breeding for one: Cooperation and competition as a callitrichine reproductive strategy. Evol. Anthropol. 5, 187–199.

Greuter, C.C., Chapais, B., Zinner, D., 2012. Evolution of multilevel social systems in nonhuman primates and humans. Int. J. Primatol. 33, 1002–1037.

Hamilton, W.D., 1971. Geometry for the selfish herd. J. Theor. Biol. 31 (2), 295–311.

Higham, J.P., 2016. Field endocrinology of nonhuman primates: Past, present, and future. Horm. Behav. 84, 145–155.

Higham, J. P., Hebets, E. A., 2013. An introduction to multimodal communication. Behav. Ecol. Sociobiol, 67, 1381–1388.

Higham, J.P., Brent, L.J.N., Dubuc, C., et al. 2010. Color signal information content and the eye of the beholder: A case study in the rhesus macaque. Behav. Ecol. 21, 739–746.

Hinde, R. A., 1976. Interactions, relationships and social structure. Man, 1–17.

Hinde, K., Milligan, L.A., 2011. Primate milk: Proximate mechanisms and ultimate perspectives. Evol. Anthropol. 20, 9–23.

Hrdy, S.B., 2009. Mothers and Others: The Evolutionary Origins of Mutual Understanding. Harvard University Press, Cambridge, Massachusetts.

Hunt, K.D., Cant, J.G.H., Gebo, D.L., et al., 1996. Standardized descriptions of primate locomotor and postural modes. Primates 37, 363–387.

Isbell, L.A., 2009. The Fruit, the Tree, and the Serpent: Why We See So Well. Harvard University Press, Cambridge, Massachusetts.

Isler, K., van Schaik, C.P., 2012. Allomaternal care, life history and brain size evolution in mammals. J. Hum. Evol. 63 (1), 52-63.

Janson, C.H., 1992. Evolutionary ecology of primate social structure In: Smith, E.A., Winterhalder, B. (Eds.), *Evolutionary Ecology and Human Behavior*. Routledge. London, pp. 95–130.

Janson, C.H., 2000. Primate socio-ecology: The end of a golden age. Evol. Anthropol. 9, 73–86.

Janson, C., 2003. Puzzles, predation and primates: Using life history to understand selection pressures. In: Kappeler, P.M., Pereira, M.E. (Eds.), Primate Life Histories and Socioecology. University of Chicago Press, Chicago, pp. 103–131.

Janson, C.H., van Schaik, C.P., 1993. Ecological risk aversion in juvenile primates: Slow and steady wins the race. In: Pereira, M., Fairbanks, L.A. (Eds.), Primates: Juvenile Life History, Development, and Behavior. University of Chicago Press, Chicago, pp. 57–74.

Janson, C.H., Byrne, R., 2007. What wild primates know about resources: Opening up the black box. Anim. Cogn. 10, 357–367.

Kappeler, P.M., 1997. Determinants of primate social organization: Comparative evidence and new insights from Malagasy lemurs. Biol. Rev. 72, 111–151.

Kappeler, P.M., Pereira, M.E., 2003. Primate Life Histories and Socioecology. University of Chicago Press, Chicago.

Kappeler, P.M., van Schaik, C., 2002. Evolution of primate social systems. Int. J. Primatol. 23, 707–740.

Kappeler, P.M., Watts, D.P. (Eds.), 2012. Long-term Field Studies of Primates. Springer, Dordrecht.

Koenig, A., Borries, C., 2009. The lost dream of ecological determinism: Time to say goodbye? Or a white queen's proposal. Evol. Anthropol. 18, 166–174.

LaFleur, M., Sauther, M., Cuozzo, F., et al. 2014. Cathemerality in wild ring-tailed lemurs (*Lemur catta*) in the spiny forest of Tsimanampetsotsa National Park: Camera trap data and preliminary behavioral observations. Primates 55, 207–217.

Lambert, J.E., Rothman, J.M., 2015. Fallback foods, optimal diets, and nutritional targets: Primate responses to varying food availability and quality. Annu. Rev. Anthropol. 44, 493–512.

Lee, P.C., 1997. The meanings of weaning: Growth, lactation, and life history. Evol. Anthropol. 5, 87–96.

Lee, P.C., Majluf, P., Gordon, I.J., 1991. Growth, weaning and maternal investment from a comparative perspective. J. Zool. Lond. 225, 99–114.

Leigh, S.R., 1994. Ontogenetic correlates of diet in anthropoid primates. Am. J. Phys. Anthropol. 94, 499–522.

Leigh, S.R., Blomquist, G.E., 2011. Life history. In: Campbell, C., Fuentes, A., MacKinnon, K., et al. (Eds.), Primates in Perspective, second ed. Oxford University Press, Oxford, pp. 418–428.

Marshall, A.J., Boyko, C.M., Feilen, K.L., et al. 2009. Defining fallback foods and assessing their importance in primate ecology and evolution. Am. J. Phys. Anthropol. 140, 603–614.

Oftedal, O.T., 1991. The nutritional consequences of foraging in primates: The relationship of nutrient intakes to nutrient requirements. Phil. Trans. R. Soc. Lond. B 334, 161–170.

Orkin, J.D., Kuderna, L.F., Marques-Bonet, T., 2021. The diversity of primates: from biomedicine to conservation genomics. Annu. Rev. Anim. Biosci. 9, 103–124.

Palagi, E., Dapporto, L., 2006. Beyond odor discrimination: demonstrating individual recognition by scent in Lemur catta. Chemical Senses 31(5), 437–443.

Palombit, R., 2009. "Friendship" with males: A female counterstrategy to infanticide in chacma baboons of the Okavango Delta. In: Muller, M., Wrangham, R.W. (Eds.), Sexual Coercion in Primates and Humans: An Evolutionary Perspective on Male Aggression Against Females. Harvard University Press, Cambridge Massachusetts, pp. 377–409.

Parga, J.A., 2011. Nocturnal ranging by a diurnal primate: Are ring-tailed lemurs (*Lemur catta*) cathemeral? Primates 52, 201–205.

Pontzer, H., Raichlen, D.A., Shumaker, R.W., et al., 2010. Metabolic adaptation for low energy throughput in orangutans. Proc. Natl. Acad. Sci. USA 107, 14048–14052.

Ramsier, M.A., Cunningham, A.J., Moritz, G.L., et al., 2012. Primate communication in the pure ultrasound. Biol Lett. 8 (4), 508–511.

Rendall, D., Di Fiore, A., 2007. Homoplasy, homology, and the perceived special status of behavior in evolution. J. Human Evol. 52, 504–521.

Ross, C., 1992. Life history patterns and ecology of macaque species. Primates 33, 207–215.

Ross, C., 1998. Primate life histories. Evol. Anthropol. 6, 54–63.

Rothman, J.M., Van Soest, P.J., Pell, A.N., 2006. Decaying wood is a sodium source for mountain gorillas. Biol. Lett. 2 (3), 321–324.

Scordato, E.S., Dubay, G., Drea, C.M., 2007. Chemical composition of scent marks in the ringtailed lemur (Lemur catta): glandular differences, seasonal variation, and individual signatures. Chemical Senses, 32(5), 493–504.

Setchell, J.M., Dixson, A.F., 2001. Changes in the secondary sexual adornments of male mandrills (*Mandrillus sphinx*) are associated with gain and loss of alpha status. Horm. Behav. 39 (3), 177–184.

Seyfarth, R. M., Cheney, D. L., & Marler, P., 1980. Monkey responses to three different alarm calls: evidence of predator classification and semantic communication. Science, 210 (4471), 801–803.

Smuts, B.B., 1985. Sex and Friendship in Baboons. Aldine, Hawthorne, New York.

Snaith, T.V., Chapman, C.A., 2007. Primate group size and socioecological models: Do folivores really play by different rules? Evol. Anthropol. 16, 94–106.

Sterck, E.H.M., Watts, D.P., van Schaik, C.P., 1997. The evolution of female social relationships in nonhuman primates. Behav. Ecol. Sociobiol. 41, 291–309.

Struhsaker, T. T., 1969. Correlates of ecology and social organization among African cercopithecines. Folia Primatol. 11, 80–118.

Swedell, L., Saunders, J., 2006. Infant mortality, paternity certainty, and female reproductive strategies in hamadryas baboons. In: Swedell, L., Leigh, S.R., (Eds.), Reproduction and Fitness in Baboons: Behavioral, Ecological, and Life History Perspectives. Springer, New York, pp. 19–51.

Tattersall, I., 1988. Cathemeral activity in primates: A definition. Folia Primatol. 49, 200–202.

Tattersall, I., 2006. The concept of cathemerality: History and definition. Folia. Primatol. 77, 7–14.

Tecot, S.R., Baden, A.L., Romine, N., Kamilar, J.M., 2012. Infant parking and nesting, not allomaternal care, influence Malagasy primate life histories. Behav. Ecol. Sociobiol. 66, 1375–1386.

Tecot, S.R., Baden, A.L., Romine, N., Kamilar, J., 2013. Reproductive strategies and infant care in the Malagasy primates. In: Clancy, K., Hinde, K., Rutherford, J. (Eds.), Building Babies: Primate Development in Proximate and Ultimate Perspective. Springer, New York, pp. 321–359.

Terborgh, J., 1983. Five New World Primates. Princeton University Press, Princeton, New Jersey.

Thierry, B., 2008. Primate socioecology, the lost dream of ecological determinism. Evol. Anthropol. 17, 93–96.

Thompson, M.E., 2017. Energetics of feeding, social behavior, and life history in non-human primates. Horm. Behav. 91, 84–96.

van Schaik, C.P., Dunbar, R.I.M., 1990. The evolution of monogamy in large primates: A new hypothesis and some critical tests. Behaviour 115, 30–62.

van Schaik, C.P., Paul, A., 1997. Male care in primates: Does it ever reflect paternity? Evol. Anthropol. 5, 152–156.

van Schaik, C.P., Janson, C.H., 2000. Infanticide by Males and Its Implications. Cambridge University Press, Cambridge.

Wheeler, B.C., 2009. Monkeys crying wolf? Tufted capuchin monkeys use anti-predator calls to usurp resources from conspecifics. Proc. R. Soc. B: Biol. Sci., 276(1669), 3013–3018.

Wrangham, R.W., 1979. On the evolution of ape social systems. Sci. Inform. 18, 335–368.

Wright, P.C., 1996. Patterns of paternal care in primates. Int. J. Primatol. 11, 89–102.

Wrogemann, D., Zimmermann, E., 2001. Aspects of reproduction in the eastern rufous mouse lemur (*Microcebus rufus*) and their implications for captive management. Zoo Biology 20 (3), 157–167.

Further Reading

Campbell, C., Fuentes, C., MacKinnon, A., et al. (Eds.), 2011. Primates in Perspective, second ed. Oxford University Press, Oxford.

Di Fiore, A., Lawler, R.R., Gagneaux, P., 2011. Molecular primatology. In: Campbell, C., Fuentes, A., MacKinnon, K., Bearder, S., Stumpf, R. (Eds.), Primates in Perspective, second ed. Oxford University Press, Oxford, pp. 390–416.

Gursky, S., Nekaris, A. (Eds.), 2007. Primate Anti-Predator Strategies. Springer Publishing, New York.

Kappeler, P.M., Pereira, M.E. (Eds.), 2003. Primate Life History and Socioecology. University of Chicago Press, Chicago.

Mitani, J.C., Call, J., Kappeler, P.M., et al. (Eds.), 2012. The Evolution of Primate Societies. University of Chicago Press, Chicago.

Strier, K.B., 2021. Primate Behavioral Ecology, sixth ed. Routledge, New York.

Sussman, R.W., Hart, D., Colquhoun, I.C. (Eds.), 2022. The Natural History of Primates: A Systematic Survey of Ecology and Behavior. Rowman & Littlefield, Lanham.

van Schaik, C.P., 1989. The ecology of social relationships among female primates. In: Standen, V., Foley, R. (Eds.), Comparative Socioecology of Mammals and Man. Blackwell Scientific, Oxford, pp. 195–218.

Wheaton, C.J., Savage, A., Lasley, B.L., 2011. Advances in the understanding of primate reproductive endocrinology. In: Campbell, C., Fuentes, A., MacKinnon, K., Bearder, S., Stumpf, R. (Eds.), Primates in Perspective, second ed. Oxford University Press, Oxford, pp. 377–389.

4

The Prosimians: Lemurs, Lorises, Galagos, and Tarsiers

The living members of the primate order are divided into two major phyletic groups (Fig. 1.2). One group is the Strepsirrhini, or strepsirrhines, composed of the speciose lemurs of Madagascar and the galagos and lorises of Africa and Asia. The other is the Haplorhini, or haplorhines, composed of the tarsiers of southeast Asia and the cosmopolitan anthropoids (monkeys, apes, and humans). In this chapter we will review the living strepsirrhines and some of their fossil relatives, as well as one haplorhine family, the bizarre tarsiers. Other haplorhines will be discussed in subsequent chapters.

Strepsirrhines

Living strepsirrhines are united by at least three specialized features of "hard anatomy" that can be identified in fossils: their unusual dental tooth comb (and associated small upper incisors), the laterally flaring talus, and the grooming claw on the second digit of their feet (Figs. 4.1 and 4.2). Their skull (Fig. 4.3) is characterized by the retention of primitive primate features such as a simple postorbital bar (without postorbital closure), a relatively small braincase, and a primitive mammalian nasal region with a sphenoid recess and a long horizontal nasolacrimal duct. Many of the distinctive soft structures of the strepsirrhine cranial region, such as the well-developed rhinarium, are primitive features found in other mammalian groups. Also like many other mammals, strepsirrhines have scent glands located on their wrists, chests, foreheads, or in their anogenital regions (Schilling, 1979), and rely heavily on olfactory communication to convey information about individual identity, genetic makeup, health, and reproductive state (Morelli et al., 2013; Drea, 2015; Harris et al., 2018; Grogan et al., 2019). The reflective tapetum lucidum in the eye is a more complicated feature. Although a tapetum is a common feature in many mammalian groups, the tapetum of strepsirrhines seems to involve different chemicals than that of other mammals, suggesting that the strepsirrhine tapetum may be a derived feature of that group (Martin, 1995).

The reproductive system of all strepsirrhines is characterized by at least two pairs of nipples, a bicornate uterus, and an epitheliochorial type of placentation. While the bicornate uterus and large number of nipples are primitive features, strepsirrhine placentation seems to be a derived feature (Wildman, 2006; Martin, 2008).

Malagasy Strepsirrhines

The greatest abundance and diversity of extant strepsirrhines occur on Madagascar, an island off the eastern coast of southern Africa. The world's fourth largest island, Madagascar has an area approximately as large as California and Oregon combined and lies totally in the southern hemisphere. There is tremendous regional diversity in the flora, with tropical rain forests along the east coast and in the northwest, mountain regions in the north, dry forests and spiny deserts in the west and south, and a heavily cultivated central plateau that has been almost totally denuded of natural vegetation (Fig. 4.4). Madagascar has been separated from the African mainland for over 100 million years and has an unusual and relatively limited mammalian fauna, with none of the large carnivores or ungulates found in other parts of the world. What is more, the only primates found in Madagascar are strepsirrhines; there are no monkeys or apes. This radiation of strepsirrhines – the lemurs – is one of the three major radiations of living primates. The extant Malagasy lemurs are usually divided into five families: cheirogaleids, lemurids, lepilemurids, indriids, and daubentoniids.

Cheirogaleids

The smallest strepsirrhines and perhaps the most primitive of the Malagasy families are the cheirogaleids (Table 4.1; Fig. 4.5). They share with lemurids the primitive

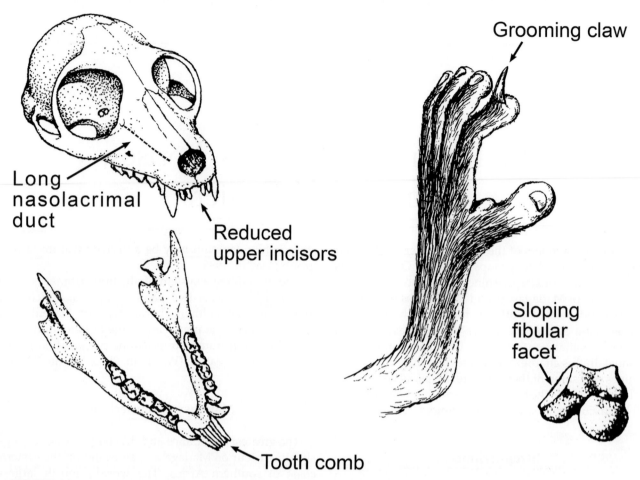

Grooming claw

Long nasolacrimal duct

Reduced upper incisors

Sloping fibular facet

Tooth comb

FIGURE 4.1   Distinctive skeletal features of strepsirrhine primates: laterally flaring talus, small upper incisors separated by a large cleft, dental tooth comb composed of lower incisors and canines, a long nasolacrimal duct, and grooming claw on the second digit of the foot.

strepsirrhine dental formula of 2.1.3.3 and, in all but one genus, they retain a primitive ear structure with the tympanic ring lying free within the bulla (Fig. 2.20). The arrangement of the cranial blood supply in cheirogaleids shows the same unique pattern as that of the lorises and galagos. In both groups, the ascending pharyngeal artery enters the skull near the center of the cranial base to form the internal carotid artery that supplies the brain (Fig. 2.13). The cheirogaleid reproductive system is unusual among primates in that females have three pairs of nipples and normally give birth to litters (Foerg, 1982; Perret, 2019). All are predominantly quadrupedal, and they show considerable dietary diversity. Cheirogaleids are all nocturnal, nest-sleeping animals that forage as solitary individuals. Nevertheless, they show a remarkable diversity of social organization in range overlap and the use of sleeping sites (Schulke and Ostner, 2005; Fig. 4.6).

The **mouse lemurs** (*Microcebus*) are among the smallest of all living primate species. They have a fairly short, pointed snout and large, membranous ears. Their limbs are short relative to the length of their trunk, and their forelimbs are slightly shorter than their hind limbs. Their hands are very humanlike in proportion, and their tail is approximately the same length as their body.

Mouse lemurs are incredibly speciose, with new taxa being discovered every year (Table 4.1). Most new species are allopatric taxa separated by rivers, but in some parts of Madagascar, there are two or more sympatric taxa. Only a few of the many species of mouse lemur have been studied, but there is nevertheless an extensive literature on many aspects of their biology.

Mouse lemurs seem to be abundant in secondary forests and in the undergrowth and lower levels of virtually all forest types, including cultivated areas, and seem to be subject to high predation pressure from a wide range of carnivores, snakes, and birds (Fig. 4.7). For example, a study at Beza Mahafaly, in the southwest of Madagascar, indicates that a population of 2000 individuals may lose as many as 500 individuals per year to predation by a single owl species (Goodman et al., 1993). Mouse lemurs

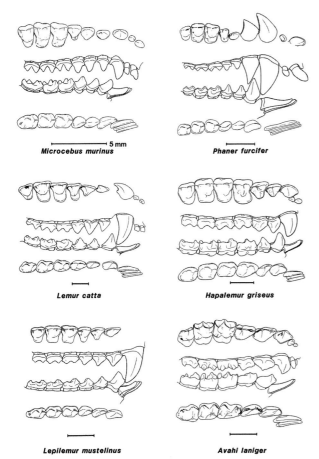

Microcebus murinus

Phaner furcifer

Lemur catta

Hapalemur griseus

Lepilemur mustelinus

Avahi laniger

FIGURE 4.2 Dentition of representative Malagasy strepsirrhines. For each species, occlusal view of upper right dentition (*above*); lateral view of upper and lower right dentition (*center*); and occlusal view of lower left dentition (*below*). (From Maier, W., 1980. Konstruktionsmorphologische Untersuchungen am Gebiß der rezenten. Prosimiae (Primates). Abh. Senckenb. Naturforsch. Ces. 538, 1–158).

are arboreal quadrupeds that move primarily by walking and running along very small branches and leaping between terminal twigs.

Mouse lemurs eat a wide range of foods, including invertebrates and small vertebrates (tree frogs, chameleons), which they catch by quick hand grasps, as well as various fruits, flowers (nectar), buds, and leaves (Fig. 4.8). Mouse lemur diets vary both geographically and seasonally, with mistletoe being a favored food at many sites. Mouse lemurs are nocturnal and seem to be most active just after nightfall and before sunrise. Activity also seems to increase with lunar phase (Deppe et al., 2016). During the day, they sleep in leaf nests, which they make among small branches or in hollow trees. Mouse lemurs undergo variable periods of torpor both daily (up to 9 hours) and seasonally (up to 24 weeks), during which they may lose considerable body mass (Schmid and Speakman, 2000).

All mouse lemurs are solitary foragers, but different populations vary considerably in the extent and pattern of range overlap among individuals and sexes and in the sleeping arrangements of individuals, which depend upon demographic factors as well as the availability of food and suitable nest sites (Fig. 4.6). In some populations, individuals sleep alone; in others, several females may sleep together; and in still others, males and females may sleep in the same nests (Schulke and Ostner, 2005). Recent work has also documented the presence of cryptic, albeit widespread social networks (Zohdy et al., 2012).

Like most Malagasy species, mouse lemurs are seasonal breeders. Individual females are receptive for as little as one day at the end of the dry season (September–October), and their birth season coincides with the wet season (November–February). Female mouse lemurs usually have litters of two or three infants, which remain in the nest while the mother forages. Infant mouse lemurs do not cling to their mother's fur. When moving them, she carries them in her mouth. Mothers will sometimes share nests with other close maternal relatives, and then cooperatively groom and nurse related offspring (Eberle and Kappeler, 2006).

Giant mouse lemurs, *Mirza*, are substantially larger than mouse lemurs. There are two species from the west coast of Madagascar: a larger southern species, *M. coquerelli* and a smaller northern species, *M. zaza* (Kappeler et al., 2005; Table 4.1). *Mirza* shares a number of dental features with the dwarf lemurs (*Cheirogaleus*) but has a very long tail and limb proportions similar to those of the mouse lemurs. Like mouse lemurs, *Mirza* has a pointed snout and large, membranous ears. *Mirza* is found sympatric with *Microcebus murinus* but seems to prefer thicker and taller forests near rivers or ponds, and is found in slightly higher parts of the canopy. Coquerel's giant mouse lemurs move mainly by quadrupedal running, with some leaping.

Mirza feeds on insects and vertebrate prey, fruits, nectar, some gums, and on secretions produced by the larvae of colonial insects (Fig. 4.8). During the dry season, these insect secretions may account for up to 60% of all feeding time. Like mouse lemurs, *Mirza* uses a variety of feeding postures, including clinging on the trunks of trees. Giant mouse lemurs construct large, elaborate, circular nests of leaves for their daytime resting. In *Mirza coquereli*, individual females and males have similar-sized, somewhat overlapping home ranges, and individuals always sleep alone. However, in *M. zaza*, several males and females sleep together in the same nest. Social interactions between males and females are rare outside its annual mating season, which is restricted to a few weeks between September and November (Stanger et al., 1995; Kraus et al., 2008). In *M. coquereli*, the male

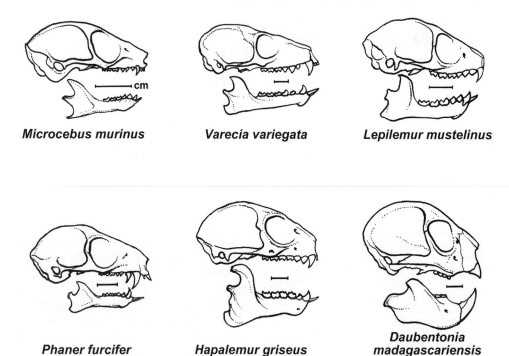

Microcebus murinus   *Varecia variegata*   *Lepilemur mustelinus*

Phaner furcifer   *Hapalemur griseus*   *Daubentonia madagascariensis*

FIGURE 4.3   Skulls of a variety of extant strepsirrhines.

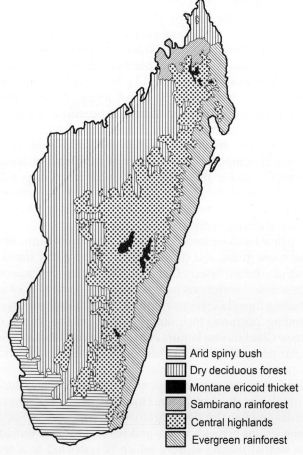

Arid spiny bush
Dry deciduous forest
Montane ericoid thicket
Sambirano rainforest
Central highlands
Evergreen rainforest

FIGURE 4.4   A map of Madagascar showing the distribution of different forest types. Map lines delineate study areas and do not necessarily depict accepted national boundaries.

testes enlarge considerably during the mating season, and the mating system of this species is best described as scramble competition polygyny (Kappeler, 1997). In the dry season, male–female interactions (including contact calls, grooming, play, and chasing) are very common and take place in the central areas of overlapping ranges. These lemurs show no indication of reduced activity during the dry season. *Mirza* females give birth to twins or triplets, which stay in the nests during the first 3 weeks of life.

The **dwarf lemurs** (*Cheirogaleus*) are currently divided into nine species (Table 4.1). All have pointed snouts, moderate-sized ears that are often partly hidden by their fur, and a tail that is slightly shorter than the body. The smaller *Cheirogaleus medius* and its sister taxa are abundant in the dry forest of the west and south, whereas the larger *C. major* and its sister taxa are found in the more humid forests of the east and on the plateau. Dwarf lemurs are arboreal quadrupeds that move more slowly than *Microcebus* or *Mirza* and are much less agile leapers. *Cheirogaleus major* is subject to predation by both carnivores and snakes (Wright and Martin, 1995).

Dwarf lemurs are predominantly frugivorous but opportunistically eat small amounts of insects, small vertebrates, gums, and nectar (Fig. 4.8). When found together, different species will feed on the same plant species and even the same trees, partitioning niches vertically within the forest strata (Lahann, 2007). Dwarf lemurs are less versatile than mouse lemurs in their feeding postures, and move almost exclusively by quadrupedal walking

FIGURE 4.5 Six genera of cheirogaleids: upper left, a mouse lemur (*Microcebus*) reaches for a dragonfly while a hairy-eared dwarf lemur (*Allocebus trichotis*) rests on a branch; to their right, Coquerel's dwarf lemur (*Mirza coquereli*) reaches for a piece of fruit; just below, a fat-tailed dwarf lemur (*Cheirogaleus medius*) also reaches for fruit; below it, a greater dwarf lemur (*Cheirogaleus major*) walks along a branch; on the right, a fork-marked lemur (*Phaner furcifer*) clings to a tree and licks exudates.

TABLE 4.1   Infraorder LEMURIFORMES

Family CHEIROGALEIDAE

Common name	Species	Intermembral index	Mass (g) M	F
Hairy-eared dwarf lemur	*Allocebus trichotis*	-	83	78
Montagne d'Ambre dwarf lemur	*Cheirogaleus andysabini*	-	253	312
Furry-eared dwarf lemur	*C. crossleyi*	-	-	-
Groves' dwarf lemur	*C. grovesi*	-	453	
Lavasoa dwarf lemur	*C. lavasoensis*	-	267	
Greater dwarf lemur	*C. major*	72	438	362
Eastern fat-tailed dwarf lemur	*C. medius*	68	283	282
Lesser iron-grey dwarf lemur	*C. minisculus*	-	-	302
Ankarana dwarf lemur	*C. shethi*	-	120	
Sibree's dwarf lemur	*C. sibreei*	-	262	
Arnhold's mouse lemur	*Microcebus arnholdi*	-	50	
Madame Berthae's mouse lemur	*M. berthae*	-	33	33
Bongolava mouse lemur	*M. bongolavensis*	-	54	54
Nosy Boraha mouse lemur	*M. boraha*	-	56.5	
Danfoss' mouse lemur	*M. danfossi*	-	61	66
Ganzhorn's mouse lemur	*M. ganzhorni*	-	-	
Gerp's mouse lemur	*M. gerpi*	-	67	69
Rufous-grey mouse lemur	*M. griseorufus*	-	48	57
Jonah's mouse lemur	*M. jonahi*	-	-	
Jolly's mouse lemur	*M. jollyae*	-	44	
Goodman's mouse lemur	*M. lehilahytsara*	-	48	45
MacArthur's mouse lemur	*M. macarthurii*	-	54	
Bemanasy mouse lemur	*M. manitatra*	-	58	
Claire's mouse lemur	*M. mamiratra*	-	61	
Margot Marsh's mouse lemur	*M. margotmarshae*	-	41	
Marohita mouse lemur	*M. marohita*	-	76.5	
Mittermeier's mouse lemur	*M. mittermeieri*	-	44	
Gray mouse lemur	*M. murinus*	72	59	63
(Peters') Pygmy mouse lemur	*M. myoxinus*	-	49	
Golden-brown mouse lemur	*M. ravelobensis*	-	52	58
Brown mouse lemur	*M. rufus*	71	43	42
Sambirano mouse lemur	*M. sambiranensis*	-	44	
Simmons' mouse lemur	*M. simmonsi*	-	76.5	
Northern rufous mouse lemur	*M. tavaratra*	-	61	
Anosy mouse lemur	*M. tanosi*	-	54	
Coquerel's giant mouse lemur	*Mirza coquereli*	70	317	299
Northern giant mouse lemur	*M. zaza*	-	287	299
Amber Mountain fork-marked lemur	*Phaner electromontis*	-	387	
Eastern fork-marked lemur	*P. furcifer*	68	460	
Pale fork-marked lemur	*P. pallescens*	70	328	351
Parienti's fork-marked lemur	*P. parienti*	-	360	

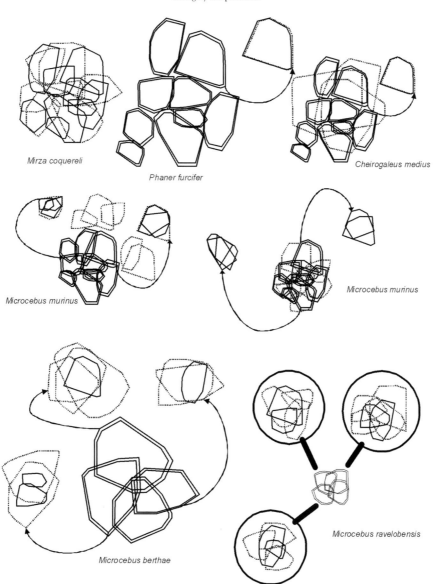

FIGURE 4.6   Diversity in the social organization and range use among nocturnal cheirogaleids. *(Courtesy of Oliver Schulke.)*

and running. During the night, they intersperse periods of activity with periods of rest.

Dwarf lemurs adapt to the dry seasons of Madagascar by hibernating for 6 to 8 months of each year. During this time, they metabolize the enormous fat reserves stored in their tails during the wet season (Blanco et al., 2018). In western Madagascar, mean adult body weight can drop by one-third between March and November. Similarly, eastern species hibernate between May and late September and may halve their weight during hibernation. There are, however, notable geographic differences in the duration of hibernation and corresponding life history variables, even within a species. For instance, *C. medius* living in rainforest habitats spend less time hibernating and have larger, more frequent litters than animals living in dry forest sites, likely as a response to

their shorter life expectancy and heightened levels of predation (Lahann and Dausman, 2011).

Lesser dwarf lemurs have been reported to nest in hollow trees during both normal daytime sleeping and hibernation, and *C. major* has been observed building nests of leaves. More recently, some species of eastern dwarf lemur have also been found to hibernate in burrows underground (Blanco et al., 2013).

Individuals of *C. medius* live in dispersed pairs of one male and one female with their offspring, though single males may live in large ranges overlapping those of several pairs (Schulke and Ostner, 2005).

The **fork-marked lemur** (*Phaner*) is one of the largest and ecologically most specialized of the cheirogaleids (Fig. 4.5). Its characteristic facial features include large, membranous ears and dark rings around the eyes that

FIGURE 4.7   In southwestern Madagascar, mouse lemurs (*Microcebus murinus*) are preyed upon extensively by barn owls.

join on the top of the skull to form a stripe down the back. There are currently four species of fork-marked lemurs. All fork-marked lemurs have relatively long hindlimbs and a very long, bushy tail. Fork-marked lemurs are widely, but discontinuously, distributed over many parts of Madagascar, although they are most common in the west. These lemurs forage in all levels of the forest and specialize on gum (Fig. 4.8).

Perhaps not surprisingly, they have a number of distinctive anatomical adaptations commonly found among primates with this unusual diet. *Phaner* have very large hands and feet with expanded digital pads, and their fingernails are keeled like claws for clinging to the trunks of trees. For obtaining gums, they have relatively long, procumbent upper and lower incisors, long canines and anterior upper premolars, and a long and narrow tongue (Fig. 4.3). Their gut is characterized by a large caecum in which the gums are chemically broken down. Their locomotion is rapid quadrupedal walking and running interspersed with leaps from branch to branch.

Like lesser dwarf lemurs, male and female fork-marked lemurs seem to live in dispersed pairs (Schulke and Kappeler, 2003). They forage one at a time at the gum sites, with the females appearing to have first choice. During the day, fork-marked lemurs normally

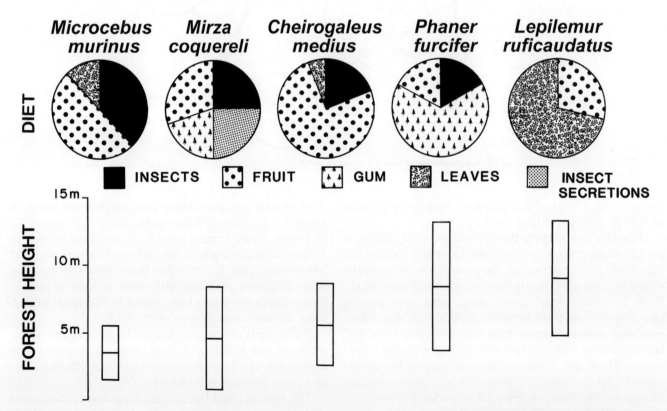

FIGURE 4.8   Diet and forest height preference for five sympatric strepsirrhines in the dry forest of western Madagascar. *(Data from) Hladik, C.M., Charles-Dominique, P., Petter, J.J., 1980. Feeding strategies of five nocturnal prosimians in the dry forest of the west coast of Madagascar. In: Charles-Dominique, P., Cooper, H.M., Hladik, A., et al. (Eds.), Nocturnal Malagasy Primates: Ecology, Physiology, and Behavior, Academic Press, New York, pp. 41–73.).*

sleep in pairs in tree holes or in nests built by *M. coquereli*. Pairs commonly groom one another.

The **hairy-eared dwarf lemur** (*Allocebus trichotis*) lives in restricted areas of the eastern rainforest. *Allocebus* is slightly larger than *M. murinus* and has ear tufts (Fig. 4.5). The most distinctive feature of the genus, which separates it from other cheirogaleids, is the construction of its auditory region. Rather than having a free tympanic ring within the auditory bulla, *Allocebus* resembles lorises and galagos in having a tympanic ring fused to the wall of the bulla. Although *Allocebus* is more closely related to mouse lemurs and dwarf lemurs than to fork-marked lemurs, it shares several similarities to *Phaner*, including aspects of its dentition and the presence of keeled nails on its digits associated with a diet of gums.

Very little is known about the behavioral ecology of the species. In one study, nearly 70% of *Allocebus* food consumption involved flying insects (Biebouw, 2012). Other food sources include fruits and leaves, as well as gum and/or insect exudates. In *Allocebus*, several individuals share large overlapping home ranges and common sleeping sites, suggesting that this species lives either in dispersed pairs with offspring or in multimale, multifemale groups (Biebouw, 2009).

Lemurids

The lemurids (Fig. 4.9; Table 4.2) are the typical Malagasy lemurs. They share the same dental formula with the cheirogaleids (2.1.3.3), and their tympanic ring lies free within the auditory bulla, as in most cheirogaleids and indriids (Fig. 2.20). Their cranial blood supply to the anterior part of the brain, however, is largely through the stapedial artery rather than through the ascending pharyngeal artery as in cheirogaleids. They are medium-sized (1–4 kg), diurnal or cathemeral, group-living prosimians that, with one exception, do not build nests. The lemurids are currently divided into five genera: *Lemur, Eulemur, Hapalemur, Prolemur*, and *Varecia*.

FIGURE 4.9   Three lemurid species from different parts of Madagascar: *above*, a pair of brown lemurs (*Eulemur fulvus*); *center*, a pair of ruffed lemurs (*Varecia variegata*); *below*, three ring-tailed lemurs (*Lemur catta*) on the ground.

TABLE 4.2   Infraorder LEMURIFORMES

Family LEMURIDAE

Common name	Species	Intermembral index	Mass (g) M	F
Ring-tailed lemur	*Lemur catta*	70	2171	2204
Black-and-white ruffed lemur	*Varecia variegata*	72	3600	3500
Red ruffed lemur	*V. rubra*	72	3550	3470
Lac Alaotra bamboo lemur	*Hapalemur alaotrensis*	-	1228	1251
Golden bamboo lemur	*H. aureus*	65	1514	1355
Gray bamboo lemur or gentle lemur	*H. griseus*	67	914	915
Southern bamboo lemur	*H. meridionalis*	60	1069	1075
Northern bamboo lemur	*H. occidentalis*	-	847	1188
Greater bamboo lemur	*Prolemur simus*	74	2238	2250
White-fronted brown lemur	*Eulemur albifrons*	-	2213	2150
White-collared brown lemur	*E. albocollaris*	-	1950	2150
Grey-headed brown lemur	*E. cinereiceps*	76	2190	2140
Red-collared brown lemur	*E. collaris*	71	2070	2217
Crowned lemur	*E. coronatus*	69	1280	1080
Blue-eyed black lemur	*E. flavifrons*	-	1880	1760
Brown lemur	*E. fulvus*	73	1867	1775
Black lemur	*E. macaco*	71	1932	1937
Mongoose lemur	*E. mongoz*	72	1140	1280
Red-bellied lemur	*E. rubriventer*	68	2067	1960
Red-fronted brown lemur	*E. rufifrons*	-	2178	2251
Rufus brown lemur	*E. rufus*	74.5	1790	1840
Sanford's brown lemur	*E. sanfordi*	-	1852	

Lemur catta, the **ring-tailed lemur** (Fig. 4.9), is probably the best-recognized Malagasy mammal. It is a gray animal with a long, striped tail; the sexes are **monomorphic**, meaning they look alike. Ring-tailed lemurs are diurnal (or perhaps cathemeral; Parga, 2011), live in the dry south of Madagascar, and feed both on the ground and in the trees. They are the most terrestrial of living strepsirrhines, spending 30% of each day and 65% of their overall traveling time on the ground. They are primarily quadrupedal walkers and runners. Their diet contains large amounts of both fruit and leaves, and varies from region to region, depending on both habitat and competition from other lemurid species.

Ring-tailed lemurs live in large social groups of about 20 individuals that contain approximately equal numbers of males and females. Groups can travel almost a kilometer a day and occupy a home range averaging between 10 and 32 ha, depending on the richness of habitats. *L. catta* societies are strikingly similar to those of many Old World monkeys, such as macaques, in that they are centered around one or more female matrilines. Males normally emigrate from their natal troop and subsequently change troops approximately every 3 years. Like other strepsirrhines, the females are dominant over the males. One or more of the adult immigrant males in a ring-tailed lemur troop usually occupies a central position and is frequently associated with the core of females, whereas other males, both natal and immigrant, occupy more peripheral positions. There are consistent dominance hierarchies among both females and males.

Scent marking plays an important role in mediating *L. catta* social lives. Both sexes have scent glands in their genital regions, and males boast two additional specialized glands on their wrists and chests. During intergroup encounters, males will sometimes anoint their tails with scent from their antebrachial (wrist) glands and then waft their tails at each other in a display known as "stink fighting" (Jolly, 1966). Males also use this anointing and tail wafting display as an honest signal of dominance for resident males, and to attract female mating partners (Walker-Bolton and Parga, 2017).

Eulemur is the most widespread and speciose genus in the family, with 10 or more currently recognized species. They are all arboreal and cathemeral. They are similar in size to *L. catta*, with many species showing sexual dichromatism, or differences in pelage coloration between males and females. There are several species groups within *Eulemur*. The species of **brown lemur** form a ring around the island of Madagascar and, as such, can be very different from one another ecologically. Despite considerable variability in chromosome number, all seem to be capable of interbreeding. Brown lemurs are totally arboreal and move primarily by quadrupedal walking, running, and leaping. Their diet consists of leaves, fruit, and flowers, with percentages varying considerably among the different species, from habitat to habitat, and from season to season. Two brown lemurs that have been well studied are the brown lemur (*Eulemur fulvus*), from the southwest, and the rufous lemur, *E. rufifrons*, from the eastern rainforest. Brown lemurs tend to live in somewhat smaller groups than *L. catta*, averaging seven to twelve individuals with roughly equal numbers of males and females. In the southwest of Madagascar, brown lemurs typically travel less than 150 m each day within their tiny (less than 1 ha) home ranges. This varies seasonally, with groups sometimes traveling as much as a kilometer outside of their range to access water during seasonal droughts (Scholz and Kappeler, 2004). In contrast, *E. rufifrons*, in the eastern rainforest, has a much larger day range of nearly 1 km and a home range of approximately 100 ha. Like *E. fulvus*, *E. rufifrons* will regularly make long foraging trips outside of its normal home range in times of food scarcity. Unlike many other strepsirrhine species, *E. rufifrons* shows no evidence of female dominance over males or male dominance over females.

The **red-bellied lemur**, *Eulemur rubriventer*, is found sympatric with *E. rufifrons* throughout Madagascar's eastern rainforests and has been extensively studied at Ranomafana (e.g., Overdorff, 1998, 1993a,b, 1996; Overdorff and Tecot, 2006; Tecot, 2008). Red-bellied lemurs are similar in size to the brown lemurs. They are also cathemeral. Red-bellied lemurs move about half the time by leaping and half by quadrupedal walking and running. Their diet varies considerably from season to season, but overall consists of about 80% fruit, with smaller amounts of leaves and nectar from flowers.

Red-bellied lemurs live in small, monogamous family groups. While extra-pair paternities do occur, they are rare and often the result of group turnover (Jacobs et al., 2018). Births peak in October, usually with only a single offspring at a time. While reproduction does occur outside of this time, infants typically do not survive to weaning (Tecot, 2010). Males and juvenile siblings assist in caring for the infant (Tecot and Baden, 2018; Tecot et al., 2023). Red-bellied lemurs have relatively small home ranges, averaging 19 ha. Day ranges average less than 450 m, with considerable seasonal variability.

The **mongoose lemur** (*Eulemur mongoz*) lives in forested areas both in the north of Madagascar and on the nearby Comoro Islands of Anjouan and Moheli. Mongoose lemurs show extreme variability in their activity pattern, both between different populations and, in at least one area, from season to season. Field observations suggest that mongoose lemurs are cathemeral throughout the year, but exhibit shifts toward nocturnal activity in dry conditions and diurnal activity in cold, wet conditions (Curtis et al., 1999). The nocturnal diet and ranging patterns of mongoose lemurs are well known. Like most *Eulemur* species, *E. mongoz* consumes a predominantly frugivorous diet supplemented mainly by leaves, flowers, and nectar (Curtis, 2004). The species consumes high-fiber foods year-round; however, mongoose lemurs switch from high-energy foods (mature fruit, nectar, and seeds) during the wet season to high-energy (mature fruit and flowers), high-protein (immature leaves) foods in the dry season. Animals supplement their diets in the dry season with mature leaves and petioles, perhaps for their mineral content. The social organization of this dichromatic species seems to be as flexible as its activity period, but the two are not clearly correlated. Most populations live in small monogamous family groups composed of an adult male, an adult female, and their offspring (Curtis and Zaramody, 1998). However, some groups seem to have more adults.

The charismatic **ruffed lemur** (Genus *Varecia*), from the east coast rainforests, is the largest living lemurid and seems to be the most primitive in many aspects of its behavior. Ruffed lemurs have a long, doglike snout, thick fur, and a long tail (Fig. 4.9). Compared to other lemurs, they have relatively short limbs. Ruffed lemurs are totally arboreal and restricted to forests with a continuous canopy of large trees. They are almost exclusively quadrupedal but frequently adopt hindlimb suspensory postures while feeding (Britt, 2000). They are almost totally frugivorous and are effective seed dispersers for many fruits (Razafindratsima and Martinez, 2012). Indeed, some tree species are now totally dependent on ruffed lemurs for seed dispersal (Federman et al., 2016). However, recent work suggests they may exhibit more significant seasonal variation in fruit consumption than once thought, consuming as much as 50% leaves during some lean season months (Beeby and Baden, 2021).

The social behavior of ruffed lemurs has been studied in a wide range of both seminatural and wild conditions (Vasey et al., 2023). Ruffed lemurs live in highly variable social organizations ranging from pairs to large communities of 18-30 individuals; however, all long-term observations of the species indicate that they are best described as exhibiting fission–fusion social dynamics (Baden et al., 2016; Holmes et al., 2016). Their loud calls may help to maintain social coordination, even when out of visual contact (Batist et al., 2022). Despite this, the highest rates of association tend to be among male–female dyads of "preferred" social partners and their offspring (Baden et al., 2020b), whose ranges cluster spatially into "neighborhoods" of social partners (Baden et al., 2020a).

In many features of their reproductive behavior, ruffed lemurs resemble the small, nocturnal cheirogaleids moreso than other lemurids. *Varecia* have three pairs of nipples and regularly give birth to twins and triplets. Mothers stash infants in nests, including communal nests (or créches), until they are able to travel on their own (Vasey et al., 2018; Baden, 2011, 2019). Infants receive considerable attention from other group members, and mating partners do not always correspond to infant care-providers (Vasey, 2007; Baden et al., 2013). Like other Malagasy species, ruffed lemurs are strict seasonal breeders, and females are fertile on only 1 to 2 days per year. They seem to have very high infant mortality, as well as great differences in fertility from year to year (Baden et al., 2013).

Bamboo lemurs (Fig. 4.10) are unusual among primates for their specialized diet. There are three types of bamboo lemurs: several small species, the **gray bamboo lemurs** or **gentle lemurs** (*Hapalemur griseus, H. occidentalis, H. meridonalis*, and *H. aloatrensis*), with a fairly broad allopatric distributions; and two larger, rarer species, the **golden bamboo lemur** (*H. aureus*) and the **greater bamboo lemur** (*Prolemur simus*), both from restricted areas in the southeastern rainforest (Fig. 4.11). All *Hapalemur* species have relatively short faces and small, hairy ears. They have short arms and long legs compared to other lemurids. All show a preference for bamboo when it exists and live almost entirely on bamboo shoots and leaves; however, each of the bamboo lemurs seems to specialize on different parts of the plant. *H. griseus* eats primarily new shoots of several bamboo lianas; the larger *P. simus* can break open and eat the pith from mature culms; and *H. aureus* specializes on the growing shoots of a species containing very high levels of cyanide (Tan, 1999). Two species, the southern gentle lemur (*H. meridonalis*) and the Alaotran gentle lemur (*H. aloatrensis*), occupy habitats that contain little or no woody bamboo, raising questions as to how these lemurs subsist within this ecological context (Eppley et al., 2011).

The locomotor and postural behavior of bamboo lemurs is mainly clinging and leaping, but they also move quadrupedally along bamboo branches when feeding. One species, the southern bamboo lemur, also spends significant time as a terrestrial quadruped (Eppley et al., 2016). Most *Hapalemur* species seem to be diurnal, and are regularly found in small (presumably family) groups and have single births. The larger-bodied *P. simus* is cathemeral and lives in much larger groups that regularly have more than one breeding female.

Lepilemurids

Lepilemur, the **sportive**, or **weasel lemurs** (Fig. 4.12; Table 4.3), comprises a very speciose group of small nocturnal animals found all over Madagascar. These small, drab-colored lemurs are characterized by a lack of permanent upper incisors (Fig. 4.2), an unusual articulation

FIGURE 4.10   A family of gentle bamboo lemurs (*Hapalemur griseus*) in a typical bamboo habitat.

FIGURE 4.11 Three sympatric bamboo lemurs from Ranomafana National Park in southeastern Madagascar: *above*, the gentle bamboo lemur (*Hapalemur griseus*); *middle*, the golden bamboo lemur (*Hapalemur aureus*); *below*, the greater bamboo lemur (*Prolemur simus*).

FIGURE 4.12 Several sportive lemurs (*Lepilemur mustelinus*) in a dry forest of Didiereaceae bush.

TABLE 4.3 Infraorder LEMURIFORMES

Family LEPILEMURIDAE

Common name	Species	Intermembral index	Mass (g) M	F
Antafia sportive lemur	*Lepilemur aeeclis*	-	868	902
Tsiombikibo sportive lemur	*L. ahmansonorum*	-	610	610
Ankarana sportive lemur	*L. ankaranensis*	-	770	
Betsileo sportive lemur	*L. betsileo*	-	1150	
Gray's sportive lemur	*L. dorsalis*	-	500	
Milne-Edwards' sportive lemur	*L. edwardsi*	60	928	934
Andoahela sportive lemur	*L. fleuretae*	-	980	
Anjiamangirana sportive lemur	*L. grewcockorum*	-	940	
Mananra–Nord sportive lemur	*L. hollandorum*	-	990	
Zombitse sportive lemur	*L. hubbardorum*	-	990	
Manombo sportive lemur	*L. jamesorum*	-	780	
White-footed sportive lemur	*L. leucopus*	-	617	594
Small-toothed sportive lemur	*L. microdon*	-	970	-
Daraina sportive lemur	*L. milanoii*	-	720	
Mittermeier's sportive lemur	*L. mittermeieri*	-	730	
Weasel sportive lemur	*L. mustelinus*	65	1000	
Ambodimahabibo sportive lemur	*L. otto*	-	938.5	
Petter's sportive lemur	*L. petteri*	-	633	600
Bemaraha sportive lemur	*L. randrianasoli*	-	788	
Red-tailed sportive lemur	*L. ruficaudatus*	63	806	803
Sahamalaza sportive lemur	*L. sahamalazensis*	-	710	679
Masoala sportive lemur	*L. scottorum*	-	888	
Seal's sportive lemur	*L. seali*	-	950	
Sahafary sportive lemur	*L. septentrionalis*	63	675	
Nosy Be sportive lemur	*L. tymerlachsonorum*	-	880	
Wright's sportive lemur	*L. wrightae*	-	1850	

between the mandible and the skull, large digital pads on the hands and feet, and a large caecum. They have relatively long legs and a long tail. There is considerable variability in the chromosomes among the widespread populations of this genus, and current studies recognize over 25 separate species.

Sportive lemurs are found in all types of forest throughout Madagascar and are often locally abundant. They prefer vertical postures and travel mainly by quadrupedal running and leaping. *Lepilemur* are predominantly nocturnal and folivorous (Fig. 4.8).

Like other folivorous primates, they are unable to digest cellulose, the main structural component of leaves, so they rely on the bacteria in their digestive tract for this task. In sportive lemurs, as in horses and rabbits, the bacteria live in the caecum, at the base of the large intestine; in this way, the cellulose is broken down very near the end of the digestive tract. Because of this, *Lepilemur* (like rabbits) have been reported to re-ingest their feces (containing the broken-down cellulose) during the day, a practice known as coprophagy. The basal metabolic rate of *Lepilemur* is one of the lowest recorded for any primate (Schmid and Ganzhorn, 1996); this is both an energy conservation strategy and perhaps an adaptation for digesting toxic chemicals.

Like many of the cheirogaleids, sportive lemurs forage as solitary individuals or, for females, with their offspring. However, based on range overlap and their use of sleeping trees, some species seem to be characterized by dispersed monogamy, while others show a noyau arrangement. Individuals have very small home ranges of about a hectare, and groups or individuals may be intensely territorial, as their common name suggests. Males give loud crow-like calls throughout the night and have been reported to engage in fisticuffs in defense of their tiny, leafy domains. In fact, some studies suggest that pairs may use loud call exchanges as a vocal display to signal territory ownership, thus limiting direct aggressive encounters between neighbors and strangers (Rasoloharijaona et al., 2006). *Lepilemur* are also known to use communal latrines (specific defecation/urination sites), which have been suggested to further facilitate communication and territory defense via olfactory signaling (Dröscher & Kappeler, 2014). Most of their evenings, however, are spent resting and guarding their territory, presumably letting their leaves digest (Nash, 1998). Sportive lemurs have single births.

Indriids

The living indriids are a relatively uniform family, consisting of three similar genera (Table 4.4) that differ most obviously in size and activity pattern (Fig. 4.13). They have a reduced dental formula, with only two premolars in each quadrant and four rather than six teeth in their tooth comb (Fig. 4.2). Indriids are remarkably uniform in their cranial morphology (Fig. 4.14) and, like lemurids, have a tympanic ring that lies free in the bulla and a large stapedial artery. All extant indriids are specialized leapers with long hindlimbs and a short, dorsally oriented ischium. Their digestive tract has an enlarged caecum and an extra loop in their large intestine. Indriids tend to have single births and are remarkable in the precocious development of their dentition. Individuals are born with the deciduous dentition fully erupted and the permanent molars ready to erupt. They are born ready to chew. However, other aspects of growth and development are slow compared with lemurids.

The smallest of the indriids are the **woolly lemurs** (*Avahi*). They are mottled brown, white, and beige,

TABLE 4.4   Infraorder LEMURIFORMES
Family INDRIIDAE

Common name	Species	Intermembral index	Mass (g) M	F
Betsileo woolly lemur	*Avahi betsileo*	-	1000	1175
Bemaraha woolly lemur	*A. cleesei*	-	1000	1058
Eastern woolly lemur	*A. laniger*	58	1106	1307
Southern woolly lemur	*A. meridionalis*	-	1100	1200
Masoala woolly lemur	*A. mooreorum*	-	920	
Western woolly lemur	*A. occidentalis*	-	814	777
Peyrieras' woolly lemur	*A. peyrierasi*	59	1080	
Manombo woolly lemur	*A. ramanantsoavanai*	-	897	1019
Sambirano woolly lemur	*A. unicolor*	-	850	
Silky sifaka	*Propithecus candidus*	64	6300	
Coquerel's sifaka	*P. coquereli*	60	3703	3757
Crowned sifaka	*P. coronatus*	-	3900	
Decken's sifaka	*P. deckenii*	-	2930	2630
Diademed sifaka	*P. diadema*	63	5940	6260
Milne–Edwards' sifaka	*P. edwardsi*	63	5500	5700
Perrier's sifaka	*P. perrieri*	-	4700	4500
Tattersall's sifaka	*P. tattersalli*	-	3390	3590
Verreaux's sifaka	*P. verreauxi*	61	3020	3200
Indri	*Indri indri*	64	5830	6840

FIGURE 4.13   Three indriids from different parts of Madagascar: *left*, the sifaka (*Propithecus verreauxi*); *right*, a pair of indris (*Indri indri*); *below*, a family of the nocturnal wooly lemurs (*Avahi laniger*) in a typical sleeping posture.

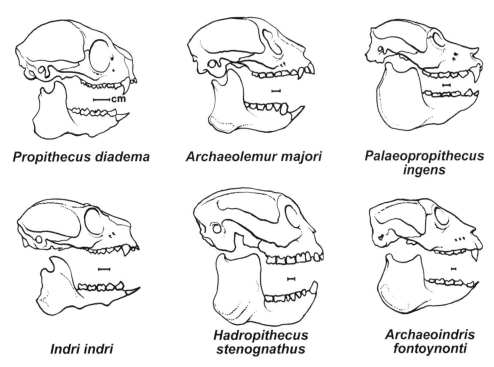

Propithecus diadema   **Archaeolemur majori**   **Palaeopropithecus ingens**

Indri indri   **Hadropithecus stenognathus**   **Archaeoindris fontoynonti**

FIGURE 4.14   Skulls of a variety of extant and recently extinct indriids.

thickly furred animals with a long tail. There are numerous mostly cryptic species. *Avahi laniger* is common in the wet forests of eastern Madagascar; the other species have more restricted distributions in moist areas of the north and west. *Avahi* are the only nocturnal indriids. Woolly lemurs move mainly by leaping between vertical supports in the understory of the forest. Like the similarly sized, equally folivorous *Lepilemur*, their diet is almost exclusively leaves, but they avoid competition by specializing in leaves containing tannins (Ganzhorn, 1988).

Woolly lemurs live in monogamous family groups that may forage either separately or together. Both males and females use three acoustically distinct call types (two loud calls and one softer call) to coordinate group spacing and coordination (Ramanankirahina et al., 2016). Home ranges are between 1 and 2 ha. Like all indriids, woolly lemurs have single births. Woolly lemurs do not build nests: during the day, they huddle together among tangles of vines and leaves in the lower parts of trees (Ramanankirahina et al., 2012; Fig. 4.13).

The **sifakas** (*Propithecus*) are much larger indriids. There are numerous species found allopatrically all around the island, including the small *P. verreauxi*, from the dry forest in the south and west; the larger *P. diadema* and *P. edwardsi* from the rainforests in the east; and the beautiful *P. tattersalli* and *P. candidus* from restricted areas in the northeast (Tattersall, 2007; Fig. 4.15). Like *Avahi*, species of *Propithecus* have very long limbs, especially long legs, and a long tail. All are diurnal. They are found

in a variety of forest types, and all are primarily vertical clingers and leapers, meaning they travel by leaping between vertical supports. When they come to the ground, they progress by means of bipedal hops. During feeding, they use a variety of suspensory postures, often hanging upside down among small branches (Fig. 4.13).

The diet of *P. verreauxi* varies from locality to locality but always includes large amounts of fruit (particularly in the wet season) and leaves (especially in the dry season). All *Propithecus* species have similar day ranges, but *P. edwardsi* has a home range 15 times the size of that used by the dry forest species *P. verreauxi*. Sifakas live in moderate-size groups (3–9 individuals), often with more than one breeding female. In some species, females actively manipulate group size by encouraging subordinate males to join and remain in social groups (Lewis, 2008). Power relationships are unambiguously biased toward females (Voyt et al., 2019). Females generally lead group progressions and are dominant over males in access to food. Males regularly change groups, and there is intense male–male competition in the breeding season. Sifakas normally give birth to one infant every 2 years. In a long-term study of *P. edwardsi* in eastern Madagascar, Wright (1995) found that 67% of the young died before the age of reproduction, and that all males emigrate from their natal group before breeding and continue to emigrate throughout their lifetimes (Morelli et al., 2009). Male infanticide has been reported in several sifaka species. Sifakas are regularly preyed on by the fossa, a large carnivorous mustelid.

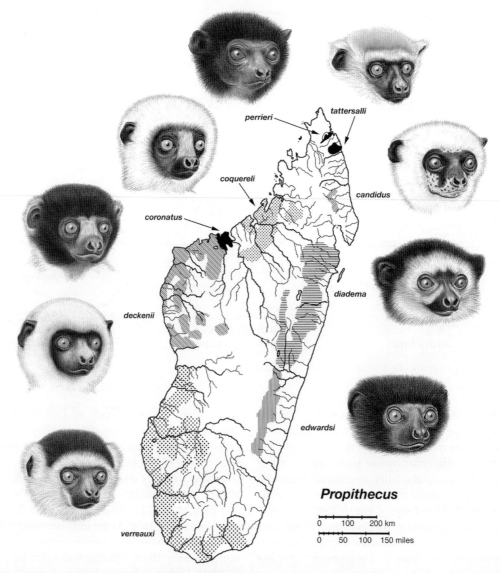

FIGURE 4.15 Map illustrating the allopatric distribution of sifaka (*Propithecus*) species around the island of Madagascar. Map lines delineate study areas and do not necessarily depict accepted national boundaries. *(Drawing by Stephen Nash.)*

The **indri** (*Indri indri*) is a diurnal species from the hilly rainforests of the east coast. Like *Propithecus*, *Indri* has a short face and thick fur. Their body weights are similar to those of the largest individuals of *P. edwardsi*. These creatures have extremely long hands, long feet, and slender arms as well as very long legs, but only a very short tail (Fig. 4.16). They move primarily by leaping in lower levels of the forest but also hang below more horizontal supports, especially during feeding. Like sifakas, indris eat both fruit and leaves, with the proportions varying seasonally. They seem to avoid mature leaves, however, and specialize on young leaves and shoots. Like *Avahi*, they seem to avoid leaves containing alkaloids. The small family groups, averaging about four individuals, defend relatively large territories from other groups. They are extremely vocal and give long,

haunting morning calls to advertise their presence to neighboring groups, which answer sequentially.

Daubentoniids

The **aye-aye** (*Daubentonia madagascariensis*) is about as improbable an animal as one could imagine (Fig. 4.17). This moderately sized (3 kg; Table 4.5), black animal with coarse, shaggy fur, enormous ears, and a large bushy tail has more extreme morphological specializations than any other living primate, but retains many features that clearly link it with other Malagasy lemurs. The aye-aye has a greatly reduced dental formula of 1.0.1.3. The most distinctive feature of its dentition – indeed of the entire skull – is the pair of large, ever-growing rodent-like incisors (Fig. 4.18). The skull has a

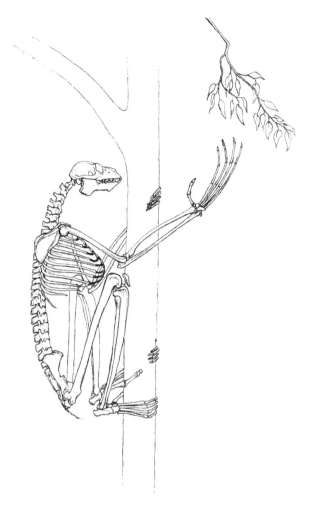

FIGURE 4.16   The skeleton of an indri (*Indri indri*). Note the extremely long hindlimbs, the slender forelimbs, and the reduced tail.

relatively large, globular braincase compared to other lemurs, but the auditory bulla and the cranial arteries are like those found in lemurids and indriids. Aye-ayes have relatively large, clawed digits on both hands and feet (except for the hallux), and the third digit of each hand is extremely long and slender. This is accompanied by a pseudothumb, which consists of both a bony component and a dense cartilaginous extension that is thought to compensate for its overspecialized third digit during grasping (Hartstone-Rose et al., 2020).

Aye-ayes are nocturnal and seem to have a limited diet consisting of fruits, seeds, plant parts, insect larvae, and flowers, most of which are obtained through the use of their gnawing teeth and elongated probing finger. They have been described as specialists on structurally defended foods, both fruits and insects, which are unavailable to other animals. Captive studies have shown that aye-ayes locate insect prey within a log by tapping with their specialized digits and listening to the response. They then obtain the food by gnawing away the bark with their large incisors and by using the

slender finger as a probe. There are no woodpeckers on Madagascar, and the aye-aye has been described as a woodpecker avatar (Cartmill, 1974).

Aye-ayes are mainly quadrupedal and are considered solitary foragers. They live in a noyau-type spatial arrangement, with very large overlapping male ranges and smaller nonoverlapping female ranges (Sterling, 1993; Sefczek et al., 2020). Both sexes have very large night ranges averaging about 1 to 1.5 km. In contrast to most Malagasy species, aye-ayes seem to be nonseasonal breeders, with females coming into estrus at different times during the year, each pursued by several males. They have single births and build large, round nests of sticks and leaves in forks of tree trunks. Aye-ayes are characterized by a very slow ontogeny. Infants remain in their mother's nest until 2 and 3 months, travel independently by about 4 months, and begin percussive foraging by 5 to 8 months. Infants can nurse for as many as 17 months, and do not begin nesting alone until that same time (Rakotondrazandry et al., 2021). Their large brain size and protracted development have been explained by their need to learn extractive foraging behaviors.

Subfossil Malagasy Strepsirrhines

Despite their diversity, the living strepsirrhines of Madagascar are only a fraction of the primate fauna that inhabited the island in the very recent past. An extensive fauna of larger genera and species is known from fossil deposits from as recently as 500 years ago (Table 4.6; Fig. 4.19), and there are reports of large lemurs still living as recently as 350 years ago (Flacourt, 1661, in Tattersall, 1982). Bones of the extinct species have been found in conjunction with human artifacts or in sites indicating human activity, suggesting that humans were largely responsible for their extinction. However, while human hunting may have triggered early megafaunal population declines, a shift toward agriculture combined with human population growth eventually led to many subfossil lemur extinctions (Godfrey et al., 2019). Nevertheless, the persistence of some species for over 1000 years after the appearance of humans indicates that it took many centuries for the lemurs to eventually become extinct. Moreover, new evidence of cut marks on elephant bird bones (*Aepyornis*) suggests human occupations as early as 10,000 years ago, long before Madagascar's largest lemurs went extinct (Hansford et al., 2018, 2020).

Most of the extinct species were large and probably diurnal; some were partly terrestrial, but many were suspensory – two locomotor behaviors that are rare on Madagascar today. In discussing the adaptive radiation of the Malagasy strepsirrhines, these subfossil taxa are most appropriately considered with the living species, since their demise is a very recent and perhaps ongoing

FIGURE 4.17   The solitary and nocturnal aye-aye (*Daubentonia madagascariensis*) probing the bark of a tree with its slender third digit. In the background is an aye-aye nest made of leaves.

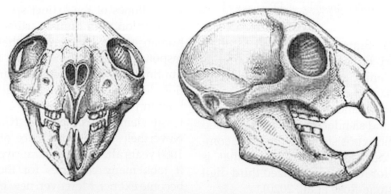

FIGURE 4.18   The skull of an aye-aye (*Daubentonia madagascariensis*) showing the large, domed neurocranium, the large incisors, and the reduced posterior dentition.

TABLE 4.5   Infraorder LEMURIFORMES
            Family DAUBENTONIIDAE

Common name	Species	Intermembral index	Mass (g)	
			M	F
Aye-aye	*Daubentonia madagascariensis*	71	2620	2490

event. The large extinct Malagasy forms are an integral part of the lemur radiation. All of the extinct species are related to the living families and are often found in association with fossil remains of living species and genera. The extinct taxa are not spread evenly among families. There are no known extinct cheirogaleids, and the fossil aye-aye species was much larger, but probably ecologically similar to the living one. The extinct members of other families greatly extend our knowledge of the adaptive radiation of Madagascar primates (Jungers et al., 2002). This is most dramatic in the case of the indriids, whose extinct relatives include six genera in two very distinct subfamilies.

Palaeopropithecines or Sloth Lemurs

Among the most remarkable of the extinct lemurs is an extensive radiation of large, folivorous suspensory indriids that inhabited the forests of Madagascar in recent millennia. The **sloth lemurs**, as they are called, share with the living indriids a reduced dental formula, with two premolars in each quadrant and four anterior teeth in a toothcomb. They are generally placed in a separate subfamily, the Palaeopropithecinae, with four genera that exhibit an increasing adaptation to sloth-like behavior with increasing size. Like extant Malagasy primates, they apparently lacked any sexual size dimorphism. Although the limb proportions and skeletal adaptations in many of these extinct lemurs were extraordinary, suspensory feeding postures are not unusual in the larger living indriids and provide evidence for a behavioral continuity between the living leapers and this extinct radiation. Like living indriids, they were characterized by precocious dental development combined with relatively slow somatic growth.

The genus *Mesopropithecus*, with three species, is the smallest and most primitive palaeopropithecine and the most similar to the living indriids. Dentally and cranially, *Mesopropithecus* is similar to *Propithecus*, but it is slightly larger, more robust, and has larger upper incisors. Like the living indriids, it seems to have been largely folivorous. *Mesopropithecus* was a slow arboreal quadruped with some suspensory abilities, and skeletal remains show that *Mesopropithecus* is a basal member of the sloth lemurs, with several of the distinctive adaptations for suspension found in the larger paleopropithecines (Simons et al., 1995).

Babakotia is from northern Madagascar and has an estimated body mass of just over 15 kg. Its dentition suggests a folivorous diet. Its limbs show more indications of suspensory habits than living indriids or *Mesopropithecus*, but less than the larger sloth lemurs. For moving about, it seems to have relied on a combination of vertical climbing and suspension.

Palaeopropithecus is a much larger genus, with an estimated weight, based on limb size, approaching 50 kg. *Palaeopropithecus* is dentally similar to *Propithecus*, with long, narrow molars and well-developed shearing crests, but differed in that it had vertical lower incisors but no real toothcomb. The skull is similar to that of living indriids but more robustly built, with a longer snout and a heavily buttressed nasal region, suggesting more prehensile lips (Fig. 4.14). The auditory region is superficially quite different from that of living indriids, in that it has a tubular meatus extending laterally from the tympanic ring, apparently related to the extreme development of the mastoid region.

Unlike the living indriids, which have relatively long legs and extraordinary leaping abilities, *Palaeopropithecus* has considerably longer forelimbs than hindlimbs (Fig. 4.20). It has very long, curved phalanges and very mobile joints. It was the most suspensory of all known strepsirrhines, with locomotor abilities that have generally been compared to those of sloths or orangutans.

The genus *Archaeoindris* is closely related to *Palaeopropithecus* but was substantially larger, with an estimated weight of roughly 160 kg, the size of a male gorilla. Dentally and cranially, *Archaeoindris* is similar to *Palaeopropithecus* and was probably folivorous (Fig. 4.14). The morphology of the few limb bones attributed to this giant indriid, together with its great size, suggest that it was probably more terrestrial (Fig. 4.18). Its limb bones suggest that the closest locomotor analogy is with ground sloths.

Archaeolemurines or Monkey Lemurs

Whereas the sloth lemurs seem to have been the Malagasy equivalents of orangutans or sloths, the other subfamily of extinct indriids, the Archaeolemurinae, evolved remarkable morphological similarities to cercopithecoid monkeys, such as macaques. Although small compared to the largest of the sloth lemurs (Table 4.6), monkey lemurs were larger than any living Malagasy genera. Archaeolemurines, like lemurs and cheirogaleids, have one more premolar than either living indriids or palaeopropithecines, and they have large upper central incisors, expanded but slightly procumbent lower incisors and canines, and a fused mandibular symphysis. Both the primitive dental formula and recent genetic data suggest that archaeolemurines are the sister group to all other indriids (Kistler et al., 2015). They were characterized by slow maturation, probably associated with learning a complex diet (Godfrey et al., 2006).

Archaeolemur is one of the best-known fossil Malagasy primates, with hundreds of bones from nearly 20 localities. Traditionally, *Archaeolemur* has been divided into two species: the larger *A. edwardsi* and the smaller *A. majori*. In *Archaeolemur*, the anterior premolar is caniniform, and the entire premolar row forms a long cutting edge. The broad molars have low, rounded cusps arranged in a bilophodont pattern similar to that characterizing Old World monkeys. The dental similarities between *Archaeolemur* and living Old World monkeys, as well as the structure of its dental enamel, suggest that *Archaeolemur* probably had a diverse diet that included fruit and hard objects such as seeds, and invertebrates. In contrast to *Palaeopropithecus*, *Archaeolemur* is

FIGURE 4.19   Artistic reconstruction of the fossil site of Ampazambazimba, Madagascar (*ca.* 8000–1000 BP), showing a variety of subfossil lemurs from that locality. At the upper left, a *Megaladapis* feeds on leaves while clinging to a trunk. Below are two individuals of *Pachylemur insignis* and to the right a family of sloth-like *Palaeopropithecus*. On the ground, an *Archaeoindris* feeds in the background while another individual of *Megaladapis* ambles along to another tree. In the foreground, a group of *Archaeolemur* feed on tamarind pods while a group of *Hadropithecus* wander in from the right.

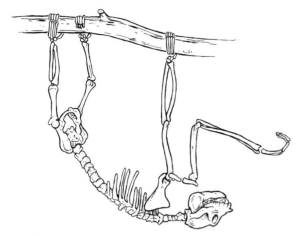

FIGURE 4.20 A skeleton of *Palaeopropithecus ingens* restored in a suspensory feeding position.

TABLE 4.6 Recently extinct Malagasy lemurs

Family LEMURIDAE	Intermembral index	Mass
Pachylemur insignis	98	11,500
P. julliyi	94	13,400
Family MEGALADAPIDAE		
Megaladapis edwardsi	120	85,100
M. grandidieri	115	74,300
M. madagascariensis	114	46,500
Family INDRIIDAE		
Subfamily ARCHAEOLEMURINAE		
Archaeolemur edwardsi	92	26,500
A. majori	92	18,200
Hadropithecus stenognathus	100	35,400
Subfamily PALAEOPROPITHECINAE		
Mesopropithecus pithecoides	99	11,000
M. globiceps	97	11,300
M. dolichobrachion	113	13,700
Babakotia radofilai	119	20,700
Palaeopropithecus ingens	138	41,500
P. maximus	144	45,800
Archaeoindris fontoynonti	~100–135	161,200
Family DAUBENTONIIDAE		
Daubentonia robusta	85	14,200

characterized by a very slow dental development relative to somatic growth. Cranially, it is similar to living indriids (Fig. 4.14), with no monkey-like features.

In skeletal anatomy *Archaeolemur* has striking similarities to Old World monkeys in its limb proportions and in the configuration of individual limb elements and joint surfaces. While many details of its limbs suggest a terrestrial habitat, the short limbs relative to trunk length are characteristic of arboreal quadrupeds. It probably exploited both arboreal and terrestrial supports.

The single species of **Hadropithecus** was the most specialized of the monkey lemurs. It has smaller incisors, reduced anterior premolars, and expanded, molarized posterior premolars compared with *Archaeolemur*. The molars have additional foldings of enamel that form a complex array of dentine and enamel crests and develop extremely flat wear. Like *Archaeolemur*, *Hadropithecus* was characterized by very slow dental development. Both functional and isotopic studies of its dentition suggest that *Hadropithecus* had a diet that was very different from that of any other lemur, living or extinct. Perhaps it specialized on underground tubers (Godfrey et al., 2010). *Hadropithecus* has a very short face and a robust skull with well-developed sagittal and nuchal crests (Fig. 4.14). Limb bones of *Hadropithecus* and *Archaeolemur* show no similarities to those of either indriids or palaeo-propithecines, but are more similar to lemurids and Old World monkeys. *Hadropithecus* had relatively short limbs for its trunk length and probably moved both on the ground and in the trees (Fig. 4.19; Godfrey et al., 2006).

Megaladapines, the Koala Lemurs

Another unusual extinct lemur was **Megaladapis**, with three species divided into two subgenera. The largest species had an estimated body size of ~85 kg, based on limb size, but had the teeth of an animal twice as large. This extinct giant shares several unusual dental and cranial features with the living *Lepilemur*, and the two have often been placed in the same family. However, studies of ancient DNA indicate that *Megladapis* is actually more closely related to the lemurids (Kistler et al., 2015; Marciniak et al., 2021). *Megaladapis* has long, narrow molars with well-developed crests, indicating a folivorous diet, and no upper incisors. The lack of incisors suggests to some authors that *Megaladapis* may have had a pad covering the premaxillary region (as in living artiodactyls) that occluded with the lower incisors for cropping herbivorous foods. In addition, the pronounced nasal region suggests the possibility of large prehensile lips, perhaps for cropping tough, thorny vegetation. The expanded mandibular condyle on *Megaladapis* is like that of *Lepilemur* and distinct from other strepsirrhines. *Megaladapis* had a fused mandibular symphysis like many of the larger fossil lemurs. This giant lemur has a long, flat cranium, a long snout, huge frontal sinuses, and a small braincase. Like the folivorous palaeopro-pithecines, *Megaladapis* showed a fast pattern of dental development relative to cranial growth.

Megaladapis has long forelimbs relative to the size of its hindlimbs, and a very long trunk. In contrast to the slender limbs of *Palaeopropithecus*, those of *Megaladapis* are extremely robust and short. The hands and feet are enormous (longer than its femur), with moderately curved phalanges and robust, divergent pollex and hallux. This genus seems to have been a vertical clinger and climber, similar to the living koala of Australia, and it probably clung to vertical trunks while cropping vegetation with its snout (Fig. 4.19). On the ground, *Megaladapis* no doubt moved quadrupedally between trees. Differences in the skeletal morphology suggest that the larger *M. edwardsi* was more terrestrial than the other species.

According to both dental and mitogenomic data, **Pachylemur** is a subfossil lemurid that is most closely related to the living ruffed lemur, *Varecia*, but is much larger (Kistler et al., 2015). There are two species. *Pachylemur* is more robustly built in its limb skeleton than *Varecia* and has forelimbs and hindlimbs that are similar in length. It was a slow, arboreal quadruped with

even less leaping ability than the living *Varecia* (Fig. 4.19). *Pachylemur* probably had a frugivorous diet.

Adaptive Radiation of Malagasy Primates

Like the Galapagos finches, the Malagasy strepsirrhines were a natural experiment in evolution. The entire radiation likely evolved from two colonization (rafting) events: one for the aye-aye, and another for the ancestor of other living lemurs (Gunnell et al., 2018). Isolated from repeated faunal invasions and from ecological competition with other primates (until humans arrived) and many other groups of mammals, this lineage evolved an extremely diverse array of species with dietary and locomotor adaptations for exploiting a wide range of ecological conditions (Fig. 4.21). In size, these animals

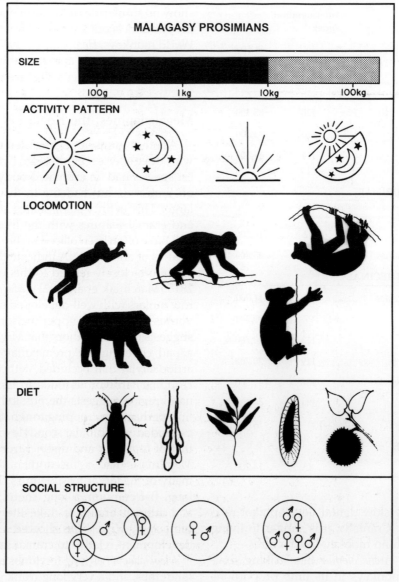

FIGURE 4.21   The adaptive diversity of Malagasy prosimians, including both extant and subfossil species.

ranged from the smallest living primate, a mouse lemur (30 g), to *Archaeoindris*, which must have weighed as much as living gorillas. In their gross dietary preferences there are species specializing on insects, on gums, on fruits, on nectar, on leaves, on bamboo plants, and perhaps on underground tubers, and numerous dental adaptations have evolved in accordance with these dietary differences. In locomotor abilities, they include prodigious leapers, arboreal and probably terrestrial quadrupeds, long-armed suspensory species, and some, such as the koala-like *Megaladapis* and the ground sloth-like *Archaeoindris*, which have no analog among other primates. Their limb proportions and many other aspects of their musculature and skeletal anatomy show considerable anatomical diversity that is functionally related to the behavioral differences.

The Malagasy radiation includes numerous diurnal, nocturnal, and cathemeral species. The remarkably flexible activity period of the mongoose lemur is not uncommon among lemurs, but such flexibility in activity periods is not known for other primates (but see Ankel-Simons and Rasmussen, 2008). Generalizations about the social organization of the Malagasy primates are more difficult, both because many aspects are poorly understood and because we have little information about the behavior of the larger (diurnal, and in some cases probably terrestrial) subfossil species (but see Godfrey et al., 1997, 2010). Although many of the nocturnal species are solitary foragers, they show considerable diversity in their ranging patterns and use of sleeping sites: some show a noyau arrangement, others are pair-living, and still others live in larger groups with numerous males and females. The more gregarious diurnal species are equally diverse. In *L. catta*, multimale and multifemale groups are structured around female matrilines, whereas the large groups in *Eulemur* have been hypothesized to be clusters of monogamous units (van Schaik and Kappeler, 1996; but see Overdorff, 1998). Overall, the diversity in patterns of social organization among the Malagasy primates exceeds that of any other primate radiation (see Kappeler, 2012).

The history and causal mechanisms that have been selected for these various grouping patterns are the subject of lively debate. In many ways the social relationships of Malagasy primates are quite different from those that characterize most living anthropoids (see Kappeler, 2012). Particularly striking is the frequency of female feeding dominance over males. Several hypotheses have been advanced to explain this phenomenon. These include the energy frugality hypothesis (Wright, 1999), which notes that Malagasy species often have low metabolic rate and eat low-quality food resources (e.g., Ganzhorn et al., 2009). In contrast, van Schaik and Kappeler (1996) have proposed the evolutionary disequilibrium hypothesis, which suggests that the social behavior of many diurnal lemurs is "in transition" from a recent nocturnal habitus. No hypothesis is completely satisfactory. Some Malagasy primates (genus *Eulemur*) are dichromatic, but in contrast to most anthropoids, there is rarely sexual dimorphism in either body mass or canine size. Both males and females have large canines, suggesting considerable intraspecific competition in both sexes. The uniformity of these features among all of the Malagasy species, despite their great adaptive diversity and the striking habitat diversity of the island, is remarkable and difficult to understand. Studies of this extraordinary radiation continue at a rapid pace (Gould and Sauther, 2009; Kappeler, 2012).

Galagos and Lorises

In addition to the diverse radiation of strepsirrhines on Madagascar, there is a mainland radiation represented by the galagos in Africa and the lorises in Africa and Asia (Figs. 4.22 and 4.23). Galagos and lorises share with the Malagasy families the strepsirrhine characteristics of a dental tooth comb, a grooming claw on the second digit, and a flared talus (Fig. 4.1), but they also have cranial features that separate them from most of the Malagasy families. Lorises and galagos share with cheirogaleids a unique blood supply to the anterior part of the brain through the ascending pharyngeal artery (Fig. 2.13) rather than through the stapedial artery. In the ear region of lorises and galagos, the tympanic ring is fused to the lateral wall, as in *Allocebus*, rather than being suspended within the bulla, as in most Malagasy forms. In both of these features, the lorises and galagos show similarities with cheirogaleids, but this must be the result of parallel evolution. Genetic studies have repeatedly confirmed that the Malagasy taxa and loris–galago radiations are separate clades (e.g., Roos et al., 2004). The overall cranial morphology of lorises and galagos is very similar and distinct from that seen in indriids and most lemurids. Galagos and lorises are nocturnal and arboreal, but the two families are extremely different in their postcranial morphology and locomotor behavior. The former are primarily leapers; the latter are slow climbers.

Galagids

The galagos, or bushbabies, are a far more diverse group than previously realized (Table 4.7), and the numbers of genera and species continue to grow with the increasing use of genetics, as well as new behavioral techniques for studying nocturnal primates (Pozzi et al., 2014, 2020). Current authorities recognize 6 genera and at least 18 species (e.g., Nekaris and Bearder, 2011; Masters et al., 2017).

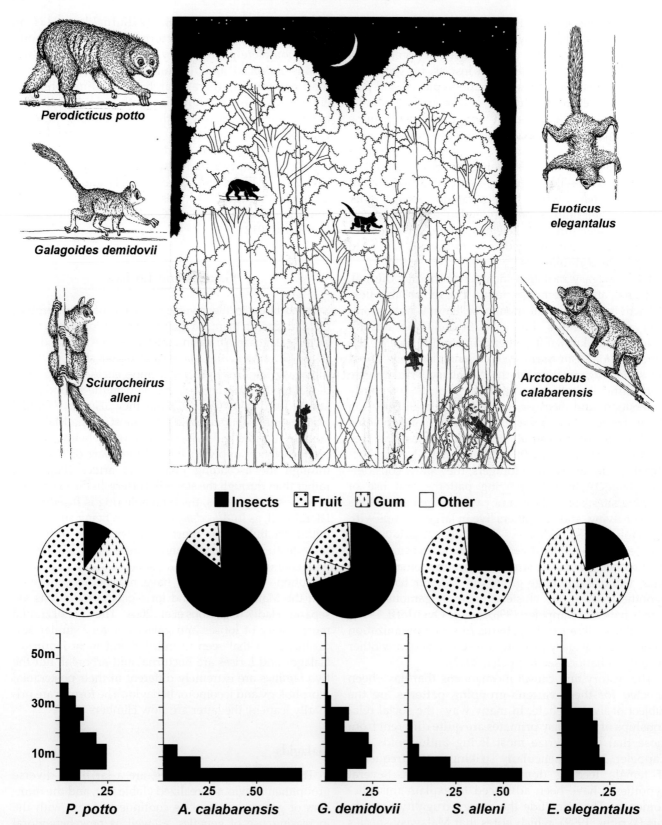

FIGURE 4.22 Diet and forest height for five sympatric lorisoids from Gabon. *(Data from Charles-Dominique, P., Martin, R.D., 1970. Evolution of lorises and lemurs. Nature 227, 257–260).*

Otolemur, the **greater galago**, weighs approximately 1 kg and is found in eastern and southern Africa. *Otolemur crassicaudatus*, the **thick-tailed greater galago**, is the largest of the galagos. A slightly smaller species, *O. garnettii*, has a more limited distribution on the coast of eastern Africa. Thick-tailed galagos have large ears, a long tail, relatively long lower limbs, and an elongated calcaneus and navicular. These proportions of both the limbs and the ankle in *Otolemur* are less extreme than in other galagos. In both of these species of *Otolemur*, males are significantly heavier than females.

Like all galagos, greater galagos are nocturnal. They are found in relatively low forests between 6 and 12 m high and move mainly by quadrupedal walking and running, and less often by leaping. Their diet consists primarily of fruits and gums and varies considerably from season to season. In contrast, the diet of *O. garnettii* consists of fruit and animal prey with no gums.

Greater galagos are solitary foragers that live in a noyau social organization with multiple males and females overlapping in range use. Females build leaf nests for their twin, or even triplet, offspring, and carry their infants in their mouth if they must move them.

There are several genera of smaller bushbabies (Nekaris and Bearder, 2011). The lesser bushbabies, genus *Galago*, are the most widespread, with four species. *Galago senegalensis*, the **Senegal bushbaby**, has the largest distribution, extending from Senegal in the west across central Africa to eastern Africa; *Galago moholi*, the

FIGURE 4.23 The two Asian lorisids: *left*, a slow loris (*Nycticebus coucang*) from southeast Asia; *right*, a slender loris (*Loris tardigradus*) from India and Sri Lanka.

TABLE 4.7 Infraorder LORISIFORMES

Family GALAGIDAE

Common name	Species	Intermembral index	Mass (g) M	Mass (g) F
Allen's bushbaby	*Sciurocheirus alleni*	-	340	
Gabon Allen's bushbaby	*S. gabonensis*	-	-	
Makande Allen's bushbaby	*S. makandensis*	-	260	
Demidoff's dwarf galago	*Galagoides demidovii*	68	60	55
Angola dwarf galago	*G. kumbirensis*	-	-	-
Thomas' dwarf galago	*G. thomasi*	67	82	75
Kenya coast dwarf galago	*Paragalago cocos*	-	150	138
Mozambique dwarf galago	*P. granti*	-	150	
Mountain dwarf galago	*P. orinus*	-	90	
Rondo dwarf galago	*P. rondoensis*	-	60	
Zanzibar bushbaby	*P. zanzibaricus*	60	136	
Northern lesser galago	*Galago senegalensis*	52	225	200
Somali lesser galago	*G. gallarum*	-	200	-
Southern lesser galago	*G. moholi*	54	187	173
Spectacled lesser galago	*G. matschiei*	-	209	
Southern needle-clawed galago	*Euoticus elegantulus*	64	300	
Northern needle-clawed galago	*E. pallidus*	-	279	241
Thick-tailed bushbaby	*Otolemur crassicaudatus*	70	1190	1110
Small-eared greater bushbaby	*O. garnetti*	69	794	734

South African lesser galago, is distributed broadly over much of southern Africa; *Galago gallarum*, the **Somali galago**, has more restricted distributions in eastern Africa; and *Galago matschiei*, the **spectacled galago**, is restricted to the Albertine Rift in Uganda, DRC, Rwanda, and Burundi. Lesser bushbabies inhabit a wide range of forests, woodlands, and vegetation thickets, but are commonly found in *Acacia* woodlands. They are all smaller than *Otolemur*, have relatively long legs and ankle bones, and are spectacular leapers that occupy the opposite extreme of the galagid locomotor spectrum from *Otolemur*. They are found in many forest levels on smaller supports and travel almost exclusively by leaping. Their diet is comprised mainly of insects, but gum is a major component in the dry season. During gum feeding, these small galagos cling to the rough bark of acacia trees by grasping.

The best-studied species of the lesser bushbabies is *G. moholi*. In this species, individuals forage separately in large, individual home ranges, but they often group together in daytime sleeping nests. They are not particular about their choice of nest sites, frequently choosing tangles as well as holes in trees. Among adult males there appears to be a social dominance hierarchy related to age and weight. Because of the harsh, unpredictable nature of their environment and high mortality rates, these primates seem to have a "boom or bust" reproductive strategy, with females capable of having up to two litters of twins per year. They will also sometimes enter brief seasonal bouts of torpor, like the cheirogaleids, but only when other methods of thermoregulation (e.g., huddling) fail (Nowack et al., 2013).

The dwarf galagos include eight named species, and several more populations remain to be described (Nekaris and Bearder, 2011). Two of the best known are **Demidoff's galago**, *Galagoides demidoff* (Fig. 4.22), and **Thomas' galago**, *G. thomasi*. Dwarf galagos resemble *Microcebus* in many behavioral features (Charles-Dominique and Martin, 1970). Their range extends in a band across central Africa from west to east. Throughout this region, they are very common in dense vegetation of either the canopy of primary forests or the understory of secondary forests. They are less specialized leapers than either *G. senegalensis* or *Sciurocheirus alleni* and move mainly by quadrupedal walking and running, with short leaps between branches. Their diet in western Africa is predominantly insects (70%), with lesser amounts of fruit (19%) and gums (10%) (Fig. 4.22). Their social structure is a noyau system with overlapping male and female ranges, and daytime sleeping nests are shared by groups of females and occasional visiting males. They seem to have single births once a year in some parts of their range, but in other areas they frequently have twins.

Allen's bushbaby, or **Allen's squirrel galago** (*S. alleni*), is one of several squirrel galagos from Gabon, Cameroon,

and Bioko (Table 4.7; Fig. 4.22). It is a medium-sized galago from West Africa whose affinities have been subject to some debate. *S. alleni* moves by leaping between small vertical supports in the understory and between these small trees and the ground. Its diet varies considerably across sites and also perhaps seasonally. In the primary forests of Gabon, Allen's bushbabies eat 25% animal matter, 75% fruit, much of which is from the ground, and some gums; but in a secondary forest locality they eat a much higher proportion of insects. Individuals forage alone, and males have large home ranges overlapping the ranges of several females. Allen's bushbabies are much less prolific than Senegal bushbabies. In their relatively stable rainforest habitat, females give birth to only a single infant per year.

The most specialized of the galagos is *Euoticus*, the **needle-clawed galago** (Fig. 4.22). Comprising two species, these medium-sized primates resemble the cheirogaleid *Phaner* in having numerous morphological specializations, such as procumbent upper incisors, caniniform upper anterior premolars, and laterally compressed, claw-like nails related to its gum-eating habits. Both *Phaner* and *Euoticus* can be found in primary and secondary forests, although both specialize in habitats with tall, large-diameter trees (Forbanka, 2018). Needle-clawed galagos use all levels of the canopy and move both quadrupedally and by leaping. They are particularly adept at clinging to large trunks and branches, which are the source of their main food – gums. Most of their foraging is solitary, and their social behavior is largely unknown.

Lorisids

In contrast to the rapid running and leaping of galagos, lorises (Table 4.8) are best known for their slow, stealthy habits. They have smaller ears than galagos, their forelimbs and hindlimbs are more similar in length, and they lack long tails. Despite their slowness, they have the broadest geographic distribution (between endpoints) of any strepsirrhine family, with two or three genera in Africa and two in Asia. The number of lorisid species is debated, and there is considerable morphological and behavioral diversity within recognized taxa (Nekaris and Bearder, 2011).

The **pottos** (*Perodicticus potto*) are the largest of the African lorises and the most widespread. Their range extends from Liberia in the west to Kenya in the east. Pottos prefer the main, continuous canopy of both primary and secondary forests and move on relatively large supports (Fig. 4.22). At some sites, their diet has been reported to be largely frugivorous, with much smaller amounts of animal matter and gums. At other sites, however, pottos eat much more gum and less fruit and animal material. These might be seasonal differences. The potto has been described as an animal that specializes in

TABLE 4.8 Infraorder LORISIFORMES

Family LORISIDAE

Common name	Species	Intermembral index	Mass (g) M	Mass (g) F
West African Potto	*Perodicticus potto*	88	830	836
Milne–Edwards' Potto	*P. edwardsi*	-	1397	1371
East African Potto	*P. ibeanus*	-	844	
Calabar Angwantibo	*Arctocebus calabarensis*	89	312	306
Golden Angwatibo	*A. aureus*	-	210	
Red Slender Loris	*Loris tardigradus*	90	163	132
Gray Slender Loris	*L. lydekkerianus*	86	264	269
Sunda Slow Loris	*Nycticebus coucang*	88	679	626
Bengal Slow Loris	*N. benegalensis*	-	1100	1020
Pygmy Slow Loris	*N. pygmaeus*	91	418	423
Philippine Slow Loris	*N. menagensis*	-	265–300	
Javan Slow Loris	*N. javanicus*	83	687	626
Kayan Slow Loris	*N. kayan*	-	410	
Bangka Slow Loris	*N. bancanus*	-	-	-
Bornean Slow Loris	*N. borneanus*	-	-	-
Sumatran Slow Loris	*N. hilleri*	-	-	-

olfactory foraging, and its most characteristic activity is slow climbing along large supports, with its nose to the branch.

Like galagos, pottos seem to be solitary foragers and to have overlapping male and female ranges. Studies of Milne-Edward's pottos (*Perodicticus edwardsi*), however, suggest these animals regularly form social networks, and they are commonly observed near conspecifics with sleeping associations between two adults and offspring (Pimley et al., 2005; Nekaris, 2006). Female pottos have single births. Pottos do not build nests but rather sleep curled up on branches. During the night, females do not carry their infant but "park" it for the evening and return for it later. The infants are white at birth and turn a drab brown within a few months.

The **angwantibos** are smaller, more slender African lorisids, with a restricted distribution in west central Africa. There are two species, *Arctocebus calabarensis* (**angwantibo**) and *A. aureus* (**golden angwantibo**). Angwantibos prefer the understory of primary and secondary forests, where they are usually found less than 5m above the ground. They move by slow quadrupedal climbing on very small branches and lianas. They are predominantly insectivorous, specializing on noxious caterpillars. Socially, they appear to be very similar to pottos.

Asian lorises (Fig. 4.23), like their African counterparts, come in two shapes, thin and plump, but the two genera are not sympatric. The **slender lorises**, *Loris*, are from Sri Lanka and southern India. The smaller red slender loris, *Loris tardigradus*, is from Sri Lanka. The larger grey slender loris, *L. lydekkerianus*, is from Sri Lanka and parts of southern India. Slender lorises have been accurately described as bananas on stilts. They are found in the understory of dry forests and in the canopy of wetter forests, where they move about among the fine twigs. Their diet consists totally of animal matter, including a diverse menu of both invertebrates and vertebrates ranging in size from ants to lizards (Nekaris and Bearder, 2011). Although often considered solitary, slender lorises spend only about half of their time alone (Nekaris, 2006). While females rarely interact with each other, males are more social, and are often responsible for maintaining proximity to both male and female conspecifics (Nekaris, 2006).

The **slow lorises** (*Nycticebus*), from Southeast Asia, are stockier animals. There are numerous species. The largest is the Bengali slow loris, *Nycticebus bengalensis* from Assam. *Nycticebus coucang* is from mainland Southeast Asia and Sumatra, and has two closely related taxa, *N. javanicus* from Java and *N. menangensis* from Borneo. The small *N. pygmaeus* from Vietnam and Laos is sympatric with *N. coucang* (Ravosa, 1998). Slow lorises are found primarily in the main canopy but seem to show no preference for primary, secondary, or deciduous forests. Slow lorises move mainly by slow quadrupedal climbing and bridging. All reports indicate that fruit, animal prey, and gums make up the diet of all species of slow lorises (Nekaris and Bearder, 2011).

Most species have been reported to sleep in either tree hollows or vine tangles, and sleeping groups include one or more females with offspring and often a male. Much like with some pottos, recent studies of Malay slow lorises (*Nycticebus coucang coucang*) have discovered that these animals are more social than previously thought, forming detectable social networks (Wiens and Zitzmann, 2003; Nekaris, 2006). They are also one of only seven known venomous mammals. In fact, their facial markings are considered by some to be a form of Muellerian mimicry of cobras (Nekaris et al., 2013). When threatened, animals can activate their venom by mixing saliva with oils secreted from their brachial gland (close to the elbow). Though its function is still not well understood, slow lorises may use their venom to defend against parasites, predators, and even conspecifics (Nekaris, 2014).

Adaptive Radiation of Galagos and Lorises

Compared to that of their Malagasy relatives, the adaptive radiation of galagos and lorises is limited. They are all relatively small and nocturnal, a likely correlate of their geographic overlap with the larger, diurnal catarrhine primates throughout their range. Nevertheless, on the basis of recent systematic and ecological studies, we are coming to appreciate more of the diversity of these nocturnal strepsirrhines (Rasmussen and Nekaris, 1998; Pozzi et al., 2014; 2020; Nekaris and Burrows, 2020). Within their nocturnal habits, the galagos and lorises have evolved sufficient adaptive differences to permit as many as five sympatric species and a wide diversity of allopatric taxa (Fig. 4.22; Nekaris and Bearder, 2011). Both families include some species that specialize on fruits, gums, or insects, as well as many that show seasonal specializations, but a more diverse diet over a full year. Even though the lorises seem to have a rather stereotyped, stealthy locomotor behavior, their different sizes permit sympatric species (in Africa) to use different types of supports in different parts of the canopy. The galagos are more speciose and show more diversity in diet and locomotor abilities, with extremely saltatory species, largely quadrupedal species, and specialized, clawed trunk scramblers. The nocturnal lorises and galagos appear to show less diversity in social organization. Most seem to be solitary foragers, and species vary in the composition of sleeping groups, but this aspect of their behavior remains very poorly documented compared to that of other primates.

Phyletic Relationships Among Strepsirrhines

Molecular studies over the past decade have yielded a consistent picture of many of the phylogenetic relationships among extant strepsirrhines (Fig. 4.24), but there remain many unresolved relationships (e.g., Roos et al., 2004; Karanth et al., 2005; Horvath et al., 2008;

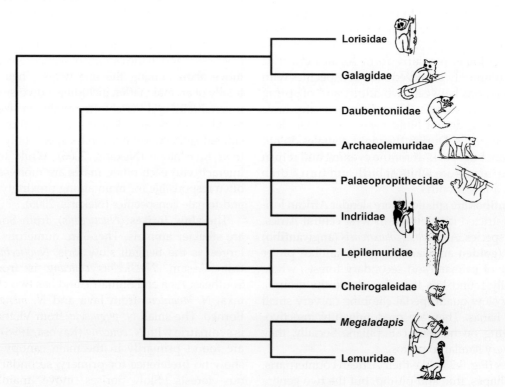

FIGURE 4.24   A phylogeny of strepsirrhines, including representative subfossil families and genera.

Orlando et al., 2008; Perelman et al., 2011; Springer et al., 2012). Despite some anatomical similarities between lorisiforms and cheirogaleids, all studies agree that the Malagasy primates are a distinct radiation from the lorises and galagos, and that cheirogaleids are nested within the Malagasy radiation. Within lorisiforms, the lorisids and galagids are distinct clades, and the African lorises and Asian lorises are each a natural group, even though each biogeographic group contains plump and slender taxa.

Among the Malagasy lemurs there is fairly general, though not universal, agreement that lemurids, indriids, lepilemurids, cheirogaleids, and the aye-aye are distinct groups of strepsirrhines, and that the aye-aye is the sister taxon to all other Malagasy primates. Relationships within lemurids seem stable: *Lemur*, *Hapalemur*, and *Prolemur* form a clade that is the sister taxon to *Eulemur*, and *Varecia* is the sister taxon to all other lemurids. On the basis of ancient DNA, *Megaladapis* groups as the sister taxon to lemurids, not with *Lepilemur* as suggested by previous morphological analyses (Marciniak et al., 2021). Within cheirogaleids, most studies find that *Mirza* is the sister taxon to *Microcebus*, with *Allocebus* and *Cheirogaleus* successive sister taxa. Indriids are a natural group and are related to archaeolemurids and palaeopropithecids, although the details among these are unresolved (Orlando et al., 2008; Kistler et al., 2015; Herrera and Davalos, 2016).

Relationships among families of extant lemurs have been difficult to resolve, but numerous molecular studies have recently converged on a stable hypothesis (Fig. 4.24). Multiple studies analyzing both nuclear and mitochondrial DNA now suggest that lemurids are the sister clade to a group including indriids, cheirogaleids, and lepilemurids. Within this latter group, indriids are the sister to a cheirogaleid plus lepilemurid clade (Perelman et al., 2011; Springer et al., 2012). These relationships are also recovered when subfossil lemurs are included (Kistler et al., 2015; Marciniak et al., 2021; but see Herrera and Davalos, 2016). Difficulty in resolving the phylogeny of Malagasy primates undoubtedly reflects the fact that the major diversification of the different families and genera took place within a short period of time.

Haplorhines

In the past, primates were frequently divided into two major groups. The prosimians included strepsirrhines and tarsiers, while the higher primates (monkeys, apes, humans) were placed in a separate order, the Anthropoidea (Fig. 4.25). It has become increasingly evident in recent decades that the major phyletic division among living primates is between the tooth-combed strepsirrhines (galagos, lorises, and Malagasy lemurs) and the haplorhines (tarsiers and anthropoids). The living haplorhines share a number of very derived cranial

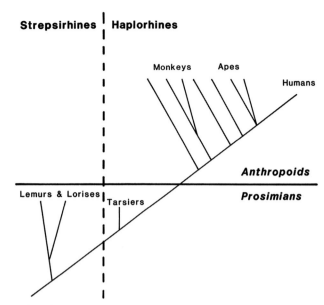

FIGURE 4.25 The gradistic division of primates into prosimians and anthropoids contrasted with the phyletic division into strepsirrhines and haplorhines.

features, including postorbital closure to some extent, a retinal fovea in their eyes, a reduced number of nasal conchae, a short, vertical nasolacrimal duct (Fig. 2.16; Rossie and Smith, 2007) and the lack of a moist rhinarium, giving them (us) the dry nose and continuous upper lip from which the term "haplorhine" derives. In addition, haplorhines all have a hemochorial placenta and an inability to synthesize vitamin D.

Tarsiers

The **tarsiers** of Southeast Asia (Table 4.9) are among the smallest and most unusual of all living primates. They show a mixture of strepsirrhine and anthropoid features. However, their similarities to strepsirrhines are primitive: an unfused mandibular symphysis, molar teeth with high cusps, grooming claws on their second (and third) toes, multiple nipples, and a bicornuate uterus. Their similarities to anthropoid primates seem to be derived specializations indicative of a close phyletic relationship. In addition, tarsiers have many distinctive features all their own (Fig. 4.26).

The most striking feature of a tarsier is the size of its eyes, each of which is actually larger than the animal's brain, not to mention its stomach. Tarsiers differ from other nocturnal primates, and resemble all diurnal primates in having a retinal fovea and lacking the reflective tapetum found in all lemurs and lorises, as well as many other groups of mammals. Their large eyes are protected by a bony socket that is partly closed posteriorly, similar to that of higher primates. The nose of tarsiers resembles that of higher primates as well, both externally in the lack

TABLE 4.9   Infraorder TARSIIFORMES

Family TARSIIDAE

Common name	Species	Intermembral index	Mass (g)	
			M	**F**
Philippine tarsier	*Carlito syrichta*	58	135	119
Western tarsier	*Cephalopachus bancanus*	52	128	117
Dian's tarsier	*Tarsius dentatus*	-	118	105
Makassar tarsier	*T. fuscus*	-	130	119
Lariang tarsier	*T. lariang*	-	118	102
Peleng tarsier	*T. pelengensis*	-	-	-
Sulawesi mountain tarsier	*T. pumilus*		49	52
Great Sangihe tarsier	*T. sangirensis*	-	135	150
Selayar tarsier	*T. tarsier*	53	136	116
Siau Island tarsier	*T. tumpara*	-	-	-
Wallace's tarsier	*T. wallacei*	-	115	99
Gursky's spectral tarsier	*T. spectrumgurskyae*	-	126	119
Jatna's tarsier	*T. supriatnai*	-	135	-
Niemitz's tarsier	*T. niemitzi*	-	-	-

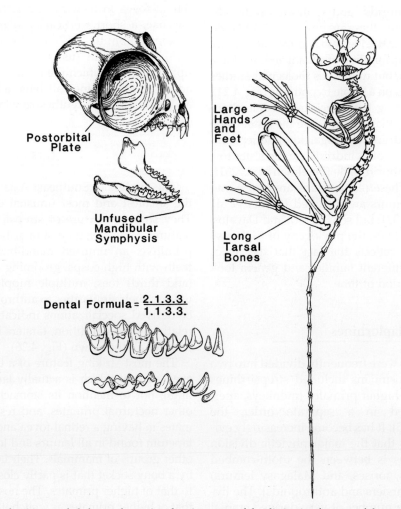

FIGURE 4.26   The skull, dentition, and skeleton of a tarsier, showing some of the distinctive features of the group.

of an attached upper lip with a median fold, and internally in the greatly reduced turbinates and in the absence of a sphenoid recess. In tarsiers, as in higher primates, the major blood supply to the brain comes through the promontory branch of the internal carotid (Fig. 2.13). The tympanic ring lies external to the auditory bulla and extends laterally to form a bony tube, the external auditory meatus. The tarsier dental formula, 2.1.3.3/1.1.3.3, is unique among primates, but tarsier teeth resemble those of anthropoids in overall proportions, with large upper central incisors, small lower incisors, and large canines.

The postcranial skeleton of tarsiers is striking in many of its proportions. The hands and feet of these tiny animals are relatively enormous, reflecting both their clinging abilities and their predatory habits (Figs. 4.26 and 4.27). They have extremely long legs and many more specific adaptations for leaping, including a fused tibia and fibula and the very long ankle region responsible for their name.

In their reproductive physiology, tarsiers show several similarities to anthropoids, as noted above. They have a hemochorial placenta rather than the epitheliochorial type found in lemurs, and they produce relatively large offspring. Female tarsiers undergo monthly sexual cycles, with swellings reminiscent of some Old World monkeys.

Living tarsiers are currently divided into 3 genera and 14 species (Table 4.9; Fig. 4.28; Groves and Shekelle, 2010). The genus *Tarsius*, with 12 species, is restricted to

FIGURE 4.27   A pair of tarsiers. *(Drawing by Stephen Nash.)*

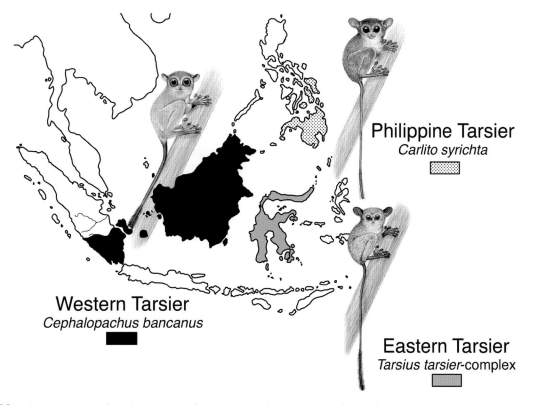

Philippine Tarsier
Carlito syrichta

Western Tarsier
Cephalopachus bancanus

Eastern Tarsier
Tarsius tarsier-complex

FIGURE 4.28   The distribution of the three genera of tarsiers in southeast Asia. Map lines delineate study areas and do not necessarily depict accepted national boundaries. *(Drawing by Stephen Nash.)*

Sulawesi and nearby small islands. The other two tarsier genera are, at present, monotypic. *Carlito syrichta* is from the Philippines, and *Cephalopachus bancanus* is from Borneo, Sumatra, and nearby small islands. The allopatric tarsier genera differ in numerous morphological features, including body size, chromosome number, limb proportions, orbit size, number of nipples, and patterns of vocalization and sociality.

Tarsiers seem to be common in many types of forests, but they are particularly abundant in secondary forests and scrub. They are totally nocturnal and spend their days sleeping in the grass or on vines in trees. Tarsiers often travel and feed very near the ground. In Sulawesi, other activities, such as calling and resting, take place higher in the canopy. All tarsiers move mainly by rapid leaps of up to 3 m. The Bornean tarsier seems to be most committed to vertical clinging and leaping, but this may reflect differences in data collection (Dagosto et al., 2001). *Tarsius pumilus*, the pygmy tarsier, is unusual in having keeled nails for clinging to moss-covered trees. All tarsiers are totally faunivorous: they eat insects, arachnids, and small vertebrates such as snakes and lizards. There is now also evidence that several species of tarsier communicate in the ultrasonic range, although the function of these calls remains poorly understood (Ramsier et al., 2012; Gursky, 2015).

There is evidence from both naturalistic and captive studies of considerable differences in social behavior within and among species (Gursky, 2011). Philippine and Bornean tarsiers have been reported to live in a noyau system, with solitary individuals living in overlapping ranges. Philippine tarsiers always sleep alone. Sulawesi tarsiers, on the other hand, live in families or perhaps small polygynous groups. These families sleep together, give complex territorial duet calls early every evening, and then forage together all night. Males and females actually chase other tarsiers out of their territories.

All tarsiers have a remarkably long, 6-month gestation period and one of the lowest rates of fetal growth among mammals (Roberts, 1994). Singleton infants weigh up to 30% of the mother's weight at birth and suckle for 2 months. During that time, they seem to develop adult prey-catching patterns. Females are largely responsible for care of the infant and usually park the infant while foraging (Gursky, 1994). However, recent evidence suggests that subadult females, and to a lesser extent, adult and subadult males also contribute to allomaternal care (Gursky, 2000).

References

Ankel-Simons, F., Rasmussen, D.T., 2008. Diurnality, nocturnality, and the evolution of primate visual systems. Yearb. Phys. Anthropol. 51, 100–117.

Baden, A.L., 2011. Communal infant care in black-and-white ruffed lemurs (*Varecia variegata*). Stony Brook University. PhD Thesis.

Baden, A.L., 2019. A description of nesting behaviors, including factors impacting nest site selection, in black-and-white ruffed lemurs (*Varecia variegata*). Ecol. Evol. 9 (3), 1010–1028.

Baden, A.L., Oliveras, J., Gerber, B.D., 2020a. Sex-segregated range use by black-and-white ruffed lemurs (*Varecia variegata*) in Ranomafana National Park, Madagascar. Folia Primatol. 92 (1), 12–34.

Baden, A.L., Webster, T.H., Bradley, B.J., 2020b. Genetic relatedness cannot explain social preferences in black-and-white ruffed lemurs, *Varecia variegata*. Anim. Behav. 164, 73–82.

Baden, A.L., Webster, T.H., Kamilar, J.M., 2016. Resource seasonality and reproduction predict fission–fusion dynamics in black-and-white ruffed lemurs (*Varecia variegata*). Am. J. Primatol. 78 (2), 256–279.

Baden, A.L., Wright, P.C., Louis, E.E., Bradley, B.J., 2013. Communal nesting, kinship, and maternal success in a social primate. Behav. Ecol. Sociobiol. 67 (12), 1939–1950.

Batist, C.H., Razafindraibe, M.N., Randriamanantena, F., Baden, A.L., 2022. Factors affecting call usage in wild black-and-white ruffed lemurs (*Varecia variegata*) at Mangevo, Ranomafana National Park. Primates 63 (1), 79–91.

Beeby, N., Baden, A.L., 2021. Seasonal variability in the diet and feeding ecology of black-and-white ruffed lemurs (*Varecia variegata*) in Ranomafana National Park, southeastern Madagascar. Am. J. Phys. Anthropol. 174 (4), 763–775.

Biebouw, K., 2009. Revealing the behavioural ecology of the elusive hairy-eared dwarf lemur (*Allocebus trichotis*). Oxford Brookes University. PhD Thesis.

Biebouw, K., 2012. Preliminary results on the behavioral ecology of the hairy-eared dwarf lemur (*Allocebus trichotis*) in Andasibe, eastern Madagascar. In: Masters, J., Gamba, M., Genin, F. (Eds.), Leaping Ahead: Advances in Prosimian Biology. Springer, New York, NY, pp. 113–120.

Blanco, M., Dausmann, K., Ranaivoarisoa, J., et al., 2013. Underground hibernation in a primate. Sci. Rep. 3, 1768. https://doi.org/10.1038/srep01768

Blanco, M.B., Dausmann, K.H., Faherty, S.L., Yoder, A.D., 2018. Tropical heterothermy is "cool": The expression of daily torpor and hibernation in primates. Evol. Anthropol.: Issues, News, and Reviews 27 (4), 147–161.

Britt, A., 2000. Diet and feeding behavior of the black-and-white ruffed lemur (*Varecia variegata*) in the Betampona Reserve, eastern Madagascar. Folia Primatol. 71, 133–141.

Cartmill, M., 1974. *Daubentonia, Dactylopsila*, woodpeckers and kinorhynchy. In: Martin, R.D., Doyle, G.A., Walker, A.C. (Eds.), Prosimian Biology. Duckworth, London, pp. 655–672.

Charles-Dominique, P., Martin, R.D., 1970. Evolution of lorises and lemurs. Nature 227, 257–260.

Curtis, D.J., 2004. Diet and nutrition in wild mongoose lemurs (*Eulemur mongoz*) and their implications for the evolution of female dominance and small group size in lemurs. Am. J. Phys. Anthropol. 124 (3), 234–247.

Curtis, D.J., Zaramody, A., 1998. Group size, home range use, and seasonal variation in the ecology of Eulemur mongoz. Int. J. Primatol. 19 (5), 811–835.

Curtis, D.J., Zaramody, A., Martin, R.D., 1999. Cathemerality in the mongoose lemur, Eulemur mongoz. Am. J. Primatol 47 (4), 279–298.

Dagosto, M., Gebo, D.L., Dolino, C., 2001. Positional behavior and social organization of the Philippine tarsier (*Tarsius syrichta*). Primates 42, 233–243.

Deppe, A.M., Baden, A., Wright, P.C., 2016. The effects of the lunar cycle, temperature, and rainfall on the trapping success of wild brown mouse lemurs (Microcebus rufus) in Ranomafana National Park, southeastern Madagascar. In: Lehman, S.M., Radespiel, U., Zimmerman, E. (Eds.), The Dwarf and Mouse Lemurs of Madagascar: Biology, Behavior and Conservation Biogeography of the Cheirogaleidae. Cambridge University Press, Cambridge, pp. 195–209.

Drea, C.M., 2015. D'scent of man: A comparative survey of primate chemosignaling in relation to sex. Horm. Behav. 68, 117–133. https://doi.org/10.1016/j.yhbeh.2014.08.001

Dröscher, I., Kappeler, P.M., 2014. Maintenance of familiarity and social bonding via communal latrine use in a solitary primate (Lepilemur leucopus). Behav. Ecol. Sociobiol. 68 (12), 2043–2058.

Eberle, M., Kappeler, P.M., 2006. Family insurance: Kin selection and cooperative breeding in a solitary primate (Microcebus murinus). Behav. Ecol. Sociobiol. 60, 582–588. https://doi.org/10.1007/s00265-006-0203-3

Eppley, T.M., Verjans, E., Donati, G., 2011. Coping with low-quality diets: A first account of the feeding ecology of the southern gentle lemur, Hapalemur meridionalis, in the Mandena littoral forest, southeast Madagascar. Primates 52, 7–13.

Eppley, T.M., Donati, G., Ganzhorn, J.U., 2016. Determinants of terrestrial feeding in an arboreal primate: The case of the southern bamboo lemur (Hapalemur meridionalis). Am. J. Phys. Anthropol. 161, 328–342.

Federman, S., Dornburg, A., Daly, D.C., et al., 2016. Implications of lemuriform extinctions for the Malagasy flora. Proceedings of the National Academy of Sciences 113, 5041–5046.

de Flacourt, E., 1661. Histoire de la grande Isle Madagascar, Avec Une Relation de ce qui s'est passe des annees 1655, 1656, & 1657, non encor veue par la premiere Impression. Paris.

Foerg, R., 1982. Reproduction in Cheirogaleus medius. Folia Primatologica 39 (1–2), 49–62.

Forbanka, D.N., 2018. Microhabitat utilization by fork-marked dwarf lemurs (Phaner spp.) and needle-clawed galagos (Euoticus spp.) in primary and secondary forests. Am. J. Primatol. 80, e22864.

Ganzhorn, J.U., 1988. Food partitioning among Malagasy primates. Oecologia 75 (3), 436–450.

Ganzhorn, J.U., Arrigo-Nelson, S., Boinski, S., et al., 2009. Possible fruit protein effects on primate communities in Madagascar and the neotropics. PLoS One 4, e8253.

Godfrey, L.R., Jungers, W.L., Schwartz, G.T., 2006. Ecology and extinction of Madagascar's subfossil lemurs. In: Gould, L., Sauther, M.L. (Eds.), Lemurs: Ecology and Adaptation. Springer, New York, pp. 41–64.

Godfrey, L.R., Jungers, W.L., Reed, K.E., et al., 1997. Subfossil lemurs: Inferences about past and present primate communities in Madagascar. In: Goodman, S.M., Patterson, B.D. (Eds.), Natural Change and Human Impact in Madagascar. Smithsonian, Washington, pp. 218–256.

Godfrey, L.R., Jungers, W.L., Burney, D.A., 2010. Subfossil lemurs of Madagascar. In: Werdelin, L., Sanders, W.J. (Eds.), Cenozoic Mammals of Africa. University of California Press, Berkley, pp. 351–367.

Godfrey, L.R., Scroxton, N., Crowley, B.E., et al., 2019. A new interpretation of Madagascar's megafaunal decline: The "Subsistence Shift Hypothesis." J. Hum. Evol. 130, 126–140.

Goodman, S.M., OConnor, S., Langrand, O., 1993. A review of predation on lemurs: Implications for the evolution of social behavior in small, nocturnal primates. In: Kappeler, P.M., Ganzhorn, J.U. (Eds.), Lemur Social Systems and Their Ecological Basis, Plenum Press, New York, pp. 51–66.

Gould, L., Sauther, M., 2009. Lemurs. Ecology and Adaptation. Springer, New York.

Grogan, K.E., Harris, R.L., Boulet, M., Drea, C.M., 2019. Genetic variation at MHC class II loci influences both olfactory signals and scent discrimination in ring-tailed lemurs. BMC Evol. Biol. 19 (1), 1–16.

Groves, C., Shekelle, M., 2010. The genera and species of Tarsiidae. Int. J. Primatol. 31, 1071–1082.

Gunnell, G.F., Boyer, D.M., Friscia, A.R., et al., 2018. Fossil lemurs from Egypt and Kenya suggest an African origin for Madagascar's aye-aye. Nat. Commun. 9, 1–12.

Gursky, S., 1994. Infant care in the spectral tarsier (Tarsius spectrum) Sulawesi, Indonesia. Int. J. Primatol. 15, 843–854.

Gursky, S., 2000. Allocare in a nocturnal primate: Data on the spectral tarsier, Tarsius spectrum. Folia Primatol. 71(1-2), 39–54.

Gursky, S., 2011. Tarsiiformes. In: Campbell, C., Fuentes, A., MacKinnon, K., et al. (Eds.), Primates in Perspective, second ed. Oxford University Press, Oxford, pp. 79–90.

Gursky, S., 2015. Ultrasonic vocalizations by the spectral tarsier, Tarsius spectrum. Folia Primatol. 86, 153–163.

Hansford, J., Wright, P.C., Rasoamiaramanana, A., et al., 2018. Early Holocene human presence in Madagascar evidenced by exploitation of avian megafauna. Sci. Adv. 4 (9). https://doi.org/10.1126/sciadv.aat6925

Hansford, J.P., Wright, P.C., Perez, V.R., et al., 2020. Evidence for early human arrival in Madagascar is robust: A response to Mitchell. J. Isl. Coast. Archaeol 15, 596–602. https://doi.org/10.1080/15564894.2020.1771482.

Harris, R.L., Boulet, M., Grogan, K.E., Drea, C.M., 2018. Costs of injury for scent signalling in a strepsirrhine primate. Sci. Rep. 8, 9882. https://doi.org/10.1038/s41598-018-27322-3

Hartstone-Rose, A., Dickinson, E., Boettcher, M.L., Herrel, A., 2020. A primate with a Panda's thumb: The anatomy of the pseudothumb of Daubentonia madagascariensis. Am. J. Phys. Anthropol. 171, 8–16.

Herrera, J.P., Davalos, L.M., 2016. Phylogeny and divergence times of lemurs inferred with recent and ancient fossils in the tree. Syst. Biol. 65, 772–791.

Holmes, S.M., Gordon, A.D., Louis, E.E., Johnson, S.E., 2016. Fission-fusion dynamics in black-and-white ruffed lemurs may facilitate both feeding strategies and communal care of infants in a spatially and temporally variable environment. Behav. Ecol. Sociobiol. 70 (11), 1949–1960.

Horvath, J.E., Weisrock, D.W., Embry, S.L., et al., 2008. Development and application of a phylogenomic toolkit: Resolving the evolutionary history of Madagascar's lemurs. Genome Res. 18, 489–499.

Jacobs, R.L., Frankel, D.C., Rice, R.J., et al., 2018. Parentage complexity in socially monogamous lemurs (Eulemur rubriventer): Integrating genetic and observational data. Am. J. Primatol. 80, e22738.

Jolly, A., 1966. Lemur Behavior. University of Chicago Press, Chicago.

Jungers, W.L., Godfrey, L.R., Simons, E.L., et al., 2002. Ecomorphology and behavior of giant extinct lemurs from Madagascar. In: Plavcan, J.M., Kay, R.F., Jungers, W.L., van Schaik, C.P. (Eds.), Reconstructing Behavior in the Primate Fossil Record. Kluwer Academic/Plenum Publishers, New York, pp. 371–411.

Kappeler, P.M., 1997. Determinants of primate social organization: comparative evidence and new insights from Malagasy lemurs. Biol. Rev. 72, 111–151.

Kappeler, P.M., 2012. Behavioral ecology of strepsirrhines and tarsiers. In: Mitani, J.C., Call, J., Kappeler, P.M., et al. (Eds.), The Evolution of Primate Societies. University of Chicago Press, Chicago.

Kappeler, P.M., Rasoloarison, R.M., Rasafimanantoa, L., et al., 2005. Morphology, behavior and molecular evolution of giant mouse lemurs (Mirza spp.). Gray, 1870, with description of a new species. Primate Report 71, 3–26.

Karanth, K.P., Delefosse, T., Rakotosamimanana, B., et al., 2005. Ancient DNA from giant extinct lemurs confirms single origin of Malagasy primates. Proceedings of the National Academy of Sciences, USA 102, 5090–5095.

Kistler, L., Ratan, A., Godfrey, L.R., et al., 2015. Comparative and population mitogenomic analyses of Madagascar's extinct, giant "subfossil" lemurs. J. Hum. Evol. 79, 45–54.

Kraus, C., Eberle, M., Kappeler, P.M., 2008. The costs of risky male behaviour: Sex differences in seasonal survival in a small sexually monomorphic primate. Proceedings of the Royal Society of London B 275, 1635–1644.

Lahann, P., 2007. Feeding ecology and seed dispersal of sympatric cheirogaleid lemurs (*Microcebus murinus, Cheirogaleus medius, Cheirogaleus major*) in the littoral rainforest of south-east Madagascar. J. Zoology 271, 88–98.

Lahann, P., Dausmann, K.H., 2011. Live fast, die young: Flexibility of life history traits in the fat-tailed dwarf lemur (*Cheirogaleus medius*). Behav. Ecol. Sociobiol. 65, 381–390.

Lewis, R.J., 2008. Social influences on group membership in *Propithecus verreauxi verreauxi*. Int. J. Primatol. 29, 1249–1270.

Marciniak, S., Mughal, M.R., Godfrey, L.R., et al., 2021. Evolutionary and phylogenetic insights from a nuclear genome sequence of the extinct, giant, "subfossil" koala lemur *Megaladapis edwardsi*. Proc. Nat. Acad. Sci. USA 118, e2022117118. https://doi.org/10.1073/pnas.2022117118

Martin, R.D., 1995. Prosimians: From obscurity to extinction? In: Alterman, L., Doyle, G.A., Izard, M.K. (Eds.), Creatures of the Dark: The Nocturnal Prosimians, Plenum Press, New York, pp. 535–563.

Martin, R.D., 2008. Evolution of placentation in primates: Implications of mammalian phylogeny. Evol. Bio. 35, 125–145.

Masters, J.C., Genin, F., Couette, S., et al., 2017. A new genus for the eastern dwarf galagos (Primates: Galagidae). Zool. J. Linnean Soc. 181, 229–241.

Morelli, T.L., King, S.J., Pochron, S.T., Wright, P.C., 2009. The rules of disengagement: Takeovers, infanticide, and dispersal in a rainforest lemur. *Propithecus edwardsi*. Behavior 146, 499–523.

Morelli, T.L., Hayes, R.A., Nahrung, H.F., et al., 2013. Relatedness communicated in lemur scent. Naturwissenschaften 100 (8), 769–777.

Nash, L.T., 1998. Vertical clingers and sleepers: Seasonal influence on the activities and substrate use of *Lepilemur leucopus* at Beza Mahafaly Special Reserve, Madagascar. Folia Primatol. 69, 204–217.

Nekaris, K.A.I., 2006. Social lives of adult Mysore slender lorises (*Loris lydekkerianus lydekkerianus*). Am. J. Primatol. 68, 1171–1182.

Nekaris, K.A.I., 2014. Extreme primates: Ecology and evolution of Asian lorises. Evol. Anthropol. 23, 177–187.

Nekaris K.A.I., Bearder, S.K., 2011. The lorisiform primates of Asia and mainland Africa: Diversity shrouded in darkness. In: Campbell, C., Fuentes, A., MacKinnon, K., et al. (Eds.), Primates in Perspective, second ed. Oxford University Press, Oxford.

Nekaris, K.A.I., Moore, R.S., Rode, E.J., Fry, B.G., 2013. Mad, bad and dangerous to know: The biochemistry, ecology and evolution of slow loris venom. J. Venom. Anim. Toxins Incl. Trop. Dis. 19 (1), 10.

Nekaris, K.A.I., Burrows, A.M. (Eds.), 2020. Evolution, Ecology and Conservation of Lorises and Pottos. Cambridge University Press, Cambridge.

Nowack, J., Mzilikazi, N., Dausmann, K.H., 2013. Torpor as an emergency solution in *Galago moholi*: Heterothermy is triggered by different constraints. J. Comp. Physiol. B 183, 547–556.

Orlando, L., Calvignac, S., Schnebelen, C., et al., 2008. DNA from extinct giant lemurs links archaeolemurids to extant indriids. BMC Evol. Biol. 8, 121.

Ostner, J., Schülke, O., 2005. Big times for dwarfs: Social organization, sexual selection, and cooperation in the Cheirogaleidae. Evol. Anthropol. 14, 170–185.

Overdorff, D.J., 1993a. Similarities, differences, and seasonal patterns in the diets of *Eulemur rubriventer* and *Eulemur fulvus rufus* in the Ranomafana National Park, Madagascar. Int. J. Primatol. 14 (5), 721–753.

Overdorff, D.J., 1993b. Ecological and reproductive correlates to range use in red-bellied lemurs (*Eulemur rubriventer*) and rufous lemurs (*Eulemur fulvus rufus*). In: Kappeler, P.M., Ganzhorn, J.U. (Eds.), Lemur Social Systems and Their Ecological Basis. Springer, New York, pp. 167–178.

Overdorff, D.J., 1996. Ecological correlates to social structure in two lemur species in Madagascar. Am. J. Phys. Anthropol. 100, 487–506.

Overdorff, D.J., 1998. Are *Eulemur* species pair-bonded? Social organization and mating strategies in *Eulemur fulvus rufus* from 1988–1995 in southeast Madagascar. Am. J. Phys. Anthropol. 105, 153–166.

Overdorff, D.J., Tecot, S.R., 2006. Social pair-bonding and resource defense in wild red-bellied lemurs (*Eulemur rubriventer*). In: Gould, L., Sauther, M.L. (Eds.), Lemurs. Springer, Boston, MA, pp. 235–254.

Parga, J.A., 2011. Nocturnal ranging by a diurnal primate: Are ring-tailed lemurs (*Lemur catta*) cathemeral? Primates 52, 201–205.

Perelman, P., Johnson, W.E., Roos, C., et al., 2011. A molecular phylogeny of living primates. PLoS Genetics 7 (3), e1001342.

Perret, M., 2019. Litter sex composition affects first reproduction in female grey mouse lemurs (*Microcebus murinus*). Physiol. Behav. 208, 112575.

Pimley, E.R., Bearder, S.K., Dixson, A.F., 2005. Social organization of the Milne-Edward's potto. Am. J. Primatol. 66, 317–330.

Pozzi, L., Disotell, T.R., Masters, J.C., 2014. A multilocus phylogeny reveals deep lineages within African galagids (Primates: Galagidae). BMC Ecol. and Evol. 14, 72. https://doi.org/10.1186/1471-2148-14-72

Pozzi, L., Penna, A., Bearder, S.K., et al., 2020. Cryptic diversity and species boundaries within the *Paragalago zanzibaricus* species complex. Mol. Phylogenet. Evol. 150, 106887. https://doi.org/10.1016/j.ympev.2020.106887

Rakotondrazandry, J.N., Ravelomandrato, F., Sefczek, T.M., et al., 2021. Developmental timeline of wild aye-aye (*Daubentonia madagascariensis*) infants in Kianjavato and Torotorofotsy, Madagascar. Int. J. Primatol. 42, 344–348.

Ramanankirahina, R., Joly, M., Scheumann, M., Zimmermann, E., 2016. The role of acoustic signaling for spacing and group coordination in a nocturnal, pair-living primate, the western woolly lemur (*Avahi occidentalis*). Am. J. Phys. Anthropol. 159 (3), 466–477.

Ramanankirahina, R., Joly, M., Zimmermann, E., 2012. Seasonal effects on sleeping site ecology in a nocturnal pair-living lemur (*Avahi occidentalis*). Int. J. Primatol. 33 (2), 428–439.

Ramsier, M.A., Cunningham, A.J., Moritz, G.L., et al., 2012. Primate communication in the pure ultrasound. Biol. Lett. 8, 508–511.

Rasoloharijaona, S., Randrianambinina, B., Braune, P., Zimmermann, E., 2006. Loud calling, spacing, and cohesiveness in a nocturnal primate, the Milne Edwards' sportive lemur (*Lepilemur edwardsi*). Am. J. Phys. Anthropol. 129 (4), 591–600.

Ravosa, M.J., 1998. Cranial allometry and geographic variation in slow lorises (*Nycticebus*). Am. J. Primatol. 45, 225–43.

Rasmussen, D.T., Nekaris, K.A., 1998. The evolutionary history of lorisiform primates. Folia Primatol. 69 (Suppl. 1), 250–285.

Razafindratsima, O.H., Martinez, B.T., 2012. Seed dispersal by red-ruffed lemurs: Seed size, viability, and beneficial effect on seedling growth. Ecotropica 18 (1), 15–26.

Roberts, M., 1994. Growth, development, and parental care in the western tarsier (*Tarsius bancanus*) in captivity: Evidence for a "slow" life-history and nonmonogamous mating system. Int. J. Primatol. 15 (1), 1–28.

Roos, C., Schmitz, J., Zischler, H., 2004. Primate jumping genes elucidate strepsirrhine phylogeny. Proc. Nat. Acad. Sci. 101, 10650–10654.

Rossie, J.B., Smith, T.D., 2007. Ontogeny of the nasolacrimal duct in Haplorhini (Mammalia: Primates): Functional and phylogenetic significance. J. Anat. 210, 195–208.

Schilling, A., 1979. Olfactory communication in prosimians. In: Doyle, G.A., Martin, R.D. (Eds.), The Study of Prosimian Behavior. Academic, New York, pp. 461–542. https://doi.org/10.1016/B978-0-12-222150-7.X5001-6

Schmid, J., Ganzhorn, J.U., 1996. Resting metabolic rates of *Lepilemur ruficaudatus*. Am. J. Primatol. 38 (2), 169–174.

Schülke, O., Kappeler, P.M., 2003. So near and yet so far: Territorial pairs but low cohesion between pair partners in a nocturnal lemur, *Phaner furcifer*. Anim. Behav. 65, 331–343.

Schmid, J., Speakman, J.R., 2000. Daily energy expenditure of the grey mouse lemur (*Microcebus murinus*): A small primate that uses torpor. J. Comp. Physiol. B 170 (8), 633–641.

Scholz, F., Kappeler, P.M., 2004. Effects of seasonal water scarcity on the ranging behavior of *Eulemur fulvus rufus*. Int. J. Primatol. 25, 599–613.

Sefczek, T.M., Hagenson, R.A., Randimbiharinirina, D.R., et al., 2020. Home range size and seasonal variation in habitat use of aye-ayes (*Daubentonia madagascariensis*) in Torotorofotsy, Madagascar. Folia Primatol. 91 (6), 558–574.

Simons, E.L., Burney, D.A., Chatrath, P.S., et al., 1995. AMS 14C dates for extinct lemurs from caves in the Ankarana Massif, northern Madagascar. Quat. Res. 43, 249–254.

Springer, M.S., Meredith, R.W., Gatesy, J., et al., 2012. Macroevolutionary dynamics and historical biogeography of primate diversification inferred from a species supermatrix. PLoS One 7 (11), e49521.

Stanger, K.F., Coffman, B.S., Kay Izard, M., 1995. Reproduction in Coquerel's dwarf lemur (*Mirza coquereli*). Am. J. Primatol. 36, 223–237.

Sterling, E.J., 1993. Patterns of range use and social organization in aye-ayes (*Daubentonia madagascariensis*) on Nosy Mangabe. In: Kappeler, P.M., Ganzhorn, J.U. (Eds.), Lemur Social Systems and Their Ecological Basis. Springer, New York, pp. 1–10.

Tan, C.L., 1999. Group composition, home range size, and diet of three sympatric bamboo lemur species (Genus *Hapalemur*) in Ranomafana National Park, Madagascar. Int. J. Primatol. 20, 547–566.

Tattersall, I., 1982. The Primates of Madagascar. Columbia University Press, New York.

Tattersall, I., 2007. Madagascar's lemurs: Cryptic diversity or taxonomic inflation? Evol. Anthropol. 16, 12–23.

Tecot, S.R., 2008. Seasonality and predictability: The hormonal and behavioral responses of the red-bellied lemur, *Eulemur rubriventer*, in southeastern Madagascar. University of Texas at Austin, PhD dissertation.

Tecot, S.R., 2010. It's all in the timing: Birth seasonality and infant survival in *Eulemur rubriventer*. Int. J. Primatol. 31, 715–735.

Tecot, S.R., Baden, A.L., 2018. Profiling caregivers: Hormonal variation underlying allomaternal care in wild red-bellied lemurs, *Eulemur rubriventer*. Physiol. Behav. 193, 135–148.

Tecot, S.T., Birr, M., Dixon, J., et al., 2023. Functional relationships between estradiol and paternal care in male red-bellied lemurs, *Eulemur rubriventer*. Horm. Behav. 150, 105324.

van Schaik, C.P., Kappeler, P.M., 1996. The social systems of gregarious lemurs: Lack of convergence with anthropoids due to evolutionary disequilibrium? Ethology Formerly Zeitschrift fuer Tierpsychologie 102 (11), 915–941.

Vasey, N., 2007. The breeding system of wild red ruffed lemurs (*Varecia rubra*): A preliminary report. Primates 48, 41–54.

Vasey, N., Baden, A.L., Ratsimbazafy, J., 2023. Varecia, ruffed or variegated lemurs, varikandana, varijatsy. In: Goodman, S.M. (Ed.), The New Natural History of Madagascar. Princeton University Press, Princeton.

Vasey, N., Mogilewsky, M., Schatz, G.E., 2018. Infant nest and stash sites of variegated lemurs (*Varecia rubra*): The extended phenotype. Am. J. Primatol. 80 (9), e22911.

Voyt, R.A., Sandel, A.A., Ortiz, K.M., Lewis, R.J., 2019. Female power in Verreaux's sifaka (*Propithecus verreauxi*) is based on maturity, not body size. Int. J. Primatol. 40, 417–434.

Walker-Bolton, A.D., Parga, J.A., 2017. "Stink flirting" in ring-tailed lemurs (*Lemur catta*): Male olfactory displays to females as honest, costly signals. Am. J. Primatol. 79e22724.

Wiens, F., Zitzmann, A., 2003. Social structure of the solitary slow loris *Nycticebus coucang* (Lorisidae). J. Zool. 261, 35–46.

Wildman, D.E., Chen, C., Erez, O., et al., 2006. Evolution of the mammalian placenta revealed by phylogenetic analysis. Proc. Nat. Acad. Sci. USA 103, 3203–3208.

Wright, P.C., 1995. Demography and life history of free-ranging *Propithecus diadema edwardsi* in Ranomafana National Park. Madagascar. Int. J. Primatol. 16, 835–854.

Wight, P.C., 1999. Lemur traits and Madagascar ecology: Coping with an island environment. Am. J. Phys. Anthropol. 110(S29), 31–72.

Wright, P.C., Martin, L.B., 1995. Predation, pollination and torpor in two nocturnal primates, *Cheirogaleus major* and *Microcebus rufus*, in the rainforest of Madagascar. In: Alterman, L., Doyle, G.A., Izard, M.K. (Eds.), Creatures of the Dark: The Nocturnal Prosimians, Springer, New York, pp. 45–60.

Zohdy, S., Kemp, A.D., Durden, L.A., et al., 2012. Mapping the social network: Tracking lice in a wild primate (*Microcebus rufus*) population to infer social contacts and vector potential. BMC Ecol. 12, 4. https://doi.org/10.1186/1472-6785-12-4

Further Reading

Asalis, S., Sussman, R.W., 2022. Ecology and life history of the nocturnal lemurs. In: Sussman, R.W., Hart, D., Colquhoun, I.C. (Eds.), The Natural History of Primates: A Systematic Survey of Ecology and Behavior. Rowman & Littlefield, Lanham.

Campbell, C., Fuentes, A., MacKinnon, K., et al., (Eds.), 2011. Primates in Perspective, second ed. Oxford University Press, Oxford.

Colquhoun, I.C., Powzyk, J., 2022. Diurnal and cathemeral lemurs. In: Sussman, R.W., Hart, D., Colquhoun, I.C. (Eds.), The Natural History of Primates: A Systematic Survey of Ecology and Behavior. Rowman & Littlefield, Lanham.

Godfrey, L.R., Jungers, W.L., 2002. Quaternary fossil lemurs. In: Hartwig, W. (Ed.), The Primate Fossil Record. Cambridge University Press, Cambridge, pp. 97–122.

Godfrey, L.G., Jungers, W.L., Burney, D.A., 2010. Subfossil lemurs of Madagascar. In: Sanders, W., Werdelin, L. (Eds.), Cenozoic Mammals of Africa. University of California Press, Berkeley, pp. 351–367.

Gould, L., Sauther, M.L., 2009. Lemurs: Ecology and Adaptation. Springer, New York.

Gould, L., Sauther, M., Cameron, A., 2011. Lemuriformes. In: Campbell, C., Fuentes, A., MacKinnon, K., et al. (Eds.), Primates in Perspective, second ed. Oxford University Press, Oxford, pp. 55–78.

Gursky, S., 2011. Tarsiiformes. In: Campbell, C., Fuentes, A., MacKinnon, K., et al. (Eds.), Primates in Perspective, second ed. Oxford University Press, Oxford, pp. 79–90.

Kappeler, P.M., 2012. Behavioral ecology of strepsirrhines and tarsiers. In: Mitani, J.C., Call, J., Kappeler, P.M., et al. (Eds.), The Evolution of Primate Societies. University of Chicago Press, Chicago.

Lehman, S.M., Radespiel, U., Zimmermann, E. (Eds.), 2016. The Dwarf and Mouse Lemurs of Madagascar: Biology, Behavior and Conservation Biogeography of the Cheirogaleidae, vol. 73. Cambridge University Press, Cambridge.

Maier, W., 1980. Konsruktion morphologishe Untersuchungen am Gebis der rezenten. Prosimiae (Primates). Abh. Senckenb. Naturforsch. Ces. 538, 1–158.

Masters, J., Gamba, M., Genin, F. (Eds.), 2013. Leaping Ahead: Advances in Prosimian Biology. Springer, New York.

Nekaris, K.A.I., Bearder, S.K., 2011. The lorisiform primates of Asia and mainland Africa: Diversity shrouded in darkness. In: Campbell, C., Fuentes, A., MacKinnon, K., et al. (Eds.), Primates in Perspective, second ed. Oxford University Press, Oxford.

Nekaris, K.A.I., Burrows, A.M. (Eds.), 2020. Evolution, Ecology and Conservation of Lorises and Pottos. Cambridge University Press, Cambridge.

Nekaris, K.A.I., Poindexter, S.A., 2022. Lorises and Galagos: The lorisiform primates. In: Sussman, R.W., Hart, D., Colquhoun, I.C.

(Eds.), The Natural History of Primates: A Systematic Survey of Ecology and Behavior. Rowman & Littlefield, Lanham.

Shekelle, M., Gursky, S.L., Achorn, A., Colquhoun, I.C., 2022. Tarsiers. In: Sussman, R.W., Hart, D., Colquhoun, I.C. (Eds.), The Natural History of Primates: A Systematic Survey of Ecology and Behavior. Rowman & Littlefield, Lanham.

Sussman, R.W., Hart, D., Colquhoun, I.C. (Eds.), 2022. The Natural History of Primates: A Systematic Survey of Ecology and Behavior. Rowman & Littlefield, Lanham.

Wright, P., Simons, E., Gursky, S., 2003. Tarsiers: Past, Present and Future. Rutgers University Press, NJ.

5

New World Anthropoids

Primate Grades and Clades

Primates have traditionally been divided into two suborders: Prosimii and Anthropoidea. This is a **gradistic** classification, because prosimians are a grade; they are identified by their retention of primitive anatomical features, not by unique derived features that distinguish them from other primates. As discussed in the previous chapter, the more appropriate, phyletic division of primates is into strepsirrhines (galagos, lorises, and Malagasy lemurs) and haplorhines (tarsiers and anthropoids) (Fig. 4.25). However, anthropoids are also a **clade**, or natural phyletic unit within haplorhines, because the features that distinguish anthropoids from other primates are unique to anthropoids and are derived with respect to other primates. Although we are certain from both morphological and molecular evidence that living anthropoids are a natural unit with a single common ancestry, there is much less agreement about where anthropoids came from, as we shall see in later chapters.

There are three major radiations of extant anthropoids, or higher primates: the platyrrhines, or New World monkeys from South and Central America; and two groups of catarrhines from Africa, Europe, and Asia – the cercopithecoids, or Old World monkeys, and the hominoids, apes and humans. The origins and evolutionary divergences of these groups are discussed in later chapters. In this chapter and in Chapters 6 and 7 we consider the evolutionary diversity of the living higher primates. After a general characterization of the features shared by all anthropoids, we begin with the platyrrhines of the New World.

Anatomy of Anthropoids

Several anatomical features, primarily in the skull, distinguish anthropoids as a group from other primates (Figs. 5.1 and 5.2). Although some of these features are found in various other mammals (and quite a few in tarsiers), the suite as a whole is diagnostic of anthropoids and presumably reflects a unique ancestry for the various groups of primates belonging to that suborder.

The dentition of anthropoids is more conservative than that of either strepsirrhines or tarsiers in both dental formula and the shape of the teeth. All higher primates have two incisors in each quadrant. Although these may be somewhat procumbent, as in pitheciines, they are never markedly so, as in strepsirrhine tooth combs. The anthropoid canine is always larger in caliber and usually taller in height than the incisors. There is considerable diversity in the size and shape of anthropoid canines, both among different species and between sexes within many species, but the canine is never absent, drastically reduced, or markedly procumbent, as in strepsirrhines, and it usually retains a large root, even when the crown is reduced, as in *Homo sapiens*.

Old World monkeys and apes have two premolars, and New World monkeys have three. The morphology of higher primate premolars varies considerably, but the anteriormost premolar is often a simple tooth with a broad buccal surface that sharpens the upper canine, and the posteriormost premolar is always a semimolariform tooth with a differentiated trigonid and talonid and two subequal cusps on the trigonid (Fig. 5.2).

All living anthropoids have three molars, except for marmosets and tamarins, which have two. Anthropoid upper molars usually have a moderate-sized hypocone, making a relatively square tooth, while their lower molars are relatively broad, with reduced trigonids (usually no paraconid) and an expanded talonid basin.

In most strepsirrhines, tarsiers, and placental mammals in general, the two halves of the mandible are joined at the front by a mobile symphysis that permits some degree of independent movement by the two halves of the jaw during chewing. In anthropoids, the two halves of the mandible are fused together. This condition seems to be functionally associated with the development of vertical lower incisors in anthropoids.

In conjunction with their diurnal habits, anthropoids as a group have reduced their reliance on smell and chemical communication and emphasized their reliance on sight. This change is reflected in several diagnostic characteristics of skull morphology (Figs. 5.1 and 5.2). In strepsirrhines, the frontal bone joins with the

Anthropoids   Prosimians

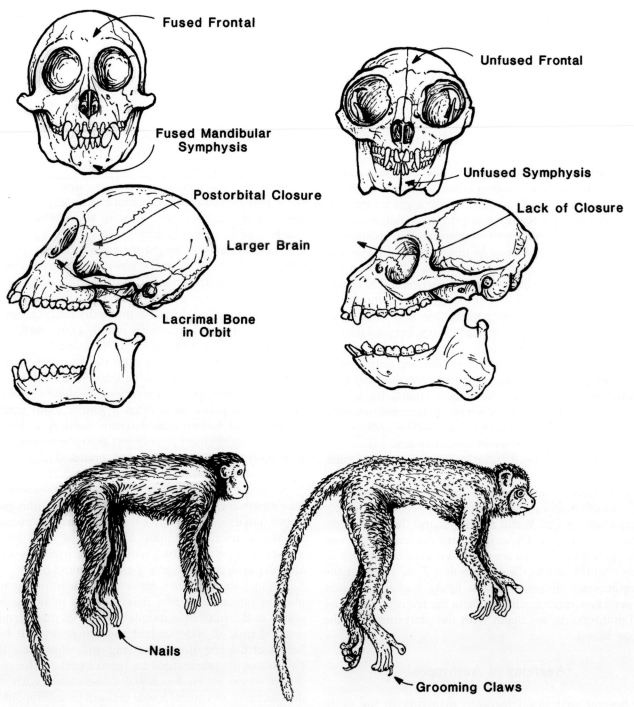

FIGURE 5.1   Anatomical features characteristic of anthropoids.

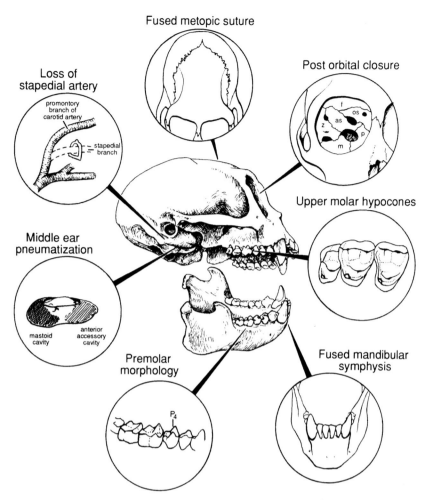

FIGURE 5.2   Cranial and dental features characteristic of extant anthropoids.

zygomatic bone to form a postorbital bar and complete a bony ring around the orbit. In anthropoids, the frontal, zygomatic, and sphenoid bones expand the postorbital bar to form a bony cup that surrounds the eye and separates it from the temporal fossa behind, a condition known as **postorbital closure**. In tarsiers, there is partial closure of the back of the orbit, but the homology between the tarsier pattern and that of anthropoids is debated.

Like tarsiers, anthropoids have forward facing, vertically oriented eyes and well-developed stereoscopic vision. They also lack a reflecting tapetum lucidum and have a retinal fovea on the posterior surface of their eyeball. Most anthropoids have some form of color vision, although there is considerable variation in sensitivity to different parts of the spectrum; Old World monkeys and apes tend to exhibit routine trichromacy, while males and some females in most New World monkeys are red-green colorblind, also known as polymorphic trichromacy. Most anthropoids also have a relatively short snout, and the lacrimal bone lies within the orbit rather than external to it on the snout (Fig. 2.16). Internally, the nasal region of anthropoids is reduced, and there are fewer conchae, and a short, vertical naso-lacrimal duct, as in tarsiers.

In anthropoids, as in tarsiers, the blood supply to the brain is primarily through the promontory branch of the internal carotid artery; the stapedial artery is usually absent in adult anthropoids (Fig. 2.13). The tympanic ring is fused to the lateral wall of the auditory bulla in all anthropoids, but the shape of the tympanic bone differs among major living groups (Fig. 2.20).

In skeletal anatomy, anthropoids share a few diagnostic features that clearly characterize them as a group. In general, most anthropoids are larger than other primates and have relatively shorter trunks. Anthropoid forelimbs and hindlimbs are more similar in length, or the forelimbs are longer. With our relatively long legs, we humans have very unusual anthropoid proportions. Anthropoids generally do not have grooming claws.

It is important to emphasize that this suite of features characterizes living anthropoids. As we will see in later chapters, they evolved in a mosaic pattern during the course of millions of years of evolution, so that the earliest anthropoids did not have the entire set of these defining traits.

Platyrrhines

The living anthropoids of the tropical areas of Central and South America, the platyrrhines (Fig. 5.3), have an evolutionary history extending back over 30 million years and are a much more diverse group than their common name, New World monkeys, suggests. They are "monkeys" only in that they are not apes (tailless, close relatives of humans). Evolving in South America in the absence of other primates (including strepsirrhines), platyrrhines have evolved some species that are strepsirrhine-like in habits, some that are ape-like, and many that have no close analogy among other groups of living primates.

Several distinctive anatomical features distinguish platyrrhines from other primates. Living New World monkeys are all small to medium-size primates, ranging from just under 100g to just over 10kg in weight. Their name is derived from the broad, flat shape of their external nose, which often – but not always – distinguishes them from the Old World anthropoids, which often have narrow, or catarrhine, nostrils. In dental and cranial anatomy (Fig. 5.4) platyrrhines maintain many primitive features that have been lost or modified in the evolution of Old World catarrhines. For example, they have three premolars, and the tympanic ring is fused to the side of the auditory bulla but does not extend laterally as a bony tube. They also share some unique specializations of their own. The first two lower molar teeth of living platyrrhines usually lack hypoconulids, and on the lateral wall of the skull (the pterion region) the parietal and zygomatic bones join to separate the frontal bone above from the sphenoid below. In many platyrrhines, the meningeal layer separating the cerebellum from the cerebral part of the brain is ossified. In addition, the cranial sutures of

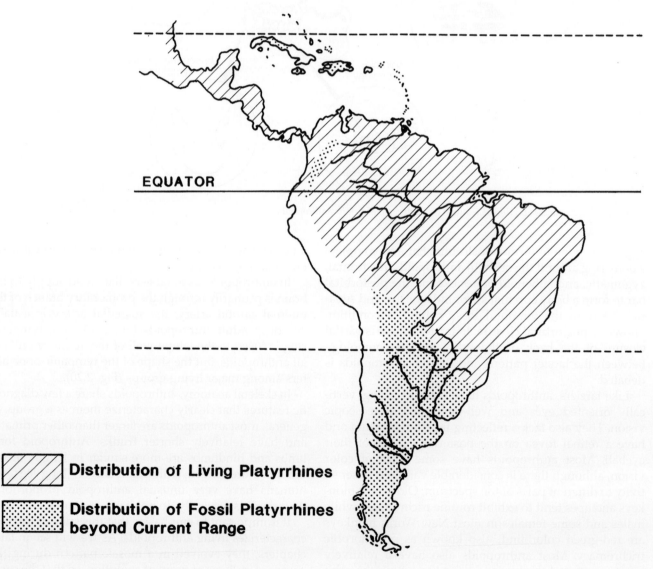

EQUATOR

////// **Distribution of Living Platyrrhines**

:::::: **Distribution of Fossil Platyrrhines beyond Current Range**

FIGURE 5.3   Geographic distribution of extant and extinct platyrrhines. Map lines delineate study areas and do not necessarily depict accepted national boundaries.

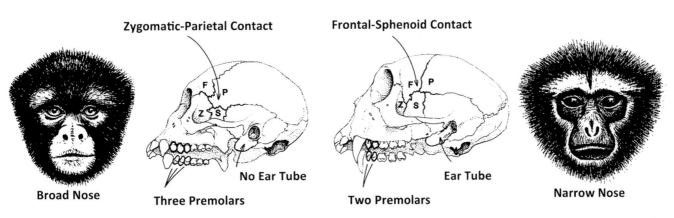

FIGURE 5.4   Skulls of a platyrrhine and a catarrhine, showing some of the features distinguishing these two major groups of anthropoids.

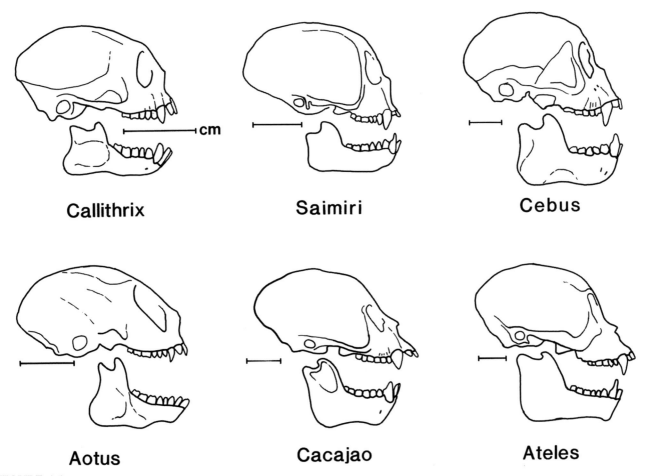

FIGURE 5.5   Skulls of several extant platyrrhines.

platyrrhines fuse relatively late, and many species have relatively long, narrow skulls (Fig. 5.5).

Platyrrhine limb proportions are relatively conservative, with intermembral indices ranging between 70 and 100. Platyrrhines lack the extremely low intermembral indices found among many strepsirrhines, or the high indices of apes. Most have a relatively short forearm, and many lack an opposable thumb. All have a tail of some sort, and in five genera the tail is a prehensile, fifth limb.

The 22 genera of extant platyrrhines are a very diverse group of species that are found from Mexico in the north to Argentina in the south. There is general agreement from studies of molecular systematics that they can be placed in three families: Pitheciidae, Atelidae, and Cebidae. Each of these families is characterized by distinctive ecological (and morphological) adaptations, and there are divisions within each of the families. Some of the ecological diversity of platyrrhines is illustrated and characterized in Fig. 5.8.

Pitheciids

The Pitheciidae are a distinctive family of platyrrhines containing two subfamilies: the Callicebinae, consisting of the genus *Callicebus* and its close relatives *Plecturocebus* and *Cheracebus*, and the Pitheciinae, which contains three very specialized genera: *Pithecia*, *Chiropotes*, and *Cacajao*.

Callicebinae

Titi monkeys (Table 5.1; Figs. 3.5 and 5.6) are the least specialized members of the Pitheciidae and lack the extreme dental specializations found in pitheciines. In many ways they are probably close to the primitive morphology of the first platyrrhines. Titi monkeys have a broad distribution throughout the Amazonian and Orinoco Basins in Southeastern Brazil, but do not extend into the Guianas or Central America. They are currently placed in 33 species, divided into 3 genera, often with distinctive chromosome numbers and features of coloration and body size (van Roosmalen et al., 2002). Species within these groups are allopatric, usually separated by rivers, but members of different species groups are sometimes found living sympatrically. Titi monkeys have short faces, fluffy bodies with long, fluffy tails, and long legs. Compared to other platyrrhines, they have very short canine teeth with little, if any, sexual dimorphism. Titi monkeys have relatively short snouts and long skulls, with simple molar teeth compared to other platyrrhines. Their skeletal anatomy is very similar to that of *Aotus*.

Species groups of titi monkeys frequently show habitat preferences. In Peru, *Cheracebus torquatus* prefers high ground and mature high forest, and is most frequently sighted in the main canopy, whereas *Plecturocebus cupreus* favors the understory and low levels in lowland forest and bamboo thickets. Titi monkeys are both quadrupeds and leapers, but *P. cupreus* leaps more frequently than does *C. torquatus*.

Titi monkeys are mainly frugivorous, but *P. cupreus* and its close relatives eat large numbers of leaves (and bamboo shoots), whereas *C. torquatus* supplements its fruity diet with less foliage and more insects. Titi

monkeys live in monogamous family groups that advertise their presence with elaborate dawn duets. In addition, they actively defend their relatively small territories. Titi monkeys generally have relatively small day ranges and seem to specialize on fruits that are found in small patches, which they can harvest on a regular basis. All species are diurnal, with feeding peaks in the early morning and late afternoon and a long resting peak in late morning. They have single births, and the young are carried by the male after their first week.

Pitheciinae

The pitheciines comprise three genera of medium-sized monkeys: *Pithecia*, *Chiropotes*, and *Cacajao* (Table 5.2; Fig. 5.7). The latter two genera are more closely related to one another than either is to *Pithecia*. Pitheciines are characterized by unusual dental specializations for processing relatively soft fruits and seeds encased in tough outer coverings that are generally too hard for other monkeys to bite through. These dental specializations include large procumbent incisors, robust canines, and relatively small, square premolar and molar teeth with low cusps. In association with their dental specializations, they have a slightly prognathic snout, a narrow, U-shaped palate, and enlarged nasal bones (see *Cacajao* in Fig. 5.5). Their pelage varies considerably from species to species. Pitheciines live in a wide variety of habitats, including rainforest, mountain savannah forest, flooded forests, and liana forest. They are found in extremely diverse social groups, ranging from small family groups to large, fluid multimale–multifemale groupings in which individuals or small groups may forage separately or together, depending on available food resources. The subfamily ranges throughout Amazonia and parts of the Guianas, but not into Central America to the north or beyond the Amazonian ecosystem to the south.

With their broad noses, bushy fur, and long fluffy tails, **sakis** (*Pithecia*) are very striking in appearance (Fig. 5.7). They are the smallest pitheciines, averaging about 2 kg. They have a slightly more gracile skull and jaw than other pitheciines, a relatively longer trunk, and longer legs. There are sixteen species of sakis, divided into two species groups based on size, pelage, and overall skeletal morphology (Marsh, 2014). The first species group is monotypic, and includes the white-faced saki, *Pithecia pithecia*, from the Guianas and northeastern Brazil, where it is frequently found sympatric with *Chiropotes* (Fig. 5.7). The remaining 15 saki species form a second Amazonian species group.

Species of *Pithecia* have been reported in a wide range of forest types, and the available data provide no clear evidence of habitat preferences. However, white-faced sakis from the Guinean shield stand out in their ecological

TABLE 5.1 Infraorder Platyrrhini
Family PITHECIIDAE
Subfamily CALLICEBINAE

Common name	Species	Intermembral index	Mass (g) M	F
Barbara Brown's titi	*Callicebus barbarabrownae*	–	–	–
Coimbra's titi	*C. coimbrai*	–	–	1215
Coastal black-handed titi	*C. melanochir*	–	–	1370
Black-fronted titi	*C. nigrifrons*	–	1350	1300
Masked titi	*C. personatus*	73	1270	1378
Golden Palace titi	*Plecturocebus aureipalatii*	–	1000	900
Baptista Lake titi	*P. baptista*	–	1165	
Prince Bernhard's titi	*P. bernhardi*	–	950	–
Brown titi	*P. brunneus*	–	854	805
Booted titi	*P. caligatus*	–	880	–
Caquetá titi monkey	*P. caquetensis*	–	–	–
Ashy black titi	*P. cinerascens*	–	740	
Coppery titi	*P. cupreus*	77	1020	1120
Red titi	*P. discolor*	–	935	1075
White-eared titi	*P. donacophilus*	75	991	909
Hershcovitz's titi	*P. dubius*	–	–	–
Hoffman's titi	*P. hoffmansi*	–	1090	1030
Milton's titi monkey	*P. miltoni*	–	1335	
Rio Beni titi	*P. modestus*	–	800	
Duksy titi	*P. moloch*	74	1020	956
Rio Mayo titi	*P. oenanthe*	–	–	–
Olalla brother's titi	*P. olallae*	–	–	–
Ornate titi	*P. ornatus*	–	1178	1163
White-coated titi	*P. pallescens*	–	800	–
Stephen Nash's titi	*P. stephennashi*	–	780	668
Toppin's titi monkey	*P. toppini*	–	–	–
Vieira's titi monkey	*P. vieirai*	–	955	–
Lucifer titi	*Cheracebus lucifer*	–	1500	–
Black titi	*Ch. lugens*	–	1150	1050
Colombian black-handed titi	*Ch. medemi*	–	1110	1310
Rio Purus titi	*Ch. purinus*	–	1350	
Red-headed titi	*Ch. regulus*	–	–	–
Widow titi	*Ch. torquatus*	79	1280	1210

tolerance for drier habitats and their distinct social systems (Van Belle et al., 2016). White-faced sakis are most commonly seen in the understory and lower canopy levels, where they move primarily by leaping (Fig. 5.8). They are among the most saltatory of all New World monkeys and frequently move by spectacular leaps. Other *Pithecia* species seem to be more quadrupedal and are found in higher parts of the forest. Like all pitheciines, sakis have the dental adaptations for feeding on seeds encased in a hard outer covering. However, compared with *Chiropotes* and *Cacajao*, *Pithecia* seems to eat fruits with relatively softer outer coverings, and has a more diverse diet.

FIGURE 5.6   The titi monkey (*Callicebus*).

TABLE 5.2   Infraorder Platyrrhini
Family PITHECIIDAE
Subfamily PITHECIINAE

Common name	Species	Intermembral index	Mass (g) M	Mass (g) F
Equatorial saki	*Pithecia aequatorialis*	–	2250	
Buffy saki	*P. albicans*	–	3000	–
Cazuza's saki	*P. cazuzai*	–	3200	2750
Golden-faced saki	*P. chrysocephala*	–	1900	1900
Hairy saki	*P. hirsuta*	–	–	–
Burnished saki	*P. inusta*	–	–	–
Bald-faced saki	*P. irrorata*	–	2250	2070
Isabel's saki	*P. isabela*	–	–	–
Miller's saki	*P. milleri*	–	2700	2177
Mittermeier's Tapajos saki	*P. mittermeieri*	–	–	–
Monk saki	*P. monachus*	77	2610	2110
Napo saki	*P. napensis*	–	–	–
Pissinatti's saki	*P. pissinattii*	–	–	–
White-faced saki	*P. pithecia*	75	1940	1580
Rylands' bald-faced saki	*P. rylandsi*	–	–	–
Vanzolini's bald-faced saki	*P. vanzolinii*	–	–	–
White-nosed bearded saki	*Chiropotes albinasus*	80	3150	2490
Bearded saki	*C. chiropotes*	–	3000	2600
Reddish-brown bearded saki	*C. sagulatus*	83	3000	2600
Black saki	*C. satanas*	–	3100	2960
Uta Hick's bearded saki	*C. utahickae*	83	3250	2750
Ayres black uacari	*Cacajao ayresi*	–	4500	3100
Bald-headed uacari	*C. calvus*	83	3450	2888
Hosom's uacari	*C. hosomi*	82	3850	2950
Black-headed uacari	*C. melanocephalus*	82	3160	2710

FIGURE 5.7 Two pitheciine primates that live sympatrically in Suriname and other areas of northeastern South America: *above*, the bearded saki (*Chiropotes sagulatus*); *below*, the white-faced saki (*Pithecia pithecia*).

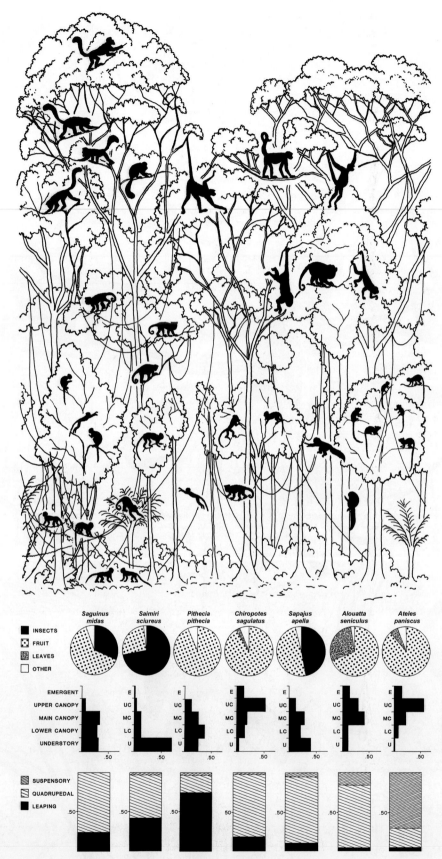

FIGURE 5.8 Seven sympatric species in a Suriname rain forest, showing typical dietary, locomotor, and postural behavior as well as use of different heights in the forest. At the highest levels are the bearded saki (*Chiropotes sagulatus*) and the black spider monkey (*Ateles paniscus*); below them are tufted capuchins (*Sapajus apella*) on the *left* and red howling monkeys (*Alouatta seniculus*) on the *right*; in the lower levels are squirrel monkeys (*Saimiri sciureus*) on the *left* and golden-handed tamarins (*Saguinus midas*) and white-faced sakis (*Pithecia pithecia*) on the *right*.

Sakis have generally been reported to live in socially monogamous family groups that travel and sleep together, but often separate while feeding during the day. There are also reports of larger groups with multiple males and females, as well as reports of small groups coming together. Interestingly, groups with more than one same-sex adult might be more characteristic of *P. pithecia* than of other saki species (Norconk, 2011; Norconk and Setz, 2013). Estimated home range sizes vary considerably, from 10ha or less for *P. pithecia* to as large as 200ha for *P. albicans*. The few documented day ranges are small. Sakis have single births, and the young seem to be cared for primarily by the females (Fernandez-Duque et al., 2013).

Bearded sakis (*Chiropotes*) are medium-sized monkeys (2–4kg) that include five species (*C. albinasus*, *C. chiropotes*, *C. sagulatus*, *C. satanas*, and *C. utahickae*) distributed throughout the Amazon and Orinoco basins of the Guianas, Suriname, Venezuela, and Brazil (Table 5.2). They are larger, and have more robust skulls and jaws than *Pithecia* (Figs. 5.7 and 5.9). The best known are *Chiropotes sagulatus*, a chocolate brown species with a black beard and bouffant hairdo from Venezuela, the Guianas, and northeastern Brazil, and a larger species, *C. albinasus*, with black fur and a white nose, from Brazil south of the Amazon.

Chiropotes prefers tall rainforest habitat, but can also be found in savannahs, swamps, and mangroves. Early work noted their absence from secondary habitats, which resulted in the widespread misconception that bearded sakis were intolerant of habitat disturbance (Johns and Ayres, 1987). However, more recent research suggests that they can be relatively abundant in isolated forest fragments (Ferrari et al., 1999).

Bearded sakis are extremely agile and fast-moving, are superb climbers and leapers, and spend most of their time in the highest forest strata (Fig. 5.8). They are primarily arboreal quadrupeds that frequently use hindlimb suspension when feeding. Bearded sakis have highly specialized craniodental anatomy, powerful masseter and temporalis muscles, and large, divergent canines, that allow them to exploit unripe fruits and seeds with very hard shells. They also occasionally ingest insects.

In contrast to the small family groups of *Pithecia*, bearded sakis live in large, fluid multimale–multifemale groups characterized by high fission–fusion dynamics (Shaffer, 2013). They have extremely large day ranges and home ranges (often well over 500ha), and groups may travel several kilometers in a day. Their mating system is poorly understood, though recent observations suggest a polygynandrous system. They have single births and show no evidence of paternal care.

Uacaris (*Cacajao*) are the largest pitheciines and are easily distinguished from the other genera by their large size and short tail. There are four species: three black uacaris from the upper reaches of the Rio Negro and Rio Orinoco in Brazil, eastern Colombia, and southern Venezuela, and the bald uacari from the upper Amazon (Solimoes) drainage. Its range is almost exclusively allopatric to that of *Chiropotes*, with the exception of a possible zone of sympatry in the northern Amazon basin (Boubli, 2002). Both subspecies of bald uacaris, *C. calvus*, have scarlet faces and bald heads, but in one, the long, shaggy fur is red; in the other, it is white.

While once regarded as flooded-forest specialists, uacaris are now known to also exploit terra firme, white-sand campinas (savannahs), and palm swamp forests. Nevertheless, the habitats used by uacaris all share the characteristic of being nutrient-restricted and often richer than average in hard-shelled large-seeded fruiting trees (ter Steege et al., 2006). Unlike *Pithecia* and *Chiropotes*, there are no reliable records of any *Cacajao* taxon inhabiting secondary, fragmented or degraded habitats for any extended period.

Like other pitheciines, uacaris specialize on fruits, and they move primarily by quadrupedal walking and running, both in the trees and on the ground. Like *Chiropotes*, uacaris often feed using hindlimb suspension.

Like *Chiropotes*, *Cacajao* lives in large multi-male, multi-female groups of 20 to 50 individuals, though groups of as many as 200 or more individuals have been reported. Many uacari populations exhibit a high degree of fission–fusion dynamics, with groups of white uacaris frequently dividing into very small foraging parties, often for several days. They also have some of the largest home ranges and daily travel distances of any Neotropical primate, a behavior that has been attributed to their reliance on unripe hard fruits, which are highly spatially dispersed and often very quickly depleted. Very little is known about their mating and reproduction. Births generally occur just after the periods of greatest fruit production by igapó trees. However, out-of-season births (November) have also been observed.

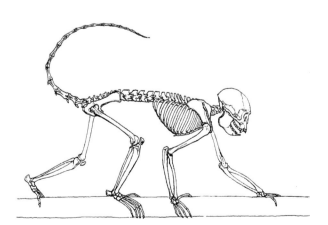

FIGURE 5.9   The skeleton of a bearded saki (*Chiropotes*).

Atelids

The four atelid genera are the largest platyrrhines; the largest individuals of each genus weigh approximately 10 kg. All atelids have a long, prehensile tail with friction ridges, similar to fingerprints, on the distal part of its ventral surface. In many aspects of limb and trunk anatomy and in their use of suspensory behavior, they show similarities to the extant apes (Erikson, 1963; Larson, 1998). In dental and cranial anatomy, as well as diet and social structure, they are quite diverse. Howling monkeys (*Alouatta*) are the most distinct genus phylogenetically and are usually placed in a separate subfamily.

Howler monkeys (*Alouatta*) (Table 5.3) are the most widespread genus of platyrrhines, with 14 recognized species, ranging from eastern Mexico to northern Argentina (Fig. 5.10). All are large (6–10 kg) and sexually dimorphic in size. Individual species vary dramatically in color, ranging from red to brown, black, or blond. In some species both sexes are the same color; in others the sexes look strikingly different.

TABLE 5.3   Infraorder Platyrrhini
Family ATELIDAE

Common name	Species	Intermembral index	Mass (g) M	F
Subfamily ALOUATTINAE				
Ursine howling monkey	*Alouatta arctoidea*	–	7000	5750
Red-handed howling monkey	*A. belzebul*	–	7270	5520
Black-and-gold howling monkey	*A. caraya*	97	6420	4330
Spix's red-handed howler	*A. discolor*	–	7250	5500
Red-and-black howling monkey	*A. guariba*	–	6730	4350
Jurua red howler monkey	*A. juara*	–	–	–
Guianan red howling monkey	*A. macconnelli*	–	7200	5550
Amazon black howling monkey	*A. nigerrima*	–	6450	
Mantled howler monkey	*A. palliata*	98	6023	4855
Black howling monkey	*A. pigra*	–	7505	5608
Purus red howler monkey	*A. puruensis*	–	–	–
Bolivian red howling monkey	*A. sara*	–	–	–
Red howling monkey	*A. seniculus*	97	6700	5725
Maranhao red-handed howling monkey	*A. ululata*	–	4700	–
Subfamily ATELINAE				
White-bellied long-haired spider monkey	*Ateles belzebuth*	109	8260	7880
Peruvian spider monkey	*A. chamek*	–	9410	9330
Brown-headed spider monkey	*A. fusciceps*	103	8890	9160
Geoffroy's spider monkey	*A. geoffroyi*	105	7780	7290
Brown spider monkey	*A. hybridus*	–	8250	9000
White-cheeked spider monkey	*A. marginatus*	–	–	–
Black spider monkey	*A. paniscus*	105	9110	8440
Peruvian yellow-tailed woolly monkey	*Lagothrix flavicauda*	–	10000	
Humboldt's woolly monkey	*L. lagotricha*	98	7280	7020
Southern muriqui	*Brachyteles arachnoides*	104	9610	8070
Northern muriqui	*B. hypoxanthus*	–	9600	8400

FIGURE 5.10   A troop of red howling monkeys (*Alouatta seniculus*).

Howler monkeys have relatively small incisors and large, sexually dimorphic canines. The lower molars have a narrow trigonid and a large talonid; the upper molars are quadrate, with well-developed shearing crests characteristic of folivorous primates. The skull of *Alouatta* is distinguished by its relatively small cranial capacity and lack of cranial flexion. The mandible is quite large and deep, and the hyoid bone is expanded into a very large, hollow resonating chamber. Howler monkeys have forelimbs and hindlimbs that are similar in length, and a long, prehensile tail. Like many platyrrhines, howler monkeys have a poorly differentiated thumb and usually hold objects and branches between their second and third digits, a grasp known as **schizodactyly** ("between fingers").

Howler monkeys are found in a variety of habitats, including primary rainforests, secondary forests, dry deciduous forests, montane forests, and llanos habitats containing patches of relatively low trees in open savannah. Their distribution ranges from sea level to altitudes above 3200 m and from Mexico to Argentina. Within these diverse forest habitats, most species seem to prefer the main canopy and emergent levels (Fig. 5.8), but several species that live in drier areas (especially *A. caraya*) regularly come to the ground and cross open areas between patches of forest. Howlers are slow, quadrupedal monkeys that rarely leap (Fig. 5.8). During feeding, and less often during travel, they use suspensory locomotion, primarily climbing, in which all five limbs grasp supports opportunistically. During feeding they frequently use their tail in suspension. Howler monkeys are the most folivorous of all New World monkeys (Fig. 5.8). There is considerable variation in their diet from month to month, but leaves, especially new leaves, usually constitute half or more of the yearly diet. Fruits and flowers are the next most common components. Shifts in the composition and activity of the gut microbiota are thought to help enable these dietary changes (Amato et al., 2015).

Most howler monkeys live in groups containing one or a few adult males, several adult females, and their offspring, but the normal group composition seems to vary by species, by habitat richness, and as a function of the age of a troop. Groups of *A. palliata* may contain from 12 to 30 individuals, but in *A. seniculus* and *A. caraya* troops are usually smaller. Howler day ranges are relatively small, usually about 500 m, because of the howlers' ability to subsist on both a diversity of food items and on relatively common foods such as leaves, and group movement is typically initiated by the most socially connected female in the group (Van Belle et al., 2013). Home ranges vary considerably depending on group size and habitat, but are usually very small for New World monkeys of their biomass. Compared with other atelids, howler monkeys spend less time traveling each day and have longer periods of resting and digesting.

As their name indicates, howler monkey groups regularly advertise their presence with low-frequency, lion-like roars given by both males and females. These loud calls are important for intergroup communication and are used to regulate space use between neighboring groups. Vocal battles are often reinforced by actual physical combat between individuals, usually males, but there is disagreement regarding the extent to which howlers actively defend their territories or merely their daily positions. The loud calls of howler monkey males, in particular, seem to be functionally linked to the access and defense of valuable resources, including food and mates, as well as to the defense of vulnerable infants. In fact, recent work has shown that howler monkey species characterized by single-male groups have relatively larger hyoids than those living in multimale groups, suggesting high levels of vocally mediated competition in these groups (Dunn et al., 2015).

While many *Alouatta* troops have several adult males, most mating is by a single male, and there is considerable competition among males for access to a troop and the females within it. Males taking over a troop have been reported to kill the dependent infants that were probably the offspring of the ousted male. An unusual feature of howling monkey demography is that in some species, both males and females regularly transfer among troops, with neither sex consistently remaining in their natal troop. Howlers have single births. Infants are frequently cared for by females other than their mother.

The two species of **woolly monkeys** (*Lagothrix*) (Table 5.3; Fig. 5.11) are restricted to the western parts of the Amazon and Orinoco basins. The very rare yellow-tailed woolly monkey, *L. flavicauda*, is restricted to the cloud forests of Peru. Like howler monkeys, woolly monkeys are very sexually dimorphic in body size, with the largest males weighing over 10 kg.

Woolly monkeys seem to inhabit primarily high rainforests, but extend into gallery forests in Colombia and Venezuela. They are primarily arboreal quadrupeds and climbers that rely less extensively on forelimb suspension or leaping. Mature fruit pulp is the major item in their diet, but there is considerable seasonal variation, and some populations rely extensively on insects, a rare dietary item for such a large monkey. In the dry season, when fruits are less available, the primary item in their diet changes to new leaves, young seeds, or legume pod secretions, depending on availability.

Woolly monkeys live in large, cohesive groups ranging between 20 and 50 individuals, depending on habitat richness and food distribution. Home ranges and day ranges are extremely large compared to those of howling monkeys. Woolly monkeys are particularly susceptible to hunting pressure.

The seven species of **spider monkeys** (*Ateles*) (Table 5.3) range from the Yucatan peninsula of Mexico through Amazonia and vary in color from beige to solid black. They are large, graceful, long-limbed monkeys with a long, prehensile tail (Fig. 5.12). Male and female spider monkeys are virtually identical in size and color in every species, and this monomorphism is augmented by the long, pendulous clitoris in the females.

The dentition of spider monkeys is characterized by relatively large, broad incisors and small molars with low, rounded cusps. The skull has large orbits, a globular braincase, and a relatively shallow mandible (Fig. 5.5). Spider monkeys have relatively long, slender limbs (Fig. 2.22) that resemble those of gibbons and other suspensory species in many features. Their fingers and toes are long and slender (Figs. 2.25 and 2.27), and most species lack an external thumb.

Spider monkeys are restricted mainly to high primary rainforests, where they prefer the upper levels of the main canopy (Fig. 5.8). They have extremely diverse locomotor abilities. During travel, they use both arboreal quadrupedalism and suspensory behavior, including brachiation and climbing. They move bipedally in the trees and occasionally leap. During feeding they are almost totally suspensory, and they can hang by combinations of all five limbs. Spider monkeys feed primarily on ripe fruit (Fig. 5.8), but in some seasons they may eat large amounts of new leaves.

Spider monkey groups are generally large, comprising a dozen or more individuals of both sexes and all ages, and exhibit a high degree of fission–fusion dynamics. During the day the large social group generally breaks down into smaller foraging units of two to five individuals which often give loud, barking contact calls. These units are most frequently either adult females and their offspring or groups of males. Males typically associate

FIGURE 5.11   A troop of woolly monkeys (*Lagothrix lagotricha*).

more than females and females tend to associate less selectively (Ramos-Fernández et al., 2009). However, both fission–fusion dynamics and association patterns can vary between seasons in response to changes in fruit availability (Chapman et al., 1995; Aguilar-Melo et al., 2020). Males generally remain in their natal group, whereas females disperse. Spider monkeys have single births, and the young are cared for by the mother.

Brachyteles arachnoides, the **woolly spider monkey**, or **muriqui** (Table 5.3; Fig. 5.13), is probably the largest nonhuman primate in the Neotropics. Reliable body weights are rare and show a tremendous range, but seem to average between 8 and 10 kg. There are two species, the northern *B. hypoxanthus* and the southern *B. arachnoides*, both found in the Atlantic coastal rainforests of southeastern Brazil. Although *Brachyteles* resembles the

spider monkeys in limb proportions and in its lack of a thumb, the dentition is superficially more like that of *Alouatta*, with numerous shearing crests on the molars. The canines of both sexes are small.

In what remains of their disappearing habitat, muriquis are totally restricted to high forest areas, where they prefer main canopy levels. Like *Ateles*, muriquis are arboreal quadrupeds that rely extensively on suspensory behavior during travel, and especially during feeding. Southern muriquis are found in more continuous forests that are less disturbed, structurally more complex, and with a higher tree density than northern muriqui habitats (Boubli et al., 2010). Southern muriqui habitats are also less seasonal, and fruits are more readily available year-round and in larger patches than where northern muriquis typically occur (de Moraes

FIGURE 5.12　A group of black spider monkeys (*Ateles paniscus*).

et al., 1998; Talebi et al., 2006; Couto-Santos 2007). This is perhaps why southern muriquis seem to preferentially consume fruits, whereas northern muriquis are considered much more folivorous.

The social organization of southern muriquis seems most comparable to that of spider monkeys, where they are consistently observed to exhibit fission–fusion dynamics. Northern muriquis, on the other hand, appear to be more variable in their grouping patterns, as researchers have observed both fluid and cohesive groups (Coles et al., 2012). Males are philopatric, form strong affiliative bonds, and spend a high proportion of time in close proximity (Strier et al., 2002). Muriquis are considered incredibly peaceful (Strier, 1994). Mating is promiscuous, with females frequently copulating with several males successively. Males have extremely large testicles, suggesting that mating competition among males is largely by sperm competition rather than by interindividual aggression. In fact, it has been shown that paternity skew within this taxon is low (Strier et al., 2011). Male muriquis remain in their natal group, whereas females disperse.

Cebids

The family Cebidae comprises three very different types of monkeys, each placed in a separate subfamily. The Cebinae contains capuchin monkeys and squirrel monkeys, the two most omnivorous platyrrhines; the Aotinae contains several species of owl monkey, the only nocturnal platyrrhines; and the Callitrichinae contains the smallest platyrrhines, marmosets, tamarins, and Goeldi's monkeys (*Callimico*).

FIGURE 5.13 A troop of muriquis, or woolly spider monkeys (*Brachyteles arachnoides*), from southeastern Brazil, one of the most endangered living primates.

Cebinae

The **capuchins** are among the most easily recognizable of New World monkeys. Formerly common at street corners and in circuses, they are now frequently seen in movies, television programs, and clothing advertisements. These medium-sized, sexually dimorphic primates all have large premolars and square molar teeth with thick enamel, which they use to open hard nuts. Their forelimbs and hindlimbs are more similar in length than those of many platyrrhines. Capuchins have a relatively short, fur-covered, prehensile tail. They have short fingers and an opposable thumb and are the most dexterous platyrrhines. For their body size, capuchins seem to have a "slow" life history schedule with long lactation periods, long interbirth intervals, and a long lifespan, with many individuals living over 40 years in captivity.

Capuchins are currently divided into 23 species, which are placed in 2 genera (Table 5.4), the tufted (*Sapajus*) and untufted capuchins (*Cebus*). *Sapajus* includes *S. apella* and related taxa (Fig. 5.14), which are relatively robust monkeys found throughout most of the Neotropics from northern South America to northern Argentina. Species of untufted capuchins, the genus *Cebus*, are sympatric with *Sapajus* throughout much of South America. *Cebus imitator*, the Panamanian white-throated capuchin, is the only species from North and Central America.

Capuchins are found in virtually all types of Neotropical forest. They seem to prefer the main canopy levels (Fig. 5.8), but they frequently descend to the understory or to the ground during both travel and feeding. This is especially true of *Sapajus libidinosus*, which lives in the open cerrado and caatinga habitats of Brazil,

TABLE 5.4   Infraorder Platyrrhini
Family CEBIDAE
Subfamily CEBINAE

Common name	Species	Intermembral index	Mass (g) M	F
Ecuadorian white-fronted capuchin	*Cebus aequatorialis*	81	2650	1700
Humboldt's white-fronted capuchin	*C. albifrons*	82	3180	2290
Brown weeper capuchin	*C. brunneus*	–	–	–
Colombian white-throated capuchin	*C. capucinus*	81	3680	2540
Chestnut capuchin	*C. castaneus*	–	–	–
Rio Cesar white-fronted capuchin	*C. cesarae*	–	–	–
Shock-headed capuchin	*C. cuscinus*	–	2900	
Panamanian white-throated capuchin	*C. imitator*	–	3800	2650
Ka'apor capuchin	*C. kaapori*	–	3050	2150
Sierra de Perija white-fronted capuchin	*C. leucocephalus*	–	–	–
Santa Marta white-fronted capuchin	*C. malitiosus*	–	–	–
Guianan weeper capuchin	*C. olivaceus*	83	3290	2520
Trinidad white-fronted capuchin	*C. trinitatis*	–	–	–
Spix's white-fronted capuchin	*C. unicolor*	–	–	–
Varied white-fronted capuchin	*C. versicolor*	–	–	–
Peruvian white-fronted capuchin	*C. yuracus*	–	3350	
Guianan brown capuchin	*Sapajus apella*	81	3610	2470
Hooded capuchin	*S. cay*	–	3215	1770
Blonde capuchin	*S. flavius*	–	2775	2115
Bearded capuchin	*S. libidinosus*	89	3100	1975
Black-horned capuchin	*S. nigritus*	–	3247	2215
Crested capuchin	*S. robustus*	–	3200	–
Yellow-breasted capuchin	*S. xanthosternos*	–	3400	2350
Black-capped squirrel monkey	*Saimiri boliviensis*	–	911	711
Humboldt's squirrel monkey	*S. cassiquiarensis*	–	888	875
Collins' squirrel monkey	*S. collinsi*	–	795	575
Central American squirrel monkey	*S. oerstedii*	80	897	680
Guianan squirrel monkey	*S. sciureus*	79	779	662
Golden-backed squirrel monkey	*S. ustus*	–	921	799
Black-headed squirrel monkey	*S. vanzolinii*	–	950	650

where the monkeys frequently walk bipedally and are adept at using tools to smash palm nuts (Fig. 5.15; Ottoni and Izar, 2008). All species are arboreal quadrupeds that use their prehensile tails mainly during feeding (Fig. 5.8). The capuchin diet includes many types of fruit and animal matter. Tufted capuchins such as *S. apella* are opportunistic monkeys that use their manipulative abilities and their strength to obtain foods that are unavailable to other species. They forage destructively for invertebrates in bark and leaf litter, and are also able to break open hard palm nuts and the hard shells covering immature flowers. In contrast, the more gracile *C. albifrons* specializes on superabundant fruit sources in the same forests as tufted species.

All capuchins live in social groups of 8 to 30 individuals, with several adult males, several adult females, and

FIGURE 5.14 Two platyrrhines that live sympatrically throughout much of South America: *above*, the tufted capuchin (*Sapajus apella*); *below*, the squirrel monkey (*Saimiri sciureus*).

FIGURE 5.15 Bipedal behavior and tool use by *Sapajus libidinosus* in Brazil. *(Photo courtesy of B. Wright.)*

offspring. Females tend to stay in their natal groups, whereas males disperse. There is greater sexual size dimorphism and a more obvious dominance hierarchy among males of *S. apella* than of *C. albifrons*. Capuchins have large home ranges and relatively long day ranges. They have single births and allomaternal nursing is common in some species.

Saimiri, the **squirrel monkeys** (Table 5.4; Fig. 5.14), resemble capuchins in their dental proportions, as well as in their omnivorous, insectivorous diet and quadrupedal locomotion. Squirrel monkeys are less than 1 kg in average body mass. They have a distinctive cranial morphology with a very long occipital region and a foramen magnum that lies well under the skull base (Fig. 5.6). The orbits are so close together that the interorbital septum is perforated by a large opening. Squirrel monkeys have relatively broad, quadrate upper molars with a large lingual cingulum, and very small last molars. Their cheek teeth have very sharp cusps, associated with an insectivorous diet. The canines are sexually dimorphic. Squirrel monkeys are very precocious in their development compared to capuchins, with more prenatal brain growth and early motor development (Hartwig, 1995). However, there is considerable variation among different populations in rates of postnatal development.

The postcranial skeleton of squirrel monkeys (Fig. 5.16) is characterized by a relatively long trunk, long hindlimbs, and a long tail that is prehensile in infants but not in adults. Their hands have comparatively short fingers and a short but relatively unopposable thumb. There is frequently a small amount of fusion between the tibia and fibula distally, making the ankle joint more restricted to simple hinge movements than in other platyrrhine species.

There are seven squirrel monkey species, many of which are allopatric throughout Panama and Costa Rica, south through Amazonian Brazil, Colombia, Ecuador, Peru, Bolivia, the Guianas, and southern Venezuela. All are small, gray to yellow monkeys with short fur and a long, thin tail. Squirrel monkeys occupy a variety of rainforest habitats but show a preference for riverine and secondary forests, where they are commonly found in the lower levels. They are arboreal quadrupeds that frequently leap, especially when traveling in lower forest levels (Fig. 5.8). During feeding they are almost totally quadrupedal, and occasionally come to the ground.

Squirrel monkeys are frugivores and insectivores (Fig. 5.8) that specialize on large fruit trees throughout the year, and there is evidence that different allopatric species specialize on distinct foods. As they travel between fruit trees, they forage for insects (for almost half of each day) and frequently catch large arthropods on the wing. South American species of *Saimiri* often travel and forage in conjunction with *S. apella* (Fig. 5.14).

Squirrel monkeys live in large, continuously moving groups that range from 20 to over 50 individuals, with numerous adults of both sexes and offspring. They communicate frequently throughout the day by means of high-pitched whistles and chatter; a group is usually heard well before it comes into view. The estimated size of the overlapping home ranges varies dramatically from troop to troop, but is often more than 200 ha.

Squirrel monkeys occupy large day ranges (2.5–4 km), in keeping with the large group sizes and extensive insect foraging. They appear to be subject to very high rates of predation, primarily from raptors.

There is considerable diversity in squirrel monkey social organization among the populations that have been studied (e.g., Mitchell et al., 1991). *Saimiri boliviensis* in Peru tend to forage in relatively larger parties than *S. oerstedii* in Costa Rica. In addition, female interactions are strikingly different in the two species, in association with the use of differently distributed fruit resources. In *S. oerstedii*, fruit resources are found in small, evenly distributed patches. There is little direct feeding competition, and females form loose associations and frequently transfer among groups. However, in Peru, *S. boliviensis* feed more frequently on larger fruit resources that generate more direct feeding competition. In this species, there is a strict dominance hierarchy among females and it is the males that disperse.

In both Costa Rican and Peruvian squirrel monkeys, as in many lemur species, adult females are dominant over males for most of the year and also spend more time feeding. Males tend to be peripheral members of troops. However, in the mating season, males increase in size through the retention of water, especially around their shoulders, and become more aggressive. Mating success seems to be correlated with male size, as a result of both intrasexual competition and female choice. Squirrel monkeys have restricted breeding seasons, with all births in a single troop occurring within a week. Females give birth to relatively large single offspring at yearly intervals. In contrast to the males of most small New World monkeys, male squirrel monkeys do not play an important role in the care of infants; rather, infants are cared for by several females in addition to their mother.

Aotines

Aotus, the **owl monkeys**, night monkeys, or douracoulis (Figs. 3.5 and 5.17), are the only nocturnal anthropoid primates, and their position in platyrrhine phylogeny has long been a subject of considerable debate. Biomolecular studies consistently group them with cebines and callitrichines in the Cebidae, although relationships among these three subfamilies are not resolved. Owl monkeys are found throughout much of South America, from Panama to northern Argentina, but they are absent from the Guianas and southeastern Brazil. Over this broad range there are at least 11 allopatric species (Table 5.5) that differ from one another in the relative size of their dentition, the coloration of their neck and tail, and often in their karyotypes. All are of medium size (about 1 kg), with no marked sexual dimorphism. They have relatively long legs and a long tail.

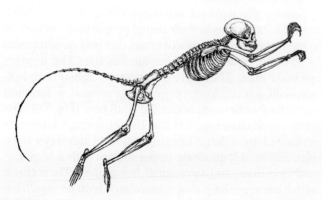

FIGURE 5.16   The skeleton of a squirrel monkey (*Saimiri sciureus*).

They have large digital pads on their hands and feet, a slightly opposable thumb, and often a compressed, claw-like grooming nail on the second digit of each foot (Fig. 2.28). They have very large upper central incisors and small third molars. Their basal metabolic rate is lower than that of other platyrrhines, a feature common among nocturnal mammals (Wright, 1989). The most distinctive feature of their cranial anatomy, associated with their nocturnal habits, is the large size of their orbits, which are the largest of any anthropoid (Fig. 5.5). Despite their nocturnal habits, *Aotus*, like other anthropoids, have no tapetum lucidum. Instead, they have a tapetum fibrosum. The retina of *Aotus* has only a single type of cone, so they lack color vision. *Aotus* almost certainly evolved from diurnal ancestors.

Owl monkeys are found in a variety of forest habitats, and there are no indications that they prefer any

particular canopy level. They are predominantly quadrupedal but are adept leapers. Their diet is primarily frugivorous and is supplemented by both foliage and insects. They feed in small, evenly dispersed trees that produce a small number of fruits on a regular basis. Unlike other small platyrrhines, they also feed in larger trees at night when more dominant species are asleep.

Owl monkeys live in monogamous families of two to four individuals that occupy home ranges of 6 to 10 ha. Night ranges are long in the wet season and very short (250 m) in the dry season, when available food is more clumped in its distribution. During the southern winters of Paraguay and northern Argentina, owl monkeys are active during both day and night. Each family has several daytime sleeping nests consisting of tree holes, vine tangles, or open branches. Solitary individuals give low, owl-like hoots, perhaps to attract mates.

In contrast to most monogamous monkeys, adult owl monkeys rarely groom one another. Nevertheless, pairs stay in close contact throughout the night and sleep huddled together. Females give birth to single offspring annually. During the first week of life the infant is increasingly entrusted to the male, who, for the remainder of the infant's dependency, carries it throughout much of the night and sleeps with it during the day. The infant returns to its mother only to nurse.

Callitrichines

Callitrichines are the smallest and most morphologically derived New World anthropoids (Table 5.6). There are three morphologically distinct groups among the callitrichines: tamarins (*Saguinus*, *Leontocebus*, and *Leontopithecus*),

FIGURE 5.17 The night monkey, or owl monkey (*Aotus trivirgatus*), the only nocturnal higher primate.

TABLE 5.5 Infraorder Platyrrhini
Family CEBIDAE
Subfamily AOTINAE

Common name	Species	Intermembral index	Mass (g) M	Mass (g) F
Azara's night monkey	*Aotus azarae*	83	1246	1243
Brumback's night monkey	*A. brumbacki*	–	875	455
Grey-legged night monkey	*A. griseimembra*	–	1009	923
Hernández-Camacho's night monkey	*A. jorgehernandezi*	–	–	–
Lemurine night monkey	*A. lemurinus*	–	921	859
Andean night monkey	*A. miconax*	–	–	–
Ma's night monkey	*A. nancymaae*	–	794	780
Black-headed night monkey	*A. nigriceps*	–	875	1040
Humboldt's night monkey	*A. trivirgatus*	74	831	751
Spix's night monkey	*A. vociferans*	–	706	698
Panamanian night monkey	*A. zonalis*	–	889	916

TABLE 5.6   Infraorder Platyrrhini
Family CEBIDAE
Subfamily CALLITRICHINAE

Common name	Species	Intermembral index	Mass (g) M	F
Goeldi's monkey	*Callimico goeldii*	69	499	468
Pied tamarin	*Saguinus bicolor*	–	428	430
Geoffroy's tamarin	*S. geoffroyi*	76	482	502
Emperor tamarin	*S. imperator*	75	474	475
Mottle-face tamarin	*S. inustus*	–	585	–
Red-bellied tamarin	*S. labiatus*	73	490	529
White-footed tamarin	*S. leucopus*	74	494	490
Martins' bare-faced tamarin	*S. martinsi*	–	–	–
White/Crandall's saddle-back tamarin	*S. malanoleucus*	–	–	–
Midas tamarin	*S. midas*	77	515	575
Moustached tamarin	*S. mystax*	76	510	539
Black-handed tamarin	*S. niger*	–	404	406
Cotton-top tamarin	*S. oedipus*	74	418	404
Eastern black-handed tamarin	*S. ursulus*	–	–	–
Cruz-Lima's saddle-back tamarin	*Leontocebus cruzlimai*	–	–	–
Spix's saddle-back tamarin	*L. fuscicollis*	79	343	358
Lessons's saddle-back tamarin	*L. fuscus*	–	375	
Illiger's saddle-back tamarin	*L. illigeri*	79	292	296
Red-mantle saddle-back tamarin	*L. lagonotus*	–	375	
Andean saddle-back tamarin	*L. leucogenys*	–	375	
Black-mantled tamarin	*L. nigricollis*	78	468	484
Geoffroy's saddle-back tamarin	*L. nigrifrons*	79	377	376
Golden-mantled saddle-back tamarin	*L. tripartitus*	80	400	
Weddell's saddle-back tamarin	*L. weddelli*	79	368	360
Golden lion tamarin	*Leontopithecus rosalia*	89	620	598
Golden-headed lion tamarin	*L. chrysomelas*	86	620	603
Black-faced lion tamarin	*L. caissara*	–	607	552
Black lion tamarin	*L. chrysopygus*	–	575	–
Common marmoset	*Callithrix jacchus*	76	318	322
Buffy-tufted-ear marmoset	*C. aurita*	74	453	432
Buffy-headed marmoset	*C. flaviceps*	–	430	430
Geoffroy's tufted-ear marmoset	*C. geoffroyi*	–	307	340
Black-tufted-ear marmoset	*C. penicillata*	76	344	307
Wied's black-tufted-ear marmoset	*C. kuhlii*	74	375	
Rio Acarí marmoset	*Mico acariensis*	–	–	420
Silvery marmoset	*M. argentatus*	76	348	355

TABLE 5.6 Infraorder Platyrrhini—Cont'd
Family CEBIDAE
Subfamily CALLITRICHINAE

Common name	Species	Intermembral index	Mass (g) M	F
Golden-white tassel-ear marmoset	*M. chrysoleucus*	–	280	310
Snethlage's marmoset	*M. emiliae*	–	313	330
Santarém marmoset	*M. humeralifer*	–	475	472
Rio Aripuanã marmoset	*M. intermedius*	–	500	–
Golden-white bare-ear marmoset	*M. leucippe*	–	500	–
Maué's marmoset	*M. mauesi*	–	345	398
Marca's marmoset	*M. marcai*	–	371	396
Black-tailed marmoset	*M. melanurus*	–	406	380
Munduruku marmoset	*M. munduruku*	–	435	311
Black-headed marmoset	*M. nigriceps*	–	335	350
Rondon's marmoset	*M. rondoni*	–	327	
Sateré marmoset	*M. saterei*	–	470	413
Western pygmy marmoset	*Cebuella pygmaea*	83	110	122
Eastern pygmy marmoset	*C. niveiventris*	83	116	133
Black-crowned dwarf marmoset	*Callibella humilis*	79	136	168

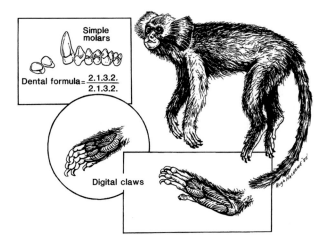

FIGURE 5.18 The unusual features of callitrichines.

marmosets (*Callithrix*, *Mico*, *Callibella*, and *Cebuella*), and Goeldi's monkey (*Callimico*). They are all small (100–750 g), often brightly colored monkeys with little if any sexual dimorphism in size or pelage coloration.

Marmosets and tamarins have a unique dentition (Fig. 5.18), with a dental formula of 2.1.3.2. They have lost their third molars from the primitive platyrrhine condition. They have simple, tritubercular upper molars, with no hypocone, that differ from the tritubercular molars of most primitive mammals in also lacking conules. *Callimico* has three molars and a tiny hypocone. All callitrichines have very short snouts and long braincases (Fig. 5.5).

Callitrichines have skeletons with relatively long trunks, tails, and legs. All digits except the great toe end in claw-like **tegulae**, rather than the flat nails characteristic of other higher primates (Fig. 5.18), an adaptation that enables them to cling to the sides of large tree trunks to feed on gums, saps, and insects. Ecologically, callitrichines seem to be characterized by the ability to exploit marginal and disturbed habitats; dietary preferences for fruit, insects, and exudates; and locomotor adaptations for quadrupedal walking and running and leaping. Many species are adept at clinging to large, vertical supports. Within these general outlines, the subfamily exhibits considerable ecological diversity.

Despite having a simple unicornuate uterus and a single pair of nipples – features usually found among mammals characterized by single births – marmosets and tamarins typically give birth to dizygotic twins, which share a common placenta but may have different fathers. *Callimico* has single births. In all callitrichines, males play a major role in the care of infants and are primarily responsible for transporting them them (Burkart et al., 2009; Erb and Porter, 2017).

Although captive callitrichines regularly live in stable monogamous groups, in natural situations they exhibit a variety of larger social groupings and mating patterns. These include monogamy, polygyny, polyandry, and polygynandry. In general, callitrichine groups are extended families. Commonly, there is a single breeding female and several breeding adult males. Most groups also contain other, nonbreeding adult males, adult females, and offspring of various ages, all of whom assist in caring for infants. Both sexes normally disperse from their natal group, but there is considerable variability in the timing and frequency of dispersion.

The unique, derived anatomical features of callitrichines have been related to their small size (e.g., Ford, 1980) or to their unusual ecological adaptations for insectivory and exudate eating (Sussman and Kinzey, 1984). Among platyrrhines, and primates in general, smaller species generally have larger infants (relative to the size of the mother), which results in considerable problems for the female in birthing and early postnatal care. Marmosets and tamarins have overcome this by giving birth to twins (rather than a single large infant) and by extensive **allomaternal care**, or care for infants by several group members (Erb and Porter, 2017). However, as many authors (e.g., Martin, 1992) have subsequently noted, twinning has such extreme ramifications for callitrichine life history and ecology that it is more than just a simple solution to an obstetric problem. It is probably also related to the heavy predation pressure suffered by callitrichines and their adaptations as colonizing species. Likewise, their small size permits both a high-energy diet and a reduced dentition, compared to that needed by larger, more folivorous or frugivorous species (see Chapter 9). Ecologically, their small size has yielded a time-minimizing foraging strategy due to the benefits of a specialization on energy-rich insects and fruits found in small patches and the cost of a high predation risk. However, the sequence in which many of the unique callitrichine features appeared and the ecological context of their evolution are far from resolved.

Tamarins

The tamarins (*Saguinus, Leontocebus, and Leontopithecus*) are the most widespread and diverse of callitrichines, with over 20 species displaying an extraordinary array of pelage color patterns and elaborations of facial hair (Table 5.6; Fig. 5.19). *Saguinus* has the widest distribution of any callitrichine genus and is found throughout Amazonia, on both banks of the Amazon, as well as the Guianas, Colombia, and Central America. Tamarins are commonly divided into three species groups, based on pelage and aspects of skeletal morphology. All are small, with relatively long trunks, legs, and tails. Their diet consists of fruit, animal prey, and exudates in varying proportions. Like most other anthropoid primates, tamarins have canines that are much larger than their incisors. Although many tamarins eat exudates, they seem to lack the marmosets' ability to gnaw holes in tree bark. Compared with other callitrichines, tamarins are characterized by smaller social groups, large home ranges, a single annual birth peak, and lower population densities. They show considerable ecological diversity.

Saguinus oedipus, the **crested** or **cotton-top tamarin** (Fig. 5.19), lives in low secondary forests of Panama and Colombia. These monkeys move primarily by quadrupedal walking and running on medium-sized supports and less frequently by leaping between vertical trunks. Fruit (40%) and animal material (40%) seem to make up the bulk of their diet, and exudates are an important third component. In foraging for fruits, they range from the middle of the canopy to the ground. They forage for insects in the shrub layer and feed on exudates by clinging to the sides of relatively large trunks.

Saguinus bicolor   ***Saguinus imperator***   ***Saguinus oedipus***   ***Saguinus labiatus***

FIGURE 5.19   The faces of four tamarins.

Cotton-top tamarins have a mean group size of six individuals, but this varies considerably. All groups contain at least one adult female and male, with some groups containing several adult males and females. The genetic relationships among individuals in these groups are not known from current studies; however, group composition changes frequently, and there is considerable immigration and emigration between groups (Savage et al., 1996).

The **golden-handed tamarin** (*Saguinus midas*) has been studied in both eastern Colombia and Suriname. In Suriname these tamarins are most common in primary forest, but they prefer the edge habitats between forest types. They spend most of their time in the middle levels of the forest, where they move primarily by quadrupedal walking and running along medium-sized supports and by leaping between the ends of branches (Fig. 5.9). They seem to be largely frugivorous; insects and exudates are less important components of their diet. Social groups of golden-handed tamarins average about six individuals, with a considerable range (2–12 individuals).

Leontocebus (Fig. 5.20), the **saddle-back tamarins** (named for the distinctive patterns on its back), are speciose and distributed throughout Amazonia. These small monkeys move and feed in the lower levels of the forest. Their locomotion involves frequent leaps between large vertical tree trunks and from branches to trunks. During the wet season they are primarily frugivorous, specializing on small, widely dispersed fruits. In the dry season, when fruits are rare, the herbivorous portion of their diet consists almost totally of nectar. Insect foraging accounts for nearly half of the daily feeding time in this species, which specializes on relatively large, cryptic insects that it locates by probing in the hollows, crevices, and bases of trees.

FIGURE 5.20 Two sympatric tamarins from Bolivia: *above*, the white-lipped tamarin (*Saguinus labiatus*); *below*, the saddle-back tamarin (*Leontocebus fuscicollis*).

Saddle-back tamarins live in small groups of three to eight individuals and actively defend their territories against neighboring groups of the same species. The most common group structure is a polyandrous mating system with a single breeding female and two breeding males. Monogamous and polygynous groups are less common, and all monogamous groups studied failed to raise their offspring successfully without a second male caretaker. Although group territories remain stable from year to year, there is considerable turnover of individuals in a group as a result of predation and births, as well as emigration and immigration. Groups often range more than 1 km a day.

Throughout their distribution, saddle-back tamarins are normally found in association with one of the moustached tamarins. They seem to be the more parasitic members of this association, for they rarely participate in defense of the common territories, and gain considerable foraging benefit from insects flushed by moustached tamarins as well as the antipredator benefits of the larger groups. There are three members of the moustached group whose behaviors are well documented: the white-lipped tamarin, the emperor tamarin, and the moustached tamarin.

Saguinus labiatus, the **white-lipped tamarin** (Figs. 5.19 and 5.20), has a relatively small distribution in the middle Amazonian region of western Brazil and eastern Bolivia, where it lives in sympatry with saddle-back tamarins. Although similar in size to the saddle-back species, white-lipped tamarins differ in several aspects of behavior and ecology. They are found at higher levels in the forest, most commonly in the middle levels of the canopy, where they move by quadrupedal running and short leaps between branches. Like saddle-backs, they eat both fruits and insects. In contrast to the actively foraging saddle-back tamarins, these monkeys spend much of their time visually scanning for prey, which they find among the leaves and terminal branches within the main canopy.

White-lipped tamarins and saddle-back tamarins have home ranges of similar size (30 ha) for their small groups (two to four individuals). Most striking is the fact that groups of the two species overlap almost completely in their daily ranging behavior, not only when traveling and feeding but also during resting and sleeping. This seems to be primarily an adaptation for predator detection.

Saguinus imperator, the **emperor tamarin** (Fig. 5.19), was named for Emperor Franz Joseph of Austria because of its sweeping moustache. This medium-sized tamarin from the upper Amazonian region of Peru and Bolivia is similar in many aspects of its ecology to the white-lipped tamarin. It relies on fruit from small trees throughout the year as its dietary staple, but it takes a substantial amount of nectar during the dry season, when fruits are less abundant. Like white-lipped tamarins, emperor tamarins forage for visible insects among the leaves and small branches of the forest canopy.

The small family groups of emperor tamarins share their moderate territories (30 ha) with a group of saddle-back tamarins of similar size, but they defend it against conspecifics. They often travel as much as 1 km a day. Twin offspring are normally born at the beginning of the rainy season.

The **moustached tamarin** (*S. mystax*) lives in the middle Amazon region of northern Peru and western Brazil. Among the available habitats, these tamarins seem to prefer the drier, upland forest and to avoid flooded forests. They travel and feed most commonly on thin, flexible supports. Moustached tamarins live in groups of three to eight individuals, which usually contain a single breeding female, up to three presumably reproductively active adult males, and several nonreproductive females and subadults. Usually more than one adult male participates in the care of infants, and census data on numerous groups show a correlation between the number of male helpers and the number of surviving infants.

Lion tamarins (*Leontopithecus*), from southeastern Brazil (Fig. 5.21), are the largest callitrichines, with an adult body mass of over 600 g (Table 5.6). There are four allopatric species: *L. rosalia*, the golden lion tamarin; *L. chrysomelas*, the golden-headed lion tamarin; *L. chrysopygus*, the golden-rumped or black lion tamarin; and *L. caissara*, the black-faced lion tamarin. All are on the verge of extinction in the wild because of extensive habitat destruction. Of the four, *L. chrysomelas* is the most distinctive in dental and cranial features, with particularly large anterior teeth.

Lion tamarins are confined largely to the lowland primary rainforest of southeast Brazil and seem to fare poorly in secondary forests. All species are found primarily in the main canopy levels, where they move in a quadrupedal fashion. Lion tamarins feed on a wide range of invertebrates and small vertebrates as well as on fruit, but they have never been observed eating exudates. They use their long fingers to extract insects from holes and crevices, beneath tree bark, and especially inside bromeliad plants. It has been suggested that the enlarged anterior teeth of *L. chrysomelas* are used to gnaw through bark to expose insects. Despite the similarities among these species, each species seems to have a distinct ecology, with the inland *L. chrysopygus* occupying the most distinctive habitat.

Lion tamarins live in relatively small social groups similar in size to those of *Saguinus*. As in all callitrichines, the social groups of lion tamarins usually have a single breeding female, but group composition and mating patterns are quite variable, with monogamy, polyandry, and polygyny all reported for wild populations. Lion tamarin females generally breed once a year and give birth to twins that are cared for by other group members. Lion tamarins have the largest home ranges

FIGURE 5.21   A group of lion tamarins (*Leontopithecus rosalia*), one of the most beautiful and most endangered primates.

of all callitrichines, often over 100 ha; likewise, day ranges are very large for most populations. All lion tamarins use holes in trees for sleeping.

Marmosets

The **marmosets** (Table 5.6) are the smallest platyrrhines and have the most specialized dentition (Fig. 5.22). They are distinguished from tamarins and other platyrrhines by their uniquely enlarged incisors, which are similar in height to their canines. Furthermore, these large incisors have only a thin layer of enamel on the lingual surface, which quickly wears away and causes the teeth to assume a chisel-like shape similar to the incisors of rodents. Marmosets use these chisel-like incisors for biting holes in trees to elicit the flow of gums, saps, and resins. Compared to tamarins, marmosets tend to inhabit more secondary, drier habitats, have large

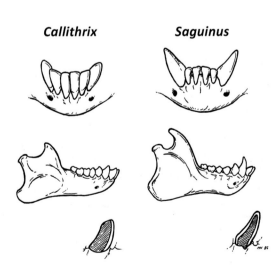

FIGURE 5.22   The lower jaw and teeth of a marmoset (*left*) and a tamarin (*right*), showing the differences in proportion of the canines and incisors and the thickness of the enamel on the lower incisors.

social groups and smaller home ranges, and to exhibit bimodal annual birth peaks.

There are several genera of marmosets. The genus *Callithrix* consists of six species, including *Callithrix jacchus*, the **common marmoset** (Fig. 5.23) from the eastern part of Brazil where they are found in edge habitats and secondary forest environments. Common marmosets forage for insects in the vine tangles of the understory and move by a combination of quadrupedal walking and running, as well as by leaping. They use clinging postures on tree trunks when eating exudates. At night *Callithrix* often sleep in tree holes. *Callithrix* species all eat fruits, insects, and exudates, with the latter being preferred, especially at the end of the wet season, when fruits are scarce.

The genus *Mico* consists of over a dozen species from wetter, forested habitats in southern Amazonia. These include *Mico argentata*, the **silvery marmoset**, and *Mico humeralifer*, the **black and white tassel-eared marmoset**. Field studies of *M. argentata* indicate that its ranging pattern is more similar to that of other marmosets that rely on exudates in the dry season than to those of tamarins that occupy similar rainforest habitats (Veracini, 2009).

Members of the genus *Cebuella*, the **pygmy marmoset** (Fig. 5.24), which includes an eastern (*C. niveiventris*) and western species (*C. pygmaea*), have an average adult body weight of approximately 100 g. It is the smallest marmoset, the smallest platyrrhine, the smallest anthropoid, and only a little larger than *Microcebus* and *Galagoides*, the smallest living primates. They are found in the Amazonian regions of Colombia, Peru, Ecuador, Brazil, and Bolivia, where their distribution and density seem to be linked to the abundance of special feeding trees.

Feeding on tree exudates occupies 67% of pygmy marmoset feeding time, part of which is devoted to actual feeding from a primary exudate tree and part of which is devoted to preparing new trees by gnawing holes in their bark. The remainder of the monkeys' feeding time consists of foraging for insects and occasional fruits. Because they are so dependent on tree exudates, pygmy marmosets tend to be found in the lower levels of the forest. Their insect foraging takes place in vine tangles. During exudate eating, they frequently adopt clinging positions on the large trunks and move by leaping between vertical supports.

Pygmy marmosets live primarily in small groups with a single adult male, a single adult female, and offspring of various ages. They have tiny home ranges centered around whatever the main food tree is at the time, and individual groups seem to be widely spaced from one another. Because these primary exudate trees change from year to year, so do the home ranges of pygmy marmoset groups. Pygmy marmosets give birth to dizygotic twins at approximately 6-month intervals. The young are carried most frequently by the adult male. During the night, pygmy marmosets sleep in vine tangles or in tree holes.

Callibella humilis, the **black-crowned dwarf marmoset** or **Amazonian dwarf marmoset**, is a recently described monkey from the Brazilian Amazon south of the Rio Madiera (van Roosmalen and van Roosmalen, 2003), and is intermediate in size between *Mico* and *Cebuella*. Adults have a dark olive brown coat with a yellow-golden belly. *Callibella* has been reported only in areas characterized by human disturbance, such as orchards and gardens. Limited observations of both wild and captive individuals indicate very unusual reproductive behavior. All known births were singletons, not twins, as in most callitrichines; the infant was carried only by the mother and was parked.

Goeldi's monkey, *Callimico goeldii* (Table 5.6; Fig. 5.25), is a tufted, silky black monkey from the upper Amazon regions of Colombia, Ecuador, Peru, Bolivia, and Brazil. This genus has a full set of three molars and a small hypocone on the upper molars. *Callimico* also has single births, the primitive condition for all New World monkeys. In other respects, such as claws and limb proportions, *Callimico* resembles tamarins and marmosets. It is slightly smaller than the squirrel monkey. Males and

Callithrix jacchus   **Mico argentatus**   **Callithrix geoffroyi**   **Callithrix aurita**

FIGURE 5.23   The faces of four marmosets, showing the diversity in facial ornamentation.

FIGURE 5.24   A family of pygmy marmosets (*Cebuella pygmaea*).

FIGURE 5.25   A group of Goeldi's monkeys (*Callimico goeldii*) in a bamboo habitat.

females are virtually indistinguishable in size and coloration. Although morphologists have traditionally placed *Callimico* as either the most primitive callitrichine or as a separate group among platyrrhines, molecular studies of platyrrhine phylogeny have consistently grouped *Callimico* with *Callithrix*, *Mico*, and *Cebuella*, indicating either that its primitive-looking features (three molars, single births) are derived from a more marmoset-like condition, or that marmosets and tamarins evolved their molar loss and twinning independently.

Goeldi's monkeys have a very limited distribution, and occur very patchily throughout Amazonia. They have been reported in primary and secondary forests, but most commonly in low bushes and bamboo thickets

of the understory. They move mainly by leaping from trunk to trunk a few meters off the ground. The diet of *Callimico* includes large amounts of invertebrates and also fruits, but fungus plays a large role in the diet at many sites. They rarely feed on exudates.

In captivity, these monkeys live in family groups of one adult male and one adult female, but under natural conditions groups contain more adults of both sexes. Goeldi's monkeys have relatively large home ranges and day ranges, and they frequently form polyspecific associations with tamarins. *Callimico* groups have an unusual ranging pattern in which they associate with numerous different tamarin groups in succession (Porter, 2004; Porter and Garber, 2004). They have single offspring twice a year. All adults in the group aid in carrying and feeding the infant.

Adaptive Radiation of Platyrrhines

Like the Malagasy lemurs, the platyrrhines of the Neotropics arrived on an island continent tens of millions of years ago and have evolved into a diverse radiation with no competition from other groups of primates. The extent of their adaptive diversity (Fig. 5.26) is indicated by the presence of numerous sympatric species throughout most of South America (e.g., Fig. 5.8), and up to 13 species at some Amazonian sites (Fig. 5.7).

Platyrrhines are small to medium in size; living species range from approximately 100 g for the pygmy marmoset to approximately 10 kg for several of the atelids. The recently extinct atelids *Protopithecus*, *Caipora*, and *Cartelles* were much larger (see Chapter 14). All platyrrhine genera but one are diurnal, but the single nocturnal genus (*Aotus*) is very widespread. Platyrrhines normally have dichromatic color vision, with trichromatic color vision present only in some females. Only in *Alouatta* is trichromatic color vision found in all or almost all individuals, as in catarrhines, including humans.

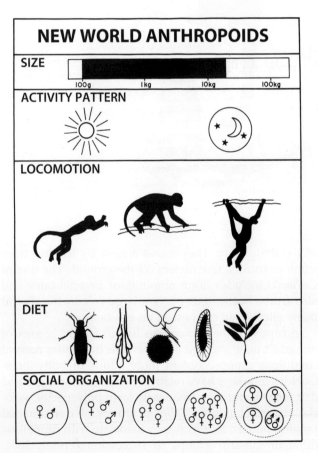

FIGURE 5.26   The adaptive diversity of the platyrrhines.

Although they lack the extremes in limb proportion or skeletal specialization seen in many other groups of primates, platyrrhines show a range of locomotor abilities. Some species are excellent leapers, many are arboreal quadrupeds, and the larger species frequently use suspensory postures. Platyrrhines are the only primates to have evolved a prehensile tail, an organ that adds considerably to the locomotor and postural abilities of five genera. Callitrichines have clawed digits which enable them to cling to vertical trunks, an adaptation for feeding on exudates and cryptic prey on trunks and in tree holes. *Sapajus libidinosus* is unique among nonhuman primates in its frequent use of bipedal posture and locomotion during terrestrial foraging.

A striking feature of the platyrrhine radiation is the absence of terrestrial species. A few genera (*Alouatta*, *Cacajao*, *Saimiri*, and *Sapajus*) occasionally forage on the ground or travel short distances between trees, but no platyrrhines, except for perhaps some species of *Sapajus*, spend a large portion of each day feeding on the ground.

The New World anthropoids include species that specialize on gums, on fruits, on leaves, and on seeds. Some of the smaller species rely heavily on nectar during the dry periods of the year, and many rely extensively on animal prey, either year-round or seasonally. There are only two predominantly folivorous genera, *Alouatta* and northern populations of *Brachyteles*. The use of tools in foraging by *Sapajus* is matched only by chimpanzees among nonhuman primates.

Because their lack of large or terrestrial species, and the limited number of folivorous and nocturnal species, the overall ecological diversity of platyrrhines is much less than that of the primates found in other biogeographical areas (Fleagle and Reed, 1999; see Chapter 8). Efforts to explain the limited ecological diversity of platyrrhines compared with primates on other continental areas have invoked patterns of phenology, competition with other mammalian groups, and limited time since divergence as possible explanations (see Chapter 8). Studies by Ganzhorn and colleagues (2009) suggest that the absence of committed folivores among Neotropical primates might be due to the fact that Neotropical fruits are generally high in protein, so platyrrhines do not need to resort to folivory to meet their protein needs.

Platyrrhine social organization is much more diverse than that found among any other major radiation of primates. Many New World monkeys live in pairs (*Aotus*, *Callicebus*, and *P. pithecia*) that seem to be stable for several years. In several genera (*Ateles*, *Brachyteles*, and *Chiropotes*), large groups exist within a fission–fusion arrangement in which many adult males and females fragment into smaller foraging units, depending on the distribution of food resources. One genus (*Alouatta*) lives in single-male groups in some environments and multi-male groups in others, or when the population density changes at different sites. *Cebus*, *Sapajus*, and *Saimiri* live in more complex groups of several adults of each sex. The social organization of *Saimiri* is reminiscent of that of Malagasy lemurs, with female dominance throughout the year and intense male–male competition for a few brief weeks in the breeding season. Social organization in the callitrichines is much more complex than the simple monogamy suggested by studies of captive monkeys. Social groups of many species in natural environments seem to be extended families with a single breeding female, but diverse mating systems have been described, including monogamous, polygynous, polyandrous, or polygynandrous arrangements. All are cooperative breeders with many group members aiding in the raising of infants. This platyrrhine diversity gives us a valuable natural laboratory for understanding the ecological causalities underlying primate social behavior.

Phyletic Relationships of Platyrrhines

The results of molecular studies over the past decade and a half have yielded a remarkably consistent picture of the relationships among extant platyrrhines (Fig. 5.27) (e.g., Disotell, 2008; Hodgson et al., 2009; Osterholtz et al., 2009; Wildman et al., 2009; Perelman et al., 2011; Springer et al., 2012; Schneider and Sampaio, 2015). There are three distinct platyrrhine clades: pitheciids, atelids, and cebids. Pitheciids are the sister taxon to the other two groups. Within pitheciids, callicebines are the sister group to the pitheciines. Among the atelids,

Alouatta is the sister taxon to the other genera. Within cebids, the relationships among *Aotus*, cebines, and callitrichines remain unresolved.

The molecular results generally agree with morphological studies of Rosenberger (e.g., 1981, 1992), but there are some striking differences. A surprising, but consistent result of the molecular studies is the grouping of *Aotus* with cebines and callitrichines rather than with callicebines. *Aotus* and *Callicebus* are extremely similar in many aspects of morphology and behavior, and it is remarkable that they are always placed in different families. This is most reasonably interpreted as an indication that the similarities shared by these two taxa represent the primitive condition for platyrrhines. Another surprising result of the molecular studies is the placement of *Callimico* well within the callitrichine clade, suggesting that many features, such as loss of the last molar and twin births, evolved independently several times, or that *Callimico* reverted to the primitive condition for these features. The position of *Brachyteles* among the atelines has been a subject of some debate, but recent results seem to suggest that it is the sister taxon of *Lagothrix* rather than *Ateles* (Perelman et al., 2011; Di Fiore et al., 2012; Springer et al., 2012). This suggests that the postcranial similarities between *Brachyteles* and *Ateles* evolved in parallel. In contrast with the situation among other major radiations of primates, there are so far no studies suggesting that hybridization has played a role in the evolution of any platyrrhine taxa.

However satisfying it is to sort out the phylogeny of platyrrhines, this is just the first step in obtaining an understanding of the evolution of the group. Using this

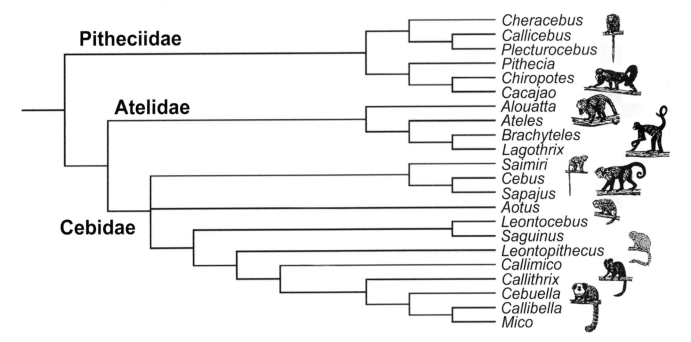

FIGURE 5.27 A summary phylogeny of New World monkeys.

phylogeny as a base, we can then start to inquire as to how and when the unique adaptations of the different clades evolved and the morphological and ecological transformations that were involved. It is the origin of adaptations that should be the goal in understanding platyrrhine phylogeny.

References

Aguilar-Melo, A.R., Calme, S., Pinacho-Guendulain, B., Smith-Aguilar, S.E., Ramos-Fernandez, G., 2020. Ecological and social determinants of association and proximity patterns in the fission–fusion society of spider monkeys (*Ateles geoffroyi*). Am. J. Primatol. 82, e23077.

Amato, K.R., Leigh, S.R., Kent, A., et al., 2015. The gut microbiota appears to compensate for seasonal diet variation in the wild black howler monkey (*Alouatta pigra*). Microb. Ecol. 69, 434–443.

Boubli, J.P., 2002. Western extension of the range of bearded sakis: A possible new taxon of *Chiropotes* sympatric with *Cacajao* in the Pico da Neblina National Park, Brazil. Neotrop. Primates 10, 1–4.

Boubli, J.P., Cuoto-Santos, F.R., Mourthe, I.M.C., 2010. Quantitative assessment of habitat differences between northern and southern muriquis (Primates, Atelidae) in the Brazilian Atlantic Forest. Ecotropica 16, 63–69.

Burkart, J.M., Hrdy, S.B., van Schaik, C.P., 2009. Cooperative breeding and human cognitive evolution. Evol. Anthrop. 18, 175–186.

Chapman, C.A., Chapman, L.J., Wrangham, R.W., 1995. Ecological constraints on group size: An analysis of spider monkey and chimpanzee subgroups. Behav. Ecol. Sociobiol. 36, 59–70.

Coles, R.C., Lee, P.C., Talebi, M., 2012. Fission–fusion dynamics in southern muriquis (*Brachyteles arachnoides*) in continuous Brazilian Atlantic forest. Int. J. Primatol. 33, 93–114.

Couto-Santos, F.R., 2007. Fenologia de especies arboreas do dossel e sub-dossel em um fragment de Mata Atlantica semi-decidua em Caratinga, Minas Gerais, Brasil. MA thesis, Universidade Federal de Minas Gerais, Belo Horizonte, Brazil.

de Moraes, P.L.R., de Carvalho, O., Strier, K.B., 1998. Population variation in patch and party size in muriquis (*Brachyteles arachnoides*). Int. J. Primatol. 19, 325–337.

Di Fiore, A., Cornejo, F., Chaves, P.B., et al., 2012. Phylogenetic position of the yellow-tailed woolly monkey based on whole mitochondrial genomes. Am. J. Phys. Anthropol. 147, 131–131.

Disotell, T.R., 2008. Primate Phylogenetics Encyclopedia of Life Sciences. John Wiley and Sons. https://doi.org/10.1002/9780470015902.a0005833.pub2

Dunn, J.C., Halenar, L.B., Davies, T.G., et al., 2015. Evolutionary trade-off between vocal tract and testes dimensions in howler monkeys. Curr. Biol. 25, 2839–2844.

Erb, W.M., Porter, L.M., 2017. Mother's little helpers: What we know (and don't know) about cooperative infant care in callitrichines. Evol. Anthrop. 26, 25–37.

Erikson, G.E., 1963. Brachiation in New World monkeys and in anthropoid apes. Symp. Zool. Soc. London 10, 135–164.

Fernandez-Duque, E., Di Fiore, A., de Luna, A.G., 2013. Pair-mate relationships and parenting in equatorial saki monkeys (*Pithecia aequatorialis*) and red titi monkeys (*Callicebus discolor*) of Ecuador. In: Veiga, L.M., Barnett, A.A., Ferrari, S.F., Norconk, M.A. (Eds.), Evolutionary Biology and Conservation of Titis, Sakis and Uacaris. Cambridge University Press, Cambridge, pp. 295–302.

Ferrari, S.F., Emidio-Silva, C., Lopes, M.A., et al., 1999. Bearded sakis in southeastern Amazonia–back from the brink? Oryx 33, 346–351.

Fleagle, J.G., Reed, K.E., 1999. Phylogenetic and temporal perspectives on primate ecology. In: Fleagle, J.G., Janson, C., Reed, K.E.

(Eds.), Primate communities, Cambridge University Press, Cambridge, pp. 92–115.

Ford, S.M., 1980. Callitrichids as phyletic dwarfs and the place of the Callitrichidae in Platyrrhini. Primates 21, 31–43.

Ganzhorn, J.U., Arrigo-Nelson, S., Boinski, S., et al., 2009. Possible fruit protein effects on primate communities in Madagascar and the Neotropics. PLoS One 4, e8253.

Hartwig, W.C., 1995. Effect of life history on the squirrel monkey (Platyrrhini *Saimiri*) cranium. Am. J. Phys. Anthropol. 97, 435–449.

Hodgson, J.A., Sterner, K.N., Matthews, L.J., et al., 2009. Successive radiations, not stasis, in the South American primate fauna. P. Natl. Acad. Sci. 106, 5534–5539.

Johns, A.D., Ayres, J.M., 1987. Bearded sakis beyond the brink. Oryx 21, 164–167.

Larson, S.G., 1998. Parallel evolution in the hominoid trunk and forelimb. Evol. Anthropol. 6, 87–99.

Marsh, L.K., 2014. A taxonomic revision of the saki monkeys, *Pithecia* Desmarest, 1804. Neotrop. Primates 21, 1–165.

Martin, R.D., 1992. Goeldi and the dwarfs: The evolutionary biology of the small New World monkeys. J. Hum. Evol. 22, 367–393.

Mitchell, C.L., Boinski, S., van Schaik, C.P., 1991. Competitive regimes and female bonding in two species of squirrel monkey (*Saimiri oerstedii* and *S. sciureus*). Behav. Ecol. Sociobiol. 28, 55–60.

Norconk, M.A., 2011. Sakis, uakaris, and titi monkeys. In: Campbell, C.J., Fuentes, A., MacKinnon, K.C., et al. (Eds.), Primates in Perspective. Oxford University Press, Oxford, pp. 122–139.

Norconk, M.A., Setz, E.Z., 2013. Ecology and behavior of saki monkeys (genus Pithecia). Cambridge Studies in. Biol. Evol. Anthropol. 1262–271.

Osterholtz, M., Walter, L., Roos, C., 2009. Retropositional events consolidate the branching order among New World monkey genera. Mol. Phylog. Evol. 50, 507–513.

Ottoni, E.B., Izar, P., 2008. Capuchin monkey tool use: Overview and implication. Evol. Anthropol. 17, 171–178.

Perelman, P., Johnson, W.E., Roos, C., et al., 2011. A molecular phylogeny of living primates. PLoS Genet. 7 (3), e1001342. https://doi.org/10.1371/journal.pgen.1001342

Porter, L.M., 2004. Differences in forest utilization and activity patterns among three sympatric callitrichines: *Callimico goeldii*, *Saguinus fuscicollis*, and *Saguinus labiatus*. Amer. J. Phys. Anthrop. 124, 139–153.

Porter, L.M., Garber, P.A., 2004. The Goeldi's monkey: A primate paradox? Evol. Anthropol. 13, 104–115.

Ramos-Fernández, G., Boyer, D., Aureli, F., Vick, L.G., 2009. Association networks in spider monkeys (*Ateles geoffroyi*). Behav. Ecol. Sociobiol. 63, 999–1013. https://doi.org/10.1007/s00265-009-0719-4

Rosenberger, A.L., 1981. Systematics: The higher taxa. In: Coimbra-Filho, A.E., Mittermeier, R.A. (Eds.), Ecology and Behavior of Neotropical Primates. Academia Brasileira de Ciencias, Rio de Janeiro, pp. 9–27.

Rosenberger, A.L., 1992. Evolution of feeding niches in New World monkeys. Am. J. Phys. Anthropol. 88, 525–562.

Savage, A., Giraldo, L.H., Soto, L.H., Snowdon, C.T., 1996. Demography, group composition, and dispersal in wild cotton-top tamarin (*Saguinus oedipus*) groups. Am. J. Primatol. 38, 85–100.

Schneider, H., Sampaio, I., 2015. The systematics and evolution of New World primates – a review. Mol. Phylogenet. Evol. 82, 348–357.

Shaffer, C.A., 2013. GIS analysis of patch use and group cohesiveness of bearded sakis (*Chiropotes sagulatus*) in the upper Essequibo Conservation Concession, Guyana. Am. J. Phys. Anthropol. 150, 235–246.

Springer, M.S., Meredith, R.W., Gatesy, J., et al., 2012. Macroevolutionary dynamics and historical biogeography of primate diversification inferred from a species supermatrix. PLoS One 7 (11), e49521. https://doi.org/10.1371/journal.pone.0049521

Strier, K.B., 1994. Brotherhoods among atelins: Kinship, affiliation, and competition. Behaviour 130, 151–167.

Strier, K.B., Chaves, P.B., Mendes, S.L., et al., 2011. Low paternity skew and the influence of maternal kin in an egalitarian, patrilocal primate. Proc. Natl. Acad. Sci. 108, 18915–18919.

Strier, K.B., Dib, L.T., Figueira, J.E.C., 2002. Social dynamics of male muriquis *Brachyteles arachnoides hypoxanthus*. Behaviour 139, 315–342.

Sussman, R.W., Kinzey, W.G., 1984. The ecological role of the Callitrichidae: A review. Am. J. Phys. Anthropol. 64, 419–449.

Talebi, M.G., Pope, T.R., Vogel, E.R., et al., 2006. Polymorphism of visual pigment genes in the muriqui (Primates, Atelidae). Mol. Ecol. 15, 551–558.

ter Steege, H., Pitman, N., Phillips, O.L., et al., 2006. Continental-scale patterns of canopy tree composition and function across Amazonia. Nature 443, 444–447.

Van Belle, S., Estrada, A., Garber, P.A., 2013. Collective group movement and leadership in wild black howler monkeys (*Alouatta pigra*). Behav. Ecol. Sociobiol. 67, 31–41.

Van Belle, S., Fernandez-Duque, E., Di Fiore, A., 2016. Demography and life history of wild red titi monkeys (*Callicebus discolor*) and equatorial sakis (*Pithecia aequatorialis*) in Amazonian Ecuador: A 12-year study. Am. J. Primatol. 78, 204–215.

van Roosmalen, M.G.M., van Roosmalen, T., 2003. The description of a new marmoset genus, *Callibella* (Callitrichinae, Primates), including its molecular phylogenetic status. Neotrop. Primates 11, 1–12.

van Roosmalen, M.G.M., van Roosmalen, T., Mittermeier, R.A., 2002. A taxonomic review of the titi monkeys, genus *Callicebus* Thomas, 1903, with the description of two new species, *Callicebus bernhardi* and *Callicebus stephennashi*, from Brazilian Amazonia. Neotrop. Primates 10 (Suppl.), 1–53.

Veracini, C., 2009. Habitat use and ranging behavior of the silvery marmoset (*Mico argentatus*) at Caxiuanã National Forest (eastern Brazilian Amazonia). In: Ford, S.M., Davis, L.C., Porter, L.M. (Eds.), The Smallest Anthropoids: The Marmoset/*Callimico* Radiation. Springer, New York, pp. 221–240.

Wildman, D.E., Jameson, N.M., Opazo, J.C., Soojin, V.Y., 2009. A fully resolved genus level phylogeny of Neotropical primates (Platyrrhini). Mol. Phylogenet. Evol. 53, 694–702.

Wright, P.C., 1989. The nocturnal primate niche in the New World. J. Hum. Evol. 18, 635–658.

Further Reading

Bezanson, M., Strier, K.B., Sussman, R.W., 2022. Chapter 11: Howler monkeys, spider monkeys, woolly monkeys, and muriquis: Behavioral ecology of the largest primates of the Americas. In: Sussman, R.W., Hart, D., Colquhoun, I.C. (Eds.), The Natural History of Primates: A Systematic Survey of Ecology and Behavior. Rowman and Littlefield, Lanham, pp. 258–287.

Bowler, M., 2007. The ecology and conservation of the red uakari monkey on the Yavarí River, Peru. PhD thesis. University of Kent, Canterbury.

Campbell, C.J., 2008. Spider Monkeys: Behavior, Ecology and Evolution of the Genus Ateles. Cambridge University Press, Cambridge.

Deluycker, A.M., 2022a. Chapter 7: Owl monkeys: Under the moonlight. In: Sussman, R.W., Hart, D., Colquhoun, I.C. (Eds.), The Natural History of Primates: A Systematic Survey of Ecology and Behavior. Rowman and Littlefield, Lanham, pp. 180–192.

Deluycker, A.M., 2022b. Chapter 8: Titi monkeys: Tail-twining in the trees. In: Sussman, R.W., Hart, D., Colquhoun, I.C. (Eds.), The Natural History of Primates: A Systematic Survey of Ecology and Behavior. Rowman and Littlefield, Lanham, pp. 193–208.

Di Fiore, A., Link, A., Campbell, C.J., 2011. Atelines: Behavioral and socioecological diversity in a New World monkey radiation. In: Campbell, C., Fuentes, A., MacKinnon, K., et al. (Eds.), Primates in Perspective, second ed. Oxford University Press, Oxford, pp. 155–188.

Digby, L.J., Ferrari, S.F., Saltzman, W., 2011. Callitrichines: The role of competition in cooperatively breeding species. In: Campbell, C., Fuentes, A., MacKinnon, K., et al. (Eds.), Primates in Perspective, second ed. Oxford University Press, Oxford, pp. 91–107.

Fernandez-Duque, E., 2011. Aotinae: Social Monogamy in the Only Nocturnal Anthropoid. In: Campbell, C., Fuentes, A., MacKinnon, K., et al. (Eds.), Primates in Perspective, second ed. Oxford University Press, Oxford, pp. 140–154.

Fernandez-Duque, E., Fiori, Di, Huck, M., A., 2012. The behavior, ecology, and social evolution of New World monkeys. In: Mitani, J.C., Call, J., Kappeler, P.M., et al. (Eds.), The Evolution of Primate Societies. University of Chicago Press, Chicago, pp. 43–64.

Ford, S.M., Porter, L.M., Davis, L.C., 2009. The Smallest Anthropoids: The Marmoset/Callimico Radiation. Springer, New York.

Garber, P.A., Estrada, A., Bicca Marques, J.-C., et al., 2009. South American Primates: Comparative Perspectives in the Study of Behavior, Ecology, and Conservation. Springer, New York.

Garber, P.A., Sussman, R.W., 2022. Chapter 6: Tamarins, callimicos, and marmosets: the evolutionary and ecological challenges of body size reduction and reproductive twinning. In: Sussman, R.W., Hart, D., Colquhoun, I.C. (Eds.), The Natural History of Primates: A Systematic Survey of Ecology and Behavior. Rowman and Littlefield, Lanham, pp. 139–179.

Jack, K.M., 2011. The Cebines: Toward and explanation of variable social structure. In: Campbell, C., Fuentes, A., MacKinnon, K., et al. (Eds.), Primates in Perspective, second ed. Oxford University Press, Oxford, pp. 107–122.

Mackinnon, K.C., Bezanson, M., 2022. Chapter 10: Ecological niches and behavioral strategies of capuchins and squirrel monkeys. In: Sussman, R.W., Hart, D., Colquhoun, I.C. (Eds.), The Natural History of Primates: A Systematic Survey of Ecology and Behavior. Rowman and Littlefield, Lanham, pp. 228–257.

Norconk, M.A., 2011. Sakis, uakaris, and titi monkeys. In: Campbell, C., Fuentes, A., MacKinnon, K., et al. (Eds.), Primates in Perspective, second ed. Oxford University Press, Oxford, pp. 122–139.

Norconk, M.A., Setz, E.Z., 2013. Ecology and behavior of saki monkeys (genus *Pithecia*). Cambridge Studies in Biol. Evol. Anthropol. 1 (65), 262–271.

Rosenberger, A.L., 2020. New World Monkeys: The Evolutionary Odyssey. Princeton University Press, Princeton, NJ.

Shaffer, C., Ormond, P.T., 2022. Chapter 9: Sakis, bearded sakis, and uakaris: Platyrrhine seed predators. In: Sussman, R.W., Hart, D., Colquhoun, I.C. (Eds.), The Natural History of Primates: A Systematic Survey of Ecology and Behavior. Rowman and Littlefield, Lanham, pp. 209–227.

Veiga, L.M., Barnett, A.A., Ferrari, S.F., Norconk, M.A. (Eds.), 2013. Evolutionary Biology and Conservation of Titis, Sakis and Uacaris. Cambridge University Press, Cambridge.

Old World Monkeys

The platyrrhine monkeys are the only primates in the Neotropics, where they fill a diverse array of ecological niches. In the Old World there are two very distinct radiations of higher primates that make up the infraorder Catarrhini – the Old World monkeys (Cercopithecoidea) and the hominoid apes (Hominoidea). As we shall see in later chapters, the evolutionary history of these two groups is quite different, as is their current diversity. The living hominoids are restricted to a few species from the tropical forests of Africa and Asia and one cosmopolitan species – humans; they are the subject of Chapter 7. Compared to hominoids, there are many more species and genera of Old World monkeys; these are the subject of the present chapter.

Catarrhine Anatomy

Catarrhines are characterized by numerous anatomical specializations that set them apart from the more primitive platyrrhines. The name is derived from the shape of their nostrils, which are often narrow and facing downward rather than round and facing laterally, as in most New World monkeys. In their dentition, catarrhines have two rather than three premolars in each quadrant, for a dental formula of 2.1.2.3. On the external surface of their skull, the frontal bone contacts the sphenoid bone and separates the zygomatic bone anteriorly from the parietal bone posteriorly. In the auditory region, the tympanic bone extends laterally to form a tubular external auditory meatus (Figs. 2.20 and 5.4). All Old World monkeys, gibbons, and some chimpanzees have expanded ischial tuberosities and well-developed sitting pads, and all lack an entepicondylar foramen on the humerus. In general, the living catarrhines are much larger than the living platyrrhines, and include more folivorous and terrestrial species.

Cercopithecoids

Old World monkeys, Cercopithecoidea, are the more taxonomically diverse and numerically successful of the living catarrhines. Although Old World monkeys have traditionally been viewed as primitive catarrhines, this is not the case. They are in fact a very specialized radiation of primates that is not only different from living apes but is also quite derived. We now know that both Old World monkeys and apes are each specialized in different ways with respect to the earliest catarrhines. We return to this issue in later chapters as we deal with early anthropoid evolution.

Cercopithecoid monkeys have several anatomical features that distinguish them from hominoids (apes and humans) (Fig. 6.1). Most characteristic are the specialized molar teeth, in which the anterior two cusps and the posterior two cusps are aligned to form two ridges, or lophs. Teeth with this structure are described as **bilophodont**. Most Old World monkeys have large, dagger-like canines in the males and smaller ones in the females. In both sexes the canines are sharpened by a narrow anterior lower premolar. In cranial anatomy, Old World monkeys have relatively narrow nasal openings and narrow tooth rows compared with apes. Most lack maxillary sinuses.

The limbs of Old World monkeys are characterized by a very narrow elbow joint with a reduced medial epicondyle and a relatively long olecranon process on the ulna. Sitting pads on the expanded ischial tuberosities are a distinctive feature of the group, and many have a long tail.

Cercopithecoid monkeys are found throughout Africa and Asia (Fig. 6.2). In Europe there is a small (probably introduced) population of Barbary macaques on the Gibraltar headland, though in the recent past they had a much more extensive distribution on that continent (Modolo et al., 2005; Alba et al., 2016). Today, Old World monkeys are found in a wider range of latitudes, climates, and vegetation types than any other group of living primates except humans. There are two very different groups of Old World monkeys, placed in two separate subfamilies: the cercopithecines, or cheek-pouch monkeys, and the colobines, or leaf-eating monkeys. Both have undergone extensive adaptive radiations and include numerous genera and species.

The two subfamilies, Cercopithecinae and Colobinae, are distinct in many aspects of their anatomy (Fig. 6.3).

Old World Monkeys   **Apes**

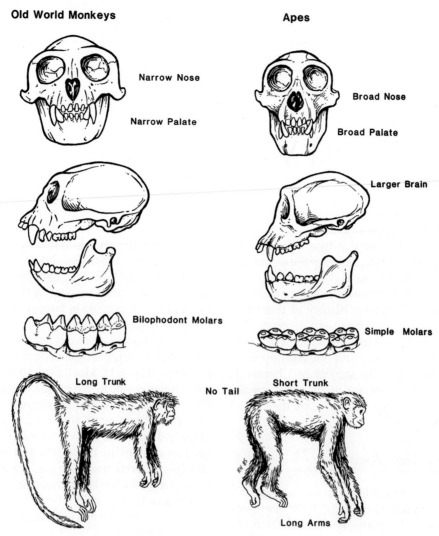

FIGURE 6.1   Characteristic features that distinguish the two groups of catarrhine primates, Old World monkeys (Cercopithecoidea) and apes (Hominoidea).

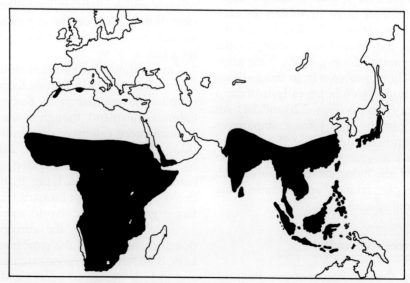

FIGURE 6.2   Geographic distribution of extant cercopithecoid monkeys. Map lines delineate study areas and do not necessarily depict accepted national boundaries.

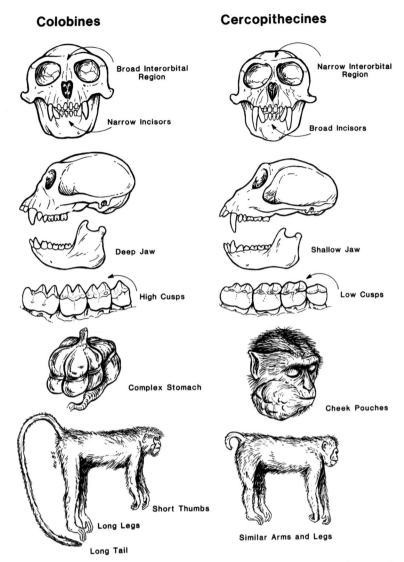

FIGURE 6.3 Characteristic features of the extant subfamilies of Old World monkeys, colobines and cercopithecines.

Many of their differences are related to basic dietary adaptations. The colobines are predominantly leaf and seed eaters, whereas the cercopithecines are predominantly fruit eaters. Cercopithecines have cheek pouches, broader incisor teeth, and molar teeth with high crowns and relatively low cusps, whereas colobines have no cheek pouches, narrower incisors, and molar teeth with high cusps. Colobines have a large, complex, ruminant-like stomach. In cranial anatomy (Fig. 6.3, Fig. 6.4), cercopithecines have a narrow interorbital region; in colobines, the interorbital region is broader. In general, cercopithecines have longer snouts and shallower mandibles than do colobines. Most cercopithecines also have longer thumbs and shorter fingers than colobines, which often lack a thumb. Cercopithecine forelimbs and hindlimbs tend to be similar in size, whereas colobines usually have longer hindlimbs.

Cercopithecines

The cercopithecines are a predominantly African group. Only a single, very successful genus, *Macaca*, is found in Asia or Europe. Cercopithecines range in size from the tiny (just over 1 kg) arboreal talapoin monkeys of western Africa to the large (as much as 50 kg), mostly terrestrial mandrills and baboons found in much of sub-Saharan Africa. There are two distinct clades of cercopithecines: the larger papionins (macaques, mangabeys, mandrills, geladas, and baboons) and the smaller cercopithecins (guenons and relatives).

Macaques

Macaques (*Macaca*) are medium-sized cercopithecines and are relatively generalized in many aspects of their anatomy compared to other members of the subfamily.

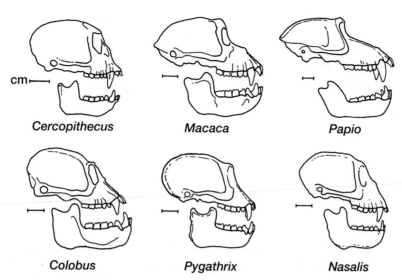

cm —

Cercopithecus   *Macaca*   *Papio*

Colobus   *Pygathrix*   *Nasalis*

FIGURE 6.4   Skulls of three cercopithecine monkeys (*above*) and three colobine monkeys (*below*).

Macaques are characterized by moderately long snouts, high-crowned molar teeth with very low cusps, and long third molars (Fig. 6.4). They share several features with other papionins, including relatively long faces and a chromosome number of 42. In general, their limbs are more slender than those of the African baboons and mangabeys and more robust than those of the smaller guenons.

Macaca has the widest distribution of any nonhuman primate genus. There are over 20 species of *Macaca* ranging from Morocco and Gibraltar in the west to Japan, Taiwan, the Philippines, Bali, and Sulawesi in the east (Table 6.1). In fact, there are seven species on the island of Sulawesi alone (Riley, 2010). *Macaca sylvanus*, the Barbary macaque, is the only living nonhuman primate in Europe, and it is unclear whether the current population in Gibraltar is native or recently introduced; *M. fuscata*, the Japanese macaque, ranges farther to the north and east than any other primate species; and *M. fascicularis*, the crab-eating macaque, from the island of Bali, extends farthest to the southeast of any nonhuman primate species. Macaques are generally divided into groups of related species (Table 6.1), with broad areas of overlap in their distributions.

Macaca species are as equally diverse in their appearance as they are in their distribution. Many macaque species have brown pelage and pale pink faces. In some species, face pigmentation has been related to rank status or sexual receptivity (Waitt et al., 2003; Gerald et al., 2007; Higham et al., 2011). Among other species, a wide variety of phenotypes exist. The pig-tailed macaque (*Macaca nemestrina*), closely resembles other macaques in its coloration, but differs in its short, almost pig-like tail. As its name suggests, the lion-tailed macaque (*Macaca silenus*) from the Western Ghats of India is named for its long black tail accented with a tuft of silver hair and a large silvery mane framing its face. The Crested macaques

(*Macaca nigra*) from the island of Sulawesi have expressive, almost human-like faces with a black tuft or mohawk of hair atop their head (Fig. 6.5).

Because macaques are common laboratory primates, the anatomy, physiology, and captive behavior of this genus are the most thoroughly studied of all nonhuman primates. However, much less is known about the natural behavior and ecology of many species. Macaques occupy a wider range of habitats and climates than any other nonhuman primate genus. Some have specific habitat preferences. Other macaques have an ability to coexist with humans and exploit a range of modified environments that surpasses that of all other nonhuman primates. This "weed" adaptation (Richard et al., 1989) is an important ecological strategy of several species, including *M. fascicularis*, *M. mulatta*, *M. radiata*, and *M. sinica*. All seem to reach their highest densities in places where they overlap with humans.

The ecological differences among macaques in nonhuman settings have been documented for several species. Two species, *M. nemestrina*, the **pig-tailed macaque**, and *M. fascicularis*, the **long-tailed** or **crab-eating macaque**, are sympatric throughout much of Southeast Asia (Fig. 6.6). The smaller (3–5 kg) *M. fascicularis* prefers lowland and secondary forests with a denser, more continuous forest structure, often near rivers, while the larger (6–10 kg) *M. nemestrina* prefers upland and more hilly environments, with a less continuous canopy and less dense undergrowth. Similarly, sympatric populations of **Assamese** (*M. assamensis*) and **rhesus macaques** (*M. mulatta*) exhibit differences in diet and habitat use that allow them to coexist. Assamese macaques are primarily terrestrial and prefer the cliffsides of seasonal limestone rainforests, whereas rhesus macaques are present mostly on hillsides, and show a preference to lower and middle canopy (Zhou et al., 2014).

TABLE 6.1 Subfamily CERCOPITHECINAE, Macaques

Common name	Species	Intermembral index	Mass (g) M	F
Sylvanus Group				
Barbary macaque	*Macaca sylvanus*	-	14,530	10,140
Silenus Group				
Northern pig-tailed macaque	*M. leonina*	-	7249	4967
Sunda pig-tailed macaque	*M. nemestrina*	94	10,094	6263
Pagai macaque	*M. pagensis*	-	-	4500
Siberut macaque	*M. siberu*	-	-	-
Lion-tailed macaque	*M. silenus*	92	8721	6100
Sulawesi Group				
Buton macaque	*M. brunnescens*	99	14,500	7500
Heck's macaque	*M. hecki*	93	11,200	6800
Moor macaque	*M. maura*	-	9720	6050
Crested macaque	*M. nigra*	84	9890	5470
Gorontalo macaque	*M. nigriscens*	96	5800	5500
Booted macaque	*M. ochreata*	99	5300	2600
Tonkean macaque	*M. tonkeana*	94	14,900	9000
Sinica Group				
Assamese macaque	*M. assamensis*	96	11,318	6851
Arunachal macaque	*M. munzala*	-	15,000	-
Bonnet macaque	*M. radiata*	-	6670	3850
Toque macaque	*M. sinica*	-	5680	3200
Tibetan macaque	*M. thibetana*	96	17,676	14,100
Fasicularis Group				
Long-tailed macaque	*M. fascicularis*	92	5335	3588
Mulatta Group				
Taiwanese macaque	*M. cyclopis*	-	10,210	6690
Japanese macaque	*M. fuscata*	-	13,320	10,348
Rhesus macaque	*M. mulatta*	93	7889	5339
Arctoides Group				
Stump-tailed macaque	*M. arctoides*	98	11,988	7925

Macaque species vary considerably in the extent to which they are arboreal or terrestrial. All species use both settings to some extent, but with different frequencies. *M. fascicularis* is primarily an arboreal species that normally feeds and travels in the trees. These macaques are most often found in the lower levels of the main canopy but utilize all levels, including the ground. *M. nemestrina* travels more on the ground, but feeds frequently in the trees. Macaque locomotion is almost totally quadrupedal walking and running, with very little leaping and no suspensory behavior aside from occasional hindlimb hanging during feeding. Macaques are extremely dexterous and have short fingers and an opposable thumb (Fig. 2.25).

All macaques are frugivores, but many consume considerable amounts of seeds, leaves, flowers, and other plant materials, as well as various animal prey (Thierry, 2011). Japanese macaques subsist on bark during the

sylvanus   *mulatta*   *nemestrina*

silenus   *nigra*   *arctoides*

FIGURE 6.5   Faces of six macaque species illustrating the diversity of facial form within the genus. *(Drawing courtesy of S. Nash.)*

cold winters. *M. sylvanus* eats lots of seeds and *M. fascicularis*, the crab-eating macaque, eats a variety of small prey – not only crabs but also termites and small vertebrates. Burmese long-tailed macaques (*M. fascicularis aurea*) living in intertidal habitats even use stones to process marine prey, focusing primarily on shelled mollusks including oysters and snails (Gumert and Malaivijitnond, 2012).

All macaques live in relatively large, multimale social groups, with troops of some species containing 50 or more individuals. During the day these groups regularly split into smaller foraging parties. Home range size and patterns of habitat use vary considerably from species to species. Groups of about 20 *M. fascicularis* have home ranges of 40 to 100 ha and day ranges of less than 1 km. Home ranges and day ranges for the larger groups of *M. nemestrina* are considerably larger because they use rapid terrestrial travel to exploit widespread, often erratically available food resources.

Social relations within macaque groups are complex. Female macaques are primarily philopatric. Matrilines and social hierarchies are particularly important to interindividual relations and troop politics, though female social styles can vary considerably. Females of some species are highly despotic (*M. mulatta*, *M. fuscata*), whereas others are highly tolerant (*M. nigra*). By contrast, males usually migrate from troop to troop several times during their lifetime. While male–male relationships are antagonistic among many of the better-known species (*M. fascicularis*, *M. fuscata*, *M. mulatta*), others show high levels

of affiliative behavior (*M. assamensis*, *M. radiata*, *M. sylvanus*, *M. thibetana*) (Riley et al., 2014).

As is typical of most cercopithecines, in many macaque species, females have sexual swellings that change in color and tumescence throughout their menstrual cycle (Dixson, 1983). In some species, these changes may reflect changes in estrogen and progesterone secretion, with ovulation usually occurring during the period of maximum swelling size (*M. sylvanus*: Brauch et al., 2007). However, they may also have limited (*M. fascicularis*: Engelhardt et al., 2005) or no apparent signaling content (*M. thibetana*: Li et al., 2005; *M. assamensis*: Fürtbauer et al., 2010). Most macaques have single births every 12 to 24 months (Kappeler and Pereira, 2003). Females provide the vast majority of infant care; however, several macaque species exhibit a form of allomaternal infant handling known as "bridging," when two handlers simultaneously pick up, hold, and/or inspect the same infant (Zhang et al., 2018). The function of bridging remains unclear, though it may act as a social buffer, and can be a predictor of social bonds, as well as future mating success in males.

Mangabeys

Mangabeys (Table 6.2; Fig. 6.7) are large, forest-living monkeys with long molars, very large incisors, relatively long snouts, and hollow cheeks. They have relatively long limbs and long tails. Formerly placed in a single genus, the living mangabeys are now widely recognized to be an unnatural group containing two distinct genera that differ in numerous aspects of their dental, cranial,

FIGURE 6.6 Two macaque species that are found sympatrically throughout Southeast Asia: *upper right*, the crab-eating or long-tailed macaque (*Macaca fascicularis*); *below*, the pig-tailed macaque (*Macaca nemestrina*).

TABLE 6.2　Subfamily CERCOPITHECINAE, Mangabeys and Kipunjis

Common name	Species	Intermembral index	Mass (g) M	F
Agile mangabey	*Cercocebus agilis*	84	8807	5405
Sooty mangabey	*C. atys*	-	10,600	6200
Golden-bellied mangabey	*C. chrysogaster*	-	-	-
Tana River mangabey	*C. galeritus*	84	9610	5260
White-naped mangabey	*C. lunulatus*	83	9900	5300
Sanje River mangabey	*C. sanjei*	-	10,300	5800
White-collared mangabey	*C. torquatus*	83	11,756	5800
White-cheeked mangabey	*Lophocebus albigena*	78	8250	6020
Northern black-crested mangabey	*L. aterrimus*	-	7840	5760
Kipunji	*Rungwecebus kipunji*	-	13,000	-

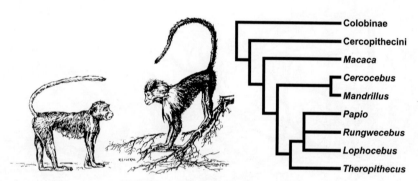

FIGURE 6.7　Molecular phylogenies indicate that the two species groups of living mangabeys have separate phylogenetic relationships. The more terrestrial mangabeys of the genus *Cercocebus*, illustrated by the agile mangabey, *Cercocebus agilis* (*left*), are more closely related to mandrills. The more arboreal mangabeys of the genus *Lophocebus*, illustrated by the gray-cheeked mangabey *Lophocebus albigena* (*right*), are more closely related to baboons and kipunjis.

and postcranial anatomy, as well as their genetics and ecology. The two genera, *Cercocebus* and *Lophocebus*, have very different phylogenetic relationships. *Cercocebus* is most closely related to mandrills and drills; *Lophocebus* is more closely related to kipunjis and baboons (Fig. 6.7).

Cercocebus is divided into seven species that are found throughout much of western and central Africa, with one relict species from the Tana River in eastern Kenya and another from the highlands of Tanzania. They are large, sexually dimorphic monkeys and can be distinguished from *Lophocebus* by a number of cranial features, including very large premolar teeth, less concave nasal bones, widely divergent temporal lines at the front of the skull, upturned nuchal crests at the back of the skull, and the lack of noticeable fossae on the lateral side of the mandible. These monkeys are found in a wide range of forest types, but seem to be dependent on swamps or forests that are flooded at least seasonally. They prefer the understory and are most commonly found on the ground during both traveling and feeding.

Cercocebus eat fruits, and especially hard nuts and seeds, which they find on the forest floor. They spend significant time foraging for invertebrates. *Cercocebus* live in groups that average 40 to over 60 individuals (Ehardt et al., 2005; Rovero et al., 2009), but these groups seem to divide regularly into subgroups and occasionally several come together in large supertroops of up to 100 individuals.

Females are commonly philopatric, and form strong linear dominance hierarchies (Range and Noe, 2002; Range et al., 2007). Males typically disperse at sexual maturity, and can either join a group to become a resident male or remain transient, periodically entering groups for several hours to several months. Once in a group, adult males, too, compete to form linear dominance hierarchies. Access to estrus females is rank dependent, though sneak copulations are common. *Cercocebus* mangabeys exhibit mating and birth seasonality (Range et al., 2007; McCabe and Emery Thompson, 2013). Females typically have singleton births every two

TABLE 6.3 Subfamily CERCOPITHECINAE, Baboons and Geladas

Common name	Species	Intermembral index	Mass (g) M	F
Olive baboon	*Papio anubis*	97	24,382	13,140
Yellow baboon	*P. cynocephalus*	96	22,545	12,394
Hamadryas baboons	*P. hamadryas*	95	17,498	9600
Kinda baboon	*P. kindae*	-	16,020	9830
Guinea baboons	*P. papio*	-	23,000	12,000
Chacma baboon	*P. ursinus*	96	29,858	14,856
Gelada	*Theropithecus gelada*	100	19,000	11,700

years (Fruteau et al., 2010; Fernandez et al., 2014). Infant handling in *Cercocebus* mangabeys is common, and females often trade grooming for access to infants, also known as "baby markets," in line with biological market theory (Fruteau et al., 2011). Infanticide in this genus has also been reported (Fruteau et al., 2011).

Lophocebus mangabeys include two species that are largely restricted to central Africa. They are smaller, less dimorphic monkeys that can be distinguished from *Cercocebus* by their smaller premolar teeth, more concave nasal bones, constricted temporal lines, downward projecting nuchal lines, and well-marked mandibular corpus fossae.

Lophocebus are strictly arboreal monkeys that prefer the main canopy levels in a variety of forests. These species also feed predominantly on fruits, seeds, and invertebrates. They live in smaller groups with usually fewer than 20 individuals.

Kipunji Monkeys

The **kipunji**, *Rungwecebus kipunji*, is a recently described monkey from the highlands of southern Tanzania. Morphologically, the kipunji is most similar to *Lophocebus* mangabeys (Gilbert et al. 2011), and it was originally described as a species within the genus *Lophocebus* (Jones et al., 2005). However, subsequent genetic studies indicate that it is more closely related to baboons (*Papio*), sharing mitochondrial DNA with yellow baboons that live in the same region of Tanzania (Burrell et al. 2009; Zinner et al. 2009 a,b). Kipunjis are present in isolated areas where baboons are no longer present, and it seems to be a stable taxon.

So far, kipunjis are only known from protected areas within two separate mountainous regions of southern Tanzania. Land outside of these areas is dominated by agriculture, and habitat models suggest limited potential for population expansion beyond these protected areas (Bracebridge et al., 2011). However, recently proposed reforestation efforts along riverine corridors stand to increase total population size by as much as 8% (Bracebridge et al., 2013). In fact, recent conservation

efforts appear to have increased the population of kipunjis by around 65% (Davenport et al., 2022).

The kipunji diet is reported to be primarily frugivorous, but includes seasonally variable amounts of leaves, pith, exudates, and bark (Davenport et al., 2010; Bracebridge et al., 2012). Average group size is between 25 and 35 individuals with multiple males (Swedell, 2011; Davenport et al., 2022). Their vocalizations are reported to be more similar to those of baboons than of mangabeys.

Baboons

Baboons (***Papio***) (Table 6.3; Figs. 6.8 and 6.9) are among the largest and perhaps the best known of all cercopithecines. They were important figures in the mythology of ancient Egypt and were well known to Greek and Roman scholars. As savannah-dwelling primates they have played an important role as models for various aspects of early human evolution, including reproductive strategies and biogeography (e.g., Jolly, 2001; Swedell and Plummer, 2019). Baboons are very large monkeys and are nearly all sexually dimorphic in body size; in many species, females are only half the size of males. Only the **kinda baboons** (*P. kindae*) of Zambia, eastern Angola, and southern DR Congo exhibit reduced cranial, canine, and body size dimorphism compared to other baboon species (Delson et al., 2000; Singleton et al., 2017; Petersdorf et al., 2019). Baboons have a long snout (Fig. 6.4), a long mandible, and pronounced brow ridges. They are characterized by long molars and broad incisors. Their canines are very sexually dimorphic, and the long anterior lower premolars form a sharpening blade for the dagger-like canines. Their limbs are nearly equal in length, and their forearm is much longer than their humerus (Fig. 6.10); they have relatively short digits on their hands and feet. Compared to other cercopithecines, baboons have relatively short tails and large ischial callosities. Nevertheless, despite these broad similarities, baboon species differ notably in their fur color and sexually selected characteristics. For instance, adult males of both Guinea and hamadryas baboons exhibit distinctive capes that are

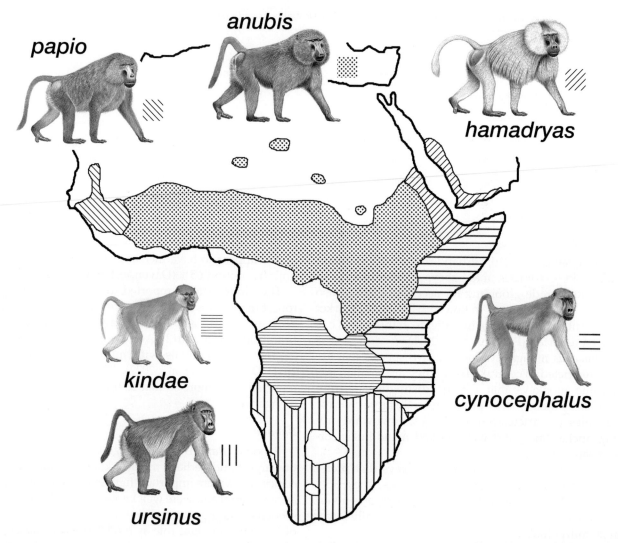

FIGURE 6.8   Distribution of baboon species around Africa and the Arabian Peninsula. Map lines delineate study areas and do not necessarily depict accepted national boundaries. *(Drawing courtesy of S. Nash.)*

only moderately present in olive baboons, and absent in yellow, chacma, and kinda baboons. Moreover, while kinda and yellow baboons exhibit similar golden-brown coloration, kindas have longer hair that is softer in texture, a distinctive pink circumorbital skin coloration, and well-developed sagittal hair crests (Jolly et al., 2011). Females have very pronounced sexual swellings during estrus, though the appearance of these swellings varies considerably among species (Petersdorf et al., 2019).

Baboons are found throughout the forests and savannahs of sub-Saharan Africa and the heel of the Arabian Peninsula (Fig. 6.8). There are between seven and ten distinct populations of baboons, commonly – but not comfortably – placed in six species: olive (*P. anubis*), yellow (*P. cynocephalus*), hamadryas (*P. hamadryas*), kinda (*P. kindae*), Guinea (*P. papio*), and chacma (*P. ursinus*) (Jolly, 1993; Frost et al., 2003; Zinner et al., 2009b, 2013; Rogers et al., 2019; Elton and Dunn, 2020; Roos et al., 2021). These six are all allopatric, with variable amounts

of interbreeding at their boundaries, and there are several additional populations that are as distinct as the commonly recognized species.

Baboon ecology and behavior has been the subject of many studies over the past six decades, including several long-term studies, the Amboseli Baboon Research Project among them. Such longitudinal projects have allowed researchers to ask detailed questions about baboon biology, life history, and generational health outcomes that have not been possible in many other primate taxa. (e.g., Alberts, 2019; Weibel et al., 2020). By contrast, the natural history and behavior of other populations, including the widespread and distinct kinda baboon (*P. kindae*), remain largely undocumented, though this has increased over the last decade (reviewed in Petersdorf et al., 2019).

Baboons are ecologically quite flexible, and can be found in semi-desert grasslands, woodland savannahs, and acacia scrubs, but also in gallery forests and some rainforest environments (Fuchs et al., 2018; Chala et al.,

FIGURE 6.9   A group of savannah baboons (*Papio anubis*) in eastern Africa.

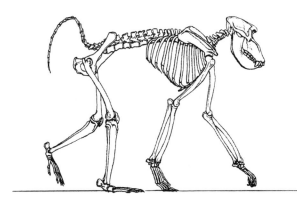

FIGURE 6.10   The skeleton of a baboon (*Papio*).

2019). They forage and travel primarily on the ground by quadrupedal walking and running, but they almost always climb trees or rocky cliffs for sleeping and often for resting. They are extremely eclectic feeders that subsist mainly on ripe fruits, roots, and tubers, as well as on grass seeds, gums, and leaves. In addition, most baboons are opportunistic faunivores and have been reported to catch and eat numerous small mammals (hares, young gazelles, vervet monkeys) as well as many invertebrates. They also eat bird eggs.

Chacma, olive, kinda, and yellow baboons (recently dubbed "COKY" baboons; Jolly, 2020) normally live in large, stable multimale–multifemale groups ranging from 40 to approximately 100 individuals, although some mountain populations of chacma baboons (*P. ursinus*) are found in one-male groups (Swedell, 2011; Markham et al., 2015). As in many Old World monkeys, these so-called COKY baboons are generally characterized by female philopatry, and males usually emigrate to other troops (reviewed in Fischer et al., 2019). These female-bonded kin groups are generally considered to form the basic structure of a baboon troop. There is usually a pronounced dominance hierarchy among both females and males, and adult males compete intensely for access to estrous females. This competition involves a whole repertoire of social maneuvers – not just simple physical prowess but also coalitions and infant care. Females frequently mate with several males during the course of their cycle. Baboon troops occupy large (4000 ha) home ranges and travel long distances (over 5 km) every day, usually as a single group.

Social organization and dispersal patterns in **hamadryas baboons** (*P. hamadryas*) are quite different from those found among the COKY baboons described above.

These handsome silver baboons from the arid scrublands of Ethiopia live in groups of a single adult male with one to four females plus their offspring. Males guard the females in their harem jealously, and actually herd them by chasing any straying females and biting them on the neck to keep the group together. They live in multilevel societies (Fig. 3.10; Kummer, 1968), as do the **guinea baboons** (*P. papio*) of west African gallery forests and adjacent woodland savannas (Fischer et al., 2017). In these species, one-male units (OMUs) form the smallest reducible, social and reproductive unit. Several OMUs, probably led by related males (Patzelt et al., 2014), regularly associate to form clans, then bands, and then troops of as many as several hundred individuals (Schreier and Swedell, 2009). Individual OMUs forage separately during the day but congregate at night on rocky cliffs in troops of 150 or more animals.

Geladas

The **gelada** (*Theropithecus gelada*) is a very distinctive monkey from the highlands of Ethiopia (Fig. 6.11). Geladas are the only surviving species of a more successful and widespread radiation during the Pliocene and Pleistocene (see Chapter 16). Like baboons, geladas are extremely sexually dimorphic in both size and appearance. Males have long, shaggy manes and pronounced facial whiskers, whereas the female pelage is shorter. Both sexes have striking red hourglass patches of skin on their chests, and in females these are outlined with white vesicles. The distinctive molar teeth of *Theropithecus* are characterized by complex enamel foldings. Male canines are very large, even by papionin standards. The snout and mandible are relatively short and deep. The hands of geladas are characterized by a relatively long thumb compared to the other digits, an adaptation for foraging for grass blades and seeds.

Geladas live in the largely treeless Ethiopian highlands, where they forage on the ground all day and sleep on rocky cliffs at night. They are the most terrestrial nonhuman primates and always move by quadrupedal walking and running. They are almost exclusively herbivorous, eating grass, seeds, and roots throughout the year, though they occasionally eat fruit, as well as some invertebrate and vertebrate prey (Fashing et al., 2014; Kifle and Bekele, 2021). Although grasses and sedges are staples, underground food items seem to be important fallback foods, particularly for populations living in disturbed habitats (Jarvey et al., 2018). Geladas

FIGURE 6.11   Geladas (*Theropithecus gelada*) from the highlands of Ethiopia.

feed by sitting upright, and plucking grass blades and seeds by hand.

Geladas live in highly fluid, multilevel societies (Snyder-Mackler et al., 2012). The primary level of organization in gelada society is either the OMU comprising a reproductive leader male, up to 12 adult females, their dependent offspring, and possibly one or more follower males, or the all-male group (AMG) comprising young adult and subadult males. These units join to form bands, which in turn form communities. Units may also join to form herds of several hundred individuals, though these are generally unstable, short-term associations. Related females form the core of gelada multilevel societies (Snyder-Mackler et al., 2014). In fact, patterns of female relatedness mirror those of their multilevel associations, such that females within units are more closely related than females within bands, and females in bands are more closely related than females in communities. Female ranks appear to be maternally inherited (Le Roux et al., 2011). By contrast, males mediate gene flow by dispersing from their natal units to other bands and, likely, to other communities.

Mandrills and Drills

Mandrills (*Mandrillus sphinx*) and **drills** (*M. leucophaeus*) are large (male mean about 30 kg) forest monkeys from western and central Africa (Table 6.4; Fig. 6.12). Although traditionally grouped with baboons, all recent molecular and morphological studies indicate that they are more closely related to terrestrial mangabeys of the genus *Cercocebus* (Fig. 6.7). They are extremely sexually dimorphic, with male mandrills averaging three times the size of females. They are also dimorphic in coloration. Males of both mandrills and drills are characterized by brightly colored faces and rumps. Both male mandrills and male drills have long muzzles with

TABLE 6.4 Subfamily CERCOPITHECINAE, Mandrills and Drills

			Mass (g)	
Common name	Species	Intermembral index	M	F
Mandrill	*Mandrillus sphinx*	95	31,050	10,678
Drill	*M. leucophaeus*	-	26,450	10,050

FIGURE 6.12 The mandrill (*Mandrillus sphinx*).

pronounced maxillary ridges and long tooth rows. Like baboons, they also have forelimbs and hindlimbs of nearly equal length. Both have very short tails.

Drills and mandrills fill similar ecological niches and tend to be shy as a result of human hunting. However, drills can be found in a wider variety of habitat types within a greater elevation range than mandrills, which occur almost exclusively within closed canopy, lowland forests (Wood, 2007). Although drills commonly occupy lowland forests, they may also occur in montane forests (Butynski and Koster, 1994; Gonzalez-Kirchner and de la Maza, 1996) and grasslands (Wild et al., 2005). Within the forest both drills and mandrills are primarily terrestrial, but females and juveniles regularly climb into trees. Their diet has been described as fruit, seeds, leaves, pith, and insects, though this appears to vary somewhat with altitude (Owens et al., 2015). They are very active foragers who rummage through leaf litter and decaying wood on the forest floor, and like *Cercocebus*, mandrills seem to specialize on hard nuts, seeds, and even woody tissue (e.g., bark and roots) when other foods are scarce (Hongo et al., 2018).

Drills and mandrills have been seen in one-male groups, multimale groups, and large congregations numbering hundreds of individuals (e.g., Hoshino et al., 1984; Abernethy et al., 2002), making their social organization difficult to identify. Reported sex ratios range from 1 male to ~14 females (Hoshino et al., 1984) to as high as 1 male to 775 females (Abernethy et al., 2002), though the latter estimate is considered by some to be an outlier. Recent work suggests that animals consistently associate in groups of one to several adult males, many females, and their dependent offspring. This is corroborated by oral reports from local Gabonese community members. Nevertheless, there appears to be an influx of breeding-aged males during periods of female receptivity (Brockmeyer et al., 2015). Historical data on this same population suggest that wild subadult (6–9 years old) and adult males emigrate at all ages, and that secondary dispersal also occurs. Female mandrills have a high degree of control over male group membership via female–female coalitions (Setchell et al., 2006) and mating partner choice (Charpentier et al., 2005).

Guenons and Relatives

The smaller cercopithecines, recognized as the Tribe Cercopithecini and often broadly referred to as guenons, consist of six genera: *Chlorocebus*, the Vervet and Grivet or Savannah monkeys; *Allenopithecus*, the Swamp Monkey; *Allochrocebus*, the terrestrial L'Hoest's monkeys and relatives; *Erythrocebus*, the Patas monkeys; *Miopithecus*, the small Talapoin monkeys; and *Cercopithecus*, the arboreal guenons, with ~20 species, divided into 6 species groups.

The majority of the guenons are restricted to the equatorial forest belt but a small number of species, belonging to three groups, range more widely: the Savanna Monkeys (aka vervet or grivet monkeys, *Chlorocebus*) range throughout the more wooded parts of sub-Saharan Africa; the Patas monkeys (*Erythrocebus*) are primarily a Sahelian group; and the Blue Monkey species group (*Cercopithecus mitis* group) have an extensive coastal, riverine and montane range, mostly through the south-eastern quarter of Africa.

In general, the guenons are predominantly frugivorous and insectivorous, but some species eat leaves during certain parts of the year. Larger species generally eat more leaves than smaller species. In contrast to many other primates, females actively defend group territories (e.g., Hill, 1994; Cords, 2000, 2007). Most species live in single-male groups, and male guenons are generally antagonistic to one another. However, the relationship between male residence and reproductive success is not clear. In many species, large numbers of males join troops and mate with the females during the mating season, and resident males of one troop may also travel to mate with females in other troops (Buzzard and Eckardt, 2007). All guenons seem to have annual birth peaks associated with the time of greatest food abundance. Infanticide has been reported in some species.

The arboreal guenons (*Cercopithecus*) are small to medium-sized forest monkeys found throughout sub-Saharan Africa. The 20 *Cercopithecus* species (Table 6.5) are remarkably diverse in color and appearance (Fig. 6.13), but relatively uniform in size and body proportions (Schultz, 1970). Arboreal guenons average about 3 to 5 kg in size, and most species exhibit a moderate amount of sexual dimorphism. All arboreal guenons have sexually dimorphic canines, relatively narrow molar teeth, and short third molars. Due in part to their smaller size, they have relatively short snouts compared to larger cercopithecines (Fig. 6.4). They also have longer hindlimbs than forelimbs, and long tails.

Arboreal guenons are all forest dwellers and show considerable variation from species to species in their forest preference and use of canopy levels. Nearly all species are basically arboreal quadrupeds, but some frequently come to the ground and some are quite good leapers. Arboreal guenons are generally divided into numerous species groups (Table 6.5), but the relationships among these are complex and almost certainly reflect a history of hybridization (e.g., Xing et al., 2007; Perelman et al., 2011; Lo Bianco et al., 2017; Jensen et al., 2023). Behavioral studies of *Cercopithecus* monkeys are limited, with the exception of a few species and populations (e.g., Tai forest monkeys, Kibale monkeys, blue monkeys).

Arboreal guenons often feed and travel in mixed-species groups, also known as **polyspecific associations**, probably as an antipredator strategy (Noe and Bshary, 1997; McGraw and Zuberbuhler, 2008). Despite the overall structural uniformity of guenons, individual species

TABLE 6.5   Subfamily CERCOPITHECINAE, Guenons and Relatives

Common name	Species	Intermembral index	Mass (g) M	F
Allen's swamp monkey	*Allenopithecus nigroviridis*	83	6142	3260
Northern talapoin monkey	*Miopithecus ogouensis*	84	1520	1280
Angolan talapoin monkey	*M. talapoin*	83	1250	775
Grivet monkey	*Chlorocebus aethiops*	83	4200	2800
Malbrouck monkey	*C. cynosurus*	86	5400	3700
Bale monkey	*C. djamdjamensis*	-	4750	3400
Vervet monkey	*C. pygerythrus*	84	4850	3400
Green monkey	*C. sabaeus*	-	6300	4400
Tantalus monkey	*C. tantalus*	81	5350	3950
Dryas monkey	*C. dryas*	79	3000	2000
Patas monkey	*Erythrocebus patas*	93	12,400	6500
Southern Patas monkey	*E. baumstarki*	-	-	-
Heuglin's Patas monkey	*E. poliophaeus*	-	-	-
L'Hoest's monkey	*Allochrocebus lhoesti*	79	6100	3500
Preuss's monkey	*A. preussi*	78	6425	3550
Sun-tailed monkey	*A. solatus*	-	6900	4000
Neglectus Group				
De Brazza's monkey	*Cercopithecus neglectus*	81	6680	3614
Diana Group				
Diana monkey	*C. diana*	81	5200	3900
Roloway monkey	*C. roloway*	-	5000	2260
Hamlyni Group				
Hamlyn's monkey	*C. hamlyni*	82	5500	3600
Lesula	*C. lomamiensis*	83	5550	3750
Nictitans Group				
Syke's monkey	*C. albogularis*	80	5700	3600
Blue monkey	*C. mitis*	82	6500	3967
Greater spot-nosed monkey	*C. nictitans*	81	5900	4100
Cephus Group				
Red-tailed monkey	*C. ascanius*	80	3700	2920
Moustached monkey	*C. cephus*	80	4178	2800
Red-bellied monkey	*C. erythrogaster*	-	4100	2400
Red-eared monkey	*C. erythrotis*	-	3950	3300
Lesser spot-nosed monkey	*C. petaurista*	80	4400	2900
Sclater's monkey	*C. sclateri*	-	5300	3900
Mona Group				
Campbell's monkey	*C. campbelli*	83	4800	2600
Dent's monkey	*C. denti*	-	4250	2830
Lowe's monkey	*C. lowei*	-	3900	2600
Mona monkey	*C. mona*	83	4700	2800
Crowned monkey	*C. pogonias*	83	4100	2850
Wolf's monkey	*C. wolfi*	-	3910	2870

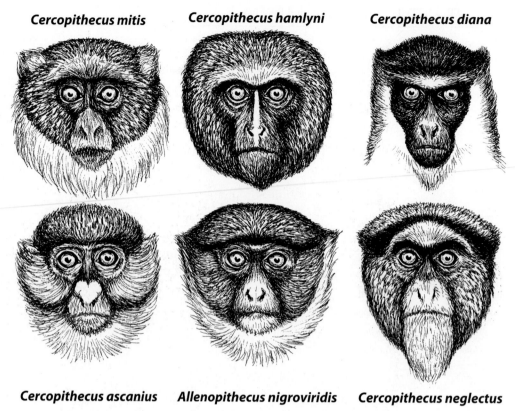

Cercopithecus mitis   **Cercopithecus hamlyni**   **Cercopithecus diana**

Cercopithecus ascanius   **Allenopithecus nigroviridis**   **Cercopithecus neglectus**

FIGURE 6.13   The faces of six guenons.

have unique foraging strategies that distinguish them from sympatric species (Gautier-Hion, 1978). Three frequently associated guenon species that have been well studied in Gabon are *C. cephus*, the **moustached monkey**; *C. pogonias*, the **crowned monkey**; and *C. nictitans*, the **greater spot-nosed monkey** (Fig. 6.14). *Cercopithecus cephus* is the smallest species, *C. pogonias* is slightly larger, and *C. nictitans* is much larger and more sexually dimorphic. All live in primary rainforests, but *C. cephus* is also very common in secondary forests. All are arboreal quadrupeds. *Cercopithecus pogonias* and *C. nictitans* are found primarily in the middle and upper levels of the main canopy, whereas *C. cephus* prefers the lower levels of the canopy and the understory and occasionally comes to the ground.

The diets of the three monkeys have been documented through examination of stomach contents. *Cercopithecus pogonias* is the most frugivorous, the most insectivorous, and the least folivorous. *C. cephus* has a diet that is intermediate, and *C. nictitans* is the most folivorous, with leaves, flowers, and other vegetable materials accounting for almost 30% of its diet. The two species that overlap most in canopy use overlap least in diet. The types of insects eaten by the species also differ considerably: *C. pogonias* appears to specialize on mobile prey, whereas *C. nictitans* eats cryptic immobile prey almost exclusively. Again, *C. cephus* is intermediate, but it occupies a different level of the forest. All three species live in

moderately small home ranges, with mean group size increasing from *C. cephus* to *C. pogonias* to *C. nictitans*. In each species there is usually a single adult male.

Two other guenon species that have been particularly well studied are *C. mitis*, the **blue monkey**, and *C. ascanius*, the **red-tailed monkey**; both are found in the Kibale forest of Uganda as well as in Kakamega forest in Kenya. *Cercopithecus ascanius*, a close relative of *C. cephus*, is smaller and less dimorphic than *C. mitis*, a relative of *C. nictitans*.

Cercopithecus hamlyni, the **owl-faced monkey**, and its close relative *C. lomamiensis*, the **lesula** (Hart et al., 2012), are two poorly known guenons from the Congo that seem to be relatively terrestrial (Arenson et al., 2020).

One of the most handsome guenons is *Cercopithecus neglectus*, the **De Brazza's monkey** (Fig. 6.15), from western and central Africa. *Cercopithecus neglectus* is one of the largest and most sexually dimorphic arboreal guenons, with males averaging over 6 kg and females less than 4 kg. De Brazza's monkeys prefer flooded forests, including islands in rivers, where they move primarily in the understory and on the ground. They are slow quadrupedal monkeys. Their diet is predominantly frugivorous and includes smaller amounts of leaves, animal matter, and mushrooms.

The most unusual feature of De Brazza's monkey is its social organization. In some parts of its range they live in polygynous groups of 8 to 12 individuals, but in Gabon,

FIGURE 6.14 Three guenon species from Gabon that frequently forage together in mixed species groups: *above*, the crowned guenon (*Cercopithecus pogonias*); *middle*, the greater spot-nosed guenon (*C. nictitans*); *below*, the moustached guenon (*C. cephus*).

where they have been quite well studied, most live in monogamous family groups. The two sexes seem to forage somewhat independently and have different strategies for dealing with predators. The small females hide in the undergrowth, whereas the large males climb trees and elicit alarm calls at the predator, apparently to distract it from the females and young. Both the day ranges and the home ranges of these family groups are very small.

The genus *Allochrocebus* contains three species of relatively poorly known guenons that are more terrestrial than members of the genus *Cercopithecus*. They are most closely related to vervets and patas monkeys and include **L'Hoest's or mountain monkeys** (*A. lhoesti*) from East and Central Africa, and two species from West Africa, **Preuss' monkey** (*A. preussi*), and the **sun-tailed monkey** (*A. solatus*).

L'Hoest's monkeys thrive in markedly disturbed areas within or adjacent to forests in Burundi, DR Congo,

Rwanda, and Uganda, and most individuals live outside protected areas (Kingdon, 2013). They are unique among guenons in their preference for terrestrial herbs (Kaplin and Moermond, 2000), though they may also serve as important agents of seed dispersal, particularly given their tendency to frequent disturbed environments (Kaplin and Moermond, 1998). They regularly raid agricultural crops. Despite this, they are not commonly hunted. L'Hoest's monkeys live in groups of 20 to 30 individuals including one adult male, several females, and immature individuals. As in most guenons, *A. lhoesti* females typically remain in their natal groups, whereas males disperse upon reaching sexual maturity (Kingdon, 2013). Polyspecific associations are uncommon in this species.

Like *A. lhoesti*, **sun-tailed monkeys** live in degraded vegetation close to villages and in heavily logged forests

FIGURE 6.15   A family of De Brazza's monkeys (*Cercopithecus neglectus*).

in Gabon. Here the dense undergrowth provides relatively abundant food and protection against predators (Gautier et al., 1992). Group sizes range from 9 to 25 individuals, including one male, several females and their offspring, though solitary individuals are common (22% of observations). Sun-tailed monkeys are cryptic, perhaps because of their proximity to humans. Animals rarely call except to convey warning (Gautier, 1988). Similarly, polyspecific associations are rare, and only under conditions where danger is not anticipated, presumably because large groups increase the risk of detection by predators.

By contrast, **Preuss' monkeys** occur in primary and secondary moist forest habitats in Nigeria, Cameroon, and Bioko Island. Shoots, herbs, and pith make up nearly half of the food items consumed by Preuss' monkeys, more than any other guenon. Group sizes are considerably smaller than in other forest-living guenons. Solitary animals occur but are uncommon. Groups often have more than one male. Unlike *A. lhoesti* and *A. solatus*, *A. preussi* are often found in polyspecific associations with Mona monkeys (*C. mona*), putty-nosed monkeys (*C. nictitans*), and red-capped mangabeys (*Cercocebus torquatus*), among others.

Chlorocebus, the **savannah monkeys** or **vervets** (Fig. 6.16), range throughout sub-Saharan Africa and are widespread cercopithecines. The genus is currently divided into six or seven distinct species, though as with baboons, they regularly hybridize at areas of contact (Table 6.5). The large range of these monkeys is undoubtedly associated with their preference for woodland savannah and gallery forests. Only the **djam-djam monkey** or **bale monkey** (*C. djamdjamensis*) makes regular use of bamboo forests (Mekonnen et al., 2010). Vervet monkeys are more terrestrial than arboreal guenons and frequently cross open areas between feeding trees and forage on the ground. Terrestrial movements account for approximately 20% to 35% of their locomotion; the remainder is arboreal. Vervets are predominantly quadrupedal, both on the ground and in the trees; leaping accounts for only 10% of their locomotor activity. Their diverse diet consists of fruits, gums, shoots, and a variety of invertebrates, and, with the exception of *C. djamdjamensis*, which specializes largely on bamboo, they are generally considered omnivores. *Chlorocebus sebaeus* in Senegal even catches crabs, a behavior reminiscent of the crab-eating macaque of Southeast Asia (Galat and Galat-Luong, 1976).

FIGURE 6.16 A troop of grivet monkeys (*Chlorocebus aethiops*) in a woodland savannah habitat.

Vervet monkeys differ from most arboreal guenons in that they live in relatively large troops with several adult males. Females remain in their natal groups throughout life, whereas males disperse from their natal groups around sexual maturity (Cheney and Seyfarth, 1983). Dispersing males almost always join adjacent groups (Henzi and Lucas, 1980; Isbell et al., 2002). As with baboons, there is a clear linear dominance hierarchy among both males and females (Struhsaker, 1967a; Whitten, 1983; Isbell and Pruetz, 1998). Female hierarchies are more stable than males, and daughters inherit their mother's rank (Cheney, 1983; Lee, 1983).

Vervets communicate via a diverse vocal repertoire (Struhsaker, 1967b), including calls that convey details about individual or group identity (Cheney and Seyfarth, 1980, 1982a,b), as well as information that elicits predator-specific responses from conspecifics (Seyfarth et al., 1980). Vervets also attend to the alarm calls of other nonprimate animals (Cheney and Seyfarth, 1990).

The little-known **dryas monkey** (*Chlorocebus dryas*) from the central Congo Basin has recently been suggested to be a close relative of vervet monkeys by molecular and morphological analyses (Van der Valk et al., 2020; Gilbert et al., 2021). Compared to vervets, dryas

monkeys are very shy and prefer the dense rainforest understory. From what little is known, they appear to be more arboreal than vervet monkeys (Gilbert et al., 2020; Alempijevic et al., 2021, 2022), and they may represent a separate genus from *Chlorocebus*.

The **patas**, or **hussier monkeys** (*Erythrocebus*), are close relatives of the vervets and mountain monkeys. They possess extreme specializations for life on the open grasslands. Patas are medium-sized, very sexually dimorphic monkeys with slender bodies, long limbs, and long tails. They have narrow hands and feet with short digits and a reduced pollex and hallux.

The three species of patas monkeys are widespread throughout the grasslands and savannahs of western, central, and eastern Africa. Although they usually sleep in trees at the edge of the forest, most of their foraging takes place in the open grass, where they move by quadrupedal walking and running. Patas monkeys are extremely agile, fast runners (55 km/h, according to Kingdon, 1971), and they frequently stand bipedally to look over the tall grass for potential predators or conspecifics. The bulk of their diet seems to be grass seeds, new shoots, and acacia gums. They also eat the beans of tamarind trees and a variety of other tough savannah fruits, seeds, and berries. They supplement their herbivorous diet with insects and various other prey items. They seem normally to eat on the move, picking up bits of food items as they walk.

Groups of patas monkeys average about a dozen individuals, usually with a single adult male. Females are philopatric. Nongroup males often live together in all-male bands, but there seems to be considerable male turnover in patas groups. Groups are not territorial but are intolerant of other groups. Aggressive intergroup encounters are led by females, and resident males are rarely involved (Chism, 1999). Within groups, agonism is rare and dominance hierarchies are weak and unstable (Isbell and Pruetz, 1998). Day ranges are extraordinarily variable, ranging from 700 to nearly 12,000 m, with a central tendency of between 2000 and 2500 m. Sometimes groups forage cohesively and other times members of a group are separated by as much as 800 m. The estimated home ranges are over 5000 ha, the largest known for any nonhuman primate species. It is not surprising that they can run so fast.

The **talapoin monkeys** (*Miopithecus*) are the smallest Old World monkeys, weighing just over 1 kg. They are also the smallest living catarrhines (Table 6.5). Many of the distinguishing features of talapoins – a relatively large head, large eyes, and a short snout – are characteristics of young animals, possibly suggesting that talapoins are a neotenic guenon (Shea, 1992). Detailed morphometric analyses suggest that allometric scaling trends across guenons are also responsible for distinctive talopoin cranial features (Cardini and Elton, 2008a,b).

Talapoins are among the least studied cercopithecins. They live in riverine, flooded, swamp, and secondary growth forests of western and central Africa. They are found most frequently in the dense undergrowth, where they move by leaping and quadrupedal walking and running. They are among the best leapers of all cercopithecines, and they also seem to be good swimmers. Their diet contains large amounts of insects and fruit, and they are among the most insectivorous of Old World monkeys (Gautier-Hion et al., 1988).

Talapoins live in large social groups averaging over sixty individuals, with roughly one adult male for every two adult females. Talapoin troops are always on the move, flushing insects as they travel. They always sleep by the water, which they leap into when frightened.

Allen's swamp monkey (*Allenopithecus nigroviridis*) is a medium-sized monkey that lives in swamps and flooded forests in western and central Africa. Compared to other guenons, this species has broader, macaque-like molar teeth. On the basis of its dental and skeletal anatomy, several authorities have suggested that this species is the most primitive guenon. Swamp monkeys are omnivores. In addition to consuming fruits, seeds, and leaves, they consume animal prey, including crustaceans,

snails, and earthworms, which make up about 20% of their diet. In flooded areas, they regularly dive and wade in rivers actively searching for aquatic prey (Gautier-Hion et al., 1988). They live in relatively large groups with several adult males (Gautier-Hion et al., 1988). Home ranges are determined by riverine boundaries. Little is known about their social or mating behaviors.

Colobines

The other subfamily of Old World monkeys are the colobines, or leaf-eating monkeys, of Africa and Asia. Colobines are easily distinguished from cercopithecines by their sharp-cusped cheek teeth and relatively narrow incisors. Their skulls have relatively short snouts (Fig. 6.4), narrow nasal openings, broad interorbital areas, and deep mandibles (Fig. 6.3). They have complex, sacculated stomachs similar to those of cattle that enable them to maintain bacterial colonies for digesting cellulose. Their skeletons are characterized by relatively long legs, long tails, and thumbs that are usually short or absent.

In general, the living colobine monkeys are more arboreal and folivorous than cercopithecines, and they are also better leapers. Colobine infants generally mature much faster than cercopithecines. Socially, many colobines live in single-male groups, but in some species, groups can be very large (see Kirkpatrick, 2011). Infants are frequently passed around among females within colobine troops, and in many species infants have a distinctive **natal coat** coloration.

There are two major groups of colobine monkeys: the colobus monkeys of Africa (Colobini) and the langurs, or leaf monkeys, of Asia (Presbytini). However, the genetic and phylogenetic relationships within and between the African and Asian colobines are very complex. Like the relationships among cercopithecines, they probably reflect a complex history of hybridization (Ting, 2008; Ting et al., 2008; Roos et al., 2011; Wang et al., 2012; Roos and Zinner, 2022; Kuderna et al., 2023).

African Colobines

The African colobus monkeys come in three color schemes: black and white, red, and olive. The three groups are quite distinct behaviorally and ecologically. The **black-and-white colobus monkeys** are the largest and most spectacular of the African colobines. These are quite robust monkeys, with considerable sexual dimorphism in size. There are six black-and-white species: *Colobus angolensis*, *C. caudatus*, *C. guereza*, *C. polykomos*, *C. satanas*, and *C. vellerosus* (Table 6.6). Black-and-white colobus (Fig. 6.17) live in a wide range of forest types throughout sub-Saharan Africa, in both primary rainforests and patchy dry forests. They are extremely hardy and can survive in a variety of habitats. They are arboreal and prefer the main canopy levels. All colobines are

TABLE 6.6 Subfamily COLOBINAE, African colobus monkeys

Common name	Species	Intermembral index	Mass (g) M	F
Angolan colobus	*Colobus angolensis*	-	9234	7299
Mt. Kilimanjaro guereza	*C. caudatus*	-	-	-
Guereza	*C. guereza*	79	10,900	8300
King colobus	*C. polykomos*	78	10,088	7935
Black colobus	*C. satanas*	-	10,700	8800
White-thighed colobus	*C. vellerosus*	-	8500	6900
Bay colobus and Temminck's red colobus	*Piliocolobus badius*	87	8391	7839
Bouvier's red colobus	*P. bouvieri*	-	-	-
Niger Delta red colobus	*P. epieni*	-	-	-
Foa's red colobus	*P. foai*	-	11,000	8000
Udzungwa red colobus	*P. gordonorum*	-	-	-
Kirk's red colobus	*P. kirkii*	-	6956	6604
Lang's red colobus	*P. langi*	-	7650	5680
Kahuzi red colobus	*P. lulindicus*	-	-	-
Oustalet's red colobus	*P. oustaleti*	-	12,200	7988
Lomami River red colobus	*P. parmentieri*	-	9200	7470
Pennant's red colobus	*P. pennantii*	-	11,000	10,000
Preuss's red colobus	*P. preussi*	-	8300	7300
Tana River red colobus	*P. rufomitratus*	-	9667	7214
Semliki red colobus	*P. semlikiensis*	-	-	-
Ashy red colobus	*P. tephrosceles*	-	9568	6264
Tshuapa red colobus	*P. tholloni*	-	-	-
Miss Waldron's red colobus	*P. waldroni*	-	6400	5750
Olive colobus	*Procolobus verus*	80	4490	4081

good leapers, but black-and-white species move more frequently by quadrupedal walking and bounding. They virtually never engage in suspensory behavior. They usually feed by sitting on branches and pulling food toward themselves.

The diet of black-and-white colobus is predominantly leaves and seeds, often of only a few tree species. They rely extensively on young and mature leaves for much of the year. However, fruits and seeds can account for nearly 50% of their diet in some seasons (Matsuda et al., 2020). For *C. angolensis*, lichens can be an important part of the diet (Miller et al., 2020).

The social groups of black-and-white colobus are extremely variable. *Colobus guereza* are usually found in small groups with a single male; *C. polykomos* and *C. vellerosus* are commonly found in larger, multimale groups. For *C. angolensis*, group sizes range from fewer than a dozen to over 300 individuals (Fashing, 2011). One subspecies from Uganda, *C. a. ruwenzorii*, exhibits a three-tiered multilevel social organization reminiscent of those exhibited by some Asian colobines (Stead and Teichroeb, 2019). Populations living in small groups can have extremely small home ranges, which they advertise with loud vocal calls called "roars," as well as very small day ranges. In some seasons black-and-white colobus can be virtually sedentary, and spend many days feeding in a single tree. The infants are white at birth and are commonly cared for by several members of the social group (Stead et al., 2021). Such instances of infant handling provide mothers with feeding opportunities that ultimately benefit their energy intake (Raboin et al., 2021).

FIGURE 6.17 Two sympatric colobines from Africa: *above* and *below*, the red colobus (*Piliocolobus badius*); center, the black-and-white colobus (*C. guereza*).

The **black colobus** (*Colobus satanas*), from western and central Africa, is quite distinct from the other black-and-white species in both anatomy and ecology. This species has larger, flatter teeth and a distinctively shaped skull. In contrast to other black-and-white species, black colobus feed predominantly on hard seeds; leaves and fruits are less important parts of their diet. As a result of this specialization on seeds, rather than leaves, as a protein source, *C. satanas* is able to thrive in areas that are uninhabitable for more folivorous colobines, because many leaves have high levels of poisonous tannins.

Black colobus live in multimale groups but have lower levels of sexual dimorphism compared to other black-and-white species. They have larger home ranges. In contrast, day ranges are, on average, quite small but vary dramatically from season to season, depending on the availability of food. When food is abundant, black colobus often feed in a few trees for several days; when food is scarce, they move long distances and feed briefly from many different trees.

Red colobus (genus *Piliocolobus*) are found sympatric with black-and-white colobus throughout most of sub-Saharan Africa. They are currently placed in many allopatric species, and they are the most threatened group of African monkeys (Linder et al., 2021). Red colobus are smaller than black-and-white colobus, and their relatively short pelage gives them a much more slender appearance (Fig. 6.17). They have broader incisors, prominent sagittal crests and supraorbital regions, robust mandibles, longer legs, and less sexual size dimorphism than many *Colobus* species, and they are strikingly different from guerezas in many aspects of their behavior and ecology as well. They seem to reach higher densities in primary forests than black-and-whites but are not found in drier forests. They range through all levels of the main canopy and in emergent trees. Compared to black-and-white species, red colobus eat fewer mature leaves, preferring fruit, young leaves, and shoots.

Many red colobus live in large troops of 40 to 90 animals, with numerous adults of both sexes. Troops can have overlapping home ranges of 100ha or more, and there seems to be direct competition and a dominance hierarchy between troops over access to specific feeding areas. Day ranges of red colobus are larger than those of black-and-white colobus. All of these ranging differences can be related to the distribution of the major food sources preferred by the two species. The more spatially and temporally homogeneous diet of mature leaves can be exploited effectively in small groups in most species of black-and-white colobus. The fruits, shoots, and new leaves preferred by red colobus are less evenly distributed, but when available they are quite abundant; thus, the size of food resources probably does not limit their group size. Furthermore, larger groups are better able to obtain and defend large food sources.

Many aspects of the social organization of red colobus are unusual for colobines. Females have a large estrus swelling, like the cercopithecines that live in multimale groups. Females, rather than males, generally emigrate from their natal group, though both sexes disperse in some species. Red colobus are commonly preyed on by chimpanzees, who can take as much as 10% of the red colobus population in a single year, and it has been suggested that the disjunct distribution and large number of subspecies of red colobus may be the result of repeated local extinctions due to chimpanzee predation. The large groups of red colobus have probably evolved in part as an antipredator strategy.

The **olive colobus** (*Procolobus verus*) is the smallest colobine and the most poorly known of the African species. It is sexually dimorphic in size and has the smallest thumb and the largest feet of all African colobines. Olive colobus live in the lowland moist forests and swamp forests of West Africa. They are cryptic animals that prefer areas with thick undergrowth and abundant lianas, such as river margins, secondary forest, and canopy gaps, where they range almost exclusively in the understory. They are the most saltatory of the African colobines (McGraw, 1998). Olive colobus forage as a very dispersed group in thick vegetation and regularly associate with groups of guenons, especially *Cercopithecus diana* (Oates and Whitesides, 1990). They eat mainly new leaves.

The social organization of olive colobus seems to be similar to that of red colobus but on a smaller scale. They live in small multimale groups of 10 to 15 animals. In contrast to red colobus, olive colobus seem to avoid predation through crypsis or by associating with large groups of *Cercopithecus* species. Like red colobus, female olive colobus show sexual swellings. Females carry young infants in their mouth, like many lemurs and lorises. Little is known of their ranging behavior.

Asian Colobines

It is in Asia that the colobine monkeys have reached their greatest diversity and abundance (Table 6.7). This diversity is largely a result of Pleistocene fluctuations in climate and sea level in the Sunda Shelf, as well as the uplift of the Tibetan Plateau and its effects on the geography and climate of Southeast Asia. Two or three sympatric species are a common pattern, and the density of leaf-eating monkeys in southern and eastern Asia exceeds that of any other forest vertebrates. The most common, most diverse, and most abundant are the langurs, formerly all placed in the genus *Presbytis*, but now divided into three separate genera. In addition, there are a number of most unusual monkeys commonly grouped as the "odd-nosed monkeys."

The systematics and phylogenetic relationships of Asian colobines are the subject of considerable debate, and they have had a complex biogeographic and

TABLE 6.7 Subfamily COLOBINAE, Langurs and leaf monkeys

Common name	Species	Intermembral index	Mass (g)	
			M	F
Black-and-white leaf monkey	*Presbytis bicolor*	-	6050	6800
Miller's leaf monkey	*P. canicrus*	-	6500	5750
Cross-marked leaf monkey	*P. chrysomelas*	-	6521	6917
Javan leaf monkey	*P. comata*	76	6680	6710
Banded leaf monkey	*P. femoralis*	-	6260	6190
White-fronted leaf monkey	*P. frontata*	76	5560	5670
Hose's leaf monkey	*P. hosei*	75	6180	6154
Black-crested Sumatran leaf monkey	*P. melalophos*	78	6590	6470
Mitred leaf monkey	*P. mitrata*	-	-	-
Natuna Islands leaf monkey	*P. natunae*	-	4536	5292
East Sumatran banded leaf monkey	*P. percura*	-	5982	6889
Mentawi leaf monkey	*P. potenziani*	-	6170	6400
Maroon leaf monkey	*P. rubicunda*	76	6290	6170
Sabah grizzled leaf monkey	*P. sabana*	-	-	6577
Pale-thighed leaf monkey	*P. siamensis*	-	6685	6877
Siberut leaf monkey	*P. siberu*	-	6500	6400
Black Sumatran leaf monkey	*P. sumatrana*	-	7371	7938
Thomas' leaf monkey	*P. thomasi*	-	6770	6690
Chamba sacred langur	*Semnopithecus ajax*	-	19,959	12,701
Hanuman langur	*S. entellus*	83	13,000	9890
Terai sacred langur	*S. hector*	-	17,237	13,608
Malabar sacred langur	*S. hypoleucos*	-	13,268	10,055
Tufted gray langur	*S. priam*	-	16,783	8845
Nepal sacred langur	*S. schistaceus*	-	19,200	14,800
John's or Nilgiri langur	*S. johnii*	80	11,709	11,192
Purple-faced langur	*S. vetulus*	-	8170	5900
Cristatus Group				
Lutung or Ebony langur	*Trachypithecus auratus*	-	6656	5841
Silvered leaf monkey	*T. cristatus*	81	6610	5760
Germain's silver leaf monkey	*T. germaini*	-	8826	-
Annamese langur	*T. margarita*	-		
West Javan langur	*T. mauritius*	-	7100	
Selangor silvered leaf monkey	*T. selangorensis*	-	-	-
Obscurus Group				
Gray leaf monkey	*T. barbei*	-	7100	5150
Indochinese gray langur	*T. crepusculus*	-	6836	6792
Phayre's leaf monkey	*T. phayrei*	-	7870	6300
Dusky leaf monkey	*T. obscurus*	83	7900	6260

TABLE 6.7   Subfamily COLOBINAE, Langurs and leaf monkeys—cont'd

Common name	Species	Intermembral index	Mass (g) M	F
Francoisi Group				
Francois' leaf monkey	*T. francoisi*	81	8450	7200
Hatinh langur	*T. hatinhensis*	79	8450	7200
White-browed black leaf monkey	*T. laotum*	-	7500	6500
White-headed langur	*T. leucocephalus*	-	8750	7820
Cat Ba Hooded or White headed black leaf monkey	*T. poliocephalus*	-	7800	7350
White-rumped black leaf monkey	*T. delacouri*	77	8600	7800
Pileatus Group				
Golden leaf monkey	*T. geei*	-	10,800	9500
Capped leaf monkey	*T. pileatus*	-	12,852	10,218
Shortridge's langur	*T. shortridgei*	-	13,155	9526

evolutionary history, likely involving varying amounts of hybridization among taxa (e.g., Ting et al., 2008; Roos et al., 2011; Wang et al., 2012; Roos and Zinner, 2022). Numerous alternative arrangements have been proposed based on studies of skulls, teeth, pelage, and molecules, with no clear resolution. In this chapter the langurs are divided into three genera: *Semnopithecus*, from South Asia and adjacent areas; *Trachypithecus*, from much of Southeast Asia and parts of South Asia and China; and *Presbytis*, restricted to Southeast Asia. The four genera of odd-nosed monkeys (*Nasalis*, *Simias*, *Pygathrix*, and *Rhinopithecus*) form a separate clade.

The **sacred, gray** or **Hanuman langurs** (*Semnopithecus*) of South Asia (Fig. 6.18) are among the most adaptable of all higher primates. They are currently divided into eight separate species (Table 6.7) found from Sri Lanka in the south, all over India, to Pakistan in the west, Bangladesh in the East, and well into the Himalayan Mountains of Nepal, Bhutan, and parts of China. Over this broad geographic range, the genus shows considerable diversity in body weight and skeletal morphology. In general, they are long-limbed, gray to brown monkeys with a very long tail, reduced thumbs, and long feet. They thrive in virtually all imaginable habitats found in South Asia, including tropical rainforests, deciduous forests, temperate dry forests, and conifer forests, as well as deserts and cities. Hanuman langurs are the most terrestrial of the colobines. In the trees they use both quadrupedal gaits and leaps, but mainly the former. They feed primarily in seated postures.

Considering the diversity of habitats they occupy, it is not surprising that the diet of Hanuman langurs is quite eclectic. Though once considered primarily folivorous, they are now known to eat leaves, fruit, flowers, and insects, as well as occasional bark, gum, and soil. In general, Hanuman langurs strongly prefer seeds, young leaves, and flowers to mature leaves (Stanford, 1991). Nevertheless, Hanuman langurs fallback on mature leaves during several resource-poor months each year (Koenig and Borries, 2001).

Three types of social groups have commonly been observed in Hanuman langurs: one-male troops, multimale troops, and all-male bands (Newton, 1988; Koenig and Borries, 2001). The ratio of one-male to multimale troops varies between populations; however, both types of social groupings can be found in varying proportions at most sites. These groups vary considerably in size, from solitary males to as many as 64 members (Newton, 1988). Home ranges often reflect group size and habitat quality, ranging from 24 to 200 ha.

Hanuman langur troops seem to be centered around groups of related adult females that aid one another and care for each other's offspring. By contrast, male residence in a troop is usually relatively short-lived. In one-male troops, the single adult male fathers all of the offspring and drives out rival males. At fairly regular intervals, bands of roving males attack a group, drive out the resident male, and kill dependent infants. One of these intruders then establishes himself as the dominant male, drives away the others, and starts the cycle anew (Hrdy, 1977). In multimale troops, male membership usually changes less drastically (Bishop, 1979; Laws and Vonder Haar Laws, 1984). A newcomer male either replaces one of the resident males or simply joins the group. Newly immigrant males may still commit infanticide. Hanuman langurs have variable patterns of

FIGURE 6.18 Two colobine species from India and Sri Lanka: *above*, the purple-faced monkeys (*Semnopithecus vetulus*); *below*, the Hanuman langur (*Semnopithecus entellus*).

reproduction. Breeding seasonality is common, though provisioned populations living in proximity to temples or cities can breed more or less year-round (Borries et al., 2001). Provisioned populations also experience earlier age at first parturition, shorter gestation periods, and reduced interbirth intervals, suggesting that nutrition is the primary factor underlying reproductive seasonality in this species. Allomothering is common in Hanuman langurs (Hrdy, 1976). Infants are regularly passed around among adult females and are often cared for by females other than their mother.

Two species whose affinities have become a source of confusion are *Semnopithecus vetulus*, the **purple-faced langur**, from Sri Lanka (Fig. 6.18), and a closely related species, the **Nilgiri langur** or **John's langur**, *Semnopithecus johnii*, from western India. Both are characterized by purple faces, white sideburns, and relatively small size. In previous decades, some authorities placed them with the Hanuman langurs in *Semnopithecus* (Brandon-Jones et al., 2003), whereas others put them with *Trachypithecus*.

Recent genetic studies have consistently found that these two species are more closely related to Hanuman langurs (Roos and Zinner, 2022). Both taxa are restricted to forested areas. They are almost totally arboreal and are excellent leapers. Both are more folivorous than sympatric Hanuman langurs and can subsist almost exclusively on mature leaves. Purple-faced langurs live in very small, one-male groups or families, with home ranges of 1 ha or less. Their day ranges are small and they often spend an entire day in one tree. Male takeovers and infanticide have also been reported in purple-faced langurs.

Closely related to *Semnopithecus* is the genus *Trachypithecus*, the **Lutongs**, which are divided into 4 species groups and over 15 individual species, ranging from India in the west to China in the east (Table 6.7). These monkeys are smaller than Hanuman langurs and are often sexually dimorphic. They are mostly silver, gray, or black as adults, but infants are bright yellow or orange.

FIGURE 6.19 Two sympatric leaf monkey species from Malaysia: *above*, the spectacled langur, or dusky leaf monkey (*Trachypithecus obscurus*); *below*, the white-thighed surili (*Presbytis siamensis*).

Dusky langurs (*T. obscurus*) (Fig. 6.19), and the closely related **Phayre's leaf monkey** (*T. phayrei*) with its striking white eye rings, belong to the *T. obscurus* species group. Members of this group prefer closed primary forests, but are also found in old-growth secondary and disturbed forests, as well as in plantations and in the jhum cultivation (shifting cultivation) areas of southeast Asia. *T. obscurus* can also be found in urban areas, and parks, suggesting a fairly high adaptive capability when compared with most colobine species. In west Malaysia, where it has been most thoroughly studied, *T. obscura* prefers the main canopy levels and, to a lesser degree, the emergent trees. These monkeys are primarily arboreal quadrupeds that leap much less frequently than the sympatric *Presbytis siamensis* (Fig. 6.20).

Both *T. pharyei* and *T. obscurus* are predominantly folivorous, although their diet is relatively varied. Animals will also consume fruit, flowers, shoots, and other items (Molur et al., 2003; Bose, 2005; Aziz and Feeroz, 2009). *T. obscurus* is able to take advantage of unripe fruit, which have chemical defenses, by the same means that they break down toxins in plant leaves, using the bacteria found in their digestive system (MacKinnon and MacKinnon, 1980).

Members of the *T. obscurus* group live in both unimale– and multimale–multifemale groups, with group sizes of 10 to 30 individuals. Groups forage together in the early morning or late afternoon, and often split up during the day into small foraging units, each of which travels a relatively short distance and may spend hours in one place (Gupta and Kumar, 1994). Home range approximates 33 ha and daily movement of 950 m/day. Dispersal is male-biased, though some males return to their natal groups to breed (Koenig and Borries, 2012). Infants are born with an orange natal coat and, like Hanuman langurs, they pass their brightly colored infants around among the females.

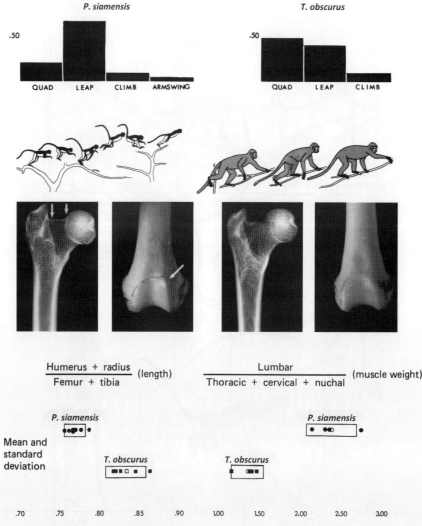

FIGURE 6.20 Locomotor and anatomical differences between *Presbytis siamensis* and *Trachypithecus obscurus*. *Presbytis siamensis* leaps more frequently, and its femur is characterized by a short, straight neck and a prominent lateral ridge on the patellar groove (*arrows*). This species also has relatively longer hindlimbs and larger back muscles.

Relatively little is known about members of the *T. cristatus* group. The **silvered langur**, *T. cristatus*, is commonly found in mangrove swamps and in flooded land along long rivers in west Malaysia and much of the Sunda Shelf; but some populations are also found inland (Furuya, 1961; MacKinnon and MacKinnon, 1987). This species is mainly folivorous, with leaves comprising 60% to 80% of the diet (Nadler et al., 2003). However, it will also feed on fruit, seeds, flowers, and young shoots (Bernstein, 1968; Furuya, 1961). *Trachypithecus cristatus* seem to live more frequently in single-male groups. They are not seasonal breeders and the average gestation period is 194.6 days with an ovarian cycle of about 24 days. Allomothering is common in this species.

There is also little known about the *T. pileatus* group, which includes three species, the **Capped langur**, *T. pileatus*, **Shortridge's langur**, *T. shortridgei*, and **Gee's golden langur**, *T. geei*. Their native habitat includes the subtropical or tropical dry forests of India, Bhutan, Bangladesh, Myanmar, and China. Among them, *Trachypithecus pileatus* are the best studied. *Trachypithecus pileatus* consume a seasonally variable diet consisting primarily of new and mature leaves, though this is supplemented by high levels of flowers and unripe fruits during certain times of year (Borah et al., 2021). Both *T. pileatus* and *T. shortridgei* live in multimale–multifemale groups that range in size from 8 to 13 individuals (Hasan et al., 2018). Mating in both species is seasonal (Solanki et al., 2007; Li et al., 2015), and like most other colobines, allomaternal care is common (Kumar and Solanki, 2014).

A complex group of *Trachypithecus* species from Indochina are those in the Francoisi group. Several of these species, including **Delacour's langur** (*T. delacouri*), the **white-headed langur** (*T. poliocephalus*), and **François langur** (*T. francoisi*) live among jagged limestone cliffs in Vietnam and China. They are among the most folivorous of all colobines, with leaves comprising over 80% of their annual diet (Workman, 2010).

The genus *Presbytis*, the **surilis**, consists of over 15 species from all over the Sunda region of Southeast Asia, including Thailand and west Malaysia, Sumatra, Borneo, the Mentawai Islands, and Java (Meijaard and Groves, 2004). There are numerous subspecies with a dazzling variability in adult coat color. By contrast, all infants show a distinct banded or cruciform pattern with a dark head, dark body and tail, and dark arms.

One of the best-known species, *P. siamensis*, the **white-thighed surili** from Peninsular Malaysia, is a relatively small colobine (6 kg) that shows little sexual size dimorphism (Fig. 6.19). Like other members of its species group, this monkey has a relatively short face. Its postcranial skeleton is characterized by relatively long legs, a long trunk, and slender arms that are associated with its leaping locomotion (Fig. 6.20).

Leaf monkeys of this genus seem to be found in a variety of inland forests, but not in swamps or montane forests. In west Malaysia the white-thighed surili is more common in secondary forests than in upland primary forests, and moves and feeds more frequently in the understory and lower canopy levels than do other sympatric langurs. They are extraordinary leapers and move less frequently by arboreal quadrupedalism. They occasionally use forelimb suspension as well (Fig. 6.19).

White-thighed surilis eat primarily young leaves, seeds, and fruit, and rarely, if ever, partake of mature leaves. They live in small groups of a dozen or fewer animals, frequently with a single adult male. Home range sizes are quite variable from locality to locality and do not seem to be defended. Troops forage as a group over day ranges that average about 700 m. They are active throughout the day and often well into early evening. Males often show wounds, scars, and other evidence of conflicts, and infanticide seems likely.

Presbytis potenziani, the **Mentawai Island leaf monkey**, exhibits flexibility in its social organization, occurring in both monogamous groups of one adult male and one female, as well as one-male–multifemale and multimale–multifemale groups (Tilson and Tenza, 1976; Watanabe, 1981; Fuentes, 1996). Their diet is predominantly fruits, seeds, and leaves (Fuentes, 1996). Like many monogamous primates, the males and females have been reported to sing together in daily vocal duets (Tilson and Tenaza, 1976).

The remaining genera of colobine monkeys from Asia are all characterized by odd-shaped noses and are referred to as the **odd-nosed monkeys** (Table 6.8). Molecular studies have shown that they are a distinct clade from other Asian colobines. The best known of these is *Nasalis larvatus*, the **proboscis monkey** of Borneo (Fig. 6.21). This large, red monkey is the most sexually dimorphic of all colobines. Males are almost twice the size of females and have an enormous, pendulous nose; females have a smaller, turned-up proboscis.

Proboscis monkeys are restricted to areas of riverine and coastal forest, where their main predators are the false gavials and clouded leopards. Proboscis monkeys practice a riverine refuging pattern, foraging widely during the day but returning to sleeping trees at the riverside each night (Matsuda et al., 2008). Groups regularly cross the river by swimming, usually at the narrowest spots (Yeager, 1991a). The majority of their diet consists of young leaves and unripe fruits. In some months proboscis monkeys eat primarily fruit, and for most of the year they eat predominantly leaves (Yeager, 1989; Matsuda et al., 2009). Like all colobines, proboscis monkeys have a complex stomach for digesting leaves. Interestingly, they have been observed to occasionally regurgitate and remasticate their food in the same way that many ungulates "chew their cud" (Matsuda et al., 2011). Individual

TABLE 6.8   Subfamily COLOBINAE, Odd-nosed Monkeys

Common name	Species	Intermembral index	Mass (g) M	Mass (g) F
Proboscis monkey	*Nasalis larvatus*	93	20,400	9820
Gray-shanked douc langur	*Pygathrix cinerea*	-	11,500	8450
Red-shanked douc langur	*P. nemaeus*	94	10,886	8165
Black-shanked douc langur	*P. nigriceps*	92	11,000	9007
Tonkin snub-nosed monkey	*Rhinopithecus avunculus*	-	14,500	8250
Yunnan snub-nosed monkey	*R. bieti*	86	15,300	9100
Guizhou snub-nosed monkey	*R. brelichi*	-	14,500	5000
Golden snub-nosed monkey	*R. roxellana*	-	17,900	11,600
Burmese snub-nosed monkey	*R. strykeri*	-	14,000	8500
Simakobu or Pig-tailed langur	*Simias concolor*	-	9148	6889

FIGURE 6.21   The proboscis monkey (*Nasalis larvatus*) in a nipa mangrove swamp.

foraging groups, mostly polygynous one-male groups and nonreproductive groups of males and immatures, have large overlapping home ranges ranging from 125 to 220 ha (Yeager, 1989; Boonratana, 2000).

Proboscis monkey social organization seems to be a two-tiered arrangement in which individual foraging groups (one-male groups and nonreproductive – mostly all-male – groups) regularly associate with adjacent groups to form a band (Yeager, 1991b). Individual groups spend about one-third of their time alone and two-thirds in association with other groups. Group interactions are not random; rather, each group seems to have preferred neighbors. These patterns seem to be driven by the density and availability of sleep sites, and differ from the patterns described for other multilevel societies (Matsuda et al., 2010). All males seem to disperse before adolescence, and many females disperse as well (Boonratana, 1999).

Simias concolor, the **Simakobu monkey** or **pig-tailed langur**, is a close relative of *Nasalis* from the Mentawai Islands, off the west coast of Sumatra. Both males and females of this species have a short, turned-up nose, much like that of a female proboscis monkey. This species has very unusual body proportions for a colobine, with similar-sized forelimbs and hindlimbs and a short tail. They are sexually dimorphic. Males are on average 30% larger than females and have much larger canines (Tenaza and Fuentes, 1995). These monkeys' daily activities, including feeding and much of their travel, take place predominately in trees, but when frightened they will sometimes descend to the ground and flee terrestrially. The bulk of their diet comprises young leaves, flowers, and unripe fruits (Erb et al., 2012b).

Simakobu monkeys live in both monogamous family groups and polygynous groups with up to five adult females (Tenaza and Fuentes, 1995; Erb et al., 2012a), with the former occurring most frequently where hunting rates are high. Adult and juvenile males typically reside in all-male groups, which, like those of proboscis monkeys, occasionally include nonreproductive females; however, solitary individuals also occur in some populations. Average group size on the Pagai Islands is 4.1 individuals (Tenaza and Fuentes, 1995), whereas in northern Siberut, groups comprise 7.9 individuals on average (Erb et al., 2012a). Home ranges are small, between 6 and 7 ha, and exhibit little overlap. Simakobu males seem to disperse prior to adulthood, and both juvenile and adult females temporarily visit and disperse to other groups (Erb et al., 2012a). Adult simakobu males produce loud, distinct vocalizations, or loud calls, which seem to play a role in male–male competition (Tenaza, 1989). Owing to hunting pressure and increasing deforestation, these monkeys are rapidly becoming extinct (Whittaker, 2006; Erb et al., 2012a).

Pygathrix, the **Douc langurs** from Vietnam, Cambodia, and Laos, are colorful monkeys with little or no sexual dimorphism. They live in mixed, partly deciduous forests and feed primarily on young leaves, fruits, and seeds (Duc et al., 2009). They are leaping monkeys, but they also engage in arm-swinging behavior as well (Byron and Covert, 2004). Black-shanked douc langurs (*Pygathrix nigripes*) typically form one-male groups with seven individuals on average, with up to 12 in the dry season and up to 20 in the wet season (Rawson, 2009). Red-shanked douc langurs (*Pygathrix nemaeus*) form multimale groups that range in size from 17 to 45, but are also seasonally variable, such that larger groups fission into smaller groups during the lean season (Phiapalath et al., 2011). Food sharing among individuals is common.

Rhinopithecus, the **snub-nosed monkeys** (Fig. 6.22), are among the largest of all colobines, with some individuals weighing 30 kg or more. One species is from the monsoon forests of Vietnam (*R. avunculus*); three others (*R. roxellana*, *R. brelichi*, and *R. bieti*) are from China (Table 6.8), and a new species (*R. strykeri*) is from Myanmar (Geissmann et al., 2011). All have a short, turned-up nose and a bluish face. They have a long tail, relatively short limbs, and short digits that are covered with long fur on the dorsal surface. All species are very dimorphic in both body size and canine size.

The behavior and ecology of *Rhinopithecus* is only just becoming known (Kirkpatrick and Grueter, 2010). These monkeys live in a wide range of habitats, from the tropical forests of Vietnam to temperate conifer forests in the mountains of southern China, including some remarkably harsh environments. Little has been reported on their locomotion, but their limb anatomy, foot structure, and some anecdotal field observations suggest that in some habitats they may be among the most terrestrial of all Asian colobines. They feed on a variety of leaves, fruits, and seeds, but are unusual in that all rely heavily on lichens, particularly at higher latitudes.

The basic social group among snub-nosed monkeys is single-male–multifemale group, but these small groups, together with AMGs, aggregate to form large bands of more than 400 individuals (Kirkpatrick and Grueter, 2010). All species of *Rhinopithecus* have very large day ranges and enormous home ranges, as large as 4000 ha in *R. roxellana*. Births seem to be strictly seasonal.

Adaptive Radiation of Old World Monkeys

Compared to either strepsirrhines or Neotropical anthropoids, Old World monkeys are a remarkably uniform group in many aspects of morphology and behavior (Fig. 6.23). This may be due partly to the recentness

FIGURE 6.22   The golden monkey (*Rhinopithecus roxellana*) from China.

of their adaptive radiation. In size, they barely range over one order of magnitude, with no very small species and no extremely large species. Sexual dimorphism in canine teeth is the rule, and body size dimorphism is common. Although their bilophodont molars vary according to dietary habits in details of cusp height, length, and breadth and in the length of shearing crests, they are remarkably similar overall (Kay, 1978).

In their locomotor habits, cercopithecoid monkeys are all primarily quadrupedal walkers and runners, and some, especially colobines, are also good leapers. Some of the Asian colobines also engage in arm-swinging behavior. When feeding in trees, they always sit above branches rather than using suspensory postures. No other primate radiation includes so many terrestrial species or is so limited in gait patterns and locomotor repertoire. Old World monkeys are primarily frugivorous or folivorous; a few are seed specialists. Although many species feed to varying degrees on insects (most cercopithecines) and gums (baboons, patas monkeys), there are no species that specialize on these foods to the degree found in many smaller strepsirrhines and platyrrhines.

Old World monkeys generally live in either single-male or multimale polygynous groups; monogamy is rare. In social organization, they are a distinctive and relatively uniform radiation compared to either platyrrhines or hominoids (Di Fiore and Rendall, 1994). Nevertheless, there are several species of Old World monkeys (hamadryas baboons, geladas, *Rhinopithecus*) that live in large multilevel societies, a feature not found in other primate radiations. In many species females tend to spend their entire lives in their natal troop, and social organization centers around matrilines and female hierarchies, with a few notable exceptions, such as red colobus, hamadryas baboons, and some populations of chacma baboons. Males regularly emigrate from their natal group and may move through several troops during their lifetime. All have single births.

Despite – or perhaps because of – this relative uniformity, Old World monkeys have successfully colonized more vegetative zones and climates than has any other group of living primates. In numbers of individuals, numbers of species, and biomass density, they are probably the most successful nonhuman primates. This success results from a variety of factors. For the colobines,

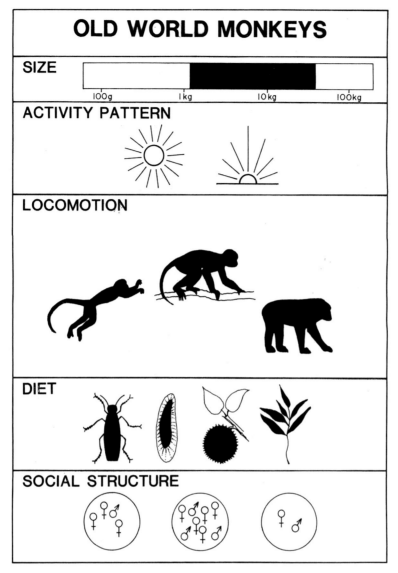

FIGURE 6.23 The adaptive radiation of Old World monkeys.

it is almost certainly their ability to digest cellulose and to exploit folivorous food sources that are unavailable to other animals. For the cercopithecines, their terrestrial locomotor potential, manipulative abilities, and intelligence enable them to exploit a wide range of foods and environments that less flexible species cannot endure. Cercopithecines are also relatively prolific breeders for higher primates, with many species giving birth on an annual basis.

Phyletic Relationships of Old World Monkeys

There are relatively few major problems concerning the phyletic relationships among cercopithecoids (Fig. 6.24). However, it has become increasingly evident in recent years that hybridization between what have been previously recognized as distinct species, and genera, is not uncommon in Old World monkeys (Zinner et al., 2009a, 2011, 2018; Roos et al., 2011; Tung and Berreiro, 2017; Roos and Zinner, 2022). Above the genus level, most cercopithecoid phylogenies are in agreement. However, because of unresolved – or perhaps unresolvable – issues concerning the phyletic relationships among individual species and genera, there is often no consensus on how to construct a nomenclature that reflects the phylogeny. Among cercopithecines, it is now well-established that *Cercocebus* mangabeys are most closely related to *Mandrillus*, and *Lophocebus* mangabeys are related to *Rungwecebus*, *Papio*, and *Theropithecus*. However, the proper phylogenetic placement of *Rungwecebus* is difficult to resolve, and there are also conflicting genetic relationships among many

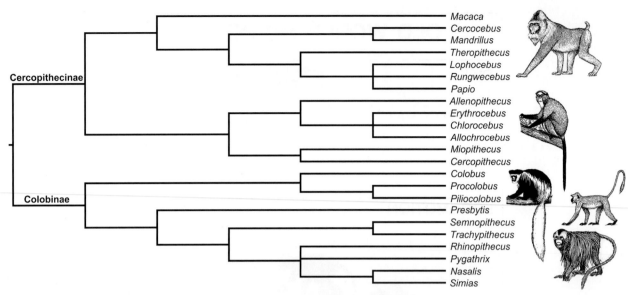

FIGURE 6.24   Phylogeny of Old World monkeys

guenon species and species groups outside of the terrestrial clade containing *Chlorocebus*, *Allochrocebus*, and *Erythrocebus* (e.g., Tosi et al., 2005; Xing et al., 2007; Perelman et al., 2011; Lo Bianco et al., 2017; Kuderna et al., 2023; Jensen et al., 2023).

Among colobines there are some questions concerning the proper taxonomy and phylogenetic relationships among the African genera and within the Asian radiation. Most analyses recover a monophyletic African clade as well as a monophyletic Asian clade (Roos et al., 2011; Wang et al., 2012; Roos and Zinner, 2022; Kuderna et al., 2023). However, there are complex genetic relationships among the Asian genera in particular as well as debates and ongoing new discoveries among the species of Asian colobines. These include the relationships among *Semnopithecus*, *Trachypithecus*, and *Presbytis*, with various species seemingly the result of varying degrees of hybridization. Fig. 6.24 provides a summary of current views of the genus-level relationships among Old World monkeys.

References

Abernethy, K.A., White, L.J.T., Wickings, E.J., 2002. Hordes of mandrills (*Mandrillus sphinx*): Extreme group size and seasonal male presence. J. Zool. 258, 131–137.

Alba, D.M., Madurell-Malapeira, J., Delson, E., et al., 2016. First record of macaques from the Early Pleistocene of Incarcal (NE Iberian Peninsula). J. Hum. Evol. 96, 139–44.

Alberts, S.C., 2019. Social influences on survival and reproduction: Insights from a long-term study of wild baboons. J. Anim. Ecol. 88 (1), 47–66.

Alempijevic, D., Boliabo, E.M., Coates, K.F., et al., 2021. A natural history of *Chlorocebus dryas* from camera traps in Lomami National Park and its buffer zone, Democratic Republic of the Congo, with notes on the species status of *Cercopithecus salongo*. Am. J. Primatol. 83, e23261. https://doi.org/10.1002/ajp.23261

Alempijevic, D., Hart, J.A., Hart, T.B., Detwiler, K.M., 2022. Using local knowledge and camera traps to investigate occurrence and habitat preference of an Endangered primate: The endemic dryas monkey in the Democratic Republic of the Congo. Oryx 56, 260–267.

Arenson, J.L., Sargis, E.J., Hart, J.A., et al., 2020. Skeletal morphology of the lesula (*Cercopithecus lomamiensis*) and the evolution of guenon locomotor behavior. Am. J. Phys. Anthropol. 172, 3–24.

Aziz, M.A., Feeroz, M.M., 2009. Utilization of forest flora by Phayre's Leaf Monkey *Trachypithecus phayrei* (Primates: Cercopithecidae) in semi-evergreen forests of Bangladesh. J. Threat. Taxa 257–262.

Bernstein, I.S., 1968. The lutong of Kuala Selangor. Behaviour. 32, 1–16.

Bishop, N.H., 1979. Himalayan langurs: Temperate colobines. J. Hum. Evol. 8, 251–281.

Boonratana, R., 1999. Dispersal in proboscis monkeys (*Nasalis larvatus*) in the Lower Kinabatangan, Northern Borneo. Trop. Biod. 6, 179–187.

Boonratana, R., 2000. Ranging behavior of proboscis monkeys (*Nasalis larvatus*) in the Lower Kinabatangan, Northern Borneo. Int. J. Primatol. 21, 497–518.

Borah, D.K., Solanki, G.S., Bhattacharjee, P.C., 2021. Feeding ecology of capped langur (*Trachypithecus pileatus*) in Sri Surya Pahar, a disturbed habitat in Goalpara District, Assam, India. Trop. Ecol. 62 (3), 492–498.

Borries, C., Koenig, A., Winkler, P., 2001. Variation of life history traits and mating patterns in female langur monkeys (*Semnopithecus entellus*). Behav. Ecol. Sociobiol. 50, 391–402.

Bose, J., 2005. An ecobehavioural study of Phayre's leaf monkey *Trachypithecus phayrei* Blyth 1847. Ph.D. thesis. Guwahati University, Guwahati, India.

Bracebridge, C.E., Davenport, T.R.B., Marsden, S.J., 2011. Can we extend the area of occupancy of the kipunji, a critically endangered African primate? Anim. Conserv. 14, 687–696.

Bracebridge, C.E., Davenport, T.R., Marsden, S.J., 2012. The impact of forest disturbance on the seasonal foraging ecology of a critically endangered African primate. Biotropica 44 (4), 560–568.

Bracebridge, C.E., Davenport, T.R., Mbofu, V.F., Marsden, S.J., 2013. Is there a role for human-dominated landscapes in the long-term conservation management of the critically endangered kipunji (*Rungwecebus kipunji*)? Int. J. Primatol. 34, 1122–1136.

Brandon-Jones, D., Eudey, A.A., Geissmann, T., et al., 2003. Asian primate classification. Int. J. Primatol. 25, 97–162.

Brauch, K., Pfefferle, D., Hodges, K., et al., 2007. Female sexual behavior and sexual swelling size as potential cues for males to discern the female fertile phase in free-ranging Barbary macaques (*Macaca sylvanus*) of Gibraltar. Horm. Behav. 52, 375–383.

Brockmeyer, T., Kappeler, P.M., Willaume, E., et al., 2015. Social organization and space use of a wild mandrill (*Mandrillus sphinx*) group. Am. J. Primatol. 77 (10), 1036–1048.

Burrell, A.S., Jolly, C.J., Tosi, A.J., Disotell, T.R., 2009. Mitochondrial evidence for the hybrid origin of the kipunji, *Rungwecebus kipunji* (Primates: Papionini). Mol. Phylo. Evol. 51, 340–348.

Butynski, T.M., Koster, S.H., 1994. Distribution and conservation status of primates in Bioko Island, Equatorial Guinea. Biodivers. Conserv. 3, 893–909.

Buzzard, P., Eckardt, W., 2007. The social systems of the guenons. In: McGraw, W.S., Zuberbühler, K., Noë, R (Eds.), Monkeys of the Taï Forest: An African Primate Community. Cambridge University Press, Cambridge, pp. 51–71.

Byron, C.V., Covert, H.H., 2004. Unexpected locomotor behavior: Brachiation by an Old World monkey (*Pygathrix nemaeus*) from Vietnam. J. Zool. 263, 101–106.

Cardini, A., Elton, S., 2008a. Variation in guenon skulls (I): Species divergence, ecological and genetic differences. J. Hum. Evol. 54, 615–637.

Cardini, A., Elton S., 2008b. Variation in guenon skulls (II): Sexual dimorphism. J. Hum. Evol. 54, 638–647.

Chala, D., Roos, C., Svenning, J.-C., Zinner, D., 2019. Species-specific effects of climate change on the distribution of suitable baboon habitats – Ecological niche modeling of current and Last Glacial Maximum conditions. J. Hum. Evol. 132, 215–226. https://doi.org/10.1016/j.jhevol.2019.05.003

Charpentier, M., Peignot, P., Hossaert-McKey, M., et al., 2005. Constraints on control: factors influencing reproductive success in male mandrills (*Mandrillus sphinx*). Behav. Ecol. 16, 614–623.

Cheney, D.L., 1983. Extrafamilial relationships among vervet monkeys. In: Hinde, R.A. (Ed.), Primate Social Relationships: An Integrated Approach. Blackwell, Oxford, pp. 278–289.

Cheney, D.L., Seyfarth, R.M., 1980. Vocal recognition in free-ranging vervet monkeys. Anim. Behav. 28, 362–367.

Cheney, D.L., Seyfarth, R.M., 1982a. How vervet monkeys perceive their grunts: field playback experiments. Anim. Behav. 30, 739–751.

Cheney, D.L., Seyfarth, R.M., 1982b. Recognition of individuals within and between groups of free-ranging vervet monkeys. Am. Zool. 22, 519–529.

Cheney, D.L., Seyfarth, R.M., 1983. Nonrandom dispersal in free-ranging vervet monkeys: social and genetic consequences. Am. Nat. 122 (3), 392–412.

Cheney, D.L., Seyfarth, R.M., 1990. How Monkeys See the World: Inside the Mind of Another Species. University of Chicago Press, Chicago, p. 377.

Chism, J., 1999. Intergroup encounters in wild patas monkeys (*Erythrocebus patas*) in Kenya. Am. J. Primatol. 49, 43.

Cords, M., 2000. The agonistic and affiliative relationships of adult females in a blue monkey group. In: Jolly, C., Whitehead., P. (Eds.), Old World Monkeys. Cambridge University Press, Cambridge, pp. 453–479.

Cords, M., 2007. Variable participation in the defense of communal feeding territories by blue monkeys in the Kakamega Forest, Kenya. Behaviour 144 (12), 1537–1550.

Davenport, T.R.B., De Luca, D.W., Bracebridge, C.E., et al., 2010. Diet and feeding patterns in the kipunji (*Rungwecebus kipunji*) in Tanzania's Southern Highlands: A first analysis. Primates 51, 213–220.

Davenport, T.R.B., Machaga, S.J., Mpunga, N.E., et al., 2022. A reassessment of the population size, demography, and status of Tanzania's endemic kipunji *Rungwecebus kipunji* 13 years on: Demonstrating conservation success. Int. J. Primatol. 43, 317–338. https://doi.org/10.1007/s10764-022-00281-3

Delson, E., Terranova, C.J., Jungers, W.L., et al., 2000. Body mass in Cercopithecidae (Primates, Mammalia): estimation and scaling in extinct and extant taxa. Anthropol. Pap. Am. Mus. Nat. Hist. 8, 1–159.

Di Fiore, A., Rendall, D., 1994. Evolution of social organization: A reappraisal for primates by using phylogenetic methods. Proc. Natl. Acad. Sci. USA 91, 9941–9945.

Dixson, A.F., 1983. Observations on the evolution and behavioral significance of "sexual skin" in female primates. Adv. Study Behav. 13, 63–106.

Duc, H.M., Baxter, G.S., Page, M.J., 2009. Diet of *Pygathrix nigripes* in Southern Vietnam. Int. J. Primatol. 30, 15–28.

Ehardt, C.L., Jones, T.P., Butynski, T.M., 2005. Protective status, ecology and strategies for improving conservation of *Cercocebus sanjei* in the Udzungwa Mountains, Tanzania. Int. J. Primatol. 26, 557–583.

Elton, S., Dunn, J., 2020. Baboon biogeography, divergence, and evolution: Morphological and paleoecological perspectives. J. Hum. Evol. 145, 102799. https://doi.org/10.1016/j.jhevol.2020.102799

Engelhardt, A., Hodges, J.K., Niemitz, C., Heistermann, M., 2005. Female sexual behavior, but not sex skin swelling, reliably indicates the timing of the fertile phase in wild long-tailed macaques (*Macaca fascicularis*). Hormones and Behavior 47, 195–204.

Erb, W.M., Borries, C., Lestari, N.S., Ziegler, T., 2012a. Demography of simakobu (*Simias concolor*) and the impact of human disturbance. Am. J. Primatol. 74, 580–590.

Erb, W.M., Borries, C., Lestari, N.S., Hodges, J.K., 2012b. Annual variation in ecology and reproduction of wild simakobu (*Simias concolor*). Int. J. Primatol. 33, 1406–1419.

Fashing, P.J., 2011. African colobine monkeys: Their behavior, ecology, and conservation. In: Campbell, C., Fuentes, A., MacKinnon, K., et al., (Eds.), Primates in Perspective, second ed. Oxford University Press, Oxford, pp. 203–229.

Fashing, P.J., Nguyen, N., Venkataraman, V.V., Kerby, J.T., 2014. Gelada feeding ecology in an intact ecosystem at Guassa, Ethiopia: Variability over time and implications for theropith and hominin dietary evolution. Am. J. Phys. Anthropol. 155 (1), 1–16.

Fernandez, D., Doran-Sheehy, D., Corries, C., Brown, J.L., 2014. Reproductive characteristics of wild Sanje mangabeys (*Cercocebus sanjei*). Am. J. Primatol. 76, 1163–1174.

Fischer, J., Kopp, G.H., Dal Pesco, F., et al., 2017. Charting the neglected West: The social system of Guinea baboons. Am. J. Phys. Anthropol. 162, 15–31.

Fischer, J., Higham, J.P., Alberts, S.C., et al., 2019. Insights into the evolution of social systems and species from baboon studies. eLife. 8, e50989. https://doi.org/10.7554/eLife.50989

Frost, S.R., Marcus, L.F., Bookstein, F.L., et al., 2003. Cranial allometry, phylogeography, and systematics of large-bodied papionins (primates: Cercopithecinae) inferred from geometric morphometric analysis of landmark data. Anat. Rec. A Discov. Mol. Cell. Evol. Biol. 275A, 1048–1072.

Fruteau, C., Range, F., Noe, R., 2010. Infanticide risk and infant defence in multi-male free-ranging sooty mangabeys, *Cercocebus atys*. Behav. Processes 83, 113–118.

Fruteau, C., van de Waal, E., van Damme, E., Noe, R., 2011. Infant access and handling in sooty mangabeys and vervet monkeys. Anim. Behav. 81, 153–161.

Fuchs, A.J., Gilbert, C.C., Kamilar, J.M., 2018. Ecological niche modeling of the genus *Papio*. Am. J. Phys. Anthropol. 166, 812–823. https://doi.org/10.1002/ajpa.23470

Fuentes, A., 1996. Feeding and ranging in the Mentawai Island langur (*Presbytis potenziani*). Int. J. Primatol. 17, 525–548.

Fürtbauer, I., Schülke, O., Heistermann, M., Ostner, J., 2010. Reproductive and life history parameters of wild female *Macaca assamensis*. Int. J. Primatol. 31, 501–517.

Furuya, Y., 1961. The social life of silvered leaf monkeys. Primates 3, 41–60.

Galat, G., Galat-Luong, A., 1976. La colonisation de la mangrove par *Cercopithecus aethiops* sabaeus au Sénégal. Revue d'Ecologie (Terre et Vie) 30, 3–30.

Gautier, J.-P., 1988. Interspecific affinities among guenons as deduced from vocalization. In: Gautier-Hion, A., Bourliere, E., Gautier, J.-P., Kingdon, J. (Eds.), A Primate Radiation Evolutionary Biology of the African Guenons, Cambridge University Press, Cambridge, pp. 194–226.

Gautier, J.-P., Moysan, F., Feistner, A.T.C., Loireau, J.-N., Cooper, R.W., 1992. The distribution of Cercopithecus (lhoesti) solatus, an Endemic Guenon of Gabon. Revue d'Écologie, 47(4), 367–381.

Gautier-Hion, A., 1978. Food niches and coexistence in sympatric primates in Gabon. In: Chivers, D.J. Herbert, J. (Eds.),. In: Recent Advances in Primatology, vol. 1. Academic Press, New York, pp. 269–286.

Gautier-Hion, A., Bourliere, F., Gautier, J.-P., Kingdon, J., 1988. A Primate Radiation: Evolutionary Biology of the African Guenons. Cambridge University Press, Cambridge.

Geissmann, T., Lwin, N., Aung, S.S., et al., 2011. A new species of snub-nosed monkey, genus *Rhinopithecus* Milne-Edwards, 1872 (Primates, Colobinae), from Northern Kachin State, Northeastern Myanmar. Am. J. Primatol. 73, 96–107.

Gerald, M.S., Corri, W., Little, A.C., Kraiselburd, E., 2007. Females pay attention to female secondary sexual color: An experimental study in *Macaca mulatta*. Int. J. Primatol. 28, 1–7.

Gilbert, C.C., Arenson, J.L., Hart, J.A., et al., 2020. New skeletons of *Cercopithecus dryas* and their implications for locomotor evolution and taxonomy within the guenon radiation. Am. J. Phys. Anthropol. 171 (Suppl. 60), 100.

Gilbert, C.C., Gilissen, E., Arenson, J.L., et al., 2021. Morphological analysis of new Dryas Monkey specimens from the Central Congo Basin: Taxonomic considerations and an emended diagnosis. Am. J. Phys. Anthropol. 176, 361–389.

Gilbert, C.C., Stanley, W.T., Olson, L.E., et al., 2011. Morphological systematics of the kipunji (*Rungwecebus kipunji*) and the ontogenetic development of phylogenetically informative characters in the Papionini. J. Hum. Evol. 60, 731–745.

Gonzalez-Kirchner, J.P., de la Maza, M.S., 1996. Preliminary notes on the ecology of the Drill (Mandrillus leucophaeus) on Bioko Island, Rep. Equatorial Guinea. Garcia de Orta Ser Zool 21, 1–5.

Gumert, M.D., Malaivijitnond, S., 2012. Marine prey processed with stone tools by Burmese long-tailed macaques (*Macaca fascicularis aurea*) in intertidal habitats. Am. J. Phys. Anthropol. 149, 447–457.

Gupta, A.K., Kumar, K., 1994. Feeding ecology and conservation of the Phayre's leaf monkey *Presbytis phayrei* in northeast India. Biol. Conserv. 69, 301–306.

Hart, J.A., Detwiler, K.M., Gilbert, C.C., et al., 2012. Lesula: A new species of *Cercopithecus* monkey endemic to the Democratic Republic of Congo and implications for conservation of Congo's Central Basin. PLoS One 7 (9), e44271.

Hasan, M.A.U., Khatun, M.U.H., Neha, S.A., 2018. Group size, composition and conservation challenges of capped langur (*Trachypithecus pileatus*) in Satchari National Park, Bangladesh. Jagannath Univ. J. Life Earth Sci. 4, 135–153.

Henzi, S.P., Lucas, J.W., 1980. Observations on the inter-troop movement of adult vervet monkeys (*Cercopithecus aethiops*). Folia Primatol. 33, 220–235.

Higham, J.P., Hughes, K.D., Brent, L.J.N., et al., 2011. Familiarity affects the assessment of female facial signals of fertility by free-ranging male rhesus macaques. Proc. Roy. Soc. B 278, 3452–3458.

Hill, C.M., 1994. The role of female diana monkeys, *Cercopithecus diana*, in territorial defence. Anim. Behav. 47 (2), 425–431.

Hongo, S., Nakashima, Y., Akomo-Okoue, E.F., Mindonga-Nguelet, F.L., 2018. Seasonal change in diet and habitat use in wild mandrills (*Mandrillus sphinx*). Int. J. Primatol. 39 (1), 27–48.

Hoshino, J., Mori, A., Kudo, H., Kawai, M., 1984. Preliminary report on the grouping of mandrills (*Mandrillus sphinx*) in Cameroon. Primates 25, 295–307.

Hrdy, S.B., 1976. Care and exploitation of nonhuman primate infants by conspecifics other than the mother. In: Rosenblatt, J.S. Hinde, R.A. Shaw, E. Beer, C. (Eds.), Advances in the Study of Behavior, Vol. 6. Academic Press, New York, pp. 101–158.

Hrdy, H.S., 1977. The Langurs of Abu. Harvard University Press, Cambridge, Massachusetts.

Isbell, L.A., Cheney, D.L., Seyfarth, R.M., 2002. Why vervet monkeys (*Cercopithecus aethiops*) live in multimale groups. In: Glenn, M.E., Cords, M. (Eds.), The guenons: Diversity and Adaptation in African Monkeys. Springer, Boston, MA, pp. 173–187.

Isbell, L.A., Pruetz, J.D., 1998. Differences between vervets (*Cercopithecus aethiops*) and patas monkeys (*Erythrocebus patas*) in agonistic interactions between adult females. Int. J. Primatol. 19, 837–855.

Jarvey, J.C., Low, B.S., Pappano, D.J., et al., 2018. Graminivory and fallback foods: Annual diet profile of geladas (*Theropithecus gelada*) living in the Simien Mountains National Park, Ethiopia. Int. J. Primatol. 39 (1), 105–126.

Jensen, A., Swift, F., de Vries, D., et al., 2023. Complex evolutionary history with extensive ancestral gene flow in an African primate radiation. Mol. Biol. Evol. 40(12):msad247. https://doi.org/10.1093/molbev/msad247

Jolly, C.J., 1993. Species, subspecies, and baboon systematics. In: Kimbel, W.H., Martin, L.B. (Eds.), Species, Species Concepts, and Primate Evolution. Plenum Press, New York, pp. 67–107.

Jolly, C.J., 2001. A proper study for mankind: Analogies from the Papionin monkeys and their implications for human evolution. Yearb. Phys. Anthropol. 44, 177–204.

Jolly, C.J., 2020. Philopatry at the frontier: A demographically driven scenario for the evolution of multilevel societies in baboons (*Papio*). J. Hum. Evol. 146, 102819. https://doi.org/10.1016/j.jhevol.2020.102819

Jolly, C.J., Burrell, A.S., Phillips-Conroy, J.E., et al., 2011. Kinda baboons (*Papio kindae*) and grayfoot chacma baboons (*P. ursinus griseipes*) hybridize in the Kafue river valley, Zambia. Am. J. Primatol. 73, 291–303.

Jones, T., Ehardt, C.L., Butynski, T.M., et al., 2005. The highland mangabey *Lophocebus kipunji*: A new species of African monkey. Science 308, 1161–1164.

Kaplin, B.A., Moermond, T.C., 1998. Variation in seed handling by two species of forest monkeys in Rwanda. Am. J. Primatol. 45 (1), 83–101.

Kaplin, B.A., Moermond, T.C., 2000. Foraging ecology of the mountain monkey (*Cercopithecus l'hoesti*): Implications for its evolutionary history and use of disturbed forest. Am. J. Primatol. 50 (4), 227–246.

Kappeler, P.M., Pereira, M.E., 2003. Primate Life Histories and Socioecology. University of Chicago Press, Chicago.

Kay, R.F., 1978. Molar structure and diet in extant Cercopithecoidae. In: Butler, P.M., Joysey, K.A. (Eds.), Development, Function and Evolution of Teeth. Academic Press, New York, pp. 309–339.

Kifle, Z., Bekele, A., 2021. Feeding ecology and diet of the southern geladas (*Theropithecus gelada obscurus*) in human-modified landscape, Wollo, Ethiopia. Ecol. Evol. 11 (16), 11373–11386.

Kingdon, J., 1971. East African Mammals, vol. 1. Academic Press, New York.

Kingdon, J. 2013. Mammals of Africa: Volume II Primates. In: Butynski, T., Kingdon, J., Kalina, J. (Eds.), Bloomsbury Publishing, London, p. 562.

Kirkpatrick, R.C., 2011. The Asian colobines: Diversity among leaf-eating monkeys. In: Campbell, C., Fuentes, A., MacKinnon, K., et al., (Eds.), Primates in Perspective, second ed. Oxford University Press, Oxford, pp. 189–202.

Kirkpatrick, R.C., Grueter, C.C., 2010. Snub-nosed monkeys: Multilevel societies across varied environments. Evol. Anthropol. 19, 98–113.

Koenig, A., Borries, C., 2001. Socioecology of Hanuman langurs: The story of their success. Evol. Anthropol. 10 (4), 122–137.

Koenig, A., Borries, C., 2012. Social organization and male residence pattern in Phayre's leaf monkeys. In: Kappeler, P.M., Watts, D.P. (Eds.), Long-term field studies of primates. Springer, Berlin, Heidelberg, pp. 215–236.

Kuderna, L.F.K., Gao, H., Janiak, M.C., et al., 2023. A global catalog of whole-genome diversity from 233 primate species. Science 380, 906–913.

Kumar, A., Solanki, G.S., 2014. Role of mother and allomothers in infant independence in capped langur Trachypithecus pileatus. J. Bombay Nat. Hist. Soc. 111 (1), 3–9.

Kummer, H., 1968. Social Organization of Hamadryas Baboons. University of Chicago Press, Chicago.

Laws, J.W., Vonder Haar Laws, J., 1984. Social interactions among adult male langurs (Presbytis entellus) at Rajaji Wildlife Sanctuary. Int. J. Primatol. 5, 31–50.

Le Roux, A., Beehner, J.C., Bergman, T.J., 2011. Female philopatry and dominance patterns in wild geladas. Am. J. Primatol. 73 (5), 422–430.

Lee, P.C., 1983. Context-specific unpredictability in dominance interactions. In: Hinde, R.A. (Ed.), Primate Social Relationships: An Integrated Approach. Blackwell, Oxford, pp. 35–44.

Li, Y.-C., Liu, F., He, X.-Y., et al., 2015. Social organization of Shortridge's capped langur (Trachypithecus shortridgei) at the Dulongjian Valley in Yunnan, China. Zool. Res. 36, 152–160.

Li, J.H., Yin, H., Wang, Q.S., 2005. Seasonality of reproduction and sexual activity in female Tibetan macaques Macaca thibetana at Huangshan, China. Acta Zool. Sinica 51(3), 365–375.

Linder, J.M., Cronin, D.T., Ting, N., et al. 2021. Red Colobus (Piliocolobus) Conservation Action Plan 2021–2026. IUCN, Gland, Switzerland.

Lo Bianco, S., Masters, J.C., Sineo, L., 2017. The evolution of the Cercopithecini: a (post)modern synthesis. Evol. Anthropol. 26, 336–349.

MacKinnon, J.R., MacKinnon, K.S., 1980. Niche differentiation in a primate community. In: Chivers, D.J. (Ed.), Malayan Forest Primates: Ten Years' Study in Tropical Rain Forest. Springer, New York, pp. 167–190.

MacKinnon, J., MacKinnon, K., 1987. Conservation status of the primates of the Indo-Chinese subregion. Primate Conserv. 8, 187–195.

Markham, A.C., Gesquiere, L.R., Alberts, S.C., Altmann, J., 2015. Optimal group size in a highly social mammal. PNAS 112, 14882–14887. https://doi.org/10.1073/pnas.1517794112

Matsuda, I., Kubo, T., Tuuga, A., Higashi, S., 2010. A Bayesian analysis of the temporal change of local density of proboscis monkeys: Implications for environmental effects on a multilevel society. Am. J. Phys. Anthropol. 142 (2), 235–245.

Matsuda, I., Murai, T., Clauss, M., et al., 2011. Regurgitation and remastication in the foregut-fermenting proboscis monkey (Nasalis larvatus). Biol. Lett. 7, 786–789.

Matsuda, I., Tuuga, A., Akiyama, Y., Higashi, S., 2008. Selection of river crossing location and sleeping site by proboscis monkeys (Nasalis larvatus) in Sabah, Malaysia. Am. J. Primatol. 70, 1097–1101.

Matsuda, I., Tuuga, A., Higashi, S., 2009. The feeding ecology and activity budget of proboscis monkeys. Am. J. Primatol. 71, 478–492.

Matsuda, I., Ihobe, H., Tashiro, Y., et al., 2020. The diet and feeding behavior of the black-and-white colobus (Colobus guereza) in the Kalinzu Forest, Uganda. Primates 61, 473–484.

McCabe, G.M., Emery Thompson, M., 2013. Reproductive seasonality in wild Sanje mangabeys (Cercocebus sanjei), Tanzania: Relationship between the capital breeding strategy and infant survival. Behaviour 150, 1399–1429.

McGraw, W.S., 1998. Comparative locomotion and habitat use of six monkeys in the Tai Forest, Ivory Coast. Am. J. Phys. Anthropol. 105, 493–510.

McGraw, W.S., Zuberbuhler, K., 2008. Socioecology, predation and cognition in a community of West African monkeys. Evol. Anthropol. 17, 254–266.

Meijaard, E., Groves, C.P., 2004. The biogeographic evolution and phylogeny of the genus Presbytis. Primate Rep. 68, 71–90.

Mekonnen, A., Bekele, A., Fashing, P.J., et al., 2010. Diet, activity patterns, and ranging ecology of the Bale monkey (Chlorocebus djamdjamensis) in Odobullu Forest, Ethiopia. Int. J. Primatol. 31, 339–363.

Miller, A., Judge, D., Uwingeneye, G., et al., 2020. Diet and use of fallback foods by Rwenzori Black-and-White Colobus (Colobus angolensis ruwenzorii) in Rwanda: Implications for supergroup formation. Int. J. Primatol. 41, 434–457.

Modolo, L., Salzburger, W., Martin, R.D., 2005. Phylogeography of Barbary macaques (Macaca sylvanus) and the origin of the Gibraltar colony. Proc. Natl. Acad. Sci. U S A 102, 7392–7397.

Molur, S., Brandon-Jones, D., Dittus, W., et al., 2003. Status of South Asian Primates. Conservation Assessment and Management Plan (C.A.M.P.) Workshop report. Coimbatore, India.

Nadler, T., Momberg, F., Dang, F., Lormee, N., 2003. Leaf Monkeys: Vietnam Primate Conservation Status Review 2002, Part 2. Fauna and Flora International, Asia Pacific Programme Office, Hanoi, Vietnam.

Newton, P.N., 1988. The variable social organization of Hanuman langurs (Presbytis entellus), infanticide, and the monopolization of females. Int. J. Primatol. 9, 59–77.

Noë, R., Bshary, R., 1997. The formation of red colobus-diana monkey associations under pressure from chimpanzees. Proc. Roy. Soc. B 264, 253–259.

Oates, J.F., Whitesides, G.H., 1990. Association between olive colobus (Procolobus verus), Diana guenons (Cercopithecus diana), and other forest monkeys in Sierra Leone. Am. J. Primatol. 21, 129–146.

Owens, J.R., Honarvar, S., Nessel, M., Hearn, G.W., 2015. From frugivore to folivore: Altitudinal variations in the diet and feeding ecology of the Bioko Island drill (Mandrillus leucophaeus poensis). Am. J. Primatol. 77 (12), 1263–1275.

Patzelt, A., Kopp, G.H., Ndao, I., et al., 2014. Male tolerance and male-male bonds in a multilevel primate society. PNAS 111, 14740–14745. https://doi.org/10.1073/pnas.1405811111

Perelman, P., Johnson, W.E., Roos, C., et al., 2011. A molecular phylogeny of living primates. PLoS Genet. 7 (3), e1001342.

Petersdorf, M., Weyher, A.H., Kamilar, J.M., et al., 2019. Sexual selection in the Kinda baboon. J. Hum. Evol. 135, 102635.

Phiapalath, P., Borries, C., Suwanwaree, P., 2011. Seasonality of group size, feeding, and breeding in wild red-shanked douc langurs (Lao PDR). Am. J. Primatol. 73, 1134–1144.

Raboin, D.L., Baden, A.L., Rothman, J.M., 2021. Maternal feeding benefits of allomaternal care in black-and-white colobus (Colobus guereza). Am. J. Primatol. 83 (10), e23327.

Range, F., Forderer, T., Storrer-Meystre, Y., et al., 2007. The structure of social relationships among sooty mangabeys in Taï. In: McGraw, W.S., Zuberbühler, K., Noë, R. (Eds.), Monkeys of the Taï Forest: An African Primate Community. Cambridge University Press, Cambridge, pp. 109–132.

Range, F., Noë, R., 2002. Familiarity and dominance relations among female sooty mangabeys in the Taï National Park. Am. J. Primatol. 56, 137–153.

Rawson, B.M., 2009. The socio-ecology of the black-shanked douc (Pygathrix nigripes) in Mondulkiri Province, Cambodia. PhD thesis. Australian National University, Canberra.

Richard, A.F., Goldstein, S.J., Dewar, R.E., 1989. Weed macaques: The evolutionary implications of macaque feeding ecology. Int. J. Primatol. 10, 569–594.

Riley, E., 2010. The endemic seven: Four decades of research on the Sulawesi macaques. Evol. Anthropol. 19, 22–36.

Riley, E., Sagnotti, C., Carosi, M., Oka, N.P., 2014. Socially tolerant relationships among wild male moor macaques (*Macaca maura*). Behaviour 151, 1021–1044.

Rogers, J., Raveendram, M., Harris, R.A., et al., 2019. The comparative genomics and complex population history of *Papio* baboons. Sci. Adv. 5. https://doi.org/10.1126/sciadv.aau6947

Roos, C., Knauf, S., Chuma, I.S., et al., 2021. New mitogenomic lineages in *Papio* baboons and their phylogeographic implications. Am. J. Phys. Anthropol. 174, 407–417.

Roos, C., Zinner, D., 2022. Molecular phylogeny and phylogeography of colobines. In: Matsuda, I., Grueter, C.C., Teichroeb, J.A. (Eds.), The Colobines: Natural History, Behaviour and Ecological Diversity. Cambridge University Press, Cambridge, pp. 32–43.

Roos, C., Zinner, D., Kubatlo, L.S., et al., 2011. Nuclear versus mitochondrial DNA: Evidence for hybridization in colobine monkeys. BMC Evol. Biol. 11, 77.

Rovero, F., Marshall, A.R., Jones, T., Perkin, A., 2009. The primates of the Udzungwa Mountains: Diversity, ecology and conservation. J. Anthropol. Sci. 87, 93–126.

Schreier, A.L., Swedell, L., 2009. The fourth level of social structure in a multi-level society: Ecological and social functions of clans in hamadryas baboons. Am. J. Primatol. 71, 948–955.

Schultz, A.H., 1970. The comparative uniformity of the Cercopithecoidea. In: Napier, J.R., Napier, P.H. (Eds.), Old World Monkeys. Academic Press, New York, pp. 39–52.

Setchell, J.M., Knapp, L.A., Wickings, E.J., 2006. Violent coalitionary attack by female mandrills against an injured alpha male. Am. J. Primatol. 68, 411–418.

Seyfarth, R.M., Cheney, D.L., Marler, P., 1980. Monkey responses to three different alarm calls: Evidence of predator classification and semantic communication. Science 210, 801–803.

Shea, B., 1992. Ontogenetic scaling of skeletal proportions in the talapoin monkey. J. Hum. Evol. 23, 283–307.

Singleton, M., Seitelman, B.C., Krecioch, J.R., Frost, S.R., 2017. Cranial sexual dimorphism in the kinda baboon (*Papio hamadryas kindae*). Am. J. Phys. Anthropol. 164 (4), 665–678.

Snyder-Mackler, N., Alberts, S.C., Bergman, T.J., 2014. The socio-genetics of a complex society: Female gelada relatedness patterns mirror association patterns in a multilevel society. Mol. Ecol. 23, 6179–6191.

Snyder-Mackler, N., Beehner, J.C., Bergman, T.J., 2012. Defining higher levels in the multilevel societies of geladas (*Theropithecus gelada*). Int. J. Primatol. 33, 1054–1068.

Solanki, G.S., Kumar, A., Sharma, B.K., 2007. Reproductive strategies of *Trachypithecus pileatus* in Arunachal Pradesh, India. Int. J. Primatol. 28 (5), 1075–1083.

Stanford, C.B., 1991. The diet of the capped langur (*Presbytis pileata*) in a moist deciduous forest in Bangladesh. Int. J. Primatol. 12, 199–216.

Stead, S.M., Badescu, I., Raboin, D.L., Sciotte, P., 2021. High levels of infant handling by adult males in Rwenzori Angolan colobus (*Colobus angolensis ruwenzorii*) compared to two closely related species, *C. guereza* and *C. vellerosus*. Primates 62, 637–646.

Stead, S.M., Teichroeb, J.A., 2019. A multi-level society comprised of one-male and multi-male core units in an African colobine (*Colobus angolensis ruwenzorii*). PLoS One 14 (10), e0217666. https://doi.org/10.1371/journal.pone.0217666

Struhsaker, T.T., 1967a. Social structure among vervet monkeys (*Cercopithecus aethiops*). Behaviour. 29, 83–121.

Struhsaker, T.T., 1967b. Auditory communication among vervet monkeys (*Cercopithecus aethiops*). In: Altma, S. (Ed.), Social Communication Among Primates. University of Chicago Press, Chicago, pp. 281–324.

Swedell, L., 2011. African papionins: diversity of social organization and ecological flexibility. In: Campbell, C., Fuentes, A., MacKinnon, K., et al., (Eds.), Primates in Perspective, second ed. Oxford University Press, New York, pp. 241–277.

Swedell, L., Plummer, T., 2019. Social evolution in Plio-Pleistocene hominins: Insights from hamadryas baboons and paleoecology. J. Hum. Evol. 137, 102667.

Tenaza, R.R., 1989. Intergroup calls of male pig-tailed langurs (*Simias concolor*). Primates 30, 199–206.

Tenaza, R.R., Fuentes, A., 1995. Monandrous social organization of pigtailed langurs (*Simias concolor*) in the Pagai Islands, Indonesia. Int. J. Primatol. 16, 295–310.

Thierry, B., 2011. The Macaques, A Double Layered Social Organization. Oxford Press, New York.

Tilson, R.L., Tenaza, R.R., 1976. Monogamy and duetting in an Old World monkey. Nature 263, 230–231.

Ting, N., 2008. Mitochondrial relationships and divergence dates of the African colobines: Evidence of Miocene origins for the living colobus monkeys. J. Hum. Evol. 55, 312–325.

Ting, N., Tosi, A.J., Li, Y., Zhang, Y.-P., Disotell, T.R., 2008. Phylogenetic incongruence between nuclear and mitochondrial markers in the Asian colobines and the evolution of the langurs and leaf monkeys. Mol. Phylogenet. Evol. 46, 466–474.

Tosi, A.J., Detwiler, K.M., Disotell, T.R., 2005. X-chromosomal window into the evolutionary history of the guenons (Primates: Cercopithecini). Mol. Phylo. Evol. 36, 58–66.

Tung, J., Berreiro, L.B., 2017. The contribution of admixture to primate evolution. Curr. Opin. Genet. Dev. 47, 61–68.

Van der Valk, T., Gonda, C.M., Silegowa, H., et al., 2020. The genome of the endangered dryas monkey provides new insights into the evolutionary history of the vervets. Mol. Biol. Evol. 37, 182–194.

Waitt, C., Little, A.C., Wolfensohn, S., et al., 2003. Evidence from rhesus macaques suggests that male coloration plays a role in female primate mate choice. Biol. Lett. 7, S144–S146.

Wang, X.P., Yu, L., Roos, C., 2012. Phylogenetic relationships among colobine monkeys revisited: New insights from analyses of complete mt genomes and 44 nuclear non-coding markers. PLoS One 7, e36274.

Watanabe, K., 1981. Variations in group composition and population density of the two sympatric Metawaian leaf monkeys. Primates 22, 145–160.

Weibel, C.J., Tung, J., Alberts, S.C., Archie, E.A., 2020. Accelerated reproduction is not an adaptive response to early-life adversity in wild baboons. Proc. Natl. Acad. Sci. U S A 117 (40), 24909–24919.

Whittaker, D.J., 2006. A conservation action plan for the Mentawai primates. Primate Conserv. 20, 95–105.

Whitten, P.L., 1983. Diet and dominance among female vervet monkeys (*Cercopithecus aethiops*). Am. J. Primatol. 5, 139–159.

Wild, C., Morgan, B.J., Dixson, A.F., 2005. Conservation of drill populations in Bakossiland, Cameroon: Historical trends and current status. Int. J. Primatol. 26, 759–773.

Wood, K., 2007. Life-history and behavioural characteristics of a semi-wild population of drills (*Mandrillus leucophaeus*) in Nigeria. Ph.D. dissertation. University of Massachusetts, Boston.

Workman, C., 2010. Diet of the Delacour's Iangur (*Trachypithecus delacouri*) in Van Long Nature Reserve, Vietnam. Am. J. Primatol. 72, 317–324.

Xing, J., Wang, H., Zhang, Y., et al., 2007. A mobile element-based evolutionary history of guenons (tribe Cercopithecini). BMC Biol. 5, 5. https://doi.org/10.1186/1741-7007-5-5

Yeager, C.P., 1989. Feeding ecology of the proboscis monkey (*Nasalis larvatus*). Int. J. Primatol. 10, 497–530.

Yeager, C.P., 1991a. Possible antipredator behavior associated with river crossing by proboscis monkeys (*Nasalis larvatus*). Am. J. Primatol. 24, 61–66.

Yeager, C.P., 1991b. Proboscis monkey (*Nasalis larvatus*) social organization: Intergroup patterns of association. Am. J. Primatol. 23, 73–86.

Zhang, D., Xia, D.-P., Wang, X., Zhang, Q.-X., Sun, B.-H., Li, J.-H., 2018. Bridging may help young female Tibetan macaques Macaca thibetana learn to be a mother. Sci. Rep. 8, 16102. https://doi.org/10.1038/s41598-018-34406-7

Zhou, Q., Wei, H., Tang, H., Huang, Z., Krzton, A., Huang, C., 2014. Niche separation of sympatric macaques, *Macaca assamensis* and *M. mulatta*, in limestone habitats of Nonggang, China. Primates 55, 125–137.

Zinner, D., Arnold, M.L., Roos, C., 2009a. Is the new primate genus *Rungwecebus* a baboon? PLoS ONE 4 (3), e4859.

Zinner, D., Chuma, I.S., Knauf, S., Roos, C., 2018. Inverted intergeneric introgression between critically endangered kipunjis and yellow baboons in two disjunct populations. Biol. Lett. 14, 20170729. https://doi.org/10.1098/rsbl.2017.0729

Zinner, D., Groeneveld, L.F., Keller, C., Roos, C., 2009b. Mitochondrial phylogeography of baboons (*Papio* spp.) – Indication for introgressive hybridization? BMC Evol. Biol. 9, 83.

Zinner, D., Arnold, M.L., Roos, C., 2011. The strange blood: Natural hybridization in primates. Evol. Anthropol. 20, 96–103.

Zinner, D., Wertheimer, J., Liedigk, R., et al., 2013. Baboon phylogeny as inferred from complete mitochondrial genomes. Am. J. Phys. Anthropol. 150, 133–140.

Further Reading

Ashmore, P.C., Hart, D., Shaffer, C.A., 2022. Chapter 17: Macaques: Ecological plasticity in primates. In: Sussman, R.W., Hart, D., Colquhoun, I.C. (Eds.), The Natural History of Primates: A Systematic Survey of Ecology and Behavior. Rowan and Littlefield, Lanham, pp. 429–480.

Cords, M., 2012. The behavior, ecology, and social evolution of cercopithecine monkeys. In: Mitani, J.C., Call, J., Kappeler, P.M., et al., (Eds.), The Evolution of Primate Societies. University of Chicago Press, Chicago, pp. 91–112.

Davies, A.G., Oates, J.F., 1994. Colobine Monkeys: Their Ecology, Behavior and Evolution. Cambridge University Press, Cambridge.

Fashing, P.J., 2011. African colobine monkeys: Their behavior, ecology, and conservation. In: Campbell, C., Fuentes, A., MacKinnon, K., et al., (Eds.), Primates in Perspective, second ed. Oxford University Press, Oxford, pp. 203–229.

Jaffe, K.E., Isbell, L.A., 2011. The guenons: Polyspecific associations in socioecological perspective. In: Campbell, C., Fuentes, A., MacKinnon, K., et al. (Eds.), Primates in Perspective, second ed. Oxford University Press, New York, pp. 277–299.

Kirkpatrick, R.C., 2011. The Asian colobines: Diversity among leaf-eating monkeys. In: Campbell, C., Fuentes, A., MacKinnon, K., et al., (Eds.), Primates in Perspective, second ed. Oxford University Press, Oxford, pp. 189–202.

Lambert, J.E., Sussman, R.W., Lyke, M.M., 2022. Chapter 14: Ecology and behavior of the arboreal African guenons. In: Sussman, R.W., Hart, D., Colquhoun, I.C. (Eds.), The Natural History of Primates: A Systematic Survey of Ecology and Behavior. Rowan and Littlefield Lanham, pp. 335–354.

Matsuda, I., Grueter, C.C., Teichroeb, J.A., 2022. The Colobines: Natural History, Behaviour and Ecological Diversity. Cambridge University Press, Cambridge.

Nystrom, P., 2022. Chapter 16: Mangabeys, mandrills, drills, baboons, and geladas. In: Sussman, R.W., Hart, D., Colquhoun, I.C. (Eds.), The Natural History of Primates: A Systematic Survey of Ecology and Behavior. Rowan and Littlefield, Lanham, pp. 380–428.

Pruetz, J.D., Sussman, R.W., Cramer, J.D., 2022. Chapter 15: Behavioral ecology of Patas monkeys and vervets. In: Sussman, R.W., Hart, D., Colquhoun, I.C. (Eds.), The Natural History of Primates: A Systematic Survey of Ecology and Behavior. Rowan and Littlefield, Lanham, pp. 355–379.

Sterck, E.H.M., 2012. The behavioral ecology of colobine monkeys. In: Mitani, J.C., Call, J., Kappeler, P.M., et al., (Eds.), The Evolution of Primate Societies. University of Chicago Press, Chicago, pp. 65–90.

Struhsaker, T.T., 2010. The Red Colobus Monkeys. Oxford University Press, Oxford.

Swedell, L., 2011. African papionins: Diversity of social organization and ecological flexibility. In: Campbell, C., Fuentes, A., MacKinnon, K., et al., (Eds.), Primates in Perspective, second ed. Oxford University Press, New York, pp. 241–277.

Swedell, L., McGraw, W.S., 2022. Chapter 19: Regarding Old World monkeys. In: Sussman, R.W., Hart, D., Colquhoun, I.C. (Eds.), The Natural History of Primates: A Systematic Survey of Ecology and Behavior. Rowan and Littlefield, Lanham, p.526.

Thierry, B., 2011. The macaques, a double layered social organization. In: Campbell, C., Fuentes, A., MacKinnon, K., Bearder, S., Stumpf, R. (Eds.), Primates in Perspective, second ed. Oxford University Press, New York, pp. 229–240.

Vandercone, R., Coudrat, C., Ormond, P.T., 2022. Chapter 18: The ecology and social structure of the Asian Colobines. In: Sussman, R.W., Hart, D., Colquhoun, I.C. (Eds.), The Natural History of Primates: A Systematic Survey of Ecology and Behavior. Rowan and Littlefield, Lanham, pp. 481–525.

Whitehead, P.F., Jolly, C.J., 2000. Old World Monkeys. Cambridge University Press, Cambridge.

Wikberg, E.C., Kelley, E., Sussman, R.W., Ting, N., 2022. Chapter 13: The African colobines: Behavioral flexibility and conservation in a changing world. In: Sussman, R.W., Hart, D., Colquhoun, I.C. (Eds.), The Natural History of Primates: A Systematic Survey of Ecology and Behavior. Rowman and Littlefield, Lanham.

7

Apes and Humans

Hominoids are the less taxonomically diverse group of living catarrhines. There are only eight genera of extant hominoids. They range in size from about 4 kg for the smallest gibbons to over 200 kg for male gorillas. With the notable exception of our own species, hominoids have a rather restricted distribution: the tropical forests of Africa and Southeast Asia (Fig. 7.1). As we will see in later chapters, in earlier times hominoids were much more taxonomically diverse and widespread. Humans are the only hominoids that occur naturally in the New World, despite persistent rumors of Sasquatch or other ape-like animals from the wooded parts of North America. However, both humans and chimpanzees have been sighted from time to time in extraterrestrial environments.

Hominoids

Hominoids are distinguished from Old World monkeys by a variety of both primitive catarrhine features and unique specializations (Fig. 6.1). Like cercopithecoids, all living hominoids have a tubular ectotympanic bone and a dental formula of 2.1.2.3. Compared to Old World monkeys, hominoids have relatively primitive molar teeth, with rounded cusps rather than the bilophodont crests of monkeys. The lower molars are characterized by an expanded talonid basin surrounded by five main cusps. The upper molars are quadrate and have a distinct trigon anteriorly and a large hypocone posteriorly. The anterior lower premolar varies in shape from an elongate shearing blade in gibbons to a bicuspid tooth in humans. Most hominoids have relatively broad incisors. Hominoid canines are much more variable than those of cercopithecoids in both shape and degree of sexual dimorphism.

Hominoids are characterized by relatively broad palates, broad nasal regions, and large brains. Hominoid skeletons show a variety of distinctive features (Fig. 7.2). The axial skeleton is characterized by a reduced lumbar region, an expanded sacrum, and the absence of a tail.

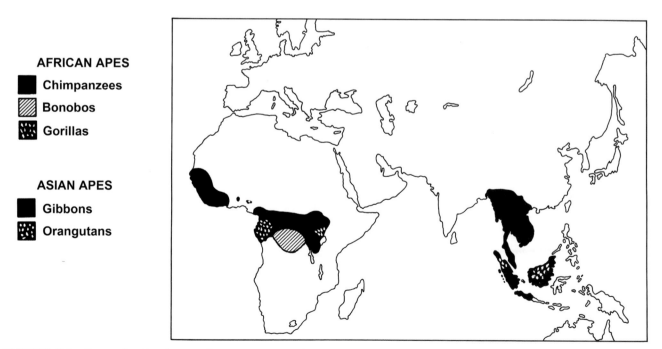

AFRICAN APES
- Chimpanzees
- Bonobos
- Gorillas

ASIAN APES
- Gibbons
- Orangutans

FIGURE 7.1 Geographic distribution of extant apes. Map lines delineate study areas and do not necessarily depict accepted national boundaries.

Primate Adaptation and Evolution
https://doi.org/10.1016/B978-0-12-815809-8.00007-2

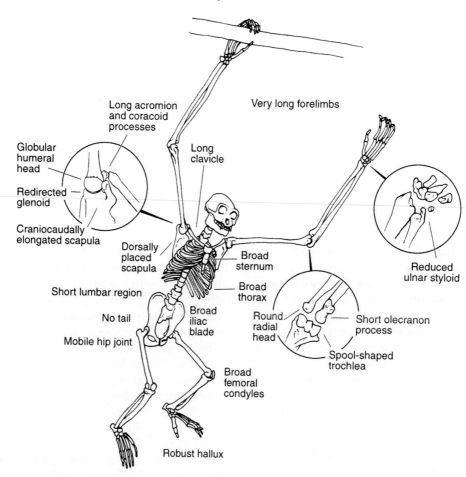

FIGURE 7.2   Characteristic skeletal features of extant apes, illustrated by a siamang.

All hominoids have a relatively broad thorax with a dorsally positioned scapula. Hominoids have relatively long upper limbs, and their elbow joint is characterized by a spool-shaped trochlea on the humerus and a short olecranon process on the ulna. Their wrist lacks an articulation between the ulna and the carpal bones; instead, a fibrous meniscus separates the two bones. The hindlimbs of hominoids are characterized by a broad ilium, broad femoral condyles, and usually a large, robust hallux. Many of the features of the hominoid skeleton are related to their more suspensory behavior compared to Old World monkeys (see Chapter 9). However, the extent to which many of the skeletal similarities among extant hominoids are descended from a common suspensory ancestor or were to some degree evolved independently is a debated issue that is impossible to resolve on current evidence (e.g., Larson, 1998; Young, 2003: Fleagle and Lieberman, 2013).

Hominoids are quite different from Old World monkeys in many aspects of their behavior and ecology, as well as their life history. However, it is unclear to what extent these are derived hominoid features or primitive anthropoid features. In general, if similar-sized animals can be compared, hominoids have longer gestation and longer time to first reproduction than cercopithecoids. Hominoid social organization is characterized by a general absence of female philopatry, and their societies are not organized around female matrilines.

The eight hominoid genera are placed in two families: hylobatids and hominids. Humans and African apes are more closely related to each other than they are to orangutans. The phyletic relationships among hominoids are discussed later in the chapter (see also Chapter 1).

Hylobatids

The gibbons, from Southeast Asia (Fig. 7.3), are the smallest, the most taxonomically diverse, and the most numerically successful of living apes (Table 7.1). These lesser apes are anatomically the most primitive of living apes and retain many monkey-like features, but in some aspects, such as their limb proportions, they are the most specialized of the living hominoids. The numerous gibbon species are relatively uniform in morphology. All are

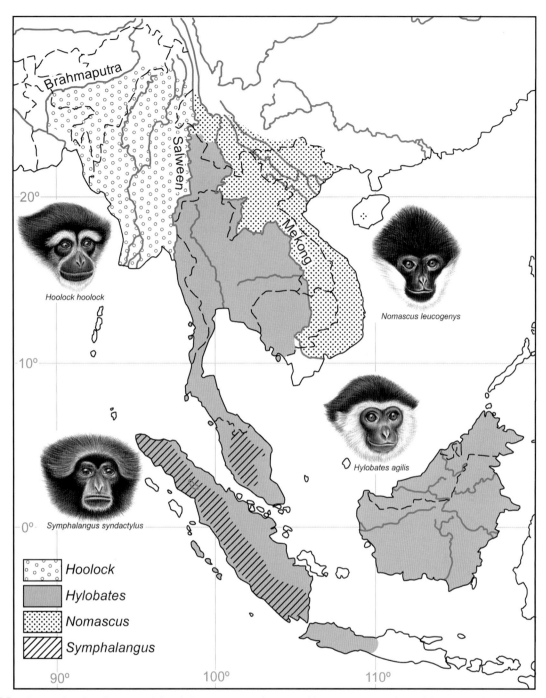

FIGURE 7.3 Geographic distribution and facial characteristics of extant gibbon genera. Map lines delineate study areas and do not necessarily depict accepted national boundaries.

relatively small (5–11 kg), with no sexual size dimorphism. They have simple molar teeth characterized by low, rounded cusps, and broad basins (Fig. 7.4). Their incisors are relatively short but broad. Both sexes have long, dagger-like canines and blade-like lower anterior premolars for sharpening the upper canine.

Gibbons have short snouts and shallow faces, large orbits with protruding rims, and a wide interorbital distance. Their braincase is globular and has no nuchal

cresting. Only occasionally do they develop a sagittal crest. The mandible is shallow and has a broad ascending ramus.

Gibbons are outstanding among living primates in their limb proportions (Figs. 7.2 and 7.5). They have the longest forelimbs relative to body size of any living primates. They have long, curved, slender digits on their hands and feet as well as a long muscular pollex and hallux. Gibbons are the only apes that consistently have

TABLE 7.1   Infraorder Catarrhini
             Family HYLOBATIDAE

Common name	Species	Intermembral index	Mass (g) M	Mass (g) F
Siamang	*Symphalangus syndactylus*	147	11,900	10,700
Western hoolock gibbon	*Hoolock hoolock*	129	6870	6880
Eastern hoolock gibbon	*H. leuconedys*	–	–	–
Skywalker hoolock gibbon	*H. tianxing*	–	–	–
Agile gibbon	*Hylobates agilis*	129	5880	5820
Bornean white-bearded gibbon	*H. albibarbis*	128	5450	6450
Kloss's gibbon	*H. klossii*	126	5670	5920
Lar gibbon	*H. lar*	130	5900	5340
Silvery gibbon	*H. moloch*	127	6580	6250
Mueller's gibbon	*H. muelleri*	129	5710	5350
Abbott's gray gibbon	*H. abbotti*	129	6150	5750
Eastern Bornean gray gibbon	*H. funereus*	129	5570	5250
Pileated gibbon	*H. pileatus*	–	5500	5440
Western black-crested gibbon	*Nomascus concolor*	140	7790	7620
Southern yellow-cheeked crested gibbon	*N. gabriellae*	144	–	5233
Hainan black-crested gibbon	*N. hainanus*	–	6610	6563
Northern white-cheeked crested gibbon	*N. leucogenys*	131	7410	7320
Eastern black-crested gibbon	*N. nasutus*	–	–	–
Southern white-cheeked crested gibbon	*N. siki*	141	6575	7867
Northern yellow-cheeked crested gibbon	*N. annamensis*	–	7000	–

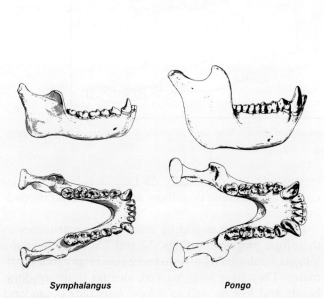

Symphalangus   **Pongo**

FIGURE 7.4   Lower jaws of a siamang (*left*) and an orangutan (*right*).

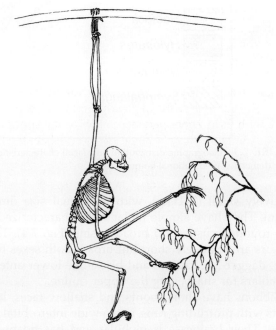

FIGURE 7.5   The skeleton of a gibbon (*Hylobates*).

ischial callosities, and females also show small sexual swellings that change shape and color during the estrous cycle.

Gibbons are found throughout the evergreen forests of Southeast Asia, from eastern India to southern China on the mainland, as well as on Borneo, Java, Sumatra, and nearby islands of the Sunda Shelf (Fig. 7.3). Gibbons are currently divided into four mostly allopatric genera, each with a different chromosome number. Phylogenetic relationships among these genera are unresolved. The **hoolock** or **white-browed gibbon** (*Hoolock*) is the westernmost genus, with a fragmented distribution ranging from the eastern parts of India and Bangladesh, through much of Myanmar, and into parts of Yunnan Province in China, with an eastern boundary at the Selween River. They are medium-sized gibbons ranging between 6 and 7 kg in size, and are sexually dichromatic. Males are all black except for a white brow, whereas females are lighter in color. Until recently, most authorities recognized two species: *Hoolock hoolock* in the west and *Hoolock leuconedys* in the east. However, the recently named skywalker hoolock gibbon (*Hoolock tianxing*) from eastern Myanmar and southwestern China has recently increased the taxonomic diversity of the genus (Fan et al., 2017).

The **crested gibbons**, genus *Nomascus*, are found east of the Selween, in southern China, including Hainan Island, Laos, Vietnam, and eastern Cambodia. Most authorities recognize between six and ten species. They are all medium-sized gibbons that are sexually dichromatic. Males are black with pale cheeks and females are buff-colored.

The genus *Hylobates*, commonly called the **lar gibbons**, range from China in the north through Thailand, southern Myanmar, eastern Cambodia, the Malay peninsula, and the islands on the Sunda Shelf, including Sumatra, the Mentawai Islands, Java, and Borneo. They are all small to medium in size. Different species have characteristic coloration patterns, but none are sexually dichromatic. However, some species may show color variations within populations. There are at least seven allopatric species with distinctive vocalizations, but there are hybrid zones between many of the species.

The **siamangs** (*Symphalangus syndactylus*) are the largest gibbons, with a body mass of over 10 kg. All individuals are completely black, but males have a scrotal tuft that females lack. Siamangs are sympatric with populations of *Hylobates* in peninsular Malaysia and on Sumatra.

The behavior and ecology of the different gibbon species is remarkably uniform considering the deep genetic differences and broad geographic range (Bartlett, 2011). However, despite this uniformity, differences in behavior and ecology have been documented between the two sympatric species, the siamang and the lar gibbon, which are found together in the forests of west Malaysia and Sumatra

(Fig. 7.6). All gibbon species show a preference for moist primary forests rather than secondary or riverine forests, but siamangs are found at higher elevations and more commonly in mountainous regions than are the sympatric lar gibbons. Gibbons move and feed mainly in the middle and upper levels of the canopy and virtually never descend to the ground. They are the most suspensory of all primates and are aerialists *par excellence*, moving almost exclusively by two-armed brachiation and by slower quadrumanous climbing (Fig. 7.7; Fleagle, 1976). The larger siamangs travel mainly by slow, pendulum-like arm-over-arm brachiation, whereas the smaller gibbon species use more rapid ricocheting brachiation in which they throw themselves from one tree to the next over gaps of 10 m or more. During feeding, all gibbons use more deliberate quadrumanous climbing when moving among small terminal branches. They use a wide variety of both seated and suspensory feeding postures.

Gibbons specialize on a diet of ripe fruit, part of which is found in small, widely scattered clumps throughout the forest and part of which occurs in large bonanzas, such as many figs (Elder, 2009). Gibbon species also eat varying amounts of new leaves and invertebrates, such as termites and arachnids. The proportions of these foods in their diet vary from season to season, from species to species, and from locality to locality. In both Malaysia and Sumatra the larger siamangs rely more on new leaves than do the smaller lar gibbons, which eat more fruit. Crested gibbons (*Nomascus*) also eat more leaves than other gibbon taxa. Kloss's gibbon, from the island of Siberut, is unusual in that it does not eat leaves, only fruit and invertebrates.

Groups of the more folivorous siamang usually forage as a unit and have smaller day ranges (1 km or less) and smaller home ranges (18–50 ha) than the smaller gibbons. The latter often forage individually, with the members of a single family separated by as many as several hundred meters, and they have larger day ranges (2 km) and home ranges. These differences in ranging patterns have been related to the different distribution of preferred foods: leaves for the siamang and scattered fruits for the gibbons. A clumped resource such as leaves can be readily exploited by a group, whereas widely scattered fruits are more easily sought out by individuals. Moreover, when feeding on clumped fruit resources, such as figs, siamangs are dominant over the smaller lar gibbons and are able to exclude them from food trees until the siamangs have had their fill.

Most gibbons live in small groups composed of one male, one female, and up to four dependent offspring. However, some populations may have more than one adult individual of either sex. Larger groups are particularly common in some siamang populations on Sumatra. In addition, extra-pair matings have been reported for many gibbon populations. Numerous hypotheses have

FIGURE 7.6 Two sympatric gibbons from west Malaysia; *upper left*, the lar gibbon (*Hylobates lar*); *below*, the siamang (*Symphalangus syndactylus*).

Brachiation

Bipedalism

Climbing

Feeding postures

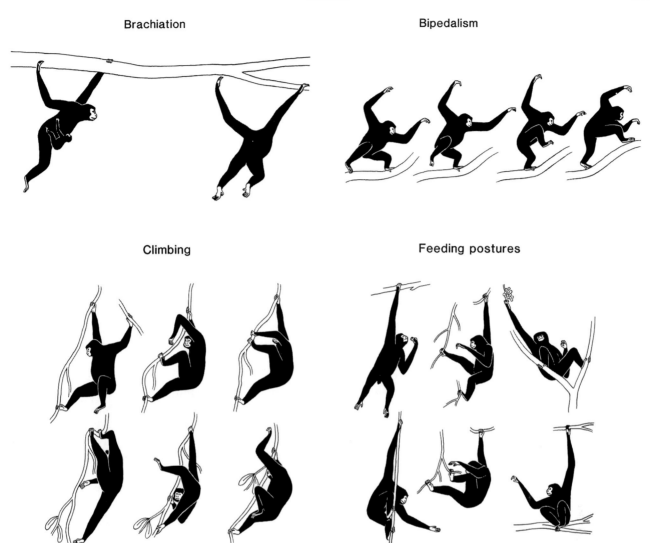

FIGURE 7.7   Locomotor behavior and feeding postures of the Malayan siamang (*Symphalangus syndactylus*).

been put forth to explain pair-bonding in gibbons, including paternal care, mate-guarding, resource defense, and infanticide defense (e.g., Bartlett, 2011; Borries et al., 2011).

All gibbons are fiercely territorial and defend their core areas with daily calling bouts and occasional intergroup conflict. Apart from vocal duetting, social interactions within a gibbon group are usually limited and consist primarily of occasional grooming bouts. Gibbons have single births every 4 or 5 years, with considerable individual variance in female reproductive success (Palombit, 1995). Siamang males carry their offspring during their second year of life. In other species, male investment is not so extensive. Young gibbons spend up to 10 years in their family group before leaving, usually after harassment by their same-sex parent. There is some evidence that gibbons may have a very long lifespan for their size. For example, one *H. muelleri* lived for 60 years in captivity (Geissmann et al., 2009).

Hominids

There are four living genera of **hominids**, the great apes and humans. Orangutans are the only great apes from Asia. Gorillas, chimpanzees, and bonobos, or pygmy chimpanzees, are from Africa, and humans are essentially cosmopolitan, although our lineage is of African origin. Great apes share many skeletal features with the lesser apes, such as relatively short trunks, the absence of a tail, a broad chest, long arms, and long hands and feet, but they are distinguished by their large size and many more detailed anatomical characteristics, including more robust canine teeth, broader premolars, a very broad ilium, and a robust fibula.

The great apes are primarily forest primates. All great apes are herbivorous and eat varying proportions of fruit and leaves. These large hominids are less committed to a suspensory lifestyle than are the lesser apes, and all are to some extent terrestrial in their habits. They are, however, more suspensory than are living cercopithecoid monkeys, and all are characterized by some use of quadrumanous climbing. Even large male gorillas regularly climb trees in some seasons to feed (Remis, 1995, 1999).

All great apes build nests for sleeping and resting, but share few unifying features in their social behavior. Each species seems to show a different pattern of social organization, sexual dimorphism, and individual grouping tendencies. In keeping with their large size, all have a relatively long lifespan and a long, slow ontogeny. All have single births and gestation periods similar to those of humans.

Orangutans

The shaggy red orangutans (*Pongo*) are the largest living Asian apes. They are currently divided into three species: *P. abelii* and *P. tapanuliensis* on Sumatra and *P. pygmaeus* on Borneo. However, orangutans had a much larger range in the past, which included Java and many parts of Southeast Asia and southern China. Orangutans are quite distinct morphologically from the African apes. Dentally, they are characterized by cheek teeth with thick enamel, low, flat cusps, and crenulated occlusal surfaces (Fig. 7.4). They have large upper central incisors and small, peg-like upper lateral incisors. The canines are large and sexually dimorphic. Orangutan crania are characterized by a relatively high, rounded braincase, poorly developed brow ridges, a deep face with small, oval orbits set close together, and a uniquely prognathic snout with a large, convex premaxilla. The mandible is deep and has a high ascending ramus.

The limbs of orangutans show extreme specializations for suspensory behavior. They have very long forelimbs and long, hook-like hands with long, curved fingers and a short pollex. Their extremely mobile hindlimbs are relatively short, and they have hand-like feet with long, curved digits and a reduced hallux. Sexual dimorphism is very odd. Females weigh approximately 40 kg, but adult males seem to come in two morphs: flanged and unflanged. Young (unflanged) males are similar to females in size, and some fully adult, reproductively active individuals retain that appearance throughout life. Other males develop broad facial flanges and grow to a body size that can be over twice that of females (Delgado and van Schaik, 2000).

Orangutans seem to prefer upland rather than lowland forest areas. Orangutan habitats in Sumatra are generally richer than those on Borneo and densities are thus higher on Sumatra (Marshall et al., 2009). Females and immature individuals are almost totally arboreal, whereas adult (especially old) males on Borneo frequently descend to the ground to travel (Ancrenaz et al., 2014; Ashbury et al., 2015). In the trees, orangutans move almost exclusively by slow quadrumanous climbing in which they use their hands and feet interchangeably as they move within tree crowns and transfer themselves from tree to tree (Fig. 7.8). On the ground they move quadrupedally, with their hands held in a fist (Tuttle, 1969). When feeding, they use both seated and suspensory postures. They frequently use their strong arms to bend or break branches to bring food to their mouth rather than change positions. Orangutans have very diverse diets and feed on twice as many plant genera as do other great apes (Knott and Kahlenberg, 2011). They eat primarily fruits (many of which contain hard seeds that they crush with their flat molars), and they also consume considerable amounts of new leaves, shoots, and bark. Choice of fruit trees in an orangutan's diet seems to be based on several variables, including the size of the available fruit crop and the amount of pulp in the fruit. They avoid tannins and seem to select fruit for its energy content rather than for its protein content (Leighton, 1993). There are differences in the diets of male and female orangutans on Borneo (Vogel et al., 2017; Schuppli et al., 2021). The ecology of fruit masts and intervening periods of fruit scarcity have a profound effect on female nutritional intake and energy balance (Knott, 1998, 1999). At one end of the spectrum, female conception is more common just after masting periods (Knott, 1999). On the other, during periods of fruit scarcity orangutan females suffer negative energy balance, which can modify female reproductive cycling (Knott, 1998). In fact, both males and females often catabolize their own fat reserves during chronic periods of fruit scarcity, despite their unusually low metabolic requirements (Vogel et al., 2012; O'Connell et al., 2021). During these lean times, orangutans are forced to fall back on low-quality foods, such as the inner cambium layer of bark (Rodman, 1977; Knott, 1999). Male orangutans eat more termites, in conjunction with their terrestrial forays. Orangutan populations show a variety of cultural differences, including aspects of tool use (van Schaik et al., 2003) and vocalizations (Wich et al., 2009).

Adult orangutans are usually solitary, but may come together in travel bands, temporary feeding aggregations, and consortships. The only consistent social group among orangutans is a female with her immature offspring. Individual females (with young) live in relatively large but stable home ranges, varying from as little as 150 ha in some forests up to 850 or more in others (Singleton and van Schaik, 2001; Singleton et al., 2009). Adult female day ranges are much smaller, approximately 500 m. The large, flanged adult males occupy much larger home ranges that overlap the ranges of several adult females and also those of other males (Singleton and van Schaik, 2001). They move much farther each day, partly in search of more food to supply their greater bulk and also to monitor the whereabouts of their female consorts and male competitors. They also interact with other orangutans through a long call,

FIGURE 7.8   The orangutan (*Pongo pygmaeus*).

which, in flanged males, is thought to play a role in mediating male development (Delgado and van Schaik, 2000). In fact, flanged males make their travel plans well in advance and use their long calls to announce them to conspecifics (van Schaik et al., 2013). Thus, even though, by other primate standards, orangutans appear to live solitary lives, it seems likely that they maintain extensive social networks of individual relationships.

The details of the relationships among adult male orangutans in the larger community, however, are not well understood. Other, unflanged adult males that have not acquired their own territory seem to have a very different ranging behavior and reproductive strategy. They forage with adult females for weeks or months at a time and forcibly mate with the usually uncooperative female. Interactions between adult male orangutans are usually aggressive, occasionally involving fierce battles, but more often only vocal exchanges. Male–female sexual encounters vary dramatically from violent, sexually coercive interactions between young males and adult females (Kunz et al., 2021) to occasional, long erotic tree-top trysts, which usually occur between older adult males and females.

The care and upbringing of young orangutans is totally the responsibility of the females. Female orangutans become sexually mature between 11 and 15 years (Galdikas, 1981; Leighton et al., 1995). Sexual maturation in male orangutans is a more variable and interesting phenomenon. They become sexually competent adults somewhere between the ages of 8 and 15 years, but individual males may or may not develop facial flanges and long calls, depending on various social factors (Dunkel et al., 2013). Orangutans can live up to 60 years in captivity, though estimates of lifespan in wild individuals are around 45 years (Leighton et al., 1995).

Gorillas

Gorillas (genus *Gorilla*) are the largest living primates and share with chimpanzees the distinction of being our closest primate relatives. Most authorities recognize two species: *Gorilla gorilla* in the west, which includes the widespread western lowland gorillas and a distinct population along the Cross River in Nigeria and Cameroon, and *Gorilla beringei* in the east, which includes the mountain gorillas and the eastern lowland gorillas of the Democratic Republic of Congo (Fig. 7.9).

FIGURE 7.9   The mountain gorilla (*Gorilla beringei*) of the Virunga volcanoes of Rwanda.

Gorillas have extreme sexual size dimorphism: females weigh 70 to 90 kg and males up to 200 kg. This dimorphism is also evident in many aspects of their skeletal anatomy, where it manifests itself in the greater general robustness of the males. The molar teeth of gorillas have a greater development of crests than those of any other hominoid, a feature associated with their high degree of folivory. They have large, tusk-like canines and relatively small incisors. Gorillas have relatively long snouts, pronounced brow ridges, and, in males, well-developed sagittal and nuchal crests.

Gorillas have relatively long forelimbs. Their hands are very broad and have a large pollex and (like all African apes) dermal ridges on the dorsal surface of their digits. Their trunk is relatively short and broad and has a wide thorax and a broad, basin-like pelvis. Their hindlimbs are relatively short, and their feet are broad. In mountain gorillas, the hallux is somewhat adducted and connected to the other digits by webbing, giving them a very human-like footprint.

Gorillas have a limited distribution in the tropical forests of sub-Saharan Africa. Mountain gorillas show a preference for secondary and herbaceous forests and are among the most terrestrial of all primates. Adult mountain gorillas rarely climb trees, and they nest on the ground (Schaller, 1963; Rothman et al., 2006a). Western gorillas are found in a wide range of forests. They are much more arboreal, especially females and youngsters, and they normally feed, rest, and build their sleeping nests both in trees and on the ground (Mehlman and Doran, 2002).

On the ground, all gorillas move by quadrupedal walking and running (Doran, 1996); like chimpanzees, gorillas have an unusual hand posture for quadrupedal standing and moving, called knuckle-walking (Fig. 7.10). Rather than support their forelimb on the palm of their hand (like most primates) or on the palmar surface of their fingers (like many baboons), they support it on the dorsal surface of the third and fourth digits of their curled hands. In the trees they are relatively good climbers, but they rarely use suspensory feeding postures. Gorillas use their great strength to good advantage when foraging, often ripping apart branches or whole trees.

Mountain gorillas have the most herbaceous diet of any living ape. They eat mainly leaves and the stems and pith of herbaceous plants in great quantities, but when available they prefer fruit (Watts, 1984, 2003; Rothman et al., 2007, 2008). Eastern lowland gorillas are much more frugivorous, and their diets are more diverse than their highland counterparts (Rogers et al., 2004; Rothman et al., 2006b), where fruit can account for more than 50% of their diet in some seasons (Masi et al., 2009). When fruits are scarce, however, lowland gorillas fall back on a wide variety of leaves and fallback herbs (Doran-Sheehy et al., 2009). Other herbs are eaten year-round and considered staples of the diet. Lowland gorillas sometimes supplement their diets with termites and ants (Cipolletta et al., 2007), though females are more likely to do so than males. Reports of gorillas eating meat are extremely rare. In their feeding activities, gorillas are extremely destructive of the vegetation, and their ranging patterns seem to involve harvesting and destroying favorite patches of rapidly regenerating vegetation on a systematic and regular basis (Remis, 1994). Nevertheless, their foraging behavior often involves a complex series of detailed manipulations of individual food items.

Gorillas live in stable social groups averaging between eight and ten individuals that usually contain several adult females and their offspring and one or more mature (silverback) adult males (Harcourt and Stewart, 2007a). Multimale groups are more common in mountain gorillas, and solitary males are found in all populations. Where silverbacks coexist, second-ranking males also sire offspring (Bradley et al., 2005). Mountain gorillas have relatively small, overlapping home ranges and travel as cohesive groups through ranges of about 500 to 1000 m per day. Day ranges are larger in lowland gorillas, and ranging seems to be correlated with fruit consumption (Doran-Sheehy et al., 2004; Cipolletta, 2004).

Although the composition of a gorilla group is similar to that of many primate groups, with a single adult male and several adult females, its formation and maintenance seem to be based on different demographic and social relationships. Most primate groups among Old World monkeys seem to be organized around groups of related females who grow up and remain in their natal group while males transfer from one troop to another. In gorilla society, both males and females emigrate from their natal groups (Harcourt and Stewart, 2007a,b); however, among western gorillas, dispersing males often do not go far and remain in the vicinity of male kin, forming neighborhoods of close male relatives (Bradley et al., 2004). Emigrating males remain solitary until they can attract females (Robbins, 2001; Parnell, 2002). Females may (secondarily) transfer more than once between groups (Sicotte, 2001; Stokes et al., 2003), though among western gorillas, females are often more related than random dispersal would suggest (Bradley et al., 2007). Most interactions are between the adult male and individual females rather than among group members of the

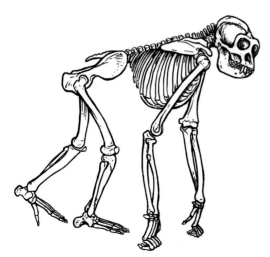

FIGURE 7.10   The skeleton of a gorilla (*Gorilla gorilla*).

same sex, although co-resident females sharing maternal kinship sometimes have close relationships (Watts, 2003). Gorillas do, however, resemble other single-male groups of primates in the intense competition between males for control of a troop, and takeovers may be accompanied by infanticide (Watts, 1989).

Chimpanzees

Pan troglodytes, the chimpanzee (Fig. 7.11), has a distribution that extends in a broad belt across much of central Africa, from Senegal in the west to Tanzania in the east. There are four commonly recognized subspecies: *Pan troglodytes verus* in West Africa; *P. t. ellioti* from Nigeria and adjacent parts of Cameroon; *P. t. troglodytes* in West-Central Africa, and *P. t. schweinfurthii* in East Africa. The western subspecies is the most distinctive. Chimpanzee subspecies differ in body weight, with *P. t. troglodytes* much larger than other subspecies. All have moderate levels of sexual dimorphism (Table 7.2).

Compared to gorillas, chimpanzees have broader incisors and cheek teeth with broader basins and lower, more rounded cusps. Their skulls are very similar in overall shape to those of gorillas, but they have shallower faces and mandibles and do not show such extensive development of sagittal and nuchal crests. The limbs of chimpanzees are more similar in length than those of gorillas and are also less robust. They have narrower hands and feet, with more slender, curved digits.

Chimpanzees occupy a variety of habitats, from rainforests to dry savannah areas with very few trees. Most of our knowledge of chimpanzee behavior comes from woodland, open forested environments. It has become strikingly clear in recent decades that chimpanzee behavior is different from site to site. Part of this can be attributed to differences in habitat and available resources (Doran et al., 2002a,b), but part of it also seems to reflect local traditions (Lycett et al., 2009). In all habitats, chimpanzees generally feed in trees much of each day (depending on the season), and travel on the ground between feeding sites using a quadrupedal, knuckle-walking locomotion. In an arboreal setting they use both quadrupedal and suspensory locomotion to move about within a feeding source, and they also use a variety of seated and suspensory feeding postures. Chimpanzees are more suspensory than gorillas, but considerably less suspensory than either gibbons or orangutans.

Chimpanzees eat primarily fruit (60%) as well as leaves (21%), with strikingly different dietary habits among different populations. For example, nuts of the oil palm are a staple for chimpanzees at Gombe but are not eaten by the chimps at Mahale, less than 20 km away. Other populations may eat different parts of the same fruit species, or process the fruits in different ways. Chimpanzees also seem to use a wide range of plants for medicinal purposes. Although primarily frugivorous, chimpanzees also regularly eat social insects and various smaller mammals, including other primates, particularly red colobus monkeys. In fact, chimpanzees eat more meat than any other nonhuman primate (Watts, 2020). Predation and hunting patterns of chimpanzees on red colobus monkeys have been well documented at many sites (Boesch, 1994; Stanford et al., 1994; Teelen, 2008).

TABLE 7.2 Infraorder Catarrhini
Family HOMINIDAE

Common name	Species	Intermembral index	Mass (g)	
			M	F
Bornean orangutan	*Pongo pygmaeus*	139	78,500	35,800
Sumatran orangutan	*P. abelli*	139	77,900	35,600
Tapanuli orangutan	*P. tapanuliensis*	–	–	–
Chimpanzee	*Pan troglodytes schweinfurthii*	103	41,700	34,400
	P. t. troglodytes	106	59,700	45,800
	P. t. ellioti	–	–	–
	P. t. verus	106	47,000	41,600
Pygmy chimpanzee, bonobo	*Pan paniscus*	104	45,000	33,200
Western lowland gorilla	*Gorilla gorilla gorilla*	116	170,400	71,500
Cross River gorilla	*G. g. diehli*	113	166,000	93,000
Mountain gorilla	*Gorilla beringei beringei*	116	162,500	97,500
Eastern lowland gorilla	*G. b. graueri*	–	175,200	71,000
Human	*Homo sapiens*	69	47,000–78,000	42,000–73,000

FIGURE 7.11   The chimpanzee (*Pan troglodytes*).

Hunting is predominantly an activity of adult males, who account for about 80% of all kills at both Gombe in Tanzania and Tai in Ivory Coast. Hunting is commonly a group activity in which several individuals participate and cooperate, but success seems to involve considerable personal abilities. Some individuals are much more interested in and adept at hunting than others. The hunting party subsequently shares food with other members of the group. In addition to these general similarities, there are also striking differences in hunting habits of chimpanzees at different sites.

Chimpanzees use many types of tools and other aspects of material culture in the course of their daily activities. Most chimpanzee populations utilize "tool kits" of approximately 20 different tool types, including using leaves as sponges, twigs as probes or digging sticks, and clubs and stones to smash hard nuts (Watts, 2008; McGrew, 2010). In fact, with the exception of humans, chimpanzees show the most diverse and complex tool-using repertoires of any extant primate (Sanz and Morgan, 2007). However, as with many aspects of chimpanzee behavior, there are striking differences between populations in the amount, nature, and outcomes of tool use. Indeed, although many populations use tools, no one behavior is universal. At Goualougo, Republic of Congo, for instance, the most commonly used tools are made from perishable materials such as twigs and branches used for extractive foraging (Sanz and Morgan, 2007); whereas at Ngogo, tools more often comprise leaves and are typically used for personal hygiene and courtship (Watts, 2008); and chimpanzees in West Africa will readily use stone tools to crack open nuts (Carvalho et al., 2009), a behavior that is most striking for students of human evolution. Although there is no evidence that the stone tools are manufactured, particularly good ones may be carried around for some time or stored for future use. These differences in tool diversity and complexity between sites are not readily attributable to phylogenetic or ecological variables (Boesch and Boesch, 1990; McGrew, 1992). With long-term studies of chimpanzee behavior demonstrating more and more details of the patterns of their material culture, the distinction between chimpanzees and humans has become increasingly harder to define.

Social groups of chimpanzees are more fluid than those of many higher primates. Their group dynamics can best be described as exhibiting "fission–fusion" (Kummer, 2006; Aureli et al., 2008), and closely resemble patterns observed in *Ateles* (Symington, 1990; Chapman et al., 1995; Spehar et al., 2010). Adults of both sexes spend large portions of their time foraging alone, but they regularly join other individuals in temporary associations or parties (Mitani et al., 2002). Chimpanzee sociality seems to vary considerably from population to population. In eastern Africa, for instance, females are less social than males: they spend more time alone (Goodall, 1986; Wrangham, 2000; Lehmann and Boesch, 2008), have shorter day ranges, and have smaller individual home ranges (e.g., Wrangham and Smuts, 1980; Doran, 1997). Female–female social interactions are relatively rare, except among close relatives such as mothers and daughters. By contrast, in western Africa, as well as

at some sites in eastern Africa, chimpanzee females are more social: they frequently forage together with males (Lehmann and Boesch, 2008; Wakefield, 2008), and exhibit more similar patterns of range use and overlap (Lehmann and Boesch, 2005, 2008).

Foraging and feeding parties are drawn from a relatively closed social community of 15 to 80 chimpanzees who share a common home range that is actively defended. Males regularly form patrol groups that monitor the boundaries of the community home range (Mitani, 2009). Interactions between adults (except some estrous females) of neighboring communities are usually aggressive. Among chimpanzees, young females usually disperse from their natal groups when they are about 11 years old; male chimpanzees tend to remain in their natal groups. Thus, long-term community structure is based on continuity of the male members rather than of the more mobile females. Nevertheless, adult males of a community are not necessarily more closely related to each other than are females (Lukas et al., 2005, but see Inoue et al., 2008).

Chimpanzee mating occurs in a number of contexts, including promiscuous mating, possessive behavior on the part of an individual male toward a fertile female, and consortships, during which an individual male and a receptive female travel together for several days at a time, apart from other individuals. Chimpanzees may live over 50 years in the wild.

Bonobos

Pan paniscus, the **pygmy chimpanzee** or **bonobo** (Fig. 7.12), is similar in adult body weight to the smallest subspecies of chimpanzee (*Pan troglodytes schweinfurthii*), but has a darker face, a more gracile skull, more slender limbs, and longer hands and feet. Although there seems to be sexual dimorphism in the body weight of bonobos, there is virtually no sexual dimorphism in either their dentition (only the canines) or their limb skeleton. Bonobos have a relatively restricted distribution in Central Africa south of the Zaire River, where they can be found in primary, secondary, and swamp forests (Terada et al., 2015), as well as in forest-savannah mosaics (Myers Thompson, 1997; Serckx et al., 2014).

Like chimpanzees, bonobos travel mainly on the ground by knuckle-walking, and feed both on the ground and in trees. Their arboreal locomotion involves quadrupedal, suspensory, quadrumanous climbing, leaping, and bipedal activities (Susman et al., 1980; Doran, 1993). They also use a variety of seated, standing, and suspensory feeding postures. Bonobos are reported to be the most suspensory of the African apes.

Like chimpanzees, bonobos eat primarily fruit and leaves, though they also rely heavily on terrestrial herbaceous vegetation and fungi (Terada et al., 2015; Lucchesi et al., 2021). Bonobos also consume occasional

FIGURE 7.12   Bonobos or pygmy chimpanzees (*Pan paniscus*).

prey, including small ungulates, monkeys, snakes, fish, and insects (Hohmann and Fruth, 2007; Surbeck et al., 2009), and unlike chimpanzees, both males and females actively participate in prey acquisition (Surbeck and Hohmann, 2008). Tool use in bonobos is more common than originally believed, though in contrast to chimpanzees, it typically occurs in nonforaging contexts (Furuichi et al., 2015; Samuni et al., 2022).

Bonobos, like chimpanzees, live in large multimale, multifemale communities characterized by female dispersal and male philopatry (Gerloff et al., 1999; Hohmann and Fruth, 2002). However, unlike chimpanzees, bonobos forage in larger, more cohesive mixed-sex and all-female subgroups (Hohmann and Fruth, 2002). Social relationships are more commonly affiliative than in chimpanzees, with both sexes relying on frequent nonagonistic displays such as genital contacts to maintain social status (Furuichi et al., 1998; Hohmann and Fruth, 2000). Moreover, females tend to develop stronger social bonds than do males (Furuichi, 1989; White, 1988; Ihobe, 1992). Bonobo mating is opportunistic and promiscuous, and occurs both within and between social communities (Gerloff et al., 1999).

Humans

Homo sapiens – humans or people – is a very odd primate species and a very unusual creature by any standard (Fig. 7.13), and attempts to identify the defining features of humans generate a great diversity of views (e.g., Calcagno and Fuentes, 2012). We are more similar to other hominoids in our dental and skeletal anatomy than our striking external and postural features would suggest, and we also have many outstanding specializations that set us apart. We are distinguished dentally by our relatively small canines, broad premolars, and reduced or absent third molars in many populations; otherwise our teeth are similar to those of many chimpanzees. Our mandible, with its protruding chin, is very unusual. Our skull (Fig. 2.3), which has a small, short

FIGURE 7.13   The skeleton of a human.

face, a large, balloon-like cranium with poorly developed crests, and a foramen magnum that lies well beneath the skull base, is extremely derived from that of our closest relatives (Fleagle et al., 2010). Relative to our body size, we have extremely large brains.

Like all hominoids, we have a relatively short trunk and long arms. Our hand is characterized by short, slender fingers and an extremely opposable thumb. The most distinctive features of our skeleton are our long lower extremities, which are associated with our upright, bipedal locomotion. Our pelvic bone is extremely short and broad, and the femur, tibia, and fibula are extremely long. Unlike all other primates, we have a hallux that is not opposable, but aligned with the other toes. We have a long heel, long metatarsals, and very short pedal phalanges. The bones of our foot form two arches, one longitudinal and one transverse. These give us our characteristic footprint, in which we land on the heel, pass our weight along the lateral border of the foot to the front, then push off with the ball of the foot and ultimately the great toe. Many of the distinctive anatomical features of living humans probably are adaptations to running rather than simply to bipedal walking (see Fig. 17.21; Bramble and Lieberman, 2004).

One of our most striking external features is the apparent lack of hair over most of our body, which contrasts with the noticeable concentrations of hair on our heads, under our arms, and in the genital region. Unlike the fur of other animals, human head hair is ever-growing, and this phenomenon plays a major role in sexual selection and many cultural activities. The development of human facial hair is an extremely variable feature that differs not only between sexes but also among different human populations. In fact, the density of hairs on the human body is not much different from that found in large apes such as chimpanzees and gorillas: human hairs are just very short and often lightly pigmented. Another striking feature that distinguishes humans from other primates is the distribution of subcutaneous body fat. Like body hair, this is not only quite different between sexes of many human populations – men tend to store fat in their abdomen, women store it in their breasts and on their hips and buttocks – but is also quite variable among major human population groups. Despite striking differences in body form and appearance among different human populations, we are remarkably uniform genetically, compared to other hominoids. Most of our genetic diversity is within rather than between populations.

As primates, we humans are uniquely cosmopolitan. Only Antarctica has steadfastly resisted permanent colonization by our species. All other habitats, including tropical rainforests, woodlands, savannahs, plains, deserts, mountains, and arctic coastlines, have supported human populations for many thousands of years. We are also the most terrestrial of all primates. Among primates in their natural habitats, only humans (and perhaps geladas) regularly live their entire lives without ever climbing a tree for food or sleep. Our bipedal gait is unique among mammals, but it does not seem to endow us with particularly striking speed or locomotor efficiency compared to other mammals, including nonhuman primates. It does, however, permit more endurance at slow speeds: only humans can run marathons (Bramble and Lieberman, 2004) (Fig. 7.14).

The "natural" human diet is probably something that exists only in television commercials and on billboards. Humans are opportunistic and omnivorous – we eat virtually anything. We lack the notable digestive specializations that characterize more vegetarian primate species and show greater similarities to faunivores. Most significantly, and unlike any other mammal, we cook our food (Wrangham and Carmody, 2010). Even more than chimpanzees, we humans regularly make use of tools of various kinds in our behavior, and we share knowledge both among unrelated individuals and from generation to generation, so that as a species we have a great store of communal knowledge (Hill et al., 2009).

FIGURE 7.14   A member of the human race.

There is no single pattern of social organization among humans: there is more variability in the social organization of human societies than is found in any other primate species. Monogamous families and single-male groups of one male with several females are the most common arrangements throughout the world, but there are human populations in which polyandry (one female with several males), or even more promiscuous multimale and multifemale groups, are common. It is generally argued that culture complicates comparisons with nonhuman primate social organizations, since human social norms are often mandated by religion or law. Comparing human sexual dimorphism with that found in other primates offers no more convincing evidence of a natural social structure for our species. Our body size dimorphism allies us with polygynous mammal species (Alexander et al., 1979). Our lack of canine dimorphism is more similar to that found among monogamous primates or among those living in fission–fusion societies. However, humans of both sexes have small canines, whereas in many monogamous primates (except *Callicebus*) both sexes have large canines. The morphological evidence thus suggests that human social structures are organized on a different morphological basis than those of other primates (e.g., Plavcan and van Schaik, 1994).

Like those of some other primates, human societies are generally multilevel structures of small subsistence units grouped into larger communities. However, humans are unusual in that both males and females tend to maintain relationships with their kin, regardless of normal dispersal (or residence) patterns. Intercommunity violence is generally the prerogative of males: human females do not form coalitions to fight females from other communities. The evolutionary and adaptive origins of these unique human social features are the subject of considerable debate.

Human communication is strikingly different from that of other primates in our extensive use of language, but there is increasing evidence of remarkable cognitive and communication abilities in chimpanzees and bonobos, indicating that, although dramatic and substantial, this is to some degree a quantitative rather than an absolute distinction. Similarly, many aspects of nonverbal communication in humans can readily be recognized in chimpanzees. Even human notions of morality have clear bases among other primates.

From a primate perspective, the most striking reproductive features of humans are aspects of our life history and patterns of childcare. Our newborn infants are extremely large compared to adult body size, despite being born at a relatively immature stage of development. Humans are also unusual among primates in the extraordinary amount of prosocial behavior that characterizes our species. Human infants are raised by

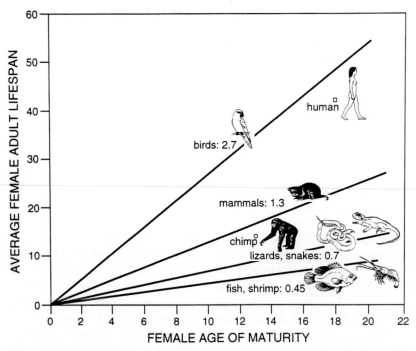

FIGURE 7.15   Humans are unusual among most other vertebrates in having a long female postreproductive lifespan. *(Redrawn from Hill, K., 1993. Life history theory and evolutionary anthropology. Ev. Anthropol. 2, 78–88).*

communities rather than just their close relatives (Hrdy, 2009; Bogin et al., 2014).

Adult mortality among humans is strikingly low, leading to our long juvenile period and very long lifespan. More unusual than our long lifespan is the fact that a very large portion of the lifespan of human females is after menopause and seems nonreproductive (Fig. 7.15). Hawkes and colleagues (1997, 1998) have argued that the long postmenopausal lifespan of human females has been selected because, in our species, older females or grandmothers make important contributions to the survival and reproductive success of their daughters and grandchildren.

Adaptive Radiation of Hominoids

Ecologically, all apes are diurnal, and all except humans are more or less restricted to forested areas (Fig. 7.16). Although all apes seem to utilize some suspensory locomotion, the African apes more commonly travel using arboreal and terrestrial quadrupedal (knuckle-walking) gaits. There are no hominoid leapers, and human bipedalism is unique among primates. Living apes are all frugivorous and folivorous: there are no seed, insect, or gum specialists. Again, humans are very unusual in our dietary diversity and our regular use of agriculture and animal domestication, and in our dependence on cooking food.

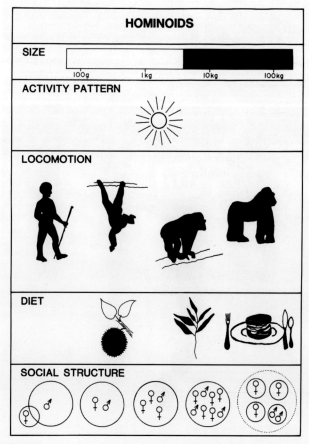

FIGURE 7.16   The adaptive radiation of living hominoids.

The most diverse aspect of hominoid behavior is their social organization. This is different for every genus, between chimpanzee species, and among populations of humans. Certainly, this diversity belies most attempts to identify an ancestral social system for humans by simple extrapolation from our nearest relatives. Several factors probably contribute to this. One is the large size of most hominoids. Associated with this large size is a considerable longevity compared with that of most primates. Our extended lifespan permits a versatility in reproductive strategies at different life stages that is not available to animals with shorter lifespans. Finally, the larger brain size and increased intelligence of apes probably permit a greater flexibility of social interactions based on memory and unique interindividual relationships.

Phyletic Relationships of Hominoids

As discussed above, with the exception of humans, which are unquestionably a single species, there is considerable debate about how many species should be recognized within almost every genus of hominoids. In contrast, it is almost universally agreed that the phylogenetic relationships among the genera of hominoids are well established based on numerous genetic studies (Fig. 7.17). The basal division is between the gibbons (Hylobatidae) and the great apes and humans (Hominidae). Within great apes and humans, orangutans are the sister taxon to the African ape and human clade. Within this latter group, humans are more closely related to chimpanzees than to gorillas. However, we are not just made-over chimps, some parts of the human genome are more similar to that of gorillas than chimpanzees (Scally et al., 2012). Thus, the question of what the last common ancestor of chimpanzees and humans, or chimpanzees, humans, and gorillas, looked like is difficult to reconstruct, and is a hotly debated issue that will be discussed further in Chapter 17.

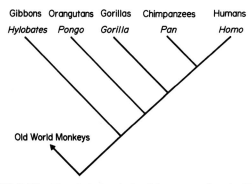

FIGURE 7.17   The phyletic relationships among hominoids.

References

Alexander, R.D., Hoogland, J.L., Howard, R.D., et al., 1979. Sexual dimorphisms and breeding systems in pinnipeds, ungulates, primates and humans. In: Chagnon, N.A., Irons, W. (Eds.), Evolutionary Biology and Human Social Behavior: An Anthropological Perspective. Duxbury Press, Boston, MA, pp. 402–435.

Ancrenaz, M., Sollmann, R., Meijaard, E., et al., 2014. Coming down from the trees: Is terrestrial activity in Bornean orangutans natural or disturbance driven? Sci. Rep. 4, 1–5.

Ashbury, A.M., Posa, M.R.C., Dunkel, L.P., et al., 2015. Why do orangutans leave the trees? Terrestrial behavior among wild Bornean orangutans (Pongo pygmaeus wurmbii) at Tuanan, Central Kalimantan. Am. J. Primatol. 77, 1216–1229.

Aureli, F., Schaffner, C.M., Boesch, C., et al., 2008. Fission-fusion dynamics: New research frameworks. Curr. Anthropol. 49, 627–654.

Bartlett, T.Q., 2011. The Hylobatidae: Small apes of Asia. In: Campbell, C.J., Fuentes, A., MacKinnon, K.C. et al., (Eds.), Primates in Perspective. Oxford University Press, New York, pp. 300–312.

Boesch, C., 1994. Chimpanzees-red colobus monkeys: A predator-prey system. Animal Behav. 47, 1135–1148.

Boesch, C., Boesch, H., 1990. Tool use and tool making in wild chimpanzees. Folia Primatol. 54, 86–99.

Bogin, B., Bragg, J., Kuzawa, C., 2014. Humans are not cooperative breeders but practice biocultural reproduction. Ann. Hum. Biol. 41, 368–380.

Borries, C., Savini, T., Koenig, A., 2011. Social monogamy and the threat of infanticide in larger mammals. Behav. Ecol. Sociobiol. 65, 685–693.

Bradley, B.J., Doran-Sheehy, D.M., Lukas, D., et al., 2004. Dispersed male networks in western gorillas. Curr. Biol. 14, 510–513.

Bradley, B.J., Doran-Sheehy, D.M., Vigilant, L., 2007. Potential for female kin associations in wild western gorillas despite female dispersal. Proc. Roy. Soc. B 274, 2179–2185.

Bradley, B.J., Robbins, M.M., Williamson, E.A., et al., 2005. Mountain gorilla tug-of-war: Silverbacks have limited control over reproduction in multimale groups. Proc. Nat. Acad. Sci. USA 102, 9418–9423.

Bramble, D.M., Lieberman, D.E., 2004. Endurance running and the evolution of Homo. Nature 432, 345–352.

Calcagno, J.M., Fuentes, A., 2012. What makes us human? Answers from evolutionary anthropology. Evol. Anthropol. 21, 182–194.

Carvalho, S., Biro, D., McGrew, W.C., Matsuzawa, T., 2009. Tool-composite reuse in wild chimpanzees (Pan troglodytes): Archaeologically invisible steps in the technological evolution of early hominins? Anim. Cogn. 12, 103–114.

Chapman, C.A., Chapman, L.J., Wrangham, R.W., 1995. Ecological constraints on group size: An analysis of spider monkey and chimpanzee subgroups. Behav. Ecol. Sociobiol. 36, 59–70.

Cipolletta, C., 2004. Effects of group dynamics and diet on the ranging patterns of a western gorilla group (Gorilla gorilla gorilla) at Bai Hokou, Central African Republic. Am. J. Primatol. 64, 193–205.

Cipolletta, C., Spagnoletti, N., Todd, A., Robbins, M.M., Cohen, H., Pacyna, S., 2007. Termite feeding by Gorilla gorilla gorilla at Bai Hokou, Central African Republic. Int. J. Primatol. 28, 457–476.

Delgado R.A., Jr., van Schaik, C.P., 2000. The behavioral ecology and conservation of the orangutan (Pongo pygmaeus): A tale of two islands. Evol. Anthropol. 9, 201–218.

Doran, D.M., 1993. Comparative locomotor behavior of chimpanzees and bonobos: The influence of morphology on locomotion. Am. J. Phys. Anthropol. 91, 83–98.

Doran, D.M., 1996. Comparative positional behavior of the African apes. In: McGrew, W.C., Marchant, L.F., Nishida, T. (Eds.), Great Ape Societies. Cambridge University Press, Cambridge, pp. 213–224.

Doran, D., 1997. Influence of seasonality on activity patterns, feeding behavior, ranging, and grouping patterns in Tai chimpanzees. Int. J. Primatol. 18, 183–206.

Doran, D.M., Jungers, W.L., Sugiyama, Y., et al., 2002a. Multivariate and phylogenetic approaches to understanding chimpanzee and bonobo behavioral diversity. In: Boesch, C., Hohmann, G., Marchant, L. (Eds.), Behavioral Diversity in Chimpanzees and Bonobos. Cambridge University Press, Cambridge, pp. 14–34.

Doran, D.M., McNeilage, A., Greer, D., et al., 2002b. Western lowland gorilla diet and resource availability: New evidence, cross-site comparisons, and reflections on indirect sampling methods. Am. J. Primatol. 58, 91–116.

Doran-Sheehy, D.M., Greer, D., Mongo, P., Schwindt, D., 2004. Impact of ecological and social factors on ranging in western gorillas. Am. J. Primatol. 64, 207–222.

Doran-Sheehy, D., Mongo, P., Lodwick, J., Conklin-Brittain, N.L., 2009. Male and female western gorilla diet: Preferred foods, use of fallback resources, and implications for ape versus old world monkey foraging strategies. Am. J. Phys. Anthropol. 140, 727–738.

Dunkel, L.P., Arora, N., van Noordwijk, M.A., et al., 2013. Variation in developmental arrest among male orangutans: A comparison between a Sumatran and a Bornean population. Front. Zoo 10, 1–11.

Elder, A., 2009. Hylobatid diets revisited: The importance of body mass, fruit availability, and interspecific competition. In: Lappan, S., Whittaker, D.J. (Eds.), The Gibbons: New Perspectives on Small Ape Socioecology and Population Biology. Springer, New York, pp. 133–159.

Fan, P.F., He, K., Chen, X., et al., 2017. Description of a new species of Hoolock gibbon (Primates: Hylobatidae) based on integrative taxonomy. Am. J. Primatol. 79, e22631.

Fleagle, J.G., 1976. Locomotion and posture of the Malayan siamang and implications for hominoid evolution. Folia Primatol. 26, 245–269.

Fleagle, J.G., Gilbert, C.C., Baden, A.L., 2010. Primate cranial diversity. Am. J. Phys. Anthropol. 142, 565–578.

Fleagle, J.G., Lieberman, D.E., 2013. The evolution of primate locomotion: Many transformations, many ways. In: Dial, K., Shubin, N., Brainerd, E. (Eds.), Great Transformations: Major Events in the History of Vertebrate Life. University of California Press, Berkeley, CA.

Furuichi, T., 1989. Social interactions and the life history of female Pan paniscus in Wamba, Zaire. Int. J. Primatol., 10173–197.

Furuichi, T., Idani, G.I., Ihobe, H., et al., 1998. Population dynamics of wild bonobos (Pan paniscus) at Wamba. Int. J. Primatol. 19, 1029–1043.

Furuichi, T., Sanz, C., Koops, K., et al., 2015. Why do wild bonobos not use tools like chimpanzees do? In: Hare, B., Yamamoto, S. (Eds.),. Bonobo Cognition and Behaviour, Brill, pp. 179–214. https://doi.org/10.1163/9789004304178_010

Galdikas, B.M.E., 1981. Orangutan reproduction in the wild. In: Graham, E.E. (Ed.), Reproductive Biology of the Great Apes. Academic Press, New York, pp. 281–300.

Geissmann, T., Geschke, K., Blanchard, B.J., 2009. Longevity in gibbons (Hylobatidae). Gibbon J. 5, 81–92.

Gerloff, U., Hartung, B., Fruth, B., et al., 1999. Intracommunity relationships, dispersal pattern, and paternity success in a wild living community of Bonobos (Pan paniscus) determined from DNA analysis of faecal samples. Proc. Roy. Soc. B 266, 1189–1195.

Goodall, J., 1986. Social rejection, exclusion, and shunning among the Gombe chimpanzees. Ethol. Sociobiol. 7, 227–236.

Harcourt, A.H., Stewart, K.J., 2007a. Gorilla society: What we know and don't know. Evol. Anthropol. 16, 147–158.

Harcourt, A.H., Stewart, K.J., 2007b. Gorilla Society: Conflict, Compromise, and Cooperation Between the Sexes. University of Chicago Press, Chicago.

Hawkes, K., OConnell, J.F., Blurton-Jones, N., 1997. Hadza women's time allocation, offspring provisioning, and the evolution of postmenopausal lifespans. Curr. Anthropol. 38, 551–577.

Hawkes, K., O'Connell, J.F., Blurton-Jones, N., et al., 1998. Grandmothering and the evolution of human life histories. Proc. Natl. Acad. Sci. 95, 1336–1339.

Hill, K., Barton, M., Hurtado, M., 2009. The emergence of human uniqueness: Characters underlying behavioral modernity. Evol. Anthropol. 18, 187–200.

Hohmann, G., Fruth, B., 2000. Use and function of genital contacts among female bonobos. Anim. Behav. 60, 107–120.

Hohmann, G., Fruth, B., 2002. Dynamics in social organization of bonobos (Pan paniscus). In: Boesch, C., Hohmann, G., Marchant, L. (Eds.), Behavioural Diversity in Chimpanzees and Bonobos. Cambridge University Press, Cambridge, pp. 138–150.

Hohmann, G., Fruth, B., 2007. New records on prey capture and meat eating by bonobos at Lui Kotale, Salonga National Park, Democratic Republic of Congo. Folia Primatol. 79, 103–110.

Hrdy, Sarah., B., 2009. Mother and Others: The Evolutionary Origins of Mutual Understanding. Harvard University Press, Cambridge.

Ihobe, H., 1992. Male–male relationships among wild bonobos (Pan paniscus) at Wamba, Republic of Zaire. Primates 33, 163–179.

Inoue, E., Inoue-Murayama, M., Vigilant, L., et al., 2008. Relatedness in wild chimpanzees: Influence of paternity, male philopatry, and demographic factors. Am. J. Phys. Anthropol. 137, 256–262.

Knott, C.D., 1998. Changes in orangutan caloric intake, energy balance, and ketones in response to fluctuating fruit availability. Int. J. Primatol. 19, 1061–1079.

Knott, C., 1999. D. Reproductive, Physiological and Behavioral Responses of Orangutans in Borneo to Fluctuations in Food Availability. PhD Thesis. Harvard University.

Knott, C.D., Kahlenberg, S., 2011. Orangutans: Understanding forced copulations. In: Campbell, C., Fuentes, A., MacKinnon, K. et al. (Eds.), Primates in Perspective, second ed. Oxford University Press, Oxford, pp. 290–305.

Kummer, H., 2006. Primate Societies: Group Techniques of Ecological Adaptation. Routledge, Abingdon-on-Thames. 160.

Kunz, J.A., Duvot, G.J., van Noordwijk, M.A., et al., 2021. The cost of associating with males for Bornean and Sumatran female orangutans: A hidden form of sexual conflict? Behav. Ecol. Sociobiol. 75, 1–22.

Larson, S.G., 1998. Parallel evolution in the hominoid trunk and forelimb. Evol. Anthropol. 6, 87–99.

Lehmann, J., Boesch, C., 2005. Bisexually bonded ranging in chimpanzees (Pan troglodytes verus). Behav. Ecol. Sociobiol. 57, 525–535.

Lehmann, J., Boesch, C., 2008. Sexual differences in chimpanzee sociality. Int. J. Primatol. 29, 65–81.

Leighton, M., 1993. Modeling dietary selectivity by Bornean orangutans: Evidence for integration of multiple criteria in fruit selection. Int. J. Primatol. 14, 257–313.

Leighton, M., Seal, U.S., Soemarna, K., et al., 1995. Orangutan life history and vortex analysis. In: Nadler, R.D., Galdikas, B.M.F., Sheeran, L.K., Rosen, N. (Eds.), The Neglected Ape. Plenum Press, New York, pp. 97–107.

Lucchesi, S., Cheng, L., Wessling, E.G., et al., 2021. Importance of subterranean fungi in the diet of bonobos in Kokolopori. Am. J. Primatol. 83, e23308.

Lukas, D., Reynolds, V., Boesch, C., Vigilant, L., 2005. To what extent does living in a group mean living with kin? Mol. Ecol. 14, 2181–2196.

Lycett, S.J., Collard, M., McGrew, W.C., 2009. Cladistic analyses of behavioural variation in wild Pan troglodytes: Exploring the chimpanzee culture hypothesis. J. Hum. Evol. 57, 337–349.

Marshall, A.J., Ancrenaz, M., Brearley, F.Q., et al., 2009. The effects of forest phenology and floristics on populations of Bornean and Sumatran orangutans. In: Wich, S.A., Atmoko, S.U., Setia, T.M., van Schaik, C.P. (Eds.), Orangutans: Geographic Variation in Behavioral Ecology and Conservation. Oxford University Press, New York, pp. 97–117.

Masi, S., Cipolletta, C., Robbins, M.M., 2009. Western lowland gorillas (*Gorilla gorilla gorilla*) change their activity patterns in response to frugivory. Am. J. Primatol. 71, 91–100.

McGrew, W.C., 1992. Chimpanzee Material Culture: Implications for Human Evolution. Cambridge University Press, Cambridge.

McGrew, W.C., 2010. Chimpanzee technology. Science 328, 579–580.

Mehlman, P.T., Doran, D.M., 2002. Influencing western gorilla nest construction at Mondika Research Center. Int. J. Primatol. 23, 1257–1285.

Mitani, J.C., 2009. Cooperation and competition in chimpanzees: Current understanding and future challenges. Evol. Anthropol. 18, 215–227.

Mitani, J.C., Watts, D.P., Muller, M.M., 2002. Recent developments in the study of wild chimpanzee behavior. Evol. Anthropol. 11, 9–25.

Myers Thompson, J.A., 1997. The history, taxonomy and ecology of the bonobo (Pan paniscus Schwarz, 1929) with a first description of a wild population living in a forest/savanna mosaic habitat. PhD Thesis. University of Oxford.

O'Connell, C.A., DiGiorgio, A.L., Ugarte, A.D., et al., 2021. Wild Bornean orangutans experience muscle catabolism during episodes of fruit scarcity. Sci. Rep. 11, 10185. https://doi.org/10.1038/s41598-021-89186-4

Palombit, R.A., 1995. Longitudinal patterns of reproduction in wild female siamang (*Hylobates syndactylus*) and white-handed gibbons (*Hylobates lar*). Int. J. Primatol. 16, 739–760.

Parnell, R.J., 2002. Group size and structure in western lowland gorillas (*Gorilla gorilla gorilla*) at Mbeli Bai, Republic of Congo. Am. J. Primatol. 56, 193–206.

Plavcan, M.J., van Schaik, C.P., 1994. Canine dimorphism. Evol. Anthropol. 2, 208–214.

Remis, M.J., 1994. Feeding ecology and positional behavior of western lowland gorillas (*Gorilla gorilla gorilla*) in the Central African Republic. PhD thesis. Yale University.

Remis, M., 1995. Effects of body size and social context on the arboreal activities of lowland gorillas in the Central African Republic. Am. J. Phys. Anthropol. 97, 413–433.

Remis, M.J., 1999. Tree structure and sex differences in arboreality among western lowland gorillas (*Gorilla gorilla gorilla*) at Bai Hokou, Central African Republic. Primates 40, 383–396.

Robbins, M.M., 2001. Variation in the social system of mountain gorillas: The male perspective. In: Robbins, M.M., Sicotte, P., Stewart, K.J. (Eds.), Mountain Gorillas: Three Decades of Research at Karisoke. Cambridge University Press, New York, pp. 29–58.

Rodman, P.S., 1977. Feeding behaviour of orangutans of the Kutai Nature Reserve, East Kalimantan. In: Clutton-Brock, T.H. (Ed.), Primate Ecology: Studies of Feeding and Ranging Behavior in Lemurs, Monkeys, and Apes. Academic Press, New York, pp. 383–413.

Rogers, M.E., Abernethy, K., Bermejo, M., et al., 2004. Western gorilla diet: A synthesis from six sites. Am. J. Primatol. 64, 173–192.

Rothman, J.M., Dierenfeld, Hintz, H.F., Pell, A.N., 2008. Nutritional quality of gorilla diets: Consequences of age, sex and season. Oecologia 155, 111–122.

Rothman, J.M., Pell, A.N., Dierenfeld, E.S., McCann, 2006a. Plant choice in the construction of night nests by gorillas in Bwindi Impenetrable National Park, Uganda. Am. J. Primatol. 68, 361–368.

Rothman, J.M., Pell, A.N., Nkurunungi, J.B., Dierenfeld, E.S., 2006b. Nutritional aspects of the diet of wild gorillas: How do Bwindi gorillas compare? In: Newton-Fisher, N.E., Notman, H., Paterson, J.D., Reynolds, V. (Eds.), Primates of Western Uganda. Springer, New York, pp. 153–169.

Rothman, J.M., Plumptre, A.J., Dierenfeld, E.S., Pell, A.N., 2007. Nutritional composition of the diet of the gorilla (*Gorilla beringei*): A comparison between two montane habitats. J. Tropical Ecol. 23, 673–682.

Samuni, L., Lemieux, D., Lamb, A., et al., 2022. Tool use behavior in three wild bonobo communities at Kokolopori. Am. J. Primatol. 84, e23342.

Sanz, C.M., Morgan, D.B., 2007. Chimpanzee tool technology in the Goualougo Triangle, Republic of Congo. J. Hum. Evol. 52, 420–433.

Scally, A., Dutheil, J.Y., Hillier, L.W., et al., 2012. Insights into hominid evolution from the gorilla genome sequence. Nature 483, 169–175.

Schaller, 1963. The Mountain Gorilla. University of Chicago Press, Chicago.

Schuppli, C., Atmoko, S., Vogel, E.R., et al., 2021. The development and maintenance of sex differences in dietary breadth and complexity in Bornean orangutans. Behav. Ecol. Sociobiol. 75, 85.

Serckx, A., Huynen, M.-C., Bastin, J.-F., et al., 2014. Nest grouping patterns of bonobos (*Pan paniscus*) in relation to fruit availability in a forest-savannah mosaic. PLoS One 9, e93742.

Sicotte, P., 2001. Female mate choice in mountain gorillas. In: Robbins, M.M., Sicotte, P., Stewart, K.J. (Eds.), Mountain Gorillas: Three Decades of Research at Karisoke. Cambridge University Press, New York, pp. 59–88.

Singleton, I., Knott, C.D., Morrogh-Bernard, H.C., et al., 2009. Ranging behavior of orangutan females and social organization. In: Wich, S.A., Utami Atmoko, S.S., Mitra Setia, T., van Schaik, C.P. (Eds.), Orangutans: Geographic Variation in Behavioral Ecology and Conservation. Oxford University Press, New York, pp. 205–213.

Singleton, I., van Schaik, C.P., 2001. Orangutan home range size and its determinants in a Sumatran swamp forest. Int. J. Primatol. 22, 877–911.

Spehar, S.N., Link, A., Di Fiore, A., 2010. Male and female range use in a group of white-bellied spider monkeys (Ateles belzebuth) in Yasuni National Park, Ecuador. Am. J. Primatol. 72, 129–141.

Stanford, C.B., Wallis, J., Matama, H., Goodall, J., 1994. Patterns of predation by chimpanzees on red colobus monkeys in Gombe National Park, 1982–1991. Am. J. Phys. Anthropol. 94, 213–228.

Stokes, E.J., Parnell, R.J., Olejniczak, C., 2003. Female dispersal and reproductive success in wild western lowland gorillas (*Gorilla gorilla gorilla*). Behav. Ecol. Sociobiol. 54, 329–339.

Surbeck, M., Fowler, A., Deimel, C., Hohmann, G., 2009. Evidence for the consumption of arboreal, diurnal primates by bonobos (*Pan paniscus*). Am. J. Primatol. 71, 171–174.

Surbeck, M., Hohmann, G., 2008. Primate hunting by bonobos at LuiKotale, Salonga National Park. Curr. Biol. 18, R906–R907.

Susman, R.L., Badrian, N.L., Badrian, A.J., 1980. Locomotor behavior of Pan paniscus in Zaire. Am. J. Phys. Anthropol. 53, 69–80.

Symington, M.M., 1990. Fission-fusion social organization in Ateles and Pan. Int. J. Primatol. 11, 47–61.

Teelen, S., 2008. Influence of chimpanzee predation on the red colobus population at Ngogo, Kibale National Park, Uganda. Primates 49, 41–49.

Terada, S., Nackoney, J., Sakamaki, T., et al., 2015. Habitat use of bonobos (*Pan paniscus*) at Wamba: Selection of vegetation types for ranging, feeding, and night-sleeping. Am. J. Primatol. 77, 701–713.

Tuttle, R.H., 1969. Quantitative and functional studies on the hands of the Anthropoidea – I. The Hominoidea. J. Morphol. 128, 309–364.

van Schaik, C.P., Ancrenaz, M., Borgen, G., et al., 2003. Orangutan cultures and the evolution of material culture. Science 299, 102–105.

van Schaik, C.P., Damerius, L., Isler, K., 2013. Wild orangutan males plan and communicate their travel direction one day in advance. PLoS One 8, e74896.

Vogel, E.R., Knott, C.D., Crowley, B.E., et al., 2012. Bornean orangutans on the brink of protein bankruptcy. Biol. Lett. 8, 333–336.

Vogel, E.R., Alavi, S.E., Utami-Atmoko, S.S., et al., 2017. Nutritional ecology of wild Bornean orangutans (Pongo pygmaeus wurmbii) in a peat swamp habitat: Effects of age, sex, and season. Am. J. Primatol. 79, e22618.

Wakefield, M.L., 2008. Grouping patterns and competition among female *Pan troglodytes schweinfurthii* at Ngogo, Kibale National Park, Uganda. Int. J. Primatol. 29, 907–929.

Watts, D.P., 1984. Composition and variability of mountain gorilla diets in the central Virungas. Am. J. Primatol. 7, 323–356.

Watts, D.P., 1989. Infanticide in mountain gorillas: New cases and a reconsideration of the evidence. Ethology 81, 1–18.

Watts, D.P., 2003. Gorilla social relationships: A comparative overview. In: Taylor, A.B., Goldsmith, M.L. (Eds.), Gorilla Biology: A Multidisciplinary Perspective. Cambridge University Press, Cambridge, pp. 302–327.

Watts, D.P., 2008. Tool use by chimpanzees at Ngogo, Kibale National Park, Uganda. Int. J. Primatol. 29, 83–94.

Watts, D.P., 2020. Meat eating by nonhuman primates: A review and synthesis. J. Hum. Evol. 149, 102882.

White, F.J., 1988. Party composition and dynamics in *Pan paniscus*. Int. J. Primatol. 9, 179–193.

Wich, S.A., Atmoko, S.S.U., Setia, T.M., van Schaik, C.P., 2009. Orangutans: Geographic Variation in Behavioral Ecology and Conservation. Oxford University Press, New York.

Wrangham, R.W., 2000. Why are male chimpanzees more gregarious than mothers? A scramble competition hypothesis. In: Kappeler, P.M. (Ed.), Primate Males, Causes and Consequences of Variation in Group Composition. Cambridge University Press, Cambridge, pp. 248–258.

Wrangham, R., Carmody, R., 2010. Human adaptation to the control of fire. Evol. Anthropol. 19, 187–199.

Wrangham, R., Smuts, B., 1980. Sex differences in the behavioural ecology of chimpanzees in the Gombe National Park, Tanzania. J. Reprod. Fertil. Supplement 28, 13–31.

Young, N.M., 2003. A reassessment of living hominoid postcranial variability: Implications for ape evolution. J. Hum. Evol. 45, 441–464.

Further Reading

Bartlett, T.Q., 2011. The Hylobatidae: Small apes of Asia. In: Campbell, C.J., Fuentes, A., MacKinnon, K.C. et al. (Eds.), Primates in Perspective. Oxford University Press, New York, pp. 300–312.

Bartlett, T.Q., Sussman, R.W., 2022. Chapter 20: Gibbons: Arboreal acrobats of Southeast Asia. In: Sussman, R.W., Hart, D., Colquhoun, I.C. (Eds.), The Natural History of Primates: A Systematic Survey of Ecology and Behavior. Rowman and Littlefield, Lanham, pp. 535–556.

Boesch, C., Hohmann, G., Marchant, L.M., 2002. Behavioral Diversity in Chimpanzees and Bonobos. Cambridge University Press, Cambridge.

Delgado, R.A., 2022. Chapter 21: The Orangutans: Asia's Endangered Great Apes. In: Sussman, R.W., Hart, D., Colquhoun, I.C. (Eds.), The Natural History of Primates: A Systematic Survey of Ecology and Behavior. Rowman and Littlefield, Lanham, pp. 557–570.

Goldsmith, M., 2022. Chapter 24: The wonderful world of apes...and the adventures we have studying them. In: Sussman, R.W., Hart, D., Colquhoun, I.C. (Eds.), The Natural History of Primates: A Systematic Survey of Ecology and Behavior. Rowman and Littlefield, Lanham, pp. 642–646.

Harcourt, A.H., 2012. Human Biogeography. University of California Press, Berkeley.

Harcourt, A.H., Stewart, K.J., 2007. Gorilla Society: Conflict, Compromise, and Cooperation Between the Sexes. University of Chicago Press, Chicago.

Knott, C.D., Kahlenberg, S., 2011. Orangutans: Understanding forced copulations. In: Campbell, C., Fuentes, A., MacKinnon, K. et al. (Eds.), Primates in Perspective, second ed. Oxford University Press, Oxford, pp. 290–305.

Lappan, S., Whittaker, D.J., 2009. The Gibbons: New Perspectives on Small Ape Socioecology and Population Biology. Springer, New York.

Morgan, D.B., Sussman, R.W., Cooksey, K.E., et al., 2022. Chapter 22: Gorillas. In: Sussman, R.W., Hart, D., Colquhoun, I.C. (Eds.), The Natural History of Primates: A Systematic Survey of Ecology and Behavior. Rowman and Littlefield, Lanham, pp. 571–600.

Reichard, U.H., Hirai, H., Barelli, C., 2016. Evolution of Gibbons and Siamang: Phylogeny, Morphology, and Cognition. Springer, New York.

Robbins, M., 2011. Gorillas: Diversity in ecology and behavior. In: Campbell, C., Fuentes, A., MacKinnon, K. et al. (Eds.), Primates in Perspective, second ed. Oxford University Press, Oxford, pp. 326–339.

Sanz, C.M., Sussman, R.W., Musgrave, S., et al., 2022. Chapter 23: Chimpanzees and bonobos. In: Sussman, R.W., Hart, D., Colquhoun, I.C. (Eds.), The Natural History of Primates: A Systematic Survey of Ecology and Behavior. Rowman and Littlefield, Lanham, pp. 601–641.

Stumpf, R.M., 2011. Chimpanzees and bonobos: Inter- and intraspecific diversity. In: Campbell, C., Fuentes, A., MacKinnon, K. et al. (Eds.), Primates in Perspective, second ed. Oxford University Press, Oxford, pp. 340–357.

Taylor, A.B., Goldsmith, M.L., 2003. Gorilla Biology: A Multidisciplinary Perspective. Cambridge University Press, Cambridge.

Watts, D.P., 2012. The apes: Taxonomy, biogeography, life histories, and behavioral ecology. In: Mitani, J.C., Call, J., Kappeler, P.M. et al. (Eds.), The Evolution of Primate Societies. University of Chicago Press, Chicago.

Wich, S.A., Atmoko, S.S.U., Setia, T.M., van Schaik, C.P., 2009. Orangutans: Geographic Variation in Behavioral Ecology and Conservation. Oxford University Press, New York.

8

Primate Communities and Biogeography

In the preceding chapters we discussed the behavior and ecology of living primates as individual species or as phyletic radiations of related taxa. This is not how primates are typically encountered in their natural habitats. Rather, if one walks through a tropical forest in South America, Africa, Madagascar, or Asia, one normally encounters a collection of distantly related, sympatric species sharing various parts of a common habitat. In this chapter we move from the earlier chapters' focus on systematic organization to a consideration of living primates from a biogeographical perspective. We will compare and contrast the numbers and ecological adaptations of primates that are found on different continents in order to examine patterns of primate diversity on a global scale.

As the previous chapters document, most living primate species have to some degree been studied under naturalistic conditions. Virtually all have been the subject of short-term observations or surveys, and many have been the subject of longer studies lasting a year or more. In some cases, these long-term studies have been part of extensive research projects comparing and contrasting a number of species in a single community. Much of our understanding about primate behavior and ecology has come from comparative studies of sympatric species at sites such as Manu and Raleighvallen-Voltzberg in South America; Kibale, Taï, and Makokou in Africa; Ranomafana, Marosalaza, Kirindy, and Beza Mahafaly in Madagascar; and Kuala Lompat, Kutai, and Ketembe in Asia. Despite the large number of comparative studies examining the similarities and differences among species within communities, there have been relatively few attempts to compare patterns of species diversity and ecology among primate communities and between communities on different continents. In this chapter we consider geographical patterns in primate ecology. What are the broad biogeographical patterns that describe the distribution of primate species on different continents (Fig. 8.1)? How similar or different are the primates and primate communities that we find in South America, Africa, Madagascar, and Asia? What are the factors responsible for differences in the composition of primate communities on different continents and at different sites within continents?

Primate Biogeography

There are over 70 genera and 500 species of living primates. What general factors determine the distribution of these species on a global basis? It has been well established for many types of organisms that the number of species in any given region is a function of the geographical area of the region. This phenomenon is known as the **species–area relationship** (Schoener, 1976). The causal mechanisms underlying this pattern are not clearly understood, though it seems reasonable that larger areas provide more habitat diversity and potential geographical barriers, thus giving rise to or maintaining greater species diversity in those areas (Lehman, 2003).

As discussed in Chapter 3, primates are primarily tropical forest animals, with only a few hardy species extending to temperate or non-forested habitats. The number of primate species found in major continental areas and on large islands is largely a function of the amount of tropical forest in those regions (Fig. 8.2). Thus, the islands of Java, Sumatra, and Borneo have increasing numbers of species in accordance with the increasing amount of tropical forest found on those islands; South America and Africa have the largest continuous forest blocks and thus the largest numbers of species. Madagascar has many more species than predicted by the size of the island. This may partly reflect the fact that the present forest area is greatly reduced from the time when the species there first evolved, but it also probably reflects the absence of many other groups of birds and mammals on Madagascar to compete with primates for resources. Thus, primates make up a larger proportion of Madagascar's species than they do in areas with a more diverse mammalian fauna.

In addition to showing a species–area relationship, primates are similar to many other organisms in showing greater diversity near the equator than at higher latitudes. This is often described as the **Forster effect** (e.g., Harcourt and Schreier, 2009). This latitudinal gradient has been demonstrated for South America (Ruggiero, 1994) and Africa (Eeley and Lawes, 1999), as well as for primates as an order (Harcourt and Schreier, 2009). Like species–area curves, increased diversity at low latitudes

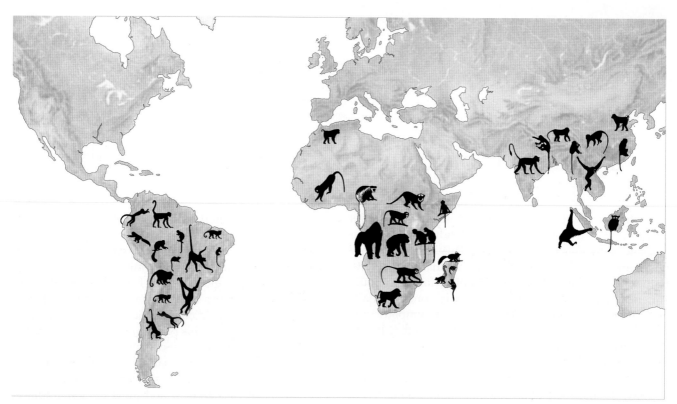

FIGURE 8.1   Living primates are found in four major biogeographical regions: the Neotropics of Central and South America, Africa, Madagascar, and Southern Asia.

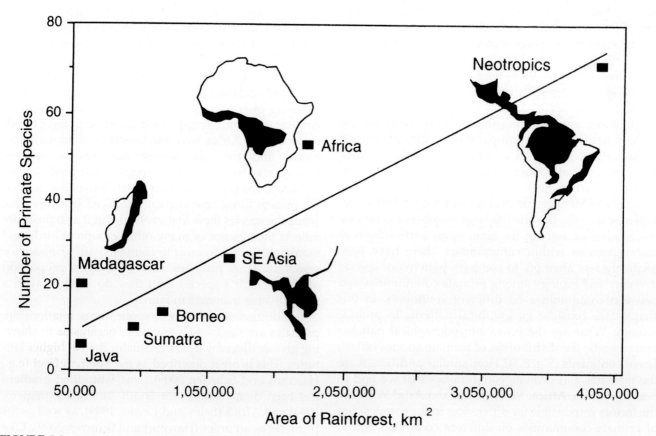

FIGURE 8.2   The number of primate species on large islands or continental areas is largely a function of the area of tropical rain forest.

is a widespread but poorly understood phenomenon. Common explanations include higher productivity near the equator associated with greater solar radiation, less climatic seasonality, and tropical niche conservatism (e.g., Wiens and Donoghue, 2004). However, others have argued that the apparent latitudinal gradients of many taxa may be heavily driven by species–area effects, since for South America and Africa the greatest area of these continents surrounds the equator.

In addition to the increased density of primate species at low latitudes, primate species found near the equator tend to have smaller geographic ranges than those found at higher latitudes (Ruggiero, 1994; Eeley and Lawes, 1999; Harcourt, 2000). This pattern is called **Rapoport's Rule**. Again, the underlying causality of this pattern is the subject of considerable debate. Most authorities feel that it reflects increasing climatic harshness with distance from the equator. Species that can live in more temperate habitats are likely to be generalists, with wider habitat tolerances than more equatorial species. Also, the greater diversity of most organisms at low latitudes may generate more competition and greater specialization on restricted resources.

Although these general patterns provide a broad guide to the number of species one is likely to find on islands and continents, in many parts of the world the number of species at any individual site seems to be closely correlated with rainfall (Fig. 8.3). The Neotropics, Africa, and Madagascar exhibit a positive relationship between

rainfall and the number of primate species found at individual sites up to rainfall levels of about 2000 mm per year (Reed and Fleagle, 1995). However, this relationship is probably not a simple linear fit; rather, at higher levels of rainfall, diversity either flattens out or decreases, especially in Asia (Kay et al., 1997), perhaps owing to leaching of soils. There is also a positive relationship between rainfall and primate species diversity in Southeast Asia, but the correlation is very low, with various relationships in different parts of this geographically complex region (Shekelle et al., 2009; Wang et al., 2013).

A number of hypotheses have been put forward to explain why primate species diversity is positively related to rainfall. Several authors have found that plant species diversity is also positively associated with rainfall, and that primate species diversity may, in turn, reflect the diversity of plant foods available (e.g., Ganzhorn et al., 1997; Stevenson, 2001). Others have argued, more generally, that primate diversity reflects overall productivity of the forests, in terms of the leaves, fruits, and flowers that primates eat. Indeed, Kay et al. (1997) have argued that productivity is the factor driving primate species diversity, and that both species diversity and productivity show the same curvilinear relationship with rainfall. In contrast to the relationships between primate species diversity and rainfall or productivity, attempts to explain diversity in primate biomass have not yielded any simple results, although differences in the nutritional content of foods seem to explain some of these patterns. Specifically,

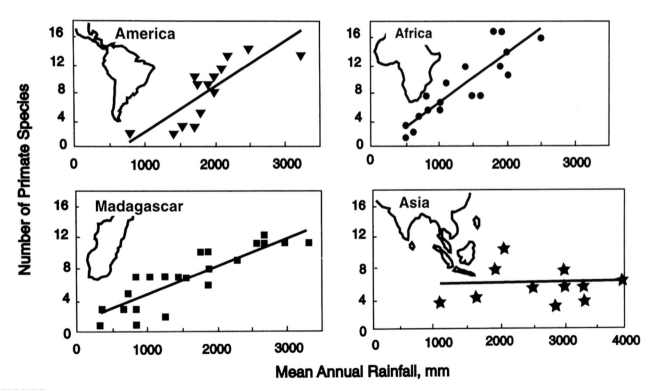

FIGURE 8.3   The number of species at individual sites in South America, Africa, and Madagascar is positively correlated with the annual rainfall. Asia exhibits no clear relationship.

seasonality in fruit availability and protein-to-fiber ratios in mature leaves explain substantial variation in frugivore and folivore biomass, respectively (Chapman et al., 2002, 2004; Hanya et al., 2011). Other biogeographic rules that are consistent with the distribution of primates include **Bergmann's Rule** – animals are larger at higher latitudes (Harcourt and Schreier, 2009) and **Gloger's Rule** – pelage tends to be darker in species that inhabit warm, wet habitats (Kamilar and Bradley, 2011).

Ecology and Biogeography

The preceding paragraphs discussed general patterns describing the distribution of numbers of primate species as a function of continental area, latitude, and climate. However, no general prediction was made about the behavior and ecology of the primate species found in different parts of the world. How similar or different are the primate fauna found in different continental areas?

Primates today are largely restricted to the tropical forested regions of Central and South America, sub-Saharan Africa, Madagascar, and southern Asia (Fig. 8.1). Each of these four major biogeographical regions has a distinctive fauna, with no common species between them, and, in the case of the Neotropics and Madagascar, no common families. Indeed, the only genus shared by any two regions is *Macaca*, which has numerous species in Asia and a single species in North Africa. Africa and Asia also share several common families.

Although these different biogeographical regions inhabited by primates are similar in that they support tropical forests of various types, there are many differences in the forest habitats, both within and between regions. These include differences in climate and seasonality, in the nature of the underlying geology and soils, in the taxonomic and structural composition of the flora, and in the animals that live in association with primates. Several studies have compared aspects of life history, ecology, and behavior of the primates found on different continents (Kappeler and Heymann, 1996; Fleagle and Reed, 1996, 1999).

In body size distributions, Neotropical primates are significantly smaller than those found on other continents, but only if the subfossil species of Madagascar are included. If the Malagasy subfossil species are not included, Madagascar shows a similar size range as primates found in the Neotropics, but the distribution of sizes is significantly different among all major faunas (Fig. 8.4). Although there are both nocturnal and diurnal species in all major primate faunal radiations, Madagascar is significantly different from other regions in having a higher proportion of nocturnal and cathemeral species (Fig. 8.5).

All primate fauna include primates with a variety of diets, including species that specialize on fauna, gums, fruits, leaves, and various combinations of these. As discussed in Chapter 3, classifying the diets of individual species is a difficult undertaking that ignores many important details of inter- and intraspecific behavior. Nevertheless, there are a number of striking patterns in

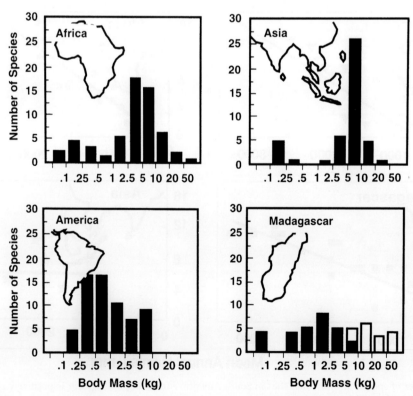

FIGURE 8.4 The distribution of primate body sizes in four biogeographical areas. White columns indicate recently extinct Malagasy species. *(Courtesy of P. Kappeler.)*

the frequency of different dietary habits among the primates of major continental areas (Fig. 8.5). Compared to other regions, Madagascar is dominated by folivorous species. This is true for the living fauna, and even more so if the subfossil species are included. In contrast, folivory is relatively rare in the Neotropics, which are dominated by frugivorous and insectivorous species (see Ganzhorn et al., 2009). In Asia, it is the frugivorous–folivorous species that are dominant. Compared to other continents, African species are unusual in their balance of different dietary habits and the lack of any dominant dietary type.

Even more so than diet, gregariousness is difficult to reduce to simple, quantitative comparisons. Nevertheless, we can say that most primates tend to forage in groups of a characteristic size. When these are plotted by continental area, we see different distributions for each of the major regions (Fig. 8.6). Most noticeable is the high frequency among Malagasy primates of species that forage solitarily, and their absence in Neotropical species.

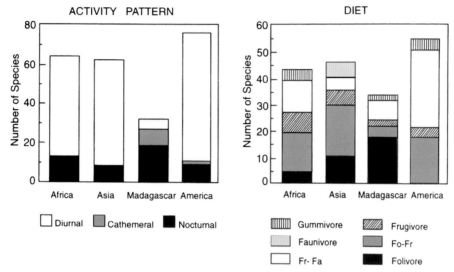

FIGURE 8.5 The distribution of activity pattern (*left*) and diet (*right*) among primate species of four major geographical areas. (*Courtesy of P. Kappeler.*)

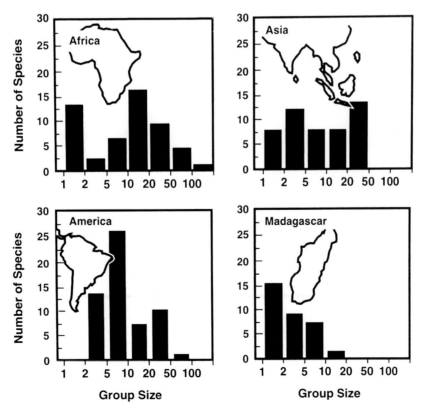

FIGURE 8.6 The distribution of group sizes in primate species from four major geographical areas. (*Courtesy of P. Kappeler.*)

Comparing Primate Communities

In addition to examining the frequency of different ecological and behavioral features among the primates of different continents, we can also look specifically at the ecological composition of individual communities. Despite their different phylogenetic histories, do the primates of South America, Africa, Madagascar, and Asia occupy roughly the same types of "ecological space"? Or are primates on different continents doing very different things? Are the ecological differences between primate communities within a single biogeographical area greater or less than those between regions?

In fact, primate communities within biogeographical regions are remarkably similar, despite differences in species number and local ecology (Fleagle and Reed 1996, 1999). For example, although rainfall and primate species diversity in Ranomafana (Fig. 8.7), an eastern rainforest habitat, are more than double or triple that of Marosalaza, a western dry forest site (Fig. 8.8; see also Fig. 4.9), the two Madagascar primate communities are strikingly similar in the overall ecology of the component species (Fig. 8.9). Both contain numerous small species, nocturnal species, and many folivorous species

FIGURE 8.8   The diurnal primate communities (*above*) and the nocturnal primate communities (*below*) from the western dry forest of Marazolaza, near Morondava. Note the dramatic differences in the typical vegetation and the smaller number of primate species compared with Ranomafana.

FIGURE 8.7   The diurnal primate communities (*above*) and the nocturnal primate communities (*below*) of the eastern rain forest of Ranomafana, Madagascar.

compared with primate communities in other parts of the world. The main differences are that Ranomafana has more medium-sized, quadrupedal frugivorous species (*Eulemur rubriventer*, *E. rufifrons*, and *Varecia variegata*) than Marosalaza, as well as three species that specialize on bamboo (*Hapalemur* and *Prolemur*).

Likewise, even though they are separated by 5000 km and have fewer than half of their primate species in common, the communities of Taï Forest in Ivory Coast (Fig. 8.10) and Kibale Forest in Uganda are very similar. The distinctive shape of the African communities (Fig. 8.9) compared to those of other continents is defined by the small, nocturnal, leaping, insectivorous galagos and the numerous larger, frugivorous, quadrupedal species (*Pan*, *Papio*, *Cercopithecus*).

The two Asian communities – Kuala Lompat, Malaysia (Fig. 8.11) and Ketambe, in Indonesia – are distinctive in the paucity of small, nocturnal, and insectivorous species and also in having the suspensory gibbons. Compared to communities from other parts of the world, Asian primate communities generally have very few species (Fig. 8.12).

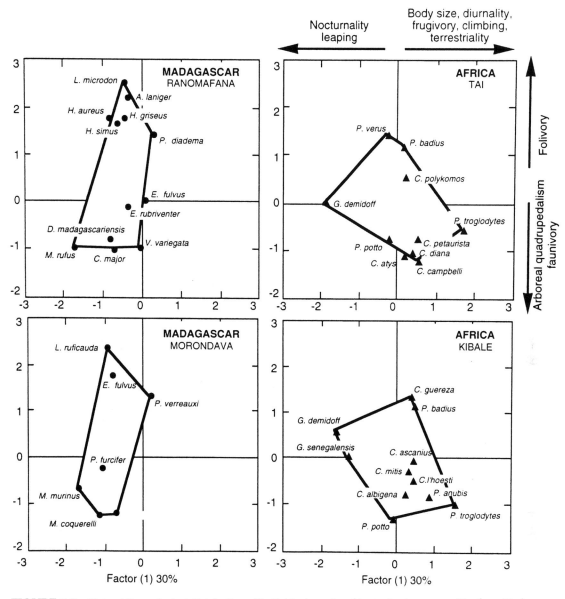

FIGURE 8.9 Plots of the ecological distribution of individual species of two primate communities from Madagascar – Ranomafana and Marazolaza (near Morondava) – and two communities from Africa – Taï Forest, Ivory Coast, and Kibale Forest, Uganda. Note the overall similarity of the communities from the same biogeographical area, despite differences in the individual species at each site. Also compare with Fig. 8.12.

The two communities from South America are virtually identical to one another, despite coming from different biogeographical provinces and having very different numbers of species (Fig. 8.12). Raleighvallen-Voltzberg (Fig. 8.13) is in Suriname on the Guianan shield, whereas Manu is in the upper reaches of the Amazon drainage in Peru. Nevertheless, these primate communities occupy the same ecological space, despite Manu's greater number of species. The most striking feature of the South American communities is the remarkable uniformity of the different species and the limited amount of

ecological space that they define. Compared with primates in other parts of the world, Neotropical monkeys are mostly diurnal, medium-sized, frugivorous quadrupeds. They lack the extreme specialists of other continents. There is only one nocturnal species (*Aotus*) and one largely folivorous species (*Alouatta seniculus*), and even the most saltatory or suspensory species (such as *Ateles*) are mostly quadrupedal.

These community comparisons accord well with other broad comparisons of primate communities on a global scale (Bourliere, 1985; Terborgh and van Schaik,

FIGURE 8.10   The diurnal primate communities (*above*) and the nocturnal primate communities (*below*) of the Taï Forest, Ivory Coast.

FIGURE 8.11   The diurnal primates of Kuala Lompat, Malaysia. Note the relatively small number of species.

1987; Kappeler and Heymann, 1996). Malagasy primates are characterized by an abundance of nocturnal and folivorous species; South American communities lack folivores; Africa has an abundance of large frugivores; and Asian communities are relatively low in species diversity, especially in small species, but are unusual in the number of suspensory taxa. The communities reflect the overall characteristics of the fauna of these different regions, and individual communities within regions are extremely similar to one another. Why are the primates on different continents so distinctive? At present there are many hypotheses and interesting observations.

Several types of arguments are commonly offered to explain the major differences among primate fauna throughout the world (Reed and Bidner, 2004). One view is that the ecological differences among the primate fauna of South America, Africa, Madagascar, and Asia reflect major differences in the nature of the forests and the available resources on these continents, perhaps associated with broad geographical patterns of soils or climate, including seasonality (Brockman and van Schaik, 2005). For example, Terborgh and van Schaik (1987) have argued that the lack of primate folivores in South America reflects

the synchronous productivity of leaves and fruits owing to the extreme seasonality of rainfall in much of South America. Thus, when fruits are not available, a species cannot switch to a diet of leaves, and vice versa (but see Heymann, 2001). Both Ganzhorn and colleagues (2009) and Donati et al. (2017) have argued that the relative abundance of folivores in Madagascar and their dearth in the Neotropics may be the result of widespread continent-specific differences in the nutritional composition of fruits on the two continents. In addition, Madagascar has relatively few plants that produce fleshy fruits.

Likewise, it has been argued that Asian forests have relatively few understory insects compared to other continents, and the low numbers of primate species in Asia have been related to a generally low level of productivity in Asian forests and to the abundance of dipterocarps, a group of trees that are not utilized extensively by primates. Also, the equatorial region of Asia (e.g., the Sunda Shelf) has less land area than South America and Africa. As this region experiences very high rainfall levels (>3000 mm/year), which are associated with a high number of cloudy days, the reduction of available sunlight may reduce plant productivity.

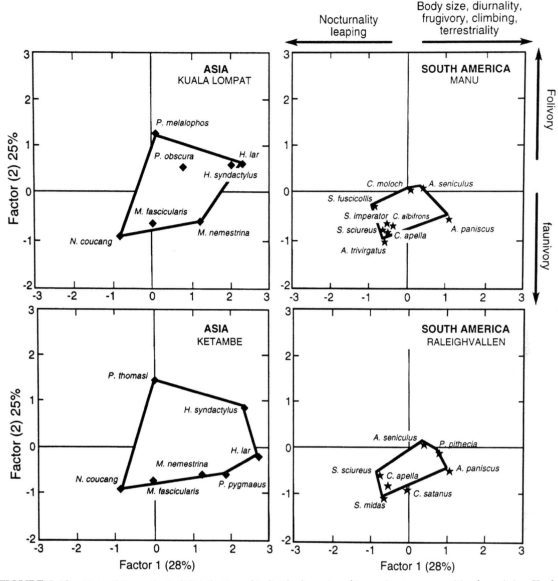

FIGURE 8.12 Plots of the ecological distribution of individual species of two primate communities from Asia – Kuala Lompat, Malaysia, and Ketambe, Sumatra – and two from South America – Raleighvallen-Voltsberg, Suriname, and Manu, Peru. Note the overall similarity of the communities from the same biogeographical area, despite differences in the individual species found at each site. Also compare with Fig. 8.9.

Another possible explanation is that the differences in primate fauna are the result of competition with other groups of vertebrates, known as **competitive exclusion**. The availability of primate niches may be constrained or influenced in many areas by the presence of other mammals or birds (e.g., Beaudrot et al., 2013a,b). Thus, the lack of primate folivores in South America has been related to an earlier presence of sloths on that continent, while the lack of nocturnal insectivorous species has been attributed to the diversity of small nocturnal marsupials. These resource availability and competition explanations are to some degree complementary and provide tests of one another (Fleagle and Reed, 2008).

For example, the presence of sloths, often in great numbers, indicates that South American forests can certainly support mammalian folivores, albeit very specialized ones. Likewise, the presence and diversity in the Neotropics of many nocturnal, frugivorous opossums, many of which are very primate-like, demonstrates that this niche (or guild) is both available and well occupied. Indeed, there were opossums in South America well before primates appeared. In contrast, many aspects of

FIGURE 8.13   The primate community of the Raleighvallen-Voltsberg Nature Reserve in Suriname.

explain. Why are there no galagos in Asia, despite the presence of lorises? Is this a historical accident? Does it reflect a scarcity of resources? Or is it due to competition from the very specialized but currently geographically restricted tarsiers? Why is the platyrrhine radiation of South America so narrow ecologically?

Despite the clear pattern of biogeographical differences in primate ecological adaptations among the primate faunas of the world, there are at present few general explanations. In most cases, the present-day patterns seem to reflect a combination of ecological and historical factors rather than a single, unitary cause (Fleagle and Reed, 2004, 2008).

Interspecific competition among different primate species may also influence community structure. For example, if two species occupy a similar ecological niche, then they may not be able to coexist in the same community (Schreier et al., 2009). This is more often the case for frugivorous primates (Kamilar and Ledogar 2011), suggesting that interspecific competition for fruit is stronger than for other food types, such as leaves, and leads to some frugivorous primates never being found together in the same community.

Interspecific competition may also be detected using a phylogenetic approach. If we assume that closely related species occupy similar ecological niches (e.g., Kamilar and Cooper, 2013; Porter et al., 2016), then we may expect that these species would be unlikely to coexist in the same community because of past competition for the same niche space. If this is true, then modern primate communities should consist of species that are more distantly related than would be expected by chance alone. However, this hypothesis is not well supported for most primate communities in Africa, Asia, and the Neotropics. In contrast, it was supported in the majority of Malagasy communities (Kamilar and Guidi, 2010; Razafindratsima et al., 2012). Although a large proportion of Malagasy primate species have become extinct in the recent past, this does not seem to be the primary cause of phylogenetically distinct species comprising modern Malagasy communities. Rather, the phylogenetic patterning of Madagascar's primate communities may be the result of the rapid diversification that likely occurred early in lemur history (e.g., Kamilar and Muldoon, 2010).

More recent research has also begun to investigate the importance of current environmental conditions, historical patterns of dispersal and vicariance, and competition for shaping primate communities (Kamilar et al., 2009; Beaudrot and Marshall, 2011; Muldoon and Goodman, 2010; Rowan et al., 2016; 2020). These so-called macroecology studies examine relationships between primates and their environments over much greater spatial and temporal scales (Keith et al. 2012). For example, within continents, communities in closer proximity tend to have more similar species compositions (Kamilar, 2009;

the vertebrate fauna of Madagascar and Asia parallel the ecological patterns seen in primates, suggesting that differences in resource availability may be more important. For example, Madagascar has both a low number of primate frugivores and very few frugivorous birds compared to other parts of the world. The dearth of small insectivorous primates in much of Asia is paralleled by a low diversity of insectivorous frogs and lizards in these forests as well (Duellman and Pianka, 1990).

Some of the continental differences in primate ecology may be the result of historical accidents of biogeography, of differences in the adaptive potential of the radiations that initially colonized the regions, or of recent extinctions. For example, it is clear that the lack of large species among the living primates of South America and Madagascar is a recent phenomenon: much larger primates inhabited both regions in the relatively recent past. Likewise, the extinct Malagasy species include numerous suspensory and more quadrupedal species than the living fauna. However, in diet they only further emphasize the distinctiveness of that fauna, for they include a preponderance of folivores. Other patterns in the biogeography of primate adaptations are difficult to

Beaudrot and Marshall, 2011). This geographic distance effect may be an artifact of historical patterns of species dispersal and vicariance. Rivers are often important barriers that limit dispersal (e.g., Ayres and Clutton-Brock, 1992; Ganzhorn et al., 2006; Goodman and Ganzhorn, 2004; Harcourt and Wood, 2012; Lehman, 2004; Meijaard and Groves, 2006), as is altitude (Wilmé et al., 2006). In Africa, Madagascar, and the Neotropics, various environmental characteristics, such as rainfall, temperature, and nutrient content also determine the presence and absence of species at a particular location. Nevertheless, environmental variables are often relatively weak predictors of primate community structure, suggesting that many primate species are ecologically flexible.

Further comparisons of primate communities from different biogeographical areas, as well as broader comparisons in both geography and time, seem unlikely to yield any simple patterns, but may provide greater insight into the many ways in which community composition is determined. In addition to being an area of major interest for understanding the origins of the patterns of biodiversity we see around us, primate biogeography and community structure are critical for conservation planning as primate habitats become increasingly reduced (Jernvall and Wright, 1998; Wright and Jernvall, 1999; Harcourt et al., 2002, 2005; Harcourt, 2006).

References

Ayres, J.M., Clutton-Brock, T.H., 1992. River boundaries and species range size in Amazonian primates. Am. Nat. 140, 531–537.

Beaudrot, L., Marshall, A.J., 2011. Primate communities are structured more by dispersal limitation than by niches. J. Anim. Ecol. 80, 332–341.

Beaudrot, L., Struebig, M.J., Meijaard, E., et al. 2013a. Co-occurrence patterns of Bornean vertebrates suggest competitive exclusion is strongest among distantly related species. Oecologia 173 (3), 1053–1062.

Beaudrot, L., Struebig, M.J., Meijaard, E., et al. 2013b. Interspecific interactions between primates, birds, bats, and squirrels may affect community composition on Borneo. Am. J. Primatol. 75 (2), 170–185.

Bourliere, F., 1985. Primate communities: Their structure and role in tropical ecosystems. Int. J. Primatol. 6, 1–26.

Brockman, D.K., van Schaik, C.P., 2005. Seasonality in Primates: Studies of Living and Extinct Human and Nonhuman Primates. Cambridge University Press, Cambridge.

Chapman, C.A., Chapman, L.J., Bjorndal, K.A., Onderdonk, D.A., 2002. Application of protein-to-fiber ratios to predict colobine abundance on different spatial scales. Int. J. Primatol. 23 (2), 283–310.

Chapman, C.A., Chapman, L.J., Naughton-Treves, L., et al. 2004. Predicting folivorous primate abundance: Validation of a nutritional model. Am. J. Primatol. 62 (2), 55–69

Donati, G., Santini, L., Eppley, T.M., et al. 2017. Low levels of fruit nitrogen as drivers for the evolution of Madagascar's primate communities. Sci. Rep. 7 (1), 1–9

Duellman, W.E., Pianka, E.R., 1990. Biogeography of nocturnal insectivores: Historical events and ecological filters. Annu. Rev. Ecol. Syst. 21, 57–68.

Eeley, H., Lawes, M., 1999. Large-scale patterns of species richness and species range size in African and South American primates. In: Fleagle, J.G., Janson, C., Reed, K.E. (Eds.), Primate Communities. Cambridge University Press, Cambridge, pp. 191–219.

Fleagle, J.G., Reed, K.E., 1996. Comparing primate communities: A multivariate approach. J. Human Evol. 30, 489–510.

Fleagle, J.G., Reed, K.E., 1999. Primate communities and phylogeny. In: Fleagle, J.G., Janson, C., Reed, K.E. (Eds.), Primate Communities. Cambridge University Press, Cambridge, pp. 92–115.

Fleagle, J.G., Reed, K.E., 2004. The evolution of primate ecology: Patterns of geography and phylogeny. In: Anapol, F., German, R.Z., Jablonski, N. (Eds.), Shaping Primate Evolution: Papers in Honor of Charles Oxnard. Cambridge University Press, Cambridge.

Fleagle, J.G., Reed, K.E., 2008. What do differences in primate communities tell us about the rest of the ecosystem? Past and Recent History of Tropical Ecosystems: ATBC Annual Meeting, 9–13 June, Surinme, Paramaribo.

Ganzhorn, J.U., Malcomber, S., Andrianantoanina, O., Goodman, S.M., 1997. Habitat characteristics and lemur species richness in Madagascar. Biotropica 29, 331–343.

Ganzhorn, J.U., Goodman, S.M., Nash, S., Thalmann, U., 2006. Lemur biogeography. In: Lehman, S.M., Fleagle, J.G. (Eds.), Primate Biogeography: Progress and Prospects. Springer, New York, pp. 229–254.

Ganzhorn, J.U., Arrigo-Nelson, S., Boinski, S., et al., 2009. Possible fruit protein effects on primate communities in Madagascar and the Neotropics. PLoS One 4 (12), e8253.

Goodman, S.M., Ganzhorn, J.U., 2004. Biogeography of lemurs in the humid forests of Madagascar: The role of elevational distribution and rivers. J. Biogeogr. 31 (1), 47–55.

Hanya, G., Stevenson, P., van Noordwijk, M., et al. 2011. Seasonality in fruit availability affects frugivorous primate biomass and species richness. Ecography 34 (6), 1009–1017.

Harcourt, A.H., 2000. Latitude and latitudinal extent: A global analysis of the Rapoport effect in a tropical mammalian taxon: Primates. J. Biogeogr. 27 (5), 1169–1182.

Harcourt, A.H., 2006. Rarity in the tropics: Biogeography and macroecology of the primates. J. Biogeogr. 33, 2077–2087.

Harcourt, A.H., Coppeto, S.A., Parks, S.A., 2002. Rarity, specialization and extinction in primates. J. Biogeogr. 29, 445–456.

Harcourt, A.H., Coppeto, S.A., Parks, S.A., 2005. The distribution-abundance (density) relationship: Its form and causes in a tropical mammal order, Primates. J. Biogeogr. 32 (4), 565–579.

Harcourt, A.H., Schreier, B.M., 2009. Diversity, body mass, and latitudinal gradients in primates. Int. J. Primatol. 30, 283–300.

Harcourt, A.H., Wood, M.A., 2012. Rivers as barriers to primate distributions in Africa. Int. J. Primatol. 33, 168–183.

Heymann, E.W., 2001. Can phenology explain the scarcity of folivory in New World primates? Am. J. Primatol. 55, 171–175.

Jernvall, J., Wright, P.C., 1998. Diversity components of impending primate extinctions. Proc. Natl. Acad. Sci. USA 95, 11279–11283.

Kamilar, J.M., 2009. Environmental and geographic correlates of the taxonomic structure of primate communities. Am. J. Phys. Anthropol. 139, 382–393.

Kamilar, J.M., Guidi, L.M., 2010. The phylogenetic structure of primate communities: Variation within and across continents. J. Biogeogr. 37, 801–813.

Kamilar, J.M., Bradley, B.J., 2011. Interspecific variation in primate coat color supports Gloger's rule. J. Biogeogr. 38, 2270–2277.

Kamilar, J.M., Cooper, N., 2013. Phylogenetic signal in primate behaviour, ecology, and life history. Philos. Trans. R. Soc. Lond. B Biol. Sci. 368, 20120341.

Kamilar, J.M., Ledogar, J.A., 2011. Species co-occurrence patterns and dietary resource competition in primates. Am. J. Phys. Anthropol. 144, 131–139.

Kamilar, J.M., Martin, S.K., Tosi, A.J., 2009. Combining biogeographic and phylogenetic data to examine primate speciation: An example using cercopithecin monkeys. Biotropica 41, 514–519.

Kamilar, J.M., Muldoon, K.M., 2010. The climatic niche diversity of Malagasy primates: A phylogenetic approach. PLoS One 5, e11073.

Kappeler, P.M., Heymann, E.W., 1996. Nonconvergence in the evolution of primate life history and socio-ecology. Biol. J. Linn. Soc. 59, 297–326.

Kay, R.F., Madden, R.H., van Schaik, C., Higdon, D., 1997. Primate species richness is determined by plant productivity: Implications for conservation. Proc. Natl. Acad. Sci. USA 94, 13023–13027.

Keith, S.A., Webb, T.J., Bohning-Gaese, K., et al., 2012. What is macroecology? Biol. Lett. 8, 904–906.

Lehman, S.M., 2003. Ecogeography of primates in Guyana: Species-area relationships and ecological specialization. Am. J. Phys. Anthropol. 120 (S36), 138.

Lehman, S.M., 2004. Distribution and diversity of primates in Guyana: Species-area relationships and riverine barriers. Int. J. Primatol. 25 (1), 73–95.

Meijaard, E., Groves, C.P., 2006. The geography of mammals and rivers in mainland Southeast Asia. In: Lehman, S.M., Fleagle, J.G. (Eds.), Primate Biogeography: Progress and Prospects. Springer, New York, pp. 305–329.

Muldoon, K., Goodman, S., 2010. Ecological biogeography of Malagasy non-volant mammals: Community structure is correlated with habitat. J. Biogeogr. 37, 1144–1159.

Porter, L.M., Gilbert, C.C., Fleagle, J.G., 2014. Diet and phylogeny in primate communities. Int. J. Primatol 35, 1144–1163. https://doi.org/10.1007/s10764-014-9794-0

Razafindratsima, O.H., Mehtani, S., Dunham, A.E., 2012. Extinctions, traits and phylogenetic community structure: Insights from primate assemblages in Madagascar. Ecography 35, 47–56.

Reed, K.E., Fleagle, J.G., 1995. Geographic and climatic control of primate diversity. Proc. Natl. Acad. Sci. USA 92, 7874–7876.

Reed, K.E., Bidner, L.R., 2004. Primate communities: Past, present, and possible future. Yearb. Phys. Anthropol. 47, 2–39.

Rowan, J., Kamilar, J.M., Beaudrot, L., Reed, K.E., 2016. Strong influence of palaeoclimate on the structure of modern African mammal communities. 2016. Proc. Roy. Soc. B: Biol. Sci. 283 (1840), 20161207.

Rowan, J., Beaudrot, L., Franklin, J., et al. 2020. Geographically divergent evolutionary and ecological legacies shape mammal biodiversity in the global tropics and subtropics. Proc. Natl. Acad. Sci. USA 117 (3), 1559–1565.

Ruggiero, A., 1994. Latitudinal correlates of the sizes of mammalian geographical ranges in South America. J. Biogeog. 21, 545–559.

Schoener, T.W., 1976. The species–area relationship within archipelagoes: Models and evidence from island birds. Proceedings of the XVI International Ornithological Congress 6, 629–642.

Schreier, B.M., Harcourt, A.H., Coppeto, S.A., Somi, M.F., 2009. Interspecific competition and niche separation in primates: A global analysis. Biotropica 41, 283–291.

Shekelle, M., Srivathsan, A., Salim, A., 2009. Asian primate species richness correlates with rainfall using GIS modeling. Am. J. Phys. Anthropol. S48, 237.

Stevenson, P.R., 2001. The relationship between fruit production and primate abundance in neotropical communities. Biol. J. Linn. Soc. 72, 161–178.

Terborgh, J., van Schaik, C.P., 1987. Convergence vs. nonconvergence in primate communities. In: Gee, J.H.R., Giller, P.S. (Eds.), Organization of Communities. Blackwell Scientific Publications, Oxford, pp. 205–226.

Wang, Y.-C., Srivathsan, A., Feng, C.-C., Salim, A., Shekelle, M., 2013. Asian primate species richness correlates with rainfall. PLoS One 8 (1), e54995.

Wiens, J.J., Donoghue, M.J., 2004. Historical biogeography, ecology, and species richness. Trends Ecol. Evol. 19, 639–644.

Wilmé, L., Goodman, S.M., Ganzhorn, J.U., 2006. Biogeographic evolution of Madagascar's microendemic biota. Science 312 (5776), 1063–1065.

Wright, P.C., Jernvall, J., 1999. The future of primate communities: A reflection of the present? In: Fleagle, J., Janson, C., Reed, K. (Eds.), Primate Communities. Cambridge University Press, Cambridge, pp. 295–309.

Further Reading

Brockman, D.K., van Schaik, C.P., 2005. Seasonality in Primates: Studies of Living and Extinct Human and Non-human Primates. Cambridge University Press, Cambridge.

Cowlishaw, G., Dunbar, R., 2000. Primate Conservation Biology. University of Chicago Press, Chicago.

Fleagle, J.G., Janson, C.H., Reed, K.E., 1999. Primate Communities. Cambridge University Press, Cambridge.

Harcourt, A.H. (Ed.), 2012. Human Biogeography. University of California Press, Berkeley.

Kamilar, J.M., Beaudrot, L., 2013. Understanding primate communities: Recent developments and future directions. Evol. Anthropol. 22 (4), 174–185.

Lehman, S.M., Fleagle, J.G., 2006. Primate Biogeography: Progress and Prospects. Springer, New York.

9

Primate Adaptations

In the preceding chapters we discussed the anatomy, behavior, and comparative ecology of the major radiations of living primates. In this chapter we examine size, diet, vision, locomotor behavior, and other aspects of comparative ecology for consistent associations between anatomical and behavioral characteristics, as well as for correlations among the different aspects of behavior and ecology. By investigating the functional relationships between morphological features such as size, tooth shape, bone shape, and behavioral habits, we can understand why and how primate species have evolved many of their anatomical differences. We can also use this information to reconstruct some aspects of the behavior and ecology of extinct species.

Effects of Size

Body size is a basic aspect of the adaptation of any primate species. An animal's size is associated with both opportunities and restrictions on its ecological options, and many of the differences between species in structure, behavior, and ecology are correlated with absolute body size. Much of this size-dependent variation in morphology, physiology, and ecology can be explained by the impact of simple mathematical considerations on basic physiological and mechanical phenomena.

As the linear dimensions of any object – including an animal – increase, so too do its areal (e.g., cross-sectional) dimensions and its volume. But linear dimensions, area, and volume do not increase at the same rates. Area increases as a function of the square of linear dimensions (L^2) and volume increases as a function of the cube of linear dimensions (L^3) (Fig. 9.1). If an animal were to double in length, breadth, and width, for example, its cross-sectional dimensions would increase fourfold, and its volume would be eight times as great. When these simple mathematical considerations are applied to animal bodies, the consequences are great. An animal's weight is a function of its volume; the strength of any of its bones is a function of the cross-sectional area of the bone. Thus, an animal whose linear dimensions double would weigh eight times as much, but its structural supports would be only four times as strong. We expect, then, that animals of greatly different size will not be similarly proportioned – and this is usually what we find. Fig. 9.2 shows the femur of a pygmy marmoset expanded to the same length as a gorilla femur. The gorilla femur is thicker – an adaptation to support the gorilla's much greater volume and weight.

Such size-related scaling, first discussed by Galileo, is apparent throughout the animal kingdom. It explains why there are no 1-ton flying birds and why humans cannot jump like grasshoppers, or carry weight like ants

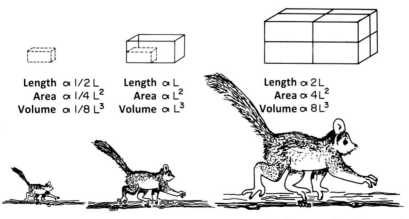

FIGURE 9.1 In a simple geometrical model, the halving or doubling of the length of an animal involves much greater changes in area or volume, which affect many physiological and structural aspects of an animal's life.

Primate Adaptation and Evolution
https://doi.org/10.1016/B978-0-12-815809-8.00009-6

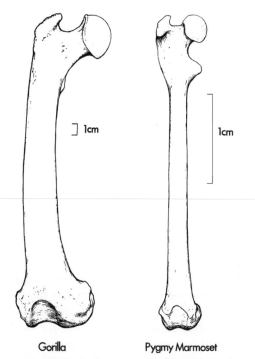

Gorilla   Pygmy Marmoset

FIGURE 9.2 The right femur of a gorilla (*Gorilla gorilla*) and a pygmy marmoset (*Cebuella pygmaea*) drawn to the same length. The bone of the gorilla is relatively much thicker than that of the small marmoset because it must support a relatively greater body mass for its length.

do. It also explains why cinematic fantasies of incredible shrinking or growing men and women are indeed fantasies. Similar considerations affect the scaling of other physiological functions, such as absorption in the digestive system. If the surface area of the digestive system increased in proportion to L^2 while the mass that must be fed increased in proportion to L^3, larger animals would have a relatively small digestive tract with which to process foods for much larger bodies (see Chivers and Hladik, 1980). If primate brains were all the same shape, larger species would have a relatively smaller brain surface area for any particular brain weight. There are many ways in which animals have dealt with such scaling issues in their evolution. Some larger primates have relatively longer intestines; some have different kinds of digestive organs; and some have different diets. Larger species tend to have more convoluted brains, so the ratio of surface area to brain weight remains approximately the same from species to species.

Metabolism is another physiological function that does not scale linearly with changes in body size, and thus it is an important consideration for understanding size-related differences in ecology and behavior. A primate's metabolic rate, or the amount of energy the individual requires for either basic body functions (basal metabolism) or daily activities (daily metabolic rate), scales in proportion to body mass raised to the power 0.75, not simply in direct proportion to body mass – or

to the surface area of the body, as had long been believed. Thus, larger animals expend relatively less energy and consequently need proportionately less food than smaller animals do. Put more simply, two 5 kg monkeys require more food than a single 10 kg monkey.

Primates have evolved in two ways to accommodate the constraints of scaling: by evolving different physical proportions (as in increasing brain convolutions) and by adapting lifestyles that capitalize on the scaling consequences. For example, whereas mechanical scaling would require that bone diameters scale in proportion to body mass$^{0.375}$ and simple geometric scaling would yield a scaling of diameters proportional to mass$^{0.333}$, most studies find that the scaling of bone diameter with body size is slightly greater than geometric scaling but not as great as a mechanical scaling would require. However, this slight positive scaling of bone thickness is often associated with postural and behavioral changes that reduce stress on bones with increasing body size (e.g., Biewener, 1990). This combination of mechanical and behavioral change thus generates a pattern of similar bone strain that has been called "dynamic similarity" (e.g., Demes and Jungers, 1993).

There is an increasingly large and sophisticated literature on size-related differences in many aspects of primate biology, including limb length, brain size, reproductive physiology, tooth size, and locomotor behavior. Altogether these studies of scaling are termed **allometry**, of which there are three general types (Fig. 9.3): **growth allometry** is the study of shape changes associated with size changes in ontogeny; **intraspecific allometry** is the study of size-related differences in adults of the same species; and **interspecific allometry** examines size-related differences across a wide range of different species for broader principles of scaling. If one variable increases faster than a baseline variable, such as the relationship of volume to area, that is described as **positive allometry**. Conversely, if one variable increases more slowly than a baseline variable, such as basal metabolism and body mass, that relationship is described as **negative allometry**.

Considerations of size are critical to our understanding of both primate adaptation and evolution, but a detailed discussion of this topic is beyond the scope of this book. In this discussion we concentrate on the role of size in ecological adaptation – i.e., the way primates of different sizes tend to have different ways of life. In considering these adaptive differences, it is often impossible to determine whether size-related differences in behavior and ecology are behavioral adaptations a species has adopted to "accommodate its size" or whether size changes themselves are better viewed as gross morphological adaptations that enable a species to better exploit a particular ecological niche. Size and adaptation are so intertwined that determining which precedes the other in evolution is akin to determining whether the

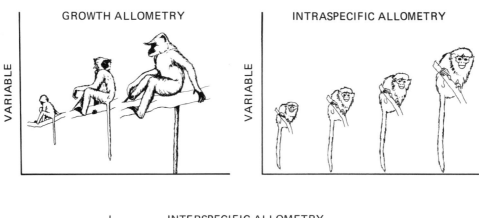

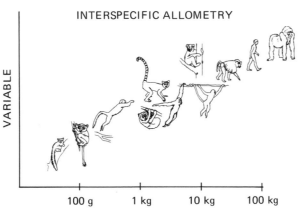

FIGURE 9.3   Graphs representative of three ways of examining the association of shape changes with size changes: growth allometry examines the shape changes associated with ontogenetic size increase; intraspecific allometry examines the shape changes associated with size differences among adults of a single species; interspecific allometry examines shape changes associated with size differences across a wide sample of different species. Allometric changes are usually plotted on a logarithmic scale so that exponential relationships appear linear.

chicken comes before the egg (e.g., Fleagle, 1985). Furthermore, any particular body size comes with both advantages and disadvantages. The best approach is to look for consistent associations between size and behavioral ecology, associations that may provide us with insight into the structure of both living and fossil primate communities.

Size and Diet

Primate diets are closely linked with body size (Fig. 9.4). Species that eat insects tend to be relatively small, whereas those that eat leaves tend to be relatively large. Fruit eaters tend to supplement their diets with either insects or leaves, depending on their size. These patterns result from the interaction of several independent, size-related phenomena. First, all primates need a balanced diet that meets not only their caloric (energy) needs but also their other nutritional requirements, such as protein and a variety of trace elements and vitamins. Although fruits are high in calories, they are usually very low in protein content (but see Ganzhorn et al., 2009); most primates must therefore turn to other sources

for their protein. The two most abundant sources of dietary protein for primates are other animals (such as insects and other invertebrates) and folivorous materials, such as leaves, shoots, and buds. Why, then, do small primates tend to eat insects and large ones folivorous material? Although these two protein strategies are in a nutritional sense complementary, the physiological and behavioral problems faced by a primate that feeds on these two dietary items are quite different.

Insects, and animal material in general, are an excellent source of nutrients (except calcium), fulfilling nearly all of a primate's requirements. Furthermore, insects are relatively high in calories per unit weight (Rothman et al., 2014). This is particularly important for small animals, which have relatively higher energy requirements than large ones (the shrew must eat several times its body weight in food every day). Insects are such a good food source that the real question is not why small primates eat them but why large ones do not. The answer seems to lie in the time normally involved in catching and handling insects. No primates have evolved the specialized abilities of anteaters to prey on large colonies of social insects; rather, they depend largely on locating and

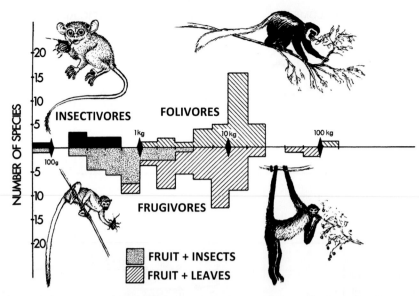

FIGURE 9.4 Primate dietary habits are correlated with body size. Insectivorous primates are relatively smaller than folivorous species. Smaller frugivorous species tend to supplement their diet with insects, and larger frugivorous species supplement their diet with leaves. *(Redrawn from Kay, R.F., 1984. On the use of anatomical features to infer foraging behavior in extinct primates. In: Cant, J., Rodman, P. (Eds.), Adaptations for Foraging in Nonhuman Primates. Columbia University Press, New York, pp. 21–53.)*

catching isolated individuals. It has been suggested – and it seems quite reasonable – that the number of insects that a primate can find and catch in a given day (or night) is likely to be relatively similar from species to species, regardless of size – assuming, of course, that they look in the appropriate places and have appropriate adaptations for insect catching, such as good grasping abilities and keen eyesight. In an 8-hour active period, any two primates might be able to ingest forty insects of one type or another. For a tiny tarsier, this catch could supply both its energy and protein requirements needed for the day; for a medium-sized monkey, however, this much food might supply all of its protein needs but not its energy requirements. Thus, although larger primates might supplement their fruity (high-energy) diet with insects, they cannot rely predominantly on insects in the way a small primate might.

Unlike insects, leaves are neither cryptic nor hard to catch, but they pose other problems for foraging primates. Although relatively high in protein (particularly young leaves, buds, and shoots), leaves also contain large amounts of less palatable components, such as cellulose or even toxins, a strategy plants have evolved to prevent or discourage predation on their leaves. Compared to insects or fruits, leaves are generally low in energy yield for their weight. Large body size helps a primate overcome some of these problems inherent in a leafy diet. First, large animals need less energy per kilogram of mass than do small animals. Thus, they can more easily afford to have a diet that is relatively lower in energy density. Second, although primates do not

have the enzymes needed to break down the cellulose in leaves, many are able to maintain colonies of microorganisms in part of their digestive tract to perform this task for them. This kind of digestion takes time, but the time it takes food to travel through an animal's digestive tract is roughly proportional to the length of the gut and hence to the animal's size. For this reason, a small primate with a short gut has less opportunity to digest plant fibers than does a larger animal with a longer gut. Furthermore, these longer, slower guts with special chambers for fermenting cellulose also seem to help detoxify some of the poisons. Thus, whereas the upper size limit of insect eaters seems to be imposed by the time required to locate and catch their prey, the lower size limit of folivores seems to be determined by metabolic and digestive parameters. In general, folivorous primates have body weights of no less than 500 g, whereas insectivores tend to weigh less than this limit. This natural physiological break at 500 g, known as **Kay's threshold**, applies throughout the order Primates (Kay, 1984).

Size and Locomotion

Like diet, locomotion shows general patterns of size-related scaling in primates. Terrestrial primates are usually larger than arboreal ones, both within taxonomic groups and for the order as a whole. Presumably, this difference reflects both the limited capability of arboreal supports to sustain large animals and perhaps also some amount of selection for large size among terrestrial species as a means of deterring potential predators.

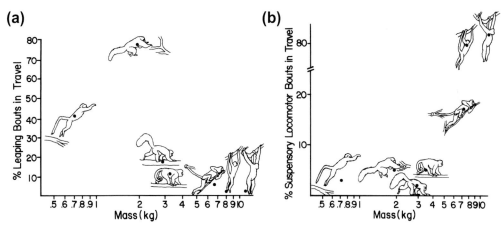

FIGURE 9.5   Primate locomotor behavior is correlated with body size. Among platyrrhine monkeys, (a) leaping is more common for smaller than for larger species and (b) suspensory behavior is more common for larger than for smaller species.

Within arboreal primates there are size-related trends in the use of different types of locomotion. Although we lack the extensive quantitative data on primate locomotion that we have for diet, the allometry of locomotor behavior has been quantitatively assessed for South American monkeys, and similar patterns seem to hold for the rest of the order (with some notable exceptions). In general, we find that leaping is more common among small primates (Fig. 9.5a), whereas suspensory behavior is more common in larger species (Fig. 9.5b). Like fruit eating, quadrupedal walking and running do not seem to show any pattern with respect to body size: there are small, medium, and large quadrupeds.

The trends we find in leaping and suspensory behavior seem to be the result primarily of simple mechanical phenomena (Fig. 9.6). Two primates, one small and one large, traveling through the forest canopy will each encounter gaps between trees that they somehow must cross to continue their journey. In the same forest, the smaller one will more frequently encounter gaps that it can cross only by leaping; the larger one will more frequently encounter gaps that can be crossed by bridging or by suspending itself between the terminal supports. Leaping, of course, involves the generation of high propulsive forces from the hindlimbs – and larger animals must generate greater forces to leap. Smaller animals will find more supports that can sustain their leaps than will larger animals. On the other hand, during both locomotion and feeding, larger animals will more frequently encounter supports too narrow or too weak to support their larger bodies and will more often need to suspend themselves below multiple branches for both support and balance (Fig. 9.7). Another relevant factor is the amount of energy a tree climber must absorb when it falls from a tree to the ground. Those animals with greater weight are likely to adopt the more cautious form of locomotion.

FIGURE 9.6   A small primate and a large primate traveling through the same forest are confronted with different locomotor problems because of the difference in their size. The small primate encounters relatively more gaps than can be crossed only by leaping, while the larger species encounters relatively more gaps that can be crossed by suspensory behavior or bridging.

All of these arguments support the scaling patterns seen in New World primates, and roughly present in the order as a whole. As with diet, there are notable exceptions, such as the small "suspensory" lorises (e.g., Terranova, 1996) or some of the larger saltatory colobines, but within taxonomic groups these broad patterns still seem to hold, so that within a community of colobines the smaller species leaps more than the larger one (McGraw, 1996).

Quadrupedal behavior seems to show no clear size restrictions: there are both large and small quadrupeds. Larger quadrupeds tend to move on larger supports, however, and the largest support is the ground. The

FIGURE 9.7   During feeding, small primates encounter more supports that can easily support their weight, while larger primates have to spread their weight over a large number of supports to feed at the same place.

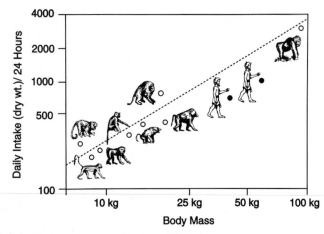

FIGURE 9.8   Total daily food intake increases with body size in living primates. *(From Barton, R.A., 1992. Allometry of food intake in free-ranging anthropoid primates. Folia. Primatol. 58, 56–59.)*

interesting exceptions to this pattern are animals that show other special adaptations, such as marmosets, which have claws for clinging to large tree trunks, or very suspensory animals, such as spider monkeys, which spread their weight over several relatively small branches.

Size and Life History

In addition to diet and locomotion, primate reproduction seems particularly closely linked to size (see Chapter 3). Although life history variables such as gestation period and lifespan increase with body size (larger primates generally take longer to grow up, and live longer), they have a negative allometric scaling, so a doubling of body size does not correspond to a doubling of gestation period or lifespan. Similarly, litter weight, or the size of the newborns, scales negatively with body size; that is, smaller primates have relatively larger babies. Obviously, the energetics of reproduction and growth are a major determinant of an individual's metabolic costs, especially for females. Although the causal relationships among size, life history variables, and ecology remain poorly understood, they are all closely integrated.

Size and Ecology

Various other aspects of primate ecology show important relationships with size. Within any habitat, smaller primates are certainly more susceptible to predation than are larger species, because they can be hunted by a larger range of predators. In some cases, size-related features of ecology may be just alternate expressions of the factors just discussed. For example, home range size for primate species increases with body size. This presumably reflects the need for larger animals to cover a wider area to support themselves, since the amount of food a primate ingests also increases with body size (Fig. 9.8; Barton, 1992). It has also been shown that primate group size increases with body size – larger species live in larger groups – but this relationship is more suspect and difficult to explain. Terrestrial species frequently aggregate into large groups, especially for sleeping and resting, as a strategy to avoid predation, but they may still forage in smaller units.

Adaptations to Diet

Diet is generally recognized as the single most important parameter underlying many of the behavioral and ecological differences among living primates, and primate diets have been more thoroughly documented than any other aspect of behavior. Food provides the energy that primates need for reproduction and seems to be the main objective of most of their daily activities. The use of hands to obtain and prepare food is a distinctive feature unifying the feeding habits of all primates, but as the previous chapters emphasize, primate species show a wide range of behavioral and morphological adaptations for obtaining and processing different types of food (Fig. 9.9).

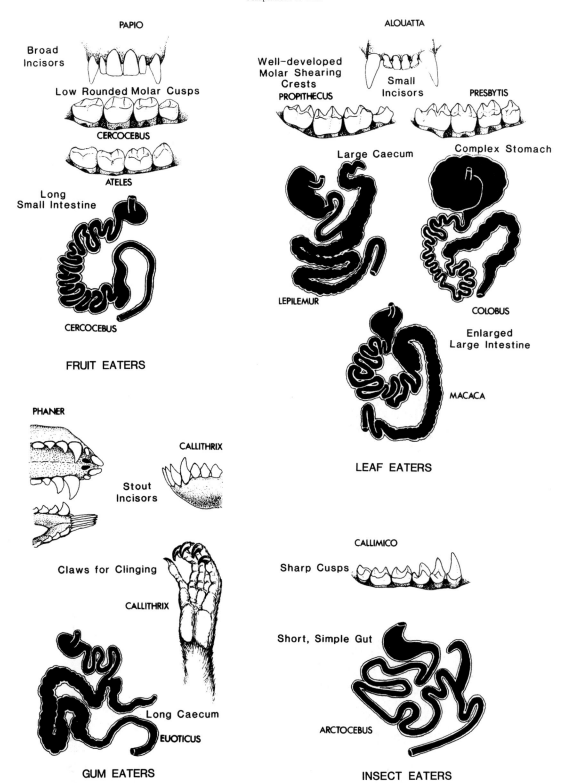

FIGURE 9.9 Morphological adaptations to diet among living primates. Fruit eaters tend to have relatively large incisors for ingesting fruits, simple molar teeth with low cusps for crushing and pulping soft fruits, and relatively simple digestive tracts without any elaboration of either the stomach or the large intestine. Leaf eaters have relatively small incisors, molar teeth with well-developed shearing crests, and an enlargement of part of the digestive tract for the housing of bacteria for the breakdown of cellulose. Gum (exudate) eaters usually have specialized incisor teeth for digging holes in bark and scraping exudates out of the holes, and claws or claw-like nails for clinging to the vertical trunks of trees. Many also have an enlarged cecum, suggesting that bacteria in the gut may function to break down the structural carbohydrates in gums or resins. Insect eaters are characterized by molar and premolar teeth with sharp cusps and well-developed shearing crests and digestive tract with a simple stomach and a short large intestine.

Dental Adaptations

The best-documented morphological adaptations to diet are those found in primate teeth, the organs primarily responsible for initial processing of food once it has been located (Ungar, 2002). Fortunately, as teeth are the parts most commonly preserved in the fossil record, they also provide us with considerable evidence for reconstructing the diets of extinct species (e.g., Ungar, 2002; Lucas, 2004). Many different aspects of the dentition have been shown to be related to diet, from measures of the overall size of teeth or their roots, to microscopic patterns of tooth wear (reviewed in Ungar, in Press).

The primary function of the anterior part of the tooth row, the incisors and canines, is ingestion, but these teeth also serve a wide range of nondietary functions, such as grooming and fighting. The role of canines and incisors in procuring or ingesting food is often not as food-specific as that of other parts of the tooth row. For example, primates that eat bark, insects, wood, or exudates may all have procumbent incisors for removing bark from trees to obtain food. Nevertheless, there are some general patterns linking incisor form with diet. Relative to the size of their molars, folivores tend to have smaller incisors than do frugivores, because leaves require less incisive preparation. Primates that feed extensively on exudates frequently have large procumbent incisors for digging holes in the bark of trees to elicit the flow of these fluids (Rosenberger, 2010).

The cheek teeth – the premolars and particularly the molars – break up food mechanically and prepare it for additional chemical processing further along the digestive system. Thus, the particular adaptations we see in molar teeth are generally not for specific foods, but for food items with particular structural properties or consistencies (e.g., Strait, 1997). There are major functional differences among primate molar teeth in the development of cusps and shearing crests or dental blades for reducing food items into small particles. Physiological experiments have demonstrated that the digestion of both insect skeletons and leaves is enhanced by reducing these food items into small pieces, thereby increasing the surface area. Thus, we find that insect eaters and folivores are characterized by molars with extensive development of these shearing crests. In folivores, this development of shearing crests is also associated with thin enamel on the tooth crown, an adaptation that creates even more shearing edges on the border between the superficial enamel and the underlying dentine once the teeth are slightly worn.

Although both insect eaters and folivores are characterized by well-developed shearing crests or blades, the optimal shapes of dental blades vary according to the physical properties of the food being cut (Strait, 1997; Fig. 9.10). For example, hard-bodied insects, which will

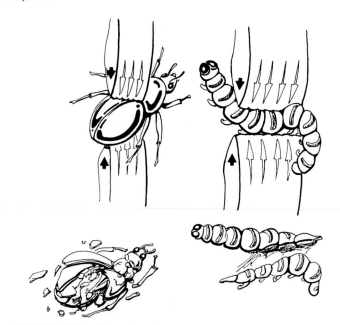

FIGURE 9.10   Short dental blades are more effective in concentrating forces to induce shattering of brittle objects, such as a hard-shelled beetle, whereas long blades are more effective in slicing ductile objects, such as a caterpillar.

shatter, are best cut with short shearing blades that concentrate force in a small area and generate self-propagating cracks. In contrast, softer insects require longer blades to separate the cut parts. Similar mechanical considerations show that the lophs of colobine teeth are effective both in propagating cracks to open up tough seeds and in cutting leaves (Lucas and Teaford, 1994). For other soft foods, such as fruits, teeth seem to be designed mainly to maximize surface area between the teeth and food items. Studies of the physical properties of primate foods, in conjunction with a better understanding of the mechanical significance of tooth shape, are providing many new insights into primate dental function.

In addition, several researchers have shown that gross measurements of overall cusp relief effectively distinguish primates in relation to major dietary categories (Boyer, 2008; Bunn et al., 2011; Kay, 1984; Evans et al., 2007). Thus, insectivores and folivores have greater cusp relief than do frugivorous taxa.

In addition to the many adaptations to diet in the gross morphology of primate teeth, differences in the microscopic structure of tooth enamel greatly affect the strength of its occlusal features, and developmental patterns of enamel structure frequently can be related to dietary habits (Lucas et al., 2008). Because teeth are in close contact with the food that a primate eats, food items (and any other material entering the mouth) frequently leave scratches on the surfaces of teeth. Although not adaptations to processing food, these scratches, or **microwear**, can provide clues to the diet of living and fossil animals. For example, primates eating hard insects show more pits on their teeth than do those eating soft insects, and primate folivores show fewer pits than do most frugivores, although their teeth lack the etching caused by acid fruits (Teaford, 1994).

Just as microwear is not an adaptation, but a feature obtained during the life of an individual, the isotopic composition of teeth and other tissues in an individual's body may reflect aspects of their diet (Crowley, 2012) and when these are measured in fossils can yield information about the dietary habits of extinct primates (Ungar and Sponheimer, 2013).

In contrast to the considerable success primatologists have had in relating dental anatomy to dietary habits, attempts to link differences in mandible shape and skull form with dietary differences have been considerably less successful, probably because cranial morphology serves so many diverse and often conflicting functions.

Other Oral Adaptations

Although teeth and jaws are the best-studied aspects of the primate mouth, other aspects of the oral cavity play important roles in the ingestion and initial preparation of foods (see Chapter 2). Some primates, especially great apes, use their lips extensively for grasping and processing food items. Chimpanzees, for example, often use their lips to extract the fluids from foods and then spit out the fibrous remains. Cercopithecine monkeys have cheek pouches that enable them to ingest large numbers of fruits rapidly without processing them immediately. Primates differ dramatically in the size and structure of the salivary glands, which play an important role in the initial digestion of foods as well as in helping control the pH balance of both the oral cavity and the initial part of the gut (e.g., Cuozzo et al., 2008).

Digestive Tract Adaptations

Although of little use to the paleontologist, the soft anatomy of the primate digestive system shows dietary adaptations as distinctive as those seen in the dentition (Fig. 9.9; e.g., Lambert, 1998). Whereas the dentition shows adaptations to relatively gross features, such as the size and mechanical characteristics of particular foods, the remainder of the digestive system shows adaptations to the chemical or nutritive properties of dietary items. Leaves and gums, for instance, which are very different in their consistency, require different dental adaptations but present similar problems for the remainder of the digestive system. Both are composed of long chains of structural carbohydrates and require extra processing chambers and the action of microorganisms.

In general, primate digestive systems show three different patterns of dietary adaptation. Faunivorous primates (mainly insect eaters, but also some omnivorous species) have a relatively short, simple digestive system with a small, simple stomach, usually a small caecum, and a very small colon relative to the size of the small intestine. In essence, the digestive system of a faunivore is devoted to absorption, the function of the small intestine. Frugivores also have relatively simple digestive systems, although large frugivores tend to have relatively large stomachs.

Folivores show the most elaborate adaptations in the visceral part of their digestive system because they must process foods containing large amounts of structural carbohydrates and also must overcome various toxins. Because primates lack the natural capability to digest the cellulose contained in the cell walls of plants, these elaborations of the visceral digestive system involve forming an enlarged pouch somewhere in the digestive tract to maintain a colony of microorganisms that can digest cellulose or other structural carbohydrates. The host primates then digest both the products of the bacterial action and the bacteria themselves. There are several possible solutions to this ranching situation, and different primate folivores grow their bacteria and break down cellulose in at least three different places.

Some folivorous strepsirrhines, such as *Lepilemur* and *Hapalemur*, have an enlarged caecum, a feature also seen in rabbits and horses. Colobine monkeys have an enlarged stomach with numerous sections, similar to, but much less elaborate than that of cows. Most other partly folivorous species, including indriids, apes (siamang and gorillas), New World monkeys (*Alouatta*), and some cercopithecine monkeys (*Macaca sylvanus*), accommodate the leafy portion of their diet by means of an enlarged colon. In addition to their role in breaking down the cellulose, it seems likely that the "fermenting" areas in the digestive systems of primate folivores help them overcome the various toxins found in many plant parts. This detoxification seems to be facilitated both directly, through actual chemical breakdown, and indirectly, by slowing down the rate at which food is processed to allow the liver more time to detoxify the absorbed food materials.

Although the visceral modifications for digestion of plant materials have been well studied in primates, there is less evidence about how and where primates break down other structural carbohydrates, such as those in gums (see Power, 2010) and the chitinous exoskeleton of invertebrates. There are anatomical indications, and a few physiological studies, suggesting that the processes used to digest these substances may be similar to those involved in cellulose digestion (e.g., Power, 2010), since primates with specialized diets of gums (*Galago*, *Cebuella*), or insects (tarsiers) are also characterized by a large caecum.

Primate Sensory Adaptations

Vision

Compared to most of our mammalian relatives, we primates are very unusual in our sensory biology. For most mammals, life tends to be dominated by smell. In contrast, vision plays a much greater role in the life of most primates than it does in many other mammals, and this is reflected in many aspects of primate anatomy that can be related to differences in visual abilities. An animal's ability to process visual information about its environment is dependent upon many factors, including the amount of ambient light available in the environment, the overall size of the eyeball (and more specifically the pupil) for admitting light into the eye, the number and nature of the cells in the retina that respond to light entering the eye, and the wiring between the retina and the brain. All of these factors vary among living primates according to differences in their behavior and visual abilities, and many can be related to bony differences in the skull, so that we can reconstruct the visual abilities and behavior of extinct taxa from the anatomy of their skull.

Eye and Orbit Size/Shape

If all other factors are equal, the quality of the image that a primate can form of the world around it is determined by the amount of light entering its eye, the absolute size of the retina, and the focal length of the eye. Larger eyes let in more light and have a larger retina for collecting that light. A longer focal length results in a more precise, higher-resolution image. Thus, in order to have visual abilities comparable to those of larger primates, smaller species need to have eyes that are relatively larger for their body size than those of larger species. In other words, the size of primate eyes, and the bony orbits that house them, shows negative allometry. Because there is less light available at night than there is during the day, nocturnal species have eyes and orbits that are larger than those of diurnal species of the same size, with cathemeral species showing an intermediate position (Kay and Kirk, 2000).

Compared to other mammals, primates have larger eyes relative to their body size, which is suggested to be an adaptation to improve visual acuity relative to other mammals by increasing the size of the retina and accompanying image (Ross and Kirk, 2007). Haplorhine primates have further improved visual acuity (see below) and image resolution, in part, by effectively increasing the focal length of their eyes relative to other mammals, including strepsirrhines (Ross and Kirk, 2007).

Visual Acuity

The sharpness of the image that a primate sees (visual acuity) is also determined by the relationship between the number of retinal cells and the number of nerves carrying information to the visual centers of the brain: a feature known as retinal summation. If one nerve carries the summed input from many retinal cells, the primate may be able to collect a visual signal in very low light, but that signal will not be very spatially precise. Alternatively, if there are many nerves carrying signals from a small number of cells, a more spatially precise signal can be produced, but more light will be required to produce a signal. Diurnal primates have less summation and greater visual acuity than nocturnal primates, and diurnal haplorhines have less summation and greater visual acuity than either nocturnal or diurnal strepsirrhines. Nerves carrying visual signals travel from the eye to the brain through the optic nerve, which travels through the optic foramen at the back of the orbit. An index of retinal summation can be measured as the ratio of the area of the optic nerve to the area of the retina, with high numbers indicating low summation and high acuity. This same relationship can be estimated in skulls by the area of the optic foramen (through which the optic nerve passes) relative to the area of the orbit, and thus estimates of retinal summation and visual acuity can be calculated for extinct primates (Kay and Kirk, 2000).

Color Vision

Primates are remarkable in the diversity of their abilities to distinguish colors, and the selective pressures underlying this diversity are subjects of considerable debate. We humans are similar to other catarrhines and Neotropical howling monkeys in having cone cells in our retina that are sensitive to three different wavelengths, giving us trichromatic color vision. In contrast, most other diurnal primates among the platyrrhines and strepsirrhines have only two types of cone cells, and thus have dichromatic color vision similar to that of "color-blind" human males. They are not able to distinguish reds and greens easily. However, color vision in several strepsirrhines and many platyrrhines is complicated by the presence of a polymorphism in the opsin gene located on the X chromosome. In those taxa, different alleles code for sensitivity to different wavelengths, and because the gene is X-linked, all males have dichromatic color vision but some females have trichromatic color vision (Jacobs, 2008).

The adaptive significance of dichromatic or trichromatic color vision in primates is a topic of considerable debate (e.g., Lucas et al., 2003). Many researchers have argued that trichromatic color vision is an adaptation to frugivory, that is, the ability to distinguish ripe from unripe fruit, or ripe fruit against a background of green foliage (e.g., Sumner and Mollon, 2000a,b; Regan et al., 2001). Experimental studies have found some support for this hypothesis (e.g., Caine and Mundy, 2000; Smith et al., 2003). However, Lucas and Dominy (e.g., Dominy and Lucas, 2001) have argued that trichromacy evolved as an adaptation for foraging on young, red leaves. This hypothesis accords well with the fact that the relatively folivorous howling monkey is the only fully trichromatic platyrrhine, as well as with some observational studies of primate foraging behavior (e.g., Lucas et al., 2003). In addition, ambient light levels and luminance contrast probably affect the utility of different types of color vision in ways that are still poorly understood (Hiramatsu et al., 2008).

Lynn Isbell (2009) offers a very different hypothesis for the evolution of color vision in primates. She argues that the regional and phylogenetic differences in the ability of primates to perceive color are adaptations to the types and coloration patterns of the most deadly snakes they are likely to encounter.

Balance, Movement, and Hearing

As a largely arboreal, and very vocal, group of mammals, primates show numerous adaptations to balance and hearing in the structure of their ear region.

Balance and Movement

In all vertebrates, movements of the head are detected by a system of three fluid-filled semicircular canals housed in the petrous part of the temporal bone. The sensitivity and other functional aspects of these canals are determined by their position in the skull and their relative dimensions and shapes (e.g., Walker et al., 2008; Malinzak et al., 2012). Thus, fast-moving or acrobatic species such as leapers (e.g., galagos or sifakas) have canals with a relatively larger radius of curvature than do slower-moving quadrupeds (such as lorises or cheirogaleids). Because the petrous portion of the temporal bone is one of the densest parts of a primate skeleton, it is often preserved in the fossil record and researchers can use semicircular canal morphology to reconstruct aspects of locomotion for extinct species.

Hearing

Vocal communication plays a large role in the behavior of most primates. The characteristics, context, and function of primate vocalizations are the subject of a large and sophisticated literature. Vocalizations play an important role in primate systematics, especially among nocturnal taxa, and in the investigation of cognitive abilities. In addition, a number of recent studies have demonstrated relationships between aspects of the bony morphology of the ear region and the sensitivity of a species to different sound frequencies (Kirk and Gosselin-Ildari, 2009; Coleman and Colbert, 2010). Because the auditory region is commonly preserved in fossil taxa, inferences can also be made about the hearing abilities of fossil primates (Coleman et al., 2010).

Locomotor Adaptations

Primates show considerable diversity in locomotion, and primate locomotor adaptations are found in many parts of the body. Most of the differences we see in the anatomy of the limbs and trunk of living primates are clearly related to differences in their locomotor and postural abilities – the way they move, hang, and sit. Locomotion and posture also affect the orientation of the head on the trunk, the shape of the thorax, and the positioning of abdominal viscera (Fleagle and Lieberman, 2013; Gebo, 2014; Larson, 2018).

Like many other adaptations, the modifications of the musculoskeletal system related to locomotor differences are influenced by the ancestry of the group being considered, and primates often have evolved different solutions to the same problem. Evolution by natural selection has worked with the available material. Thus, quadrupedal lemurs, quadrupedal monkeys, and quadrupedal apes all show similarities related to their quadrupedal habits, but they also show affinities to other lemurs, monkeys, and apes. For the paleontologist, this is a real advantage: it means that bones can provide information about both phylogeny and adaptation – if the two can be accurately distinguished.

Because locomotor adaptations may have different expressions in different species, our best approach is to examine the mechanical problems that different types of locomotion present. Then we can consider how living primate species have evolved musculoskeletal differences to meet these mechanical demands. We will concentrate on features of the skeleton that can be related to different postures and methods of progression because these are the best-documented aspects of primate locomotor anatomy and those that are most useful in reconstructing the locomotor habits of fossils. It is important to realize that such correlations between bony morphology and locomotor behavior are constantly being tested and refined by experimental studies that permit a clearer understanding of the biomechanical and physiological mechanisms of primate locomotion.

Arboreal Quadrupeds

Arboreal quadrupedalism is the most common locomotor behavior among primates, and most radiations of primates include arboreal quadrupeds. In many respects, arboreal quadrupeds show a generalized skeletal morphology that can easily be modified into any of the more specialized locomotor types, and it is likely that this type of locomotor behavior characterized both the earliest mammals and the earliest primates (Fig. 9.11).

Quadrupeds, by definition, use four limbs in locomotion. Experimental evidence suggests that in primates, as in most mammals, the hindlimbs play a greater role in

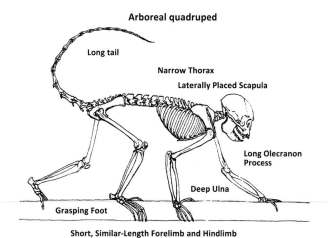

Arboreal quadruped

Long tail

Narrow Thorax

Laterally Placed Scapula

Long Olecranon Process

Deep Ulna

Grasping Foot

Short, Similar-Length Forelimb and Hindlimb

FIGURE 9.11 The skeleton of a primate arboreal quadruped, illustrating some of the distinctive anatomical features associated with that type of locomotion.

propulsion than the forelimbs (Demes et al., 1994; Granatosky et al., 2018). The major problem arboreal quadrupeds face in their locomotion is providing propulsion on an inherently unstable, uneven support that is usually very small compared with the size of the animal. Thus, stability and balance are their major concerns.

The overall body proportions of arboreal quadrupeds are adapted in several ways to meet these challenges of balance and stability. These primates have forelimbs and hindlimbs that are more similar in length than are those of either leapers, which have relatively long hindlimbs, or climbers, which have relatively long forelimbs. In addition, the forelimbs and hindlimbs of arboreal quadrupeds are usually short, to bring the center of gravity closer to the arboreal support. Most arboreal primates may also bring the center of gravity closer to the support by using more flexed limbs when they walk on arboreal supports than when on terrestrial supports (Schmitt, 1994). In addition, they tend to move more slowly. Finally, many have a long tail, which aids in balancing. The grasping hands and feet of most arboreal quadrupedal primates provide both a firm base for propulsion and a guard against falling.

The forelimbs of arboreal quadrupeds show a number of distinctive osteological features related to their typical postures and method of progression. The relatively mobile shoulder joint is characterized by an elliptically shaped glenoid fossa on the scapula and a broad humeral head surrounded by relatively large tubercles below the head for the attachment of the scapular muscles that control the position of the head of the humerus. The humeral shaft is usually moderately robust, as the forelimb plays a major role in both support and propulsion. The elbow region of an arboreal quadruped is particularly diagnostic. On the distal end of the humerus, the medial epicondyle is large and directed medially. This process, where the major flexors of the wrist and some

of the finger flexors originate, provides leverage for these muscles when the hand and wrist are in different degrees of pronation and supination. The olecranon process of the ulna is long, to provide leverage for the triceps muscles when the elbow is in the flexed position characteristic of arboreal quadrupeds. Because the elbow rarely reaches full extension, the olecranon fossa of the humerus is shallow. The ulnar shaft is relatively robust and often more bowed and deep in arboreal quadrupeds than in many other locomotor types. At the wrist, arboreal quadrupeds are characterized by a relatively broad hamate, presumably for weightbearing, and a midcarpal joint that seems to permit extensive pronation.

As a group, primates are characterized by relatively long digits and grasping hands. Among primates, however, arboreal quadrupeds usually have digits of moderate length – longer than those of terrestrial quadrupeds but shorter than those of suspensory species. They show a wide range of grasps.

The most distinctive features of the hindlimb joints of an arboreal quadruped reflect the characteristic abducted posture of that limb. The femoral neck is set at a moderately high angle relative to the shaft, enhancing abduction at the hip. At the knee, the abduction of the hindlimb is expressed in the asymmetrical size of the femoral condyles and their articulating facets on the top of the tibia. At the ankle, the tibiotalar joint is also asymmetrical. The lateral margin of the proximal talar surface is higher than the medial margin, reflecting the normally inverted posture of the grasping foot. Arboreal quadrupeds all have a large hallux and moderately long digits.

Terrestrial Quadrupeds

Compared to their abundance among other orders of mammals, terrestrial quadrupeds are relatively rare among primates, and none show the striking morphological adaptations found in runners such as cheetahs and antelopes. The main group of primate terrestrial quadrupeds is the larger Old World monkeys: baboons, mangabeys, mandrills, some macaques, and the patas monkey. These species show a number of distinctive anatomical features that separate them from more arboreal species (e.g., Frost, 2001; Fleagle and McGraw, 2002; Sargis et al., 2008; Arenson et al., 2020). Most of these features relate to the use of more extended, adducted limb postures on a broad, flat surface. Because balance is not a problem, these primates have a narrow, deep trunk and relatively long limbs, designed for long strides and speed, and their tails are often short or absent (Fig. 9.12).

The limbs of terrestrial quadrupeds seem designed for speed and simple fore-and-aft movements rather than for power, and more complex rotational movements at the joints. At the shoulder joint, the articulating surfaces of the scapula and head of the humerus provide

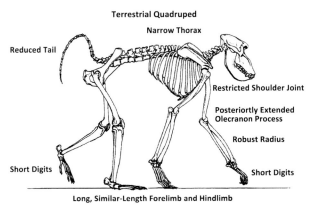

FIGURE 9.12 The skeleton of a primate terrestrial quadruped, illustrating some of the distinctive anatomical features associated with that type of locomotion.

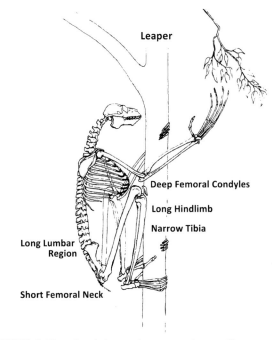

FIGURE 9.13 The skeleton of a primate leaper, illustrating some of the distinctive anatomical features associated with that type of locomotion.

only a limited anterior–posterior motion, and the greater tuberosity of the humerus is high and positioned above the humeral head in front of the shoulder joint to stabilize that joint during the support phase of locomotion (Larson and Stern, 1989).

Terrestrial quadrupeds have an elbow joint that reflects their more extended limb postures. Instead of being long and extending proximally, as in arboreal quadrupeds, the olecranon process extends dorsally to the long axis of the ulna, an orientation that maximizes the leverage of the elbow-extending muscles when the elbow is nearly straight rather than flexed. A related feature is the deep olecranon fossa on the posterior surface of the humerus, with an expanded articular surface on the lateral aspect of the fossa for articulation with the ulna. The articulation of the ulna with the humerus is relatively narrow, whereas the head of the radius is relatively large, oval in shape, and flattened proximally (Rose, 1988), suggesting that the latter bone plays a more important role in transmitting weight from the elbow to the wrist in terrestrial quadrupeds than in other primates. The medial epicondyle of the humerus is short and directed posteriorly, an orientation that facilitates the use of the wrist and hand flexors when the forearm is pronated, the normal position for terrestrial species.

The carpal bones of terrestrial quadrupeds are relatively short and broad, more suitable for weightbearing and less adapted for rotational movements. Their hands are characterized by robust metacarpals and short, straight phalanges.

The hindlimbs of terrestrial quadrupeds, like their forelimbs, are long. Their feet have robust tarsal elements, robust metatarsals, and short phalanges.

Leapers

Many primates are excellent leapers, and leaping adaptations have almost certainly evolved independently in many primate groups. Although there are

many differences among primate leapers, there are also a number of similarities resulting from the mechanical demands of such movement (Fig. 9.13). In leaping, most of the propulsive force comes from a single rapid extension of the hindlimbs, with little or no contribution from the forelimbs. The leaper's takeoff speed, and hence the distance the animal can travel during a leap, is proportional to the distance over which the propulsive force is applied – the length of its hindlimbs. Longer legs thus enable a longer leap from the same locomotor force. Relative to the length of the hindlimb, leapers have relatively short, slender forelimbs. Although the forelimbs are certainly used in landing after leaps, in clinging between leaps, and for various other tasks, including feeding, their most important role during leaping is probably for control of rotational forces during flight (Demes et al., 1996). The anatomy of the vertebral column varies considerably among leapers, in association with trunk posture. More quadrupedal species seem to have relatively long and flexible lumbar regions; however, indriids, which leap from a vertical posture, have a relatively short and stiff lumbar region (Shapiro, 1995).

There are many skeletal adaptations for leaping to be found in the hindlimb. Because hip extension is a major source of propulsive force in leaping, primate leapers usually have a long ischium, which increases the leverage of the hamstring muscles. The direction in which the ischium is extended depends on the postural habits of the species. In primates that leap from a quadrupedal position, the ischium extends distally, in line with the blade of the ilium, enhancing hip extension when the hindlimb is at a right-angle to the trunk. In prosimians that normally leap from a vertical clinging posture, the ischium is usually extended posteriorly rather than distally, increasing the moment arm of the hamstrings when the limb is near full extension, a common situation for vertical clingers (Fleagle and Anapol, 1992).

Whereas arboreal quadrupeds use abducted limbs for balancing on small supports, leapers restrict their limb excursions to simple, hinge-like flexion and extension movements, both for greater mechanical efficiency and to avoid twisting and damaging joints during the powerful takeoff. In this regard, leapers resemble swift quadrupedal mammals. Many features of the hindlimbs of leapers seem related to this alignment of movement and to increasing the range of flexion and extension. For example, the neck of the femur is very short and thick in leapers, and in many species the head of the femur has a cylindrical shape for simple flexion–extension movements, rather than the ball-and-socket joint found at the hips of most primates. At the knee joint, the femoral condyles are very deep, to permit an extensive range of flexion and extension, and they are symmetrical because of the adducted limb postures. The patellar groove has a pronounced lateral lip to prevent displacement of the patella during powerful knee extension. The tibia is usually very long and laterally compressed, reflecting the emphasis on movement in an anterior–posterior plane, and the attachments for the hamstring muscles on the tibial shaft are relatively near the proximal end so that when these muscles extend the hip they do not flex the knee as well. In many leapers the fibula is very slender and bound to the tibia distally, so that the ankle joint becomes a simple hinge joint for flexion and extension. The morphology of the tarsal region varies considerably among leapers. In many small leapers, the calcaneus and navicular are extremely long, providing a long load arm for rapid leaping. The digits of leapers seem to reflect postural habits rather than adaptations directly related to leaping.

Suspensory Primates

Many living primates hang below arboreal supports by various combinations of arms and legs. Because of the acrobatic nature of such behavior, the skeletons of suspensory primates show features that enhance their abilities to reach supports in many directions (Fig. 9.14). In their body proportions, suspensory primates have long limbs, especially forelimbs. Their trunks are relatively short and have a broad thorax, a broad fused sternum, and a very short lumbar region to reduce bending of the trunk during hanging and reaching.

The relatively deep, narrow scapula of suspensory primates is positioned on the dorsal rather than the

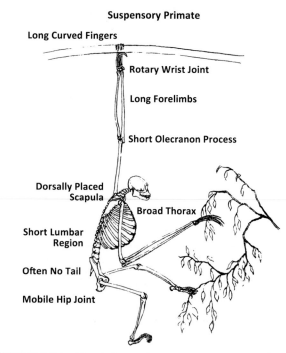

FIGURE 9.14   The skeleton of a suspensory primate, illustrating some of the distinctive anatomical features associated with that type of locomotion.

lateral side of the broad thorax, enhancing their reach in all directions. The shoulder joint, which faces upward to aid reaching above the head, is composed of a relatively small, round glenoid fossa and a very large, globular humeral head with low tubercles placed well-below the head, a combination that permits a wide range of movement. Because elbow extension is important but does not need to be powerful, the olecranon process on the ulna is short. The medial epicondyle of the humerus is large and medially oriented to enhance the action of the wrist flexors at all ranges of pronation and supination. Both the ulna and the radius are usually relatively long and slender.

Suspensory primates show numerous features of the wrist that seem to increase the mobility of that joint. In many species, the ulna does not articulate with the carpals, and the distal and proximal rows of carpal bones form a ball-and-socket joint with increased rotational ability. Suspensory species have long fingers with curved phalanges for grasping a wide range of arboreal supports. Like the forelimb, the hindlimb of suspensory primates is characterized by very mobile joints. Mobility at the hip joint is increased by a spherical head of the femur set on a highly angled femoral neck to permit extreme degrees of abduction. The knee joint is characterized by broad, shallow femoral condyles and a shallow patellar groove. There is very little bony relief on the talus at the ankle joint, a condition that allows movement in many directions rather than restricting it to one plane. In most species, the calcaneus has a short lever arm for the calf muscles that extend the ankle; there is, however, an additional process for the origin of the short flexor muscle of the toes, to enhance grasping. The feet of suspensory primates, like their hands, have long, curved phalanges for grasping branches.

Bipeds

One of the most distinctive types of primate locomotion is the bipedalism that characterizes humans. The mechanics and dynamics of human locomotion have been more thoroughly studied than those of any other type of animal movement, but many aspects of human locomotion are still poorly understood. Compared to other types of primate locomotion, bipedalism is unusual in that only one living species habitually moves in this way.

The major bony features associated with bipedalism (Fig. 9.15) are found in the trunk and lower extremity. The upper extremity of humans, like that of leapers, does not normally play a role in locomotion and is adapted for other functions. The major mechanical problems faced by a bipedal primate are balance, particularly from side to side, and the difficulty of supporting all of the body weight on a single pair of limbs. One of the most striking correlates of our upright posture is the dual curvature of our spine, with a dorsal convexity (kyphosis) in the thoracic region and a ventral convexity (lordosis) in the lumbar region. In most other primates the kyphosis extends the entire length of the spine; the unique human lumbar curvature moves the center of mass of the trunk forward and also brings it closer to the hip joint. In keeping with our vertical posture, the size of each vertebra increases dramatically from the cervical region to the lumbar region, for each successive vertebra must support a greater part of the body mass (Fig. 9.15).

The human pelvis is the most unusual in the entire primate order. It has a very short, broad iliac blade that serves to lower the center of gravity and to provide better balance and stability. This arrangement also places many of the large hip muscles on the side of the lower limb rather than behind it; in this position, they can act to balance the trunk over the lower limbs during walking and running. The human ischium, where the hip extensors originate, is shortened and extended posteriorly (as in vertical leapers) rather than inferiorly (as in most other primates). This position provides greater leverage for the major hip extensors to move the lower limbs behind the trunk.

The human femur is characterized by a very large head, which must support the weight of the entire body during much of the locomotor cycle. Unlike most other primates, humans are naturally knock-kneed: our femur is normally aligned obliquely, with the proximal ends much further apart than the distal ends. This alignment

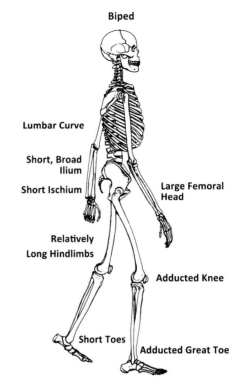

FIGURE 9.15 The skeleton of a bipedal primate, illustrating some of the distinctive anatomical features associated with that type of locomotion.

of the femur (called a valgus position) has the effect of placing the knees – as well as the legs, ankles, and feet – directly beneath the body rather than at the sides. As a result, successive footsteps involve less lurching from side to side, and during those parts of the walking cycle when only one limb is on the ground, that limb is always near the midline of the body (its center of gravity). This oblique orientation of the femur is reflected in many of its bony details, such as the long, oblique neck and the angle between the distal condyles and the shaft. A disadvantage of this oblique alignment of our femur is that it predisposes us to patellar dislocation, because the muscles extending the knee are now located lateral to the knee itself. To keep the small patella in place, we have developed a very large bony lip on the lateral side of the patellar groove.

In contrast with the grasping, hand-like foot of most primates, our foot has been transformed into a rather rigid lever for propulsion. The long tuberosity on the calcaneus forms the lever arm, while the stout metatarsals and the large hallux aligned with the other digits provide a firm load arm. The phalanges on our toes are extremely small, because they are not used for grasping, only for pushing off. The strong ligaments on the sole of the foot bind the tarsals and metatarsals together to form two bony arches that act to some degree as spring-like shock absorbers. In addition, they direct the body weight through the outside of the foot during each stride, providing us with our characteristic human footprint.

Humans are not just bipeds but are also more adept at endurance running than any other mammals. With only one living species, it is difficult to distinguish adaptations for running in particular from features more broadly related to bipedalism (Bramble and Lieberman, 2004). However, as discussed in Chapter 17, the fossil record provides a much broader comparison of bipedal hominids where some of these distinctions can be teased out.

Locomotor Compromises

In the previous sections we have portrayed primates that are somewhat hypothetical and idealistic – primates adapted for a single type of locomotion. But, as discussed in earlier chapters, most primates habitually use many types of locomotion, just as they eat many types of food. For example, many arboreal quadrupeds often leap, and some leapers are also suspensory. Nevertheless, it is reassuring that the features discussed earlier seem to distinguish not only primates that always leap from primates that always move quadrupedally, but also those species that leap more and are less quadrupedal from those that leap less and are more quadrupedal (Fig. 6.18). We can therefore have confidence that these features are likely to be useful in reconstructing the habits of extinct primates known only from bones.

There are other factors to consider when trying to understand how primate skeletons are related to locomotor habits. The same parts of the body that are used in locomotion play other roles in the animal's life. Hands are used both in locomotion and for obtaining food, perhaps catching insects, picking leaves, or opening seed pods. The bony pelvis is an anchor for the hindlimb and a site for the origin of many hip muscles, but it also supports the abdominal viscera and serves as the birth canal in females. Such multiple, often conflicting functional demands are present throughout the body, and often complicate attempts to identify features that are uniquely related to one type of movement, or to reconstruct the locomotor abilities of an extinct primate from bits of the skeleton. Still, many of the bony features discussed earlier, as well as numerous others (which can be found in more technical treatises), have proved to be generally characteristic of animals with particular locomotor habits and should provide useful evidence for reconstructing fossils.

Locomotion, Posture, and Ecology

Why do primates show such diverse locomotor and postural abilities and all the morphological specializations that accompany them? One factor is certainly size.

As discussed earlier in this chapter, within the same habitat, large and small primates are likely to face very different problems in terms of balance and the availability of strong-enough supports. Thus, larger species are more likely to be suspensory or terrestrial.

Apart from size, the major adaptive significance of different locomotor habits seems to be the access they provide to different parts of a forest habitat. In different types of forests and at different vertical levels within a forest, the density and the arrangement of available supports for a primate to move on are often quite different. Primates that live in open areas are best adapted to terrestrial walking and running. Even within a tropical rainforest, the available supports in the understory are different from those higher in the canopy, and species that travel and feed in different levels have different methods of moving (e.g., Cannon and Leighton, 1994). The lowest levels of most forests are characterized by many vertical supports, such as tree trunks and lianas, but there are few pathways that are continuous in a horizontal direction (Fig. 3.3). Primates that feed and travel in the understory are often leapers that can move best between discontinuous vertical supports. Higher, in the main canopy levels, the forest is usually more continuous horizontally and suitable for other methods of progression, such as quadrupedal walking and running or suspensory behavior.

We know remarkably little about the relationship between postural activities and aspects of primate ecology. Although most postural behavior elicits lower levels of bone strain and muscle activity than does locomotor behavior, it certainly occupies a major portion of an individual's activity budget. Clearly, there are particular types of posture that seem related to specific habitats or food sources. Primates that regularly eat gums or other tree exudates often have claws or claw-like nails so that they can cling to large tree trunks (Fig. 9.9). Among Old World monkeys, colobines tend to sit while feeding, whereas more insectivorous cercopithecines tend to stand (McGraw, 1998). It is likely that different postural abilities enable species to exploit different parts of the same resource.

However, except for special cases such as gum eaters, there are very few general associations between the patterns of locomotion and posture used by primates and their dietary habits. Primates that live in bamboo forests (*Hapalemur, Callimico*) are almost always leapers because of the predominance of vertical supports. However, it is more frequently the case that, among sympatric species, those with the most similar diets show the greatest locomotor differences; at the same time, those with the most similar locomotion show the greatest dietary differences. This suggests that primates have often evolved locomotor differences for exploiting similar foods in different parts of their environment and vice versa (Fleagle and Mittermeier, 1980; Walker, 1996).

Polygynous Social System
Dimorphic Canines

Monogamous Social System
Monomorphic Canines

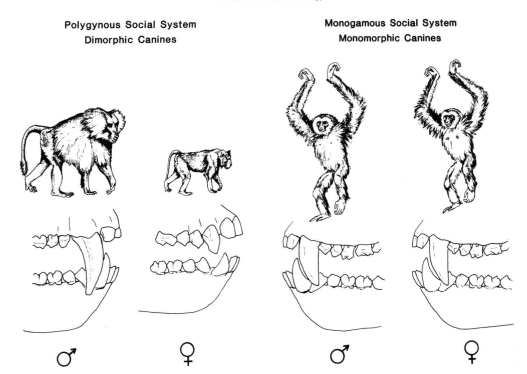

FIGURE 9.16   Canine differences between monogamous gibbons (*Hylobates*) and polygynous baboons (*Papio*).

It is also likely that many of the ways in which loco-motion contributes to a species' foraging habits have not been properly studied. As noted in Chapter 3, we nor-mally categorize foods into fruits, leaves, and insects, a classification that accords well with the mechanical and nutritional properties of dietary items. But for under-standing locomotion, we should perhaps classify foods according to their distribution in the forest, the shapes of the trees in which they are found, or the size of the branches from which they can best be harvested. Locomotor habits are certainly an integral part of pri-mate feeding strategies, and the subtle nature of this relationship deserves more study.

Anatomical Correlates of Social Organization

As we have discussed in previous chapters, primates live in many different types of social groups, and the reproductive strategies of individuals of different ages and sexes vary dramatically from species to species. There are a number of general anatomical and physio-logical features that seem to characterize species that live in particular types of social groups (Nunn and van Schaik, 2002; Plavcan, 2013). Among higher primates, the degree of canine dimorphism is closely associated with the type and amount of intrasexual competition normally found among the males and females of that species (e.g., Plavcan, 2002, 2013). Those species charac-terized by intense male–male competition, compared to

the amount of female–female competition, have greater canine dimorphism than those in which intrasexual competition is less or equal (Figs. 9.16–9.18). Body size dimorphism shows a strong correlation with patterns of intrasexual competition, but other variables, such as pre-dation and phylogeny, seem to be important as well. Not surprisingly, soft tissues of the reproductive system also show strong associations with patterns of social behav-ior. Testicular size shows a close association with the amount of intrasexual competition among males (Fig. 9.18). Monogamous and single-male species show little mating competition within a group, so the testes are relatively small. On the other hand, multimale and polyandrous groups exhibit considerable male–male competition for mating success, and males have rela-tively large testes (Harvey and Harcourt, 1984; Harcourt, 1995). Concomitantly, female catarrhines that live in multimale groups usually have sexual swellings that advertise their reproductive status throughout the men-strual cycle. There are, however, several very different hypotheses as to why these swellings evolved (Hrdy and Whitten, 1986; Sillén-Tullberg and Møller, 1993).

Brains, Behavior, and Ecology

Probably the most intriguing relationship between anatomy and behavior is the question of why primates have relatively large brains. Discussion about the adap-tive significance of primate brain size has largely

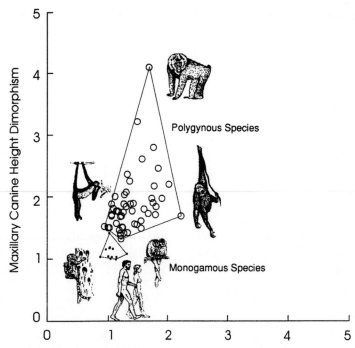

FIGURE 9.17   Relationship between relative canine dimorphism, body size dimorphism, and mating system. *(Courtesy of M. Plavcan.)*

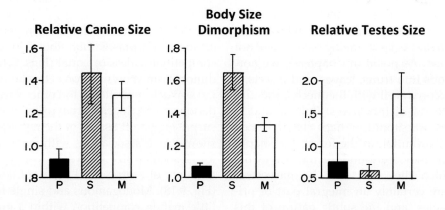

P - Polyandrous or Monogamous   S - Single male   M - Multimale

FIGURE 9.18   Morphology and social organization. Canine dimorphism and body size dimorphism separate monogamous and polyandrous species from polygynosus species; relative testes size separates primate species living in multimale groups. *(Adapted from Harvey, P.H., Harcourt, A.H., 1984. Sperm competition, testes size, and breeding systems in primates. In: Smith, R.L. (Ed.), Sperm Competition and the Evolution of Animal Mating Systems. Academic Press, London, pp. 589–600.)*

involved two types of causal variables: diet and foraging; or aspects of social behavior, including group size and patterns of social interaction. In the course of their foraging behavior, primates may consume hundreds of different plant parts of many different species with different patterns of temporal availability, spread over many hectares. The cognitive rules that primates follow in their foraging behavior are only now being investigated, but they seem to involve consideration of many different variables (e.g., Milton, 2000; Janson, 2000, 2007). Many researchers have linked interspecific differences in

primate brain size to differences in diet and foraging requirements. It has been suggested that the cognitive demands of keeping track of patterns of fruit availability are more difficult than those of obtaining a more folivorous diet. Thus, the frugivorous spider monkeys have relatively larger brains than their folivorous relatives, howling monkeys, and frugivorous chimpanzees have relatively larger brains than the more folivorous gorillas. In addition, some of the primates with the relatively largest brains, capuchin monkeys and the aye-aye, are best described as extractive foragers.

In contrast to the arguments for the importance of diet and foraging on brain size, others, notably Dunbar and colleagues (e.g., Dunbar, 1998) have argued that differences in brain size are primarily driven by differences in the number and nature of social interactions. Dunbar has shown that brain size, and especially neocortex size, is closely correlated with the average social group size for a species. Thus, primate neocortex size reflects the need to keep track of complex social relationships in an overall life history context as individuals in a primate group are born, change roles and status, immigrate and emigrate (Dunbar and Shultz, 2007).

Most recent analyses looking at ecological versus social variables in primate brain size evolution find the strongest association between diet and brain size, suggesting that ecological variables are likely more important than social ones (DeCasien et al., 2017; Chambers et al., 2021; Lopez-Aguirre et al., 2022). In particular, an omnivorous diet seems consistently correlated with larger overall brain size in primates (DeCasien et al., 2017; Chambers et al., 2021). However, as Barton (2000, 2006) has argued, the actual mechanisms of brain and brain size evolution are certainly more complex. The neocortex and the overall brain are composed of many different parts, with both different and interconnected functions. Thus, in one sense, foraging abilities and social interactions depend greatly on the ability to process and interpret visual signals, as well as information from other senses. Thus, current approaches to understand the history of brain evolution involve consideration of the interactions of many different aspects of brain function in an evolutionary and phylogenetic context (e.g., Dunbar and Schultz, 2007; Montgomery et al., 2010; DeCasien et al., 2017; Powell et al., 2017; Chambers et al., 2021; Grabowski et al., 2022).

Adaptation and Phylogeny

Although primates show tremendous diversity in morphology and behavior related to diet, locomotion, and social organization, it is largely true in any overall assessment of either morphology or behavior that closely related species are more similar to one another than to distantly related species. The reason for this is that evolution by natural selection does not begin with a blank slate for every species; rather, evolution, by definition, is the modification of ancestral forms through descent. Thus, despite showing many anatomical differences related to their locomotor adaptations, leaping and quadrupedal langurs are, in general, more similar to one another than they are to leaping and quadrupedal platyrrhines, respectively. Likewise, the teeth of frugivorous and folivorous Old World monkeys can be readily distinguished from one another, but they are more similar

to each other in many respects than they are to the teeth of frugivorous and folivorous lemurs.

Not only does natural selection work with the raw material available locally, but it also does not necessarily generate identical optimal solutions to similar ecological problems. In many cases, the adaptations found in a species or group of species may be constrained or guided by the ancestral condition from which the species or group has evolved, as well as by other aspects of the species' anatomy or behavior. Thus, indriids show very different types of adaptations to both the mastication and digestion of leaves than do colobine monkeys. Similar patterns of phylogenetic similarity are evident in many aspects of social behavior. Despite their tremendous ecological success in many different habitats, all Old World monkeys share many features of their social behavior that distinguish them from other primates (Di Fiore and Rendall, 1994).

The fact that closely related species may be more similar to one another for historical reasons rather than strictly adaptive ones is a potentially very serious complication for identifying adaptations, and it is essential that comparative studies make some effort to take phylogeny into account (Purvis and Webster, 1999; Nunn and Barton, 2001; Ross et al., 2002). However, adaptation and phylogeny should not be viewed as alternative explanations for primate diversity. Phylogenetic evolution is largely the result of adaptation, and all adaptations have evolved in a phylogenetic context. Indeed, the relationship of adaptation and phylogeny is at the heart of evolutionary biology, and we can never properly study one without the other (Nunn, 2011).

References

Arenson, J.A., Sargis, E.J., Hart, J.A., et al. 2020. Skeletal morphology of the lesula (*Cercopithecus lomamiensis*) and the evolution of guenon locomotor behavior. Am. J. Phys. Anthropol. 172, 3–24

Barton, R.A., 1992. Allometry of food intake in free-ranging anthropoid primates. Folia Primatol. 58, 56–59.

Barton, R.A., 2000. Ecological and social factors in primate brain evolution. In: Boinski, S., Garber, P. (Eds.), On the Move: How and Why Animals Travel in Groups. Chicago University Press, Chicago.

Barton, R.A., 2006. Primate brain evolution: Integrating comparative, neurophysiological and ethological data. Evol. Anthropol. 15, 224–236.

Biewener, A.A., 1990. Biomechanics of mammalian terrestrial locomotion. Science 250, 1097–1103.

Boyer, D.M., 2008. Relief index of second mandibular molars is a correlate of diet among prosimian primates and other euarchontan mammals. J. Human Evol. 55, 1118–1137.

Bramble, D.M., Lieberman, D.E., 2004. Endurance running and the evolution of *Homo*. Nature 432, 345–352.

Bunn, J.M., Boyer, D.M., Lipman, Y., et al. 2011. Comparing Dirichlet normal surface energy of tooth crowns, a new technique of molar shape quantification for dietary inference, with previous methods in isolation and in combination. Am. J. Phys. Anthropol. 145, 247–261.

Caine, N.G., Mundy, N.I., 2000. Demonstration of a foraging advantage for trichromatic marmosets (*Callithrix geoffroyi*) dependent on food color. Proc. R. Soc. Lond. B 267, 439–444.

Cannon, C.B., Leighton, M., 1994. Comparative locomotor ecology of gibbons and macaques: Selection of canopy elements for crossing gaps. Am. J. Phys. Anthropol. 93, 505–524.

Chambers, H.R., Heldstab, S.A., O'Hara, S.J., 2021. Why big brains? A comparison of models for both primate and carnivore brain size evolution. PLoS One 16, e0261185. https://doi.org/10.1371/journal.pone.0261185

Chivers, D.J., Hladik, C.M., 1980. Morphology of the gastrointestinal tract in primates: comparisons with other mammals in relation to diet. J. Morphol. 166, 337–386.

Coleman, M.N., Colbert, M.W., 2010. Correlations between auditory structures and hearing sensitivity in non-human primates. J. Morphol. 271, 511–532.

Coleman, M.N., Kay, R.F., Colbert, M.W., 2010. Auditory morphology and hearing sensitivity in fossil New World monkeys. Anat. Rec. 293, 1711–1721.

Crowley, B.E., 2012. Stable isotope techniques and applications for primatologists. Int. J. Primatol. 33, 673–701.

Cuozzo, F.P., Sauther, M.L., Yamashita, N., et al., 2008. A comparison of salivary pH in sympatric wild lemurs (*Lemur catta* and *Propithecus verreauxi*) at Beza Mahafaly Special Reserve, Madagascar. Am. J. Primatol. 70, 363–371.

DeCasien, A.R., Williams, S.A., Higham, J.P., 2017. Primate brain size is predicted by diet but not sociality. Nat. Ecol. Evol. 1, 0112. https://doi.org/10.1038/s41559-017-0112

Demes, B., Jungers, W.L., 1993. Long bone cross-sectional dimensions, locomotor adaptations and body size in prosimian primates. J. Human Evol. 25, 57–74.

Demes, B., Larson, S.G., Stern Jr., J.T., et al. 1994. The kinetics of primate quadrupedalism: "Hindlimb drive" reconsidered. J. Human Evol. 26, 353–374.

Demes, A.B., Jungers, W.L., Fleagle, J.G., et al. 1996. Body size and leaping kinematics in Malagasy vertical clingers and leapers. J. Human Evol. 31, 367–388.

Di Fiore, A., Rendall, D., 1994. Evolution of social organization: A reappraisal for primates by using phylogenetic methods. Proc. Natl. Acad. Sci. USA 91, 9941–9945.

Dominy, N.J., Lucas, P.W., 2001. Ecological importance of trichromatic vision to primates. Nature 410, 363–366.

Dunbar, R.I.M., 1998. The social brain hypothesis. Evol. Anthropol. 6, 178–190.

Dunbar, R.I.M., Shultz, S., 2007. Evolution in the social brain. Science 317, 1344–1347.

Evans, A.R., Wilson, G.P., Fortelius, M., Jernvall, J., 2007. High-level similarity of dentitions in carnivorans and rodents. Nature 445, 78–81.

Fleagle, J.G., 1985. Size and adaptation in primates. In: Jungers, W.L. (Ed.), Size and Scaling in Primate Biology. Springer, New York, pp. 1–19.

Fleagle, J.G., Anapol, F.C., 1992. The indriid ischium and the hominid hip. J. Human Evol. 22, 285–305.

Fleagle, J.G., Lieberman, D.E., 2013. The evolution of primate locomotion: Many transformations, many ways. In: Dial, K., Shubin, N., Brainerd, E. (Eds.), Great Transformations: Major Events in the History of Vertebrate Life. University of California Press, Berkeley.

Fleagle, J.G., Mittermeier, R.A., 1980. Locomotor behavior, body size and comparative ecology of seven Suriname monkeys. Am. J. Phys. Anthropol. 52, 301–322.

Fleagle, J.G., McGraw, W.S., 2002. Skeletal and dental morphology of African papionins: Unmasking a cryptic clade. J. Hum. Evol. 42, 267–292.

Frost, S.R., 2001. Fossil Cercopithecidae of the Afar Depression, Ethiopia: Species systematics and comparison to the Turkana Basin. Ph.D. Dissertation. The City University of New York, New York.

Ganzhorn, J.U., Arrigo-Nelson, S., Boinski, S., et al., 2009. Possible fruit protein effects on primate communities in Madagascar and the Neotropics. PLoS One 4, e8253.

Gebo, D.L., 2014. Primate Comparative Anatomy. Johns Hopkins University Press, Baltimore.

Grabowski, M., Kopperud, B.T., Tsuboi, M., Hansen, T.F., 2022. Both diet and sociality affect primate brain-size evolution. Syst. Biol. 72, 404–418. https://doi.org/10.1093/sysbio/syac075

Granatosky, M.C., Fitzsimons, A., Zeininger, A., Schmitt, D., 2018. Mechanisms for the functional differentiation of the propulsive and braking roles of the forelimbs and hindlimbs during quadrupedal walking in primates and felines. J. Exp. Biol. 221, jeb162917.

Harcourt, A.H., 1995. Sexual selection and sperm competition in primates: What are male genitalia good for? Evol. Anthropol. 4, 121–129.

Harvey, P.H., Harcourt, A.H., 1984. Sperm competition, testes size, and breeding systems in primates. In: Smith, R.L. (Ed.), Sperm Competition and the Evolution of Animal Mating Systems. Academic Press, London, pp. 589–600.

Hiramatsu, C., Melin, A.D., Aureli, F., et al., 2008. Importance of achromatic contrast in short-range fruit foraging of primates. PLoS One 3, e3356.

Hrdy, S.B., Whitten, P.L., 1986. Patterning of sexual activity. In: Smuts, B.B., Cheney, D.L., Seyfarth, R.M., et al. (Eds.), Primate Societies. University of Chicago Press, Chicago, pp. 370–384.

Isbell, L.A., 2009. The Fruit, the Tree, and the Serpent: Why We See So Well. Harvard University Press, Cambridge.

Jacobs, G.H., 2008. Primate color vision: A comparative perspective. Visual Neurosci. 25, 619–633.

Janson, C.H., 2000. Spatial movement strategies: theory, evidence, and challenges. In: Boinski, S., Garber, P.A. (Eds.), On the Move: How and Why Animals Travel in Groups. University of Chicago Press, Chicago, pp. 165–203.

Janson, C.H., 2007. Experimental evidence for route integration and strategic planning in wild capuchin monkeys. Anim. Cogn. 10, 341–356.

Kay, R.F., 1984. On the use of anatomical features to infer foraging behavior in extinct primates. In: Cant, J., Rodman, P. (Eds.), Adaptations for Foraging in Nonhuman Primates. Columbia University Press, New York, pp. 21–53.

Kay, R.F., Kirk, E.C., 2000. Osteological evidence for the evolution of activity pattern and visual acuity in primates. Am. J. Phys. Anthropol. 113, 235–262.

Kirk, E.C., Gosselin-Ildari, A.D., 2009. Cochlear labyrinth volume and hearing abilities in primates. Anat. Rec. 292, 765–776.

Lambert, J.E., 1998. Primate digestion: Interactions among anatomy, physiology, and feeding ecology. Evol. Anthropol. 7, 8–20.

Larson, S.G., 2018. Nonhuman primate locomotion. Am. J. Phys. Anthropol. 165, 705–725.

Larson, S.G., Stern Jr., J.T., 1989. The role of supraspinatus in the quadrupedal locomotion of vervets (*Cercopithecus aethiops*): Implications for interpretation of humeral morphology. Am. J. Phys. Anthropol. 79, 369–377.

Lopez-Aguirre, C., Lang, M.M., Silcox, M.T., 2022. Diet drove brain and dental morphological evolution in strepsirrhine primates. PLoS One 17 (6), e0269041.

Lucas, P.W., 2004. How Teeth Work. Cambridge University Press, Cambridge.

Lucas, P.W., Teaford, M.F., 1994. Functional morphology of colobine teeth. In: Davies, A.G., Oates, J.F. (Eds.), Colobine Monkeys: Their Ecology, Behavior, and Evolution. Cambridge University Press, Cambridge, pp. 173–204.

Lucas, P.W., Dominy, N.J., Riba-Hernandez, P., et al., 2003. Evolution and function of routine trichromatic vision in primates. Evolution 57, 2636–2643.

Lucas, P., Constantino, P., Wood, B., Lawn, B., 2008. Dental enamel as a dietary indicator in mammals. BioEssays 30, 374–385.

Malinzak, M.D., Kay, R.F., Hullar, T.E., 2012. Locomotor head movements and semicircular canal morphology in primates. Proc. Natl. Acad. Sci. USA 109 (44), 17914–17919.

McGraw, W.S., 1996. Cercopithecoid locomotion, support use, and support availability in the Tai Forest, Ivory Coast. Am. J. Phys. Anthropol. 100, 507–522.

McGraw, W.S., 1998. Posture and support use of Old World monkeys (Cercopithecidae): The influence of foraging strategies, activity patterns and the spatial distribution of preferred food items. Am. J. Primatol. 46, 229–250.

Milton, K., 2000. Quo vadis? Tactics of food search and group movement in primates and other animals. In: Boinski, S., Garber, P. (Eds.), On the Move: How and Why Animals Travel in Groups. University of Chicago Press, Chicago, pp. 375–418.

Montgomery, S.H., Capellini, I., Barton, R.A., Mundy, N.I., 2010. Reconstructing the ups and downs of primate brain evolution: Implications for adaptive hypotheses and *Homo floresiensis*. BMC Biology 8, 9.

Nunn, C.L., 2011. The Comparative Approach in Evolutionary Anthropology and Biology. University of Chicago Press, Chicago.

Nunn, C.L., Barton, R.A., 2001. Comparative methods for studying primate adaptation and allometry. Evol. Anthropol. 10, 81–98.

Nunn, C.L., van Schaik, C.P., 2002. A comparative approach to reconstructing the socioecology of extinct primates. In: Plavcan, J.M., Kay, R.F., Jungers, W.L., van Schaik, C.P. (Eds.), Reconstructing Behavior in the Primate Fossil Record. Plenum, New York, pp. 159–215.

Plavcan, J.M., 2002. Reconstructing social behavior from dimorphism in the fossil record. In: Plavcan, J.M., Kay, R.F., Jungers, W.L., van Schaik, C.P. (Eds.), Reconstructing Behavior in the Primate Fossil Record. Plenum, New York, pp. 297–338.

Plavcan, J.M., 2013. Reconstructing social behavior from fossil evidence. In: Begun, D.R. (Ed.), A Companion to Paleoanthropology. Wiley-Blackwell, Hoboken, pp. 226–243.

Powell, L.E., Isler, K., Barton, R.A., 2017. Re-evaluating the link between brain size and behavioural ecology in primates. Proc. R. Soc. B 284, 20171765. http://doi.org/10.1098/rspb.2017.1765

Power, M.L., 2010. Nutritional and digestive challenges to being a gum-feeding primate. In: Burrows, A.M., Nash, L.T. (Eds.), The Evolution of Exudativory in Primates. Springer, New York, pp. 25–44.

Purvis, A., Webster, A.J., 1999. Phylogenetically independent comparisons and primate phylogeny. In: Lee, P.C. (Ed.), Comparative Primate Socioecology. Cambridge University Press, Cambridge, pp. 44–70.

Regan, B.C., Julliot, C., Simmen, B., et al. 2001. Fruits, foliage and the evolution of color vision. Phil. Trans. R. Soc. Lond. B 356, 229–283.

Rose, M.D., 1988. Another look at the anthropoid elbow. J. Human Evol. 15, 333–367.

Rosenberger, A.L., 2010. Adaptive profile versus adaptive specialization: Fossils and gummivory in early primate evolution. In: Burrows, A.M., Nash, L.T. (Eds.), The Evolution of Exudativory in Primates. Springer, New York, pp. 273–295.

Ross, C.F., Lockwood, C.A., Fleagle, J.G., Jungers, W.L., 2002. Adaptation and behavior in the primate fossil record. In: Plavcan, J.M., Kay, R.F., Jungers, W.L., van Schaik, C.P. (Eds.), Reconstructing Behavior in the Primate Fossil Record. Plenum, New York, pp. 1–41.

Ross, C.F., Kirk, E.C., 2007. Evolution of eye size and shape in primates. J. Hum. Evol. 52, 294–313.

Rothman, J.M., Raubenheimer, D., Bryer, M.A.H., et al. 2014. Nutritional contributions of insects to primate diets: Implications for primate evolution. J. Hum. Evol. 71, 59–69.

Sargis, E.J., Terranova, C.J., Gebo, D.L., 2008. Evolutionary morphology of the guenon postcranium and its taxonomic implications. In: Sargis, E.J., Dagosto, M. (Eds.), Mammalian Evolutionary Morphology: A Tribute to Frederick S. Szalay. Springer, Dordrecht, pp. 361–372.

Schmitt, D., 1994. Forelimb mechanics as a function of substrate type during quadrupedalism in two anthropoid primates. J. Human Evol. 26, 441–457.

Shapiro, L., 1995. Functional morphology of indrid lumbar vertebrae. Am. J. Phys. Anthropol. 98, 323–342.

Sillén-Tullberg, B., Møller, A.P., 1993. The relationship between "concealed ovulation" and mating systems in anthropoid primates: A phylogenetic analysis. Am. Nat. 141, 1–25.

Smith, A.C., Buchanan-Smith, H.M., Surridge, A.K., et al. 2003. The effect of colour vision status on the detection and selection of fruits by tamarins (*Saguinus* spp.). J. Exp. Biol. 206, 3159–3165.

Strait, S.G., 1997. Tooth use and the physical properties of food. Evol. Anthropol. 5, 199–211.

Sumner, P., Mollon, J.D., 2000a. Catarrhine photopigments are optimized for detecting targets against a foliage background. J. Exp. Biol. 203, 1963–1986.

Sumner, P., Mollon, J.D., 2000b. Chromaticity as a signal of ripeness in fruits taken by primates. J. Exp. Biol. 203, 1987–2000.

Teaford, M.F., 1994. Dental microwear and dental function. Evol. Anthropol. 3, 17–30.

Terranova, C.J., 1996. Variation in the leaping of lemurs. Am. J. Phys. Anthropol. 40, 145–166.

Ungar, P.S., 2002. Reconstructing the diets of fossil primates. In: Plavcan, J.M., Kay, R.F., Jungers, W.L., van Schaik, C.P. (Eds.), Reconstructing Behavior in the Primate Fossil Record. Plenum, New York, pp. 261–296.

Ungar, P.S., Sponheimer, M., 2013. Hominin diets. In: Begun, D.R. (Ed.), A Companion to Paleoanthropology. Wiley-Blackwell, Hoboken, pp. 165–182.

Ungar, P.S., in press. Reconstructing fossil primate diets: Dental-dietary adaptations and foodprints for thought. In: Lambert, J.E., Bryer, M.A.H., Rothman, J.M. (Eds.), How Primates Eat: A Synthesis of Nutritional Ecology Across a Mammal Order. University of Chicago Press, Chicago.

Walker, S.E., 1996. The evolution of positional behavior in the saki-uakaris (Pithecia, Chiropotes, and Cacajao). In: Norconk, M.A., Rosenberger, A.L., Garber, P.A. (Eds.), Adaptive Radiations of Neotropical Primates. Plenum, New York, pp. 335–367.

Walker, A., Ryan, T., Silcox, M., et al. 2008. The semicircular canal system and locomotion: The case of extinct lemuroids and lorisoids. Evol. Anthropol. 17, 135–145.

Further Reading

Boinski, S., Garber, P.A., 2001. On the Move: How and Why Animals Travel in Groups. University of Chicago Press, Chicago.

Gebo, D.L., 1993. Postcranial Adaptation in Nonhuman Primates. Northern Illinois University Press, DeKalb.

Lucas, P.W., 2004. How Teeth Work. Cambridge University Press, Cambridge.

Plavcan, J.M., Kay, R.F., Jungers, W.L., van Schaik, C.P., 2002. Reconstructing Behavior in the Primate Fossil Record. Plenum, New York.

10

The Fossil Record

In the previous chapters we discussed the anatomy, behavior, and ecology of extant primates, with only a passing mention of their evolutionary history. In the following chapters we discuss primate adaptation and evolution from a paleontological perspective. Although most of our understanding of the relationships among living organisms is based on the study of living species themselves, the fossil record provides us with many types of information about the biology of primates that we could never know from the extant taxa alone.

The unique aspect of the fossil record is that it establishes a temporal framework for evolution. It provides a crude dating for individual events, such as the first appearance of particular taxonomic groups or particular anatomical features. It also provides evidence for the patterns and rates of evolutionary change, whether it was gradual or occurred in fits and starts.

We can extract several kinds of information from the fossil record that are valuable for understanding the phylogeny of living primate groups. It often shows us intermediate or primitive forms that link more anatomically distinct living groups, and it demonstrates how the living species came to be the way they are by documenting the sequence of evolutionary changes that led to their present morphologies.

The fossil record also enables us to examine adaptive changes through time. Knowledge of past adaptations can help us understand how the adaptive characteristics of extant radiations came to be the way they are, and can also suggest tests for examining causal changes between morphology and environment.

Most importantly, the fossil record provides us with a record of life in the past. It is our only evidence of extinct primates – in most cases, animals whose existence we could never have predicted or even imagined, had we not been confronted with their bones. As we shall see, some groups of primates were far more diverse in morphology, ecology, and biogeography during the very recent past than they are today, and other successful radiations from previous epochs have no living representatives.

The information available from the fossil record is quite different from what we can obtain about living species, most noticeably in its incompleteness. Time extracts its price, and our insights into the past are, alas, more often glimpses than panoramas. If we hurry, we can still observe living primates in the forest as they go about their daily activities and we can record their behavior in scientific papers, books, photographs, and films. We can examine their pelage, measure and dissect their bodies, and study their physiology, communication, and learning abilities, in addition to measuring their bones and teeth and sequencing their DNA. For fossils we have mostly only bones and teeth – mainly the latter. The occasional impression of the bushy tail of an archaic primate or the footprint of an early hominin is, unfortunately, a rare and remarkable occurrence. So far, DNA analyses have only been possible for fossils from the past few hundred thousand years (e.g., Meyer et al., 2016), and even then only under unusual types of preservation. As a result, our discussions of the behavior of extinct primates, and even the identification of different sexes and species, require a much larger dose of guesswork than our descriptions of living species.

Our greatest tool is our ability to extrapolate from the consistent patterns between morphology and behavior that we see among living species to the more poorly known animals in the fossil record. We must keep an open mind, however: the fossil record is likely to be full of unique events. Thus, before we discuss primate evolution, we must consider briefly the special attributes of the fossil record and the types of information that are available for understanding primate history.

Geological Time

The evolution of primates has taken place on a timescale that is virtually impossible to comprehend in anything but a comparative sense (Fig. 10.1). As individuals, 100 years is the most we are ever likely to

Primate Adaptation and Evolution
https://doi.org/10.1016/B978-0-12-815809-8.00010-2

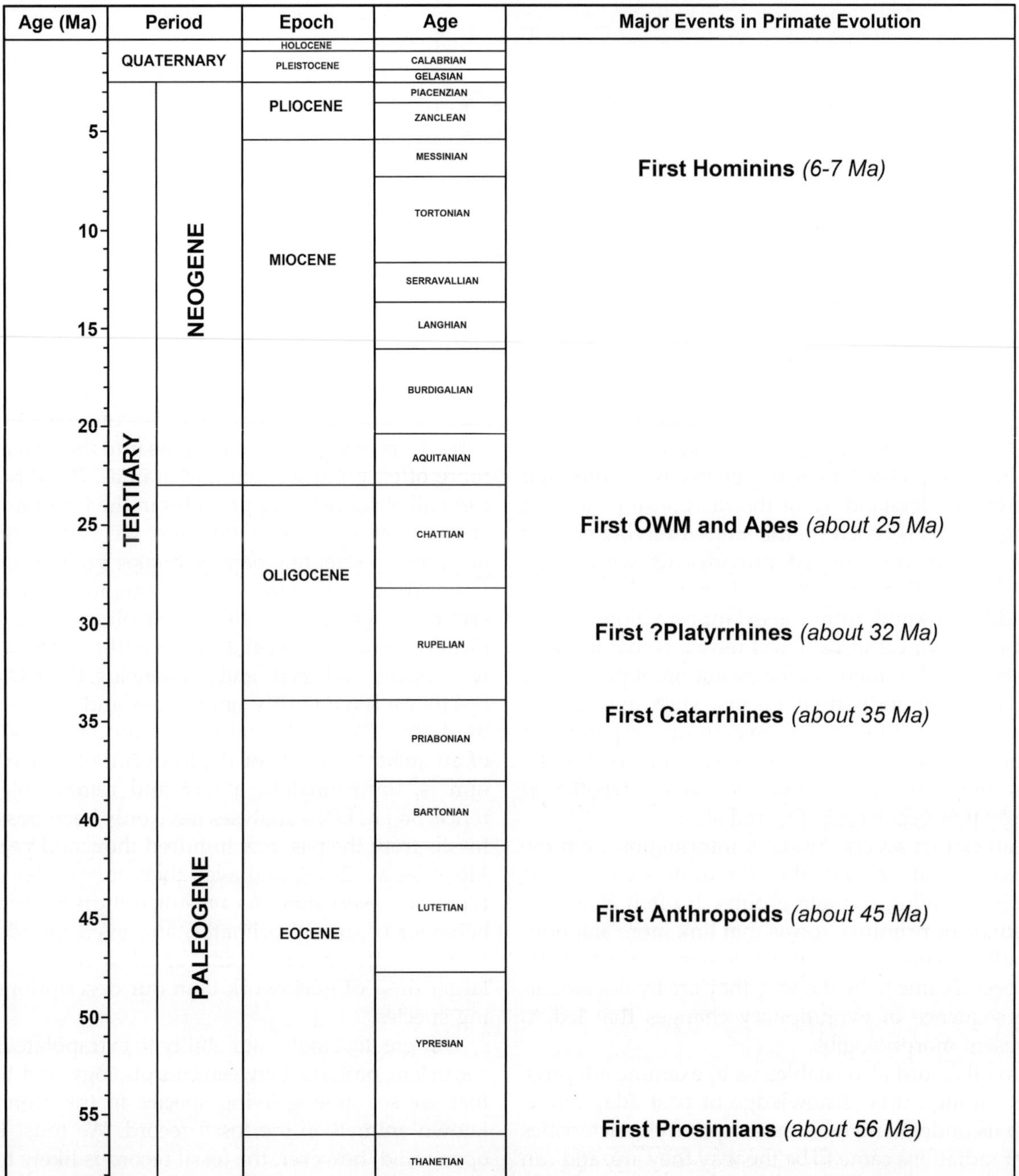

Age (Ma)	Period	Epoch	Age	Major Events in Primate Evolution
	QUATERNARY	HOLOCENE		
		PLEISTOCENE	CALABRIAN	
			GELASIAN	
		PLIOCENE	PIACENZIAN	
5			ZANCLEAN	
	NEOGENE		MESSINIAN	**First Hominins** (6-7 Ma)
10		MIOCENE	TORTONIAN	
			SERRAVALLIAN	
15			LANGHIAN	
			BURDIGALIAN	
20			AQUITANIAN	
25	TERTIARY		CHATTIAN	**First OWM and Apes** (about 25 Ma)
30		OLIGOCENE	RUPELIAN	**First ?Platyrrhines** (about 32 Ma)
35			PRIABONIAN	**First Catarrhines** (about 35 Ma)
40	PALEOGENE		BARTONIAN	
45		EOCENE	LUTETIAN	**First Anthropoids** (about 45 Ma)
50			YPRESIAN	
55				**First Prosimians** (about 56 Ma)
60		PALEOCENE	THANETIAN	
			SELANDIAN	
65			DANIAN	**First ?Plesiadapiforms** (about 66 Ma)

FIGURE 10.1   A geological time scale for the Cenozoic era, showing the epoch series, major ages within each epoch, and first appearance of major primate phyletic groups.

experience, yet few events in primate evolution can be dated to within 1 million or even 5 million years. The scale of events is more commonly on the order of tens of millions of years.

The fossil record of primates, like that of most other groups of modern mammals, is known almost totally from the Cenozoic era – the Age of Mammals – roughly the last 66 million years (Fig. 10.1). Paleontologists have traditionally divided this period into smaller units (epochs and land mammal ages) on the basis of animals commonly contained in the sediments. Through **faunal correlation**, sediments from different places and the fossils in them can be placed in a relative timescale. For example, the fossil record of mammals in Europe has been divided into a sequence of chrons based on the fossil species that are typically found within them. However, faunal correlation is a method of **relative dating**. It can enable a researcher to compare different sequences of rocks and faunas, but provides no absolute age for fossils and geological deposits. These relative scales, such as the epoch series, must be calibrated by some method of absolute dating that can assign a specific age to either geological deposits containing fossils or the fossils themselves.

A variety of techniques are available for calibrating the fossil record (Ludwig and Renne, 2000). These techniques vary considerably in the time range over which they are applicable, usually depending on the half-life of the elements being used, and in the materials that can be dated (Fig. 10.2). In general, **radiometric dating** techniques for absolute dating of geological deposits rely on the normal decay patterns of radioactive elements. Many elements on earth are naturally unstable and change to more stable elements at a characteristic, known rate. By examining the percentages of the parent isotope and of the product, it is possible to calculate how long ago the rock was formed. Thus, **carbon-14 dating** is based on the rate of radioactive decay of carbon isotopes found in many organic materials, such as wood and bone; unfortunately, it can only be used for dating events that occurred during the last 70,000 to 90,000 years (Taylor, 1996). **Potassium/argon dating** and **argon/argon dating** can be applied to a much longer range of

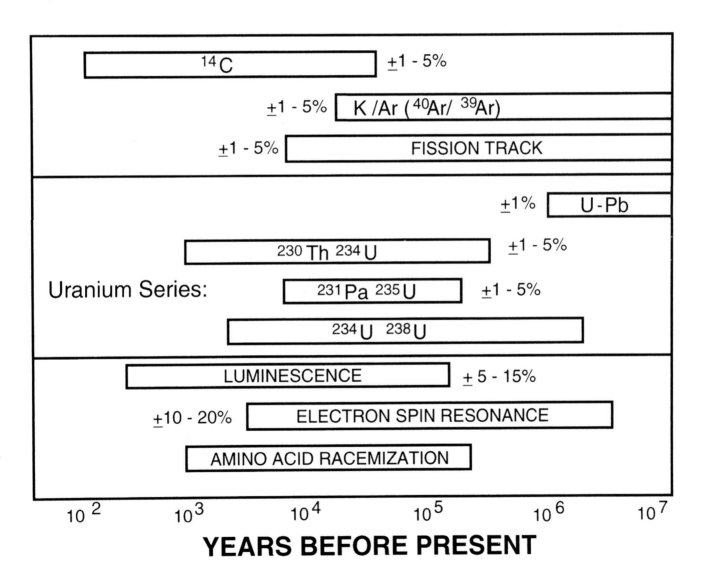

FIGURE 10.2 Time ranges for common methods of geological dating used in paleontology and anthropology. *(Modified from Schwarcz, H.P., 1992. Uranium series dating in paleoanthropology. Evol. Anthropol. 1, 56–62.)*

ages, but these dating techniques are limited to material containing these elements, usually volcanic rocks (Deino et al., 1998). However, these represent the most precise methods currently available, and are responsible for most of the absolute dates used to calibrate primate and early human evolution.

Uranium series dating can be applied to other types of rocks over a much younger time range (Schwarcz, 1992). In **fission-track dating**, the investigator measures the number of scars left in a crystal by nuclear particles given off during radioactive decay (Wagner, 1996). **Electron spin resonance** (Grun, 1993), **thermoluminescence** (Feathers, 1996; Jacobs and Roberts, 2007), and **amino acid racemization** are valuable techniques when one has some estimates of the environmental history of the materials being dated.

Radiometric dating of geological sediments in absolute numbers of years is generally possible only for certain types of rocks; for example, relatively pure volcanic ashes or lava flows that were clearly formed at a particular point in time. However, some techniques can measure time elapsed since a particular "re-setting event." For example, thermoluminescence or fission tracks may be reset by high temperatures and can be used to date hearths or fired ceramics in the archeological record, or the time since grains of sand were exposed to sunlight.

Paleomagnetism is the study of the magnetic orientation of geological deposits laid down in earlier times. One of the many startling geological discoveries of the twentieth century was that the earth's magnetic field has reversed its north and south poles many times during the past. These reversals have taken place approximately once every 700,000 years, but not at regular intervals. How and why these changes have occurred is not well understood, but geologists are continually compiling a history of magnetic reversals over the past 500 million years through combined studies of paleomagnetism and radiometric dating techniques. For this reason, paleomagnetism provides another method of correlating the ages of sediments and fossils (Kappelman, 1993). At many fossil sites that may lack rocks suitable for absolute dating, geologists can use the sequence of magnetic reversals in conjunction with faunal correlation to determine the position of the rocks in the geological timetable and to estimate their absolute age.

Tephrostratigraphy is another way of correlating sediments in different areas (Feibel, 1999). Each volcanic eruption produces ashes with a characteristic chemical signature. By careful examination of these pyroclastic deposits, it is often possible to identify a unique chemical signature for volcanic ashes from a single eruption event. Thus, geologists can identify ashes from the identical eruption in several different places and correlate geographically separated stratigraphic sections. For example, it has been possible to identify and correlate

individual ash falls across many parts of East Africa, and even into ocean cores from the Red Sea. Although tephracorrelation is a method of relative dating, it is a very precise one because individual eruptions usually take place within a very limited timespan. Moreover, it is often possible to obtain absolute dates from these same volcanic deposits.

Determining the age of particular events in primate evolution usually requires a combination of both relative and absolute dating methods. Fig. 10.1 summarizes determinations of the age of geological epochs and ages relevant to primate evolution, together with major events in the history of primates.

Molecular Dating

The methods discussed in the previous paragraphs are aimed at determining the age of paleontological and archeological deposits and the various materials they contain. The importance of dating particular fossils for our understanding of the evolutionary history and divergence times for groups of living primates depends on our ability to correctly place fossil taxa relative to extant species. However, it is also possible to date the time when extant species or groups of species were separated from one another based on the amount of genetic difference between them. These molecular dating techniques, often collectively referred to as the **molecular clock**, are generally based on various assumptions about the rate at which genetic mutations accumulate within a lineage, and have been widely applied to all types of living organisms, including primates (e.g., Chatterjee et al., 2009; Kuderna et al., 2023; Perelman et al., 2011; Pozzi et al., 2014; Springer et al., 2012). In general, the ages of primate divergence times based on a genetic molecular clock are much older than the dates suggested by fossils.

Theoretically, this is to be expected for several reasons (Steiper and Young, 2008). Genetic divergences record the polymorphisms in a gene that appear within a population before parts of that population may be segregated into anatomically and/or reproductively distinct lineages. In contrast, the fossil record is a very serendipitous sample of past life, and it is extremely unlikely, indeed virtually impossible, that the fossil record would sample the initial appearance of a distinct clade with a precise date. Rather, fossil divergences should appear after the real lineage divergence (e.g., Tavaré et al., 2002; Soligo et al., 2007). Nevertheless, most molecular divergence dates are vastly older than seem reasonable; for example, estimated dates for the divergence between tarsiers and anthropoids precede any evidence in the fossil record of modern primates or most other orders of modern mammals (e.g., Perelman et al., 2011). This is a major conflict in evolutionary biology, but one that is too rarely addressed. Rather, more often than not, molecular biologists and paleontologists continue to go their separate ways regarding the dating of primate divergence times.

However, there are several recent efforts aimed at bringing these two sources of evidence about primate evolutionary history into some type of accord. Steiper and Seiffert (2012) found that evolutionary rates are highly correlated with body size and measures of brain size, and if primate divergence times are recalculated using fossil evidence of these features in extinct taxa, the estimated divergence times are much closer to the fossil dates. Similarly, dos Rios and colleagues (2012, 2018) combined a large set of genetic data with flexible fossil calibrations in large Bayesian analyses that yielded divergence dates for many orders of mammals that accord well with the fossil record. Hopefully some synthesis of the molecular clock and fossil record is in progress.

Plate Tectonics and Continental Drift

Far from being a stable, unchanging sphere, the earth is a very dynamic body. Its surface is made up of a patchwork of individual plates that form the different continents and ocean floors. These plates are constantly in motion with respect to one another. Thus, the sizes, orientations, and connections of the continents and the positions of the oceans surrounding them have changed considerably in the past, just as they are changing today (Fig. 10.3). These geographic rearrangements have greatly influenced the routes of migration and dispersal available to plants and animals. In addition, the relative positions of land masses have had major effects on ocean currents and climate – effects with global consequences for primates and all other living things. Many of the most dramatic changes in the earth's surface took place well before the first appearance of primates, and so have little bearing on the subject of this book. Nevertheless, during the past 66 million years a number of changes in continental positions and connections have influenced primate evolution (Zachos et al., 2001; Fleagle and Gilbert, 2006).

Paleoclimate

Through studies of geological deposits, fossil plants and animals, as well as marine organisms, geologists and paleontologists have been able to reconstruct many changes in the earth's climate during evolutionary time. These studies have documented many

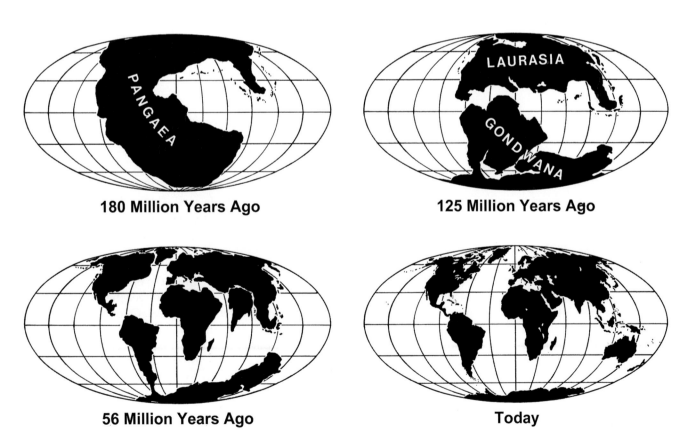

180 Million Years Ago   **125 Million Years Ago**

56 Million Years Ago   **Today**

FIGURE 10.3   Positions of the continents at various times during the past 180 million years.

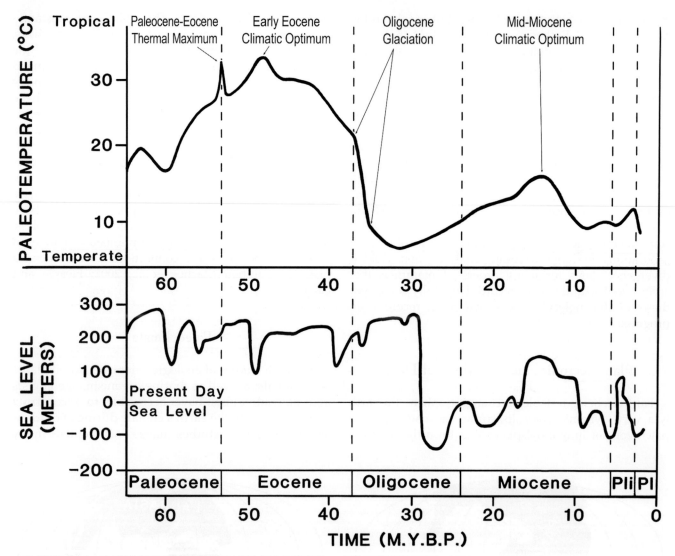

FIGURE 10.4   Temperatures and sea levels during the Cenozoic Era. *Above,* global climate changes based on North Sea foraminifera; *below,* relative sea levels based on seismic reflections.

dramatic climatic events and patterns of climate change over the past 66 million years (Fig. 10.4) that have undoubtedly been important in primate evolution (Zachos et al., 2001). For example, at the Paleocene–Eocene boundary, about 56 million years ago, there was a brief spike in global temperatures that enabled primates, and several other groups of mammals, to expand into high latitudes where there were intercontinental connections, and thus to disperse between Eurasia and North America (Fig. 12.1). It is important to remember, however, that climatic changes in any restricted area are quite likely to show different patterns from those that may characterize the earth as a whole. Likewise, our knowledge of climatic changes is well documented in some places, but it is quite crude for many parts of the world.

The formation of glaciers at polar latitudes is one of the most far-reaching global climatic events. In addition to dramatically altering regional climates and landforms, glaciers profoundly affect sea levels by changing the distribution of water on the earth's surface. In turn, these changes in sea level can affect the erosional and depositional rates of rivers, streams, and beaches, as well as the connections between land masses. Over the past 66 million years, several dramatic changes in global sea levels have taken place. Many of the most dramatic of these have been associated with the development of glaciers at the poles. Like the positions of continents, changes in sea level can have important effects on plant and animal dispersal. In the following chapters we attempt to relate these changes in continental position, climate, and sea level to the major events in primate evolution.

Fossils and Fossilization

Fossils are any remains of life preserved in rocks. We most commonly think of fossils as petrified bones and teeth, but fossils also include such things as impressions, natural molds of brains or even bodies, and traces of life such as footprints, worm burrows, and termite nests (Fig. 10.5).

Although fossils often preserve shapes of bones or teeth very accurately, most fossils are usually formed by replacement of the original biological materials with minerals derived from the sedimentary environments in which they are buried. In many cases, however, this replacement takes place at a molecular level, so that even microscopic details of morphology, such as muscle attachments, fine tooth-wear scratches, dental enamel prisms, or delicate internal bone structures, are preserved and can be analyzed with many of the same tools we apply to the study of living primate skeletons including isotopic analyses (Cerling et al., 2011).

The type of remains available to a scientist today from an animal that lived sometime in the geological past is determined by many events and processes. The study of the factors that determine which animals become fossils, what parts of their bodies are preserved, how they are preserved, and how they appear to scientists many millions of years later is **taphonomy**. Taphonomists seek to reconstruct as well as possible everything that has happened to a bone between the time it was, say, climbing a tree 35 million years ago in the body of an early fossil monkey, until the time it was discovered along with other fossils in a Kenyan sandstone (Fig. 10.6). They want to know such things as why teeth and ankle bones are commonly found as fossils but other parts may not be, or why some fossils are found as whole bodies and others as fragments. In pursuit of answers to such questions, taphonomists engage in many unusual activities, such as staking out dead antelopes on the Serengeti Plains to see what happens to them (e.g., Blumenschine, 1986; Pobiner, 2015), or placing bones in cement mixers to simulate the effects of rolling down a rocky stream. Taphonomy is a diverse science that provides a more nuanced view of the way in which fossils relate to the living animals from the past.

Taphonomic studies enable paleontologists to determine whether the remains of animals found at a particular locality or site have been transported to the site by the action of streams or perhaps predatory birds, or whether the animals are likely to have lived and died where their bones are recovered. Studies of the proportions of different skeletal elements recovered, the absence of abrasion, and the abundance of bite marks on bones of fossil primates and other mammals from the Early Eocene of Wyoming, for example, indicate that the

Termite Nest

Sectioned Tooth

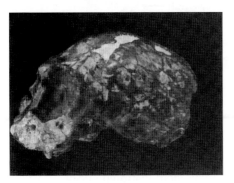

Skull

Impression of Tail

FIGURE 10.5   Different kinds of fossils.

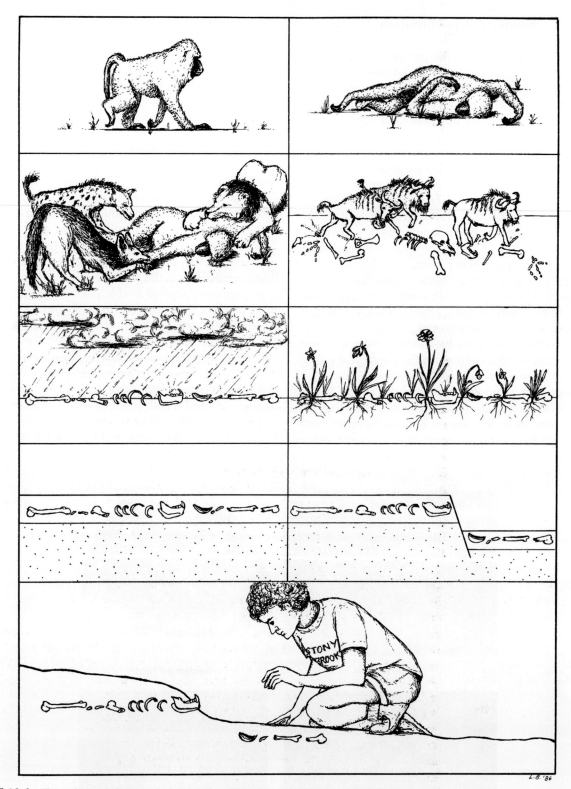

FIGURE 10.6   The taphonomic history of a fossil. (*Adapted from Shipman, P., 1981. Life History of a Fossil. Harvard University Press. Cambridge, MA.*)

fossils are the result of long-term accumulations on the surface and were not transported long distances and concentrated by stream action. Thus, the species in the fossil record at this site probably represent the species living in this area 50 million years ago. Other studies have demonstrated specific patterns of damage indicating that fossils are individuals that were killed by large raptors (McGraw et al., 2006; Gilbert et al., 2009).

Studies of the proportions of bony elements found in archeological sites have been widely used as evidence for different patterns of carcass transport associated with different strategies of procurement, such as hunting or scavenging. Although such inferences are intriguing, they are not without potential pitfalls arising from differential destruction of different body parts by carnivores, as well as simple errors due to collecting bias.

Paleoenvironments

A primate fossil is usually found along with other fossils, both plant and animal, and within a particular geological setting, all of which can yield useful information about the environment in which the animal lived and died. Whether a fossil primate is found associated with forest rodents or savannah rodents, for example, can provide clues to its habitat preferences. Land snails seem to have narrow habitat preferences, and fossil snails have proved very useful in determining the extent to which a particular fossil locality represents a forested or an open habitat. Similarly, fossil plants can yield information about both local habitat and climate.

The sediments containing fossils can provide many kinds of information about the fossils' origin. They can tell us whether a fossil deposit was preserved on a floodplain, on a river delta, in a stream channel, or on the shores of a lake. This information about where an animal's bones were preserved provides clues to where it lived or died. In addition, sediments can provide detailed information about the climatic regime during which they were formed. Was it hot, cold, wet, or dry? Was the weather relatively uniform, or was it seasonal? In addition to the information they convey about a particular fossil site, sedimentary deposits can tell us about climatic trends in a particular region. Many fossil primates are found in ancient soils, and these soils usually contain considerable information about the conditions under which they were formed (e.g., Retallack, 2007).

A very important tool for reconstructing past environments inhabited by primates comes from the study of stable isotopes in soils and in fossils. Most of the common elements on earth come in different forms, called isotopes, which have the same number of electrons and protons, and thus the same electrical properties. However, isotopes differ in their numbers of neutrons, and consequently, different isotopes have different weights. Isotopes of different weights can behave differently in various biochemical and geochemical processes. For example, the two most common isotopes of oxygen are ^{16}O (the most common) and ^{18}O. Water (H_2O) made from the lighter isotope (^{16}O) tends to evaporate more readily than water made from the heavier isotope, so that rain and snow are enriched in the lighter isotope.

Thus, in colder times, when much of the earth's water is bound up in glacial ice with a higher proportion of ^{16}O, the oceans have a relatively higher proportion of ^{18}O to ^{16}O than they would when there was less water bound up in glaciers. The oxygen isotope ratio in the shells of marine organisms is proportional to that of the water in which they were formed, so that by measuring the ratio of ^{18}O to ^{16}O in marine fossils, paleontologists can identify temperature changes in the past.

In reconstructing fossil environments, there are obvious limits to the amount of detailed information we can infer about the life of an extinct primate. There are also potential pitfalls in extrapolating from the events surrounding fossilization to the habits of an animal. For example, most fossil primates are found in sediments that were originally deposited by streams, rivers, or lakes; often channels within a stream or along floodplains of rivers that overflowed their banks during floods. One early worker, finding fossil lemur bones mixed with the bones of turtles and crocodiles, argued that the lemurs must have been aquatic. Obviously, finding lemur bones in deposits formed by water need not imply aquatic lemurs; rather, the bones of many different animals were more likely just buried together in stream or lake deposits. While the crocodiles may have lived in the river, the lemurs probably lived in trees overhanging the water, or perhaps their bodies were washed into the river during a rainstorm.

However, by careful comparisons of the ecological characteristics of animals found in different habitats today, we can often reconstruct the nature of ancient habitats by the ecological characteristics of the animals preserved as fossils. For example, comparisons of the percentage of arboreal to terrestrial mammals in a fauna usually provide a good estimate of the amount of tree cover in a habitat, and can distinguish rainforests from woodlands and savannahs (Fig. 10.7). This is true for several continents, regardless of the particular species of mammals being compared (e.g., Reed, 1997). Thus, fossil faunas containing high numbers of species with arboreal adaptations most likely represent forested environments.

Reconstructing Behavior

Generally, the best and most reliable information about the habits of an extinct primate is obtained by comparing details of its dental and skeletal anatomy with those of living primates. Sediments may tell us where it died, and taphonomy may tell us how and why it was preserved, but its teeth and bones can tell us how it lived – what it ate, how it moved, and possibly in what kind of social group it lived. In the previous chapter we discussed many of the associations between behavior and anatomy among living primates that form the basis

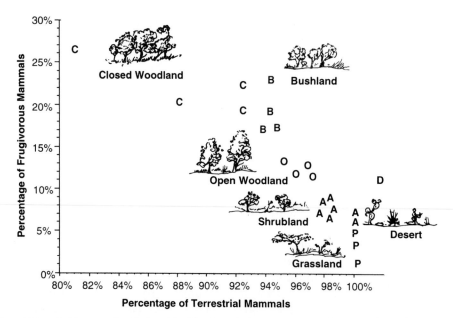

FIGURE 10.7 Modern African habitats can be distinguished by the relative proportions of terrestrial and frugivorous mammals. *C*, Closed Woodland; *B*, Bushland; *O*, Open Woodland, *A*, Shrubland; *D*, Desert; *P*, Grassland/Plains. The same criteria can be used to interpret the habitats represented by assemblages of fossil bones in paleontological sites. *(From Reed, K.E., 1997. Early hominid evolution and ecological change through the African Plio-Pleistocene. J. Hum. Evol. 32, 289–322.)*

for our interpretations of fossil behavior. Our ability to reconstruct the habits of an extinct primate from its bones is intimately linked to our understanding of how the shape of bones in living primates varies with their behavior. Associations between bony morphology and behavior that are true among living primates only "some of the time" cannot be expected to yield reliable reconstructions when applied to fossils (see Ross et al., 2002).

Furthermore, we have to always remain aware that uniformitarianism has its limits: the present is our best key to the past, but the past was not necessarily just like the present. We know, for example, that tooth size and many aspects of behavior are highly correlated with body size among living primates, but we cannot necessarily extrapolate these relationships based on a finite sample of living species to a fossil primate whose teeth are considerably larger or smaller than those of any living species. Likewise, many fossil primates had anatomical features that were quite different from anything we find among living species. We are sure to have problems interpreting such structures, and may need to compare the fossil primates with another type of mammal for an analogy.

We commonly find that fossil primates differ from living species in the combinations of anatomical features they exhibit. A fossil ape may have a humerus that resembles that of a howling monkey in some features, that of a variegated lemur in others, and that of a macaque in still others. In such a case we must examine closely the mechanical implications of the individual features rather than simply look for a living species that matches the fossil in all respects. Our reconstructions of

the behavior of extinct primates from their bones and teeth must be based not on simple analogy, but on an understanding of the physiological and mechanical principles underlying the associations between bony structure and behavior. (Ross et al., 2002).

Just as stable isotopes can be used as a tool to help reconstruct paleoenvironments, they can also provide many insights into the behavior of extinct animals, especially regarding their diet (Cerling et al., 2011; Crowley, 2012; Crowley et al., 2016). For example, the ratio of carbon isotopes in the teeth of both fossil and living mammals is correlated with the chemical composition of their diet. Thus, it is possible to distinguish species that ate predominantly plants using a C_3 photosynthetic pathway (most trees and herbs) from those that ate predominantly plants using a C_4 pathway (grasses and many tubers) (Sponheimer and Lee-Thorp, 2007). Nitrogen isotopes can distinguish carnivores from herbivores because the ratio of ^{15}N to ^{14}N increases with every step of the food chain, so that herbivores have higher values than plants and carnivores have higher values than herbivores. Other isotopes can help distinguish animals living in closed habitats from those in open habitats (e.g., Schoeninger et al., 1999; Loudon et al., 2007).

Paleobiogeography

It is a common tale that primate fossils are rare because primates typically live in jungles, which have acidic soils that destroy their bones before they can be preserved, whereas animals such as horses live on savannahs, where bones are more easily saved for posterity. Although different soils may well affect the chances of fossilization in different environments, there are many examples of tropical environments in the Cenozoic fossil record indicating that the tree-dwelling habits of primates are not primarily responsible for the gaps in our knowledge of primate evolution. In fact, the primate fossil record is, overall, more complete than that of most other groups of mammals.

The large gaps in the primate fossil record are more directly the result of a remarkably meager geological record from those parts of the world in which primates have almost certainly been most successful for tens of millions of years: the Amazon Basin in South America, the Congo Basin in central Africa, and the tropical forests of Southeast Asia. For huge amounts of time and space we lack not just fossils, but even rocks from critical places and ages. Thus, the seemingly poor fossil record of primates compared with, for example, that of horses is likely due to the fact that primates have evolved in places with virtually no fossil record, or with one that is still covered with forests and recent sediments, whereas horses were evolving in temperate areas of Europe, Asia, and western North America, which have excellent fossil records and miles of well-exposed sediments resulting from recent climatic events. Much of our knowledge of extinct primates comes from places that today are too dry and poorly vegetated to support living primates, e.g., Wyoming, North Africa, and northern Kenya. This terrain is excellent for geological and paleontological research; however, all of the paleoenvironmental evidence tells us that, during the earlier epochs when primates were abundant, many of these places were lush forests.

Because so much of our understanding of major events in primate evolution is based on a limited sampling of past life in both time and space, no aspect of primate evolution is open to more surprises than biogeography. As new fossils are discovered from parts of the world that were previously poorly known, such as Africa and Asia, many of our notions about the evolution, diversity, and biogeography of primates will be dramatically revised. For example, it now seems most likely that platyrrhines arrived in South America about 35 million years ago, because we have no record of earlier primates on that continent. But an unsuspected discovery of fossil prosimians from Brazil would dramatically change our view of the evolutionary history and biogeography of higher primates. Similarly, our current view that hominins originated in Africa is based on a lack of early hominins from other continents. We must keep in mind that our knowledge of primate evolution will continue to change with new finds and new interpretations. In the following chapters we try to evaluate the nature of the

evidence for our present understanding of primate evolution, with an eye toward particular issues that are currently unresolved.

References

Blumenschine, R.J., 1986. Early hominid scavenging opportunities: Implications of carcass availability in the Serengeti and Ngorongoro ecosystems. British Archaeological Reports International Series. Archaeopress, Oxford.

Cerling, T.E., Levin, N.E., Passey, B.H., 2011. Stable isotope ecology in the Omo-Turkana Basin. Evol. Anthropol. 20, 217–237.

Chatterjee, H.J., Ho, S.Y.W., Barnes, I., Groves, C., 2009. Estimating the phylogeny and divergence times of primates using a supermatrix approach. BMC Evol. Biol. 9, 259. https://bmcecolevol.biomedcentral.com/articles/10.1186/1471-2148-9-259

Crowley, B.E., 2012. Stable isotope techniques and applications for primatologists. Int. J. Primatol. 33, 673–701.

Crowley, B.E., Reitsema, L.J., Oelze, V.M., Sponheimer, M., 2016. Advances in primate stable isotope ecology – Achievements and future prospects. Am. J. Primatol. 78, 995–1003.

Deino, A.L., Renne, P.R., Swisher III, C.C., 1998. 40Ar/39Ar dating in paleoanthropology and archeology. Evol. Anthropol. 6, 63–75.

dos Rios, M., Inoue, J., Hasegawa, M., et al., 2012. Phylogenomic datasets provide both precision and accuracy in estimating the timescale of placental mammal phylogeny. Proc. R. Soc. B 279, 3491–3500.

dos Rios, M., Gunnell, G.F., Barba-Montoya, J., et al., 2018. Using phylogenomic data to explore the effects of relaxed clocks and calibration strategies on divergence time estimation: Primates as a test case. Sys. Biol. 67, 594–615.

Feathers, J.M., 1996. Luminescence dating and modern human origins. Evol. Anthropol. 5, 25–36.

Feibel, C.S., 1999. Tephrostratigraphy and geological context in paleoanthropology. Evol. Anthropol. 8, 87–100.

Fleagle, J.G., Gilbert, C.C., 2006. Biogeography and the primate fossil record: The role of tectonics, climate, and chance. In: Lehman, S., Fleagle, J.G. (Eds.), Primate Biogeography. Springer, New York, pp. 375–418.

Gilbert, C.C., McGraw, W.S., Delson, E., 2009. Plio-Pleistocene eagle predation on fossil cercopithecids from the Humpata Plateau, southern Angola. Am. J. Phys. Anthropol. 139, 421–429.

Grun, R., 1993. Electron spin resonance dating in paleoanthropology. Evol. Anthropol. 2, 172–181.

Jacobs, Z., Roberts, R.G., 2007. Advances in optically stimulated luminescence dating of individual grains of quartz from archaeological deposits. Evol. Anthropol. 16, 210–223.

Kappelman, J., 1993. The attraction of paleomagnetism. Evol. Anthropol. 2, 89–99.

Kuderna, L.F.K., Gao, H., Janiak, M.C., et al., 2023. A global catalog of whole-genome diversity from 233 primate species. Science 380, 906–913.

Loudon, J.E., Sponheimer, M., Sauther, M.L., Cuozzo, F., 2007. Intraspecific variation in hair δ13C and δ15N values of ringtailed lemurs (Lemur catta) with known individual histories, behavior, and feeding ecology. Am. J. Phys. Anthropol. 133, 978–985.

Ludwig, K.R., Renne, P.R., 2000. Geochronology on the paleoanthropological time scale. Evol. Anthropol. 9, 101–110.

McGraw, W.S., Cooke, C., Shultz, S., 2006. Primate remains from crowned eagle (Stephanoaetus coronatus) nests in Ivory Coast's Tai Forest: Implications for South African cave taphonomy. Am. J. Phys. Anthropol. 131, 151–165.

Meyer, M., Arsuaga, J.-L., de Filippo, C., et al., 2016. Nuclear DNA sequences from the Middle Pleistocene Sima de los Heusos hominins. Nature 531, 504–507.

Perelman, P., Johnson, W.E., Roos, C., et al., 2011. A molecular phylogeny of living primates. PLoS Genet. 7, e1001342.

Pobiner, B.L., 2015. New actualistic data on the ecology and energetics of scavenging opportunities. J. Hum. Evol. 80, 1–16.

Pozzi, L., Hodgson, J.A., Burrell, A.S., et al., 2014. Primate phylogenetic relationships and divergence dates inferred from complete mitochondrial genomes. Mol. Phylogenet. Evol. 75, 165–183.

Reed, K.E., 1997. Early hominid evolution and ecological change through the African Plio-Pleistocene. J. Hum. Evol. 32, 289–322.

Retallack, G., 2007. Paleosols. In: Henke, W., Tattersall, I. (Eds.), Handbook of Paleoanthropology, vol. 1. Springer-Verlag, Berlin, pp. 383–408.

Ross, C.F., Lockwood, C.A., Fleagle, J.G., Jungers, W.L., 2002. Adaptation and behavior in the primate fossil record. In: Plavcan, J.M., Kay, R.F., Jungers, W.L., van Schaik, C.P. (Eds.), Reconstructing Behavior in the Primate Fossil Record. Plenum, New York, pp. 1–41.

Schoeninger, M.J., Moore, J., Sept, J., 1999. Subsistence strategies of two "savanna" chimpanzee populations: Stable isotope evidence. Am. J. Primatol. 49, 297–314.

Schwarcz, H.P., 1992. Uranium series dating in paleoanthropology. Evol. Anthropol. 1, 56–62.

Shipman, P., 1981. Life History of a Fossil. Harvard University Press, Cambridge, MA.

Soligo, C., Will, O., Tavaré, S., et al., 2007. New light on the dates of primate origins and divergence. In: Ravosa, M.J., Dagosto, M. (Eds.), Primate Origins: Adaptations and Evolution. Springer, New York, pp. 29–49.

Sponheimer, M., Lee-Thorp, J., 2007. Hominin paleodiets: The contribution of stable isotopes. In: Henke, W. Tattersall, I. (Eds.), Handbook of Paleoanthropology, vol. 3. Springer-Verlag, Berlin, pp. 555–586.

Springer, M.S., Meredith, R.W., Gatesy, J., et al., 2012. Macroevolutionary dynamics and historical biogeography of primate diversification inferred from a species supermatrix. PLoS One 7 (11), e49521. https://doi.org/10.1371/journal.pone.0049521

Steiper, M.E., Seiffert, E.R., 2012. Evidence for a convergent slowdown in primate molecular rates and its implications for the timing of early primate evolution. Proc. Natl. Acad. Sci. USA 109, 6006–6011.

Steiper, M.E., Young, N.M., 2008. Timing primate evolution: Lessons from the discordance between molecular and paleontological estimates. Evol. Anthropol. 17, 179–188.

Tavaré, S., Marshall, C.R., Will, O., et al., 2002. Using the fossil record to estimate the age of the last common ancestor of extant primates. Nature 416, 726–729.

Taylor, R.E., 1996. Radiocarbon dating: The continuing revolution. Evol. Anthropol. 2, 169–181.

Wagner, G.A., 1996. Fission-track dating in paleoanthropology. Evol. Anthropol. 5, 164–171.

Zachos, J., Pagani, M., Sloan, L., et al., 2001.Thomas, E., Billups, K., Trends, rhythms, and aberrations in global climate 65 Ma to present. Science 292, 686–693.

Further reading

Deino, A.L., 2013. Geochronology. In: Begun, D.R. (Ed.), A Companion to Paleoanthropology. Wiley-Blackwell, Hoboken, pp. 244–264.

Henke, W., Tattersall, I., 2007. Handbook of Paleoanthropology, vol. 1–3. Springer-Verlag, Berlin.

Reed, K.E., 2013. Multiproxy paleoecology: Reconstructing evolutionary context in paleoanthropology. In: Begun, D.R. (Ed.), A Companion to Paleoanthropology. Wiley-Blackwell, Hoboken, pp. 204–225.

CHAPTER

11

Primate Origins

The Paleocene, the first epoch in the Age of Mammals, documents the first major radiation of many orders of placental mammals. At the end of the Cretaceous, the non-avian dinosaurs and their relatives that had dominated terrestrial faunas for the previous 160 million years had all disappeared, most probably due to the effects of a gigantic meteor impact in the Yucatan. Thus, beginning about 66 million years ago, some of the most abundant terrestrial vertebrates in the fossil record are mammals of various sorts.

Geologically, the Late Cretaceous and Paleocene were relatively active times in earth history and were marked by the rise of several major mountain groups, including the American Rockies. Geographically, the world looked different than it does today (Fig. 11.1). The North Atlantic was considerably narrower than at present, particularly in the vicinity of Greenland. The intermittent occurrence of land connections between North America and Europe is indicated by the similarity of the Paleocene faunas of the two continents. There is also faunal evidence of occasional connections between North America and Asia, presumably across the Bering Strait.

South America, Africa, and India were all island continents, except that South America and Antarctica were connected until the Oligocene. The South Atlantic was an open ocean, although somewhat narrower than it is today, and perhaps there were land surfaces lying between South America and Africa. The Panama land bridge, which currently connects North and South America, would not come into being for more than 60 million years. Africa was separated from Europe by the great Tethys Seaway, extending from China on the east to southern France on the west. India was adrift in the Indian Ocean and had not yet collided with the Asian mainland.

Paleocene climates were relatively cooler than those of either the preceding Late Cretaceous or the succeeding Eocene, but temperatures fluctuated throughout the epoch (Fig. 10.4). The flora of western North America, which has been carefully studied, was characterized by deciduous trees and conifers rather than the more tropical plants characteristic of immediately earlier and later epochs.

Euarchontans – Primates and Other Mammals

The earliest primates likely evolved sometime in the later part of the Cretaceous period or during the Paleocene. There is a general consensus from paleontology, comparative anatomy, and molecular systematics that, among living mammals, primates, treeshrews, and colugos are more closely related to one another than to other mammals. These groups are frequently placed in the superorder Euarchonta. In addition to these extant mammals, the fossil plesiadapiforms also lie near the origin of primates, either within the order Primates or as another order of closely related euarchontans. The temporal, geographic, and phylogenetic origins of primates and relationships among euarchontans are topics of considerable debate and discussion, with no clear answers. However, they involve many of the major issues confronting paleoanthropologists of all sorts, including the identification of homology and homoplasy, the relationship between morphological and molecular data, differences in the study of extant and fossil taxa, the relative completeness or incompleteness of the fossil record, methods of estimating divergence times, and basic questions about phylogeny reconstruction and systematics.

Treeshrews

Treeshrews, order Scandentia, are mostly small squirrel-like mammals from Asia. The three diurnal, partly terrestrial genera – *Tupaia* (Fig. 11.2), *Anathana*, and *Dendrogale* – are placed in the family Tupaiidae; and the nocturnal, arboreal *Ptilocercus* is placed in its own family, the Ptilocercidae (Roberts et al., 2011). Treeshrews are insectivorous and frugivorous. For many decades, following the early studies of Le Gros Clark (e.g., 1926, 1934), treeshrews were considered primitive primates. In the late 1960s and early 1970s, however, numerous studies demonstrated that many of the similarities between treeshrews and primates in brain morphology and reproductive anatomy were not homologous features,

Primate Adaptation and Evolution
https://doi.org/10.1016/B978-0-12-815809-8.00011-4

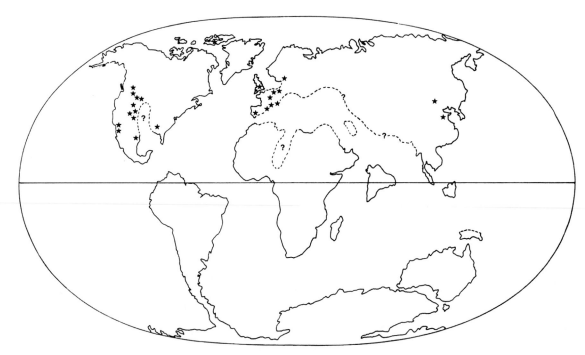

FIGURE 11.1   Map of the world during the Middle Paleocene (60 million years ago), with locations (∗) of plesiadapiform fossil sites. Map lines delineate study areas and do not necessarily depict accepted national boundaries.

and treeshrews are no longer classified with primates (Martin, 1968). Studies of cranial anatomy have pointed to other features (including a postorbital bar, the position of the maxillary artery, the bony canals for the arteries of the middle ear, and the dual jugular foramen) that link treeshrews and primates (Wible and Covert, 1987), but their homology is sometimes unclear (MacPhee, 1981; Wible, 2009, 2011). Postcranially, *Ptilocercus*, the pen-tailed tree shrew from many parts of Southeast Asia, shows a number of arboreal adaptations that link Scandentia to living primates. Inclusion of *Ptilocercus* in analyses of euarchontan relationships is important for our reconstructions of primate origins (Sargis, 2004).

Colugos (Flying Lemurs)

The colugos (sometimes called flying lemurs even though they are not lemurs and do not fly), order Dermoptera, are among the most unusual and interesting living mammals. They are gliding mammals from Southeast Asia that, like strepsirrhine primates, possess a "tooth comb". However, even that apparent similarity is misleading. In Dermoptera, the tooth comb does not comprise many separate incisors and canines as it does in many lemurs, but rather, each of the lower incisors contains numerous tines arranged like a comb (Fig. 11.3). Thus, the "tooth combs" in strepsirrhines and dermopterans are clearly not homologous. There are two living genera of dermopterans, *Cynocephalus* and *Galeopterus* (Fig. 11.2), each containing a single species. The limbs of

dermopterans show numerous specializations associated with their gliding habits, including unusual phalangeal proportions related to the webbing between their fingers (Beard, 1993a). Molecular studies have consistently found Dermoptera to be closely related to primates, and numerous researchers have argued that Primates and Dermoptera are sister taxa on the basis of genetic similarities (Janecka et al., 2007; Mason et al., 2016). The proposed clade including Dermoptera, Primates, and their fossil relatives (e.g., various plesiadapiforms) is named Primatomorpha (Beard, 1989).

Plesiadapiforms

The plesiadapiforms were an extremely successful group of primate-like mammals that flourished in the Paleocene and early Eocene of North America and Europe (Fig. 11.1), and are also known from Asia. In North America, they are the most common mammals in many Paleocene faunas. Their known taxonomic diversity (more than 55 genera and over 150 species) is much greater than that of living strepsirrhines, and their diversity in size is comparable to that of either living strepsirrhines or New World platyrrhines. However, the phylogenetic relationships among plesiadapiforms, Dermoptera, Scandentia, and living primates have been the subject of ongoing debate for many decades. Although plesiadapiforms have long been known primarily from teeth and jaws, this has all changed in recent decades with the discovery and description of many

FIGURE 11.2   The superorder Euarchonta includes Primates and two other orders of extant mammals, Scandentia (treeshrews) shown in the *lower part* of the picture, and Dermoptera (colugos) in the *upper part* of the picture. *(Modified from drawing by William Yee.)*

very complete specimens preserving almost the entire skeleton (Bloch and Boyer, 2002, 2007; Bloch et al., 2007; Boyer and Gingerich, 2019; Chester et al., 2017).

The families of plesiadapiforms show many distinctive dental specializations (Figs. 11.3 and 11.4) as well as dental features that link them with later primates. These include molar teeth with relatively low cusps (compared to contemporary or extant insectivores), lower molars with low trigonids and basin-shaped talonids, and elongated lower third molars with an extended talonid. Their upper molars often have prominent conules, a poorly developed or absent stylar shelf, and a well-developed post-protocingulum (nannopithex fold) or comparable wear facet distal to the protocone. The primitive dental formula for plesiadapiforms is 3.1.4.3. Most later members of all lineages show reduction and loss of teeth, especially incisors and the anterior premolar. Since

almost all plesiadapiforms have a dental formula with three or fewer premolars, they are too specialized to have given rise to the earliest prosimians, many of which have four premolars. In addition, most plesiadapiforms have extremely large and procumbent upper and lower central incisors (Rose et al., 1993), which are more derived than those teeth in the earliest primates (Figs. 11.3 and 11.4; see also Fig. 12.2).

The sharp cusps on the teeth of many species, as well as their small size, suggest that many plesiadapiform species were largely insectivorous. Nevertheless, many have teeth that indicate they were capable of more crushing and thus more omnivory and herbivory/frugivory than contemporary insectivores, and some show adaptations for harvesting exudates (Beard, 1991) and leaves (Boyer et al., 2010).

Plesiadapiforms have a low, flat skull with a long snout, large infraorbital foramina, a small brain, large

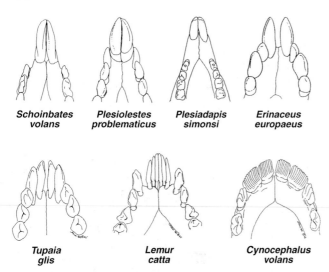

FIGURE 11.3 The anterior dentitions of *Schoinobates volans*, a folivorous marsupial; *Plesiolestes problematicus* and *Plesiadapis simonsi*, two plesiadapiforms; *Erinaceus europaeus*, a hedgehog; *Tupaia glis*, a treeshrew; *Lemur catta*, a ring-tailed lemur; and *Cynocephalus volans*, a colugo, or flying lemur.

zygomatic arches, and no bony ring surrounding their laterally directed orbits (Fig. 11.5). In these features, they are more primitive than all later primates. There has been considerable debate about the likely arterial circulation to the braincase, and whether plesiadapiforms had an auditory bulla made up of the petrosal bone (e.g., MacPhee et al., 1983; Kay et al., 1992; Bloch et al., 2007, 2016; Boyer et al., 2016). New evidence indicates that, in contrast to primates, plesiadapiforms commonly possess a reduced internal carotid circulatory system (Boyer et al., 2016). The composition of the auditory bulla in plesiadapiforms, when present, is nonpetrosal in some taxa and is debated in others (Silcox et al., 2017). Thus there are very few, if any, cranial features clearly linking plesiadapiforms with extant primates.

Analyses of the limb and trunk skeleton, particularly the foot, the elbow, and the wrist, have suggested several features that link plesiadapiforms with primates and others that link them with Dermoptera (Beard, 1993a,b; but see further below). Other studies have pointed out that the pen-tailed treeshrew, *Ptilocercus*, exhibits many of these features as well, and thus there is no special link in postcranial anatomy between colugos, plesiadapiforms, and primates (Sargis, 2002). Such analyses are limited by the available material in the fossil record. However, attributable postcrania demonstrate that most plesiadapiforms were almost certainly arboreal (Kirk et al., 2008; Chester et al., 2015, 2017).

There is debate among current authorities regarding the composition of the order Plesiadapiformes. The taxonomic scheme adopted here (Tables 11.1–11.3) is based on the work of Silcox and colleagues (e.g., Silcox, 2001; Bloch et al., 2007; Silcox and Gunnell, 2008; Silcox et al., 2017). The order Plesiadapiformes is paraphyletic

and contains 11 families. Three families – plesiadapids, carpolestids, and saxonellids – are placed in the superfamily Plesiadapoidea. Three other families – palaechthonids, paromomyids, and picrodontids – are placed in a separate superfamily, Paromomyoidea. The remaining five families – purgatoriids, microsyopids, picromomyids, micromomyids, and toliapinids, as well as two unusual genera – are left unclassified at any superfamily level.

Purgatoriids

Purgatorius, a tiny mammal from the earliest part of the Paleocene of North America, is often regarded as the most primitive plesiadapiform (Table 11.1). With a lower dental formula of 3.1.4.3 (Clemens, 2004) and very primitive molars, *Purgatorius* is generalized enough to be ancestral to both all other plesiadapiforms and later primates. However, ankle bones attributed to *Purgatorius* are derived relative to other primitive mammals, indicating arboreal behavior and placing it within Plesiadapiformes (Chester et al., 2015). It is known from at least five Early Paleocene species. A single tooth is sometimes recognized as another species, *P. ceratops*, but its provenence and identification are questionable. Regardless, the recent discovery of multiple *Purgatorius* species at the base of the Paleocene suggests that purgatoriids first arose even earlier, perhaps in the latest Cretaceous (Wilson Mantilla et al., 2021). The recently discovered *Ursolestes* is very similar to *Purgatorius* in nearly all known dental features but is much larger (Fox et al., 2015; Hovatter et al., 2024).

Microsyopids

The Microsyopidae (Table 11.1) was a very successful family, with over 20 species from the Late Paleocene through Middle Eocene of both North America and Europe. Microsyopids are relatively diverse in appearance and include both very small species such as *Uintasorex parvulus*, smaller than any living primate, and large species (*Megadelphus*), the size of a small raccoon. Microsyopids have distinctive upper incisors ("can-opener" shaped in primitive taxa, caniniform in more derived taxa) and a narrow, lanceolate (spearhead-shaped), specialized lower central incisor (Fig. 11.5).

Cranially, microsyopids have most of the primitive mammalian features characteristic of other plesiadapiforms, such as a long snout, small brain, lateral facing orbits, and lack of a postorbital bar (Fig. 11.5). The best-known genus, *Microsyops*, has a cranial blood supply that more closely resembles the cranial arterial pattern of living primates than does that of any other plesiadapiform, but it also has an auditory structure that is more

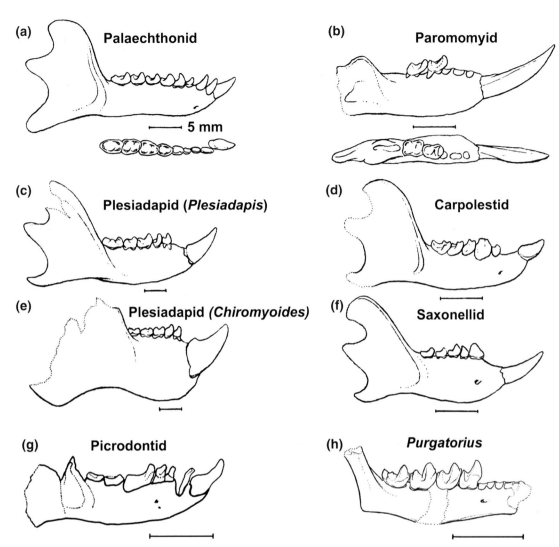

FIGURE 11.4 Mandibles of several plesiadapiforms, showing the diversity in shape and size of the dentition (a) *Plesiolestes problematicus*, lateral and occlusal views; (b) *Elwynella oreas*, lateral and occlusal views; (c) *Plesiadapis rex*, lateral view; (d) *Elphidotarsius florencae*, lateral view; (e) *Chiromyoides campanicus*, lateral view; (f) *Saxonella crepaturae*, lateral view; (g) *Picrodus silberlingi*, lateral view; (h) *Purgatorius pinecreeensis*, lateral view.

primitive than that of other plesiadapiforms in that it lacks a bony bulla. There are few limb bones attributed to microsyopids.

Micromomyids

This family has five genera from North America (Table 11.1) and possibly an undescribed taxon from Asia (Chester and Bloch, 2013; Rose et al., 1993; Tong and Wang, 1998). Micromomyids are all very tiny plesiadapiforms: they most probably weighed under 30 g. Their lower central incisors are enlarged, but their molars are primitive and resemble those of many purgatoriids. A number of species have multicusped upper central incisors similar to those of plesiadapids and carpolestids, and a large posterior lower premolar. From its small size and its molars with very acute cusps, *Foxomomys* appears to have been almost totally insectivorous. *Tinimomys* has more rounded molar cusps.

Several micromomyids are known from relatively complete skeletons (Beard, 1993a,b; Bloch et al., 2007). Although they have been suggested to show morphological features indicative of gliding habits (Beard, 1993a,b), recovery of partial skeletons and more extensive analyses indicate that this is most unlikely, and that they were arboreal, clawed climbers, similar in locomotor habits to extant *Ptilocercus* and probably *Purgatorius* as well (Bloch et al., 2007; Boyer and Bloch, 2008; Chester et al., 2015).

Picromomyids

Picromomyids are very tiny animals known only from teeth. They possess low-crowned molars and may

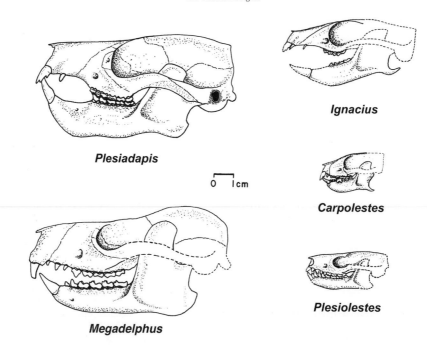

Plesiadapis

Ignacius

Carpolestes

0   1 cm

Plesiolestes

Megadelphus

FIGURE 11.5   Skulls of five plesiadapiforms.

have subsisted on a combination of soft-bodied insects, larvae, and exudates (Rose and Bown, 1996). Their relationships to other plesiadapiforms are not resolved, but their incisor morphology suggests possible relationships with micromomyids (Rose and Bown, 1996; Silcox, 2001; Silcox et al., 2002).

Paromomyids

The paromomyids (Table 11.2) are a very diverse and speciose group of plesiadapiforms. They are among the most long-lived and geographically widespread families. They ranged in time from the Early Paleocene through the Middle Eocene, and have been found in North America as far north as the Arctic Circle, as well as in Europe and Asia. Paromomyids were small to medium-size plesiadapiforms with long, slender lower central incisors and mitten-like upper central incisors with three distal cusps and one basal cusp (Silcox et al., 2017; Scott et al., 2023). The posterior lower premolar is usually tall and pointed. Paromomyids have relatively flat, low-crowned lower molars with short, squared trigonids and broad, shallow talonid basins; the upper molars are square, with expanded basins. Both upper and lower posterior molars are conspicuously elongated.

The function of the incisors is uncertain. Some authors note similarities to the incisors of shrews and suggest that they functioned in procuring insects; others suggest that paromomyid incisors were adapted for exudate eating (e.g., Boyer and Bloch, 2008). However, as noted by Rosenberger (2010), the incisors of paromomyids are structurally very different from those of marmosets and

seem unlikely to have functioned as tools for gouging bark. Rather, the delicate occlusal relationships between the lower and upper incisors suggest some sort of foraging (nipping) of herbivorous foods of some kind. The pointed P_4 seems to be adapted for puncturing food during initial preparation, and the broad, flat, lower molars suggest an herbivorous rather than an insectivorous diet, probably including fruits, gum, or nectar, for most paromomyids.

Partial skulls are known for several species. In all, the face is long and narrow and has a large infraorbital foramen, suggesting a richly innervated snout with tactile vibrissae. They have small brains and broad zygomatic arches. The bony auditory bulla is formed by the entotympanic bone.

Numerous skeletal remains of *Phenacolemur* were argued by Beard (1990, 1993a,b) to show similarities to colugos in having relatively long, straight intermediate phalanges as well as many other osteological similarities in the remainder of the skeleton, including the wrist, the femur, and the pelvis. However, analyses by Runstead and Ruff (1995) and Bloch et al. (2007) have demonstrated that the limbs of *Phenacolemur* are far more robust than those of any extant gliding or flying mammals, and more detailed analyses of paromomyid skeletons by Boyer and Bloch (2008) have largely refuted any evidence of colugo-like gliding behavior in paromomyids. They seem to have been adapted for climbing and clinging to vertical supports.

Paromomyids are the only plesiadapiforms with a geographic range that extends above the Arctic Circle. During the Early Eocene, two large species of *Ignacius* thrived on Ellesmere Island (Miller et al., 2023), which

TABLE 11.1 Order PLESIADAPIFORMES

Species	Estimated mass (g)
Family INCERTAE SEDIS	
Pandemonium (Early Paleocene, North America)	
P. hibernalis	275
P. dis	224
Asioplesiadapis (Early Eocene, Asia)	
A. youngi	54
Russellodon (Early Paleocene, Europe)	
R. haininense	621
Family PURGATORIIDAE	
Purgatorius (Early Paleocene, North America)	
P. unio	63
P. ceratops?	57
P. pinecreeensis	51
P. coracis	57
P. janisae	68
P. mckeeveri	85
Ursolestes (Early Paleocene, North America)	
U. perpetior	944
U. blissorum	4594
Family MICROSYOPIDAE	
Navajovius (Late Paleocene–Early Eocene, North America)	
N. kohlhaasae	32
Berruvius (Late Paleocene–Early Eocene, Europe)	
B. lasseroni	16
B. gingerichi	22
Subfamily UINTASORICINAE	
Uintasorex (Middle–Late Eocene, North America)	
U. parvulus	17
U. montezumicus	10
Niptomomys (Early Eocene, North America)	
N. doreenae	20
N. favorum	10
N. thelmae	41
Choctawius (Early Eocene, North America)	
C. foxi	17
C. mckennai	–
Nanomomys (Early Eocene, North America)	
N. thermophilus	9
Bartelsia (Early Eocene, North America)	
B. pentadactyla	10
Subfamily MICROSYOPINAE	
Arctodontomys (Early Eocene, North America)	
A. simplicidens	327
A. nuptus	476
A. wilsoni	248

TABLE 11.1 Cont'd

Microsyops (Eocene, North America)	
M. elegans	689
M. angustidens	455
M. annectens	1350
M. cardiorestes	297
M. knightensis	557
M. latidens	405
M. scottianus	790
M. kratos	2120
Megadelphus (Early Eocene, North America)	
M. lundeliusi	2723
Craseops (Late Eocene, North America)	
C. sylvestris	2015
Family PICROMOMYIDAE	
Alveojunctus (Middle Eocene, North America)	
A. minutus	50
A. bowni	114
Picromomys (Early Eocene, North America)	
P. petersonorum	14
Family MICROMOMYIDAE	
Micromomys (Late Paleocene–Early Eocene, North America)	
M. silvercouleei	–
Foxomomys (Late Paleocene, North America)	
F. fremdi	12
F. vossae	11
F. gunnelli	–
Chalicomomys (Early Eocene, North America)	
C. antelucanus	11
Tinimomys (Early Eocene, North America)	
T. graybulliensis	16
T. tribos	13
Dryomomys (Late Paleocene, North America)	
D. szalayi	16
D. millennius	14
D. willwoodensis	–
D. dulcifer	15
Family TOLIAPINIDAE	
Avenius (Early Eocene, Europe)	
A. amatorum	8
Toliapina (Early Eocene, Europe)	
T. vinealis	8
T. lawsoni	14

TABLE 11.2 Order PLESIADAPIFORMES
Superfamily Paromomyoidea

Species	Estimated mass (g)
Family PALAECHTHONIDAE	
Subfamily PALAECTHONINAE	
Palaechthon (Middle–Late Paleocene, North America)	
P. alticuspis	94
P. woodi	36
Palenochtha (Middle–Late Paleocene, North America)	
P. minor	18
P. weissae	22
Premnoides (Middle Paleocene, North America)	
P. douglassi	75
Subfamily PLESIOLESTINAE	
Phoxomylus (Late Paleocene, North America)	
P. puncticuspis	–
Plesiolestes (Middle–Late Paleoence, North America)	
P. problematicus	148
P. nacimienti	172
Talpohenach (Middle Paleocene, North America)	
T. torrejonius	300
Torrejonia (Middle Paleocene, North America)	
T. wilsoni	441
T. sirokyi	563
Anasazia (Paleocene, North America)	
A. williamsoni	304
Family PAROMOMYIDAE	
Edworthia (Early Paleocene, North America)	
E. lerbekmoi	40
E. greggi	48
Paromomys (Middle Paleocene, North America)	
P. farrandi	144
P. maturus	305
P. depressidens	85
P. libedianus	225
Ignacius (Middle Paleocene–Middle Eocene, North America)	

TABLE 11.2 Cont'd

I. frugivorus	111
I. fremontensis	46
I. glenbowensis	68
I. mceknnai	1147
I. dawsonae	632
I. graybullianus	176
I. clarkforkensis	242
Phenacolemur (Late Paleocene–Middle Eocene, North America, Europe)	
P. archus	165
P. praecox	272
P. citatus	207
P. fortior	292
P. jepseni	121
P. pagei	181
P. simonsi	85
P. willwoodensis	121
Acidomomys (Late Paleocene, North America)	
A. hebeticus	173
Elwynella (Middle Eocene, North America)	
E. oreas	153
Arcius (Early–Middle Eocene, Europe)	
A. rougieri	77
A. fuscus	111
A. lapparenti	168
A. zbyszewskii	56
A. moniquae	110
Family PICRODONTIDAE	
Draconodus (Middle Paleocene, North America)	
D. apertus	–
Zanycteris (Late Paleocene, North America)	
Z. paleocenus	–
Z. honeyi	–
Picrodus (Middle–Late Paleocene, North America)	
P. silberlingi	33
P. calgariensis	40
P. canpacius	22
P. lepidus	25

has been located near the Arctic Circle at 78° north latitude during most of the Cenozoic Era. Because there are several months of total darkness at that latitude today, it seems likely that the fauna was composed of cathemeral or crepuscular mammals.

Palaechthonids

Palaechthonids are a diverse and possibly paraphyletic family from the Paleocene of North America with eight genera placed in two subfamilies. In size, they are comparable to the smallest living primates (20–500 g). Most species have a dental formula of 2.1.3.3, but the canine and anteriormost premolar are very small in many species and probably absent in some. The enlarged, lanceolate lower first incisors of *Plesiolestes* form a scoop-like apparatus for cutting, an adaptation that suggests a partly herbivorous diet (Figs. 11.3 and 11.4), yet the molars have relatively acute cusps compared to those of many living primate species, suggesting that insects were a major part of the diet as well. Other taxa with more rounded cusps (e.g., *Torrejonia*) were perhaps more frugivorous (Silcox et al., 2017).

For one of these small Middle Paleocene species, *Plesiolestes nacimienti*, there is a relatively complete but crushed skull (Fig. 11.5) with relatively small, laterally directed orbits, suggesting limited stereoscopic abilities; a broad interorbital region, suggesting a large olfactory fossa and greater reliance on a sense of smell; and a large infraorbital foramen, suggesting the presence of a richly innervated snout bearing sensitive facial vibrissae.

Kay and Cartmill (1977) suggested that the small size and cranial features of *P. nacimienti* indicate that it was probably a terrestrial forager that hunted for concealed insects and other animal prey by "nosing around the ground," foraged more by hearing, smell, and its sensitive snout than by vision, and that it was probably nocturnal. New skeletal material of the early palaechthonid *Torrejonia*, however, suggests that it was an arboreal animal, broadly similar to other plesiadapiforms, with adaptations for climbing and clinging to vertical supports (Chester et al., 2017).

Picrodontids

The three genera of the family Picrodontidae (Table 11.2) are tiny and known primarily by dental and gnathic remains from the Middle and Late Paleocene of western North America. Their relationship to other plesiadapiforms is unclear, and recent analyses of *Zanycteris* note primitive features of the petrosal suggesting that they may not be plesiadapiforms at all (Crowell et al., 2024). They resemble plesiadapiforms in their incisor morphology (Fig. 11.3), but their cheek teeth are quite unusual.

The first upper and lower molars are enlarged and oddly shaped. The lower molars have very small trigonids and large, shallow talonids with crenulated enamel. Because of notable similarities between the molars of picrodontids and those of some frugivorous bats, we can infer that they probably had a diet of fruit and nectar.

Plesiadapids

The best known of the Plesiadapiformes, plesiadapids, were very diverse and abundant in the Paleocene of North America and the Paleocene through Early Eocene of Europe (Table 11.3). There are five genera. The smallest species were comparable in size to a marmoset (150 g); the largest were the size of a small guenon (2.5 kg).

Compared to many other plesiadapiforms, plesiadapids generally have molar and premolar teeth that often resemble those of primates (Fig. 11.4). The most primitive genus in the family, *Pronothodectes*, has a dental formula of 2.1.3.3, but later genera show considerable reduction and loss of incisors, canines, and premolars. All plesiadapids have relatively broad, procumbent lower incisors that occlude in a pincer-like fashion with the multicusped upper central incisors (Figs. 11.4 and 11.5). Like primates, plesiadapids have premolars and molars with low, bulbous cusps. The posterior two upper premolars are short and broad. The lower molars have a relatively low trigonid and broad talonid. These low-crowned cheek teeth, together with the relatively large size of most species, suggest that plesiadapids were predominantly herbivorous with a diversity of adaptations, including some more insectivorous species, such as the odd *Chiromyoides* with a deep aye-aye-like jaw, and *Pronothodectes* with broad chisel-like incisors (Boyer et al., 2010) (Fig. 11.4).

There are several skulls of *Plesiadapis*. The cranium of *Plesiadapis* (Fig. 11.5) has a long snout with a large premaxillary bone and a diastema between the large incisors and the cheek teeth in both the upper and the lower jaws. The auditory bulla has been suggested to be composed of the petrosal bone, a feature that this family could share with extant primates, but this has proved difficult to confirm and specimens of *Pronothodectes* seem to suggest differently (Boyer et al., 2012). The tympanic ring (tympanic bone) is fused to the bulla and extends laterally to form a bony tube. Considerable skeletal material is known for *Plesiadapis*. The short, robust limbs, the long, laterally compressed claws, and the long, bushy tail (known from a delicate limestone impression; Fig. 10.5) indicate that it was an arboreal quadruped making use of large diameter supports (Bloch and Boyer, 2007; Boyer and Bloch, 2008; Boyer, 2009; Boyer and Gingerich, 2019). The phylogenetic relationships among plesiadapids have been extensively studied in recent years (e.g., Bloch et al., 2007; Boyer et al., 2012; Boyer and Gingerich, 2019).

Carpolestids

The carpolestids (Table 11.3) are a North American family of small, mouse-sized (20–150 g) plesiadapiforms, characterized by the enlargement of their last lower premolar and last two upper premolars. Relatively complete material of *Carpolestes* has been described in recent years, including much of the postcranial skeleton (Fig. 11.6). In contrast to the clawed digits found in all other plesiadapiforms, *Carpolestes* had a divergent hallux with a mediolaterally flattened and expanded distal phalanx, suggesting that it resembled extant primates in having a nail-like structure (Bloch and Boyer, 2002). However, the hallux of *Carpolestes* also retains an overall dorsoventrally deep claw-like shape, and it is very different from the hallux of any primate.

The three most common North American genera more or less follow one another in time: *Elphidotarsius*, from the Middle and Late Paleocene; *Carpodaptes*, from the Late Paleocene; and *Carpolestes*, from the Latest Paleocene (Fig. 11.7). Because of their distinctive morphological specializations and short species durations, these North American genera have been useful as biostratigraphic indicators in Early Tertiary sediments. However, the Asian *Carpocristes* indicates a much more complex phylogenetic and biogeographical history for this family (Beard and Wang, 1995; Bloch et al., 2001). *Chronolestes*, a tiny animal from the Early Eocene of China, was originally considered to be a basal carpolestid, but other analyses place it near the base of all plesiadapoids.

Saxonellids

Saxonella (Fig. 11.4; Table 11.3) is a relative of the plesiadapids from the Late Paleocene of North America and Germany. Like carpolestids, *Saxonella* seems to be a derivative of the plesiadapids that evolved a very large lower premolar. In contrast to carpolestids, however, which enlarged the last premolar, *Saxonella* enlarged P_3. Thus, although probably adaptively similar, the two groups are not sister taxa.

Toliapinids

Toliapinids are a poorly known group of very small European plesiadapiforms represented by two genera, *Toliapina* and *Avenius*. Surprisingly, for their size toliapinids have relatively bunodont molars and were probably omnivorous, incorporating soft foods into their diet (Silcox et al., 2017).

Adaptive Radiation of Plesiadapiforms

The plesiadapiforms were a very successful group of early Cenozoic mammals that evolved a range of body

TABLE 11.3   Order PLESIADAPIFORMES
Superfamily PLESIADAPOIDEA

Species	Estimated mass (g)
Family INCERTAE SEDIS	
Chronolestes (Early Eocene, Asia)	
C. simul	35
Family PLESIADAPIDAE	
Pronothodectes (Middle–Late Paleocene, North America)	
P. matthewi	153
P. gaoi	315
P. jepi	208
Nannodectes (Late Paleocene, North America)	
N. gazini	188
N. gidleyi	396
N. simpsoni	329
N. intermedius	221
Plesiadapis (Late Paleocene–Early Eocene, North America, Europe)	
P. tricuspidens	1283
P. walbeckensis	389
P. remensis	753
P. russelli	–
P. anceps	427
P. churchilli	728
P. cookei	2560
P. dubius	376
P. fodinatus	540
P. gingerichi	1735
P. praecursor	316
P. rex	506
P. simonsi	1217
P. insignis	235
P. ploegi	1344
Chiromyoides (Late Paleocene–Early Eocene, North America, Europe)	
C. gingerichi	–
C. campanicus	329
C. caesor	227
C. major	489
C. minor	125

TABLE 11.3   Cont'd

C. potior	–
C. mauberti	189
Platychoerops (Early Eocene, Europe)	
P. daubrei	1977
P. richardsonii	–
P. georgei	1462
P. antiquus	1815
P. russelli	–
P. boyeri	1575

Family CARPOLESTIDAE

Elphidotarsius (Middle–Late Paleocene, North America)	
E. florencae	48
E. russelli	42
E. shotgunensis	48
E. wightoni	26
Carpodaptes (Late Paleocene, North America)	
C. aulacodon	48
C. cygneus	26
C. hazelae	54
C. hobackensis	48
C. rosei	26
C. stonleyi	35
Carpomegodon (Late Paleocene, North America)	
C. jepseni	113
Carpolestes (Late Paleocene–Early Eocene, North America)	
C. nigridens	60
C. dubius	110
C. simpsoni	56
Carpocristes (Early Eocene, Late Paleocene, America, Asia)	
C. oriens	18
Subengius (Late Paleocene, Asia)	
S. mengi	22

Family SAXONELLIDAE

Saxonella (Late Paleocene, North America, Europe)	
S. crepaturae	82
S. naylori	57

FIGURE 11.6   A reconstruction of *Carpolestes* showing its grasping hallux and frugivorous habits. *(Courtesy of D. Boyer.)*

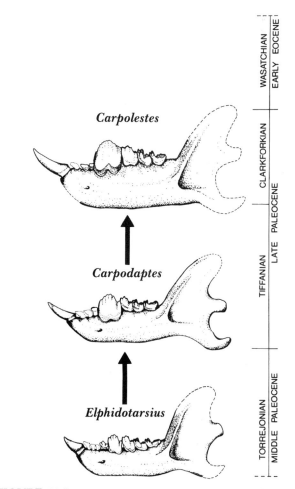

FIGURE 11.7   Differences in premolar shape among three genera of North American carpolestids. Note the increasing size of the last premolar and first premolar. *(From Rose, K.D., 1975. The Carpolestidae: Early tertiary primates from North America. Bull. Mus. Comp. Zool. 147, 1–74.)*

sizes as well as dental and postcranial adaptations (Figs. 11.3–11.5, 11.8, and 11.9). They include several species that were as large as living lemurs or New World monkeys, as well as several species that were much smaller than any living primate. Their cranial structure is so different from that of any living primate that we have little evidence of whether most taxa were diurnal or nocturnal. Their great diversity in dental morphology (Figs. 11.3 and 11.4) suggests considerable diversity in dietary adaptations (e.g., Boyer et al., 2010). It seems likely that many species specialized on insects, and many others relied on fruit, leaves, seeds, or other plant parts. In addition, it seems likely that some relied on nectar or gums. The size and shape differences in their incisor and mandible structures indicate that plesiadapiform feeding habits were probably quite different from anything found among living primates. This is especially true for picrodontids and carpolestids, which had very odd dental specializations by any standards.

The limb skeletons of plesiadapiforms have become increasingly well known in the past decade. Most taxa seem to have had digital proportions of their clawed hands that resemble arboreal rather than terrestrial mammals, including primates (Kirk et al., 2008), and one taxon, *Carpolestes simpsoni*, had a flattened nail-like structure on its divergent big toe. Other details of the limbs, including overall body proportions, indicate that all were primarily arboreal quadrupeds with no indication of leaping (Chester et al., 2017), a reconstruction that is supported by the morphology of their semicircular canals (Silcox et al., 2009). However, many were probably capable of clinging to large supports with their claws. Suggestions that paromomyids and micromomyids had gliding abilities similar to extant colugos have been largely disproved.

The social habits of plesiadapiforms are certainly beyond our ken, but we can speculate. If they were nocturnal, they probably were mostly solitary foragers, like many primitive mammals. Diurnal species may have lived in larger groups.

The radiation of plesiadapiforms was largely during the Paleocene, although several groups, including microsyopids, paromomyids, and picromomyids, survived well into the Eocene and others, including plesiadapids and carpolestids, had at least one Eocene taxon. Several explanations are commonly offered for the decline, in the beginning of the Eocene, of this once very successful group. The most common view has been that plesiadapiform decline and extinction resulted from competition with some other mammals, including rodents, early prosimians, and possibly bats. It is also likely that changes in the diversity of some plesiadapiforms were linked with climatic changes, and that their decline and extinction at the beginning of the Eocene related, either directly or indirectly, to the

dramatic changes in global climate of that epoch (Fig. 10.4). In a review of this problem, Maas et al. (1988) found that the changes in climate during the Late Paleocene and Early Eocene did not correlate well with changes in the diversity of plesiadapiforms, and that the radiation of early prosimians (see Chapter 12) came after the extinction of most plesiadapiforms. The increasing diversity and abundance of early rodents is, however, correlated with the decline of the plesiadapiforms. Moreover, functional comparisons show that certain plesiadapiforms and rodents were likely to have been similar in many aspects of their ecological adaptations, so it is not unlikely that they were competitors to some extent (but see also Prufrock et al., 2016).

Plesiadapiforms and Primates

Our knowledge of plesiadapiforms has expanded manyfold in the past two decades, so that they are now known from many nearly complete skeletons and have been the subject of numerous analyses that address their role in primate evolution. However, the relationship of plesiadapiforms to primates and to primate origins, and how this should be expressed in systematics, remains a difficult issue with many alternative views (e.g., Silcox, 2007; Silcox et al., 2017). Are some or all plesiadapiforms the close relatives of crown primates? If so, what does this tell us about the pattern and possibly the timing of primate origins? How should the phylogenetic relationships between primates and plesiadapiforms be expressed in the systematics of the two groups?

In their comprehensive phylogenetic analysis of plesiadapiforms and their relationship to crown primates, Bloch and colleagues (2007) place all plesiadapiforms within the order Primates and identify a series of nested clades within the paraphyletic plesiadapiforms, with plesiadapoids as the sister group of crown primates (Euprimates) in the group Euprimateformes following the taxonomic guidelines of Phylocode (de Queiroz and Gauthier, 1990). As noted by Silcox (2007) in her very articulate discussion of this issue, these are primarily stem-based clades rather than apomorphy-based groups, although each clade does share some – mostly dental – features. Some characteristic features of crown primates, such as an opposable hallux and possibly a nail-like structure on the hallux, seem to be found in at least one species of carpolestids, but do not characterize the whole group. As such, they would seem to be examples of homoplasy. While this phylogeny suggests an evolutionary mosaic in which digital proportions of the hand suitable for grasping (with or without claws), and perhaps a petrosal bulla, appeared in some plesiadapiforms before the evolution of crown primates, it also requires that the number of premolars must increase and the size

FIGURE 11.8   Reconstruction of a scene from the Late Paleocene of North America showing several plesiadapiforms. A small group of *Plesia-dapis rex* feeds in a tree, and *Ignacius frugivorous* feeds on exudates from the trunk. A small *Picrodus silberlingi* feeds on nectar in a bush.

and complex morphology of the incisors must decrease dramatically in the origin of Euprimates.

Thus, although plesiadapiforms are the most primate-like mammals from the Paleocene, almost all have various morphological specializations that would seem to preclude them from the ancestry of the early prosimians that appear in the beginning of the Eocene epoch, as well as from any direct relationship with other later primates (Fig. 11.9). For example, most of the earliest and most primitive Eocene crown primates have four lower premolars and simple incisors. Most plesiadapiforms, on the other hand, have very derived dental features, such as a reduced dental formula (particularly the premolars

and anterior dentition), enlarged remaining premolars, and large procumbent incisors. While some plesiadapoids and micromomyids have been suggested to have a petrosal bulla, as in extant primates, this is not the case for other taxa or families in which the cranium is known. Indeed, only the most primitive plesiadapiform, *Purgatorius*, is generalized enough in its dental formula and dentition to be a suitable ancestor for all later primates, but it is so generalized and poorly known that very few features would link it with later primates.

Any phylogeny is probably going to involve some amount of homoplasy and reversal of characters, but the

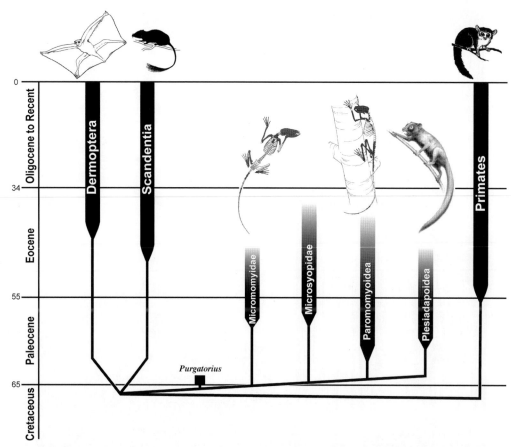

FIGURE 11.9   Phyletic relationships among Euarchonta and a phylogeny of Plesiadapiformes. *(Modified from figures by Jon Bloch and Stephen Chester. Plesiadapiform phylogeny based on analyses in Bloch, J.I., Silcox, M.T., Boyer, D.M., Sargis, E.J., 2007. New Paleocene skeletons and the relationship of plesiadapiforms to crown-clade primates. Proc. Nat. Acad. Sci. USA 104, 1159–1164.)*

scenario in which crown primates are nested within plesiadapiforms as the sister taxon of plesiadapoids involves an evolutionary reversal of increasing tooth number and dramatic reduction of the procumbent incisors that seems unlikely (e.g., Godinot, 2007; but see Boyer et al., 2010). Thus, while it is intriguing that within the radiation of plesiadapiforms there may be evidence for the evolution of several features that also characterize crown primates, it is unlikely that plesiadapiforms are stem primates in more than the most general sense. It seems more likely that, albeit primate-like in some of their adaptations, this extensive radiation of arboreal Euarchontans was collateral to, rather than ancestral to, the origin of crown primates, and that plesiadapiforms are best kept as a separate order of mammals (Fig. 11.9).

Primates Among the Euarchonta

As already noted, Primates are usually grouped with Plesiadapiformes, Scandentia (treeshrews), and Dermoptera, in a single superorder, the Euarchonta. The older term Archonta included bats (Chiroptera) within

this group, and virtually all molecular studies indicate that they are more distantly related. However, although most authorities accept the concept of a superorder Euarchonta, studies based on different anatomical systems (e.g., genes, basicranial anatomy, postcranial anatomy, dentition) and often including different taxa, have reached different results regarding the relationships among these groups, and specifically which is the sister taxon of crown primates (Silcox, 2007). Thus, on the basis of basicranial anatomy and/or the anatomy of the foot, some authors (Wible and Covert, 1987; Wible and Martin, 1993; Godinot, 2007) have found treeshrews (Scandentia) to be the sister taxon to Primates. Most molecular studies (e.g., Janecka et al., 2007; Perelman et al., 2011; Mason et al., 2016), which cannot include plesiadapiforms, have found Dermoptera to be the sister taxon to Primates, and a broad morphological study (Bloch et al., 2007) based heavily on dental features of fossils as well as extant dermopterans, treeshrews, and some crown primates found that some plesiadapiforms were the sister taxon of crown primates. More specifically, they found that crown primates are a clade within a larger radiation of plesiadapiforms. Yet other molecular studies find that dermopterans and scandentians are most closely related within Euarchonta, with primates being their sister group (Upham et al., 2019). Despite some morphological evidence, there is little molecular support for Scandentia and Primates as sister taxa.

Thus, the supraordinal relationships of Primates are certainly not resolved to everyone's satisfaction (Fig. 11.9). There are many reasons for this (Silcox et al., 2017). Undoubtedly, it partly reflects the long time between the separation of primates from other orders of living mammals over 60 million years ago, and the numerous evolutionary specializations all groups have accumulated in the meantime that may mask true relationships and/or suggest false ones. In addition, we have little fossil evidence for the evolution of Dermoptera and Scandentia. Morphological studies are the only ones that can potentially include all of the relevant groups, and the most thorough analyses yield different results regarding the relationships of Primates to plesiadapiforms, Dermoptera and Scandentia (Bloch et al., 2007; Ni et al., 2011, 2013, 2016; Chester et al., 2015; Seiffert et al., 2020). Thus, at present there is no clear resolution of the phyletic relations of Primates to other orders of Euarchonta.

The Adaptive Origin of Primates

Extant primates are distinguished from other mammals, including most plesiadapiforms, by numerous anatomical specializations, including more convergent orbits with postorbital bars, and grasping extremities with nails rather than claws on most digits, as well as details of their dentition and basicranial anatomy. There are a number of hypotheses or scenarios concerning the specific details of primate ancestry and the behaviors that were most critical in the evolution of this suite of primate adaptations (e.g., Cartmill, 1992; Bloch and Boyer, 2002; Soligo and Martin, 2006; many papers in Ravosa and Dagosto, 2007). These scenarios often emphasize different aspects of primate anatomy and are sometimes complementary.

In the **"Grasp-Leaping" Hypothesis** of Frederick Szalay and Marian Dagosto (e.g., Szalay and Delson, 1979; Szalay and Dagosto, 1980, 1988; Dagosto, 2007; Szalay, 2007), the arboreal habits of primates represent a primitive euarchontan feature, and the specific behavior that distinguished crown primates from our euarchontan ancestry, including plesiadapiforms, was a more acrobatic grasp-leaping type of locomotion, which accounts for the grasping, nailed digits and other limb features, including "enlarged feet," deep knee joints, and a "speed-adapted ankle" that characterized the earliest primates from the Early Eocene. Szalay (2007) notes that in view of the diversity of dental morphology found among both plesiadapiforms and early crown primates, it is not possible to identify a particular dietary habit involved in the origin of crown primates.

In contrast, Matt Cartmill (e.g., 1970, 1992) places particular emphasis on the visual adaptations in primate origins as an explanation for the cranial features characterizing primates, including a postorbital bar, forward-facing orbits, and reduced snouts, as well as grasping hands. Noting that most arboreal mammals do not look much like primates, because they possess neither orbital convergence nor grasping digits, he argues that we must look beyond arboreality per se to account for the distinctive features of our order. Stereoscopic vision, he notes, is particularly characteristic of nocturnal predators, such as cats or owls, that rely on vision to detect their prey. Similarly, he notes that the hands of most arboreal mammals are equipped with claws rather than nails, and that the nailed, grasping hands of primates are probably adaptations for grasping prey rather than arboreal supports. Thus, in Cartmill's **"Nocturnal Visual Predation" Hypothesis**, the ancestral primate was specifically a nocturnal visual predator that stalked and grasped its prey in "the dense tangle of small twigs and vines, which characterizes the canopy and forest margins" (Cartmill, 1970, 1992; Cartmill et al., 2007). Many of the earliest and most primitive known fossil primates are very small and almost certainly included insects in their diet, consistent with Cartmill's theory (Rothman et al., 2014).

In Bob Sussman's (1991, also Sussman et al., 2013) **"Primate/Angiosperm Coevolution" Hypothesis**, primate evolution is best viewed as something that took place in conjunction with the radiation of angiosperm plants to exploit the products of flowering plants (fruit, flowers, and nectar) in a small branch setting. Although he admits that his theory does not specifically account for primate visual adaptations, he argues that visual predation is a rare behavior among primates, who often use sound to locate prey, and is unlikely to have been important in the earliest primates. He also notes that although the convergent orbits of primates are similar to those of visual predators such as cats and owls, the mammals that seem to show the greatest resemblance to primates in their visual system are fruit bats, a totally herbivorous group. In addition, compared to many other primitive mammals, the earliest fossil primates have relatively bunodont molars indicating the inclusion of plant material in their diets. Nevertheless, the evolution of angiosperms was also accompanied by a radiation of insects that depend on these plants.

Robin Crompton (1995) has emphasized the role of locomotion, especially leaping behavior in a nocturnal animal, as the important selective factor for the evolution of primate stereoscopic vision, as well as the postcranial adaptations of early primates. In his **"Nocturnal Leaping" Hypothesis** it was locomotion rather than a particular aspect of diet that selected for primate visual and manipulative abilities.

Finally, many authors have addressed the question of whether the ancestral primate was likely diurnal or nocturnal, and most evidence supports the view that the ancestral primates were nocturnal (Heesy and Ross, 2001; Ross et al., 2007).

As many authors have emphasized, the key to reconstructing the history of any seemingly integrated suite of evolutionary adaptations lies in having a record of the sequence in which the features were acquired. Did the primate visual adaptations precede or follow the grasping abilities and the reduced molar cusps? The appearance of primate-like features is scattered among other euarchontans. Both Dermoptera and Scandentia, but not plesiadapiforms, often have orbits ringed by bone, as in Primates, but not facing forward. Many plesiadapiforms have molar teeth resembling those of later primates, but they have very specialized anterior teeth. All Euarchonta seem to have hands suitable for arboreal behavior, but none show the features of the hindlimbs and ankle related to leaping that characterize the earliest crown primates. As discussed above, it is difficult to envision the evolution of primates from any other group of euarchontans without drastic changes in anatomy and adaptation.

In the absence of a clear sequence of intermediate forms in early primate evolution, several authors have looked to other groups of mammals, specifically marsupials, to provide evolutionary models of early primate adaptations (Rasmussen and Sussman, 2007). Thus, Rasmussen (1990) conducted a field study of the Neotropical marsupial *Caluromys derbianus*, a didelphid that shows striking anatomical similarities to primates in having a relatively large brain, a nearly complete postorbital bar, a relatively short snout, and primate-like proportions in its (clawed) digits. Similarly, in studies of the hand morphology and grasping abilities as well as the gaits of marsupials and prosimian primates, *Caluromys* is strikingly similar to small prosimians such as *Cheirogaleus* and *Microcebus* (Lemelin, 1999; Lemelin and Schmitt, 2007). Since *Caluromys* is more arboreal than any of its relatives and forages for both fruits and insects among terminal branches, study of this marsupial does not offer unequivocal support for either predation or frugivory as critical for the evolution of primate features. Nevertheless, Rasmussen speculated that primate visual adaptations may have evolved for visual predation in an animal that was already adapted to foraging for fruit in terminal branches, a scenario that supports each of the different adaptive explanations for primate origins, but at different times. Unfortunately, until we have a better fossil record from the Late Cretaceous and Paleocene and of intermediate forms preceding the first appearance of crown primates, the details of primate origins will remain hidden in the mists of time and phylogeny.

References

Beard, K.C., 1989. Postcranial anatomy, locomotor adaptations, and paleoecology of Early Cenozoic Plesiadapidae, Paromomyidae, and Micromomyidae (Eutheria, Dermoptera). PhD dissertation. Johns Hopkins University.

Beard, K.C., 1990. Gliding behaviour and palaeoecology of the alleged primate family Paromomyidae (Mammalia, Dermoptera). Nature 345, 340–341.

Beard, K.C., 1991. Vertical postures and climbing in the morphotype of Primatomorpha: Implications for locomotor evolution in primate history. In: Coppens, Y., Senut, B. (Eds.), Origines de la Bipédie chez les Hominidés. Editions du CNRS (Cahiers de Paléoanthropologie), Paris, pp. 79–87.

Beard, K.C., 1993a. Origin and evolution of gliding in Early Cenozoic Dermoptera (Mammalia, Primatomorpha). In: MacPhee, R.D.E. (Ed.), Primates and Their Relatives in Phylogenetic Perspective. Plenum Press, New York, pp. 63–90.

Beard, K.C., 1993b. Phylogenetic systematics of the Primatomorpha, with special reference to Dermoptera. In: Szalay, F.S., Novacek, M.J., McKenna, M.C. (Eds.), Mammal Phylogeny: Placentals. Springer-Verlag, New York, pp. 129–150.

Beard, K.C., Wang, J., 1995. The first Asian Plesiadapoids (Mammalia: Primatomorpha). Ann. Carnegie Mus. 64, 1–33.

Bloch, J.I., Boyer, D.M., 2002. Grasping primate origins. Science 298, 1606–1610.

Bloch, J.I., Boyer, D.M., 2007. New skeletons of Paleocene-Eocene Plesiadapiformes: A diversity of arboreal positional behaviors in early primates. In: Ravosa, M., Dagosto, M. (Eds.), Primate Origins: Adaptations and Evolution. Springer, New York, pp. 535–581.

Bloch, J.I., Chester, S.G.B., Silcox, M.T., 2016. Cranial anatomy of Paleogene Micromomyidae and implications for early primate evolution. J. Hum. Evol. 96, 58–81.

Bloch, J.I., Fisher, D.C., Rose, K.D., Gingerich, P.G., 2001. Stratocladistic analysis of Paleocene Carpolestidae (Mammalia, Plesiadapiformes) with description of a new late Tiffanian genus. J. Vert. Paleontol. 21, 119–131.

Bloch, J.I., Silcox, M.T., Boyer, D.M., Sargis, E.J., 2007. New Paleocene skeletons and the relationship of plesiadapiforms to crown-clade primates. Proc. Nat. Acad. Sci. USA 104, 1159–1164.

Boyer, D.M., 2009. New cranial and postcranial remains of late Paleocene Plesiadapidae ("Plesiadapiformes," Mammalia) from North America and Europe: Description and evolutionary implications. PhD dissertation. Stony Brook University.

Boyer, D.M., Bloch, J.I., 2008. Evaluating the mitten-gliding hypothesis for Paromomyidae and Micromomyidae (Mammalia, "Plesiadapiformes") using comparative functional morphology of new Paleogene skeletons. In: Sargis, E.J., Dagosto, M. (Eds.), Mammalian Evolutionary Morphology: A Tribute to Frederick S. Szalay. Springer, Dordrecht, pp. 233–284.

Boyer, D.M., Evans, A.R., Jernvall, J., 2010. Evidence of dietary differentiation among late Paleocene-Early Eocene Plesiadapids (Mammalia, Primates). Am. J. Phys. Anthropol. 142, 194–210.

Boyer, D.M., Gingerich, P.D., 2019. Skeleton of late Paleocene *Plesiadapis cookei* (Mammalia, Euarchonta): Life history, locomotion, and phylogenetic relationships. University of Michigan Papers on Paleontology 38, 1–269.

Boyer, D.M., Kirk, E.C., Silcox, M.T., et al., 2016. Internal carotid arterial canal size and scaling in Euarchonta: Re-assessing implications for arterial patency and phylogenetic relationships in early fossil primates. J. Hum. Evol. 97, 123–144.

Boyer, D.M., Scott, C.S., Fox, R.C., 2012. New craniodental material of *Pronothodectes gaoi* Fox (Mammalia, "Plesiadapiformes") and relationships among members of Plesiadapidae. Am. J. Phys. Anthropol. 147, 511–550.

Cartmill, M., 1970. The orbits of arboreal mammals: A reassessment of the arboreal theory of primate evolution. PhD dissertation. University of Chicago.

Cartmill, M., 1992. New views on primate origins. Evol. Anthropol. 1, 105–111.

Cartmill, M., Lemelin, P., Schmitt, D., 2007. Primate gaits and primate origins. In: Ravosa, M.J., Dagosto, M. (Eds.), Primate Origins: Adaptations and Evolution. Springer, New York, pp. 403–435.

Chester, S.G.B., Bloch, J.I., 2013. Systematics of Paleogene Micromomyidae (Euarchonta, Primates) from North America. J. Hum. Evol. 65, 109–142.

Chester, S.G.B., Bloch, J.I., Boyer, D.M., Clemens, W.A., 2015. Oldest known euarchontan tarsals and affinities of Paleocene *Purgatorius* to Primates. P. Natl. Acad. Sci. USA 112, 1487–1492.

Chester, S.G.B., Williamson, T.E., Bloch, J.I., Silcox, M.T., 2017. Oldest skeleton of a plesiadapiform provides additional evidence for an exclusively arboreal radiation of stem primates in the Palaeocene. R. Soc. Open Sci. 4, 170329. https://doi.org/10.1098/rsos.170329

Clemens, W.A., 2004. *Purgatorius* (Plesiadapiformes, Primates?, Mammalia), a Paleocene immigrant into northeastern Montana: Stratigraphic occurrences and incisor proportions. Bull. Carnegie Mus. Nat. Hist. 36, 3–13.

Crompton, R.H., 1995. Visual predation, habitat structure, and the ancestral primate niche. In: Alterman, L., Doyle, G.A., Kay Izard, M. (Eds.), Creatures of the Dark: The Nocturnal Prosimians. Plenum Press, New York, pp. 11–30.

Crowell, J.W., Wible, J.R., Chester, S.G.B., 2024. Basicranial evidence suggests picrodontid mammals are not stem primates. Biol. Lett. 20, 20230335. https://doi.org/10.1098/rsbl.2023.0335.

Dagosto, M., 2007. The postcranial morphotype of primates. In: Ravosa, M.J., Dagosto, M. (Eds.), Primate Origins: Adaptations and Evolution. Springer, New York, pp. 489–534.

de Queiroz, K., Gauthier, J., 1990. Phylogeny as a central principle in taxonomy: Phylogenetic definitions of taxon names. Syst. Zool. 39, 307–322.

Fox, R.C., Scott, C.S., Buckley, G.A., 2015. A "giant" purgatoriid (Plesiadapiformes) from the Paleocene of Montana, USA: Mosaic evolution in the earliest primates. Palaeontology 58, 277–291.

Heesy, C.P., Ross, C.F., 2001. Evolution of activity patterns and chromatic vision in primates: Morphometrics, genetics and cladistics. J. Hum. Evol. 40, 111–149.

Hovatter, B.T., Chester, S.G.B., Wilson Mantilla, G.P., 2024. New records of early Paleocene (earliest Torrejonian) plesiadapiforms from northeastern Montana, USA, provide a window into the diversification of stem primates. J. Hum. Evol. https://doi.org/10.1016/j.jhevol.2024.103500.

Godinot, M., 2007. Primate origins: A reappraisal of historical data favoring tupaiid affinities. In: Ravosa, M.J., Dagosto, M. (Eds.), Primate Origins: Adaptations and Evolution. Springer, New York, pp. 83–142.

Janecka, J.E., Miller, W., Pringle, T.H., et al., 2007. Molecular and genomic data identify the closest living relative of primates. Science 318, 792–794.

Kay, R.F., Cartmill, M., 1977. Cranial morphology and adaptations of *Palaechthon nacimienti* and other Paromomyidae (Plesiadapoidea, Primates), with a description of a new genus and species. J. Hum. Evol. 6, 19–53.

Kay, R.F., Thewissen, J.G.M., Yoder, A., 1992. Cranial anatomy of *Ignacius graybullianus* and the affinities of Plesiadapiformes. Am. J. Phys Anthropol. 89, 477–498.

Kirk, E.C., Lemelin, P., Hamrick, M.W., et al., 2008. Intrinsic hand proportions of euarchontans and other mammals: Implications for the locomotor behavior of plesiadapiforms. J. Hum. Evol. 55, 278–299.

Le Gros Clark, W.E., 1926. On the anatomy of the pen–tailed tree shrew *(Ptilocercus lowii)*. Proc. Zool. Soc. Lond. 1926, 1179–1309.

Le Gros Clark, W.E., 1934. Early Forerunners of Man. Bailliere Tindall & Cox, London.

Lemelin, P., 1999. Morphological correlates of substrate use in didelphid marsupials: Implications for primate origins. J. Zool. 247, 165–175.

Lemelin, P., Schmitt, D., 2007. Origins of grasping and locomotor adaptations in primates: Comparative and experimental approaches using an opossum model. In: Ravosa, M.J., Dagosto, M. (Eds.), Primate Origins: Adaptations and Evolution. Kluwer Academic Publishers, New York, pp. 329–380.

Maas, M.C., Krause, D.W., Strait, S.G., 1988. Decline and extinction of plesiadapiforms in North America: Displacement or replacement? Paleobiology 14, 410–431.

MacPhee, R.D.E., 1981. Auditory regions of primates and eutherian insectivores: Morphology, ontogeny, and character analysis. Contrib. Primatol. 18, 1–282.

MacPhee, R.D.E., Cartmill, M., Gingerich, P.D., 1983. New Palaeogene primate basicrania and the definition of the order Primates. Nature 301, 509–511.

Martin, R.D., 1968. Towards a new definition of primates. Man 3, 377–401.

Mason, V.C., Li, G., Minx, P., Schmitz, J., et al., 2016. Genomic analysis reveals hidden biodiversity within colugos, the sister group to primates. Sci. Adv. 2, e1600633. https://doi.org/10.1126/sciadv.1600633

Miller, K., Tietjen, K., Beard, K.C., 2023. Basal Primatomorpha colonized Ellesmere Island (Arctic Canada) during the hyperthermal conditions of the early Eocene climatic optimum. PLoS One 18, e0280114. https://doi.org/10.1371/journal.pone.0280114

Ni, X., Gebo, D.L., Dagosto, M., et al., 2013. The oldest known primate skeleton and early haplorrhine evolution. Nature 498, 60–64.

Ni, X., Li, Q., Li, L., Beard, K.C., 2016. Oligocene primates from China reveal divergence between African and Asian primate evolution. Science 352, 673–677.

Ni, X., Meng, J., Beard, K.C., et al., 2011. Phylogeny of the primates and their relatives: An analysis based on a large data matrix. In: Lehmann, T., Schaal, S.F.K. (Eds.), The World at the Time of Messel: Puzzles in Palaeobiology, Palaeoenvironment, and the History of Early Primates. Senckenberg Gesellschaft für Naturforschung, Frankfurt am Main.

Perelman, P., Johnson, W.E., Roos, C., et al., 2011. A molecular phylogeny of living primates. PLoS Genetics 7, e1001342.

Prufrock, K.A., Boyer, D.M., Silcox, M.T., 2016. The first major primate extinction: An evaluation of palaeoecological dynamics of North American stem primates using a homology free measure of tooth shape. Am. J. Phys. Anthropol. 159, 683–697.

Rasmussen, D.T., 1990. The phylogenetic position of *Mahgarita stevensi*: Protoanthropoid or lemuroid? Int. J. Primatol. 1B1, 439–469.

Rasmussen, D.T., Sussman, R.W., 2007. Parallelisms among primates and possums. In: Ravosa, M.J., Dagosto, M. (Eds.), Primate Origins: Adaptations and Evolution. Kluwer Academic Publishers, New York, pp. 775–803.

Ravosa, M., Dagosto, M., 2007. Primate Origins: Adaptations and Evolution. Springer, New York.

Roberts, T.E., Lanier, H.C., Sargis, E.J., Olson, L.E., 2011. Molecular phylogeny of treeshrews (Mammalia: Scandentia) and the timescale of diversification in Southeast Asia. Mol. Phylogenet. Evol. 60, 358–372.

Rose, K.D., 1975. The Carpolestidae, early tertiary primates from North America. Bull. Mus. Comp. Zool. 147, 1–74.

Rose, K.D., Beard, K.C., Houde, P., 1993. Exceptional new dentitions of the diminutive plesiadapiforms Tinimomys and Niptomomys, with comments on the upper incisors of Plesiadapiformes. Ann. Carnegie Museum 62(4), 351–361.

Rose, K.D., Bown, T.M., 1996. A new plesiadapiform (Mammalia: Plesiadapiformes) from the early Eocene of the Bighorn Basin, Wyoming. Ann. Carnegie Mus. 65, 305–321.

Rosenberger, A.L., 2010. Adaptive profile versus adaptive specialization: Fossils and gumivory in primate evolution. In: Burrows, A., Nash, L. (Eds.), The Evolution of Exudativory in Primates. Springer, New York, pp. 273–296.

Ross, C.F., Hall, I.M., Heesy, C.P., 2007. Were basal primates nocturnal? Evidence from eye and orbit shape. In: Ravosa, M., Dagosto, M. (Eds.), Primate Origins: Adaptations and Evolution. Springer, New York, pp. 233–256.

Rothman, J.M., Raubenheimer, D., Bryer, M.A.H., et al., 2014. Nutritional contributions of insects to primate diets: Implications for primate evolution. J. Hum. Evol. 71, 59–69.

Runstead, J.A., Ruff, C.B., 1995. Structural adaptations for gliding in mammals with implications for locomotor behavior in paromomyids. Am. J. Phys. Anthropol. 98, 101–119.

Sargis, E.J., 2002. The postcranial morphology of *Ptilocercus lowii* (Scandentia, Tupaiidae): An analysis of Primatomorphan and Volitantian characters. J. Mammalian Evol. 9, 137–160.

Sargis, E.J., 2004. New views on treeshrews: The role of tupaiids in primate supraordinal relationships. Evol. Anthropol. 13, 56–66.

Scott, C.S., Lopez-Torres, S., Silcox, M.T., Fox, R.C., 2023. New paromomyids (Mammalia, Primates) from the Paleocene of southwestern Alberta, Canada, and an analysis of paromomyid interrelationships. J. Paleontol. 97, 477–498.

Seiffert, E.R., Tejedor, M.F., Fleagle, J.G., et al., 2020. A parapithecid stem anthropoid of African origin in the Paleogene of South America. Science 368, 194–197.

Silcox, M.T., 2001. A phylogenetic analysis of Plesiadapiformes and their relationship to euprimates and other archontans. PhD dissertation. Johns Hopkins University.

Silcox, M.T., 2007. Primate taxonomy, plesiadapiforms, and approaches to primate origins. In: Ravosa, M., Dagosto, M. (Eds.), Primate Origins: Adaptations and Evolution. Springer, New York, pp. 143–178.

Silcox, M.T., Bloch, J.I., Boyer, D.M., et al., 2009. Semicircular canal system in early primates. J. Hum. Evol. 56, 315–322.

Silcox, M.T., Bloch, J.I., Boyer, D.M., et al., 2017. The evolutionary radiation of plesiadapiforms. Evol. Anthropol. 26, 74–94.

Silcox, M.T., Gunnell, G.F., 2008. Plesiadapiformes. In: Janis, C.M., Gunnell, G.F., Uhen, M.D. (Eds.), Evolution of Tertiary Mammals of North America, vol. 2. Small Mammals, Xenarthrans, and Marine Mammals. Cambridge University Press, Cambridge, pp. 207–238.

Silcox, M.T., Rose, K.D., Walsh, S., 2002. New specimens of picromomyids (Plesiadapiformes, Primates) with description of a new species of *Alveojunctus*. Ann. Carnegie Mus. 71, 1–11.

Soligo, C., Martin, R.D., 2006. Adaptive origins of primates revisited. J. Hum. Evol. 50, 414–430.

Sussman, R.W., 1991. Primate origins and the evolution of angiosperms. Am. J. Primatol. 23, 209–223.

Sussman, R.W., Rasmussen, D.T., Raven, P.H., 2013. Rethinking primate origins. Amer. J. Primatol. 75, 95–106.

Szalay, F.S., 2007. Ancestral locomotor modes, placental mammals, and the origin of Euprimates: Lessons from history. In: Ravosa, M.,

Dagosto, M. (Eds.), Primate Origins: Adaptations and Evolution. Springer, New York, pp. 457–487.

Szalay, F.S., Dagosto, M., 1980. Locomotor adaptations as reflected in the humerus of Paleogene primates. Folia Primatol. 34, 1–45.

Szalay, F.S., Dagosto, M., 1988. Evolution of hallucial grasping in the primates. J. Hum. Evol. 17, 1–33.

Szalay, F.S., Delson, E., 1979. Evolutionary History of the Primates. Academic Press, New York.

Tong, Y., Wang, J., 1998. A preliminary report on the Early Eocene mammals of the Wutu fauna, Shandong Province, China. Bull. Carnegie Mus. Nat. Hist. 34, 186–193.

Upham, N.S., Esselstyn, J.A., Jetz, W., 2019. Inferring the mammal tree: Species-level sets of phylogenies for questions in ecology, evolution, and conservation. PLoS Biol. 17, e3000494. https://doi.org/10.1371/journal.pbio.3000494

Wible, J.R., 2009. The ear region of the pen-tailed treeshrew Ptilocercus lowii Gray, 1848 (Placentalia, Scandentia, Ptilocercidae). J. Mammalian Evol. 16, 199–233.

Wible, J.R., 2011. On the treeshrew skull (Mammalia, Placentalia, Scandentia). Ann. Carnegie Mus. 79, 149–230.

Wible, J.R., Covert, H.H., 1987. Primates: Cladistic diagnosis and relationships. J. Hum. Evol. 16, 1–22.

Wible, J.R., Martin, J., 1993. Ontogeny of the tympanic floor and roof in Archontans. In: MacPhee, R.D.E. (Ed.), Primates and Their Relatives in Phylogenetic Perspective. Plenum Press, New York, pp. 111–148.

Wilson Mantilla, G.P., Chester, S.G.B., Clemens, W.A., et al., 2021. Earliest Palaeocene purgatoriids and the initial radiation of stem primates. R. Soc. Open Sci. 8 (2), 210050.

Further Reading

Bloch, J.I., Silcox, M.T., Boyer, D.M., Sargis, E.J., 2007. New Paleocene skeletons and the relationship of plesiadapiforms to crown-clade primates. Proc. Natl. Acad. Sci. USA 104, 1159–1164.

Ravosa, M.J., Dagosto, M. (Eds.), 2007. Primate Origins: Adaptations and Evolution. Springer, New York.

Sargis, E.J., Dagosto, M. (Eds.), 2008. Mammalian Evolutionary Morphology: A Tribute to Frederick S. Szalay. Springer, Dordrecht.

Silcox, M.T., 2013. Primate origins. In: Begun, D.R. (Ed.), A Companion to Paleoanthropology. Wiley-Blackwell, Hoboken, NJ, pp. 341–357.

Silcox, M., Sargis, E.J., Bloch, J.I., Boyer, D.M., 2015. Primate origins and supraordinal relationships: Morphological evidence. In: Henke, W. Tattersall, I. (Eds.), Handbook of Paleoanthropology, vol. 2. Springer-Verlag, Berlin, pp. 1053–1082.

CHAPTER

12

Fossil Prosimians

In North America and Europe, the Eocene epoch (56–34 million years ago) was marked by a major change in faunas. Many modern types of mammals, including the earliest artiodactyls, perissodactyls, and rodents, replaced more archaic types of mammals. In primate evolution, the beginning of this epoch is marked by the first appearance of primates that resemble living prosimians. These faunal changes took place in a series of waves rather than in a single, broad sweep and seem to be the result of remarkable climatic changes coincident with connections between continents or major continental areas.

Eocene paleogeography was strikingly different from that of today (Fig. 12.1). The continents were in more or less familiar positions, but the connections were very different. Near the beginning of the Eocene, North America and Europe were connected at high latitudes but became increasingly separated and distinct throughout the epoch, resulting in an increasing distinctiveness in their mammalian faunas. Additionally, there is faunal evidence for intermittent connections between North America and Asia. The Tethys Seaway ran though the Mediterranean region and western Asia, separating Africa from Eurasia. In Africa during this time, a large seaway separated the northwest corner from the rest of the continent. India was coming into contact with the Asian "mainland," much of which was then a cluster of islands. South America remained isolated from all other continents except Antarctica.

Eocene climates in Europe and North America were also very different from modern conditions. Most significant was a dramatic spike in global temperatures at the Paleocene–Eocene boundary, the Paleocene–Eocene Thermal Maximum (PETM), which had a major effect on plant and animal distributions, including primate biogeography (Gingerich, 2006; Smith et al., 2006). Overall, the Eocene was warmer and more equable than the preceding Paleocene epoch (Fig. 10.4). Climates were so warm during the Early–Middle Eocene that there was a relatively diverse fauna of mammals, including a paromomyid, living well within the Arctic Circle.

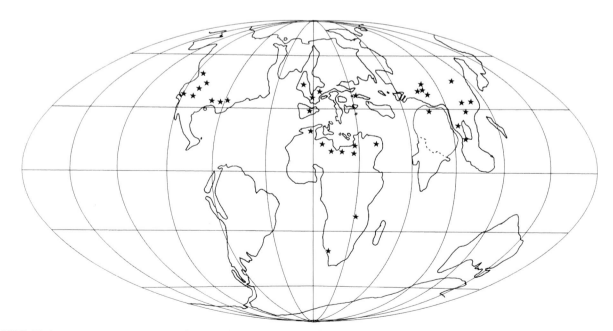

FIGURE 12.1 Geographic distribution of Eocene fossil sites yielding fossil strepsirrhines and tarsiiforms, shown on an Early Eocene paleogeography. Map lines delineate study areas and do not necessarily depict accepted national boundaries.

Primate Adaptation and Evolution
https://doi.org/10.1016/B978-0-12-815809-8.00012-6

The First Modern Primates

The primates that made their debut in the Early Eocene were quite different from the plesiadapiforms of the preceding Paleocene Epoch. They had all the anatomical features characteristic of living primates. They had a shorter snout, a smaller infraorbital foramen, forward-facing eyes, and a postorbital bar completing the bony ring around their orbits (Fig. 12.2). They had a larger, more rounded braincase, and their auditory regions and cerebral blood supplies were like those of living primates. Their skeletons had more slender limbs; hind limbs often adapted for leaping; and feet with

divergent, grasping halluces; they also possessed nails, rather than claws, on their digits (Dagosto, 1988, 2007).

All of these morphological differences indicate that the Eocene primates practiced a very different way of life from the plesiadapiforms and other related mammals. Many of the cranial differences indicate an increased reliance on vision rather than smell and tactile vibrissae. The postcranial changes suggest an increased importance of manipulative abilities, with the replacement of claws by nails, and the locomotor skeletons of many species suggest leaping abilities and more acrobatic locomotion. In several species, there are indications of canine sexual dimorphism, as in later anthropoids.

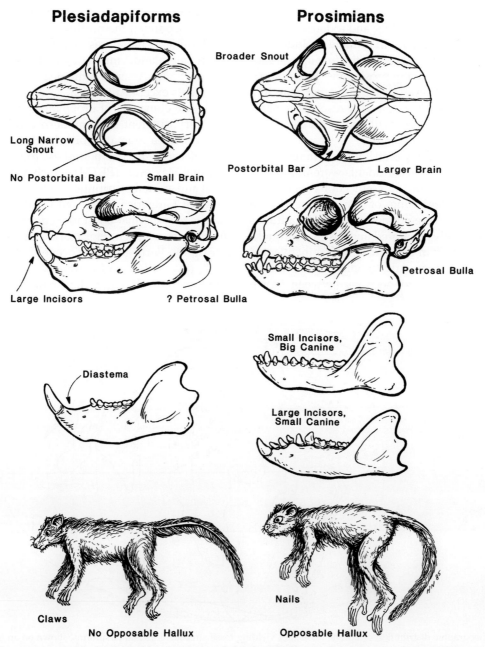

FIGURE 12.2   Comparison of early fossil primates and more archaic plesiadapiforms, showing major anatomical contrasts.

They are "primates of modern aspect," or crown primates. For those who place plesiadapiforms in Primates, they are often labeled Euprimates; in this book, they are simply Primates.

Like plesiadapiforms, the early prosimians were among the most abundant mammals of their day, but they are not equally well documented on all continents. They are common in mammalian faunas of North America and Europe and are becoming better known from Asia and Africa; there is no evidence of Eocene prosimian primates from South America (Fig. 12.1). From their first appearance in the Early Eocene, the primates of North America and Europe are commonly divided into two distinct groups: the adapoids, usually linked with later strepsirrhines, and the omomyoids, usually linked with tarsiers and anthropoids. The earliest members of the two superfamilies (*Donrussellia*, *Cantius*, and *Teilhardina*) are very similar in their dentition (Figs. 12.3 and 12.4), suggesting a relatively recent common ancestor. However, important cranial and

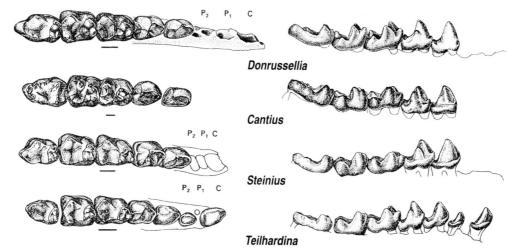

FIGURE 12.3   Occlusal and lingual views of the lower left dentitions of early omomyoids and adapoids. Scale = 1mm. *(Courtesy of K. D. Rose.)*

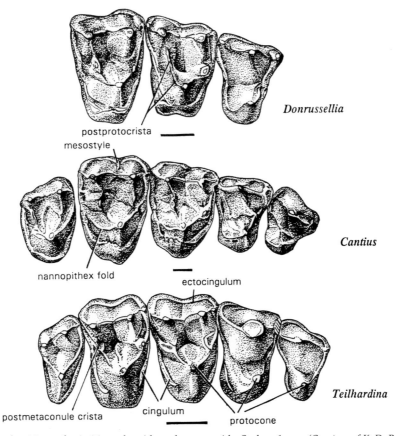

FIGURE 12.4   Upper right dentitions of primitive adapoids and omomyoids. Scale = 1mm. *(Courtesy of K. D. Rose.)*

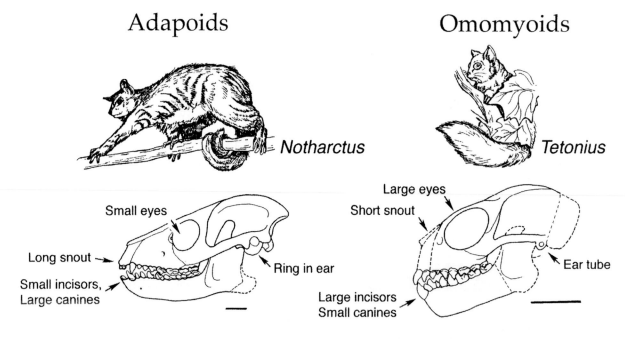

Adapoids   # Omomyoids

Notharctus   *Tetonius*

Small eyes   Large eyes
                                              Short snout
Long snout   Large eyes
Small incisors,   Ear tube
Large canines   Ring in ear

                                              Large incisors
                                              Small canines

Many premolar and molar shearing crests   **Small species with sharp pointed molar cusps**
                                              Large, later species with flat molar teeth

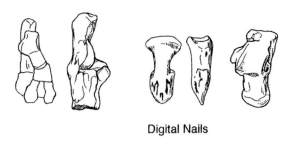

Digital Nails

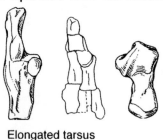

Elongated tarsus

FIGURE 12.5   Comparison of adapoids and omomyoids. *(Courtesy of K. D. Rose.)*

postcranial distinctions between adapoids and omomy-oids are already present in the earliest members of each group (Fig. 12.5; but see Dunn et al., 2016). Both super-families produced adaptive radiations of species that flourished throughout the epoch.

The first appearance of Primates seems to be at approximately the same time in Eurasia and North America at the beginning of the epoch. Early Eocene faunas of North America and Europe were extremely similar and enabled by a lingering connection between Europe and North America in high latitudes combined with the Paleocene–Eocene spike in global temperatures that enabled primates to survive and disperse at those latitudes (Smith et al., 2006; Hooker, 2007).

As we noted in Chapter 11, there are no clear phyletic ancestors for early primates among the plesiadapiforms or among any other group of early mammals. However, there are two poorly known species from Africa and Asia that may lie close to the origins of all later primates.

TABLE 12.1   Order PRIMATES
Family INCERTAE SEDIS

Species	Estimated mass (g)
Altanius (Early Eocene, Asia)	
A. orlovi	18
Altiatlasius (Late Paleocene, Africa)	
A. koulchii	88

Altanius orlovi (Table 12.1; Fig. 12.6) is a tiny primate (10–20 g) from the Early Eocene of Mongolia that is now known from many specimens preserving much of the dentition. It has elevated trigonids and tall premolars. Originally described as an early omomyoid, *Altanius* has also been considered as a possible carpolestid plesi-adapiform. On the basis of this limited dental evidence, *Altanius* is widely considered a basal primate, near the common ancestry of adapoids and omomyoids.

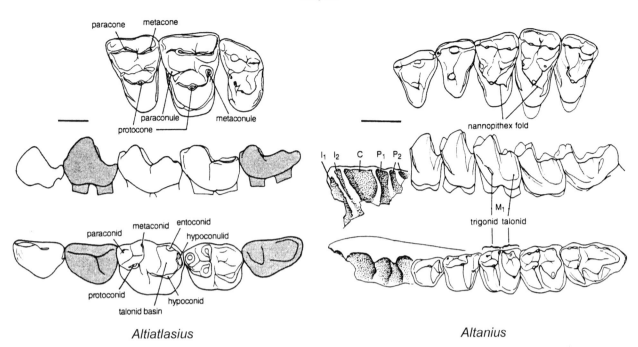

FIGURE 12.6   *Altiatlasius koulchii (left),* from the Paleocene of Morocco, and *Altanius orlovi (right),* from the Early Eocene of Mongolia, are generally considered the earliest fossil primates. *(Courtesy of K. D. Rose.)*

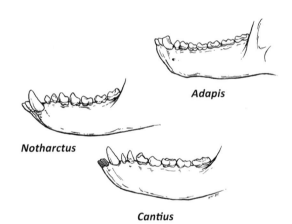

FIGURE 12.7   Mandibles of several adapoid primates.

Altiatlasius koulchii (Table 12.1; Fig. 12.6) is a larger African primate (50–100 g) from the Paleocene of Morocco that is known from ten isolated teeth. *Altiatlasius* has been variously considered as an early omomyoid, a basal primate, and even an early anthropoid. It is the oldest fossil primate, but its precise affinities within the order are uncertain.

Adapoids

In many aspects of their dental anatomy, adapoids are the most primitive of all known primates, fossil or living. Most of the dental specializations found among later primates could easily be derived from an early adapoid morphology. As we discuss later, such a basic and primitive morphology poses interesting difficulties in ascertaining the phyletic relationships of adapoids with later primate groups.

Compared to the earlier plesiadapiforms and the contemporaneous omomyoids, most adapoids were rather large primates, comparable in size to living lemurids. The primitive adapoid dental formula, retained by many relatively late members of the family, is 2.1.4.3 (Fig. 12.7). Adapoids differ from plesiadapiforms and many omomyoids and resemble living anthropoids in the proportions of their anterior dentition (Figs. 12.5 and 12.7). The lower incisors are small and positioned more or less vertically in the mandible; the upper incisors are relatively broad, but short, and are usually separated by a median gap. Both upper and lower canines are larger than the incisors. In some taxa, canines are sexually dimorphic, but the dimorphism seems to have a different developmental basis from canine dimorphism in anthropoids (Schwartz et al., 2005). The anterior premolars are often caniniform, and the posterior ones are often molariform. The upper molars are broad, and at least two major lineages of adapoids evolved a hypocone independently. Lower molars are relatively long and narrow in most taxa. Numerous shearing crests, presumably an adaptation to folivory, along with fusion of the two halves of the mandible, appear to have evolved independently in many adapoid lineages.

Adapoids have a relatively long but broad snout with small infraorbital foramina (Figs. 12.5 and 12.8). As in all living primates, each orbit is encircled by a complete bony ring. They have a large ethmoid recess with

Smilodectes   **Magnadapis**

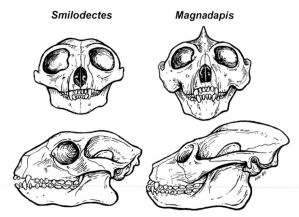

FIGURE 12.8   Reconstructed skulls of two adapoid primates.

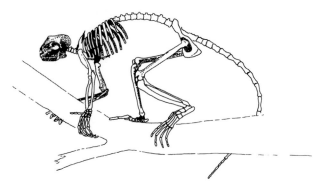

FIGURE 12.9   Reconstructed skeleton of *Smilodectes gracilis*.

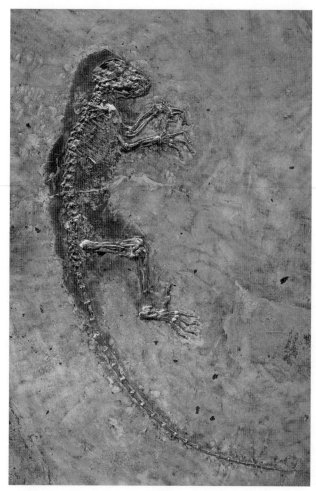

FIGURE 12.10   The crushed skeleton of "Ida," the type specimen of *Darwinius masillae* from the Eocene shale deposits in Messel, Germany. *(Courtesy of J. Hurum.)*

numerous ethmoturbinates, as in lemurs and primitive mammals generally. They also have a long nasolacrimal canal suggesting that they had a "wet nose" and a functioning vomeronasal organ as in extant strepsirrhines (Rossie and Smith, 2007; Rossie et al., 2018). The braincase is larger than that of the plesiadapiforms but smaller than in extant lemurs or anthropoids. The tympanic ring is suspended within the inflated bony bulla, much as in extant lemurs. The bony canals for stapedial and promontory branches of the internal carotid artery are apparently quite variable.

The skeletal anatomy, which is well known for several North American and European genera (Figs. 12.9 and 12.10), shows that adapoid limbs are similar to those of living strepsirrhines but are more robust. These Eocene prosimians have relatively long legs, a long trunk, and a long tail. The tarsal region is relatively short, and the talus shows distinctive features characteristic of extant strepsirrhines (e.g., Gebo et al., 2012a,b). The hands and feet of some fossils have distal phalanges suggesting the presence of nails rather than claws (Maiolino et al., 2012), and they have a grasping foot with a divergent hallux.

The systematics of adapoids have been studied by many workers and not without disagreement. Adapoids can be divided into six families that are largely, but not completely, distinct biogeographically. The notharctids are a predominantly Early–Middle Eocene group from North America, and the biostratigraphy of notharctid species from the western United States is particularly well documented. Cercamoniids (also called protoadapids) are a predominantly Early–Middle Eocene European radiation, closely related to notharctids. The caenopithecids include several European genera (e.g., *Caenopithecus*) as well as a widespread array of genera from North America, Africa, and Asia. The adapids are a European group that appeared abruptly in the Middle Eocene, probably from Asia, and became extinct by the Early Oligocene. The fifth family of adapoids, the sivaladapids, is an Asian radiation that began in the Eocene and survived into the Late Miocene. In addition, recently described adapoids from the Early Eocene of India and Pakistan have been placed in another family, the asiadapids (Rose et al., 2009). These groupings will be used here, although relationships among adapoids are almost certainly more complex and far from being resolved (Godinot, 1998; Gebo, 2002; Seiffert et al., 2009; Godinot et al., 2018).

Notharctids

The notharctids (Table 12.2) were among the most common mammals in the Early and Middle Eocene faunas of western North America; there are numerous time-successive species, but they have limited diversity in size and adaptations (Fig. 12.11). There were never more than two or three synchronic species and only a total of six genera from the Early and Middle Eocene. The earliest notharctid, and one of the earliest adapoids, is **Cantius**, with numerous species from North America (Gingerich, 1986) and two from Europe. *Cantius* was a small to medium-sized primate ranging from about 0.5 kg in the earliest and smallest species to 4 kg in the latest.

Cantius has a dental formula of 2.1.4.3 (Fig. 12.7). The lower molars have a simple trigonid with three cusps and a broad-basined talonid (Fig. 12.3); the upper molars (Fig. 12.4) are simple tritubercular teeth in the early species, but later species (in North America) developed a hypocone from the postprotocingulum (or nannopithex fold) and a mesostyle. All species have four premolars, prominent canines, and two small, relatively vertical incisors. The mandibular symphysis is unfused in this early genus. *Cantius* was probably largely frugivorous.

The partial skulls and few skeletal remains of *Cantius* resemble those of the better-known, later genera *Notharctus* and *Smilodectes* in most respects. They indicate a diurnal species that moved primarily by arboreal quadrupedal running and leaping (Fig. 12.9). Some species of *Cantius* were sexually dimorphic in canine size, similar to many extant higher primates.

Cantius is the most common Early Eocene primate taxon in the northern parts of the American West (Wyoming). In contrast, two related genera, **Pelycodus** (with large, broad teeth) and **Copelemur** (with small, narrow teeth), are more common in the southern parts (New Mexico) and are rare in northern localities. The North American climate showed a considerable warming from the Early to the Middle Eocene. Associated with this climate change was a change in faunas, including the appearance of new primates. The Middle Eocene faunas of Wyoming document two new genera, **Notharctus** and **Smilodectes**, both of which show numerous dental specializations for folivory.

Notharctus is larger (2 kg up to 4.5 kg) than most *Cantius* taxa and has larger hypocones and mesostyles on the upper molars, reduced paraconids on the lower molars, and a fused mandibular symphysis. Because the transition from *Cantius* to *Notharctus* was gradual and essentially continuous, this last feature is used arbitrarily to delineate the two genera (Fig. 12.11). The cheek teeth of *Notharctus* have well-developed shearing crests, and the genus was likely folivorous (Covert, 1986, 1995). *Notharctus* also had sexually dimorphic canines, suggesting that they lived in relatively large groups where males competed over females.

TABLE 12.2   Superfamily ADAPOIDEA
Family NOTHARCTIDAE

Species	Estimated mass (g)
Cantius (Early Eocene, North America, Europe)	
C. abditus	1768
C. angulatus	703
C. eppsi	550
C. frugivorus	1260
C. mckennai	959
C. nunienus	1192
C. ralstoni	746
C. savagei	1063
C. simonsi	4300
C. torresi	550
C. trigonodus	1247
Copelemur (Early Eocene, North America)	
C. australotutus	1404
C. consortutus	885
C. feretutus	1109
C. praetutus	888
C. tutus	2324
Notharctus (Middle Eocene, North America)	
N. pugnax	3482
N. robinsoni	2924
N. robustior	4414
N. tenebrosus	2152
N. venticolis	2262
Smilodectes (Middle Eocene, North America)	
S. gracilis	1009
S. mcgrewi	1443
S. sororis	1234
Pelycodus (Early Eocene, North America)	
P. danielsae	5579
P. jarrovii	2831
Hesperolemur (Middle Eocene, North America)	
H. actius	1800

Notharctus is similar to *Lemur* in both overall cranial proportions and in details of its basicranial anatomy. The Eocene genus is more robustly built and has a smaller braincase with more pronounced sagittal and nuchal crests. There is a moderately long snout with a large

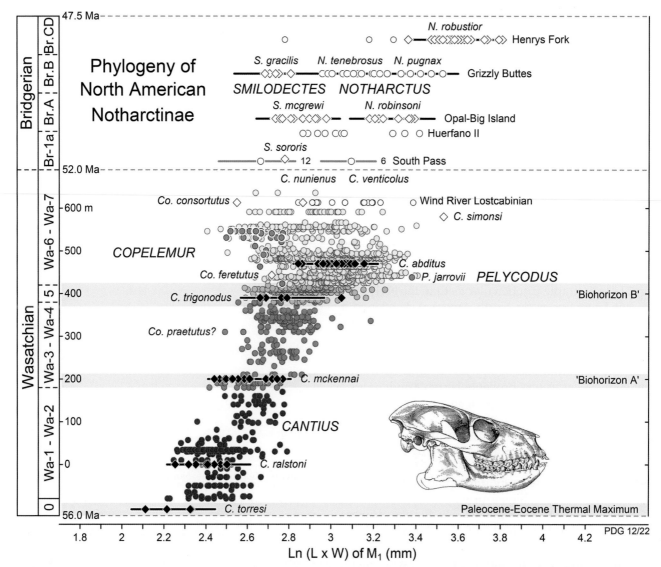

FIGURE 12.11 A phylogeny of notharctids from northern Wyoming showing a dense fossil record with gradual changes in dental dimensions through time. *(Courtesy of P. Gingerich.)*

premaxillary bone. The lacrimal bone is at the edge of the orbit rather than anterior to it, as in extant lemurs. The auditory region has a free tympanic ring lying within the bulla. The canal for the stapedial artery was generally smaller than that for the promontory artery. Although the size and position of these canals are widely used to reconstruct patterns of cranial circulation in fossil mammals, it is important to keep in mind that there is not necessarily a one-to-one correspondence between bony canals and arteries in living primates.

Several virtually complete skeletons are known for *Notharctus*. Gregory (1920) found that the Eocene genus is similar in skeletal proportions and details of limb architecture to the extant genera *Lemur*, *Varecia*, *Lepilemur*, and *Propithecus* but has relatively more robust bones. *Notharctus* has extremely long hind limbs (intermembral index = 60);

a long, flexible trunk; and a long tail. The ilium is sickle-shaped, as in extant lemurs, and the ischium is rather long, not short as in vertical-clinging and leaping indriids (Fleagle and Anapol, 1992). The pollex and hallux are large and opposable; the digits are long and tipped with nails, plus a grooming digit on the second toe (Maiolino et al., 2012). In most, but not all, details of muscle attachment that can be reconstructed, it is similar to living strepsirrhine quadrupedal leapers. The calcaneus is rather short, as in *Varecia*. There is little doubt that *Notharctus* was an adept leaper and quadrupedal runner, but it was not so restricted to vertical supports as living indriids.

Smilodectes (Figs. 12.8, 12.9, 12.11), a smaller (1–2 kg) Middle Eocene contemporary of *Notharctus*, was characterized by narrower teeth, a shorter snout, and a more rounded frontal bone. Like *Notharctus*, it was diurnal and folivorous.

TABLE 12.3 Superfamily ADAPOIDEA
Family CERCAMONIIDAE

Species	Estimated mass (g)
Donrussellia (Early Eocene, Europe)	
D. gallica	114
D. louisi	694
D. magna	432
D. provincialis	159
D. russelli	376
D. lusitanica	212
Panobius (?Early–Middle Eocene, Asia)	
P. afridi	59
P. russelli	86
P. amplior	229
Protoadapis (Early–Middle Eocene, Europe)	
P. angustidens	–
P. brachyrhynchus	2689
P. curvicuspidens	895
P. ignoratus	1128
P. muechelnensis	1096
P. reticuspidens	1004
P. weigelti	2741
P. andrei	2416
Barnesia (Middle Eocene, Europe)	
B. hauboldi	1636
Periconodon (Middle Eocene, Europe)	
P. helveticus	57
P. huerzeleri	331
P. helleri	–
P. jaegeri	437
P. lemoinei	510
Buxella (Middle Eocene, Europe)	
B. prisca	234
B. magna	350
Agerinia (Middle Eocene, Asia, Europe)	
A. roselli	453
A. marandati	420
A. smithorum	425
Agerinia. sp.	–
Anchomomys (Middle to Late Eocene, Europe)	
A. crocheti	91
A. gaillardi	54

TABLE 12.3 Superfamily ADAPOIDEA—Cont'd
Family CERCAMONIIDAE

Species	Estimated mass (g)
A. pygmaeus	73
A. quercyi	97
A. frontanyensis	85
Mazateronodon (Late Eocene, Europe)	
M. endemicus	227
Nievesia (Late Eocene, Europe)	
N. sossisensis	117
Pronycticebus (Middle to Late Eocene, Europe)	
P. gaudryi	534
P. neglectus	1071
P. cosensis	656

Its external brain morphology is known from several endocasts. Compared to other Eocene mammals, *Smilodectes* had an expanded visual cortex and reduced olfactory bulbs; its brain was larger than that of most contemporaneous mammals but smaller than that of extant strepsirrhines (Harrington et al., 2016; Gilbert and Jungers, 2017). The phylogenetic origin of *Smilodectes* is the subject of some debate. Some authors have argued that the two Middle Eocene taxa are both derived from *Cantius* (Fig. 12.11; Gingerich, 1984; Covert, 1990); however, Beard (1988) has argued that, whereas *Notharctus* is derived from *Cantius* in the north, *Smilodectes* is a descendant of the *Copelemur* lineage, which is more common in the South, but appeared in the northern part of the continent in conjunction with the climatic warming. A study based on additional intermediate fossils supports the *Copelemur–Smilodectes* phylogeny (Gunnell, 2002). Both *Notharctus* and *Smilodectes* apparently became extinct early in the Middle Eocene. *Hesperolemur* is a large notharctid from the slightly younger Uintan Land Mammal Age of southern California.

Cercamoniids

Cercamoniids (Table 12.3) are an Old World radiation of early adapoids closely related to notharctids. Cercamoniids had a much more diverse evolutionary radiation than the relatively uniform notharctids. They ranged in size from tiny, presumably insectivorous, species the size of a pygmy marmoset (55g) to larger (2500+ g), more frugivorous or partly folivorous species. Most taxa are known primarily from their dentition. Cranial and associated skeletal remains are rare. The phylogenetic relationships among the diverse and widespread cercamoniids are complex, and this group likely includes

TABLE 12.4   Superfamily ADAPOIDEA
Family ASIADAPIDAE

Asiadapis (Early Eocene, Asia)	
A. cambayensis	272
A. tapiensis	–
Marcgodinotius (Early Eocene, Asia)	
M. indicus	109

several phyletic radiations (Thalmann et al., 1989; Godinot, 1998; Rose et al., 2009; Seiffert et al., 2009; Boyer et al., 2010).

Donrussellia (Figs. 12.3 and Fig. 12.4) is the earliest and most primitive cercamoniid and is probably close to the origin of all adapoids. This tiny genus has a full dental formula of 2.1.4.3, simple tritubercular upper molars, and lower molars with a simple trigonid and a broad talonid. It is known postcranially from a single astragalus, which is also primitive and does not show the sloping fibular facet characteristic of many other adapoid taxa (Boyer et al., 2017).

Cercamoniids evolved numerous dental adaptations, indicative of considerable dietary diversity within the subfamily (Ramdarshan et al., 2012). The small (55 g) *Anchomomys gaillardi*, for example, has extremely simple upper molars, not unlike those of marmosets. Judging from its sharp molar cusps and tiny size, this species was almost certainly insectivorous. A similar species, *A. frontanyensis*, is known from Spain, and its ankle bones suggest more leaping than most other adapoids, most similar to extant *Mirza* (Marigó et al., 2016). The larger *Pronycticebus gaudryi* has relatively simple molar teeth with sharp cusps, a robust, tusk-like upper canine, and a long row of sharp premolars, suggesting a carnivorous diet (Szalay and Delson, 1979). *Periconodon* has molars with broader, more bulbous cusps, suggestive of fruit eating. In contrast to the North American notharctids, many cercamoniids developed a hypocone from the lingual cingulum rather than from the protocone.

Asiadapids

The coal mines of Western India have yielded several very small adapoid primates and an unnamed larger species (Rose et al., 2018), placed in a separate family, Asiadapidae (Table 12.4). Asiadapids appear to lie near the base of the adapoid radiation, but also show some features suggesting that they are possibly related to sivaladapids (Rose et al., 2009). *Asiadapis* (Fig. 12.12; dental formula 2-1-3-3) and *Marcgodinotius* (dental formula 2-1-4-3) are both bushbaby-sized primates with simple premolars and simple upper molars with no nannopithex fold. Hypocones and lingual cingula are either

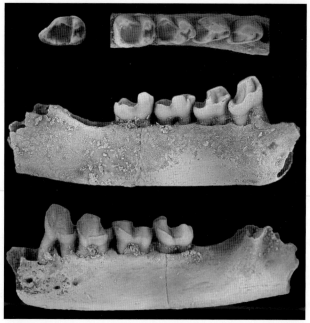

FIGURE 12.12   Asiadapid primates from the Early Eocene Vastan coal mine of India. *(Courtesy of K.D. Rose.)*

weak or absent. The postcranial elements attributed to asiadapids suggest they were primarily characterized by agile quadrupedalism rather than leaping or slow climbing (Dunn et al., 2016). *Panobius* is another adapoid from Pakistan that is likely related to asiadapids (Gunnell et al., 2008; Rose et al., 2009).

Caenopithecids

The caenopithecids (Table 12.5) are a cosmopolitan collection of adapoids that have traditionally been a part of the cercamoniids but are probably more closely related to the adapids discussed below (Thalmann, 1994; Godinot, 1998; Seiffert et al., 2009; Godinot et al., 2018).

Some of the most complete skeletal remains of any fossil primates are from two localities in Germany, the Geiseltal brown coals near Halle and the Messel Bone Pit near Frankfurt. *Godinotia neglecta* is known from a nearly complete skeleton that has been crushed flat (Thalmann et al., 1989). *Godinotia* was probably a quadrupedal, leaping and climbing form. A particularly unusual fossil from Messel is a half-skeleton of a small primate. Because only the lower half of the skeleton has been found, it cannot be confidently assigned to any genus or species. The hind limb suggests it was a leaper. This species has been reported to have a "grooming claw" on the second digit of its foot as in extant strepsirrhines. It also has a very large baculum (penis bone) for an animal of its size.

One of the most spectacular and best known of the fossil primates from Messel is *Darwinius masillae*

TABLE 12.5 Superfamily ADAPOIDEA
Family CAENOPITHECIDAE

Species	Estimated mass (g)
Caenopithecus (Late Eocene, Europe)	
C. lemuroides	2563
Europolemur (Middle to Late Eocene, Europe)	
E. dunaifi	778
E. klatti	922
E. koenigswaldi	1290
Godinotia (Eocene, Europe)	
G. neglecta	1317
Darwinius (Eocene, Europe)	
D. masillae	948
Mahgarita (Late Eocene, North America)	
M. stevensi	527
Mescalerolemur (Late Eocene, North America)	
M. horneri	305
Aframonius (Late Eocene, Africa)	
A. dieides	933
Afradapis (Late Eocene, Africa)	
A. longicristatus	2446
Masradapis (Late Eocene, Africa)	
M. tahai	744
Namadapis (Middle Eocene, Africa)	
N. interdictus	150
Notnamaia (Middle Eocene, Africa)	
N. bogenfelsi	–
Adapoides (Eocene, Asia)	
A. troglodytes	196

(Fig. 12.10; Franzen et al., 2009). This specimen of a young female individual preserves almost the entire skeleton (estimated adult body size ~650–1000 g), as well as impressions of fur and remains of its stomach contents. Dentally, it is very similar to *Godinotia* and *Europolemur*; the morphology of the teeth and the remains of fruit and leaves preserved in its stomach indicate an herbivorous diet. *Darwinius* had relatively large orbits, suggesting it was nocturnal. It possessed very long hind limbs with a low intermembral index, suggesting leaping abilities. Unfortunately, most of the diagnostic features of the cranium and ankle are crushed and impossible to reconstruct with any certainty. Although it has been argued frequently that

Darwinius is specifically related to the origin of anthropoids (and humans!) (Franzen et al., 2009; Gingerich et al., 2010), it is undoubtedly an adapoid similar to other European caenopithecids (Seiffert et al., 2009; Williams et al., 2010). Indeed, half of the skeleton was previously placed in the genus *Godinotia*.

Five adapoids similar to the European caenopithecids have been described from the Eocene of Africa (Seiffert, 2012; Godinot et al., 2018). *Aframonius dieides* (Table 12.5) is from the Late Eocene of the Fayum, Egypt (Simons et al., 1995). The morphologically similar *Masradapis tahai* is known from slightly older deposits in the Fayum along with *Afradapis longicristatus*, which is larger and more derived in having lost its anterior premolar to give it a catarrhine-like dental formula of 2.1.2.3 (Fig. 12.13) (Seiffert et al., 2009; Seiffert et al., 2017). All were almost certainly folivorous, and the talus of *Afradapis* is most similar to that of lorises, suggesting slow, quadrupedal climbing (Boyer et al., 2010). *Namadapis interdictus* and *Notnamaia bogenfelsi* are older species from Namibia and demonstrate that caenopithecids were once distributed from northern to southern Africa (Pickford et al., 2008; Godinot et al., 2018). It seems most likely that the African caenopithecids dispersed from Asia.

Adapoides troglodytes, from the Eocene of southern China, is a medium-sized adapoid that has been suggested as a possible ancestor of the adapids, indicating an Asian origin for the group (Beard et al., 1994). It is also similar to many caenopithecids. Like the African *Afradapis*, *Adapoides* has a talus suggesting slow quadrupedal climbing (Gebo et al., 2008). Similar to European caenopithecids, *Adapoides* likely had a grooming claw on the second toe as well (Gebo et al., 2017).

Mahgarita stevensi, from the latest Middle Eocene (Duchesnean) of Texas, is a North American caenopithecid and occurs after the apparent extinction of the notharctids at the end of the Middle Eocene. *Mahgarita* has relatively small premolars, and the hypocone on the upper molars is derived from the lingual cingulum. The mandibular symphysis is fused. The strong development of crests on the molar teeth, as well as its moderate size (>500 g), suggests that it was at least partly folivorous. Although *Mahgarita* has been mentioned as a possible anthropoid ancestor because of its deep snout, fused mandibular symphysis, and enlarged canal for the promontory artery (Rasmussen, 1990; but see Ross, 1994), it shows no evidence of postorbital closure and is similar to other adapoids in all other aspects of its cranial anatomy.

Mescalerolemur horneri is a related species from slightly older Uintan deposits of Texas (Kirk and Williams, 2011). The relationships of *Mescalerolemur* and *Mahgarita* to other caenopithecids, as well as how they came to be in Texas, are unresolved issues (Gunnell et al., 2008; Rose et al., 2009; Kirk and Williams, 2011).

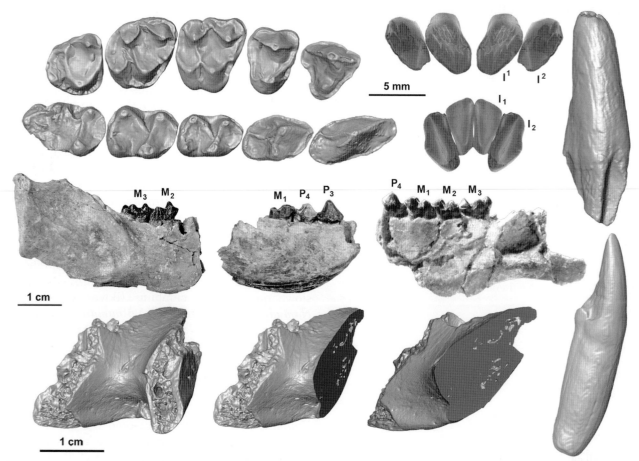

FIGURE 12.13   The mandible and dentition of *Afradapis longicristatus*, a caenopithecine adapoid from the Late Eocene of Egypt. Note the large canine and reduced number of premolars. *(Courtesy of E. Seiffert.)*

Adapids

The genus *Adapis* was the first fossil nonhuman primate named in 1822. However, it was so unusual that its describer, Cuvier, did not recognize it as a primate, and its primate affinities were not recognized until some 50 years later. The family Adapidae contains numerous genera (Table 12.6), all characterized by molarized last premolars, well-developed shearing crests, and many postcranial peculiarities (Dagosto, 1983, 1993; Godinot, 1998; Marigó et al., 2019). They are extremely common mammals in the Late Eocene of France, and there are many crania.

The best-known adapid is **Adapis parisiensis** (Fig. 12.7), a medium-sized species from several Late Eocene deposits in France. The latest occurring adapid in Europe, *A. parisiensis*, disappeared during the major European faunal turnover known as the Grande Coupure, which coincided with a major drop in temperature near the Eocene–Oligocene boundary.

A. parisiensis has a full primate dental formula of 2.1.4.3 (Fig. 12.7). Like most adapoids and living lemurs, it has upper central incisors that are small, relatively broad, and spatulate, with a gap between their bases, presumably for a Jacobson's organ. The upper lateral incisors are smaller and positioned behind the upper centrals.

The lower anterior dentition of *A. parisiensis* is unusual in that the lower incisors and canines form a single cutting edge. *Adapis* has long, narrow molars and premolars with well-developed shearing crests. They are strikingly similar to the molars of *Hapalemur*, suggesting a folivorous diet for *Adapis*.

Adapis has a very low, broad skull with flaring zygomatic arches and a small braincase (see closely related *Magnadapis*, Fig. 12.8). Prominent sagittal and nuchal crests are found on the larger individuals, suggesting that they are probably males. The orbits are relatively small, suggesting diurnal habits, and are oriented slightly upward rather than directly forward. The snout is moderately short. From the robust zygomatic arches and the extremely large temporal fossa, it is clear that *Adapis* had extremely large chewing muscles, concordant with the extensive shearing abilities seen in its dentition.

The auditory region of the *Adapis* skull has an inflated bulla with a free tympanic, as in extant strepsirrhines.

TABLE 12.6   Superfamily ADAPOIDEA
             Family ADAPIDAE

Species	Estimated mass (g)
Adapis (Late Eocene to Early Oligocene, Europe)	
A. bruni	–
A. collinsonae	1654
A. parisiensis	2500
A. stintoni	2182
A. sudrei	1256
Cryptadapis (Late Eocene, Europe)	
C. tertius	2500
C. laharpei	1700
Microadapis (Late Eocene, Europe)	
M. lynnae	–
M. sciureus	598
Leptadapis (Late Eocene to Early Oligocene, Europe)	
L. filholi	–
L. leenhardti	3742
L. magnus	4674
Palaeolemur (Late Eocene, Europe)	
P. betillei	–
Magnadapis (Late Eocene, Europe)	
M. quercyi	5663
M. fredi	?10,000
M. laurenceae	?10,000
M. intermedius	15,050
Paradapis (Late Eocene)	
P. rutimeyeri	–
P. priscus	1000

The stapedial artery appears to have been enlarged relative to the promontory artery in *Adapis*, but the relative sizes of these conduits vary within close relatives such as *Leptadapis* (Boyer et al., 2016). The brain is relatively small compared to that of extant strepsirrhines and has large olfactory bulbs.

There are numerous relatively complete limb bones of *A. parisiensis*. Initial analyses of these bones suggested that *Adapis* was most similar to the living lorises *Nycticebus* and *Perodicticus* — slow arboreal quadrupeds (Fig. 12.14; Dagosto, 1983; Spoor et al., 1998). In addition, the joint between the ulna and the wrist in *Adapis* shows features linking it with extant lemurs and lorises (Beard et al., 1988). More recent studies indicate a greater diversity of species and locomotor adaptations among the fossils attributed to *Adapis*, including generalized arboreal quadrupeds and more active, agile climbing species as well (Bacon and Godinot, 1998; Marigó et al., 2019).

In addition to *Adapis*, there are numerous species of small adapines in related genera, including **Microadapis**, **Cryptadapis**, and **Palaeolemur**.

Leptadapis and **Magnadapis** (Figs. 12.8 and 12.15) are much larger relatives of *Adapis*, each known from several species with distinctive cranial and dental morphology (Godinot and Couette, 2008). Like *Adapis*, they were probably diurnal folivores (Ramdarshan et al., 2012) that moved by quadrupedal climbing.

The phyletic and geographic origins of the adapids are unknown; they cannot be easily derived from any of the earlier European cercamoniids or the notharctids of North America. Their appearance in the Late Eocene of Europe was just as abrupt as their extinction shortly thereafter, suggesting an immigration from some other continental region. Several analyses suggest that they have a common ancestry with the widespread caenopithecids, and the Asian *Adapoides* seems to show similarities to both groups.

Sivaladapids

Long after the notharctids, cercamoniids, caenopithecids, and adapids disappeared from North America and Europe, there were relatively large adapoid primates thriving alongside fossil apes in the Late Miocene of Asia (Table 12.7). The best known of these, **Sivaladapis**, with two species from the Late Miocene of India and Pakistan, was fairly large (2.5–3.5 kg) with a dental formula of 2.1.3.3 (Fig. 12.16). The sharp crests on its molars and premolars suggest a folivorous diet. Unlike the latest members of either the European adapids or the North American notharctids, *Sivaladapis* has simple upper molars with no hypocone, which has been argued to be a secondarily derived condition (Qi and Beard, 1998). **Indraloris**, containing several species, is from the same deposits, and both *Sivaladapis* and *Indraloris* seem to have been sexually dimorphic in size (Flynn and Morgan, 2005). **Sinoadapis**, from the Latest Miocene site of Lufeng in China, is similar to *Sivaladapis* (Wu and Pan, 1985).

Sivaladapids first appear in the Eocene, and they are now known from many parts of Asia. *Hoanghonius* and *Rencunius* are from China and represent some of the earliest and most primitive sivaladapids. Similar to the condition in notharctids and caenopithecids, *Hoanghonius* possessed a grooming digit on its second toe (Gebo et al., 2015), suggesting that this feature was probably present in other sivaladapids and widespread among adapoids.

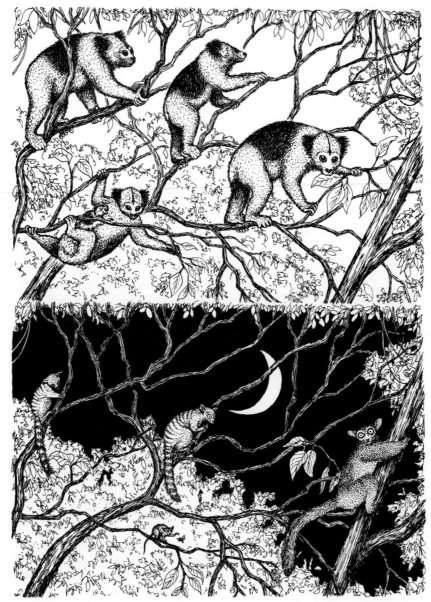

FIGURE 12.14   Scene from the Late Eocene of the Paris basin. *Above,* the diurnal *Adapis parisiensis* feed on leaves. *Below* are several nocturnal microchoerines: the tiny *Pseudoloris* attempts to catch an insect while *Necrolemur* (*left*) and *Microchoerus* (*right*) cling to branches.

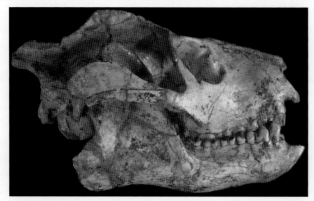

FIGURE 12.15   The skull of *Magnadapis intermedius* from the Eocene of France. *(Courtesy of J. Perry.)*

Paukkaungia and ***Kyitchaungia*** are another group of early sivaladapids, known from isolated teeth in the Eocene Pondaung Formation of Myanmar (Beard et al., 2007). Numerous postcranial elements have also been attributed to them. In general, these seem similar to the limbs of extant Malagasy lemurs and suggest active quadrupedalism and leaping (Marivaux, 2008a,b). However, there is some debate about the validity of the sivaladapids described from isolated teeth and whether the isolated limb elements in the Pondaung Formation belong to sivaladapids or amphipithecids.

Guangxilemur, from the Late Eocene of southern China and the Oligocene of Pakistan, seems to link the Eocene hoanghoniine and Miocene sivaladapine taxa (Qi and

TABLE 12.7   Superfamily ADAPOIDEA
Family SIVALADAPIDAE

Species	Estimated mass (g)
Subfamily SIVALADAPINAE	
Indraloris (Late Miocene, Asia)	
I. himalayensis	2779
I. kamlialensis	948
I. sp. LARGE	3113
Sivaladapis (Late Miocene, Asia)	
S. nagrii	2648
S. palaeindicus	3426
Sinoadapis (Late Miocene, Asia)	
S. carnosus	4449
S. shihuibaensis	4701
Siamoadapis (Middle Miocene, Asia)	
S. maemohensis	510
Ramadapis (Middle Miocene, Asia)	
R. sahnii	1290
Subfamily HOANGHONIINAE	
Hoanghonius (Eocene, Asia)	
H. stehlini	684
Rencunius (Middle Eocene, Asia)	
R. zhoui	734
Laomaki (Early Oligocene, Asia)	
L. yunnanensis	188
Kyitchaungia (Middle Eocene, Asia)	
K. takaii	1040
Paukkaungia (Middle Eocene, Asia)	
P. parva	483
Subfamily WAILEKIINAE	
Wailekia (Late Eocene, Asia)	
W. orientale	1012
Guangxilemur (Eocene to Oligocene, Asia)	
G. tongi	4800
G. singsilai	1526
Yunnanadapis (Early Oligocene, Asia)	
Y. folivorus	1228
Y. imperator	-
Subfamily INCERTAE SEDIS	
Lushius (Late Eocene, Asia)	
L. qinlinensis	1450

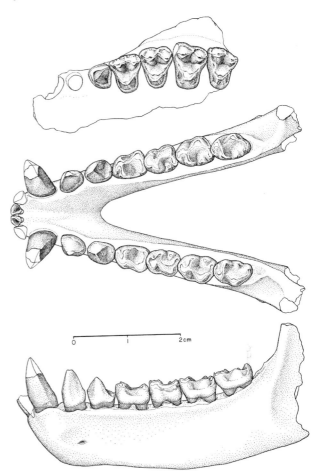

FIGURE 12.16   Upper and lower dentition of *Sivaladapis nagrii*. *(Courtesy of P. Gingerich.)*

Beard, 1998). *Wailekia*, from the Eocene of Thailand, and *Yunnanadapis*, from the Oligocene of China, seem to be a part of this intermediate group of taxa as well (Ducrocq et al., 1995; Ni et al., 2016; Gilbert et al., 2017).

As mentioned above, some researchers suggest that sivaladapids are derived from the Early Eocene asiadapids (Rose et al., 2009; Godinot, 2015), but Sivaladapidae may be a paraphyletic "wastebasket" group linked by their Asian geographic distribution more than shared derived features. Thus, their position relative to other adapoid groups is still unclear.

Other Asian Adapoids

In addition to the many genera and species that have been placed in the families and subfamilies described above, there are several adapoids from Asia that cannot be placed confidently in any of these families, and may instead represent a family of their own. (Table 12.8). *Bugtilemur* is a genus based on isolated teeth from the Oligocene of Pakistan that was originally described as part of the radiation of Malagasy lemurs and specifically

TABLE 12.8   Superfamily ADAPOIDEA
Family Ekgmowechashalidae

Species	Estimated mass (g)
Family EKGMOWECHASHALIDAE	
Bugtilemur (Oligocene, Asia)	
B. mathesonae	105
Muangthanhinius (Eocene, Asia)	
M. siami	300
Sulaimanius (Eocene, Asia)	
S. arifi	25
Gatanthropus (Early Oligocene, Asia)	
G. micros	161
Palaeohodites (Late Eocene, Asia)	
P. naduensis	969
Ekgmowechashala (Oligocene, North America)	
E. philotau	932
E. zancanellai	1216

as the sister taxon of *Cheirogaleus* (Marivaux et al., 2001). However, the subsequent discovery of a similar taxon, *Muangthanhinius* from the Late Eocene of Thailand, which lacks a dental tooth comb, suggests that neither is closely related to the Malagasy radiation (Marivaux et al., 2006). Instead, these taxa appear to form a group with *Gatanthropus* from China and the enigmatic *Ekgmowechashala* from North America, suggesting intermittent biogeographic connections and dispersals between Asia and North America into the Oligocene (Seiffert, 2007; Samuels et al., 2015; Ni et al., 2016; Rust et al., 2023). *Sulaimanius*, from the late Early Eocene of Pakistan, shares some features with *Bugtilemur* and may be related to it and other ekgmowechashalines (Gunnell et al., 2008). How these taxa relate to the other adapoids is not at all clear, as all are based on very few dental remains, but a link with sivaladapids has been recently suggested (Rust et al., 2023). Interestingly, *Ekgmowechashala* is similar to notharctids in developing a hypocone from the postprotocingulum (i.e., a pseudohypocone), but its teeth are otherwise so uniquely derived that it is difficult to place too much emphasis on any single feature. In any case, these taxa provide further evidence that the radiation of adapoids was extremely diverse, widespread, and long lived in Asia (e.g., Gunnell et al., 2008).

Amphipithecids

The amphipithecids are a small radiation of primates from the Eocene of Myanmar (Burma) and Thailand.

Since their initial discovery nearly 100 years ago, they have been the subject of ongoing debate over whether they are adapoids or primitive anthropoids. This debate is still not resolved. However, they will be discussed in the next chapter dealing with early anthropoids.

Azibiids, Djebelemurids, and *Plesiopithecus*: North African Stem Strepsirrhines?

In addition to the larger adapoids from North Africa placed in the Caenopithecidae, there are several tiny primates from North Africa and the Arabian Peninsula (Table 12.9) that show dental similarities to both Eocene adapoids and to extant lemurs, galagos, and lorises, suggesting an ancient African origin for extant strepsirrhines (Seiffert et al., 2005; Godinot, 2006; Tabuce et al., 2009). *Azibius trerki* is a tiny mammal from the Eocene of Algeria with very unusual premolars whose affinities have been debated for decades. However, new material of *Azibius* and another tiny primate, *Algeripithecus* (previously thought to be an early anthropoid based on its upper molars), indicates that these are near the ancestry of extant strepsirrhines. The anterior teeth are not known for either genus, but it has been suggested from the alveolus that *Algeripithecus* may have had a procumbent canine suggestive of a tooth comb (Tabuce et al., 2009). In addition, these North African primates resemble extant strepsirrhines in having premolars and molar crowns that overlap, and a talus attributed to an azibiid shows clear strepsirrhine affinities (Marivaux et al., 2012).

Djebelemur (Fig. 12.17) is a tiny (~70 g) adapoid from the Early Eocene site of Chambi in Tunisia that seems to be another stem strepsirrhine. The same may well be the case for *Omanodon* and *Shizarodon* from the Early Oligocene of Oman. A very similar species from the Late Eocene of Egypt, *"Anchomomys" milleri*, seems to belong in the same group. *"A" milleri* is almost certainly a djebelemurid rather than a member of the genus *Anchomomys* and needs to be given a new generic name. However, the fact that this species was originally allocated to the European genus *Anchomomys* suggests dental similarities between these North African stem strepsirrhines and European cercamoniines, particularly to the genus *Anchomomys*. Unfortunately, these tiny African and Arabian primates are known only from limited dental remains. The anterior dentition, as well as cranial and postcranial remains, is needed to confirm their phyletic position relative to extant strepsirrhines.

Plesiopithecus teras is a very unusual primate from the latest Eocene of the Fayum region of Egypt that is known from an almost complete skull and several mandibles (Fig. 12.18). The dental formula of *Plesiopithecus* is debated. The upper canines are very large, and there is

TABLE 12.9 Suborder STREPSIRRHINI

Species	Estimated mass (g)
Family AZIBIIDAE	
Azibius (Eocene, Africa)	
A. trerki	115
A. sp.	650
Algeripithecus (Eocene, Africa)	
A. minutus	56
Family DJEBELEMURIDAE	
Djebelemur (Eocene-Oligocene, Africa, Arabia)	
D. martinezi	68
"Anchomomys" (Eocene, Africa)	
"A." milleri	74
Omanodon (Oligocene, Arabia)	
O. minor	35
Shizarodon (Oligocene, Arabia)	
S. dhofarensis	84
Family DAUBENTONIIDAE	
Family INCERTAE SEDIS	
Plesiopithecus (Eocene, Africa)	
P. teras	628
Propotto (Miocene, Africa)	
P. leakeyi	910

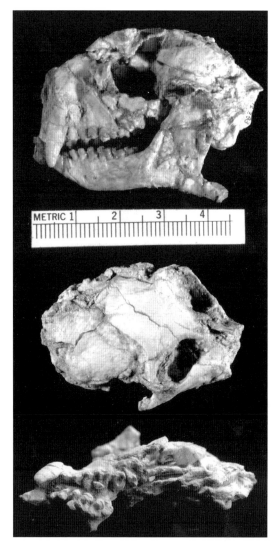

FIGURE 12.18   The cranium and mandible of *Plesiopithecus teras*, a fossil strepsirrhine from the Eocene of Egypt with large procumbent front teeth.

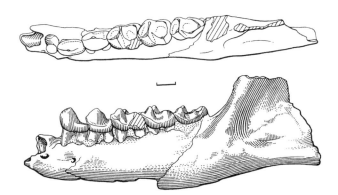

FIGURE 12.17   *Djebelemur martinezi*, a tiny primitive strepsirrhine from the Eocene of Tunisia. *(Courtesy of J.L. Hartenberger.)*

a large procumbent lower tooth that has most recently been argued to be an incisor with a tiny, reduced canine behind it (Godinot, 2010; Gunnell et al., 2018). The premaxilla is missing, so it is not known what the upper incisors were like. The upper molars are simple tritubercular teeth. The lower premolars have overlapping crowns like those of many extant strepsirrhines and the stem strepsirrhines of North Africa and Oman discussed

above. The presence of a lower first premolar (or lower canine, depending on interpretation) seems variable, but in one mandible, this tooth is angled forward over the base of the procumbent lower incisor. *Plesiopithecus* has a postorbital bar like extant strepsirrhines, and the large orbits suggest a nocturnal activity pattern. The braincase is relatively small, and the temporal lines form a small sagittal crest posteriorly.

The phylogenetic position of *Plesiopithecus* is very difficult to determine because it is so different from any other primate. Some have suggested a relationship which places it with lorisoids (Simons and Rasmussen, 1996; Rasmussen and Nekaris, 1998). However, it now seems most likely that *Plesiopithecus* and the Miocene *Propotto* are stem daubentoniids, implying an independent rafting event to Madagascar for aye-ayes apart from other lemurs (Gunnell et al., 2018).

Fossil Lorises and Galagos

In addition to the likely stem strepsirrhines and possible stem lemuroids from North Africa mentioned above, there are numerous fossil lorisoids from the Eocene through Pleistocene of Africa. The earliest records of this group come from the Late Eocene of the Fayum in Egypt, where three taxa have been identified as lorisoids. There are also fossil lorises from the Miocene of Asia (Table 12.10).

Saharagalago and *Karanisia* are two tiny primates from the Late Eocene locality of BQ-2 that show dental features clearly linking them with extant strepsirrhines. *Saharagalago* is very similar to extant galagids in its dentition and is considered a stem galagid or stem lorisoid (Seiffert et al., 2003). Originally identified as a fossil loris (Seiffert et al., 2003), *Karanisia* is more likely a stem lorisoid or some sort of stem strepsirrhine (Seiffert et al., 2005; Seiffert 2007, 2012). An isolated incisor attributed to *Karanisia* shows that it had a tooth comb like extant strepsirrhines. *Wadilemur elegans* (Fig. 12.19), from the slightly younger L-41 locality, has a dentition similar to that of extant galagids, and an isolated femur attributed to *Wadilemur* has many of the distinct adaptations for leaping found in the femur of extant galagids.

There are several genera and many species of lorisoids from the Early and Middle Miocene of Kenya and Uganda (Table 12.9; Harrison, 2010). *Mioeuoticus* (Fig. 12.20) and *Progalago* seem to be related to the lorises, and *Komba* seems to be closer to living galagos (see McCrossin, 1992; Seiffert, 2007; Harrison, 2010).

These Miocene lorisoids are very similar to living African genera in their dental and cranial anatomy, and the shape of the incisor roots indicates that they had tooth combs. Although they can generally be identified (not without debate) as lorises or galagos, none can be positively linked to any living genus or species. The dental remains indicate a size range comparable to that of modern lorises and galagos (~100–1000 g), as well as considerable dietary diversity, including frugivores and faunivores. The skulls and facial fragments indicate large orbits, which are suggestive of nocturnal habits. Although there are no postcranial bones directly associated with cranial material, limbs have been assigned to taxa based on relative size and presence at specific sites. The limb elements attributed to galagos are long, but their tarsals are not as elongated as those of living galagos. They are more similar to the tarsals of cheirogaleids (Gebo, 1989; Harrison, 2010), suggesting that these Miocene taxa lie outside the clade of extant galagids. Thus far, there is only a single humerus of the African Miocene lorisoids indicating slow climbing habits (Gebo et al., 1997). However, studies of the semicircular canals of *Mioeuoticus* show similarities to those of slow-climbing extant lorises, whereas the canals of *Komba* resemble those of leaping galagos (Walker et al., 2008).

TABLE 12.10 Suborder STREPSIRRHINI
Superfamily LORISOIDEA

Species	Estimated mass (g)
Family GALAGIDAE	
Saharagalago (Late Eocene, Africa)	
S. misrensis	122
Wadilemur (Late Eocene, Africa)	
W. elegans	100
Namaloris (Middle Eocene, Africa)	
N. rupestris	-
Progalago (Early Miocene, Africa)	
P. dorae	1200
P. songhorensis	800
Komba (Early to Middle Miocene, Africa)	
K. robusta	288
K. minor	129
K. walkeri	325
K. winamensis	840
Galago (Miocene to Recent, Africa)	
G. farafraensis	37
Laetolia (Pliocene, Africa)	
L. sadimanensis	114
Otolemur (Pliocene to Recent, Africa)	
O. howelli	700
Paragalago (Pliocene to Recent, Africa)	
P. zanzibaricus	136
Family LORISIDAE	
Mioeuoticus (Early Miocene, Africa)	
M. bishopi	600
M. shipmani	1050
M. kichoti	700
Nycticeboides (Late Miocene, Asia)	
N. simpsoni	275
Microloris (Late Miocene, Asia)	
M. pilbeami	67
?Nycticebus (Miocene, Asia)	
N. linglom	-
Family INCERTAE SEDIS	
Karanisia (Late Eocene, Africa)	
K. clarki	230
K. arenula	120
Orogalago (Early Oligocene, Africa)	
O. saintexuperyi	189
Superfamily INCERTAE SEDIS	
Orolemur (Early Oligocene, Africa)	
O. mermozi	

Galago remains from the Late Miocene of Egypt and younger fossil galagos from 2 to 4 million years ago in Ethiopia and Tanzania are similar to the living *Galago* and *Otolemur*.

The earliest fossil record of Asian lorises comes from the Middle to Late Miocene of Pakistan. Numerous isolated teeth from 14 to 7 million years ago and one relatively complete skeleton have been attributed to a single species, **Nycticeboides simpsoni**. This species seems closely related to the living slow loris, *Nycticebus*, in both cranial and postcranial anatomy (MacPhee and Jacobs, 1986). **Microloris pilbeami** is a tiny fossil loris (50 g) from similar-aged deposits in Pakistan, where there are isolated teeth suggesting other species from both younger and older deposits (Flynn and Morgan, 2005). **?Nycticebus linglom** is a possible fossil loris based on a single tooth from the Miocene of Thailand (Mein and Ginsburg, 1997) and there are isolated teeth of a possible fossil loris from the Late Miocene of India as well (Bhandari et al., 2021).

Adapoids and Strepsirrhines

Since adapoids were first identified as primates, virtually all authors have noted their many anatomical

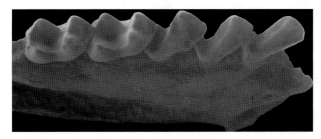

FIGURE 12.19   The lower jaw of *Wadilemur elegans*, a primitive lorisoid from the Eocene of Egypt, showing overlapping premolar crowns and procumbent lower canine. *(Courtesy of E. Seiffert.)*

similarities to living strepsirrhines, particularly to lemurs. Adapoids are lemur-like in their cheek teeth, in the overall configuration of their skull with its simple postorbital bar and moderately long snout, and in the morphology of the nasal region. The auditory region is also lemur-like, with an inflated bulla and a free ectotympanic ring. The carotid circulation is more similar to that of lemurs than to either haplorhines or lorises in that most individuals have a stapedial canal of moderate size. However, in virtually all of these features, adapoids and strepsirrhines retain the primitive primate condition found in many other mammals, rather than sharing unique specializations. Furthermore, adapoids lack a tooth comb, the derived feature that most clearly distinguishes living strepsirrhines from other primates, and they also seem to have retained more primitive hands and feet than many Malagasy lemurs.

Adapoids and living strepsirrhines share only a few anatomical features that may be unique specializations linking the two and also precluding ancestral relations to other primates. Eocene adapoids and strepsirrhines share several unusual features of the ankle: a flaring fibular surface on the talus, a long talar shelf, and the arrangement of the cuneiform facets of the navicular (Dagosto, 1988; Boyer et al., 2010). Adapids (but not notharctids) have been linked with extant strepsirrhines by a unique articulation between the ulna and the carpus (Beard et al., 1988). In addition, a grooming claw on the second toe, which is found in all extant strepsirrhines, has been identified in several taxa across several adapoid families (e.g., Notharctidae, Caenopithecidae, Sivaladapidae), although tarsiers have two such grooming digits and it is not clear that the strepsirrhine condition is derived. The overall anatomical similarity between adapoids and strepsirrhines clearly demonstrates that living strepsirrhines have retained many aspects of an adapoid-like morphology for nearly 60 million years, and most phylogenetic analyses identify adapoids as the

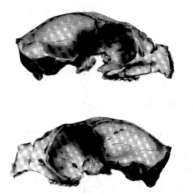

FIGURE 12.20   The skull of a fossil loris, *Mioeuoticus*, from the Miocene of eastern Africa. *(From LeGros Clark, W.E., 1956. A Miocene lemuroid skull from East Africa. In: Fossil Mammals of Africa 9, British Museum of Natural History, London, p. 6).*

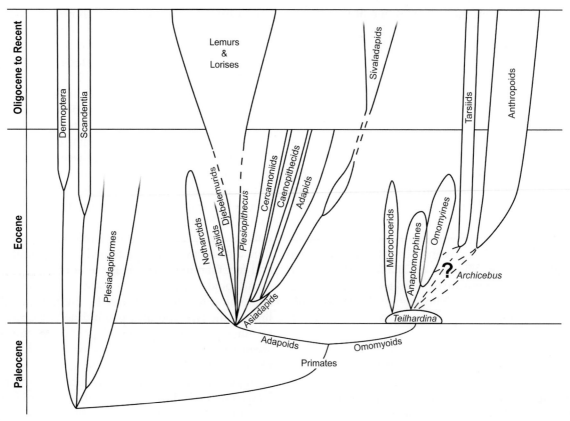

FIGURE 12.21   A schematic diagram showing the relationships among the families of fossil prosimians and later groups of primates.

sister group to extant strepsirrhines (Seiffert et al., 2009; Williams et al., 2010). However, although the two groups are very similar in overall morphology and are most probably sister taxa, there is very little evidence demonstrating a unique phyletic relationship between strepsirrhines and any particular group of Eocene adapoids (Seiffert et al., 2005; Godinot, 2006), except possibly the North African azibiids and djebelemurids. Moreover, if the Eocene azibiids and djebelemurids from North Africa are indeed stem strepsirrhines, their common ancestry with adapoids from Eurasia and North America was probably very early in that epoch or even earlier (Fig. 12.21). It seems most likely that the Eurasian and North American adapoids and the stem strepsirrhines from North Africa are successive nodes in strepsirrhine evolution.

In addition to their traditional link with strepsirrhines, adapoids have frequently been proposed as the ancestors of higher primates (Rasmussen, 1994), or even as haplorhines (Franzen et al., 2009; Gingerich et al., 2010; Gingerich, 2012; but see Williams et al., 2010). This suggestion has been based on their anthropoid-like anterior dentition, fused mandibular symphysis, and common size range, as well as postcranial features on the Messel adapoid *Darwinius*. However, these analyses are mostly based on a comparison of Eocene adapoids to living primates. Analyses that consider *Darwinius* and

other adapoids in conjunction with other fossil primates place them clearly among the adapoids and show that many of the similar features, such as the fused mandibular symphysis, clearly evolved in parallel in many groups of strepsirrhines and in early anthropoids. We will discuss the relationship between adapoids and anthropoids again in the next chapter.

Omomyoids

Like the adapoids, the tarsier-like omomyoids first appeared in the earliest Eocene of Asia, Europe, and North America (Ni et al., 2004, 2005; Smith et al., 2006; but see Beard, 2008; Morse et al., 2019). Omomyoids, like adapoids, had very different evolutionary histories in North America and in Europe. There are no omomyoids known from Africa, except possibly *Altiatlasius*. They are just becoming known from Asia. In North America, omomyoids were very diverse taxonomically throughout the Eocene. There are three widely recognized groups of omomyoids: Anaptomorphinae and Omomyinae, both predominantly North American, and the European Microchoeridae. Thus, it seems best to recognize two separate families: Omomyidae, for the diverse, predominantly North American anaptomorphines and omomyines (each divided into many tribes),

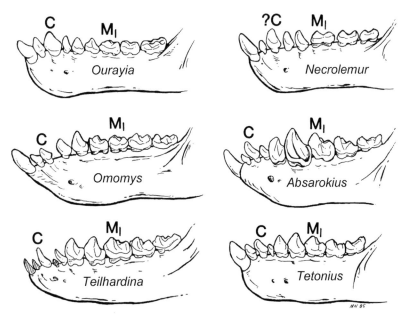

FIGURE 12.22 Mandibles of representative omomyoid primates from North America and Europe. The positions of the canine (C) and the first molar (M1) are indicated.

and Microchoeridae, for the smaller, but distinct, radiation in the European Eocene (Fig. 12.21).

The most primitive omomyoid, *Teilhardina*, is very similar to early adapoids such as *Donrussellia* and *Cantius* in dental morphology and in retaining a primitive dental formula of 2.1.4.3 (Figs. 12.3 and 12.4; Bown and Rose, 1991). However, all subsequent taxa (except *Steinius*) are characterized by reduction and reorganization of the antemolar dentition; these changes occurred independently in several lineages (Fig. 12.22). Most omomyoids have a relatively large, procumbent lower central incisor and a smaller lateral one, and the canines are usually small – never large as in adapoids or absent as in some plesiadapiforms (Figs. 12.2 and 12.22). The premolars are reduced to three or fewer in all but a few species and vary considerably in shape among subfamilies and tribes. In some, they are tall and pointed; in others, they are broad and molariform. The lower molars usually have relatively small, low, mesiodistally compressed trigonids and broad-basined talonids. The upper molars are usually broad. Many early species have a prominent postprotocingulum (nannopithex fold) joining the protocone distally, and later species developed a hypocone from the lingual cingulum. The mandibular symphysis of omomyoids is unfused.

The skulls of most omomyoids resemble those of extant tarsiers and galagos in their relatively short, narrow snout, posteriorly broadening palate, and large orbits (Figs. 12.6, 12.23, 12.24) (Rosenberger, 2011). The auditory region of some species has an inflated auditory bulla and a tympanic ring that is fused to the bullar wall and extends laterally to form a bony tube. The internal carotid circulation is known in only a few genera. In *Tetonius*, *Shoshonius*, *Necrolemur*, and *Rooneyia*, both the

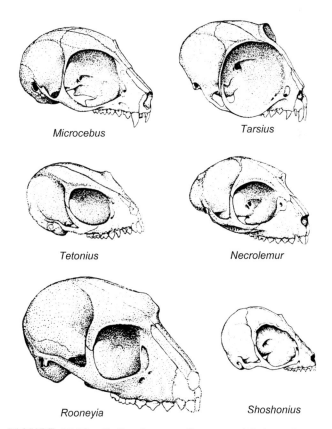

FIGURE 12.23 Skulls of two small, nocturnal living primates, *Microcebus murinus* and *Tarsius* sp., compared with reconstructed skulls of several omomyoids, *Tetonius homunculus*, *Necrolemur antiquus*, *Rooneyia viejaensis*, and *Shoshonius cooperi*. Note that *Tarsius* and *Shoshonius* have relatively larger eyes than the other primates.

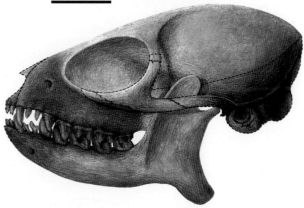

FIGURE 12.24 The skull of *Teilhardina asiatica*. *(Courtesy of X. Ni.)*

stapedial and promontory canals are present. In *Tetonius*, they are similar in size, but in the latter three, the promontory is larger (Ross, 1994; Boyer et al., 2016).

There are only a few partial skeletons known for omomyoids (Dagosto, 1993), but in all of them, the calcaneus is moderately elongated, as in extant cheirogaleids (Figs. 12.3 and 12.4). In both North American omomyids and European microchoerids, the distal tibia and fibula are either joined by an extensive fibrous joint or show evidence of some fusion, as in extant *Tarsius*. Most known skeletal elements indicate leaping habits, but not clinging, for these early prosimians, and they show greater overall similarities to the skeletons of cheirogaleids than to those of the extant tarsiers. Likewise, the morphology of omomyoid semicircular canals indicates agile or jerky locomotion, as in extant leapers (Silcox et al., 2009).

Teilhardina, Archicebus, and Baataromomys

The genus ***Teilhardina*** (Figs. 12.3, 12.4, 12.24) as normally recognized, with one species each from Asia and Europe and several species from North America, seems to lie at the base of the entire omomyoid radiation, and separate individual species may have given rise to microchoerids, and omomyids, as it dispersed from Asia to Europe and subsequently to North America at the beginning of the Eocene (Fig. 12.25; Smith et al., 2006;

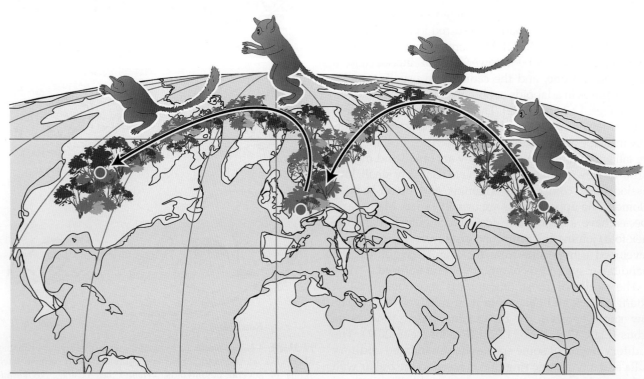

FIGURE 12.25 Morphological and phylogenetic analyses indicate that populations of the genus *Teilhardina* dispersed rapidly at the Paleocene–Eocene boundary from Asia to Europe and then to North America in conjunction with the Paleocene–Eocene Thermal Maximum while Eurasia and North America were still connected at high latitudes. Map lines delineate study areas and do not necessarily depict accepted national boundaries. *(Courtesy of P. Gingerich.)*

TABLE 12.11 Family OMOMYIDAE
Subfamily ANAPTOMORPHINAE

Species	Estimated mass (g)
Teilhardina (Early Eocene, Asia, Europe, North America)	
T. belgica	66
T. gingerichi	116
T. brandti	72
T. demissa	78
T. tenuicula	68
T. asiatica	58
T. magnoliana	63
*Bownomomy*s (Early Eocene, North America)	
B. americanus	83
B. crassidens	86
Archicebus (Early Eocene, Asia)	
A. achilles	20–30 g
Baataromomys (Early Eocene, Asia)	
B. ulaanus	75
Anaptomorphus (Middle Eocene, North America)	
A. aemulus	150
A. westi	282
Gazinius (Middle Eocene, North America)	
G. amplus	687
G. bowni	432
Tetonius (Early Eocene, North America)	
T. homunculus	50
T. mckennai	103
T. matthewi	180
Pseudotetonius (Early Eocene, North America)	
P. ambiguus	161
Absarokius (Early to Middle Eocene, North America)	
A. abbotti	206
A. gazini	150
A. metoecus	190
Tatmanius (Early Eocene, North America)	
T. szalayi	87
Strigorhysis (Middle Eocene, North America)	
S. bridgeriensis	155
S. huerfanensis	279
S. rugosus	155
Aycrossia (Middle Eocene, North America)	
A. lovei	204

TABLE 12.11 Family OMOMYIDAE—Cont'd
Subfamily ANAPTOMORPHINAE

Species	Estimated mass (g)
Trogolemur (Middle Eocene, North America)	
T. amplior	110
T. fragilis	44
T. leonardi	128
T. myodes	65
T. storeri	41
Walshina (Middle Eocene, North America)	
W. esmaraldensis	72
W. shifrae	40
W. mcgrewi	115
Sphacorhysis (Middle Eocene, North America)	
S. burntforkensis	78
Anemorhysis (Early Eocene, North America)	
A. natronensis	64
A. pattersoni	98
A. pearcei	63
A. savagei	60
A. sublettensis	50
A. wortmani	89
Arapahovius (Early Eocene, North America)	
A. advena	81
A. gazini	164
Chlororhysis (Early Eocene, North America)	
C. knightensis	117
C. incomptus	-
Artimonius (Middle Eocene, North America)	
A. witteri	148
A. nocerai	102
A. australis	227

Rose et al., 2012; but see Beard, 2008; Morse et al., 2019). Thus, *Teilhardina* is likely a paraphyletic taxon (Hooker, 2007; Rosenberger, 2011), and some species have been described as a separate genus, ***Bownomomys*** (Morse et al., 2019). However, *Teilhardina* is usually placed in the Anaptomorphinae as it is most similar to members of that subfamily (Table 12.11). The oldest and most primitive species, *T. asiatica*, is known from a crushed but nearly complete skull from the earliest Eocene of China (Fig. 12.24) (Ni et al., 2004). *Teilhardina asiatica* was a tiny primate with an estimated body size of ~28–58 g and an insectivorous diet. It has a primitive dental formula with four premolars and three molars. The canine is a large, conical tooth. An isolated incisor referred to this species is small and suggests that basal omomyoids had anterior dental proportions like those of adapoids and anthropoids. The cranium has moderate-sized orbits, which the authors of the original description suggested indicated diurnal habits. However, because *T. asiatica* probably lies outside the size range of extant primates, others have noted that it is difficult to extrapolate a regression of orbit size and tooth size (Ross and Martin, 2007; Heesy, 2009). Postcranial remains of *Teilhardina* are known from both Europe and North America. In general, they resemble the limbs of other omomyids, but there is a suggestion that the European *Teilhardina belgica* had enormous hands, like the extant tarsiers (Gebo et al., 2012a). Most recently, grooming claws have been identified among *Teilhardina* foot bones, but it is unclear whether they are restricted to the second digit as in adapoids and extant strepsirrhines or whether they are present in both the second and third digits as in living tarsiers (Boyer et al., 2018). A tiny omomyoid similar to *Teilhardina* has been recently found in Virginia (Rose et al., 2021).

Archicebus achilles is a tiny (20–30 g) basal haplorhine from the earliest Eocene of China that is known from much of a crushed skeleton (Ni et al., 2013). While its teeth closely resemble those of early anaptomorphines such as *Teilhardina*, the skeleton is unusual in having a relatively short ankle, suggesting more quadrupedal habits than other tarsiiforms. It seems to lie near the base of the tarsiiform clade near the divergence of tarsiiforms and anthropoids and may be an early representative of a *Teilhardina* clade (Fig. 12.21; Morse et al., 2019).

Baataromomys ulaanus is a tiny omomyoid from Inner Mongolia, known from a single lower molar (Ni et al., 2007). It is similar to *Teilhardina brandti* from the earliest Eocene of Wyoming and *Steinius*, a basal omomyine. Given that it falls within the variation found among the various species of *Teilhardina* and *Bownomomys* (Rose et al., 2011), *Baataromomys* may be best placed within one of these genera, close to the base of all omomyoids.

Omomyids

North American omomyids are traditionally placed in two separate subfamilies, anaptomorphines and omomyines. However, the composition of the subfamilies is unstable, with some taxa, notably the washakiins, being regularly placed in different subfamilies by different authorities (e.g., Williams, 1994; Gunnell and Rose, 2002; Gunnell et al., 2007; Tornow, 2008; Ni et al., 2010). The distribution of individual genera into distinct tribes also varies among authorities. The presence of omomyids with North American affinities in Asia further emphasizes the breadth and complexity of this radiation (e.g., Ni et al., 2010).

Anaptomorphines

Anaptomorphines (Table 12.11) include over a dozen genera; they are often regarded as the most primitive omomyoids and are probably ancestral to omomyines. Apart from the paraphyletic *Teilhardina*, which is also found in Asia and Europe, this subfamily is known only from the Early and Middle Eocene of North America. Anaptomorphines are very common and speciose in the Early Eocene, and also the early Middle Eocene, but only one taxon (**Trogolemur**) is known from the late Middle Eocene (Gunnell et al., 2007; Williams and Kirk, 2008). The early evolution of anaptomorphines in Wyoming is one of the most detailed records of population and species changes through time in the entire fossil record (Bown and Rose, 1987). Evolutionary analyses of this radiation have provided remarkable documentation of transitions between paleospecies (Rose and Bown, 1993) and evidence of considerable parallelism.

Despite their systematic diversity, anaptomorphines are all relatively similar in many aspects of their morphology. All were small, probably ranging from about 50 to 500 g. Later members of the subfamily are usually characterized by a tall, pointed P_4 and a reduced M_3. Many species have only two premolars. Their lower molars have relatively low trigonids with bulbous cusps and shallow talonids. Based on alveolus size, the lower incisors are small relative to the canines in the earliest taxa (i.e., *Teilhardina*), but incisor enlargement with canine reduction is common in many later genera. Studies of the dental morphology of the early anaptomorphines indicate that most were frugivorous, with a few species showing adaptations for processing invertebrates that have hard shells (Strait, 1991).

The skull of **Tetonius homunculus** (Fig. 12.23), from the Early Eocene of Wyoming, was recovered over 140 years ago. It has a short snout, large eyes, and a relatively globular braincase. Unfortunately, the auditory region is extremely damaged. The teeth of *Tetonius* suggest that it was probably largely insectivorous (Fig. 12.26). Its orbits are similar in size to those of a living cheirogaleid or a small galago, suggesting that it was nocturnal. Because the orbits are relatively smaller than those of tarsiers, it seems likely that it had a tapetum lucidum-like living strepsirrhines.

Very little is known of the postcranial anatomy of anaptomorphines other than *Teilhardina*. One of the best-known taxa, **Absarokius**, shows adaptations for quadrupedal leaping, including an extensive fibrous connection between the tibia and the fibula, but no indications of vertical clinging. Similar to *Teilhardina*, grooming claws have recently been identified in anaptomorphines such as *Tetonius*, *Anemorhysis/Tetonoides*, and *Arapahovius* (Boyer et al., 2018).

Omomyines

The omomyines (Table 12.12), a predominantly North American group, were almost certainly derived from an anaptomorphine-like ancestor similar to *Steinius*. The composition of the subfamily is under considerable flux: several genera (**Stockia**, **Uintanius**, **Utahia**) are frequently placed in different tribes by various researchers, and the washakiins (including **Washakius** and **Shoshonius**) are sometimes placed in the anaptomorphines. The major adaptive radiation of omomyines was later than that of the more primitive anaptomorphines: omomyines were most abundant in the Middle Eocene. Many authorities have suggested that the replacement of anaptomorphines by omomyines in the late Middle Eocene is associated with the general climatic warming and a northward movement of southern faunas (Gunnell, 1997). A few omomyines are also known from China (Beard et al., 1991, 1994).

Omomyines ranged in size from about 100 g to over 1 kg. Despite their similar taxonomic diversity, omomyines show a far greater range of dental adaptations than the anaptomorphines (Figs. 12.26 and 12.27; Gilbert, 2005). Their molars often have lower cusps, and the trigonid cusps are less inflated; the last molar is usually elongated. Some later members of the subfamily developed very flat molars with accessory cusps and crenulated enamel (Fig. 12.26). Omomyines probably occupied a variety of dietary niches (Fig. 12.26). Like anaptomorphines, the earlier, smaller species included both frugivores and species with adaptations for processing hard invertebrates such as beetles. Later taxa included folivores, such as **Macrotarsius**, the largest omomyine (Fig. 12.26).

The Early Eocene **Shoshonius cooperi** (Fig. 12.23) is one of the best-known omomyids. There are now several skulls and many parts of the postcranial skeleton (Beard et al., 1991; Dagosto et al., 1999). It was a relatively small (~120–150 g), nocturnal, insectivorous primate. *Shoshonius* shows many striking cranial similarities to the living tarsiers, including very large orbits, overlap of basicranial structures on the enlarged auditory bullae, and possibly some features of cranial circulation. In addition, it has a vertical nasolacrimal duct as in extant haplorhines (Rossie et al., 2006, 2018). The limbs, however, are more generalized and lack the distinctive features of extant tarsiers related to vertical clinging and leaping. Rather, they are more similar to primates such as cheirogaleids

TABLE 12.12 Family OMOMYIDAE
Subfamily OMOMYINAE

Species	Estimated mass (g)
Omomys (Middle Eocene, North America)	
O. carteri	220
O. lloydi	121
Chumashius (Middle to Late, North America)	
C. balchi	169
Steinius (Early Eocene, North America)	
S. annectens	298
S. vespertinus	156
Uintanius (Middle Eocene, North America)	
U. ameghini	70
U. rutherfurdi	94
Jemezius (Early Eocene, North America)	
J. szalayi	140
Macrotarsius (Middle Eocene, North America, Asia)	
M. jepseni	937
M. montanus	1659
M. roederi	980
M. seigerti	1175
M. macrorhysis	975
Hemiacodon (Middle Eocene, North America)	
H. casamissus	320
H. engardae	750
H. gracilis	658
Yaquius (Middle Eocene, North America)	
Y. travisi	1311
Ekwiiyemakius (Middle Eocene, North America)	
E. walshi	107
Gunnelltarsius (Middle Eocene, North America)	
G. randalli	298
Brontomomys (Middle Eocene, North America)	
B. cerutti	765
Ourayia (Middle Eocene, North America)	
O. uintensis	1032
O. coverti	742
Mytonius (Middle Eocene, North America)	
M. hopsoni	690
M. williamsae	591
Nesomomys (Middle Eocene, Asia)	
N. bunodens	572
Saskomomys (Middle Eocene, North America)	
S. lindsayorum	140

(Continued)

TABLE 12.12   Family OMOMYIDAE—Cont'd
Subfamily OMOMYINAE

Species	Estimated mass (g)
Diablomomys (Middle Eocene, North America)	
D. dalquesti	323
Wyomomys (Middle Eocene, North America)	
W. bridgeri	182
Ageitodendron (Middle Eocene, North America)	
A. matthewi	493
Utahia (Middle Eocene, North America)	
U. carina	38
U. kayi	94
Stockia (Middle Eocene, North America)	
S. powayensis	287
Asiomomys (Middle Eocene, Asia)	
A. changbaicus	256
Chipetaia (Middle Eocene, North America)	
C. lamporea	601
Washakius (Middle Eocene, North America)	
W. insignis	155
W. izetti	96
W. laurae	130
W. woodringi	140
Shoshonius (Early to Middle Eocene, North America)	
S. bowni	132
S. cooperi	123
Dyseolemur (Middle Eocene, North America)	
D. pacificus	144
Loveina (Early Eocene, North America)	
L. minuta	78
L. sheai	94
L. wapitiensis	116
L. zephyri	130
Subfamily INCERTAE SEDIS	
Rooneyia (Late Eocene, North America)	
R. viejaensis	350
Kohatius (Early to Middle Eocene, Asia)	
K. coppensi	130
Vastanomys (Early Eocene, Asia)	
V. gracilis	120
V. major	323

and the larger galagos that rely on both quadrupedal and leaping behaviors. This accords with the morphology of the semicircular canals (Silcox et al., 2009).

Omomys carteri, a larger species (220 g) from the Middle Eocene of Colorado, is also known from both cranial (Burger, 2010) and postcranial (Anemone and Covert, 2000) remains. *Omomys* was nocturnal and insectivorous. Based on limited evidence, *Omomys* may have retained an oblique nasolacrimal canal, a primitive olfactory configuration apparently retained in numerous other omomyoid taxa as well (see below and Rossie et al., 2018). Both the morphology of the limb bones and the semicircular canals indicate that *Omomys*, like *Shoshonius*, had a generalized quadrupedal and leaping locomotor behavior. Similar to many anaptomorphines, *Omomys* possessed grooming claws (Boyer et al., 2018).

Rooneyia viejaensis (Figs. 12.23 and 12.26), from the Late Eocene of Texas, is known from only one specimen: a nearly complete cranium. *Rooneyia* has a relatively broad, short snout and moderate-sized orbits surrounded by a complete postorbital bar and a small flange of the frontal bone behind the upper part of the orbit. *Rooneyia* had a body mass of ~350–400 g. On the basis of orbit size, it seems most likely that *Rooneyia* was diurnal, and the molar teeth with multiple rounded cusps suggest a frugivorous diet. Although there are no limb elements known for *Rooneyia*, the morphology of the semicircular canals suggests agile locomotor abilities similar to those of *Shoshonius* or *Omomys*.

The braincase is relatively large, in the range of that of extant prosimians. The auditory region has an uninflated bulla, with a tubular, bony ectotympanic partly enclosed by the bulla. The promontory branch of the internal carotid was large compared to the reduced stapedial branch (Boyer et al., 2016). However, *Rooneyia* retained relatively large olfactory bulbs and a primitive configuration of the olfactory apparatus. The nasal fossa housed a full complement of four ethmoturbinals, and the nasolacrimal canal was obliquely oriented, suggesting *Rooneyia* probably retained a wet nose as in living strepsirrhines (Kirk et al., 2014; Rossie et al., 2018; Lundeen and Kirk, 2019). It has traditionally been placed in the omomyines, but given its primitive nasal anatomy, more recent analyses suggest that it is either a late-occurring basal primate or, more likely, a primitive stem haplorhine (Ross et al., 1998, 2018; Kirk et al., 2014; Lundeen and Kirk, 2019). It is here regarded as Omomyoidea, *Incertae Sedis* (Table 12.13).

Microchoerids

The microchoerids (Table 12.14) were a diverse group of omomyoids from the Early Eocene through the

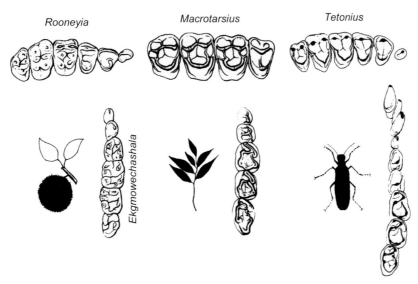

FIGURE 12.26   Dentitions of several omomyoids with different dietary adaptations.

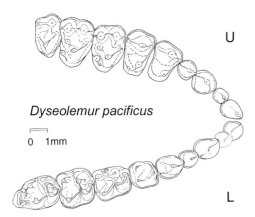

FIGURE 12.27   Upper (U) and lower (L) right dentitions of the North American omomyine, *Dyseolemur pacificus*. *(Courtesy of D.T. Rasmussen.)*

beginning of the Oligocene of western Europe (Gunnell and Rose, 2002; Hooker and Harrison, 2008), with one possible taxon from Asia. They appear to have derived from a European population of *Teilhardina* similar to *T. belgica* (Fig. 12.25; Hooker, 2007). The earliest and most primitive microchoerid is **Melaneremia**, reconstructed from isolated teeth from the Early Eocene deposits in England about one million years younger than those yielding *T. belgica* (Hooker, 2007). The presence of a primitive microchoerid in the earliest Eocene indicates that the European microchoerids are not far removed from the earliest omomyids, with *Teilhardina* as a root taxon for all omomyoids.

Microchoerids range in size from tiny **Pseudoloris** (25–65 g) to the medium-sized **Microchoerus** (325–1200 g), and many are relatively abundant in the fossil record. Microchoerids generally have large central incisors, small lateral incisors, a moderate-sized canine, two or three premolars that increase in size posteriorly, and

three molars. However, occlusal relations at the front of the jaw are often difficult to reconstruct, so the dental formula for microchoerids has been the subject of some debate (Fig. 12.28; Schmid, 1983; Thalmann, 1994) but was clearly different from that in tarsiids. The cheek teeth of microchoerids vary considerably among the genera. *Nannopithex* is a tiny genus from the Middle Eocene with an enlarged, pointed premolar and anaptomorphine-like molars with a high trigonid and deep, narrow talonid; it is most comparable to living frugivorous primates. The tiny **Pseudoloris**, however, probably had a faunivorous diet. This long-lived genus survived into the Oligocene in Spain. The larger genera, **Necrolemur** and *Microchoerus*, from the Middle and Late Eocene, have molars with low, rounded cusps and elaborate crenulations of the enamel; these suggest a more frugivorous diet or, considering their anterior dentition, perhaps a diet supplemented by gums (Ramdarshan et al., 2012). *Vectipithex* is a newly diagnosed genus of several small to medium-sized species with large procumbent lower incisors and a wide range of reconstructed dietary habits, including frugivory, insectivory, and seed-eating (Thalmann, 1994; Hooker and Harrison, 2008).

There are many nearly complete, usually crushed, skulls of *Necrolemur* (Fig. 12.23) and cranial fragments of *Microchoerus*, *Pseudoloris*, and *Nannopithex*. All have a relatively short, narrow snout with a bell-shaped palate, a gap between the upper central incisors, large eyes, and a moderately large infraorbital foramen. The olfactory bulb apparently passes above the orbits as in all extant haplorhines, but the back of the orbit is not walled off from the temporal fossa, as in tarsiers and anthropoids. *Microchoerus* and probably *Pseudoloris* possessed obliquely oriented nasolacrimal canals, as in extant strepsirrhines (Tabuce et al., 2009; Rossie et al., 2018). In the ear region, the ectotympanic forms a ring within the bulla, but extends

TABLE 12.13   Superfamily OMOMYOIDEA
Family MICROCHOERIDAE

Species	Estimated mass (g)
Nannopithex (Early to Middle Eocene, Europe)	
N. filholi	104
N. humilidens	76
N. zuccolae	102
N. aberhaldeni	93
N. barnesi	118
Pseudoloris (Middle Eocene to Early Oligocene, Europe)	
P. crusafonti	52
P. godinoti	65
P. isabenae	41
P. parvulus	27
P. pyrenaicus	54
P. saalae	31
P. cuestai	58
Necrolemur (Late Eocene, Europe)	
N. antiquus	172
N. zitteli	82
N. anadoni	140
Microchoerus (Late Eocene, Europe)	
M. creechbarrowensis	452
M. edwardsi	591
M. erinaceus	1109
M. ornatus	325
M. wardi	326
M. hookeri	563
Vectipithex (Middle to Late Eocene, Europe)	
V. quaylei	363
V. raabi	109
V. smithorum	771
V. ulmensis	1058
Melaneremia (Early Eocene, Europe)	
M. bryanti	58
M. schrevei	67
Indusomys (Early Eocene, Asia)	
I. kaliae	22
Quercyloris (Middle Eocene, Europe)	
Q. eloisae	46

Necrolemur antiquus

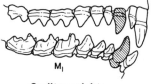
Carlito syrichta

FIGURE 12.28   A lateral view of the anterior dentition of *Necrolemur antiquus* and *Carlito syrichta*, showing the dental proportions. Various authorities have identified each of the first three teeth as the canine. It seems most likely that the shaded tooth is the canine and the teeth anterior to it are incisors. Note that regardless of how the dental formula of *Necrolemur* is interpreted, the dental proportions are very different from those of extant tarsiers. *(Adapted from Schmid, P., 1983. Front dentition of the Omomyiformes (Primates). Folia Primatol. 40, 1–10).*

laterally to form a bony tube. This unique condition makes them resemble strepsirrhines in the position of the ring but tarsiers and some anthropoids in presence of a tube. The canal for the stapedial artery and the groove for the promontory artery are similar in size. There is extensive inflation of the mastoid region behind the middle and inner ear. The large eyes of microchoerines suggest that they were all nocturnal animals, but the orbits are more like those of strepsirrhines, and the relative size of the optic foramen in *Necrolemur* and *Microchoerus* suggests that they did not possess a high-acuity retina (Kirk and Kay, 2004). They probably had a tapetum lucidum.

Although there are no complete skeletons for microchoerids, numerous isolated hind limb elements have been attributed to species of this family. These include a nearly complete femur, a partly fused tibia–fibula, a talus, and a calcaneus for *Necrolemur*, and isolated tarsal bones probably attributable to *Microchoerus*. All of these postcranial elements indicate leaping abilities. In their elongation, however, the calcanei of microchoerids are more like those of cheirogaleids than those of tarsiers (Dagosto, 1993). In several aspects of femoral morphology, microchoerines show striking similarities to anthropoids, especially parapithecids (Dagosto and Schmid, 1996).

Asian Omomyoids

The fossil evidence of omomyoids in Asia has expanded considerably in recent years. In addition to *Altanius*, generally regarded as a basal primate, and earliest Eocene omomyoids from China (see above), there are several other taxa from Pakistan, India, and China. ***Kohatius*** is a moderate-sized (130 g) species from the Early to Middle Eocene of Pakistan known from only a few teeth that has been attributed broadly to omomyoids (Gunnell et al., 2008). ***Indusomys kaliae*** is a tiny primate known from two molar teeth from the Early Eocene of Pakistan. It shows similarities to the microchoerid ***Melaneremia*** (Gunnell et al., 2008). ***Vastanomys*** from the Early Eocene Vastan Lignite coal mine in Gujarat,

TABLE 12.14   Family TARSIIDAE

Species	Estimated mass (g)
Afrotarsius (Early Oligocene, Africa)	
A. chatrathi	188
A. libycus	119
cf. *Tarsius* (Eocene to Recent, Asia)	
T. eocaenus	66
T. sirindhornae	355
Hesperotarsius (Miocene South Asia)	
H. sindhensis	170
H. thailandicus	161
Xanthorhysis (Eocene, Asia)	
X. tabrumi	144
Afrasia (Middle Eocene, Asia)	
A. djijidae	83
Oligotarsius (Early Oligocene, Asia)	
O. rarus	82

India (Bajpai et al., 2005, 2007; Rose et al., 2009; Dunn et al., 2016) is known from both dental and skeletal remains. The attributed limb elements lack the distinctive deep knee indicative of leaping that is found in most other omomyoids, suggesting that *Vastanomys* may retain a primitive skeletal anatomy close to the common ancestry of omomyoids and adapoids (Dunn et al., 2016; Martin et al., 2022). All of these taxa from South Asia are relatively poorly known at present, and further material will be needed to understand their affinities with omomyoids from other regions.

Asiomomys from the Middle Eocene of China is strikingly similar to the Late Eocene omomyine *Stockia* from California (Beard and Wang, 1991). *Nesomomys* from the Middle Eocene of Turkey shows similarities to North American Middle Eocene omomyines such as *Mytonius* (Beard et al., 2021). In addition, Beard and colleagues (1994) have described specimens from China that are very similar to the North American omomyine *Macrotarsius* and also a relative of the poorly understood *Tarka* (Ni et al., 2010). These new discoveries clearly document a radiation of omomyoids in Asia and some continuity between North American and Asian primate communities during the Eocene.

Tarsiids

Numerous fossil primates have been described that are attributable to the same family as the living tarsiers

(Table 12.14). *Tarsius eocaenus* is from the Middle Eocene Shanghuang fissure-fills and is known from numerous teeth and a small cranial fragment that is virtually identical to extant tarsiers in orbital and nasal morphology (Beard et al., 1994; Beard, 1998; Rossie et al., 2006). There are also many fossil tarsal bones from Shanghuang that have been attributed to tarsiids (Dagosto et al., 1996; Gebo et al., 1996, 2008, 2012b). *Xanthorhysis tabrumi* is another Eocene tarsier based on a well-preserved mandible from the Yanqu Basin in Central China (Beard, 1998). Most recently, *Oligotarsius rarus* has been described from isolated teeth in the early Oligocene of China (Ni et al., 2016). These remains appear to document the presence of tarsiids in Asia for at least 50 million years, an observation that fits well with their morphological and biomolecular distinctiveness from other primates (Beard, 1998).

From Miocene deposits in Thailand, there are two fossil tarsiids. *Hesperotarsius thailandicus* (Ginsburg and Mein, 1986) is known from a single lower molar, and *Tarsius srindhornae* is known from numerous jaws (Chaimanee et al., 2012). A small tarsiid *Hesperotarsius sindhensis* is known from the Miocene of southern Pakistan (Zijlstra et al., 2013). In view of the generic diversity currently recognized among extant tarsiers (Chapter 4), many of these fossil species should probably be recognized as distinct genera.

Afrotarsius chatrathi is from the Oligocene Fayum of Egypt, and another species has been described from the Eocene of Libya (Jaeger et al., 2010). Because this genus is known only from limited dental material, it has been debated whether *Afrotarsius* is more closely related to the living tarsiers, to the European microchoerids, or to early anthropoids, all of which it resembles to some degree. *Afrasia* (Chaimanee et al., 2012), from the Eocene of Burma, is known from a few isolated teeth. It is very similar to *Afrotarsius* and has also been identified by some authorities as a stem anthropoid (Chaimanee et al., 2012) and by others as a tarsiid (Seiffert, 2012).

Omomyoids, Tarsiers, and Haplorhines

As small prosimians with large eyes, elongate calcanei, and in some species, a fused tibia–fibula, omomyoids have been traditionally linked with the extant tarsiers, just as their contemporaries, the adapoids, have been allied with extant strepsirrhines (Fig. 12.21). Several authorities have even placed one or more of the European microchoerids into the family Tarsiidae. The omomyoid–tarsier connection continues to be debated extensively in light of new anatomical information, new analytical techniques, and new fossils (Rossie et al., 2018; Kirk and Lundeen, 2020), and it is clear that omomyids and microchoerids are not simply Eocene tarsiers. Some of their supposed tarsier-like resemblances are superficial similarities or features common to other Eocene prosimians as well; moreover, all known omomyoids clearly lacked many of the distinguishing features of the ear, orbit, nose, and skeleton that characterize the living tarsiers. Even more confusing are the different patterns of similarities and differences between Eocene taxa and the extant tarsiids, indicating that any phylogenetic scenario involved considerable mosaic evolution and parallelism. There are Eocene fossils from China that are more clearly placed in the family Tarsiidae, suggesting that true tarsiids were older than or contemporary with most omomyoids. Nevertheless, omomyoids, tarsiers, and anthropoids all share a number of features that lead almost all researchers to group them together in the semiorder Haplorhini. It appears that in all fossil omomyoids, as in tarsiers and anthropoids, the olfactory bulb lies above the interorbital septum, and there is no extensive sphenoethmoid recess. The anatomy of the talus, likewise, groups omomyoids, tarsiers, and anthropoids, as do various other features of the postcranial skeleton (e.g., Dagosto et al., 1999, 2008; Gebo et al., 2012a). The focus of most current debate concerns the relationships and the branching pattern among these three haplorhine groups (Fig. 13.21). Are tarsiers more closely related to omomyoids than they are to anthropoids, or are tarsiers more closely related to anthropoids than they are to omomyoids?

Although omomyoids have been generally considered as tarsier-like for over a century, there have been extensive and detailed arguments for an omomyoid–tarsier relationship, based specifically on the cranial morphology of the microchoerid *Necrolemur* (Rosenberger, 1985), the omomyine *Shoshonius* (e.g., Beard and MacPhee, 1994), and the anaptomorphine *Strigorhysis* (Rosenberger, 2011). There are a number of features that link *Shoshonius* and *Necrolemur* with tarsiers, including overlap of the pterygoid plates and the enlarged auditory bullae and a jaw articulation with a narrow gutter-like shape. However, the pattern of similarities and differences between the Eocene fossils and extant tarsiers is very complex (Fig. 12.28). Much of the tarsier-like appearance of omomyoids derives from their large orbits, and Rosenberger (2011) has argued that although there is considerable diversity in relative orbit size among omomyoids, all have a characteristic structure of their facial region reflecting the need to support large eyes. Nevertheless, the orbits of all omomyoids are structurally more similar to those of strepsirrhines than to tarsiers (or anthropoids) in lacking any postorbital closure.

Because all living haplorhines lack a tapetum, the light-catching efficiency of their eyes is less than that of strepsirrhines. Thus, both nocturnal haplorhines (tarsiers and *Aotus*) have eyes that are much larger than those of similar-sized nocturnal strepsirrhines or omomyoids, and the optic foramen is relatively large compared to orbit size because of the large number of individual nerve fibers leaving the high-acuity retina. The size of the optic foramen relative to the orbits in the few omomyoids in which these features can be measured suggests that some omomyoids (*Necrolemur* and *Microchoerus*) were like strepsirrhines in having an eye with low acuity and probably a tapetum lucidum, rather than a high-acuity retina without a tapetum, as in tarsiers and anthropoids (Kay and Kirk, 2000; Kirk and Kay, 2004). However, at least two omomyoids, *Shoshonius* and *Omomys*, approach tarsiers in relative orbit size. Unfortunately, there are no analyses of the relative size of the optic foramen in these taxa.

The nasal anatomy of many omomyoids (including microchoerids) indicates that they retain a primitive strepsirrhine configuration of the nasolacrimal canal, which is lost in all tarsiers and anthropoids. In fact, among all examined omomyoids, only *Shoshonius* exhibits the derived crown haplorhine condition shared by tarsiers and anthropoids (Rossie et al., 2018). Thus, if omomyoids are a monophyletic group linked to tarsiers, the modern haplorhine configuration of the nose likely evolved numerous times. Alternatively, if omomyoids are a paraphyletic mix of both stem and crown haplorhines, it is possible that some omomyoids such as *Shoshonius* are more closely related to tarsiers (and anthropoids) than others and that the haplorhine nasal condition evolved only once.

The postcranial similarities linking omomyoids and tarsiers are very general. Most omomyoids have a calcaneus that is elongated compared with the ankles of anthropoids and many strepsirrhines, but they are all shorter than those of extant tarsiers and lack specific features of extant tarsier ankle bones. Similarly, some omomyoids (and many small anthropoids) show varying degrees of distal connection between the tibia and fibula, but only in the microchoerine *Necrolemur* does this approach the extent seen in tarsiers. In some omomyoids, the femur shows similarities to tarsiers; in others, it does not (Dagosto and Schmid, 1996). Overall, omomyoid postcranial bones are more similar to those of cheirogaleids or more quadrupedal galagos (Dagosto et al., 1999; Anemone and Covert, 2000).

Thus, in some features of the orbits, eyes, nose, and limbs, tarsiers are far more specialized and derived than nearly all omomyoids as far as is known. Even taxa such as *Shoshonius*, *Omomys*, or *Necrolemur*, which seem to approach tarsiers in some features, are much more primitive and generalized in others. For Rosenberger and colleagues (2008, 2011), the overall cranial similarities between omomyoids and tarsiers, as well as the diversity among omomyoids in aspects of their teeth, crania, and limbs, indicate that tarsiers are best viewed as one very derived (and long-lived) branch of a monophyletic omomyoid radiation. This argument is supported to some extent by the indications that Chinese "tarsiers" are extremely similar to extant tarsiers in dental and cranial anatomy (Rossie et al., 2006), but like omomyoids, they are less derived than extant tarsiers in their skeletal anatomy (e.g., Gebo et al., 2001).

In addition to the features that link omomyoids, tarsiers and anthropoids together as haplorhines, there are also a number of features of the cranium that tarsiers seem to share uniquely with anthropoids, leading many researchers to argue that tarsiers and anthropoids share a unique common ancestry to the exclusion of omomyoids (e.g., Ross et al., 1998; Kay et al., 2004; Williams et al., 2010). Specifically, some degree of postorbital closure and an anterior accessory cavity of the middle ear are found in both extant tarsiers and anthropoids. If these features are indeed shared only by anthropoids and tarsiers (e.g., Ross, 1994; but see DeLeon et al., 2016) and are lacking in all omomyoids, then omomyoids would be best considered as stem haplorhines, preceding the subsequent divergence of anthropoids and tarsiers. Alternatively, the unique cranial features linking tarsiers and anthropoids could have evolved independently.

Finally, given recent evidence from omomyoid nasal anatomy, it is also possible that omomyoids represent a paraphyletic grouping of stem haplorhines or even stem and crown haplorhines. If the haplorhine condition is a true synapomorphy that only evolved once, then most omomyoids would appear to be stem haplorhines retaining an oblique nasolacrimal canal. In this scenario, *Shoshonius* would either be considered a derived stem haplorhine more closely related to the tarsier + anthropoid clade or it could be a possible crown haplorhine related to living tarsiers. However, a close relationship between *Shoshonius* and tarsiers would imply that the cranial similarities uniting tarsiers and anthropoids evolved independently (see above).

Any solution likely implies parallel evolution within these three groups. The presence of tarsier dental and cranial material in Asia contemporary with or slightly younger than omomyids and microchoerids in North America, Europe, and Asia indicates that any phylogenies linking one or two Eocene omomyoids with either tarsiers or anthropoids, beyond indicating a broad haplorhine cranial organization, is likely to be overly simplistic. Moreover, the presence of the most primitive omomyoid (*Teilhardina asiatica*), primitive tarsiers, and very primitive anthropoids all from the Eocene of China suggests that sorting out this trichotomy is likely to be very difficult, but further

fossils from that region should help in its resolution. This issue will be addressed again in the subsequent discussions on anthropoid origins.

Adaptive Radiations and Biogeography of Fossil Prosimians

Our understanding of the evolution of the earliest modern primates is currently in an exciting but very awkward state of complexity. Both adapoids and omomyoids have been well known in North America and Europe for over a century, and the broad patterns of their taxonomic and adaptive diversity on those continents are well established. At the same time, our appreciation of the evolution and biogeography of these early primates in Africa and Asia are relatively recent and expanding rapidly. Still, we have only the poorest knowledge of their likely diversity in either taxonomic or adaptive realms on these continents.

In North America and Europe, the adapoids and omomyoids were a diverse group of primates that occupied a wide range of ecological niches. There seem to be clear temporal trends in the adaptive radiations of these early prosimians on both continents. Throughout the Eocene, adapoids seem to have occupied adaptive niches that to some degree characterize extant higher primates (larger size, diurnality, frugivory, and folivory), whereas early omomyoids were perhaps more comparable to galagos. Only in the later part of the Eocene do the omomyoids appear to have expanded into the adaptive zones of large size and some folivory. Equally striking are the phyletic and adaptive differences between the Eocene prosimian faunas on the two continents from which they are well known.

In North America (Fig. 12.29), the omomyoids of the Early and Middle Eocene were taxonomically diverse, but all were relatively small (most less than 500 g). Their teeth suggest diets that were predominantly frugivorous, with some specializing in hard insects. Available skulls indicate nocturnal habits. In contrast, the North American notharctines from the Early and Middle Eocene were much less taxonomically diverse, with only five or six genera, and all were considerably larger (0.5–6 kg), frugivorous or folivorous, and most were probably diurnal. Only after the near disappearance of notharctid adapoids in the Late Middle Eocene do we find larger, probably frugivorous and folivorous omomyoids in North America. The locomotor adaptations of Eocene prosimians are poorly known, but most remains indicate quadrupedal and leaping abilities for both omomyoids and adapoids, rather than the specialized vertical clinging and leaping that characterizes many extant strepsirrhines and tarsiers.

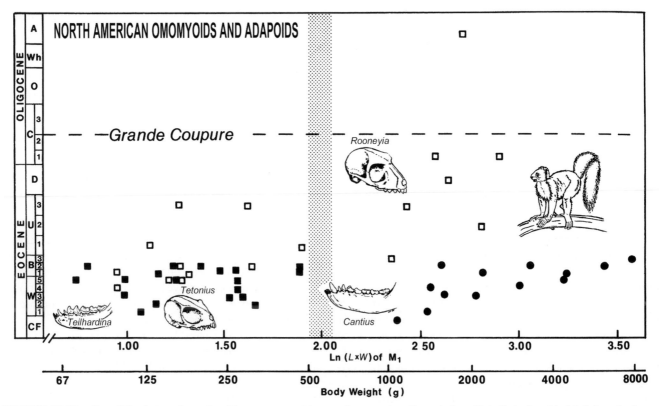

FIGURE 12.29 Size of North American adapoids, anaptomorphines, and omomyines through time. Note that adapoids (*circles*) are far larger than contemporary anaptomorphines (*filled squares*) and that the radiation of larger omomyines (*open squares*) takes place after the extinction of most of the adapoids. Cross-hatching indicates Kay's threshold.

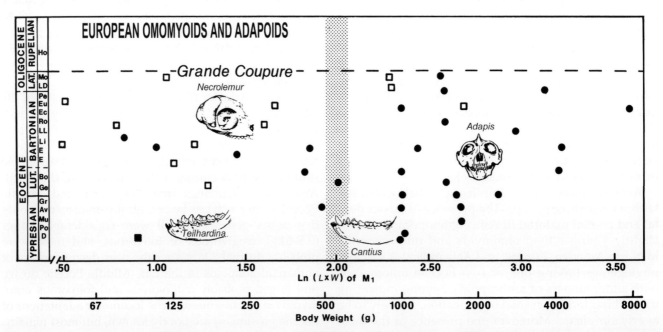

FIGURE 12.30 Size of European adapoids (*circles*) and omomyoids (*squares*) through time. Note that there is considerable overlap in body size in the two radiations and that the adapoids are more diverse. Compare with Figure 12.29.

In Europe (Fig. 12.30), the cercamoniids were more diverse, and the microchoerids were limited to only a few genera after the basal omomyoid *Teilhardina*. The Early and Middle Eocene adaptive diversity of both adapoids and omomyoids is limited (Gilbert, 2005). In the Latest Eocene, there were numerous medium to large adapoids. Although the European adapoids were generally larger than synchronous microchoerids, the size range of the two groups overlapped somewhat in the late Eocene, with the evolution of very small cercamoniids such as *Anchomomys gailardi* and large microchoerines such as *Microchoerus*. Associated with their size diversity was dietary diversity among the cercamoniids, caenopithecids, and adapids (Ramdarshan et al., 2012). There seem to have been insectivorous, frugivorous, folivorous, and possibly carnivorous (*Pronycticebus*) adapoids. The microchoerids, although less diverse, included small, insectivorous species and other species that probably specialized on fruits or gums. One ecological parameter that seems to have separated the two radiations was their activity cycle. Most microchoerids seem to have been nocturnal, and many European adapoids were probably diurnal, judging from orbit size. Furthermore, the microchoerids seem to have been leapers or cheirogaleid-like arboreal runners, whereas the skeletal remains from adapoids suggest slower quadrupedal climbing for some and leaping for others.

In Africa, where the pre-Miocene fossil record is mostly from the northern part of the continent, current evidence suggests two different radiations of strepsirrhines in the Eocene. There were numerous small species that included both stem strepsirrhines and primitive lorisoids (e.g., Tabuce et al., 2009; Godinot, 2010). These were presumably frugivorous and insectivorous and include some leaping species (Seiffert et al., 2005). Contemporary with these were larger, folivorous, slow-climbing adapoids related to the caenopithecids of Europe (Seiffert et al., 2009; Boyer et al., 2010). There is no reliable evidence of any omomyoid primates from the fossil record of Africa.

In Asia, the record of adapoids and omomyoids (and tarsiers) is more complex systematically, adaptively, and biogeographically. From the early part of the Eocene of India and Pakistan, there are several small, poorly known adapoids and omomyoids that seem to have some affinities with European taxa or later Asian groups (Gunnell et al., 2008; Rose et al., 2009, 2018). The limbs of these early adapoids resemble those of North American and European notharctids. From Myanmar and Thailand, the Eocene record is dominated by two groups of putative early anthropoids, the small eosimiids and the larger amphipithecids, along with limited evidence of sivaladapids. The Early Eocene of China contains the early omomyoid *Teilhardina* as well as *Archicebus*, both tiny,

insectivorous taxa argued to be close to omomyoid origins, and later in the Eocene and Oligocene, there is evidence of numerous small haplorhines, including tarsiers, anthropoids, sivaladapids, and various others known only from partial jaws, isolated teeth, and isolated skeletal elements (e.g., Gebo et al., 2001, 2008; Ni et al., 2016). In addition, there are a few larger, likely folivorous adapoids, some with affinities to sivaladapids, others to European and North American taxa, and one large folivorous omomyoid with North American affinities. The Late Miocene sivaladapids were large folivores.

Our only information about the social organization of fossil prosimians is the sexual dimorphism in canine size that seems to characterize some adapoids. This suggests that some type of polygynous social system was common in this radiation (Gingerich, 1995).

Origin and Early Evolution of Primates

As our knowledge of the phylogenetic, adaptive, and biogeographical diversity of primates from the Eocene has expanded, the question of the origin of the order Primates has become more confused and difficult to reconstruct with any confidence, as different sources of evidence yield very different answers. On the one hand, new fossils indicate that both omomyoids and adapoids from the earliest Eocene of China (Ni et al., 2004, 2005, 2013), India (Gunnell et al., 2008; Rose et al., 2009), Europe (Hooker, 2007), and North America (Beard, 2008) are all very similar to the early European adapoid *Donrussellia* or the cosmopolitan *Teilhardina* (e.g., Smith et al., 2006; Morse et al., 2019), and that dentally and postcranially, the earliest and most primitive adapoids and omomyoids were similar (Rose, 1994; Dunn et al., 2016; Martin et al., 2022), suggesting a temporal origin very near the Paleocene–Eocene border. Nevertheless, the earliest crania from the two groups are clearly distinct (Fig. 12.3), and although early members of both superfamilies seem to have been quadrupedal leapers, their limbs usually show anatomical differences characteristic of later haplorhines and strepsirrhines (Gebo et al., 2012a,b; but see Ni et al., 2013; Dunn et al., 2016). While some authors argue strongly for an origin of modern primates in the Paleocene, approximately 62 million years ago, from among the diverse plesiadapiforms (Bloch et al., 2007), all of the known plesiadapiforms from the Paleocene and Eocene seem too specialized to give rise to the Early Eocene primates. Thus, apart from the poorly known *Altiatlasius* from the Late Paleocene of North Africa, or *Altanius* from Asia, there are no suitable candidates for primate origins. Primates seem to just pop up suddenly in the Early Eocene of Asia, Europe, and North America. In the case of *Teilhardina*, this seems to involve a rapid dispersal from Asia at the Paleocene–Eocene boundary (Smith et al., 2006).

In contrast to this abrupt appearance of primates in the fossil record in earliest Eocene, studies of molecular systematics consistently place the origin of the order and the split between strepsirrhines and haplorhines in the Late Cretaceous, between 80 and 90 million years ago, or over 25 million years before they appear in the fossil record (e.g., Springer et al., 2007; Perelman et al., 2011; but see Steiper and Seiffert, 2012). Similar estimates of a very early origin come from estimates based on sampling probabilities of the fossil record (e.g., Tavare et al., 2002; Soligo et al., 2007). If the estimated gap of more than 25 million years between the actual origin of modern primates and their first appearance in the fossil record is accurate, it is difficult to imagine what the missing aspects of our early history may have looked like and what relationship the fossils from the Early Eocene actually bear to primate origins.

Efforts to reconstruct the geographical origins of Primates and the modern suborders are equally complex and confusing, but there are patterns. If plesiadapiforms are considered primates, or if they include the sister taxa of Primates, the origin of the group is alternately reconstructed as North America (where the best records of plesiadapiforms are found), Asia, or Africa (Bloch et al., 2007). Other biogeographical analyses identify Asia as the most likely area for primate origins because that is where the two extant sister groups of Primates, Scandentia (treeshrews) and Dermoptera (colugos), are found (Heesy et al., 2006; Beard, 2006). It is also where the basal omomyoids *Teilhardina* and *Archicebus* have been discovered (Ni et al., 2005). Only the presence of *Altiatlasius* in the Late Paleocene of Africa keeps that continent as a possibility.

An intriguing possibility that seems to unite many of the conflicts was initially proposed by Krause and Maas (1990) and more recently by others (Miller et al., 2005; Martin, 2006). These authors suggest that primates originated on part of Gondwana (probably the India–Madagascar landmass) sometime in the Cretaceous, and much of their early evolution took place while India was adrift in the Indian Ocean. The abrupt appearance of modern primates at the beginning of the Early Eocene is then coincident with, and the result of, the collision of the Indian subcontinent with the Asian mainland. At first glance, the discovery of primitive adapoids and some omomyoids in the Early Eocene of India would seem to support this hypothesis (Rose et al., 2009). However, the relationship between the age of the Early Eocene primates from India and Pakistan and the initial contact between the Indian and Asian landmasses is not at all clear. Most of the early Eocene Indian fossils seem to have similarities to Eurasian taxa, suggesting that primates and other mammals migrated into India via an ephemeral connection in the late Paleocene or earliest Eocene (Smith et al., 2016). Moreover, as noted above, all of the mammalian taxa most closely related to Primates seem to be endemic Asian orders (Scandentia, Dermoptera, Rodentia), some with a solid fossil record from the Paleocene of that region. Finally, Cretaceous mammals of Madagascar do not include any modern orders of mammals. Thus, the idea of the origin of primates taking place on the "Noah's Ark" subcontinent of India is an intriguing idea, but at present, there is little evidence to support it.

Although adapoids and omomyoids have traditionally been identified as Eocene lemurs and tarsiers, respectively, both Eocene superfamilies are decidedly more primitive in some respects than recent prosimians. Thus, adapoids are best considered primitive stem strepsirrhines and omomyoids as stem haplorhines (Fig. 12.21). Current evidence suggests that the subsequent evolution of crown strepsirrhines was initially an African event, and the evolution of crown haplorhines as represented by tarsiers was an Asian event. However, the geographical location and the morphology of the last common ancestor of basal strepsirrhines and basal haplorhines remain to be discovered.

References

Anemone, R.L., Covert, H.H., 2000. New skeletal remains of *Omomys* (Primates, Omomyidae): functional morphology of the hindlimb and locomotor behavior of a Middle Eocene primate. J. Hum. Evol. 38, 607–633.

Bacon, A.M., Godinot, M., 1998. Analyse morphofonctionnele des femurs et des tibias des "*Adapis*" du Quercy: mise en evidence de cinq types morphologiques. Folia Primatol. 69, 1–24.

Bajpai, S., Kapur, V.V., Thewissen, J.G.M., Das, D.P., et al., 2005. Early Eocene primates from Vastan Lignite Mine, Gujarat, western India. J. Palaeontol. Soc. India. 50 (2), 43–54.

Bajpai, S., Kapur, V.V., Das, D.P., Tiwari, B.N., 2007. New Early Eocene primate (Mammalia) from Vastan Lignite Mine, District Surat (Gujarat), western India. J. Palaeontol. Soc. India. 52 (2), 231–234.

Beard, K.C., 1998. A new genus of Tarsiidae (Mammalia: Primates) from the Middle Eocene of Shanxi Province, China, with notes on the historical biogeography of tarsiers. Bulletin of Carnegie Museum of Natural History. 34, 260–277.

Beard, K.C., 1988. New notharctine primate fossils from the Early Eocene of New Mexico and southern Wyoming and the phylogeny of Notharctinae. Am. J. Phys. Anthropol. 75, 439–469.

Beard, K.C., 2006. Mammalian biogeography and anthropoid origins. In: Lehman, S.M., Fleagle, J.G. (Eds.), Primate Biogeography: Progress and Prospects. Springer, New York, pp. 439–468.

Beard, K.C., 2008. The oldest North American primate and mammalian biogeography during the Paleocene-Eocene thermal maximum. Proc. Natl. Acad. Sci. USA 105 (10), 3815–3818.

Beard, K.C., Wang, B., 1991. Phylogenetic and biogeographic significance of the tarsiiform primate *Asiomomys changbaicus* from the Eocene of Jilin Province, People's Republic of China. Am. J. Phys. Anthropol. 85, 159–166.

Beard, K.C., MacPhee, R.D.E., 1994. Cranial anatomy of *Shoshonius* and the antiquity of the Anthropoidea. In: Fleagle, J.G., Kay, R.F. (Eds.), Anthropoid Origins. Plenum Press, New York, pp. 55–97.

Beard, K.C., Metais, G., Ocakoglu, F., Licht, A., 2021. An omomyid primate from the Pontide microcontinent of north-central Anatolia: Implications for sweepstakes dispersal of terrestrial mammals during the Eocene. Geobios 66-67, 143–152.

Beard, K.C., Dagosto, M., Gebo, D.L., Godinot, M., 1988. Interrelationships among primate higher taxa. Nature 331, 712–714.

Beard, K.C., Krishtalka, L., Stucky, R.K., 1991. First skulls of the Early Eocene primate *Shoshonius cooperi* and the anthropoid-tarsier dichotomy. Nature. 349, 64–67.

Beard, K.C., Tao, Q., Dawson, M.R., Wang, B., Chuanhuei, L., 1994. A diverse new primate fauna from the Middle Eocene fissure-fillings in southeastern China. Nature. 368, 604–609.

Beard, K.C., Marivaux, L., Tun, S.T., et al., 2007. New sivaladapid primates from the Eocene Pondaung formation of Myanmar and the anthropoid status of Amphipithecidae. Bulletin Carnegie Museum of Natural History 39, 67–76.

Bhandari, A., Bajpai, S., Flynn, L.J., Tiwari, B.N., Mandal, N., 2021. First Miocene rodents from Kutch, western India. Historical Biology 33, https://doi.org/10.1080/08912963.2020.1870970.

Bloch, J.I., Silcox, M.T., Boyer, D.M., Sargis, E.J., 2007. New Paleocene skeletons and the relationship of "plesiadapiforms" to crown-clade primates. Proc. Natl. Acad. Sci. USA 104, 1159–1164.

Bown, T.M., Rose, K.D., 1987. Patterns of dental evolution in Early Eocene anaptomorphine primates (Omomyidae) from the Bighorn Basin, Wyoming. Paleontol. Soc. Mem. 23, 1–162.

Bown, T.M., Rose, K.D., 1991. Evolutionary relationships of a new genus and three new species of omomyid primates (Willwood Formation, lower Eocene, Bighorn Basin, Wyoming). J. Hum. Evol. 20, 465–480.

Boyer, D.M., Seiffer, E.R., Simons, E.L., 2010. Astragalar morphology of *Afradapis*, a large adapiform primate from the Earliest Late Eocene of Egypt. Am. J. Phys. Anthropol. 143, 383–402.

Boyer, D.M., Kirk, E.C., Silcox, M.T., Gunnell, G.F., Gilbert, C.C., Yapuncich, G.S., Allen, K.L., Welch, E., Bloch, J.I., Gonzalez, L., Kay, R.F., Seiffert, E.R., 2016. Internal carotid arterial canal size and scaling in Euarchonta: Re-assessing implications for arterial patency and phylogenetic relationships in early fossil primates. J. Hum. Evol. 97, 123–144.

Boyer, D.M., Toussaint, S., Godinot, M., 2017. Postcrania of the most primitive euprimate and implications for primate origins. J. Hum. Evol. 111, 202–215.

Boyer, D.M., Maiolino, S.A., Holroyd, P.A., Morse, P.E., Bloch, J.I., 2018. Oldest evidence for grooming claws in euprimates. J. Hum. Evol. 122, 1–22.

Burger, B.J., 2010. The skull of the Eocene Primate *Omomys carteri* from western North America. Paleontol. Contrib 2, 1–19.

Chaimanee, Y., Chavasseau, O., Beard, K.C., et al., 2012. A new Middle Eocene primate from Myanmar and the initial anthropoid colonization of Africa. Proc. Natl. Acad. Sci. USA 109, 10293–10297.

Covert, H.H., 1986. Biology of Early Cenozoic primates. In: Swindler, D.R., Erwin, J. (Eds.), Comparative Primate Biology, vol. 1: Systematics, Evolution, and Anatomy. Alan R. Liss, New York, pp. 335–359.

Covert, H.H., 1990. Phylogenetic relationships among Notharctinae of North America. Am. J. Phys. Anthropol. 81, 381–398.

Covert, H.H., 1995. Locomotor adaptations of Eocene primates: adaptive diversity among the earliest prosimians. In: Altermann, L., Doyle, G.A., Izard, M.K. (Eds.), Creatures of the Dark. Plenum Press, New York, pp. 495–509.

Dagosto, M., 1983. Postcranium of *Adapis parisiensis* and *Leptadapis magnus* (Adapiformes, Primates): Adaptational and phylogenetic signals. Folia Primatol. 41, 49–101.

Dagosto, M., 1988. Implications of postcranial evidence for the origin of euprimates. J. Hum. Evol. 17, 35–56.

Dagosto, M., 1993. Postcranial anatomy and locomotion behavior in Eocene primates. In: Gebo, D.L. (Ed.), Postcranial Adaptation in Nonhuman Primates. Northern Illinois University Press, DeKalb, pp. 199–219.

Dagosto, M., 2007. The postcranial morphotype of primates. In: Ravosa, M.J., Dagosto, M. (Eds.), Primate Origins: Adaptations and Evolution. Springer, New York, pp. 489–534.

Dagosto, M., Schmid, P., 1996. Proximal femoral anatomy of omomyiform primates. J. Hum. Evol. 30 (1), 29–56.

Dagosto, M., Gebo, D.L., Beard, C., Qi, T., 1996. New primate postcranial remains from the middle Eocene Shanghuang fissures, southeastern China. Am. J. Phys. Anthropol. 22, 92–93.

Dagosto, M., Gebo, D.L., Beard, K.C., 1999. Revision of the Wind River faunas, Early Eocene of central Wyoming. Part 14. Postcranium of *Shoshonius cooperi* (Mammalia: Primates). Ann. Carnegie Mus. 68 (3), 175–211.

Dagosto, M., Gebo, D.L., Ni, X., Qi, T., Beard, K.C., 2008. Primate tibiae from the Middle Eocene Shanghuang fissure-fillings of eastern China. In: Sargis, E.J., Dagosto, M. (Eds.), Mammalian Evolutionary Morphology: A Tribute to Frederick S. Szalay. Springer, Dordrecht, pp. 315–324.

DeLeon, V.B., Smith, T.D., Rosenberger, A.L., 2016. Ontogeny of the Postorbital Region in Tarsiers and Other Primates. Anat. Rec. 299, 1631–1645.

Ducrocq, S., Jaeger, J.-J., Chaimanee, Y., Suteethorn, V., 1995. New primate from the Palaeogene of Thailand and the biogeographical origin of anthropoids. J. Hum. Evol. 28, 477–485.

Dunn, R.H., Rose, K.D., Rana, R.S., Kumar, K., Sahni, A., Smith, T., 2016. New euprimate postcrania from the early Eocene of Gujarat, India, and the strepsirrhine-haplorhine divergence. J. Hum. Evol. 99, 25–51.

Fleagle, J.G., Anapol, F.C., 1992. The indriid ischium and the hominid hip. J. Hum. Evol. 22, 285–305.

Flynn, L.J., Morgan, M.E., 2005. New lower primates from the Miocene Siwaliks of Pakistan. In: Lieberman, D.E., Smith, R.J., Kelley, J. (Eds.), Interpreting the Past: Essays on Human, Primate, and Mammal Evolution. Brill Academic Publishers, Boston, pp. 81–102.

Franzen, J.L., Gingerich, P.D., Habersetzer, J., Hurum, J.H., von Koenigswald, W., Smith, H.B., 2009. Complete primate skeleton from the Middle Eocene of Messel in Germany: Morphology and paleobiology. PLoS One 4, e5723.

Gebo, D.L., 1989. Postcranial adaptation and evolution in Lorisidae. Primates 30, 347–367.

Gebo, D.L., 2002. Adapiformes: phylogeny and adaptation. In: Hartwig, W.C. (Ed.), The Primate Fossil Record. Cambridge University Press, Cambridge, pp. 21–44.

Gebo, D.L., Dagosto, M., Beard, C., Qi, T., 1996. New primate tarsal remains from the Middle Eocene Shanghuang fissures, southeastern China. Am. J. Phys. Anthropol. Suppl. 22, 113.

Gebo, D.L., MacLatchy, L., Kityo, R., 1997. A new lorisid humerus from the Early Miocene of Uganda. Primates 38 (4), 423–427.

Gebo, D.L., Dagosto, M., Beard, K.C., Qi, T., 2001. Middle Eocene primate tarsals from China: Implications for haplorhine evolution. Am. J. Phys. Anthropol. 116, 83–107.

Gebo, D.L., Dagosto, M., Beard, K.C., Ni, X., 2008. New primate hind limb elements from the Middle Eocene of China. J. Hum. Evol. 55, 999–1014.

Gebo, D.L., Smith, T., Dagosto, M., 2012a. New postcranial elements for the earliest Eocene fossil primate *Teilhardina belgica*. J. Hum. Evol. 63, 205–218.

Gebo, D.L., Dagosto, M., Ni, X., Beard, K.C., 2012b. Species diversity and postcranial anatomy of Eocene primates from Shanghuang, China. Evol. Anthropol 21, 224–238.

Gebo, D.L., Beard, K.C., Ni, X., Dagosto, M., 2015. Distal phalanges of *Eosimias* and *Hoanghonius*. J. Hum. Evol. 86, 92–98.

Gebo, D.L., Dagosto, M., Ni, X., Beard, K.C., 2017. Phalangeal morphology of Shanghuang fossil primates. J. Hum. Evol. 113, 38–82.

Gilbert, C.C., 2005. Dietary ecospace and the diversity of euprimates during the Early and Middle Eocene. Am. J. Phys. Anthropol. 126, 237–249.

Gilbert, C.C., Jungers, W.L., 2017. Comment on relative brain size in early primates and the use of encephalization quotients in primate evolution. J. Hum. Evol. 109, 79–87.

Gilbert, C.C., Patel, B.A., Singh, N.P., Campisano, C.J., Fleagle, J.G., Rust, K.L., Patnaik, R., 2017. New sivaladapid primate from Lower Siwalik deposits surrounding Ramnagar (Jammu and Kashmir State), India. J. Hum. Evol. 102, 21–41.

Gingerich, P.D., 1984. Primate evolution. In: Broadhead, T.D. (Ed.), Mammals: Notes for a Short Course. University of Tennessee, Dept. of Geological Sciences, Knoxville, pp. 167–181.

Gingerich, P.D., 1986. Early Eocene *Cantius torresi* – oldest primate of modern aspect from North America. Nature. 319, 319–321.

Gingerich, P.D., 1995. Sexual dimorphism in earliest *Cantius torresi* (Mammalia, Primates, Adapoidea). Contributions to the Paleontological. Museum, University of Michigan 29 (8), 185–199.

Gingerich, P.D., 2006. Environment and evolution through the Paleocene-Eocene thermal maximum. Trends Ecol. Evol. 20 (5), 246–253.

Gingerich, P.D., 2012. Primates in the Eocene. In: Lehmann, T. Schaal, S.F.K. (Eds.), Messel and the terrestrial Eocene – Proceedings of the 22nd Senckenberg Conference. Palaeobiodiversity and Palaeo–environments. 92, 649–663.

Gingerich, P.D., Franzen, J.L., Habersetzer, J., Hurum, J.H., Smith, B.H., 2010. *Darwinius masillae* is a haplorhine – Reply to Williams et al. 2010. J. Hum. Evol. 59, 574–579.

Ginsburg, L., Mein, P., 1986. *Tarsius thailandica* nov. sp., Tarsiidae (Primates, Mammalia) fossile d'Asie. C. R. Acad. Sci 19, 1213–1215.

Godinot, M., 1998. A summary of adapiform systematics and phylogeny. Folia Primatol. 69, 218–249.

Godinot, M., 2006. Lemuriform origins as viewed from the fossil record. Folia Primatol. 77, 446–464.

Godinot, M., 2010. Paleogene prosimians. In: Werdelin, L., Sanders, W.J. (Eds.), Cenozoic Mammals of Africa. University of California Press, Berkeley, pp. 319–333.

Godinot, M., 2015. Fossil record of the primates from the Paleocene to the Oligocene. In: Henke, W., Tattersall, I. (Eds.), Handbook of Paleoanthropology, 2nd Edition. Springer, Heidelberg, pp. 1137–1260.

Godinot, M., Couette, S., 2008. Morphological diversity in the skulls of large adapines (Primates, Adapiformes) and its systematic implications. In: Sargis, E.J., Dagosto, M. (Eds.), Mammalian Evolutionary Morphology: A Tribute to Frederick S. Szalay, Springer, Dodrecht, pp. 285–314.

Godinot, M., Senut, B., Pickford, M., 2018. Primitive Adapidae from Namibia sheds light on the early primate radiation in Africa. Comm. Geol. Survey Namibia 18, 140–162.

Gregory, W.K., 1920. On the structure and relation of *Notharctus*, an American Eocene primate. Mem. Am. Mus. Nat. Hist. 351, 243.

Gunnell, G.F., 1997. Wasatchian-Bridgerian (Eocene) paleoecology of the western interior of North America: changing paleoenvironments and taxonomic composition of omomyid (Tarsiiformes) primates. J. Hum. Evol. 32, 105–132.

Gunnell, G.F., 2002. Notharctine primates (Adapiformes) from the early to middle Eocene (Wasatchian-Bridgerian) of Wyoming: Transitional species and the origins of *Notharctus* and *Smilodectes*. J. Hum. Evol. 43, 353–380.

Gunnell, G.F., Rose, K.D., 2002. Tarsiiformes: Evolutionary history and adaptation. In: Hartwig, W.C. (Ed.), The Fossil Primate Record. Cambridge University Press, Cambridge, pp. 45–82.

Gunnell, G.F., Rose, K.D., Rasmussen, D.T., 2007. Euprimates. In: Janis, C.M. Gunnell, G.F. Uhen, M.D. (Eds.), Evolution of Tertiary Mammals of North America, vol. 2. Cambridge University Press, Cambridge, pp. 239–261.

Gunnell, G.F., Gingerich, P.D., Ul-Haq, M., Bloch, J.I., Khan, I.H., Clyde, W.C., 2008. New primates (Mammalia) from the Early and Middle Eocene of Pakistan and their paleobiogeographical implications. Contributions from the Museum of Paleontology, University of Michigan 32 (1), 1–14.

Gunnell, G.F., Boyer, D.M., Friscia, A.R., Heritage, S., Manthi, F.K., Miller, E.R., Salam, H.M., Simmons, N.B., Stevens, N.J., Seiffert, E.R., 2018. Fossil lemurs from Egypt and Kenya suggest an African origin for Madagascar's aye-aye. Nat. Commun. 9 (3793), 1–12.

Harrington, A.R., Silcox, M.T., Yapuncich, G.S., Boyer, D.M., Bloch, J.I., 2016. First virtual endocasts of adapiform primates. J. Hum. Evol. 99, 52–78.

Harrison, T., 2010. Later Tertiary Lorisiformes (Strepsirrhini, Primates). In: Werdelin, L., Sanders, W.J. (Eds.), Cenozoic Mammals of Africa. University of California Press, Berkeley, pp. 333–349.

Heesy, C.P., 2009. Seeing in stereo: The ecology and evolution of primate binocular vision and stereopsis. Evol. Anthropol. 18, 21–35.

Heesy, C.P., Stevens, N.J., Samonds, K.E., 2006. Biogeographic origins of primate higher taxa. In: Lehman, S.M., Fleagle, J.G. (Eds.), Primate Biogeography: Progress and Prospects. Springer, New York, pp. 419–438.

Hooker, J.J., 2007. A new microchoerine omomyid (Primates, Mammalia) from the English Early Eocene and its paleobiogeographical implications. Paleontology. 50 (3), 739–756.

Hooker, J.J., Harrison, D.L., 2008. A new clade of omomyid primates from the European Paleogene. J. Vert. Paleontol. 28 (3), 826–840.

Jaeger, J.-J., Beard, K.C., Chaimanee, Y., et al., 2010. Late middle Eocene epoch of Libya yields earliest known radiation of African anthropoids. Nature 467, 1095–1098.

Kay, R.F., Kirk, E.C., 2000. Osteological evidence for the evolution of activity pattern and visual acuity in primates. Am. J. Phys. Anthropol. 113, 235–262.

Kay, R.F., Williams, B.A., Ross, C.F., Takai, M., Shigehara, N., 2004. Anthropoid origins: A phylogenetic analysis. In: Ross, C.F., Kay, R.F. (Eds.), Anthropoid Origins. New Visions, Kluwer Academics/Plenum Publishers, New York, pp. 91–135.

Kirk, E.C., Williams, B.A., 2011. New adapiform primate of Old World affinities from the Devil's Graveyard Formation of Texas. J. Human Evol. 61, 156–168.

Kirk, E.C., Kay, R.F., 2004. The evolution of high visual acuity in the Anthropoidea. In: Ross, C.F., Kay, R.F. (Eds.), Anthropoid Origins: New Visions, Kluwer Academic Press, New York, pp. 539–587.

Kirk, E.C., Daghighi, P., Macrini, T.E., Bullar, B.-A.S., Rowe, T.B., 2014. Cranial anatomy of the Duchesnean primate *Rooneyia viejaensis*: New insights from high resolution computed tomography. J. Hum. Evol. 74, 82–95.

Kirk, E.C., Lundeen, I., 2020. New observations of the nasal fossa and interorbital region of *Shoshonius cooperi* based on microcomputerized tomography. J. Hum. Evol. 141, 102748.

Krause, D.W., Maas, M.C., 1990. The biogeographic origins of Late Paleocene-Early Eocene mammalian immigrants to the Western Interior of North America. In: Bown, T.M. Rose, K.D. (Eds.), Dawn of the Age of Mammals in the Northern Part of the Rocky Mountain Interior, 243. Geol. Soc. Amer. Spec. Paper, North America, pp. 71–105.

Lundeen, I., Kirk, E.C., 2019. Internal nasal morphology of the Eocene primate *Rooneyia viejaensis* and extant Euarchonta: using μCT scan data to understand and infer patterns of nasal fossa evolution in primates. J. Hum. Evol. 132, 137–173.

Martin, C.J., Rose, K.D., Sylvester, A.D., 2022. A morphometric analysis of early Eocene euprimate tarsals from Gujarat, India. J. Hum. Evol. 164, 103141.

MacPhee, R.D.E., Jacobs, L.L., 1986. *Nycticeboides simpsoni* and the morphology, adaptations, and relationships of Miocene Siwalik Lorisidae. In: Flanagan, K.M., Lillegraven, J.A. (Eds.), Vertebrates,

Phylogeny, and Philosophy. University of Wyoming Contrib. Geol, Special Papers 3, pp. 131–162.

Maiolino, S., Boyer, D.M., Bloch, J.I., Gilbert, C.C., Groenke, J., 2012. Evidence for a grooming claw in a North American adapiform primate: Implications for anthropoid origins. PLoS One 7, e29135.

Marigó, J., Roig, I., Seiffert, E.R., Moyà-Sola, S., Boyer, D.M., 2016. Astragalar and calcaneal morphology of the middle Eocene primate *Anchomomys frontanyensis* (Anchomomyini): Implications for early primate evolution. J. Hum. Evol. 91, 122–143.

Marigó, J., Verrière, N., Godinot, M., 2019. Systematic and locomotor diversification of the *Adapis* group (Primates, Adapiformes) in the late Eocene of the Quercy (Southwest France), revealed by humeral remains. J. Hum. Evol. 126, 71–90.

Marivaux, L., Welcomme, J.L., Antoine, P.O., Metais, G., Baloch, I.M., Benammi, M., Chaimanee, Y., Ducroq, S., Jaeger, J.J., 2001. A fossil lemur from the Oligocene of Pakistan. Science 294, 587–591.

Marivaux, L., Chaimanee, Y., Tafforeau, P., Jaeger, J.J., 2006. New strepsirrhine primate from the Late Eocene of Peninsular Thailand (Krabi Basin). Am. J. Phys. Anthropol. 130, 425–434.

Marivaux, L., Beard, K.C., Chaimanee, Y., et al., 2008a. Anatomy of the bony pelvis of a relatively large-bodied strepsirrhine primate from the Late Middle Eocene Pondaung Formation (central Myanmar). J. Hum. Evol. 54, 391–404.

Marivaux, L., Beard, K.C., Chaimanee, Y., et al., 2008b. Proximal femoral anatomy of a sivaladapid primate from the Late Middle Eocene Pondaung Formation (Central Myanmar). Am. J. Phys. Anthropol. 137, 263–273.

Marivaux, L., Tabuce, R., Lebrun, R., Ravel, A., Adaci, M., Mahboubi, M., Bensalah, M., 2012. Talar morphology of azibiids, strepsirhine-related primates from the Eocene of Algeria: Phylogenetic affinities and locomotor adaptation. J. Hum. Evol. 61 (4), 447–457.

Martin, R.D., 2006. New light on primate evolution. Berichte und Abhandlungen der Berlin-Bradenburgischen Akademie der Wissenschaften 11, 379–405.

McCrossin, M., 1992. New species of bushbaby from the Middle Miocene of Maboko Island, Kenya. Am. J. Phys. Anthropol. 89, 215–234.

Mein, P., Ginsburg, L., 1997. Les mammifères du gisement miocène inférieur de Li Mai Long, Thailand: systématique, biostratigraphie et paléoenvironment. Geodiversitas 19, 783–844.

Miller, E.R., Gunnell, G.F., Martin, R.D., 2005. Deep time and the search for anthropoid origins. Ybk. Phys. Anthropol. 48, 60–95.

Morse, P.E., Chester, S.G.B., Boyer, D.M., Smith, T., Smith, R., Gigase, P., Bloch, J.I., 2019. New fossils, systematics, and biogeography of the oldest known crown primate *Teilhardina* from the earliest Eocene of Asia, Europe, and North America. J. Hum. Evol. 128, 103–131.

Ni, X., Wang, Y., Hu, Y., Li, C., 2004. A euprimate skull from the Early Eocene of China. Nature 427, 65–68.

Ni, X., Hu, Y., Wang, Y., Li, C., 2005. A clue to the Asian origin of euprimates. Anthropol. Sci. 113, 3–9.

Ni, X., Beard, K.C., Meng, J., Wang, Y., Gebo, D.L., 2007. Discovery of the first early Cenozoic euprimate (Mammalia) from Inner Mongolia. Am. Mus. Nov. 3571 (1), 11.

Ni, X., Meng, J., Beard, K.C., Gebo, D.L., Wang, Y., Li, C., 2010. A new tarkadectine primate from the Eocene of Inner Mongolia, China: Phylogenetic and biogeographic implications. Proc. R. Soc. B 277, 247–256.

Ni, X., Gebo, D.L., Dagosto, M., Meng, J., Tafforeau, P., Flynn, J.J., Beard, K.C., 2013. The oldest known primate skeleton and early haplorine evolution. Nature 498, 60–64.

Ni, X., Li, Q., Li, L., Beard, K.C., 2016. Oligocene primates from China reveal divergence between African and Asian primate evolution. Science 352, 673–677.

Perelman, P., Johnson, W.E., Roos, C., et al., 2011. A molecular phylogeny of living primates. PLoS Genet 7, e1001342.

Pickford, M., Senut, B., Morales, J., et al., 2008. Mammalia from the Lutetian of Namibia. Mem. Geol. Surv. Namibia 20, 465–514.

Qi, T., Beard, K.C., 1998. Late Eocene sivaladapid primate from Guangxi Zhuang Autonomous Region, People's Republic of China. J. Hum. Evol. 35, 211–220.

Ramdarshan, A., Merceron, G., Marivaux, L., 2012. Spatial and temporal diversity amongst Eocene primates of France: Evidence from teeth. Am. J. Phys. Anthropol. 147, 201–216.

Rasmussen, D.T., 1990. The phylogenetic position of *Mahgarita stevensi*: Protoanthropoid or lemuroid? Int. J. Primatol. 1B1, 439–469.

Rasmussen, D.T., 1994. The different meaning of a tarsioid-anthropoid clade and a new model of anthropoid origins. In: Fleagle, J.G., Kay, R.F. (Eds.), Anthropoid Origins. Plenum Press, New York, pp. 335–360.

Rasmussen, D.T., Nekaris, K.A., 1998. Evolutionary history of lorisiform primates. Folia Primatol. 69, 250–285.

Rose, K.D., 1994. The earliest primates. Ev. Anthropol. 3, 159–173.

Rose, K.D., Bown, T.M., 1993. Species concepts and species recognition in Eocene primates. In: Kimbel, W.H., Martin, L.B. (Eds.), Species, Species Concepts, and Primate Evolution. Plenum Press, New York, pp. 299–330.

Rose, K.D., Rana, R.S., Sahni, A., et al., 2009. Early Eocene primates from Gujarat, India. J. Hum. Evol. 56, 366–404.

Rose, K.D., Chester, S.G.B., Dunn, R.H., Boyer, D.M., Bloch, J.I., 2011. New fossils of the oldest North American euprimate *Teilhardina brandti* (Omomyidae) from the Paleocene-Eocene thermal maximum. Am. J. Phys. Anthropol. 146, 281–305.

Rose, K.D., Chew, A.E., Dunn, R., Kraus, M.J., Fricke, H., Zack, S.P., 2012. Earliest Eocene mammalian fauna from the Paleocene-Eocene Thermal Maximum at Sand Creek Divide, southern Bighorn Basin, 36. University of Michigan Papers on Paleontology, Wyoming. 1–122.

Rose, K.D., Dunn, R.H., Kumar, K., Perry, J.M.G., Prufrock, K.A., Rana, R.S., Smith, T., 2018. New fossils from Tadkeshwar Mine (Gujarat, India) increase primate diversity from the early Eocene Cambay Shale. J. Hum. Evol. 122, 93–107.

Rose, K.D., Perry, J.M.G., Prufrock, K.A., Weems, R.E., 2021. Early Eocene Omomyid from the Nanjemoy formation of Virginia: First fossil primate from the Atlantic Coastal Plain. J. Vert. Paleontology 41(1). https://doi.org/10.1080/02724634.2021.1923340.

Rosenberger, A.L., 1985. In favor of the *Necrolemur-Tarsier* hypothesis. Folia Primatol. 45, 179–194.

Rosenberger, A.L., 2011. The face of *Strigorhysis*: Implications of another tarsier-like, large-eyed Eocene North American tarsiiform primate. Anat. Rec. 294, 797–812.

Rosenberger, A.L., Hogg, R., Wong, S.M., 2008. *Rooneyia*, postorbital closure, and the beginnings of the age of Anthropoidea. In: Sargis, E.J., Dagosto, M. (Eds.), Mammalian Evolutionary Morphology: A Tribute to Frederick S. Szalay. Springer, Dordrecht, pp. 325–346.

Ross, C., 1994. The craniofacial evidence for anthropoid and tarsier relationships. In: Fleagle, J.G., Kay, R.F. (Eds.), Anthropoid Origins. Plenum Press, New York, pp. 469–547.

Ross, C.F., Martin, R.D., 2007. The role of vision in the origin and evolution of primates. In: Preuss, T.M. Kaas, J. (Eds.), Evolution of Nervous Systems, vol. 4. The Evolution of Primate Nervous Systems, Elsevier, Oxford, pp. 59–78.

Ross, C., Williams, B., Kay, R.F., 1998. Phylogenetic analysis of anthropoid relationships. J. Hum. Evol. 35, 221–306.

Rossie, J.B., Smith, T.D., 2007. Ontogeny of the nasolacrimal duct in primates: Functional and phylogenetic implications. J. Anat. 210, 195–208.

Rossie, J.B., Ni, X., Beard, K.C., 2006. Cranial remains of an Eocene tarsier. Proc. Natl. Acad. Sci. USA 103, 4381–4385.

Rossie, J.B., Smith, T.D., Beard, K.C., Godinot, M., Rowe, T.B., 2018. Nasolacrimal anatomy and haplorhine origins. J. Hum. Evol. 114, 176–183.

Rust, K., Ni, X., Tietjen, K., Beard, K.C., 2023. Phylogeny and paleobiography of the enigmatic North American primate Ekgmowechashala illuminated by new fossils from Nebraska (USA) and Guangxi Zhuang Autonomous Region (China). J. Hum. Evol. 185, 103452. https://doi.org/10.1016/j.jhevol.2023.103452

Samuels, J.X., Albright, L.B., Fremd, T.J., 2015. The last fossil primate in North America, new material of the enigmatic Ekgmowechashala from the Arikareean of Oregon. Am. J. Phys. Anthropol. 158, 43–54.

Schmid, P., 1983. Front dentition of the Omomyiformes (Primates). Folia Primatol. 40, 1–10.

Schwartz, G.T., Miller, E.R., Gunnell, G.F., 2005. Developmental processes and canine dimorphism in primate evolution. J. Hum. Evol. 48, 97–103.

Seiffert, E.R., 2007. Evolution and extinction of Afro-Arabian primates near the Eocene-Oligocene boundary. Folia Primatol. 78, 314–327.

Seiffert, E.R., 2012. Early primate evolution in Afro-Arabia. Evol. Anthropol. 21, 239–253.

Seiffert, E.R., Simons, E.L., Attia, Y., 2003. Fossil evidence for an ancient divergence of lorises and galagos. Nature 422, 421–424.

Seiffert, E.R., Simons, E.L., Ryan, T.M., Attia, Y., 2005. Additional remains of Wadilemur elegans, a primitive stem galagid from the Late Eocene of Egypt. Proc. Natl. Acad. Sci. USA 102 (32), 11396–11401.

Seiffert, E.R., Perry, J.M., Simons, E.L., Boyer, D.M., 2009. Convergent evolution of anthropoid-like adaptations in Eocene adapiform primates. Nature 461, 1118–1121.

Seiffert, E.R., Boyer, D.M., Fleagle, J.G., Gunnell, G.F., Heesy, C.P., Perry, J.M.G., Sallam, H.M., 2017. New adapiform primate fossils from the late Eocene of Egypt. Historical. Biology 30 (1–2), 204–226.

Silcox, M.T., Bloch, J.I., Boyer, D.M., et al., 2009. Semicircular canal system in early primates. J. Hum. Evol. 56, 315–327.

Simons, E.L., Rasmussen, D.T., 1996. A remarkable cranium of Plesiopithecus teras (Primates, Prosimii) from the Eocene of Egypt. Proc. Natl. Acad. Sci. USA 91, 9946–9950.

Simons, E.L., Rasmussen, D.T., Gingerich, P.D., 1995. New cercamonine adapid from Fayum, Egypt. J. Hum. Evol. 29, 577–589.

Smith, T., Rose, K.D., Gingerich, P.D., 2006. Rapid Asia-Europe-North America geographic dispersal of earliest Eocene primate Teilhardinia during the Paleocene-Eocene thermal maximum. Proc. Natl. Acad. Sci. USA 103, 11223–11227.

Smith, T., Kumar, K., Rana, R.S., Folie, A., Sole, F., Noiret, C., Steeman, T., Sahni, A., Rose, K.D., 2016. New early Eocene vertebrate assemblage from western India reveals a mixed fauna of European and Gondwana affinities. Geosci. Front. 7, 969–1001.

Soligo, C., Will, O.A., Tavare, S., Marchall, C.R., Martin, R.D., 2007. New light on the dates of primate origins and divergence. In: Ravosa, M.J., Dagosto, M. (Eds.), Primate Origins: Adaptations and Evolution. Springer, New York, pp. 29–50.

Spoor, F., Walker, A., Lynch, J., Liepins, P., Zonneveld, F., 1998. Primate locomotion and vestibular morphology, with special reference to Adapis, Necrolemur and Megaladapis. Am. J. Phys. Anthropol. Suppl. 26, 207.

Springer, M.S., Murphy, W.J., Eizirik, E., et al., 2007. A molecular classification for the living orders of plancental mammals and the phylogenetic placement of primates. In: Ravosa, M.J., Dagosto, M. (Eds.), Primate Origins: Adapations and Evolution. Springer, New York, pp. 1–28.

Steiper, M.E., Seiffert, E.R., 2012. Evidence for a convergent slowdown in primate molecular rates and its implications for the timing of early primate evolution. Proc. Natl. Acad. Sci. USA 109, 6006–6011.

Strait, S., 1991. Dietary Reconstruction of Small-Bodied Fossil Primates. Ph.D. Dissertation. Stony Brook University.

Szalay, F.S., Delson, E., 1979. Evolutionary History of the Primates. Academic Press, New York.

Tabuce, R., Marivaux, L., Lebrun, R., et al., 2009. Anthropoid versus strepsirhine status of the African Eocene primates Algeripithecus and Azibius: Craniodental evidence. Proc. R. Soc. B 276, 4087–4094.

Tavare, S., Marshall, C.R., Will, O., Soligo, C., Martin, R.D., 2002. Using the fossil record to estimate the age of the last common ancestor of extant primates. Nature 416, 726–729.

Thalmann, U., 1994. Die Primaten aus dem eozanen Geiseltal bei Halle/Saale (Deutschland). Courier Forschungsinstitut Senckenberg 175, 1–161.

Thalmann, U., Haubold, H., Martin, R.D., 1989. Pronycticebus neglectus – an almost complete adapid primate specimen from the Geiseltal (GDP). Palaeovertebrata 19 (3), 115–130.

Tornow, M.A., 2008. Systematic analysis of the Eocene primate family Omomyidae using gnathic and postcranial data. Bulletin of the Peabody Museum of Natural History 49 (1), 43–129.

Walker, A., Ryan, T.M., Silcox, M.T., Simons, E.L., Spoor, F., 2008. The semicircular canal system and locomotion: the case of extinct lemuroids and lorisoids. Evol. Anthropol. 17 (3), 135–145.

Williams, B.A., 1994. Phylogeny of the Omomyidae and Implications for Anthropoid Origins. Ph.D. Dissertation. University of Colorado.

Williams, B.A., Kirk, E.C., 2008. New Uintan primates from Texas and their implications for North American patterns of species richness during the Eocene. J. Hum. Evol. 55, 927–941.

Williams, B.A., Kay, R.F., Kirk, E.C., Ross, C.F., 2010. Darwinius masillae is a strepsirrhine – a reply to Franzen et al. (2009). J. Hum. Evol. 59 (5), 567–573.

Wu, R., Pan, Y., 1985. A new adapid primate from the Lufeng Miocene. Yunnana. Acta Anthropol. Sinica 4 (1), 1–6.

Zijlstra, J.S., Flynn, L.J., Wessels, W., 2013. The Westernmost tarsier: A new genus and species from the Miocene of Pakistan. J. Human Evol. 65, 544–550.

Further reading

Covert, H.H. The earliest fossil primates and the evolution of prosimians: introduction. In: Hartwig, W.C. (Ed.), The Primate Fossil Record, Cambridge University Press, Cambridge, pp. 13–20.

Gebo, D.L., 2002. Adapiformes: Phylogeny and adaptation. In: Hartwig, W.C. (Ed.), The Primate Fossil Record. Cambridge University Press, Cambridge, pp. 21–44.

Godinot, M., 1998. A summary of adapiform systematics and phylogeny. Folia Primatol. 69 (Suppl.1), 218–249.

Godinot, M., 2010. Paleogene prosimians. In: Werdelin, L., Sanders, W.J. (Eds.), Cenozoic Mammals of Africa. University of California Press, Berkeley, pp. 319–333.

Gunnell, G.F., Rose, K.D., 2002. Tarsiiformes: Evolutionary history and adaptation. In: Hartwig, W.C. (Ed.), The Fossil Primate Record. Cambridge University Press, Cambridge, pp. 45–82.

Gunnell, G.F., Rose, K.D., Rasmussen, D.T., 2007. Euprimates. In: Janis, C.M. Gunnell, G.F. Uhen, M.D. (Eds.), Evolution of Tertiary Mammals of North America, vol. 2. Cambridge University Press, Cambridge, pp. 239–261.

Harrison, T., 2010. Later tertiary lorisiformes (Strepsirrhini, Primates). In: Werdelin, L., Sanders, W.J. (Eds.), Cenozoic Mammals of Africa. University of California Press, Berkeley, pp. 333–349.

Phillips, E.M., Walker, A., 2002. Fossil lorisoids. In: Hartwig, W.C. (Ed.), The Fossil Primate Record. Cambridge University Press, Cambridge, pp. 83–95.

Rasmussen, D.T., 2007. Fossil record of primates from the Paleocene to the Oligocene. In: Henke, W. Tattersall, I. (Eds.), Handbook of Paleoanthropology, vol. 2. Springer-Verlag, Berlin, pp. 889–920.

Ravosa, M.J., Dagosto, M., 2007. Primate Origins: Adaptations and Evolution. Springer, New York.

Sargis, E.J., Dagosto, M., 2008. Mammalian Evolutionary Morphology: A Tribute to Frederick S. Szalay. Springer, Dordrecht.

13

Early Anthropoids

The Oligocene

Although the Oligocene Epoch proper is currently dated to between approximately 34 million and 23 million years ago, the Eocene-to-Oligocene transition involved a long series of geological, climatic, and paleontological changes that took place over about 10 million years. Some of these changes took place in the Late Eocene, others in the early part of the Oligocene, and many in a series of steps. By the Oligocene, the continents were beginning to look as they do today, except for the lack of a connection between North America and South America. Arabia was a part of Africa, and Asia was apparently a collection of islands. India had joined the Asian mainland, closing off the Tethys Seaway on the east, and both South America and Australia were separated from Antarctica. These last events made possible the first deep water currents around Antarctica. As a result, the Eocene-to-Oligocene transition was marked by a major drop in global temperatures from the more tropical climates of the Early and Middle Eocene. The Early Oligocene also saw a dramatic lowering of sea level as a result of glaciations at the poles. These climatic changes led to equally dramatic changes in the primate fossil record. In the northern hemisphere, the prosimians that had been abundant in the Eocene disappeared by the beginning of the Oligocene in Europe, and they became increasingly rare in North America, leaving them nearly extinct on northern continents by the Oligocene. Anthropoid primates first appear in the Eocene of Africa and Asia. In the Middle and Late Eocene, they are often found alongside various prosimians on both continents. However, by the beginning of the Oligocene, anthropoids had become dominant in Africa, where we have the most extensive record of early anthropoid evolution (Fig. 13.1).

Early Anthropoids From Africa and Arabia

For over a century, early fossil anthropoids have been known from an area in Egypt known as the Fayum Depression. That region has yielded an abundance of exciting and often relatively complete fossil material, documenting a tremendous diversity of evolving forms from the Late Eocene and Early Oligocene (e.g., Seiffert et al., 2010a,b; Seiffert, 2012). Here, in an expanse of eroded badlands on the eastern edge of the Sahara Desert (Fig. 13.2), is a sequence of highly fossiliferous sedimentary deposits (Fig. 13.3) that preserve the best

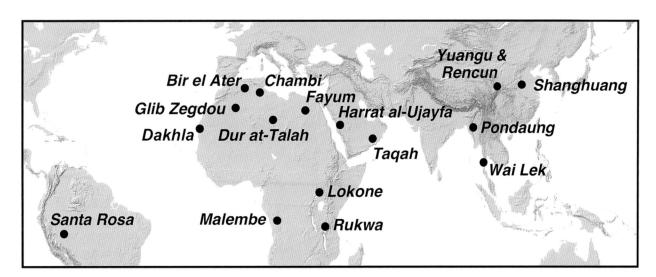

FIGURE 13.1 Geographic distribution of early fossil anthropoids. Map lines delineate study areas and do not necessarily depict accepted national boundaries.

FIGURE 13.2   The desert landscape of the Fayum Depression of Egypt.

record of early mammalian evolution in Africa (e.g., Fleagle and Gilbert, 2008; Werdelin and Sanders, 2010).

We know a great deal about the environment under which the sediments and the fossil primates within them were deposited (Fig. 13.3). From the sediments, we know that the climate was warm, wet, and somewhat seasonal. The fossil plants are most similar to species currently found in the tropical forests of Southeast Asia. The quarries yielding the primate fossils are very diverse. Some were laid down as sandbars in river channels and showed repeated sequences of standing water (probably oxbow lakes) as well as roots of mangrove-like plants. This evidence, together with abundant fossil remains of water birds, indicates a swampy environment at the time of deposition; others were probably quicksands that were dry in some seasons.

Almost all of the mammals that are found with the primates are different from Eocene or Oligocene mammals in other parts of the world and from the well-known Miocene African faunas from Kenya and Uganda. The rodents are the earliest members of the cane rat suborder (hystricognaths), which includes the guinea pig-like rodents of South America as well as African porcupines. There were also opossums as well as tenrecs, bats, primitive carnivorous mammals called hyaenodonts, and anthracotheres, an archaic group of artiodactyls that might be related to the hippopotamus. In addition, the Fayum provides the first substantial record of several African groups of mammals, such as hyraxes, elephants, and elephant-shrews, as well as the numerous groups of fossil primates.

As a result of research by Elwyn Simons, Erik Seiffert, and colleagues over the past six decades, the Fayum has yielded one of the most diverse fossil primate faunas from anywhere in the world (Fig. 13.4). There are numerous genera of prosimians, including caenopithecine adapoids, stem strepsirrhines, several lorisoids, and the tarsier-like *Afrotarsius* (see Chapter 12). In addition, there are four major groups of higher primates: parapithecoids, proteopithecids, propliopithecids, and oligopithecids, as well as a few taxa whose affinities are less clear (Seiffert et al., 2010a,b). This diverse array provides many insights into the initial radiation and diversification of higher primates (Seiffert, 2012).

Primate fossils come from two geological formations: the Birket Qarun Formation, which is early Late Eocene in age, approximately 37 Ma, and the younger Jebel Qatrani Formation, which spans the Latest Eocene through Early Oligocene (Fig. 13.3; Seiffert et al., 2005, 2010a,b; Seiffert 2006, 2007, 2012). The fossil primates have been recovered from three levels within the Jebel Qatrani Formation. In the uppermost level (Quarries M, I), primates are the most common mammals, and one species, *Apidium phiomense*, is known from hundreds of fossils. The middle level (Quarries G, V) has yielded relatively few primates, but these may contain some crucial intermediate species. The lower sequence of the Jebel Qatrani Formation, dated to the Latest Eocene and Earliest Oligocene, contains Quarries E and L-41, which have yielded different genera from the higher levels, many known from cranial material.

More recently, the Birket Qarun Formation, Quarry BQ-2, approximately 37 Ma (e.g., Seiffert et al., 2008) has produced many new species of both prosimians and anthropoids, and many more are waiting to be described and analyzed.

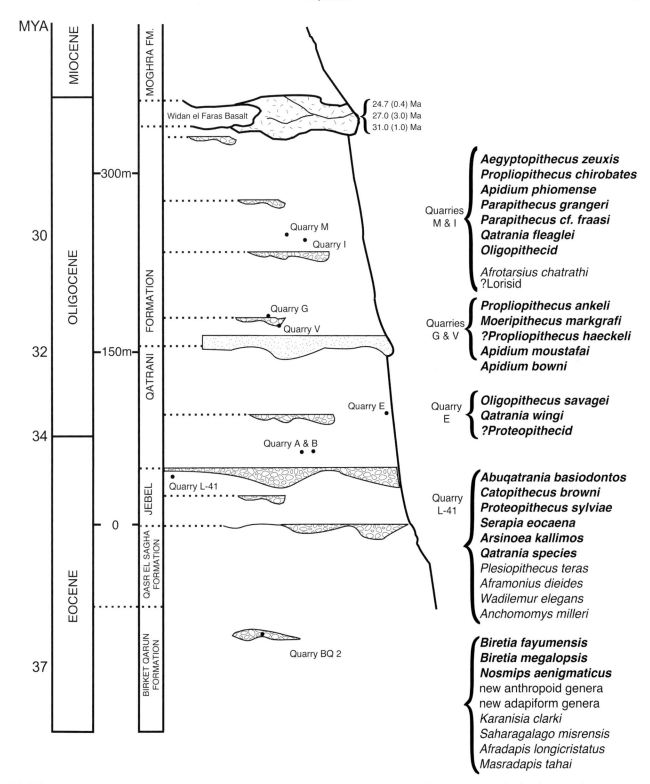

FIGURE 13.3 Stratigraphic section of the fossil-bearing sediments of the Fayum region of Egypt showing the fossil taxa and major quarries. Anthropoids are in bold type.

Parapithecoids

Although parapithecoids were first discovered near the turn of the twentieth century, an appreciation of the diversity of this group has come only in recent decades. There are over ten species of parapithecoids from the Fayum, one from Algeria, one from Morocco, two from Libya, one from Kenya, perhaps one from Tanzania, and

FIGURE 13.4   A reconstruction of the Early Oligocene environment and fauna from the Fayum.

several undescribed parapithecoids from Oman (Seiffert et al., 2010a,b; Seiffert, 2012; Mattingly et al., 2021). In addition, there is a recently recovered parapithecoid from Peru, in South America. Parapithecoid species range in size from tiny, marmoset-sized species such as *Qatrania*, which are among the smallest Old World higher primates, to the guenon-sized *Parapithecus grangeri* (Table 13.1). These Late Eocene and Early Oligocene anthropoids are among the most primitive of all known higher primates, and they have a number of anatomical features that distinguish them from all other Old World anthropoids.

Most parapithecoids have a primitive dental formula of 2.1.3.3, as in New World monkeys. This is probably the primitive dental formula for all higher primates. It has been retained by most New World monkeys, but it is not found in any living catarrhines. In the best-known species, *Apidium phiomense*, the lower incisors are small and spatulate (Fig. 13.5), but one species, *Parapithecus grangeri*, lost its permanent lower incisors altogether (Fig. 13.7). Upper incisors are poorly known. The canines in *Apidium* are similar to those of most platyrrhines, but in *P. grangeri*, they are large and tusk-like. The three lower premolars increase in size and complexity from front to back, but in all species, the last premolar is

relatively simple compared to that of extant anthropoids, with a metaconid that is smaller and distally positioned relative to the protoconid. The upper premolars of parapithecoids are broad, usually with three cusps rather than two as in other higher primates.

Parapithecoid molars (Figs. 13.6 and 13.7) are generally characterized by low, rounded cusps. The upper molars are quadrate, with well-developed conules and a large hypocone. The lower molars have a small trigonid (sometimes with a paraconid on M_1) and a broad talonid basin. In some species, accessory cusps are common, and often, there is a buccolingual alignment of the molar cusps and a narrowing in the center of the tooth, giving parapithecoid molars a "waisted" shape, superficially similar to that seen in cercopithecoid monkeys. The mandible is fused at the symphysis in most species but apparently not in *Biretia*.

The skull of parapithecoids is known from only a few relatively complete remains, but these clearly show higher primate features, such as fused frontal sutures and postorbital closure. The auditory region in parapithecoids is poorly known but seems to be characterized by a large promontory artery and anterior accessory cavity, as in anthropoids (and tarsiers), and the lack of a tubular ectotympanic.

TABLE 13.1 Suborder ANTHROPOIDEA
Superfamily PARAPITHECOIDEA

Species	Estimated mass (g)
Family PARAPITHECIDAE	
Abuqatrania (Late Eocene, Egypt)	
A. basiodontos	341
Qatrania (Late Eocene to Early Oligocene, Egypt)	
Q. wingi	242
Q. fleaglei	510
Ucayalipithecus (Late Eocene to Early Oligocene, Peru)	
U. perdita	320–365
Apidium (Early Oligocene, Egypt and Libya)	
A. phiomense	1687
A. moustafai	859
A. bowni	525
A. zuetina	1000
Lokonepithecus (Late Oligocene, Kenya)	
L. manai	695
Parapithecus (Early Oligocene, Egypt)	
P. fraasi	1768
P. grangeri	3161
P. harujensis	1000–1500
Family INCERTAE SEDIS	
Biretia (Eocene, Egypt, Libya, Algeria)	
B. piveteaui	464
B. fayumensis	273
B. megalopsis	381
Arsinoea (Late Eocene, Egypt)	
A. kallimos	552
Superfamily INCERTAE SEDIS **Family PROTEOPITHECIDAE**	
Proteopithecus (Late Eocene, Egypt)	
P. sylviae	800
Serapia (Late Eocene, Egypt)	
S. eocaena	1029
Superfamily INCERTAE SEDIS **Family INCERTAE SEDIS**	
Nosmips (Late Eocene, Egypt)	
N. aenigmaticus	?2500

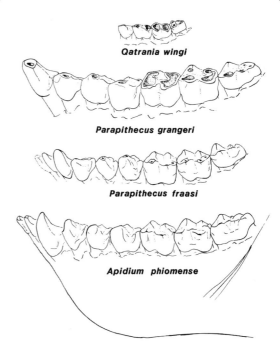

FIGURE 13.5 Mandibles of parapithecids showing proportions of anterior and posterior teeth. Note that *Parapithecus grangeri* lacks lower incisors.

Many elements of the limb skeleton have been recovered for one species, *Apidium phiomense* (Fig. 13.9). In many features of their limbs, parapithecoids are more primitive than any later Old World higher primates and resemble platyrrhines and omomyoids, or are unique among primates (Fleagle and Simons, 1995). In *Apidium*, the tibia and fibula are joined for approximately 40% of their length, a similarity to some microchoerines, some platyrrhines, and tarsiers.

Biretia is a tiny, primitive parapithecoid known from the Late Eocene of Egypt, Libya, and Algeria. The type species, *B. piveteaui*, is based on a single lower molar from the site of Bir el Ater in northern Algeria (de Bonis et al., 1988), but additional specimens have been attributed to this species from the Eocene of Libya (Jaeger et al., 2010). Two other species, *B. fayumensis* and *B. megalopsis*, are from the Late Eocene BQ2 quarry in the Fayum. Based on the Fayum material, it is evident that *Biretia* is more primitive dentally than other parapithecoids. In the larger *B. megalopsis*, the mandibular symphysis was unfused and the floor of the orbit and the palate were fused, a feature otherwise only known in tarsiers. It is not possible to determine whether *Biretia* had postorbital closure or whether it was diurnal or nocturnal. The cusps on the teeth of *Biretia* are less bulbous than those of later parapithecoids (Figs. 13.6 and 13.7), suggesting they were probably more insectivorous than later members of the superfamily.

Abuqatrania basiodontos is a small parapithecid best known from Quarry L-41 in the lower part of the Jebel Qatrani Formation (Simons et al., 2001a). It is mainly represented by lower cheek teeth (Fig. 13.6). The lower dentition is more primitive than that of most later

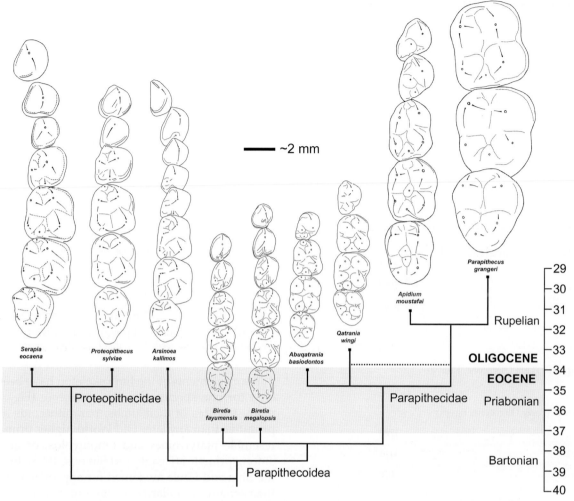

FIGURE 13.6   Lower dentitions of parapithecoids and proteopithecids in a phylogenetic context. *(Courtesy of Erik Seiffert.)*

parapithecids in having a small paraconid on P₄–M₃. The bulbous molars suggest a diet of fruit and/or gums (Kirk and Simons, 2001).

Qatrania wingi, from Quarry E in the lower part of the Jebel Qatrani Formation, is another very early and primitive parapithecid (Simons and Kay, 1983; Figs. 13.5 and 13.6). This tiny primate (less than 300 g) is known from only two lower jaws and a few isolated teeth. *Qatrania wingi* is more derived than *Abuqatrania* in having a paraconid only on the first molar. The absence of shearing crests on the teeth indicates that its diet was probably fruits or gums rather than insects (Kirk and Simons, 2001). A second species of *Qatrania*, *Q. fleaglei*, is from the upper part of the Jebel Qatrani Formation, indicating a long temporal range for this genus.

Ucayalipithecus is a *Qatrania*-like parapithecid taxon from Peru, in South America (Seiffert et al., 2020) (Fig. 13.8), demonstrating that parapithecids had a range that extended across the Atlantic Ocean to another continent.

There are four species of *Apidium*, the best-known parapithecid. Two smaller species, **A. bowni** (Quarry V) and **A.** *moustafai* (Quarry G), are from the intermediate levels of the Jebel Qatrani Formation; the larger **A.** *phiomense* is from the upper part of the formation. There is also a medium-sized species, *A. zuetina*, from Libya at a site estimated to be equivalent to Quarries G and V. *Apidium phiomense* is known from hundreds of specimens, including jaws, limb bones, and a partial cranium. *Apidium* has tiny incisors; moderate-sized, sexually dimorphic canines; and molars with numerous low, rounded cusps and few shearing crests (Figs. 13.5–13.7). The molars of *A. bowni* are more primitive than those of other *Apidium* species in having paraconids on m1 and m2. All species of *Apidium* have a fused mandibular symphysis. Functionally, the teeth indicate a predominantly frugivorous diet, but the very thick enamel on the molars suggests that seeds may also have been an important dietary component. The canine dimorphism, unusual in a primate this small, suggests that *Apidium* lived in polygynous social groups.

The few cranial remains of *Apidium* show a short snout, a small infraorbital foramen, and relatively small eyes (Fig. 13.9). It was a diurnal monkey. The most

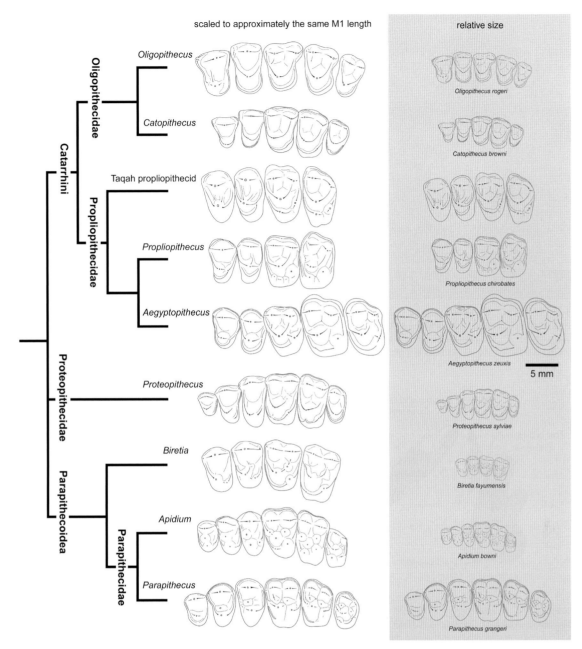

FIGURE 13.7 Upper dentitions of parapithecoids, proteopithecids, oligopithecids, and propliopithecids in a phylogenetic context. *(Courtesy of Erik Seiffert.)*

complete cranial remains are from *A. bowni* (Simons, 1995). In this species, the orbits seem to be less convergent than in other anthropoids (Simons, 2008).

The many postcranial bones attributed to *Apidium* show it was an excellent leaper (Fig. 13.10; but see Ryan et al., 2012). The hindlimb is relatively long compared with the forelimb (intermembral index = 70), the ischium is extremely long, the femoral neck is oriented at a right-angle to the shaft, and the distal femoral condyles are very deep, more so than in any other higher primate. The tibia is extremely long and laterally compressed, and the fibula is tightly attached to it distally. The ankle joint is hinged for rapid flexion and extension. *Apidium* probably had a divergent hallux. The scapula is similar to that seen in many living anthropoid quadrupedal leapers, such as *Saimiri*, and the short forelimb bones indicate quadrupedal rather than clinging habits. In many details of limb structure, *Apidium* shows greater similarities to platyrrhines and to Eocene prosimians (including omomyids and microchoerids) than to later Old World anthropoids (Fleagle and Simons, 1995).

Lokonepithecus manai is a parapithecid from the Oligocene of the Turkana Basin in Northern Kenya (Ducrocq et al., 2011). *Lokonepithecus* was small, with an

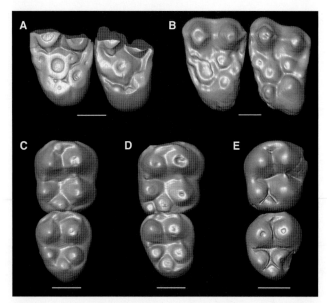

FIGURE 13.8 Upper and lower dentition of *Ucayalipithecus perdita* compared to other parapithecids from the Fayum. A, *U. perdita* M¹-M², B, *Apidium bowni* M¹-M², C, *Abuqatrania basiodontos* M₂-M₃, D, *Qatrania wingi* M₂-M₃, E, *U. perdita* M₂-M₃. Scale bars = 1mm. *(Courtesy of Erik Seiffert.)*

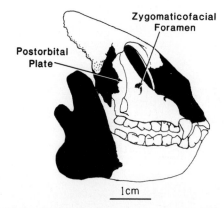

FIGURE 13.9 A reconstructed facial skeleton of *Apidium phiomense*. *(After Simons, E.L., 1971. A current review of the interrelationships of Oligocene and Miocene Catarrhini. In: Dahllery, A.A. [Ed.], Dental Morphology and Evolution, University of Chicago Press, Chicago, pp. 193–208.)*

estimated mass under 700 g, and probably fed mainly on fruits. This species, along with other, unnamed parapithecid fossils from Tanzania (Stevens et al., 2005), demonstrates that parapithecoids were not limited to North Africa but were also present in sub-Saharan Africa during the Paleogene.

The most unusual parapithecid, and also the largest, is **Parapithecus grangeri** with an estimated body size around 3000 g. Like *Apidium phiomense*, *P. grangeri* is from the upper levels of the Jebel Qatrani Formation. This species is sometimes placed in a separate genus, *Simonsius*. Like other parapithecoids, this species has three premolars and three molars. There is no paraconid on any of the molars, and the anterior and posterior

cusps, on the lower molars are superficially similar to the condition found in cercopithecoid monkeys (Fig. 13.6). The lower premolars are short with bulbous cusps, and the upper premolars have three prominent cusps, although the upper P² is very small (Fig. 13.7). It is evident from preserved parts of the premaxilla that *P. grangeri* had an upper canine and two upper incisors on each side. A most unusual feature of this species is the anterior dentition of the mandible, which demonstrates large, tusk-like canines and no permanent incisors (Kay and Simons, 1983; Simons, 1986; Fig. 13.5). The function of this tusk-like arrangement is unclear, but it seems likely that *P. grangeri* was partly folivorous or perhaps a seed predator.

There is a nearly complete cranium for *P. grangeri* (Fig. 13.11; Simons, 2001). This large parapithecoid had a prominent snout with relatively small orbits that were more laterally directed than in other anthropoids. The size of the orbits suggests diurnal habits, and the relative size of the optic foramen indicates that *Parapithecus* had high visual acuity like extant anthropoids (Kirk and Kay, 2004). *P. grangeri* had a relatively small brain for its body size, more similar to extant strepsirrhines than anthropoids. There are few limb elements attributed to *P. grangeri*. *Parapithecus* (or *Simonsius*) *harujensis* is a similar, smaller species from Libya (Mattingly et al., 2021).

Although it was one of the first parapithecids named, **Parapithecus fraasi** remains a poorly understood species. The type specimen was described early in the 1900s from an unknown site in the Fayum, and this medium-sized species is still definitively known only from the type specimen and a few jaws. The dental formula of *P. fraasi* has been debated since its initial discovery. The type specimen has only a small lateral incisor on each side. However, since *P. grangeri* lacks permanent incisors altogether, it is possible that *P. fraasi* also lacked permanent incisors and that the tiny anterior teeth preserved in the type specimen are deciduous incisors. More complete fossils are needed to resolve this question. *P. fraasi* has distinct, rounded cusps on its molars, suggesting a frugivorous diet, relatively simple premolars, and a reduced third molar.

Arsinoea (Fig. 13.6) is a Late Eocene genus from quarry L-41 in the lowest level of the Jebel Qatrani Formation. It has a dental formula of 2.1.3.3 and is known only from lower teeth. The premolar morphology of *Arsinoea* is similar to that of parapithecids, but some features of its molar morphology are not found in other parapithecoids, including the primitive *Biretia* (Seiffert et al., 2010a,b). The cheek teeth of *Arsinoea* are extremely flat and crenulated, suggesting a diet of fruit (Kirk and Simons, 2001). Overall, based on the limited material currently available, *Arsinoea* does not fit well

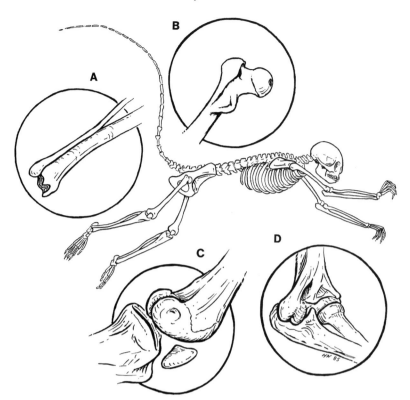

FIGURE 13.10   A restored skeleton of *Apidium phiomense* showing many of the distinctive features of the species: (A), the tibia and fibula nearly fused for the distal-most 40% of their length; (B), the large lesser trochanter of the femur; (C), the deep distal condyles of the femur; (D), the entepicondylar foramen and elongate capitulum of the humerus. In these features, *Apidium* is more like omomyoid prosimians and small platyrrhines than modern catarrhines.

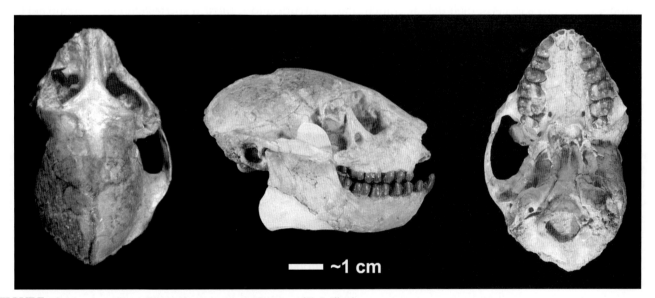

FIGURE 13.11   A cranium of *Parapithecus grangeri*. (Courtesy of E. Seiffert.)

with any of the other Fayum anthropoids but is best placed *Incertae sedis* among the parapithecoids.

Parapithecoids are among the earliest and most primitive fossil higher primates. Compared to extant anthropoids, they have many primitive features in their dentition, including three relatively simple lower premolars and occasional paraconids on their lower molars. Many skeletal features of *Apidium*, such as the lack of expanded ischial tuberosities, a large lesser trochanter, deep condyles on the femur, and the retention of an entepicondylar foramen on the humerus, are also primitive features not found in most later Old World higher primates. The arrangement of the cranial bones on the skull wall in *Apidium*, with the zygomatic bone contacting the temporal bone at pterion, and the morphology of the ear region seem to be similar to those in platyrrhines. Other parapithecids (e.g., *P. grangeri*) display the frontal-sphenoid condition at pterion, which has been argued to be the primitive condition for anthropoids (Simons, 2001, 2004; Fulwood et al., 2016).

Many authorities have noted that parapithecids share features with the platyrrhines of South America. Although most of the similarities between parapithecids and platyrrhines are probably primitive anthropoid features, the recent discovery of the *Qatrania*-like *Ucayalipithecus* in Peru demonstrates that parapithecids had crossed the South Atlantic to South America. At present, it isn't clear how broad their radiation on that continent was and what relationship, if any, they may have had in the origin of platyrrhines (also discussed in Chapter 14).

Proteopithecids

Proteopithecidae (Table 13.1) consists of two genera from the Late Eocene L-41 Quarry in the Fayum that share a primitive dental formula of 2.1.3.3 and some features of their postcranial skeleton with parapithecoids, and are probably near the origin of that group, but clearly differ from parapithecoids and from other early anthropoids in many aspects of their dental, cranial, and postcranial anatomy (e.g., Seiffert et al., 2010a,b, 2020).

Proteopithecus sylviae is known from numerous dentitions, two skulls, and parts of the skeleton (Miller and Simons, 1997; Simons, 1997; Simons et al., 1999; Simons and Seiffert, 1999). Dentally, *Proteopithecus* is characterized by very small incisors relative to the cheek teeth, sexually dimorphic canines, a relatively short P_4 similar to that of extant anthropoids, and moderately bunodont molars with relatively tall trigonids. There is a paraconid on M_1 and occasionally on the posterior molars (Fig. 13.6). The upper dentition is characterized by premolars that increase dramatically in size posteriorly and upper molars that are broad with prominent hypocones on M^1 and M^2 (Fig. 13.7). Upper and lower third molars are reduced in size. It probably had a diet of fruit and insects (Kirk and Simons, 2001). The cranium of *Proteopithecus* has relatively small orbits that suggest diurnal activity and are more convergent than those of parapithecoids. There is complete postorbital closure, and the optic foramen suggests that *Proteopithecus* had high visual acuity like extant anthropoids (Kirk and Kay, 2004).

The postcranial skeleton of *Proteopithecus* is known from a humerus, a femur, a tibia, and a talus. In most morphological features, the limbs of *Proteopithecus* are similar to those of a small platyrrhine. It probably moved by quadrupedal running and leaping in the trees, but *Proteopithecus* lacked many of the extreme adaptations for leaping found in *Apidium*, such as the very deep distal femoral condyles and the extensive apposition of the tibia and fibula.

Serapia eocaena is also from the Late Eocene Quarry L-41. Only known from lower jaws, *Serapia* has a dental formula of ?.1.3.3. Its molar teeth are broader and more bunodont than those of *Proteopithecus* but are overall very similar (Fig. 13.6), suggesting that the two taxa are closely related (e.g., Seiffert, 2012). It was predominantly frugivorous (Kirk and Simons, 2001).

Propliopithecids

The best-known group of early anthropoids from the Fayum, the propliopithecids, were as large as, or larger than, the largest parapithecid. They have a dental formula of 2.1.2.3 and a dental morphology more like that of later apes than cercopithecoid monkeys in that they lack bilophodont molars (Fig. 13.12). However, in details of their dental, cranial, and postcranial anatomy, they are more primitive than any living catarrhines. There are three genera (Table 13.2).

The first fossil "ape" described from the Fayum was **Propliopithecus**. There are three species from Egypt: *P. haeckeli*, the type species, described early in the twentieth century; *P. chirobates*, from the uppermost levels of the Fayum; and *P. ankeli*, a species with large premolars from the middle levels. *Propliopithecus chirobates*, the best-known species, is a medium-sized (4 kg) anthropoid. *Propliopithecus* has relatively broad, spatulate lower incisors with large, sexually dimorphic canines and anterior premolars (Figs. 13.12 and 13.13). As in most living anthropoids, the anterior lower premolar shears against the posterior surface of the upper canine to sharpen it, and this tooth appears to be sexually dimorphic in propliopithecids in association with the canine dimorphism (Fig. 13.12); the posterior premolar is semimolariform, with protoconids and metaconids of equal size. The lower molars resemble those of later apes in that they are formed by a broad talonid basin surrounded by five rounded cusps. There is usually no paraconid, and the trigonid is small. The three lower molars are similar in size. The upper premolars are bicuspid, and the upper molars are broad and quadrate, with a small hypocone connected to a pronounced lingual cingulum (Fig. 13.7). There are no conules or stylar cusps on the upper molars. The simple molars with low, rounded cusps and the broad incisors suggest that *Propliopithecus* was frugivorous.

There are no described cranial remains of *Propliopithecus*. Several isolated limb elements indicate that *Propliopithecus* was an arboreal quadruped with a strong grasping foot and was probably capable of some hindlimb suspension.

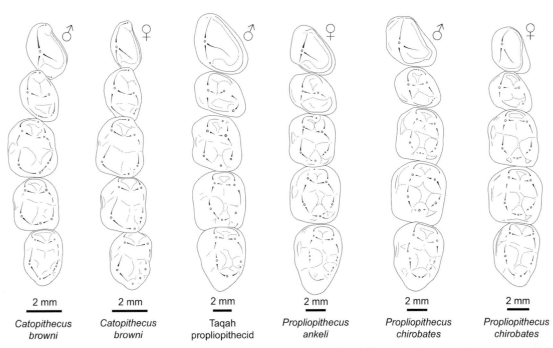

2 mm   **2 mm**   **2 mm**   **2 mm**   **2 mm**   **2 mm**

Catopithecus browni   *Catopithecus browni*   Taqah propliopithecid   *Propliopithecus ankeli*   *Propliopithecus chirobates*   *Propliopithecus chirobates*

FIGURE 13.12 Lower dentitions of propliopithecids and oligopithecids. Note the difference in the size of the anterior premolar between likely male and female specimens.

TABLE 13.2   Infraorder CATARRHINI
Superfamily PROPLIOPITHECOIDEA

Species	Estimated mass (g)
Family PROPLIOPITHECIDAE	
Propliopithecus (Early Oligocene, Egypt, Oman)	
P. haeckeli	4064
P. chirobates	4259
P. ankeli	5685
Moeripithecus (Early Oligocene, Egypt)	
M. markgrafi	4000
Aegyptopithecus (Early Oligocene, Egypt)	
A. zeuxis	6716
Family OLIGOPITHECIDAE	
Oligopithecus (Late Eocene, Egypt, Oman)	
O. savagei	1602
O. rogeri	1745
Catopithecus (Late Eocene, Egypt)	
C. browni	1088
Talahpithecus (Eocene, Libya)	
T. parvus	?370

A closely related taxon is **Moeripithecus markgrafi**, first described over 100 years ago and sometimes placed in *Propliopithecus*. It is known from a single specimen with two lower teeth from an unknown locality in the Fayum.

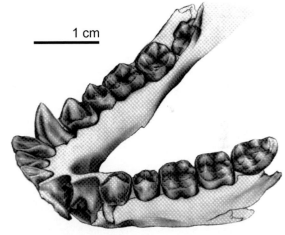

1 cm

FIGURE 13.13 A mandible of *Propliopithecus chirobates*. Note the dental formula of 2.1.2.3, the ape-like arrangement of cusps on the lower molars, the elongate anterior premolar that shears against the upper canine, and the fused mandibular symphysis. *(Courtesy of R. F. Kay.)*

In addition to the many fossils of *Propliopithecus* from the Fayum, there is a large specimen from Taqah in Oman (Fig. 13.12). It was identified by its discoverers as *Moeripithecus* (Thomas et al., 1991) but is more likely close to *Propliopithecus ankeli* (Seiffert et al., 2010a,b; Seiffert, 2012).

Aegyptopithecus zeuxis, from the uppermost quarries in the Fayum, was a much larger animal (6–8 kg) than *P. chirobates* and is one of the best known of all fossil anthropoids (Figs. 13.6, 13.14–13.16). Dentally,

Aegyptopithecus differs from *Propliopithecus* in having narrower incisors, lower molars with larger cusps and a more restricted talonid basin, and upper molars with better-developed conules and stylar cusps. In contrast to *Propliopithecus*, in which the three lower molars are similar in length, the molars of *Aegyptopithecus* increase in size posteriorly. Overall, the dental differences suggest that *A. zeuxis* was largely frugivorous but probably more folivorous than *Propliopithecus*. Like *Propliopithecus*, *Aegyptopithecus* has sexually dimorphic canines and lower anterior premolars; it probably lived in polygynous social groups.

The cranial anatomy of *Aegyptopithecus* is more primitive than that of any later Old World anthropoid but more advanced than that of any prosimian (Fig. 13.14). The skull resembles other anthropoids in that the lacrimal bone lies within the orbit, and the relatively small orbits (indicating diurnal habits) are completely walled-off posteriorly, with a bony configuration similar to that found in extant catarrhines rather than in platyrrhines. *Aegyptopithecus* has a premaxillary bone that is very large for an anthropoid. There was considerable sexual dimorphism in cranial morphology in *Aegyptopithecus*,

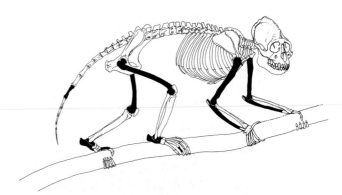

FIGURE 13.15 A reconstructed skeleton of *Aegyptopithecus zeuxis* showing, in black, the bones that have been recovered for this species.

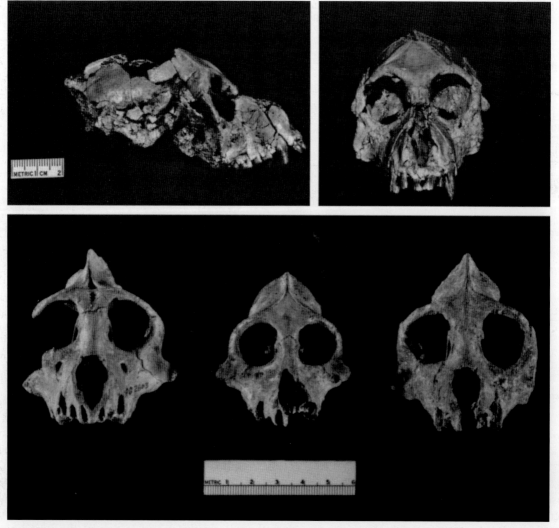

FIGURE 13.14 Cranial remains of *Aegyptopithecus zeuxis*. Note the long snout, small orbits, sagittal and nuchal crests, and converging temporal lines in older individuals. *(Courtesy of E. L. Simons.)*

FIGURE 13.16 Three Fayum anthropoids: *above*, the propliopithecids *Aeygtopithecus zeuxis* (*left*) and *Propliopithecus chirobates* (*right*); *below*, the parapithecid *Apidium phiomense*.

with females having a much more gracile cranium than males (Simons et al., 2001b). In males, the superficial cranial morphology changes dramatically with age (Simons, 1987; Fig. 13.14). In young individuals, there is a broad frontal bone. Older individuals develop a pronounced sagittal crest that divides anteriorly and extends over the brow ridges to form an increasingly narrow frontal trigon. There is also a large nuchal crest along the posterior border of the occiput. This suggests a pattern of differing cranial maturation in males and females (Sanders and Gunnell, 2012). The auditory region is like that of other anthropoids in having an anterior accessory cavity and in lacking a stapedial artery. As with all of the Fayum anthropoids, the ectotympanic is most similar to that in platyrrhines, with a bony ring fused to the lateral surface of the bulla and no bony tube. However, the venous drainage of the cranial cavity appears to be like that of later catarrhines (Kay et al., 2008).

The brain of *Aegyptopithecus* was relatively small compared to the brains of living anthropoids and more like a prosimian brain. However, compared with contemporaneous Oligocene mammals or Eocene prosimians, it was relatively large, with an expanded parietal region (Simons, 1993).

The forelimb of *Aegyptopithecus* is known from the humerus and ulna, and the hindlimb is known from the femur, talus, calcaneus, a first metatarsal, and some phalanges (Fig. 13.15). All of these elements indicate that *Aegyptopithecus* was a robust arboreal quadruped (Fig. 13.16). The foot bones indicate that it had a grasping hallux and was capable of considerable inversion of the foot. In many anatomical details, the limb elements of *Aegyptopithecus* are more similar to those of platyrrhines and prosimians than to those of either living apes or cercopithecoid monkeys. This early anthropoid retained many primitive features of its limbs which were lost in later catarrhines, such as an entepicondylar foramen on the humerus, a very long olecranon on the ulna, and a large third trochanter on the femur.

Propliopithecids have gross dental similarities to living hominoids and to later fossil apes from Europe and Africa, and for much of the twentieth century, they were considered to be primitive apes. The similarities to living apes are, however, primitive anthropoid features rather than specializations, and increasing knowledge of their cranial and postcranial anatomy has shown that these early anthropoids are stem catarrhines. They have all of the characteristic features of anthropoids (fused mandibular symphysis, postorbital closure, lacrimal bone within the orbit) and are linked with living catarrhines by their dental formula of 2.1.2.3 and a few postcranial features. However, in the anatomy of their auditory region and limbs, they lack common specializations found both in living apes and in living Old World monkeys, and they retain the more primitive, platyrrhine-like

morphology. Thus, the Fayum propliopithecids preceded the evolutionary divergence and subsequent radiations of both living groups of extant catarrhines (Fig. 13.18).

Oligopithecids

Oligopithecids (Table 13.2) are a widespread group of early anthropoids known from Egypt, Libya, Morocco, and Oman whose affinities with other anthropoids have been a source of confusion and debate since their initial discovery and continue to be so today. One of the earliest of the Fayum primates found was **Oligopithecus savagei**, from Quarry E in the lower part of the Jebel Qatrani Formation (Simons, 1962). A second, larger species, **O. rogeri**, is from the Ashawq Formation in Oman, and there are suggestions of a considerable radiation of related species in the same fauna (Gheerbrant et al., 1995). There is also an unnamed, very small oligopithecid from Quarry M in the upper part of the Jebel Qatrani Formation of Egypt (Seiffert and Simons, 2013).

Oligopithecus is about the size of a titi monkey (*Callicebus*) and has a deep mandible. Its dental formula is 2.1.2.3, as in propliopithecids, but its teeth show an odd mixture of features quite different from those of other Fayum primates (Figs. 13.7 and 13.12). The canine is small and mesiodistally compressed, and the simple P_3 is narrow with a honing blade on its mesial edge. The last premolar is strikingly similar to the same tooth in propliopithecids. The molars are very primitive compared to those of other anthropoids in having a relatively high trigonid and a small paraconid on the first molar. On both the first and second lower molars, there is a large hypoconulid near the entoconid and no posterior fovea.

Catopithecus browni (Fig. 13.17), best known from Quarry L-41, is a close relative of *Oligopithecus*. *Catopithecus* is known from numerous jaws, several skulls, and some limb bones, and has provided an excellent picture of the comparative anatomy of the oligopithecids. Despite affirming the anthropoid affinities of this group, this new material has not resolved where oligopithecids fit into anthropoid phylogeny. *Catopithecus* is a medium-size primate with an estimated body mass of between 700 and 1000 g. It has a dental formula of 2.1.2.3 with relatively broad upper and lower incisors like anthropoids. The canines are sexually dimorphic (Simons et al., 1999). The lower premolars resemble those of propliopithecids and other catarrhines, with a honing blade on P_3 and a P_4 with a similar-sized protoconid and metaconid. The upper molars have a small hypocone, and the lower molars have a small paraconid on the trigonid of the first molar and a small hypoconulid adjacent to the entoconid. The dental morphology shows a high shearing quotient, suggesting a diet of fruit and insects or leaves, although it is

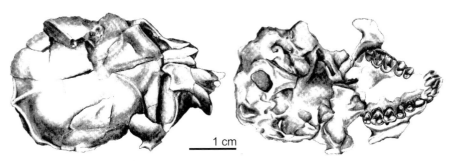

FIGURE 13.17   The cranium and dentition of *Catopithecus browni*, an oligopithecid from the late Eocene of Egypt. *(Drawing by Jennifer Reig.)*

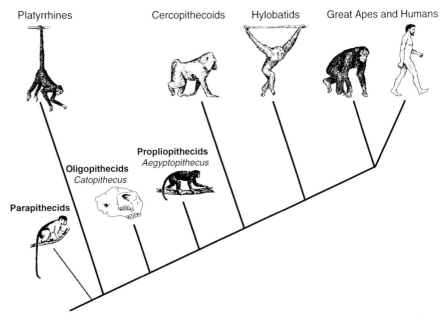

FIGURE 13.18   Phyletic relationships of early anthropoids relative to extant anthropoids.

small for a folivorous primate (Kirk and Simons, 2001). Unlike all extant anthropoids, and propliopithecids, *Catopithecus* has an unfused mandibular symphysis (Ravosa, 1999).

The cranial anatomy of *Catopithecus* is well known from several skulls. Overall, the skull is clearly anthropoid and very much like that of a small platyrrhine, with a fused frontal bone, complete postorbital closure, a large promontory artery and anterior accessory cavity of the ear, and a platyrrhine-like ectotympanic ring fused to the bulla wall (Simons, 1995). The orbit size indicates diurnal habits. Like *Aegyptopithecus*, *Catopithecus* has a relatively large premaxillary bone.

A few limb bones have been attributed to *Catopithecus*. These suggest arboreal quadrupedal habits, with no indication of either leaping or clinging behavior, and they also share features with later propliopithecids and other catarrhines, rather than with parapithecids (Seiffert et al., 2000; Seiffert and Simons, 2001).

Another possible oligopithecid, **Talahpithecus**, has been described from the Eocene of Libya (Jaeger et al., 2010). *T. parvus* is much smaller than either *Oligopithecus*

or *Catopithecus* and has an estimated size of between 225 and 370 g. *Perupithecus*, a potential stem platyrrhine from the early Oligocene of Peru, is very similar to *Talahpithecus* (Bond et al., 2015).

The phyletic affinities of oligopithecids have been much debated since the discovery of *Oligopithecus* in the early 1960s. *Oligopithecus* was initially considered a primitive propliopithecid related to *Propliopithecus* and *Aegyptopithecus* because of the similar dental formula (Simons, 1962). Others argued that the dentition is more suggestive of adapid affinities. The material of *Catopithecus* demonstrated that oligopithecids have undoubted anthropoid features of postorbital closure and a fused frontal as well as an anthropoid anterior dentition. However, the relationship of this group to other anthropoids is debated. On the basis of the unusual combination of dental features, some authorities (e.g., Kay et al., 1997; Williams et al., 2010a,b) have placed oligopithecids as basal anthropoids that precede the platyrrhine–catarrhine divergence, implying that the premolar loss in catarrhines and oligopithecids was independent. The alternative view, adopted here (Fig. 13.18; Table 13.2), is that oligopithecids are related to propliopithecids (Seiffert et al., 2010a,b; Seiffert, 2012; Seiffert et al., 2020). Indeed, Rasmussen and Simons (1992) have suggested that *Moeripithecus markgrafi* is intermediate in molar morphology between oligopithecids and *Propliopithecus* or *Aegyptopithecus*. In addition, oligopithecids are similar to propliopithecids and distinct from parapithecids in features of their humerus and talus (Seiffert et al., 2000).

Nosmips

Nosmips aenigmaticus (Seiffert et al., 2010a,b) is a fossil primate from the Late Eocene BQ2 quarry in the Fayum with relatively large, unusually shaped premolars. The phyletic affinities of *Nosmips* are unclear. The species is known from numerous isolated teeth, and most analyses identify it as a primitive anthropoid (Seiffert et al., 2020). More complete, associated material is needed to clarify its relationship to other African fossil primates.

Afrotarsiids

Afrotarsius is best known from a single lower jaw with four teeth from the upper part of the Jebel Qatrani Formation (*Afrotarsius chatrathi*) and from upper and lower molar teeth from the Late Eocene of Libya (*Afrotarsius libycus*). It was discussed in the previous chapter along with other fossil tarsiers. However, some authors consider that the affinities of *Afrotarsius* and the related Asian taxon *Afrasia* are with early anthropoids rather than tarsiers (e.g., Beard, 2002, 2016; Jaeger et al., 2010; Chaimanee et al., 2012).

Altiatlasius

Although the Paleocene-aged *Altiatlasius koulchii* from Morocco (Fig. 12.3) is considered by many to be the most likely candidate for earliest primate, some authorities have argued that it may actually be a basal anthropoid (Godinot, 1994; Beard, 2006). Since this taxon is only known from a series of isolated teeth, its phyletic position is based on very little. However, if *Altiatlasius* turns out to be a basal anthropoid, it would greatly expand the date of anthropoid origins and have important implications for the biogeography of anthropoid origins.

Eocene Stem Anthropoids From Asia

Eosimiids

Eosimiids (Fig. 13.19; Table 13.3) are a group of very primitive stem anthropoids from numerous, mostly Eocene sites in southern and eastern Asia. The first genus to be described was *Eosimias*, from the Middle Eocene Shanghuang fissure fills of southern China, where many dental and postcranial remains of early haplorhines have

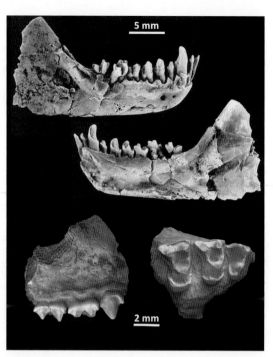

FIGURE 13.19   Eosimiids from the Eocene of China. *(Courtesy of K. C. Beard.)*

been recovered (Beard et al., 1994, 1996; Table 13.3). There are three species of *Eosimias* from China: *E. sinensis* from Shanghuang, as well as *E. centennicus* and a larger species *E. dawsonae* from the Yuanqu Basin (Beard and Wang, 2004). There is also one poorly known species from Myanmar (Takai et al., 2005).

E. centennicus is known from a nearly complete lower jaw that preserves all of the lower teeth on one side (Fig. 13.19). A second genus, **Phenacopithecus**, with two species, is also known from the Yuanqu Basin. For one of these, *P. krishtalkai*, there is a maxilla preserving part of the face (Fig. 13.19). *Eosimias* has been identified as a very primitive anthropoid on the basis of many dental features, including small, spatulate incisors and enlarged canines; broad posterior premolars with oblique roots; and several lower molar features, including relatively broad trigonids as well as an anthropoid-like mandible. The facial region of *Phenacopithecus* was relatively deep compared to that of most omomyoids or tarsiers. However, no cranial remains have been described that indicate whether *Eosimias* and *Phenacopithecus* were diurnal or nocturnal, or whether they had postorbital closure like the Fayum early anthropoids. A petrosal bone attributed to *Eosimias* (MacPhee et al., 1995) is omomyid-like with no definitively anthropoid features. Although there are no limb elements definitely associated with any of the Chinese eosimiids, there are numerous anthropoid-like limb bones, including tarsals, tibiae,

TABLE 13.3 Suborder cf. ANTHROPOIDEA

Species	Estimated mass (g)
Family EOSIMIIDAE	
Eosimias (Middle Eocene, China, Myanmar)	
E. sinensis	131
E. centennicus	125
E. dawsonae	206
E. paukkaungensis	205
Phenacopithecus (Middle Eocene, China)	
P. krishtalkai	149
P. xueshii	311
Bahinia (Middle Eocene, Myanmar)	
B. pondaungensis	652
B. banyueae	288
?Anthrasimias (Early Eocene, India)	
A. gujaratensis	131
Family AMPHIPITHECIDAE	
Amphipithecus (Middle to Late Eocene, Myanmar)	
A. mogaungensis	6743
Pondaungia (Middle to Late Eocene, Myanmar)	
P. cotteri	6132
P. savagei	9255
Ganlea (Late Middle Eocene, Myanmar)	
G. magacanina	2364
Myanmarpithecus (Middle Eocene, Myanmar)	
M. yarshensis	2220
Siamopithecus (Late Eocene, Thailand)	
S. eocaenus	9238
Bugtipithecus (Early Oligocene, Pakistan)	
B. inexpectans	549
Krabia (Late Eocene, Thailand)	
K. minuta	230
Family INCERTAE SEDIS	
Phileosimias (Early Oligocene, Pakistan)	
P. brahuiorum	285
P. kamali	408
Amamria (Middle Eocene, Tunisia)	
A. tunisiensis	100
Aseanpithecus (Middle Eocene, Myanmar)	
A. myanmarensis	3000

part of a femur, and a fragment of the bony pelvis from the same fissure fills as *Eosimias* (e.g., Gebo et al., 1996, 2001, 2008, 2012; Dagosto et al., 1996, 2008).

Bahinia pondaungensis is a larger eosimiid (300–600 g) from the Pondaung Formation of Myanmar. It is known from cranial bones, including a pair of upper jaws and part of a mandible (Jaeger et al., 1999). Its teeth are similar to those of *Eosimias* and *Phenacopithecus*. The size of its orbit indicates a diurnal activity pattern (Heesy and Ross, 2004). A second species, *B. banyueae*, from the Oligocene of China is known from several teeth.

There is also a very small calcaneus from the Pondaung Formation that has been identified as an eosimiid, but it is smaller than any of the named eosimiids there (Gebo et al., 2002).

Phileosimias is a tiny primate based on numerous isolated teeth from the Oligocene Bugti deposits in central Pakistan that would greatly extend the geographic and temporal ranges of the family (Marivaux et al., 2005). It is similar in size to *Eosimias* and *Phenacopithecus* and has been placed in the Eosimiidae, but this taxonomic position is very uncertain (see Gunnell et al., 2008; Seiffert et al., 2009).

Anthrasimias, from the Early Eocene Vastan Coal Mine in Gujarat, India, has been described as the earliest Asian anthropoid and has also been placed in the Eosimiidae (Bajpai et al., 2008). *Anthrasimias* is based on several unassociated teeth, and its identity as a taxon has been questioned by several authorities (e.g., Gunnell et al., 2008; Rose et al., 2009). However, there are isolated postcranial remains from the site that resemble bones attributed to eosimiids (Rose et al., 2009).

Thus, numerous eosimiid taxa have been described, many based on isolated teeth and fragmentary remains, from various sites in Asia ranging from Early Eocene to Early Oligocene in age. Most phylogenetic analyses place eosimiids as basal stem anthropoids because of their very primitive dental features (e.g., Kay et al., 1997; Williams et al., 2010a). There is no evidence of crown anthropoid features in the few cranial remains that have been described, and dentally, they are much more primitive than most of the (younger) anthropoids from the Fayum. However, the Shanghuang deposits that yielded the type specimens of *Eosimias sinensis* contain many limb elements likely attributable to *Eosimias* that show distinctive anthropoid features (e.g., Gebo et al., 2000, 2001, 2012; Gebo and Dagosto, 2004). In addition, analysis of these limb bones indicates a considerable diversity of other small anthropoids in the Eocene of China, including some of the smallest primates that ever lived (Gebo et al., 2001).

Nevertheless, some authors have questioned the anthropoid status of eosimiids. Gunnell and Miller (2001) found that eosimiids did not conform to their reconstructed anthropoid morphotype based on the early anthropoids from Africa and that many showed combinations of dental features unknown among other primates. In their analysis, they found that eosimiids were related to tarsiers. Rasmussen (2002, 2007) has similarly argued that eosimiids are not anthropoids but tarsiers.

These authors are certainly correct that eosimiids show great similarities to tarsiers (see also Cartmill and Smith, 2009), a group that is also common in the Asian Eocene faunas (see Chapter 12). However, given the close relationship between tarsiers and anthropoids, similarities between eosimiids and tarsiers are what one would expect in the most primitive anthropoids.

Amphipithecids

The second group of potential anthropoids from the Eocene and Oligocene of Asia is the amphipithecids (Table 13.3). The phylogenetic position of **Amphipithecus mogaungensis** and **Pondaungia cotteri** (Fig. 13.20) from the Middle Eocene Pondaung Formation of Myanmar (Burma) has been debated since their initial discoveries early in the twentieth century (see reviews by Beard, 2002; Ciochon and Gunnell, 2002; Beard et al., 2009). In the past decade, considerable additional fossil material has been described for amphipithecids, including several new taxa and numerous isolated postcranial elements that expand and either clarify or confuse our understanding of the group and its place in primate evolution.

Both *Pondaungia cotteri* and *Amphipithecus mogaungensis* were large primates, over 5000 g, with broad, low-crowned molars and deep mandibles and an unfused, but rugose interdigiting symphysis (Chaimanee et al., 2000; Kay et al., 2004). In addition, they have relatively prominent canines and lower premolars that increase in complexity posteriorly. Upper molars are quadrate with

a large hypocone. As more and more specimens have been recovered, it appears that the most obvious difference between *Pondaungia* and *Amphipithecus* is size, and some authors have suggested they (and several other species of *Pondaungia*) represent one very sexually dimorphic species, *Pondaungia cotteri,* with "*Amphipithecus*" being a male of that taxon (Jaeger et al., 2004; Kay et al., 2004; Beard et al., 2009; Ramdarshan et al., 2010). The attribution of cranial and postcranial remains to *Amphipithecus* and *Pondaungia* has been a subject of considerable confusion and debate. A frontal bone found in association with specimens of *Amphipithecus* was described and discussed by several researchers (e.g., Gunnell et al., 2002; Takai et al., 2003; Kay et al., 2004), who noted that it showed considerable postorbital constriction and no evidence of postorbital closure – a key feature of extant anthropoids. However, subsequent study by another group of researchers (Beard et al., 2005) led them to claim that the bone was not that of a primate and thus reveals nothing about the cranial morphology of amphipithecids (see also Williams et al., 2010a).

Ciochon and colleagues (Ciochon et al., 2001; Gunnell et al., 2002; Ciochon and Gunnell, 2002) have described some relatively large, associated postcranial bones (a humerus, ulna, and calcaneus) from the Pondaung Formation near Pankan and attributed them to amphipithecids because they were appropriate in size and because amphipithecids are the most common primates from Myanmar. The postcranial bones are very similar to the same elements in living strepsirrhines and in adapoids, and were important in reconstructing the locomotor behavior and phylogenetic affinities of amphipithecids (e.g., Kay et al., 2004). Subsequently, Marivaux and colleagues (2003) described a large talus from another site in the same formation that showed clear affinities with anthropoids. The anthropoid allocation of

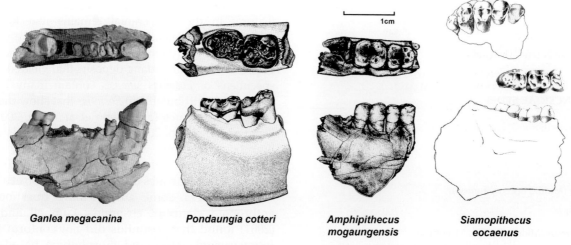

| Ganlea megacanina | Pondaungia cotteri | Amphipithecus mogaungensis | Siamopithecus eocaenus |

FIGURE 13.20   Amphipithecids from the Eocene of Asia. *Pondaungia, Amphipithecus, Ganlea,* and *Siamopithecus. (Courtesy of K. C. Beard, R. L. Ciochon, and S. Ducrocq.)*

the talus was questioned by Gunnell and Ciochon (2008), but further analyses confirmed its anthropoid affinities (Dagosto et al., 2010; Marivaux et al., 2010). Furthermore, Marivaux and colleagues showed that the talus does not seem appropriate in shape to articulate with the large calcaneus, suggesting that there are two very different postcranial ankle morphologies present in the Eocene of Myanmar, one anthropoid and one strepsirrhine. Since none of these elements can be directly associated with dental remains, it is unclear which, if any, actually represent *Pondaungia*. Recent fieldwork has suggested the presence of sivaladapids in the Pondaung Formation, and authors who support anthropoid affinities for amphipithecids have argued that the large, lemur-like bones are not from *Pondaungia* but from a large sivaladapid (e.g., Marivaux et al., 2008); however, the sivaladapids that have been described so far are too small for the postcranial remains.

Myanmarpithecus yarshensis is a much smaller primate from the same deposits in Myanmar (Takai et al., 2001) that is generally considered to be a small amphipithecid (Beard et al., 2009; but see Gunnell et al., 2008). It had an estimated size of around 2000 g and a higher shearing quotient than *Pondaungia*, suggesting that it probably had a diet that included more leaves or insects than other amphipithecids from Myanmar (Kay et al., 2004; Egi et al., 2004; Ramdarshan et al., 2010).

Ganlea megacanina, also from Myanmar, is slightly larger than *Myanmarpithecus*, and, as the name suggests, has very large canine teeth (Beard et al., 2009). It is known from dental, cranial, and postcranial remains (Jaeger et al., 2020). The molar structure of *Ganlea* suggests a diet of fruit and seeds, and it has been suggested that it used its large canine for opening hard seeds, like the New World pitheciines. However, microwear on the teeth suggests a diet of leaves (Ramdarshan et al., 2010). Cranial remains show it lacked the postorbital closure characterizing crown anthropoids, but the lacrimal bone was within the orbit. An ulna is similar to that of early anthropoids.

Siamopithecus eocaenus (Fig. 13.19), from the Late Eocene Krabi Basin (Wailek) in Thailand (Chaimanee et al., 1997), was a large primate with an estimated body mass approaching 10 kg (Ramdarshan et al., 2010). The fossils of *Siamopithecus* have been found in an abandoned (and flooded) coal mine, and most of the fossils are crushed. It is known from teeth, lower jaws, and part of a face. The mandibular and maxillary molars are broad, bunodont teeth, and the mandibular corpus is very deep. A crushed maxilla of *Siamopithecus* has been reconstructed by Zollikofer and colleagues (2009), who argue that it is morphologically similar to the face of living anthropoids (but see Rosenberger et al., 2011), and is anthropoid-like in the frontation and convergence of the orbits. There is no evidence regarding possible postorbital closure in the specimen, and there are no published postcranial remains of *Siamopithecus*.

Phylogenetic Relationships of Amphipithecidae

The phylogenetic position of amphipithecids has long been a subject of considerable debate. Some authorities have argued that amphipithecids are probably adapoids and any dental (and adaptive) similarities to anthropoids are convergent (Gunnell and Miller, 2001; Ciochon and Gunnell, 2002; Gunnell et al., 2002; Gunnell and Ciochon, 2008). However, most current analyses place amphipithecids somewhere among anthropoids, rather than among adapoids or omomyoids, although their placement relative to other anthropoid groups varies. Some place them as basal or stem anthropoids along with eosimiids (Williams et al., 2010a); others place them as sister taxa to platyrrhines (Beard et al., 2009; Seiffert et al., 2010a,b), crown anthropoids (Seiffert, 2012; Seiffert et al., 2020; Fig. 13.21), or parapithecoids (Seiffert et al., 2010a,b). Each of these placements has different implications about the biogeography of anthropoid evolution as well as the likely morphology of the earliest crown anthropoids. Any resolution of the phylogenetic position of amphipithecids will only come about through the recovery of more complete, associated cranial and postcranial material.

Other Early Anthropoids

Bugtipithecus and *Phileosimias* are primates from the Bugti Hills of Pakistan (Marivaux et al., 2005). Both are known only from collections of isolated teeth. *Bugtipithecus inexpectans* was described as an amphipithecid, and *Phileosimias kamali* was described as an eosimiid. Most authorities agree that they are probably primitive anthropoids but do not fit easily into those families.

Aseanpithecus is a recently named primate from the Pondaung Formation in Myanmar that is known by a small collection of fossils, including a maxilla, a partial mandible, and a lower third molar (Jaeger et al., 2019). *Aseanpithecus* was described as bridging the gap between the Asian stem Primates and crown anthropoids. It shares features with *Bugtipithecus*.

Amamria tunisiensis is a taxon from the Eocene of Tunisia that was described by Marivaux et al. (2014) as another primate that is intermediate between the Asian stem anthropoids and the crown anthropoids from Africa. It is known from a single tooth.

As the researchers who described them realized, these taxa are difficult to fit into any of the known families of early anthropoids, in most cases because they are known from very limited material. However, each could have important implications for the biogeographic history of crown anthropoids if they become better known.

The Biogeography of Early Anthropoid Evolution

Early Anthropoid Adaptations

For many decades, the fossil primates from the Early Oligocene of Egypt provided our only record of Old World higher primate evolution. These Oligocene anthropoids were all small to medium-sized, comparable to extant platyrrhines. Their dentitions indicate that they ate fruits, seeds, and perhaps gums, but there is no evidence of predominantly folivorous species. From the available limb bones, they seem to have been arboreal quadrupeds and leapers; there is no evidence of either terrestrial quadrupeds or suspensory species. Overall, the adaptive breadth of the Early Oligocene primates from the Fayum is more like that of extant platyrrhines than that found among later catarrhine primates of the Old World. It thus seemed likely that these platyrrhine-like morphologies represented the primitive anthropoid adaptations.

However, the new discoveries from the older Eocene deposits from the Fayum, as well as the new fossils from other parts of North Africa and Asia, have greatly expanded our knowledge of early anthropoids and also enlarged our view of the adaptive diversity of this group. With all of the diversity among early anthropoids, it is not easy to characterize the basal anthropoid adaptations, but the "large platyrrhine" morphology of the Oligocene anthropoids seems to have been a later phenomenon that greatly postdated the origin of higher primates as a group. Rather, Eocene anthropoids are generally small (100–300 g), with a range of dietary habits, with some likely more frugivorous and others more insectivorous. Interestingly, in the Eocene of North Africa, it is the adapoids that are the larger folivorous taxa (Seiffert et al., 2009). Most of the Egyptian early anthropoids that are known from skulls (*Catopithecus*, *Apidium*, and *Aegyptopithecus*) were probably diurnal.

The adaptive pattern among the stem anthropoids of Asia is quite dichotomous. The eosimiids are all small, probably largely insectivorous taxa. The one taxon for which there are orbital remains (*Bahinia*) was likely diurnal. In contrast, the amphipithecids were relatively large primates with low, rounded molar cusps indicating predominantly frugivorous diets (Ramdarshan et al., 2010), although it has been suggested that they may have been seed predators like modern pitheciins (Kay et al., 2004; Beard et al., 2009). Only the smallest amphipithecid, *Myanmarpithecus*, has a dentition suggesting insects or leaves in its diet.

Several features of the Eocene anthropoids have important implications for reconstructing adaptive scenarios to account for anthropoid origins (Kay et al., 1997; Williams et al., 2010a). Ross (1996) has hypothesized that the distinctive anthropoid feature of postorbital closure evolved in conjunction with a shift to diurnality in a small primate that was probably a visual predator. As small primates, often less than 500 g, the earliest anthropoids were probably partly insectivorous. Moreover, all anthropoids have a retinal fovea, a pit in the retina that provides greater visual acuity in part of the field of vision; in other vertebrates, this feature is usually associated with diurnal visual predation. The shift to a diurnal activity pattern in the ancestors of anthropoids would have resulted in increased convergence of the orbits. In addition, anthropoid skulls have increased frontation because of enlarged frontal lobes of the brain. This increase in the size of the frontal lobes may have been associated with living in larger social groups, a characteristic feature of diurnal primates. All of these changes in orbit shape would lead to a need for a bony septum between the orbit and the temporal fossa so that contraction of the temporal muscles during chewing would not move the eyeball and disturb an animal's vision.

Although little is known of the postcranial skeleton of the very early anthropoids, their small size suggests that they were predominantly quadrupedal but may well have included more leaping in their locomotor repertoire than many later, larger species. As a group, anthropoids are characterized by fewer adaptations to leaping in their hindlimbs than most groups of non-anthropoids. In particular, the isolated postcranial material from the Eocene of China, some of which is probably attributable to eosimiids, suggests less extreme leaping adaptations in early anthropoids than in the contemporary omomyoids and tarsiers (e.g., Gebo et al., 2012).

Phyletic Relationships of Early Anthropoids

The early anthropoids from the Fayum and elsewhere in Africa, Arabia, and Asia are all more primitive than later Old World higher primates. In general, the major groups of fossil anthropoids from the Late Eocene and Early Oligocene of North Africa seem to be intermediate forms that fill morphological gaps between the major radiations of extant anthropoids (Figs. 13.18 and 13.21). In Africa, the parapithecoids are the most primitive and the closest to the origin of crown anthropoids, and they may run the risk of becoming a wastebasket group of primitive anthropoids. They share some postcranial features with omomyoids as well as with platyrrhines, suggesting that the extant radiation of higher primates probably originated in Africa. *Proteopithecus* is likely

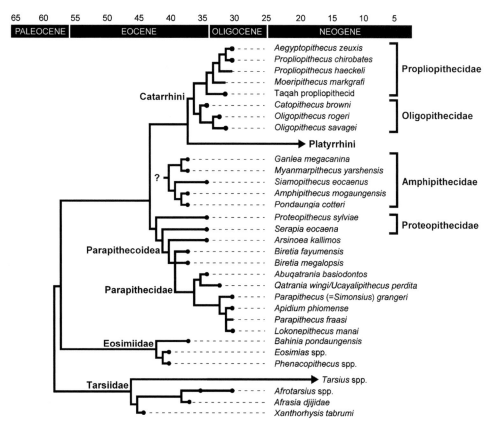

FIGURE 13.21   A broad composite phylogenetic analysis of relationships of early anthropoids. *(Modified from Seiffert, E.R., 2012. Early Primate Evolution in Afro-Arabia. Evol. Anthropol, 21, 239–253.)*

closer to crown anthropoids than the parapithecids and has even been identified as a platyrrhine ancestor by some authors. *Aegyptopithecus* and *Propliopithecus* are more advanced than platyrrhines but more primitive than later catarrhines. Although oligopithecids are clearly anthropoids and more primitive than propliopithecids, their phyletic position relative to parapithecids and platyrrhines is debated. They are most likely stem catarrhines.

As discussed above, the phyletic relationships of the Asian eosimiids and amphipithecids are much more difficult to resolve on current evidence. All show anthropoid-like features in some aspects of their anatomy, but other features suggest that they may not be anthropoids at all. Since there are no good skulls for any of these taxa, inclusion of them in Anthropoidea requires a definition of anthropoids based solely on dental and a few postcranial features, most notably ankle morphology. Eosimiids are most likely basal, stem anthropoids, and it is not difficult to derive the Late Eocene–Oligocene anthropoids from North Africa from an eosimiid-like ancestor. The position of the amphipithecids is more confusing. Despite their Middle Eocene age, most analyses group them with later anthropoids from Africa, including propliopithecids, or with platyrrhines, based on dental

similarities, rather than with earlier, more primitive anthropoids. Other analyses suggest that they are adapoid strepsirrhines and that the anthropoid features are convergent. It is difficult to envision an evolutionary and biogeographic scenario that would accommodate the Fayum primates plus eosimiids and amphipithecids in early anthropoid evolution without many faunal exchanges between Africa and Asia (see Miller et al., 2005; Beard, 2006; Seiffert, 2012). Details of cranial morphology, especially several major anthropoid features such as postorbital closure, remain unknown for these two families, and future finds will surely provide a firmer base for evaluating phylogenetic analyses of these taxa.

There has undoubtedly been considerable homoplasy in the evolution of key anthropoid features of the dentition and the postcranial skeleton. What we understand now that was less evident only a decade ago is that anthropoids were a diverse and successful group well before the Early Oligocene, with an inordinate number of tiny species, the likes of which are not found in the suborder today (e.g., Gebo et al., 2000). As the geographical and morphological scope of this Eocene diversity becomes clearer, our views of early anthropoid evolution will continue to evolve.

Prosimian Origins of Anthropoids

For most of the past decade, the enlarging fossil record of early anthropoids and increasing numbers of detailed phylogenetic analyses appeared to have settled earlier debates over the broader question of the phyletic origins of anthropoids, or, more specifically, which group of prosimians gave rise to higher primates (Fleagle and Kay, 1994; Kay et al., 1997; Williams et al., 2010b). There is widespread agreement among primatologists and mammalogists that, among living primates, anthropoids seem to be more closely related to the living tarsiers than to living lemurs and lorises. When we consider fossil prosimians, however, the possible phyletic relationships between anthropoids (and tarsiers) and various groups of living and fossil prosimians are less settled. Thus, there are three distinct views regarding which (extant or fossil) clade is the sister group of Anthropoidea (Fig. 13.22).

Tarsier Origin

As noted in Chapter 4, tarsiers share with anthropoids many similarities in reproductive anatomy, eye structure, and cranial anatomy, as well as biochemical similarities not found in other living primates. Moreover, two features that unite tarsiers and anthropoids, postorbital closure and the development of an anterior accessory chamber of the middle ear, are unique among primates and even among mammals, rather than being similar features that appear to have evolved in numerous groups. Although the cranial similarities linking tarsiers with anthropoids have been questioned (Simons and Rasmussen, 1989; Beard and MacPhee, 1994; Smith et al., 2013), there seems little reason to do so on morphological grounds (Cartmill, 1994; Ross, 1994). Among all known fossil and extant prosimians, only tarsiers share any evidence of either of these traits with anthropoids. The extensive phylogenetic analyses of anthropoid origins by Kay, Ross, and Williams (e.g., Kay et al., 1997; Williams et al., 2010b) support a sister group relationship between tarsiers and anthropoids as the most parsimonious relationship. However, although the dentition of early anthropoid taxa such as *Eosimias* could be interpreted as supporting tarsier–anthropoid affinities, the absence of common tarsier–anthropoid features in the ear region attributed *Eosimias* suggests that these features may have evolved in parallel (Kay et al., 1997). Furthermore, if *Ganlea* and amphipithecids are early anthropoids, then it is more likely that postorbital closure arose independently in tarsiers and anthropoids as well.

Omomyoid Origin

Many authorities have argued that the sister taxon of anthropoids is not the genus *Tarsius* specifically, but

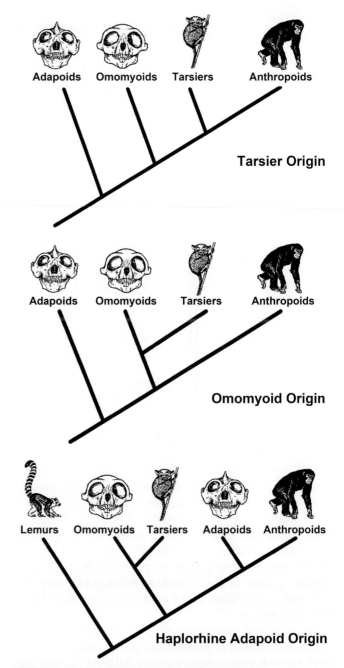

FIGURE 13.22   Hypotheses of anthropoid origins.

some omomyoid or, more, broadly, the clade including omomyoids and *Tarsius*. In their view, tarsiers are a branch within omomyoids, and thus both tarsiers and anthropoids are descended from some common tarsiiform or omomyoid ancestry (e.g., Rosenberger and Szalay, 1980; Rosenberger, 2011). In this view, the features uniting tarsiers and anthropoids are either primitive features common to most omomyoids or, in the case of postorbital closure, convergent and not really homologous. As with the tarsier–anthropoid hypothesis, the omomyoid origins hypothesis accords with notions of a haplorhine–strepsirrhine dichotomy among primates, with molecular studies linking tarsiers with anthropoids, and with studies that argue for the origin of tarsiers from particular omomyid taxa. The biggest weakness of this scenario is that the cranial features that most clearly link tarsiers with anthropoids (postorbital closure and the anterior accessory chamber of the middle ear) are not present in any omomyoids, including the taxa that share various other cranial features with tarsiers, such as *Shoshonius*. There are a few derived cranial features (probably an apical interorbital septum and reduced nasal region), and a few postcranial characters that link some omomyoids with anthropoids and tarsiers, but if tarsiers are indeed derived from any of the taxa most frequently proposed (*Tetonius*, *Shoshonius*, and *Necrolemur*), then the unique tarsier–anthropoid cranial features have evolved independently in tarsiers and anthropoids. At present, the relationships among omomyoids, tarsiers, and anthropoids is unresolved.

Adapoid Origins

Largely on the basis of similarities in the anterior dentition (small incisors, large canines, and frequently, a fused mandibular symphysis), numerous workers have historically supported an adapid–anthropoid relationship (e.g., Rasmussen, 1994). Supporters of such a relationship have argued that omomyids tend to be too specialized dentally (but see Covert and Williams, 1994; Ni et al., 2004) and that many of the distinctive features of haplorhines do not necessarily exclude adapoids because adapoids are not clearly strepsirrhines (Rasmussen, 1986). Although generally out of favor, this view has recently been resurrected and expanded by Franzen and colleagues (2009) (also Gingerich et al., 2010; Gingerich, 2012), who have argued that the exceptionally complete adapoid fossil *Darwinius masillae* is, in fact, a haplorhine and that adapoids are the sister taxon of Anthropoidea (contrary to the views of Seiffert et al., 2009; 2012; Williams et al., 2010b; and many others).

The major weakness of the adapoid–anthropoid hypothesis is that adapoids share with strepsirrhines a number of likely derived features (Williams et al., 2010b). In addition, many of the features that adapoids share with anthropoids are almost certainly primitive primate features, whereas the anthropoid features shared by tarsiers (postorbital closure) and some omomyoids (apical interorbital septum, vertical nasolacrimal duct, reduced nasal region) are almost certainly derived features. Most significant, especially in the case of *Darwinius*, is that many of the anthropoid features shared by some adapoids and extant anthropoids, such as a fused mandibular symphysis, reduction of premolar numbers in catarrhines, and development of a hypocone on upper molars, were not present in early fossil anthropoids. Rather, they have clearly evolved within the

evolutionary history of anthropoids in parallel with their evolution in adapoids. Finally, as with omomyoids, there is no indication among the known fossils that any adapoid ever approximated the diagnostic anthropoid or tarsier features of the orbit and ear. There is very little support for the view that anthropoids evolved from an adapoid ancestry or that adapoids are haplorhines. However, as noted by Rosenberger and colleagues (2011), the phylogenetic position of amphipithecids bears on the question of the relationship between adapoids and anthropoids. If amphipithecids turn out to be early anthropoids, it would provide some support for an adapoid-like morphology very early in anthropoid evolution. However, all detailed analyses support the views that anthropoids are more closely related to omomyoids and tarsiers and that adapoids are stem strepsirrhines (Fig. 12.21).

References

Bajpai, S., Kay, R.F., Williams, B.A., Das, D.P., Kapur, V.V., Tiwari, B.N., 2008. The oldest Asian record of Anthropoidea. Proc. Natl. Acad. Sci. USA 105, 11093–11098.

Beard, K.C., 2002. Basal anthropoids. In: Hartwig, W.C. (Ed.), The Primate Fossil Record. Cambridge University Press, Cambridge, pp. 133–149.

Beard, K.C., 2006. Mammalian biogeography and anthropoid origins. In: Lehman, S.M., Fleagle, J.G. (Eds.), Primate Biogeography: Progress and Prospects. Springer, New York, pp. 429–468.

Beard, K.C., 2016. Out of Asia: Anthropoid origins and the colonization of Africa. Ann. Rev. Anthropol. 45, 199–213.

Beard, K.C., MacPhee, R.D.E., 1994. Cranial anatomy of *Shoshonius* and the antiquity of the Anthropoidea. In: Fleagle, J.G., Kay, R.F. (Eds.), Anthropoid Origins. Plenum Press, New York, pp. 55–97.

Beard, K.C., Wang, B., 2004. The eosimiid primates (Anthropoidea) of the Heti Formation, Yuanqu Basin, Shanxi and Henan Provinces, People's Republic of China. J. Hum. Evol. 46, 401–432.

Beard, K.C., Tao, Q., Dawson, M.R., Wang, J., Chauanhuei, L., 1994. A diverse new primate fauna from the Middle Eocene fissure-fillings in southeastern China. Nature 386, 604–609.

Beard, K.C., Tong, Y.S., Dawson, M.R., Wang, J., Huang, X., 1996. Earliest complete dentition of an anthropoid primate from the late Middle Eocene of Shanxi Province, China. Science 272, 82–85.

Beard, K.C., Jaeger, J.-J., Chaimanee, Y., et al., 2005. Taxonomic status of purported primate frontal bones from the Eocene Pondaung Formation of Myanmar. J. Hum. Evol. 49, 468–481.

Beard, K.C., Marivaux, L., Chaimanee, Y., et al., 2009. A new primate from the Eocene Pondaung Formation of Myanmar and the monophyly of Burmese amphipithecids. Proc. R. Soc. B 276, 3285–3294.

Bond, M., Tejedor, M.F., Campbell Jr., K.E., Chornogubsky, L., Novo, N., Goin, F., 2015. Eocene primates of South America and the African origins of New World monkeys. Nature 520, 538–541.

Cartmill, M., 1994. Anatomy, antinomies, and the problem of anthropoid origins. In: Fleagle, J.G., Kay, R.F. (Eds.), Anthropoid Origins. Plenum Press, New York, pp. 549–566.

Cartmill, M., Smith, F.H., 2009. The Human Lineage. Foundations of Human Biology. Wiley-Blackwell, Hoboken.

Chaimanee, Y., Suteethorn, V., Jaeger, J.-J., Ducrocq, S., 1997. A new Late Eocene anthropoid primate from Thailand. Nature 385, 429–431.

Chaimanee, Y., Thein, T., Ducrocq, S., et al., 2000. A lower jaw of *Pondaungia cotteri* from the late Middle Eocene Pondaung formation (Myanmar) confirms its anthropoid status. Proc. Natl. Acad. Sci. USA 97, 4102–4105.

Chaimanee, Y., Chavasseau, O., Beard, K.C., et al., 2012. A new Middle Eocene primate from Myanmar and the initial anthropoid colonization of Africa. Proc. Natl. Acad. Sci. USA 109, 10293–10297.

Ciochon, R.L., Gunnell, G.F., 2002. Eocene primates from Myanmar: historical perspectives on the origin of Anthropoidea. Evol. Anthropol. 11, 156–168.

Ciochon, R.L., Gingerich, P.D., Gunnell, G.F., Simons, E.L., 2001. Primate postcrania from the late Middle Eocene of Myanmar. Proc. Natl. Acad. Sci. USA 98, 7672–7677.

Covert, H.H., Williams, B.A., 1994. Recently recovered specimens of North American Eocene omomyids and adapids and their bearing on debates about anthropoid origins. In: Fleagle, J.G., Kay, R.F. (Eds.), Anthropoid Origins. Plenum Press, New York, pp. 29–54.

Dagosto, M., Gebo, D.L., Beard, K.C., Qi, T., 1996. New primate postcranial remains from the Middle Eocene Shanghuan fissures, southern China. Am. J. Phys. Anthropol Supp. 22, 113.

Dagosto, M., Gebo, D.L., Beard, K.C., Ni, X., Qi, T., 2008. Primate tibiae from the Middle Eocene Shanghuang Fissure-fillings of eastern China. In: Sargis, E.J., Dagosto, M. (Eds.), Mammalian Evolutionary Morphology: A Tribute to Frederick S. Szalay. Springer, Dordrecht, pp. 315–324.

Dagosto, M., Marivaux, L., Gebo, D.L., et al., 2010. The phylogenetic affinities of the Pondaung tali. Am. J. Phys. Anthropol. 143, 223–234.

de Bonis, L., Jaeger, J.-J., Coiffait, P.-E., 1988. Decouverte du plus ancien primate Catarrhinien connu dans l'Eocene superieur d'Afrique du Nord. C. R. Acad. Sci. 306, 929–934.

Ducrocq, S., Manthi, F.K., Lihoreau, F., 2011. First record of a parapithecid primate from the Oligocene of Kenya. J. Hum. Evol. 61, 327–331.

Egi, N., Takai, M., Shigehara, N., Tsubamoto, T., 2004. Body mass estimates for Eocene eosimiid and amphipithecid primates using prosimian and anthropoid scaling models. Int. J. Primatol. 25, 211–236.

Fleagle, J.G., Gilbert, C.C., 2008. Elwyn Simons: A Search for Origins. Springer, New York.

Fleagle, J.G., Kay, R.F., 1994. Anthropoid Origins. Plenum Press, New York.

Fleagle, J.G., Simons, E.L., 1995. Skeletal anatomy of Apidium phiomense, an early anthropoid from Egypt. Am. J. Phys. Anthropol. 97, 235–289.

Franzen, J.L., Gingerich, P.D., Habersetzer, J., Jurum, J.H., von Koenigswald, W., Smith, B.H., 2009. Complete primate skeleton from the Middle Eocene of Messel in Germany: Morphology and paleobiology. PLoS One 4, e5723.

Fulwood, E.L., Boyer, D.M., Kay, R.F., 2016. Stem members of Platyrrhini are distinct from catarrhines in at least one derived cranial feature. J. Hum. Evol. 100, 16–24.

Gebo, D.L., Dagosto, M., 2004. Anthropoid origins: Postcranial evidence from the Eocene of Asia. In: Ross, C.F., Kay, R.F. (Eds.), Anthropoid Origins: New Visions, Kluwer Academic. Plenum Press, New York, pp. 369–380.

Gebo, D.L., Dagosto, M., Beard, K.C., Ni, X., 2008. New hindlimb elements from the Middle Eocene of China. J. Human Evol. 55, 999–1014.

Gebo, D.L., Dagosto, M., Beard, K.C., Qi, T., 1996. New primate tarsal remains from the Middle Eocene Shanghuang fissures, southeastern China. Am. J. Phys. Anthropol Supp. 22, 92–93.

Gebo, D.L., Dagosto, M., Beard, K.C., Qi, T., 2000. The smallest primates. J. Hum. Evol. 38, 585–594.

Gebo, D.L., Dagosto, M., Beard, K.C., Qi, T., 2001. Middle Eocene tarsals from China: Implications for haplorhine evolution. Am. J. Phys. Anthropol. 116, 83–107.

Gebo, D.L., Dagosto, M., Beard, K.C., Qi, T., Wang, J., 2000. The oldest anthropoid postcranial fossils and their bearing on the early evolution of higher primates. Nature 404, 276–278.

Gebo, D.L., Dagosto, M., Ni, X., Beard, K.C., 2012. Species diversity and postcranial anatomy of Eocene primates from Shanghuang. China. Evol. Anthropol. 21, 224–235.

Gebo, D.L., Gunnell, G.F., Ciochon, R.L., Takai, M., Tsubamoto, T., Egi, N., 2002. New eosimiid primate from Myanmar. J. Hum. Evol. 43, 549–553.

Gheerbrant, E., Thomas, H., Sen, S., Al-Sulaimani, Z., 1995. Noveau primate Oligopithecinae (Simiiformes) de l'Oligocene inferiur de Taqah, Sultinat d'Oman. C. R. Acad. Sci. Paris 321, 425–432.

Gingerich, P.D., 2012. Primates in the Eocene. In: Lehmann, T.M., Schaal, S.F.K. (Eds.), Messel and the terrestrial Eocene- Proceedings of the 22nd Senckenberg Conference. Palaeobiodiversity and Palaeoenvironments. 92, 649–663.

Gingerich, P.D., Franzen, J.L., Habersetzer, J., Hurum, J.H., Smith, B.H., 2010. Darwinius masillae is a haplorhine – reply to Williams et al. J. Hum. Evol. 59, 574–579.

Godinot, M., 1994. Early North American primates and their significance for the origins of Simiiformes (=Anthropoidea). In: Fleagle, J.G., Kay, R.F. (Eds.), Anthropoid Origins. Plenum Press, New York, pp. 235–296.

Gunnell, G.F., Ciochon, R.L., 2008. Revisiting primate postcrania from the Pondaung Formation of Myanmar. In: Fleagle, J.G., Gilbert, C.C. (Eds.), Elwyn Simons: A Search for Origins. Springer, New York, pp. 211–228.

Gunnell, G.F., Ciochon, R.L., Gingerich, P.D., Holroyd, P.A., 2002. New assessment of Pondaungia and Amphipithecus (Primates) from the later Middle Eocene of Myanmar, with a comment on Amphipithecidae. Contributions from the Museum of Paleontology. University of Michigan. 30, 337–372.

Gunnell, G.F., Gingerich, P.D., Haq, M., Bloch, J.I., Khan, I.H., Clyde, W.C., 2008. New primates (Mammalia) from the Early and Middle Eocene of Pakistan and their paleobiogeographical implications. Contributions from the Museum of Paleontology. University of Michigan. 32, 1–14.

Gunnell, G.F., Miller, E.R., 2001. Origin of Anthropoidea: Dental evidence and recognition of early anthropoids in the fossil record, with comments on the Asian anthropoid radiation. Am. J. Phys. Anthropol. 114, 177–191.

Heesy, C., Ross, C.F., 2004. Mosaic evolution of activity pattern, diet, and color vision in haplorhine primates. In: Ross, C.F., Kay, R.F. (Eds.), Anthropoid Origins: New Visions, Kluwer Academic/ Plenum Press, New York, pp. 665–698.

Jaeger, J.-J., Thein, T., Benammi, M., et al., 1999. A new primate from the Middle Eocene of Myanmar and the Asian early origin of anthropoids. Science 285, 528–530.

Jaeger, J.-J., Chaimanee, Y., Tafforeau, P., et al., 2004. Systematics and paleobiology of the anthropoid primate Pondaungia from the late Middle Eocene of Myanmar. Comptes Rendus Palevol 3, 243–255.

Jaeger, J.-J., Beard, K.C., Chaimanee, Y., et al., 2010. Late Middle Eocene epoch of Libya yields earliest known radiation of African anthropoids. Nature 467, 1095–1098.

Jaeger, J.-J., Chavasseau, O., Lazzari, V., Soe, A.N., Sein, C., Le Maitre, A., Shwe, H., Chaimanee, Y., 2019. New Eocene primate from Myanmar shares dental characters with African Eocene crown anthropoids. Nature Comm. 10, 3531 https://doi.org/10.1038/s41467-019-11295-6.

Jaeger, J.-J., Sein, C., Gebo, D.L., Chaimanee, Y., Nyein, M.T., Oo, T.Z., Aung, M.M., Suraprasit, K., Rugbumrung, M., Lazzari, V., Soe, A.N., Chavasseau, O., 2020. Amphipithecine primates are stem anthropoids: Cranial and postcranial evidence. Proc. Roy. Soc. B 287, 20202129.

Kay, R.F., Simons, E.L., 1983. Dental formulae and dental eruption patterns in Parapithecidae (Primates, Anthropoidea). Am. J. Phys. Anthropol. 62, 363–375.

Kay, R.F., Ross, C.F., Williams, B.A., 1997. Anthropoid origins. Science 275, 797–804.

Kay, R.F., Williams, C.F., Ross, J., Takai, M., Shigehara, N., 2004. Anthropoid origins: A phylogenetic analysis. In: Ross, C.F., Kay, R.F. (Eds.), Anthropoid Origins: New Visions. Kluwer Academic/Plenum Press, New York, pp. 91–135.

Kay, R.F., Simons, E.L., Ross, J., 2008. The basicranial anatomy of African Eocene/Oligocene anthropoids. Are there any clues for platyrrhine origins? In: Fleagle, J.G., Gilbert, C.C. (Eds.), Elwyn Simons: A Search for Origins. Springer, New York, pp. 125–158.

Kirk, E.C., Kay, R.F., 2004. The evolution of high visual acuity in the anthropoidea. In: Ross, C.F., Kay, R.F. (Eds.), Anthropoid Origins: New Visions, Kluwer Academic/Plenum Press, New York, pp. 539–602.

Kirk, E.C., Simons, E.L., 2001. Diets of fossil primates from the Fayum Depression of Egypt: A quantitative analysis of molar shearing. J. Hum. Evol. 40, 203–229.

MacPhee, R.D.E., Horovitz, I., Arredondo, O., Vasquez, O.J., 1995. A new genus for the extinct Hispaniolan monkey *Saimiri bernensis* (Rimoli, 1977), with notes on its systematic position. Am. Mus. Nov. 3134, 1–21.

Marivaux, L., Chaimanee, Y., Ducrocq, S., et al., 2003. The anthropoid status of a primate from the late Middle Eocene Pondaung Formation (Central Myanmar): Tarsal evidence. Proc. Natl. Acad. Sci. USA 100, 13173–13178.

Marivaux, L., Antoine, P.O., Barqri, S.F.H., et al., 2005. Anthropoid primates from the Oligocene of Pakistan (Bugti Hills): Data on early anthropoid evolution and biogeography. Proc. Natl. Acad. Sci. USA 102, 8336–8441.

Marivaux, L., Antoine, P.O., Baqri, S.R.H., et al., 2008. Anatomy of the bony pelvis of a relatively large-bodied strepsirrhine primate from the late Middle Eocene Pondaung Formation (central Myanmar). J. Hum. Evol. 54, 391–404.

Marivaux, L., Beard, K.C., Chaimanee, Y., et al., 2010. Talar morphology, phylogenetic affinities, and locomotor adaptation of a large-bodied amphipithecid primate from the late Middle Eocene of Myanmar. Am. J. Phys. Anthropol. 143, 208–222.

Marivaux, L., Essid, E.M., Marzougui, W., Anmar, H.K., Adnet, S., Marandat, B., et al., 2014. A morphological intermediate between eosimiiform and simiiform primates from the late Middle Eocene of Tunisia: macroevolutionary and paleobiogeographic implications of early anthropoids. Am. J. Phys. Anthropol. 154, 387–401.

Mattingly, S.G., Beard, K.C., Coster, P.M.C., Salem, M.J., Chaimanee, Y., Jaeger, J.-J., 2021. A new parapithecine (Primates: Anthropoidea) from the early Oligocene of Libya supports parallel evolution of large body size among parapithecids. J. Hum. Evol. 153, 1–11.

Miller, E.R., Simons, E.L., 1997. Dentition of *Proteopithecus sylviae*, an archaic anthropoid from the Fayum, Egypt. Proc. Natl. Acad. Sci. USA 94, 13760–13764.

Miller, E.R., Gunnell, G.F., Martin, R.D., 2005. Deep time and the search for anthropoid origins. Yearb. Phys. Anthropol. 48, 60–95.

Ni, X., Wang, Y., Hu, Y., Li, C., 2004. A euprimate skull from the early Eocene of China. Nature 427, 65–68.

Ramdarshan, A., Merceron, G., Tafforeau, P., Marivaux, L., 2010. Diet reconstruction of the Amphipithecidae (Primates, Anthropoidea) from the Paleogene of South Asia and paleoecological implications. J. Hum. Evol. 59, 96–108.

Rasmussen, D.T., 1986. Anthropoid origins: A possible solution to the Adapidae-Omomyidae paradox. J. Hum. Evol. 15, 1–12.

Rasmussen, D.T., 1994. The different meaning of a tarsioid-anthropoid clade and a new model of anthropoid origins. In: Fleagle, J.G., Kay, R.F. (Eds.), Anthropoid Origins. Plenum Press, New York, pp. 335–360.

Rasmussen, D.T., 2002. Early catarrhines of the African Eocene and Oligocene. In: Hartwig, W.C. (Ed.), The Primate Fossil Record. Cambridge University Press, Cambridge, pp. 203–220.

Rasmussen, D.T., 2007. Fossil record of the primates from the Paleocene to the Oligocene. In: Henke, W., Tattersall, I. (Eds.), Handbook of Paleoanthropology. Springer, Berlin, pp. 889–919.

Rasmussen, D.T., Simons, E.L., 1992. Paleobiology of the oligopithecines, the earliest known anthropoid primates. Int. J. Primatol. 13, 477–508.

Ravosa, M.J., 1999. Anthropoid origins and the modern symphysis. Folia Primatol. 70, 65–78.

Rose, K.D., Rana, R.S., Sahni, A., Kumar, K., Missiaen, P., Singh, H., Smith, T., 2009. Early Eocene primates from Gujarat, India. J. Hum. Evol. 56, 366–404.

Rosenberger, A.L., 2011. Functional morphology, fossils, and the origins of the tarsier and anthropoid clades. In: Lehmann, T., Schaal, S.K. (Eds.), The World at the Time of Messel: Puzzles in Paleobiology, Paleoenvironment, and the History of Early Primates, 22nd International Senckenberg Conference, pp. 147–148.

Rosenberger, A.L., Szalay, F.S., 1980. The tarsiiform origins of Anthropoidea. In: Ciochon, R.L., Chiarelli, A.B. (Eds.), Evolutionary Biology of the New World Monkeys and Continental Drift. Plenum Press, New York, pp. 139–157.

Rosenberger, A.L., Gunnell, G.F., Ciochon, R.L., 2011. The anthropoid-like face of *Siamopithecus*: cherry picking trees, phylogenetic corroboration, and the adapiform-anthropoid hypothesis. Anat. Rec. 294, 1783–1786.

Ross, C.F., 1994. The craniofacial evidence for anthropoid tarsier relationships. In: Fleagle, J.G., Kay, R.F. (Eds.), Anthropoid Origins. Plenum Press, New York, pp. 469–548.

Ross, C.F., 1996. Adaptive explanation for the origins of the Anthropoidea (Primates). Am. J. Primatol. 40, 205–230.

Ryan, T.M., Silcox, M.T., Walker, A., et al., 2012. Evolution of locomotion in Anthropoidea: The semicircular canal evidence. Proc. Roy. Soc. B279, 3467–3475.

Sanders, W.J., Gunnell, G.G., 2012. Ontogenetic, behavioral, and evolutionary considerations of cranial polymorphism in Early Oligocene *Aegyptopithecus zeuxis* (Catarrhine, Primates), 72nd Annual Meeting of the Society of Vertebrate Paleontology, 165.

Seiffert, E.R., 2006. Revised age estimates for the Later Paleogene mammal faunas of Egypt and Oman. Proc. Natl. Acad. Sci. USA 103, 5000–5005.

Seiffert, E.R., 2007. Evolution and extinction of Afro-Arabian primates near the Eocene-Oligocene boundary. Folia Primatol. 78, 314–327.

Seiffert, E.R., 2012. Early Primate Evolution in Afro-Arabia. Evol. Anthropol. 21, 239–253.

Seiffert, E.R., Simons, E.L., 2001. Astragalar morphology of Late Eocene anthropoids from the Fayum Depression (Egypt) and the origin of catarrhine primates. J. Hum. Evol. 41, 577–606.

Seiffert, E.R., Simons, E.L., 2013. Last of the oligopithecids? A dwarf species from the youngest primate-bearing level of the Jebel Qatrani formation, northern Egypt. J. Hum. Evol. 64, 211–215.

Seiffert, E.R., Simons, E.L., Fleagle, J.G., 2000. Anthropoid humeri from the Late Eocene of Egypt. Proc. Natl. Acad. Sci. USA 97, 10062–10067.

Seiffert, E.R., Simons, E.L., Clyde, W.C., et al., 2005. Basal anthropoids from Egypt and the antiquity of Africa's higher primate radiation. Science 310, 300–304.

Seiffert, E.R., Bown, T.M., Clyde, W.C., Simons, E.L., 2008. Geology, paleoenvironment, and age of Birket Qarun Locality 2 (BQ-2), Fayum Depression, Egypt. In: Fleagle, J.G., Gilbert, C.C. (Eds.), Elwyn Simons: A Search for Origins. Springer, New York, pp. 71–86.

Seiffert, E.R., Simons, E.L., Perry, J.G.M., Boyer, D.M., 2009. Convergent evolution of anthropoid-like adaptations in Eocene adapiform primates. Nature 461, 1118–1121.

Seiffert, E.R., Simons, E.L., Boyer, D.M., Perry, J.M.G., Ryan, T.M., Sallam, H.M., 2010a. A peculiar primate of uncertain affinities from the earliest Late Eocene of Egypt. Proc. Natl. Acad. Sci. USA 107, 9712–9717.

Seiffert, E.R., Simons, E.L., Fleagle, J.G., Godinot, M., 2010b. Paleogene anthropoids. In: Sanders, W.J., Werdelin, L. (Eds.), Cenozoic Mammals of Africa. University of California Press, Berkeley, pp. 369–391.

Seiffert, E.R., Tejedor, M.F., Fleagle, J.G., Novo, N.M., Cornejo, F.M., Bond, M., De Vries, D., Campbell, K.E., 2020. A parapithecid stem anthropoid of African origin in the Paleogene of South America. Science 368, 194–197.

Simons, E.L., 1962. Two new primate species from the African Oligocene. Postilla 64, 1–12.

Simons, E.L., 1986. *Parapithecus grangeri* of the African Oligocene: An archaic catarrhine without lower incisors. J. Hum. Evol. 15, 205–213.

Simons, E.L., 1987. New faces of *Aegyptopithecus* from the Oligocene of Egypt. J. Hum. Evol. 16, 273–289.

Simons, E.L., 1993. New endocasts of *Aegyptopithecus*: Oldest well-preserved record of the brain in Anthropoidea. Am. J. Sci. 293A, 383–390.

Simons, E.L., 1995. Egyptian Oligocene primates: A review. Yearb. Phys. Anthropol. 38, 119–238.

Simons, E.L., 1997. Preliminary description of the cranium of *Proteopithecus sylviae*, and Egyptian Late Eocene anthropoidean primate. Proc. Natl. Acad. Sci. USA 94, 14970–14975.

Simons, E.L., 2001. The cranium of *Parapithecus grangeri*, an Egyptian Oligocene anthropoidean primate. Proc. Natl. Acad. Sci. USA 98, 7892–7897.

Simons, E.L., 2004. Crania of *Apidium*: primitive anthropoidean (Primates, Parapithecidae) from the Egyptian Oligocene. Am. Mus. Novit. 3124, 1–10.

Simons, E.L., 2008. Convergence and frontation in Fayum anthropoid orbits. In: Vinyard, C.J., Ravosa, M.J., Wall, C.E. (Eds.), Primate Craniofacial Function and Biology. Springer, New York, pp. 407–429.

Simons, E.L., Kay, R.F., 1983. *Qatrania*, new basal anthropoid primate from the Fayum, Oligocene of Egypt. Nature 304, 624–626.

Simons, E.L., Rasmussen, D.T., 1989. Cranial morphology of *Aegyptopithecus* and *Tarsius* and the question of the Tarsier-Anthropoidean clade. Am. J. Phys. Anthropol. 79, 1–23.

Simons, E.L., Seiffert, E.R., 1999. A partial skeleton of *Proteopithecus sylviae* (Primates, Anthropoidea): first associated dental and postcranial remains of an Eocene anthropoidean. C. R. Acad. Sci. Ser. IIA 329, 921–927.

Simons, E.L., Plavcan, J.M., Fleagle, J.G., 1999. Canine sexual dimorphism in Egyptian anthropoid primates: *Catopithecus* and *Proteopithecus*. Proc. Natl. Acad. Sci. USA 96, 2559–2562.

Simons, E.L., Seiffert, E.R., Chatrath, P., Attai, Y., 2001a. Earliest record of a parapithecid anthropoid from the Jebel Qatrani Formation, northern Egypt. Folia Primatol. 72, 316–331.

Simons, E.L., Seiffert, E.R., Ryan, T.M., Attia, Y., 2001b. A remarkable female cranium of the Oligocene anthropoid *Aegyptopithecus zeuxis* (Catarrhini, Propliopithecidae). Proc. Natl. Acad. Sci. USA 104, 8731–8736.

Smith, T.D., Deleon, V.B., Rosenberger, A.L., 2013. At birth, tarsiers lack a postorbital bar or septum. Anat. Rec. 296, 365–377.

Stevens, N.J., O'Connor, P.M., Gottfried, M.D., Roberts, E.M., Ngasala, S., 2005. An anthropoid primate humerus from the Rukwa Rift Basin, Paleogene of southwestern Tanzania. J. Vert. Paleontol. 25, 986–989.

Takai, M., Aung, A.K., Egi, N., et al., 2001. Phylogenetic positions of Pondaung primates (latest Middle Eocene, Myanmar). Anthropol. Sci. 109, 94.

Takai, M., Shigehara, N., Egi, N., Tsubamoto, T., 2003. Endocranial cast and morphology of the olfactory bulb of *Amphipithecus mogaungensis* (latest Middle Eocene of Myanmar). Primates 44, 137–144.

Takai, M., Egi, N., Maung, M., Sein, C., Shigehara, N., Tsubamoto, T., 2005. A new eosimiid from the latest Middle Eocene in Pondaung, central Myanmar. Anthropol. Sci. 113, 17–25.

Thomas, H., Sen, S., Roger, J., Al-Sulaimani, Z., 1991. The discovery of *Moeripithecus markgrafi* Schlosser (Propliopithecidae, Anthropoidea, Primates), in the Ashawq Formation (Early Oligocene of Dhofar Province, Sultanate of Oman). J. Hum. Evol. 20, 33–49.

Werdelin, L., Sanders, W.J., 2010. Cenozoic Mammals of Africa. University of California Press, Berkeley.

Williams, B.A., Kay, R.F., Kirk, E.C., 2010a. New perspectives on anthropoid origins. Proc. Natl. Acad. Sci. USA 107, 4797–4804.

Williams, B.A., Kay, R.F., Kirk, E.C., Ross, C.F., 2010b. *Darwinius masillae* is a strepsirhine – a reply to Franzen et al. (2009). J. Hum. Evol. 59, 567–573.

Zollikofer, C.P.E., Ponce de Leon, M.S., Chaimanee, Y., et al., 2009. The face of *Siamopithecus*: new geometric-morphometric evidence for its anthropoid status. Anat. Rec. 292, 1734–1744.

Further reading

Beard, K.C., 2013. Anthropoid origins. In: Begun, D.R. (Ed.), A Companion to Paleoanthropology. Wiley-Blackwell, Hoboken, pp. 358–375.

Dagosto, M., Gebo, D.L., 1994. Postcranial anatomy and the origin of Anthropoidea. In: Fleagle, J.G., Kay, R.F. (Eds.), Anthropoid Origins. Plenum Press, New York, pp. 567–593.

Kay, R.F., 2012. Evidence for an Asian origin of stem anthropoids. Proc. Natl. Acad. Sci. USA 109, 10132–10133.

Simons, E.L., 1971. A current review of the interrelationships of Oligocene and Miocene Catarrhini. In: Dahllery, A.A. (Ed.), Dental Morphology and Evolution. University of Chicago Press, Chicago, pp. 193–208.

Steiper, M.E., Young, N.M., 2006. Primate molecular divergence dates. Mol. Phylogenet. Evol. 41, 384–394.

14

Fossil Platyrrhines

For most of the Cenozoic Era, South America was an island with connections to no continent other than possibly Antarctica. Much of Central America, including what is now Panama, was much farther west and North and South America became connected only at the end of the Miocene or Early Pliocene. The biogeography of the Caribbean during most of the Cenozoic remains a mystery.

The early mammalian fossil record of South America reflects its isolation. It contains many unusual mammals unique to that continent, such as armadillos, sloths, many types of marsupials, and a large radiation of endemic ungulates, rather than the artiodactyls, perissodactyls, rodents, and prosimian primates common to the Eocene of North America and Europe. The first appearances of primates and rodents mark novel additions to the South American fauna. There are no similar appearances of exogenous mammals until the joining of the Northern and Southern Hemispheres in the latest Miocene or Early Pliocene. Until recently, the earliest appearance of both primates and rodents was in Oligocene deposits of Bolivia and Argentina. However, rodents have now been recovered from latest Eocene deposits in Chile and Peru, suggesting that monkeys may also be found at a similarly early date. Where New World monkeys came from and how they got there have long been a source of debate and speculation in primate evolution. Before we tackle these questions, we examine the fossil record.

The Platyrrhine Fossil Record

Fossil New World monkeys are relatively scarce (Table 14.1), considering the great diversity of living primates found in the Neotropics today and the relatively good fossil record for other South American mammals. Until fairly recently, a large shoe box would have been sufficient to contain the primate fossils of South America and the Caribbean from the last 30 million years. Fortunately, the situation has improved greatly in recent decades. The paucity of primates among the well-documented mammalian faunas of South America suggests that much of the evolution of this group took place in areas from which there are very few fossil mammals at all, such as the vast Amazonian Basin. Although it is not extensive, the platyrrhine fossil record is expanding rapidly and provides us with a broad overview and many tantalizing hints about the evolutionary history of the group.

On the basis of geography (Fig. 14.1) and age (Fig. 14.2), fossil platyrrhines can conveniently be divided into several distinct groups: early platyrrhine fossils from a single Late Oligocene locality in Bolivia; several difficult-to-interpret genera from the Early and Middle Miocene of southern Argentina and Chile; a diversity of relatively modern genera from the Late Miocene of Colombia; an increasing number of exciting new discoveries from the Amazon basin in Peru and Brazil; and some very unusual species from Pleistocene and Recent deposits in the Caribbean and Brazil.

The Earliest Fossil Primates in South America

The earliest occurrence of primates in the fossil record of South America comes from the early Oligocene of Peru (Bond et al., 2015; Seiffert et al., 2020; Campbell et al., 2021). Although South America has an extensive record of Paleocene and Eocene deposits, these deposits have not yielded any primate fossils. A fossil locality called Santa Rosa, in eastern Peru, has recently yielded two genera of fossil primates, both with connections to similar aged primates from Africa. *Perupithecus ucayaliensis* is a small primate, based on an upper molar that is very similar to a fossil primate in Libya called *Talahpithecus* that has been identified as an oligopithecid (Bond et al., 2015). A phylogenetic analysis suggests that *Perupithecus* could be ancestral to later platyrrhines.

Another fossil primate from Santa Rosa is *Ucayalipithecus perdita*, known from several upper and lower molars. *Ucayalipithecus* is extremely similar to the small parapithecid *Qatrania* from the Fayum deposits in Egypt (Seiffert et al., 2020). Both *Perupithecus* and

Primate Adaptation and Evolution
https://doi.org/10.1016/B978-0-12-815809-8.00014-X

FIGURE 14.1   Map of the Caribbean and Central and South America showing fossil platyrrhine localities. Map lines delineate study areas and do not necessarily depict accepted national boundaries.

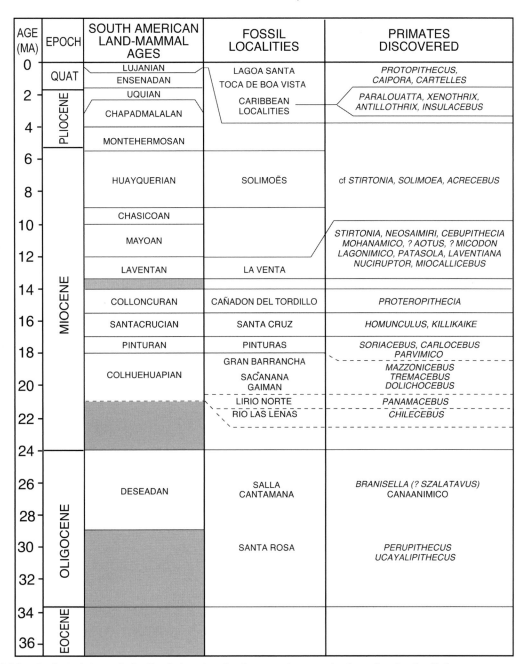

AGE (MA)	EPOCH	SOUTH AMERICAN LAND-MAMMAL AGES	FOSSIL LOCALITIES	PRIMATES DISCOVERED
0	QUAT	LUJANIAN / ENSENADAN	LAGOA SANTA / TOCA DE BOA VISTA	*PROTOPITHECUS, CAIPORA, CARTELLES*
2	PLIOCENE	UQUIAN	CARIBBEAN LOCALITIES	*PARALOUATTA, XENOTHRIX, ANTILLOTHRIX, INSULACEBUS*
4	PLIOCENE	CHAPADMALALAN		
4	PLIOCENE	MONTEHERMOSAN		
6	MIOCENE	HUAYQUERIAN	SOLIMOÕES	cf *STIRTONIA, SOLIMOEA, ACRECEBUS*
8	MIOCENE			
10	MIOCENE	CHASICOAN		*STIRTONIA, NEOSAIMIRI, CEBUPITHECIA MOHANAMICO, ? AOTUS, ? MICODON LAGONIMICO, PATASOLA, LAVENTIANA NUCIRUPTOR, MIOCALLICEBUS*
10	MIOCENE	MAYOAN		
12	MIOCENE	LAVENTAN	LA VENTA	
14	MIOCENE	COLLONCURAN	CAÑADON DEL TORDILLO	*PROTEROPITHECIA*
16	MIOCENE	SANTACRUCIAN	SANTA CRUZ	*HOMUNCULUS, KILLIKAIKE*
18	MIOCENE	PINTURAN	PINTURAS	*SORIACEBUS, CARLOCEBUS PARVIMICO*
18	MIOCENE	COLHUEHUAPIAN	GRAN BARRANCHA	*MAZZONICEBUS TREMACEBUS DOLICHOCEBUS*
20	MIOCENE	COLHUEHUAPIAN	SACANANA GAIMAN	
20	MIOCENE	COLHUEHUAPIAN	LIRIO NORTE	*PANAMACEBUS*
22	MIOCENE		RIO LAS LENAS	*CHILECEBUS*
24	OLIGOCENE			
26	OLIGOCENE	DESEADAN	SALLA CANTAMANA	*BRANISELLA (? SZALATAVUS)* CANAANIMICO
28	OLIGOCENE			
30	OLIGOCENE		SANTA ROSA	*PERUPITHECUS UCAYALIPITHECUS*
32	OLIGOCENE			
34	EOCENE			
36	EOCENE			

FIGURE 14.2 Geological timescale for South American land-mammal ages and primate-bearing fossil sites.

Ucayalipithecus strongly suggest that African anthropoids crossed the South Atlantic ocean to South America while it was an island continent.

The Earliest Platyrrhines

The next earliest primate fossils, and probably the earliest platyrrhines, come from the Late Oligocene (Deseadan) locality of Salla in Bolivia. Two genera have been described: **Branisella** and **Szalatavus** (Fig. 14.3). They are small monkeys, the size of an owl monkey, with three premolars and three molars. Both have four-cusped upper molars with a small hypocone and a well-developed lingual cingulum. They differ only in the shape of the molars: *Szalatavus* has narrower, more triangular molars than *Branisella*, and some authorities do not recognize *Szalatavus* as a distinct taxon (e.g., Takai and Anaya, 1996). The small P^2 and the shape of the mandible suggest short-faced monkeys. The low,

TABLE 14.1   Infraorder PLATYRRHINI

Species	Estimated mass (g)
Family HOMUNCULIDAE	
Soriacebus (Early to Middle Miocene, Argentina)	
S. ameghinorum	1481
S. adrianae	852
Mazzonicebus (Early Miocene, Argentina)	
M. almendrae	2033
Canaanimico (Late Oligocene, Peru)	
C. amazonensis	2000
Carlocebus (Early to Middle Miocene, Argentina)	
C. carmenensis	3302
C. intermedius	2534
Homunculus (Early to Middle Miocene, Argentina)	
H. patagonicus	2150
H. vizcainoi	1714
H. sp. nov.	–
Killikaike (Early to Middle Miocene, Argentina)	
K. blakei	1000
Dolichocebus (Early Miocene, Argentina)	
D. gaimanensis	1495
Family PITHECIIDAE	
Subfamily PITHECIIDAE	
Cebupithecia (Middle to Late Miocene, Colombia)	
C. sarmientoi	1792
Nuciruptor (Middle to Late Miocene, Colombia)	
N. rubricae	2044
Proteropithecia (Middle Miocene, Argentina)	
P. neuquensis	1587
Subfamily CALLICEBINAE	
Miocallicebus (Middle to Late Miocene, Colombia)	
M. villaviejai	2000
Subfamily XENOTRICHINAE	
Xenothrix (Recent, Jamaica)	
X. mcgregori	5746
Antillothrix (Recent, Dominican Republic)	
A. bernensis	4219
Insulacebus (Recent, Haiti)	
I. toussaintiana	4805
Family CEBIDAE	
Subfamily AOTINAE	
Tremacebus (Early Miocene, Argentina)	
T. harringtoni	1350

TABLE 14.1   Cont'd

Species	Estimated mass (g)
Aotus (Middle Miocene to Recent, Colombia)	
A. dindensis	1066
Subfamily CEBINAE	
Chilecebus (Early Miocene, Chile)	
C. carrascoensis	583
Neosaimiri (Middle to Late Miocene, Colombia)	
N. fieldsi	783
Laventiana (Middle to Late Miocene, Colombia)	
L. annectens	887
Acrecebus (Late Miocene, Brazil)	
A. fraileyi	10,000
Panamacebus (Early Miocene, Panama)	
P. transitus	2700
Subfamily cf. CALLITRICHINAE	
Micodon (Middle to Late Miocene, Colombia)	
M. kiotensis	441
Patasola (Middle to Late Miocene, Colombia)	
P. magdalena	480
Lagonimico (Middle to Late Miocene, Colombia)	
L. conclutatus	595
Family ATELIDAE	
Subfamily ALOUATTINAE	
Stirtonia (Middle to Late Miocene, Colombia, Brazil)	
S. tatacoensis	8757
S. victoriae	12,644
S. sp.	-
Cartelles (Pleistocene, Brazil)	
C. coimbrafilhoi	26,500
Paralouatta (Pleistocene, Cuba)	
P. varonai	8444
P. marianae	-
Alouatta (Holocene to Recent, North and South America)	
A. mauroi	8000
Subfamily ATELINAE	
Caipora (Pleistocene, Brazil)	
C. bambuiorum	20,500
Protopithecus (Pleistocene, Brazil)	
P. brasiliensis	22,500
Solimoea (Late Miocene, Brazil)	
S. acrensis	5376

(Continued)

TABLE 14.1 Cont'd

Species	Estimated mass (g)
Family INCERTAE SEDIS	
Branisella (Late Oligocene, Bolivia)	
B. boliviana	759
Szalatavus (Late Oligocene, Bolivia)	
S. attricuspis	418
Mohanamico (Middle to Late Miocene, Colombia)	
M. hershkovitzi	759
Parvimico (Early Miocene, Peru)	
P. materdei	235
Perupithecus (Early Oligocene, Peru)	
P. ucayaliensis	480

rounded cusps of both taxa suggest frugivorous diets, and *Branisella* has very high-crowned lower molars, suggesting possible semiterrestrial habits (Kay et al., 2002). There have been few indications of any special phylogenetic relationship between *Branisella* and any modern platyrrhine subfamily, and in phylogenetic analyses by Kay and colleagues (2008; Kay, 2015), Marivaux and colleagues (2016), and Tejedor and Novo (In Press), *Branisella* is a stem platyrrhine. However, some authors have proposed callitrichine affinities for *Szalatavus* based on its triangular upper molars (Rosenberger et al., 1991b) or *Branisella* because of the small, single-rooted P^2 (Takai et al., 2000).

The Patagonian Platyrrhines

The southern parts of Argentina and Chile, a region informally known as Patagonia, have yielded over a half dozen genera and species of platyrrhines from Early and Middle Miocene deposits (Table 14.1; Figs. 14.1 and 14.2). The primates were part of a rich fauna dominated by rodents, endemic ungulates, sloths, armadillos, and marsupials. After many decades of fossil collecting, monkeys are now known from hundreds of fossils, including numerous skulls, as well as isolated dental and postcranial remains. Like the earlier fossil monkeys from Bolivia, these Patagonian fossils are broadly related to New World monkeys as a group, but the relationships to particular living subfamilies are difficult to identify and are the subject of considerable debate.

Dolichocebus gaimanensis (Fig. 14.4) is from sediments of the Colhuehuapian Land Mammal Age (Earliest Miocene) near Gaiman, in Chubut Province, Argentina. It is known from a nearly complete but damaged skull, a couple of mandibular fragments, numerous isolated teeth, and a talus. *Dolichocebus* was a medium-sized platyrrhine; the dentition suggests a body weight of about 1500 g (Kay et al., 2008), the size of *Callicebus*, the titi monkey. *Dolichocebus* has dimorphic canines, three premolars, and three broad upper molars with a moderate-sized hypocone and a broad lingual cingulum. The molar morphology resembles that of *Saimiri*, *Callicebus*, or *Aotus*, but is more primitive than these genera in such features as broad upper molars with a paraconule and lower molars with hypoconulids. The molar morphology of *Dolichocebus* suggests a frugivorous diet.

The skull of *Dolichocebus* has a narrow, posteriorly widening maxilla, complete postorbital closure, moderate-sized orbits with a very narrow interorbital dimension, and relatively large tooth roots. The relative brain size is similar to that of extant platyrrhines. The distortion of the cranium suggests that in *Dolichocebus*, as in many living platyrrhines, the cranial sutures fused late in adulthood. The cranial morphology of *Dolichocebus* has been the subject of considerable debate. Rosenberger has argued that *Dolichocebus* had an interorbital foramen linking the right and left orbits, an unusual cranial feature found only in *Saimiri* among living primates; other researchers have questioned whether the opening is natural or due to breakage (Kay et al., 2008). The talus of *Dolichocebus* is most similar to that of *Cebus* or *Saimiri*, suggesting it was either a rapid arboreal quadruped or a leaper.

On the basis of the presumed interorbital foramen and several other aspects of the cranial morphology of *Dolichocebus*, Rosenberger (1979, 2019, 2020; also Tejedor and Novo, In Press) has argued that this genus is uniquely related to the living squirrel monkey. In a phylogenetic review of the cranial and dental morphology of *Dolichocebus*, Kay and colleagues (2008; also Kay and Fleagle, 2010; Kay, 2015; Marivaux et al., 2016) found that *Dolichocebus* was a stem platyrrhine that lay outside the radiation of extant New World monkeys.

Tremacebus harringtoni (Fig. 14.4) is from the Colhuehuapian (Early Miocene) locality of Sacanana, also in Chubut, Argentina. It was similar in size to *Dolichocebus*. The type specimen and only fossil clearly attributable to this species is a nearly complete but broken skull with a broad, posteriorly facing nuchal plane. *Tremacebus* has relatively small canines, three premolars, and three molars. The broken upper molars on the skull are quadrate with a large hypocone and a broad lingual cingulum. They are most similar to the teeth of *Callicebus* or *Aotus*. In addition to a relatively short, broad snout, *Tremacebus* has larger orbits than most diurnal platyrrhines but smaller ones than the nocturnal *Aotus*, suggesting to Hershkovitz (1974) that the species was possibly crepuscular. The posterior wall of the orbit is not completely

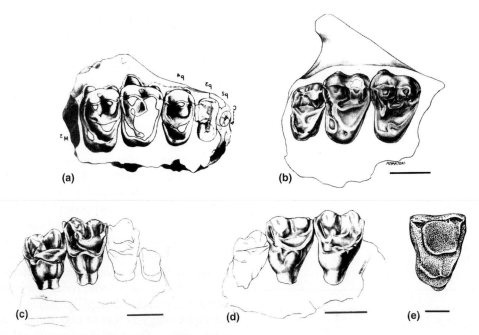

FIGURE 14.3 Maxillary remains attributed to *Branisella boliviana* (a, c), *Szalatavus attricuspis* (b, d), and *Perupithecus ucayaliensis* (e). (*Images of Branisella and Szalatavus courtesy of A. L. Rosenberger.*)

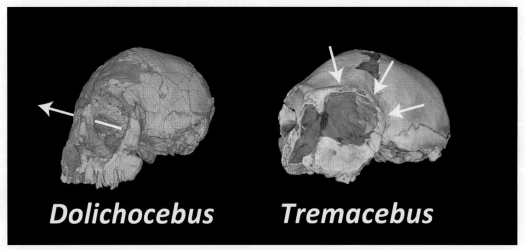

FIGURE 14.4 Crania of *Dolichocebus gaimanensis* and *Tremacebus harringtoni* from Chubut Province, Argentina. Note the opening between the orbits in *Dolichocebus* and the large orbits in *Tremacebus*. (*Images courtesy of Richard Kay.*)

walled off in the type specimen. Hershkovitz (1974) has argued from this evidence that *Tremacebus* was more primitive than any known anthropoids, for which full postorbital closure is a defining feature, but it is much more likely that the large opening in the back of the orbit is due to breakage of the fossil and that *Tremacebus* is similar to living platyrrhines in its postorbital wall.

Tremacebus shows greatest dental and cranial similarities to the extant platyrrhines *Callicebus* and *Aotus*. Rosenberger (1984, 2020) has suggested that it is an ancestor of the living owl monkey; however, in the phylogenetic analysis of Kay et al. (2008, 2015) and Marivaux and colleagues (2016), *Tremacebus*, like *Dolichocebus*, is reconstructed as a stem platyrrhine.

Chilecebus carrascoensis is a fossil platyrrhine from the Early Miocene (20 million years ago) of Chile (Flynn et al., 1995) that is known from a complete cranium. It is a small monkey with an estimated body mass of 583 g (Sears et al., 2008). Its dentition resembles that of *Saimiri* in the relatively large upper premolars. The cranium is similar to that of *Tremacebus* in the posteriorly directed occipital region. The morphology of the semicircular canals indicates moderate agility (Ni et al., 2010). Based on an endocast, the relative size of the brain (EQ) and relative size of the olfactory nerves are small but within the range of extant platyrrhines (Ni et al., 2019).

In the phylogenetic analyses of Kay (2015) and Marivaux et al. (2016), *Chilecebus* is closely related to *Dolichocebus*.

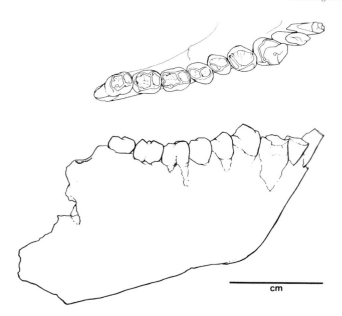

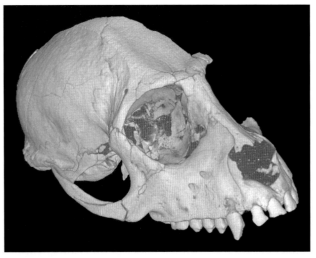

FIGURE 14.6 A cranium of *Homunculus patagonicus* from Santa Cruz Province, Argentina. *(Image courtesy of Richard Kay.)*

Soriacebus ameghinorum

FIGURE 14.5 Mandibular and maxillary remains of *Soriacebus ameghinorum* from the Pinturas Formation (Early Miocene) of southern Argentina. Note the deep mandible and large anterior teeth.

There are several genera of fossil platyrrhines from slightly younger deposits of the Pinturan and Santacrucian Land Mammal ages (early Middle Miocene) in Santa Cruz Province, Argentina (Fig. 14.1). These fossil monkeys come from two main geographical and geological areas: the slightly older Pinturas Formation in the northwest and the younger Santa Cruz Formation, primarily in the east.

The Pinturas Formation, preserved in the foothills of the Andes of southern Argentina, is approximately 17.5 to 16.5 million years old (Fleagle et al., 1995, 2012; Perkins et al., 2012) and has yielded an abundant fauna of fossil birds, reptiles, and mammals, including four primate species (Fleagle, 1990). Evidence from the sediments themselves – fossil pollen, fossil birds, and abundant nests of fossil insects – indicates that the Pinturas primates lived in a forested habitat in what must have been a time of climatic fluctuations, with periods of relative wetness separated by periods of desiccation (Bown and Larriestra, 1990). The fossil monkeys, which are known from over 300 specimens, including several facial skeletons, mandibles, and postcranial elements as well as isolated teeth, are found at two separate levels within the formation (Fleagle and Tejedor, 2002).

The best-known and most unusual of the Pinturas primates is **Soriacebus** (Fig. 14.5). There are two species: the saki-sized *S. ameghinorum* from the lowest levels of the formation and the tamarin-sized *S. adrianae* from younger levels. *Soriacebus* has a dental formula of 2.1.3.3, with large procumbent lower incisors that form a continuous arcade with the large canine; a tall P$_2$; tiny posterior premolars; and three narrow, marmoset-like molars. The jaw deepens posteriorly as in extant pitheciines. The upper teeth have large, dagger-like canines; broad premolars with a hypocone on P^3 and P^4; and small, triangular molars. The facial skeleton is very deep. *Soriacebus* was probably frugivorous and used its large front teeth for some type of gnawing. The few postcranial elements of *Soriacebus* suggest quadrupedal running and leaping habits, with some clinging. The affinities of *Soriacebus* have been debated since its initial discovery: it has been considered a primitive pitheciine on the basis of the large incisors, the morphology of the canines, the broad upper premolars, and the deep mandible (Tejedor, 2005), or as part of a radiation that preceded the radiation of extant platyrrhines (Kay et al., 2008; Kay, 2015; Marivaux et al., 2016). **Mazzonicebus almendrae** is a similar monkey from slightly older deposits at the Gran Barranca (Kay, 2010), as is **Canaanimico amazonensis** from the late Oligocene of Peru.

The other platyrrhine genus from Pinturas is **Carlocebus**, also with two species: the saki-sized *C. intermedius* from the lower levels and the larger *C. carmenensis* from throughout the formation. The lower dentition of *Carlocebus* is more generalized than that of *Soriacebus*, with small vertical incisors, a small canine, a moderate-sized anteriormost premolar, and relatively larger posterior premolars and molars. As in *Soriacebus*, the mandible is relatively deep. The upper dentition is characterized by very broad premolars and molars with large hypocones. Dentally, *Carlocebus* appears to have been frugivorous and folivorous. Its dentition is most comparable to that of *Callicebus*, the titi monkey, but much larger. Postcranial remains of *Carlocebus* suggest arboreal quadrupedal habits. Like *Soriacebus*, it has been considered a basal pitheciid or a stem platyrrhine.

Homunculus patagonicus (Fig. 14.6), from the early Middle Miocene (~16.5 million years ago) Santa Cruz

Formation on the Atlantic coast of southern Argentina, was one of the earliest fossil platyrrhines discovered (Ameghino, 1891; see also Tejedor and Rosenberger, 2008), and for many years, all fossil platyrrhines were placed in this genus. It was a medium-sized monkey, with an estimated mass of 2.0 to 2.5 kg. The dental formula is 2.1.3.3. The lower incisors are narrow and spatulate; the canines are relatively small, and probably sexually dimorphic. The molars are characterized by relatively small cusps connected by long shearing crests, suggesting a diet of fruit and leaves (Kay et al., 2012). *Homunculus* has a relatively short snout with procumbent incisors and moderate-sized orbits (indicating diurnal habits) with complete postorbital closure. The cranium is relatively gracile with no sagittal crest, and contact between the cranial bones at pteryion is variable (Fulwood et al., 2016). The brain of *Homunculus* was relatively smaller than that of extant platyrrhines. The morphology of the semicircular canals suggests agile locomotion similar to that of some leaping strepsirrhines. The limb elements of *Homunculus* are unusually robust for a small platyrrhine and indicate that it was a quadrupedal monkey with leaping abilities. The limbs have numerous primitive features reminiscent of early anthropoids from Egypt, rather than extant New World monkeys (Fleagle et al., 2022).

In addition to *H. patagonicus*, there are undescribed species from the coast of Argentina and a newly described species, **Homunculus vizcainoi**, from the valley of the Santa Cruz River (Kay et al., 2012; Kay and Perry, 2019).

As the name *Homunculus* indicates, Ameghino (1891) originally thought the genus was in the ancestry of humans; it is not. Many studies have noted dental similarities to *Aotus* and *Callicebus* (Bluntschli, 1931; Bond et al., 2015; Rosenberger, 2020). In describing a facial skeleton attributed to *Homunculus*, Tauber (1991) noted many similarities to pitheciines. Phylogenetic analyses by Kay and colleagues (2008, 2015) and Marivaux and colleagues (2016) place *Homunculus* as a stem platyrrhine, outside the radiation of extant New World monkeys.

Killikaike blakei is from the same part of the Santa Cruz Formation in southernmost Argentina as the original type specimen of *Homunculus*. The anterior part of the cranium is similar in size and morphology to *Homunculus*. In the original description (Tejedor et al., 2006), it was identified as an early cebid, but others have suggested it should be included in *Homunculus* (Kay et al., 2012).

From the slightly younger deposits of the Canadon del Tordillo Formation (Middle Miocene) in northern Patagonia, **Proteropithecia neuquenensis** is known from a few isolated teeth. It has been identified as an early pitheciid on the basis of several dental features (Kay et al., 1998, 1999), although one recent analysis suggested that it may be a stem platyrrhine (Beck et al., 2023).

The relationship of many of the Patagonian primates is a topic of much discussion, often described as a debate over whether the modern platyrrhine fauna is the result of a pattern of long lineages. All of the Patagonian platyrrhines except the poorly known *Proteropithecia* show unusual combinations of primitive anthropoid features not found in any extant platyrrhines and derived features linking them to distinct clades of extant platyrrhines. This confusion in their phylogenetic placement is perhaps not surprising, as these fossils come from the time at which molecular studies indicate that the extant subfamilies were differentiating and would be unlikely to show the suite of features found in later members of the clades (see Fig. 14.11). One possibility is that the Patagonian primates are collateral to all later evolution of platyrrhines (Kay et al., 2008; Kay, 2015; Marivaux et al., 2016), a conclusion that would be compatible with both their age and their geographic distinctiveness (Hodgson et al., 2009). Alternatively, modern clades may have lost the primitive features independently. Unusual combinations of "modern" features are often found among fossil taxa and are what make fossils especially valuable in resolving the evolutionary history of a group. There is no doubt that part of our inability to clearly resolve the phylogenetic position of these monkeys is that we lack a full understanding of the polarity of the features characterizing extant platyrrhines – that is, which features are primitive retentions and which are derived specializations in different groups. In addition, Patagonia has a long history as a separate biogeographical area within South America, with a distinct flora and invertebrate fauna, and some have suggested that these Miocene monkeys from the southernmost end of South America are not just primitive but are a geographically isolated radiation of platyrrhines separate from the modern clades that were evolving further north. However, without similarly aged primates from elsewhere in the continent, this hypothesis cannot be tested.

It is likely that the unusual combinations of features seen in these early platyrrhines reflect their proximity to the initial split of modern clades, before most acquired the larger suite of features that characterize their living members. Many of the primates from Santa Cruz Province seem to show features related to modern pitheciids (including *Callicebus*), and it is noteworthy that most molecular phylogenies (Fig. 1.2) identify extant pitheciids as the first branch in the modern platyrrhine radiation. As more complete fossils of these early platyrrhines continue to be found, they should provide further insight into the origins and early radiations of New World anthropoids.

A More Modern Community

Fossil platyrrhines are known from very few places in the vast land mass of Central and South America, and there are few fossil sites as old as the Patagonian deposits from more tropical areas of South America. The

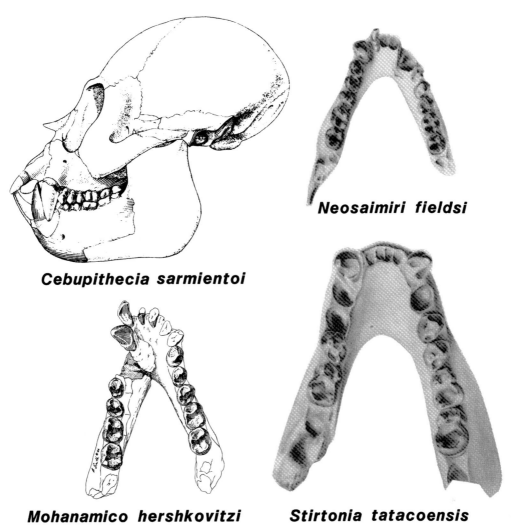

Cebupithecia sarmientoi

Neosaimiri fieldsi

Mohanamico hershkovitzi

Stirtonia tatacoensis

FIGURE 14.7 Cranium and dentition representing four primates from the Late Miocene of La Venta, Colombia.

best-known fossil platyrrhines from more tropical parts of the continent are those from La Venta in the Magdalena Valley of Colombia (Fleagle et al., 1997; Kay et al., 1997). Compared to the Patagonian fossil platyrrhines, which are difficult to place confidently in extant platyrrhine subfamilies, many of the fossil monkeys from the later Miocene of La Venta are strikingly similar to modern platyrrhines and clearly belong in living subfamilies or even genera. The fossil platyrrhines from La Venta are found in two different geological formations, which preserve a very brief period of time from approximately 13 million to 12 million years ago. Comparison of the La Venta fauna with modern South American faunas indicates a tropical forest environment for this region in the Late Miocene. Compared with the earlier Patagonian faunas, the La Venta fauna contains some of the same typical South American mammals, including caenolestid marsupials and astrapotheres, but some not found in Patagonia, such as anteaters. There are nearly a dozen species of fossil primates described from this site.

Cebupithecia sarmientoi (Fig. 14.7) was similar in size (2–3 kg) and in many aspects of morphology to the living saki, *Pithecia pithecia*. The fossil is known from parts of two skeletons, including most of the dentition, the mandible, most of the axial skeleton, and the ends of limb bones first discovered over 60 years ago. In many aspects of its dental morphology, such as the stout canines, procumbent incisors, and flat cheek teeth with little cusp relief, this Miocene genus is very similar to the living pitheciines. Like living pitheciines, *Cebupithecia* probably ate mainly fruit and used its large anterior dentition for opening hard seeds. The *Cebupithecia* skeleton shows more similarities to the saltatory *Pithecia* than to the more quadrupedal sakis, such as *Chiropotes* (Meldrum and Lemelin, 1991). In contrast to its uniquely pitheciine dentition, the skeleton retains many features found in other platyrrhine subfamilies while lacking some shared derived features of living pitheciines. There are also indications of vertical clinging habits in the morphology of the elbow. It is probably very near the base of modern pitheciine radiation.

Nuciruptor rubricae is a smaller, more primitive pitheciine from La Venta that differs from *Cebupithecia* in having a smaller canine and no canine–incisor diastema (Meldrum and Kay, 1997).

Mohanamico hershkovitzi (Fig. 14.7) is a small (1 kg) fossil monkey from La Venta that is known from a single mandible (Luchterhand et al., 1986). It has also been placed near the base of the evolutionary radiation of the pitheciines on the basis of its large lateral incisor and the structure of the canine and anterior premolar. The molar teeth indicate that it was probably frugivorous. However, *Mohanamico* is much less clearly a pitheciine than are *Nuciruptor* or *Cebupithecia*, and others have linked it with *Callimico* (Rosenberger et al., 2009). One recent phylogenetic analysis hypothesized that *Mohanamico* may instead be a stem callitrichine (Beck et al., 2023)

Miocallicebus villaviejai is known from a single maxillary fragment that is very similar to that of the extant *Callicebus* but is much larger (Takai et al. 2001).

Aotus dindensis is virtually identical in molar and premolar morphology to the extant *Aotus*, but it has narrower lower incisors. A small facial fragment suggests that the Miocene species also had large orbits similar to those of the nocturnal owl monkey. There has been debate about the affinities and similarities of *Aotus dindensis* and *Mohanamico*, and the phyletic relationships of both are by no means certain. This debate illustrates mostly the conservative nature of the mandibular dentition in small frugivorous platyrrhines.

A fossil genus that is clearly closely related to a modern subfamily is **Stirtonia** (Fig. 14.7), the largest monkey in the La Venta fauna (6–10 kg). There are two species from different stratigraphic levels. The older, larger species is *Stirtonia victoriae*; the smaller, younger species is *S. tatacoensis*. *Stirtonia* has many similarities in its upper and lower dentition to the living howling monkey (*Alouatta*). Like *Alouatta*, *Stirtonia* has long molars with a relatively small trigonid and large talonid, and very large upper molars with well-developed shearing crests and styles. It was a folivore. Isolated molars that resemble those of *Stirtonia* (and *Alouatta*) have also been recovered from Late Miocene deposits along the Rio Acre in western Brazil (Kay and Frailey, 1993).

Neosaimiri fieldsi (Fig. 14.7) is virtually identical to the living squirrel monkey in size and in all known details of dental anatomy (Rosenberger et al., 1991a), differing most clearly in molar proportions and the relative size of the lateral incisor. Like *Saimiri*, it was insectivorous and frugivorous. An isolated humerus from the same deposits is indistinguishable from the same bone in *Saimiri*. Although there is but a single genus of squirrel monkey today, with numerous allopatric species, it seems that there may have been a larger taxonomic radiation of squirrel monkeys in the Miocene. *Neosaimiri* was just one of several squirrel monkey-like taxa in the later Miocene. Marivaux and colleagues (2020) have described teeth similar to those of *Neosaimiri* from deposits of comparable age in central Peru.

Laventiana annectens (Rosenberger et al., 1991c) is another fossil species from La Venta that is similar to *Saimiri* and *Neosaimiri*, and some authorities place *L. annectens* in the same genus as *Neosaimiri* or *Dolichocebus* (Kay, 2015; Rosenberger, 2019, 2020). It is distinguished by a postentoconid notch on the lower molar, a characteristic not seen in any living platyrrhines. A talus recovered with the *Laventiana* mandible also resembles that of *Saimiri* and lacks the clearly derived traits characterizing the smaller callitrichines and larger atelines.

One of the most obvious gaps in the platyrrhine fossil record for many years has been the absence of any clear fossil evidence of callitrichines (marmosets, tamarins, and *Callimico*), even though they are the most diverse subfamily of living New World monkeys. Their evolutionary origin has been a source of endless debate, but molecular studies place them as the sister group either to *Aotus* or to the cebines, *Cebus*, *Sapajus*, and *Saimiri*. Several putative fossil callitrichines have been described from La Venta, each with different morphologies and different implications for the origin of the group.

Micodon kiotensis is a poorly known species from La Venta that is based on three small, isolated teeth (Setoguchi and Rosenberger, 1985). It has been described as a fossil marmoset, primarily on the basis of size. The type specimen, an upper molar, lacks any marmoset features and resembles that of a small pitheciine in occlusal morphology. Its describers noted that *Micodon* demonstrates that the small size of callitrichines preceded their distinctive dental features, such as loss of a hypocone. Any determination regarding either the validity of the species or its affinities must await more fossil remains.

Patasola magdalena is based on a single mandible from La Venta that is slightly smaller than that of the living squirrel monkey. It shares features with both *Saimiri* and callitrichines and was identified by its describers as a callitrichine that is closely related to marmosets and tamarins, on the basis of a simplified premolar morphology and the shape of the molar trigonid and cristid obliqua (Kay and Meldrum, 1997).

Lagonimico conclutatus (Fig. 14.8) is another possible callitrichine species from La Venta. It was roughly the size of an owl monkey and has been described by Kay (1994) as a giant tamarin. Based on a different set of features, mainly upper molar shape, he placed it in exactly the same phyletic position as *Patasola*. The absence of upper molar hypocones in a monkey the size of *Lagonimico* would suggest that acquisition of the distinctive marmoset and tamarin molar morphology did not necessarily evolve in conjunction with small size.

The presence of several putative fossil callitrichines at La Venta is exciting but clearly demands further analysis since each offers a somewhat different picture of the origin and early evolution of the group. It is almost certain that the origin of this group involved mosaic

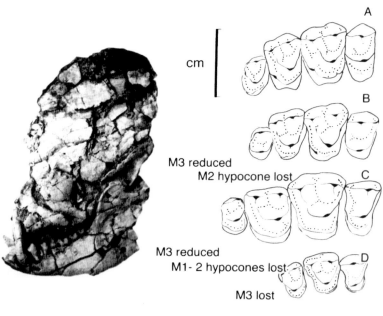

cm

M3 reduced
M2 hypocone lost

B

C

M3 reduced
M1- 2 hypocones lost

D

M3 lost

M1- 2 hypocones lost

FIGURE 14.8 *Lagonimico conclutatus*, a giant tamarin from La Venta, Colombia. (A) *Saimiri*, (B) *Callimico*, (C) *Lagonimico*, (D) *Saguinus*. Note that *Lagonimico* has a reduced M³ like *Callimico* and has lost the M¹⁻² hypocones like *Saguinus*. *(Courtesy of R. F. Kay.)*

evolution and a rather bushy phylogeny, as evidenced by the parallel and convergent features found in marmosets, tamarins, *Leontopithecus*, *Lagonimico*, and *Callimico*. However, we are still far from understanding many details of the origin and radiation of these most successful platyrrhines.

Despite debate over the validity of a few taxa and the conflicting phylogenetic relationships attributed to a few others, the fauna from La Venta clearly indicates that many groups of extant platyrrhines were well differentiated along modern lines and present in northern Colombia by 12 million years ago. Perhaps most notable is the absence at La Venta of any putative relatives of *Cebus*, one of the most widespread of modern genera, and the absence of the large spider or woolly monkeys.

The La Venta primate fauna is clearly modern compared with the fossil platyrrhines from Patagonia, both taxonomically and adaptively (Wheeler, 2010; Cooke, 2011). This modernity may reflect its relatively late age or the geographical location of La Venta closer than other fossil localities to the Amazon Basin, where living New World monkeys are most abundant today. Most probably, both of these relationships contribute to the modern appearance of this fauna. La Venta records the most diverse fauna of fossil New World monkeys from any place or period in the entire fossil record of the group. Thus, although this one site provides documentation of the major diversification of subfamilies and major clades, until recently, we had no real appreciation of platyrrhine species diversity or biogeography in other parts of the continent. This has changed in recent years.

A Fossil Monkey From Panama

Recent expansion of the Panama Canal has exposed sediments containing fossils commonly found in North America dating to the early Miocene. Among these North American mammals are dental remains of a fossil platyrrhine that Bloch and colleagues (2016) have described as **Panamacebus transitus**. *Panamacebus* appears to be closely related to living and fossil cebines.

Fossil Primates From Amazonia

The greatest diversity of extant platyrrhines is found in the central area of South America in the broad basin of the Amazon River. However, for most of its history, the fossil record of primate evolution in South America has been notably sparse from that part of the continent. Recent paleontological expeditions in the Peru Amazon Basin have resulted in the discovery of numerous fossil primates from several localities. Most surprising, and exciting, are the two new genera from the site of Santa Rosa discussed above, that are very similar to fossil primates from Africa.

In addition, as mentioned above, from late Oligocene deposits in the area of Contamana, also in Eastern Peru, Marivaux and colleagues (2016) have described *Canaanimico*, a fossil platyrrhine that is similar to *Soriacebus* from the Miocene of Patagonia. From late Miocene deposits in the same area, Marivaux and colleagues (2016) have also described isolated fossil teeth of two extant platyrrhine genera, *Cebus* and *Cebuella*.

From another site in eastern Peru, along the Madre de Dios River, Kay and colleagues (2019) have described a tiny primate upper molar similar in size to a marmoset. The fossil, named **Parvimico materdei**, is from the early Miocene and differs from all other known platyrrhines, and its phylogenetic position is uncertain.

Although all are known only from a few teeth and a talus (e.g., see Marivaux et al., 2012), these recently described fossils from the Oligocene and Miocene of Peru clearly show that the early evolution of New World monkeys took place throughout the continent of South America. In addition, there are a few fragmentary fossils from the Late Miocene (9–6 Ma) of Acre in Brazil (Kay and Cozzuol, 2006). They are placed in three separate genera.

A broken lower molar, similar to *Alouatta* was placed by Kay and Frailey (1993) in the genus *Stirtonia*, better known from the slightly older deposits in La Venta. A very large upper molar, similar to the genus *Cebus*, but much larger than any living platyrrhine, was given the name **Acrecebus fraileyi**. It was most likely a frugivorous monkey.

Solimoea acrensis is based on a well-preserved lower molar and a fragment of maxilla with very worn P^4 through M^2. It had an estimated body mass of about 5 kg, and its teeth suggest a frugivorous diet. *Solimoea* was identified by Kay and Cozzuol (2006) as an ateline.

Pleistocene Platyrrhines

In many respects, the most surprising fossil platyrrhines are the youngest: those from Pleistocene and Recent caves of the Caribbean and Brazil. One of the earliest fossil primates ever recovered, and the first that was recognized to be unlike any living species, was found in a Brazilian cave in the 1830s by the Danish naturalist Peter Wilhelm Lund. The numerous cave deposits of the Brazilian cave Lagoa Santa are Late Pleistocene and/or Recent in age and contain a mixture of extinct and extant fauna. In 1836, Lund found a proximal femur and distal humerus of an ateline-like primate that was larger than any living platyrrhine and had a body weight probably two and a half times the size of the largest living platyrrhine (Hartwig, 1995a,b; Halenar, 2011a). Although Lund's fossils were commonly placed in the same genus and species as the living muriqui, *Brachyteles arachnoides*, it is clearly a different monkey and goes by the name Lund gave it over 150 years ago, **Protopithecus brasiliensis**.

Skeletons of two other large atelids have been recovered from another Pleistocene cave in northeastern Brazil, Toca de Boa Vista, farther North than Lagoa Santa (Hartwig and Cartelle, 1996; Halenar and Rosenberger, 2013). The locomotor skeleton of **Cartelles** resembles that of a large suspensory ateline such as *Ateles* or

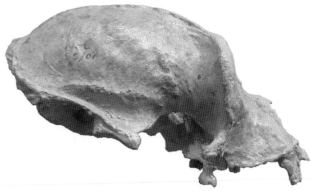

Cartelles coimbrafilhoi

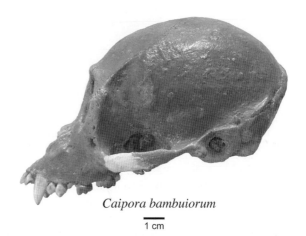

Caipora bambuiorum

1 cm

FIGURE 14.9 Crania of *Cartelles coimbrafilhoi* and *Caipora bambuiorum* from the Pleistocene of Brazil.

Brachyteles (Halenar, 2011b). However, the skull more closely resembles that of *Alouatta*, the howling monkey, with an upward flexed face (airorhynchy), a relatively small brain (Krupp et al., 2012), prominent temporal and nuchal crests, and a posteriorly facing foramen magnum (Fig. 14.9). This combination of cranial and postcranial features in *Cartelles* suggests considerable parallelism within the evolution of atelids.

A second species from the same cave as *Cartelles* in northeastern Brazil, **Caipora bambuiorum** (Fig. 14.9), is also larger than any living platyrrhine, with an estimated mass of 20 kg. In both cranial and skeletal morphology, *Caipora* most closely resembles the living spider monkey, *Ateles* (Cartelle and Hartwig, 1996), including having a very reduced thumb.

More recently, another large fossil platyrrhine was discovered from the Engrunado Cave south of the cave that yielded *Cartelles* and *Caipora*, also in the state of Bahia. It is known from a skull and some relatively complete parts of the postcranial skeleton (Dantas et al., 2019) and is currently being studied.

The Brazilian cave deposits also contain remains of a fossil species of *Alouatta* that differs from any extant howling monkey (Tejedor et al., 2008).

Caribbean Primates

There are no platyrrhine primates on any of the Caribbean islands today. When primate remains from these islands were originally described, they were generally considered not endemic species but "pets" brought over from the mainland by humans. Ironically, it now seems most likely that this region once harbored a diverse fauna of unusual endemic primates (Fig. 14.1; Table 14.1) and that humans were either directly or indirectly responsible for their extinction rather than their introduction.

Xenothrix mcgregori (Fig. 14.10) is from latest Pleistocene to Recent deposits on the island of Jamaica. It quite likely survived as late as the colonization of this island by Europeans in the last few centuries (MacPhee and Fleagle, 1991). *Xenothrix* is known from several mandibles, a face, and numerous postcranial bones (MacPhee and Horovitz, 2004; MacPhee and Meldrum, 2006). Dental dimensions suggest a mass of 5 to 6 kg, while postcranial bones give a lower estimate of approximately 2 to 4 kg. It had a dental formula of 2.1.3.2, as in marmosets and tamarins. However, it was much larger than marmosets and tamarins, and the molars, with large, bulbous cusps, are very different in both cusp morphology and proportions from any modern callitrichines, or any other platyrrhines for that matter. *Xenothrix* was probably a frugivorous species, or it may have specialized on insect larvae, like the aye-aye of Madagascar. The postcranial remains that have been attributed to *Xenothrix* evidence an unusual type of slow quadrupedal locomotion that has no counterpart among living primates. The relationship of *Xenothrix* to other platyrrhines has been uncertain for several decades. Rosenberger has suggested possible relationships to pitheciines and *Aotus*, while MacPhee and Horovitz (e.g., 2004) consider all Caribbean primates as part of a single endemic radiation most closely related to the extant titi monkeys, *Callicebus*. A recent study of ancient DNA from *Xenothrix* (Woods et al., 2018) placed the Caribbean fossil within the titi monkey radiation, most closely related to *Cheracebus*, and found that it split from the extant titi monkeys on South America approximately 11 million years ago.

Antillothrix bernensis (Fig. 14.10) is a fossil monkey from Recent cave deposits, some of which are underwater, in the Dominican Republic (MacPhee et al., 1995; Kay et al., 2011; Rosenberger et al., 2011; Halenar et al., 2017). Originally described from a maxilla with three teeth, it is now known from several complete skulls and most of a skeleton that have been discovered by Scuba divers in underwater caves. From dental and cranial remains, *Antillothrix* has an estimated body mass of 2 to 3 kg. The cranium is characterized by a posteriorly facing nuchal plane as in *Tremacebus* or *Dolichocebus*, but not in most extant platyrrhines. The dentition suggests a diet of fruit and insects, rather than either hard seeds or leaves. The limb bones associated with the cranium show unusual proportions, with the forearm longer than the robust femur, which resembles the same bone in *Xenothrix*. Based on the cranium and dentition, Rosenberger has argued that *Antillothrix* shows evidence of a phylogenetic relationship with titi monkeys and *Aotus*, the owl monkey. One recent analysis places it in a clade with *Xenothrix* and *Paralouatta* as titi monkey relatives (Beck et al., 2023). Kay (2015; Kay et al., 2011) has suggested that it is most likely a stem platyrrhine, not clearly related to any extant clade.

Insulacebus toussaintiana (Fig. 14.10) is a large fossil primate from Haiti that is based on a virtually complete, unworn dentition, a few gnathic remains, possibly a femur, and a humerus. It has an estimated body mass of around 5 kg. *Insulacebus* has a very large upper central incisor compared with the upper lateral, a feature that is found in *Aotus* and, to a lesser degree, in *Saimiri*. The cheek teeth are unusual among platyrrhines in having cusps that are closely approximated relative to the broad base.

Paralouatta varonai (Fig. 14.10) is a Pleistocene platyrrhine from Cuba known from a nearly complete skull, a mandible, numerous isolated teeth, and numerous limb elements. *Paralouatta* was a large monkey with an estimated body mass of 9 kg. The unflexed cranium has a large palate, a posteriorly facing nuchal region, a small brain, and possibly large orbits. The limbs of *Paralouatta* are unlike those of any extant platyrrhine in showing numerous adaptations in the humerus, ulna, and digits for terrestrial quadrupedalism, similar to that found in cercopithecoids (MacPhee and Meldrum, 2006). Despite its name and many similarities between the skull of *Paralouatta* and that of the living howling monkey, there is considerable debate regarding the phylogenetic relationships of *Paralouatta*. Many authorities agree with the original describers that *Paralouatta* is related to *Alouatta* (Kay et al., 2011; Rosenberger et al., 2011). Others place it within a clade containing other Caribbean monkeys, more closely related to titi monkeys than howler monkeys (Beck et al., 2023). While *Paralouatta varonai*, like most of the other Caribbean monkeys, seems to date to the end of the Pleistocene or Recent time periods, a talus similar to that of *P. varonai* and designated as a separate species, *P. marianae*, has been dated to approximately 17 Ma (MacPhee and Iturralde-Vinent, 1995), making it by far the oldest primate from the Caribbean and comparable in age to the fossil platyrrhines from southern Argentina.

The recovery of numerous fossil primates that are strikingly different from anything known elsewhere in the New World, often from sites that predate human colonization of the islands, demonstrates that there was an endemic primate fauna in the Caribbean until quite recently. As a group, the Caribbean primates also seem to be larger than most fossil platyrrhines from the mainland Neotropics, suggesting the possibility of island gigantism among the monkeys as documented for other Caribbean taxa. Interestingly, most seem to fall in the 4 to 5 kg size range, which is not seen among most extant

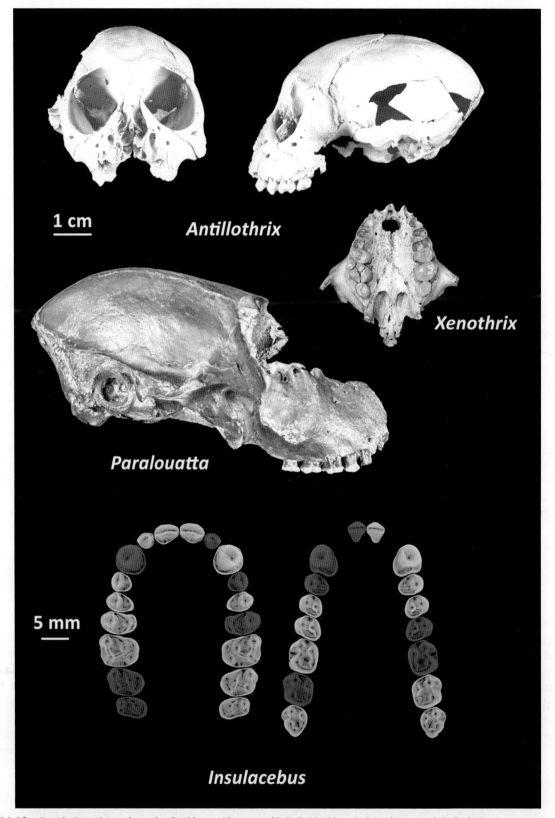

FIGURE 14.10   Fossil platyrrhines from the Caribbean. *(Courtesy of R.D.E. MacPhee, A. Rosenberger, and S. Cooke.)*

platyrrhine faunas, where the monkeys are either much smaller or larger (Cooke et al., 2011). The recovery of many Caribbean fossils, often in excellent condition, has raised interesting issues about the origin and ultimate extinction of these diverse primates.

The simplest explanation for the origin of the Caribbean fauna is over-water dispersal from nearby parts of Central and South America. Cuba is only 50 miles from the Yucatan Peninsula, where *Alouatta*, *Ateles*, and *Cebus* are found today. In addition, the Lesser Antilles are close to Venezuela, where there is an even more extensive fauna. However, morphological evidence demonstrates that the fossil primates from the Caribbean are quite different from extant platyrrhines on the "mainland." This is compatible with the view that the Caribbean primate may have been separated from other platyrrhines for many millions of years. This hypothesis was supported by the Miocene primate talus from Cuba noted above. The distinctiveness of many other aspects of the Caribbean mammal fauna has long been noted as well (Hedges, 1996). For example, the Caribbean sloths have been reported to be more similar to the Miocene sloths of Argentina than to the extant sloths of Central and South America.

There have been several theories regarding the history and relationships of the numerous Caribbean primates. Some authorities (Rosenberger et al., 2009, 2011; Kay et al., 2011) have argued that the Caribbean primates comprise a diverse group with phylogenetic relationships to several separate extant clades, whereas Kay (Kay et al., 2011; Kay, 2015) has argued that they are a diverse group of stem platyrrhines. MacPhee and colleagues have argued that all known Caribbean primates are part of a single endemic clade related to *Callicebus* (MacPhee and Horovitz, 2002, 2004; see also Tallman and Cooke, 2016). The genetic evidence that *Xenothrix* is related to extant titi monkeys and diverged from them 11 million years ago falsifies the view that all the Caribbean monkeys are stem platyrrhines, and the presence of a *Paralouatta* talus in Cuba several million years before the divergence of *Xenothrix* from titi monkeys falsifies the view that all of the Caribbean taxa are a single radiation (discussed in Woods et al., 2018). These fossils indicate that the extinct primate fauna from the Caribbean is likely the result of several over-water colonizations. The details of the history and origin of the Caribbean platyrrhines are far from resolved, and with our knowledge of Caribbean monkeys increasing at a rapid pace, these details are likely to be a hot topic for many years to come (Cooke et al., 2011; Woods et al., 2018).

Summary of Fossil Platyrrhines

Despite the relative paucity of fossils from Central and South America, compared to the record from the Eocene of North America or the Miocene through Pleistocene of Africa, the remains of fossil platyrrhines provide a number of insights into the history and timing of appearance of many modern groups of New World monkeys. Perhaps the most striking feature of the platyrrhine fossil record is the overall similarity of many extinct species to modern lineages. Although it is important to remember that our knowledge of fossil New World monkeys is based largely on fragmentary dental remains, it seems that many lineages of extant platyrrhines have been distinct since at least the Middle Miocene (Fig. 14.11). Fossil species related to the extant owl monkey (*Aotus*), squirrel monkey (*Saimiri*), titi monkey (*Callicebus*) and other pitheciids, and the howling monkey (*Alouatta*) were most likely present in the Late Miocene of Colombia (Wheeler, 2010). Evidence suggesting that some of these lineages can be traced back to the Late Oligocene (Colhuehuapian) or Early Miocene (Pinturan and Santacrucian) times is the subject of ongoing debate (e.g., Rosenberger, 2002, 2010, 2019, 2020; Kay et al., 2008; Kay, 2015; Marivaux, 2016; Beck et al., 2023; Tejedor and Novo, In Press). There is some suggestive evidence of relationships to a few extant clades, particularly for pitheciids, but also evidence from other analyses suggesting that the fossil monkeys from the Miocene of Patagonia are stem platyrrhines that precede the modern radiation (Kay, 2015; Kay et al., 2008; Hodgson et al., 2009; Marivaux et al., 2016; Beck et al., 2023).

The fossil record also provides evidence that the extant platyrrhine fauna is very impoverished from that in the Pleistocene. *Protopithecus*, *Cartelles*, and *Caipora* were much larger than any extant taxa, roughly twice the size of the largest living platyrrhines. The Recent primates from the Caribbean show a remarkable diversity in morphology and are all very different from any extant taxa in both their anatomy and their size distribution. The New World, like Madagascar, has clearly suffered dramatic Pleistocene extinctions of its primate fauna. We can only imagine what other discoveries may come from further fieldwork.

Platyrrhine Origins

Where did platyrrhines come from, and how did platyrrhines get to South America? The issue is a particularly complex one involving not only paleontological information about fossil platyrrhines but also information about paleogeography and the faunas of other continental areas. The origins question integrates our understanding of the position and interconnections among continents through time as well what we can reconstruct about the relationships and divergence times among modern primate clades.

South America was an island continent throughout most of the Early Cenozoic, separated from Africa by the South Atlantic and from North America by the Caribbean Sea. The Central American connection between the two continents did not come into place until the Late Miocene or Pliocene. Debate over the origin of platyrrhine primates has, for many decades, focused on whether North

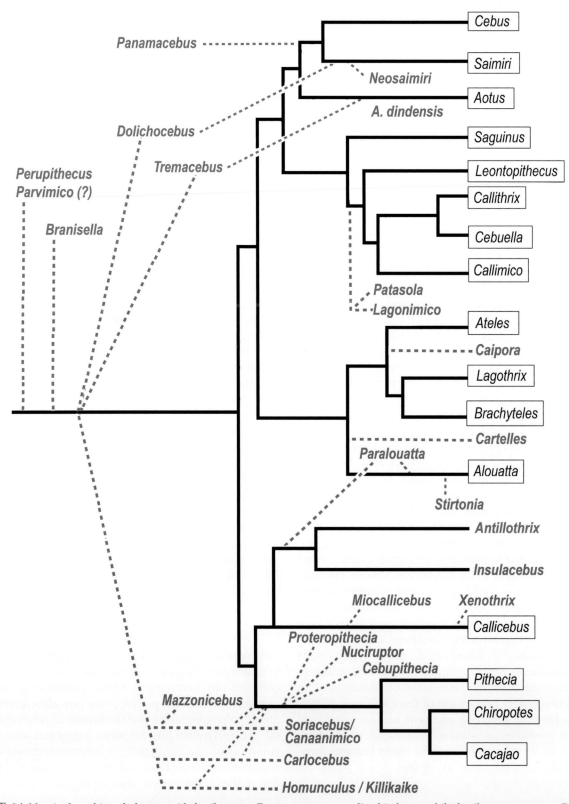

FIGURE 14.11   A platyrrhine phylogeny with fossil genera. Extant genera are outlined in boxes while fossil genera are gray. Dotted gray lines indicate possible phylogenetic positions for the fossil taxa; note that some fossil taxa have competing phylogenetic interpretations.

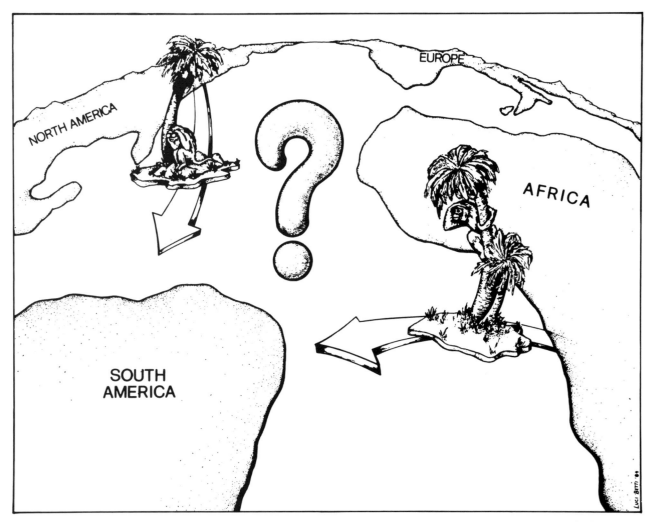

FIGURE 14.12   How did ancestral platyrrhines reach South America? Map lines delineate study areas and do not necessarily depict accepted national boundaries.

America or Africa is the most likely source of the immigrating primates (Fig. 14.12).

Most geophysical studies indicate that the positions of North and South America and Africa relative to one another were much the same in the Eocene and Oligocene as they are now; the rifting of the South Atlantic had taken place much earlier, during the Mesozoic era. There was, then, a considerable body of water for migrating primates to cross, from either North America or Africa. The geological history of the Caribbean region is, however, very poorly known.

During the Early Cenozoic, though, there were probably large areas of relatively shallow water in the South Atlantic, due to crustal uplift, and possibly a series of islands in the areas of the Walvis Ridge and the Sierra Leone Rise. In periods of low sea level, such as the Early Oligocene, these areas and the continental shelves of Africa were probably dry land, which would appreciably shorten the open-water distances between the continents. The reconstructed currents for this time also seem to favor a crossing from Africa to South America (Houle, 1999; Campbell et al., 2021).

Because all available evidence indicates that the immigrant primates that first appeared in South America and gave rise to living platyrrhines were anthropoids rather than prosimians, we must also consider the nature of the fossil primates on potential source continents. North America, Africa, or Antarctica could be a source area for platyrrhines only if there were suitable primates on those continents before or around the time that platyrrhines arrived in South America. In this respect, Africa is unquestionably the most likely source of early platyrrhines. There are numerous Eocene and Oligocene anthropoids from Africa. Moreover, there are many similarities between the Eocene and Oligocene anthropoids from Africa and extant and fossil platyrrhines. In addition, South American rodents, which appear in the fossil record slightly earlier than the primates (Wyss et al., 1993; Antoine et al., 2012), are most closely related to the rodents known from the Eocene through to the present of Africa, providing further evidence of a faunal connection between South America and Africa.

By contrast, there is no evidence of either Eocene or Oligocene anthropoids in North America. The only anthropoid to appear in North America prior to humans 20,000 years ago is *Panamacebus* at the very southern edge of the continent in the early Miocene. However, its similarity to other crown platyrrhines from South America strongly suggests that, like the Caribbean fossil platyrrhines, it is an immigrant to North America rather than an endemic taxon. Thus, arguments for a North American origin for platyrrhines must postulate either a separate prosimian ancestry for platyrrhines and catarrhines, which conflicts with all morphological and molecular evidence, or colonization of South America by a group of still-unknown North American or Central American anthropoids. Certainly, the presence of prosimians with European affinities (the caenopithecid adapoid *Mahgarita*, or the enigmatic *Rooneyia*) in the latest Eocene or earliest Oligocene of Texas demonstrates the existence of North American primates that do not seem to be related to most of the Eocene primates from the Rocky Mountain region. Nevertheless, they are not anthropoids (but see Rosenberger, 2020), and on the basis of the known North American Eocene primate fauna, that continent is a very unlikely source of early platyrrhines. In addition, it should be noted that there is no evidence for the appearance in South America of the many groups of nonprimates that are well documented and abundant in North American faunas of the Eocene through Oligocene.

The presence of numerous possible anthropoids or even catarrhines in the Eocene of Asia suggests to some the possibility of an Asian origin of platyrrhines. However, an Asian origin would presumably require a North American path of dispersal that seems belied by the lack of any fossil anthropoids in North America.

The one continent that was actually connected to South America during much of the early part of the Cenozoic is Antarctica, and there is evidence of faunal connections between the two continents as well as with Australia. However, the mammals they have in common are marsupials. There is no evidence of any primates on either Antarctica or Australia prior to the relatively recent arrival of humans.

Finally, if, as all evidence indicates, primates (and rodents) arrived in South America sometime by the Oligocene from Africa (e.g., Hartwig, 1994), they must have crossed a large expanse of open ocean. Many authorities have, quite reasonably, questioned whether long-distance rafting between any continents is a likely method for biogeographic dispersal of animals, given the dietary and climatic requirements of primates. Anthropoids seem less suited to dispersal on floating masses of vegetation than animals that can hibernate or have long periods of inactivity (such as cheirogaleids or

rodents). But regardless of how unlikely rafting may seem, it is currently the only suggested mechanism for transporting terrestrial animals between continents separated by open ocean. If South America was indeed an island continent during the period in question, we must assume that primates rafted from some other continental area (e.g., Houle, 1999). Only a revision of the paleocontinental maps or a dramatic revision of our understanding of the timing of primate evolution could eliminate the need for rafting in the origin of platyrrhines.

While researchers have argued over the way in which primates and rodents got to South America, the recent discovery of fossil primates virtually identical to those of Africa in the early Oligocene deposits of Eastern Peru serves to confirm that platyrrhines came from Africa by rafting across the South Atlantic (Bond et al., 2015; Seiffert et al., 2020; Campbell et al., 2021).

References

Ameghino, E., 1891. Nuevos restos de mamiferos fosiles descubiertos por Carlos Ameghino en al Eoceno inferior del la Patagonia Austral. Rev. Argentina. Hist. Nat. 1, 289–328.

Antoine, P.-O., Marivaux, L., Croft, D.A., Billet, G., Ganerød, M., Jaramillo, C., Martin, T., Orliac, M.J., Tejada, J., Altamirano, A.J., Duranthon, F., Fanjat, G., Rousse, S., Gismondi, R.S., 2012. Middle Eocene rodents from Peruvian Amazonia reveal the pattern and timing of caviomorph origins and biogeography. Proc. Roy. Soc. B 219, 1319–1326.

Beck, R.M.D., de Vries, D., Janiak, M.C., Goodhead, I.B., Boubli, J.P., 2023. Total evidence phylogeny of platyrrhine primates and a comparison of undated and tip-dating approaches. J. Hum. Evol. 174, 103293. https://doi.org/10.1016/j.jhevol.2022.103293.

Bloch, J.I., Woodruff, E.D., Wood, A.R., Rincon, A.F., Harrington, A.R., Morgan, G.S., Foster, D.A., Montes, C., Jaramillo, C.A., Jud, N.A., Jones, D.S., MacFadden, B.J., 2016. First North American fossil monkey and early Miocene tropical biotic interchange. Nature. 533, 243–246.

Bluntschli, H., 1931. Homunculus patagonicus und die ihm zugereihten Fossil funde aus den Santa-Cruz-Schichten Patagoniens. Morphol. Jahr. 67, 811–892.

Bond, M., Tejedor, M.F., Campbell Jr., K.E., Chornogubsky, L., Novo, N., Goin, F., 2015. Eocene primates of South America and the African origins of New World monkeys. Nature. 520, 538–541.

Bown, T.M., Larriestra, C.N., 1990. Sedimentary paleoenvironments of fossil platyrrhine localities, Miocene Pinturas Formation, Santa Cruz Province, Argentina. J. Hum. Evol. 19, 87–119.

Campbell Jr., K.E., O'Sullivan, P.B., Fleagle, J.G., Seiffert, E., 2021. An early Oligocene age for the oldest known monkeys and rodents of South America. Proc. Natl. Acad. Sci. 118, e2105956118.

Cartelle, C., Hartwig, W.C., 1996. A new extinct primate among Pleistocene megafauna of Bahia, Brazil. Proc. Natl. Acad. Sci. USA. 93, 6405–6409.

Cooke, S., 2011. Paleodiet of extinct platyrrhines with emphasis on the Caribbean forms: Three-dimensional geometric morphometrics of mandibular second molars. Anat. Rec. 294, 2073–2091.

Cooke, S., Rosenberger, A.L., Turvey, S., 2011. An extinct monkey from Haiti and the origins of the Greater Antillean primates. Proc. Natl. Acad. Sci. USA. 108, 2699–2708.

Dantas, M.A., Araujo, A.V., Ell Tink, E., Leal, L.A., Araujo-Junior, H., Liparini, A., Cozzuol, M., Fleagle, J.G., 2019. A new record of a late quaternary megaprimate from Bahia. XXV Brazilian Congress of Paleontology, vol. 2, 118548.

Fleagle, J.G., 1990. New fossil platyrrhines from the Pinturas Formation, southern Argentina. J. Hum. Evol. 19, 61–86.

Fleagle, J.G., Tejedor, M.F., 2002. Early platyrrhines of southern South America. In: Hartwig, W.C. (Ed.), The Primate Fossil Record. Cambridge University Press, Cambridge, pp. 161–174.

Fleagle, J.G., Bown, T.M., Swisher, C., Buckley, G., 1995. Age of the Pinturas and Santa Cruz Formations. VI Cong. Argentin de Paleont. Y Bioestrat. Actas, 129–135.

Fleagle, J.G., Kay, R.F., Anthony, M.R.L., 1997. Fossil New World monkeys. In: Kay, R.F., Madden, R.L., Cifelli, R.L., Flynn, J.L. (Eds.), Vertebrate Paleontology in the Neotropics: The Miocene Fauna of La Venta, Colombia. Smithsonian Institution Press, Washington, D.C.

Fleagle, J.G., Perkins, M., Heitzler, M., et al., 2012. Tephrochronology and paleontology of the Santa Cruz and Pinturas Formations, Argentina. In: Vizcaino, S.F., Kay, R.F., Bargo, S.M. (Eds.), Early Miocene Paleobiology in Patagonia: High-latitude paleocommunities of the Santa Cruz Formation. Cambridge University Press, Cambridge, pp. 41–58.

Fleagle, J.G., Gladman, J.T., Kay, R.F., 2022. A new humerus of *Homunculus patagonicus*, a stem platyrrhine from the Santa Cruz Formation (Late Early Miocene), Santa Cruz Province, Argentina. Ameghiniana. 59, 78–96.

Flynn, J.J., Wyss, A.R., Swisher, C.C., 1995. An Early Miocene anthropoid skull from the Chilean Andes. Nature. 373, 603–607.

Fulwood, E.L., Boyer, D.M., Kay, R.F., 2016. Stem members of Platyrrhini are distinct from catarrhines in at least one derived cranial feature. J. Hum. Evol. 100, 16–24.

Halenar, L.B., 2011a. Reconstructing the locomotor repertoire of *Protopithecus brasiliensis* I. Body size. Anat. Rec. 294, 2024–2047.

Halenar, L.B., 2011b. Reconstructing the locomotor repertoire of *Protopithecus brasilensis* II. Forelimb morphology. Anat. Rec. 294, 2048–2063.

Halenar, L.B., Rosenberger, A.L., 2013. A closer look at the "Protopithecus" fossil assemblages: New genus and species from the Pleistocene of Minas Gerais. Brazil. J. Human Evol. https://doi.org/10.1016/j.jhevol.2013.07.008.

Halenar, L.B., Cooke, S.B., Rosenberger, A.L., Rimoli, R., 2017. New cranium of the endemic Caribbean platyrrhine, *Antillothrix bernensis*, from La Altagracia Province, Dominican Republic. J. Hum. Evol. 106, 133–153.

Hartwig, W.C., 1994. Patterns, puzzles, and perspectives on platyrrhine origins. In: Corruccini, R.S., Ciochon, R.L. (Eds.), Integrative Paths to the Past: Paleoanthropological Advances in Honor of F. Clark Howell. Prentice Hall, New York, pp. 69–93.

Hartwig, W.C., 1995a. A giant New World monkey from the Pleistocene of Brazil. J. Hum. Evol. 28, 189–196.

Hartwig, W.C., 1995b. Protopithecus: Rediscovering the first fossil primate. Hist. Philos. Life Sci. 17, 447–460.

Hartwig, W.C., Cartelle, C., 1996. A complete skeleton of the giant South American primate. Protopithecus. Nature. 381, 307–311.

Hedges, S.B., 1996. Historical biogeography of West Indian vertebrates. Ann. Rev. Ecol. Syst. 27, 163–196.

Hershkovitz, P., 1974. A new genus of Late Oligocene monkey (Cebidae, Platyrrhine) with notes on postorbital closure and platyrrhine evolution. Folia. Primatol. 21, 1–35.

Hodgson, J.A., Sterner, K.N., Matthews, L.J., et al., 2009. Successive radiations, not stasis, in the South American primate fauna. Proc. Natl. Acad. Sci. USA. 106, 5534–5539.

Houle, A., 1999. The origin of platyrrhines: An evaluation of the Antarctic scenario and the floating island model. Am. J. Phys. Anthropol. 109, 541–559.

Kay, R.F., 1994. Giant tamarin from the Miocene of Colombia. Am. J. Phys. Anthropol. 95, 333–353.

Kay, R.F., 2010. A new primate from the Early Miocene of Gran Barranca, Chubut Province, Argentina: Paleoecological implications. In: Madden, R.L., Vucetich, G., Carlini, A.A., Kay, R.F. (Eds.), The Paleontology of Gran Barranca: Evolution and Environmental Change Through the Middle Cenozoic of Patagonia. Cambridge University Press, Cambridge, pp. 220–239.

Kay, R.F., 2015. Biogeography in deep time – what do phylogenetics, geology, and paleoclimate tell us about early platyrrhine evolution? Mol. Phylogenet. Evol. 82, 358–374.

Kay, R.F., Frailey, C.D., 1993. Large fossil platyrrhines from the Rio Acre local fauna, Late Miocene, western Amazonia. J. Hum. Evol. 25, 319–327.

Kay, R.F., Meldrum, D.J., 1997. A new small platyrrhine and the phyletic position of Callitrichidae. In: Kay, R.F., Madden, R.H., Cifelli, R.L., Flynn, J.L. (Eds.), Vertebrate Paleontology in the Neotropics: The Miocene Fauna of La Venta, Colombia. Smithsonian Institution Press, Washington, D.C, pp. 453–458.

Kay, R.F., Johnson, D.D., Meldrum, D.J., 1998. A new pitheciin primate from the Middle Miocene of Argentina. Am. J. Primatol. 45, 317–336.

Kay, R.F., Johnson, D.D., Meldrum, D.J., 1999. Proteropithecia, new name for. Propithecia. Am. J. Primatol. 47, 347.

Kay, R.F., Cozzuol, M.A., 2006. New platyrrhine monkeys from the Solimoes Formation (Late Miocene, Acre State, Brazil). J. Hum. Evol. 50, 673–686.

Kay, R.F., Fleagle, J.G., 2010. Stem taxa, homoplasy, long lineages, and the phylogenetic position of *Dolichocebus*. J. Hum. Evol. 59, 218–222.

Kay, R.F., Madden, R.H., Cifelli, R.L., Flynn, J.J., 1997. Vertebrate Paleontology in the Neotropics: The Miocene Fauna of La Venta, Colombia. Smithsonian Institution Press, Washington, D.C.

Kay, R.F., Williams, B.A., Anaya, F., 2002. Adaptations of *Branisella boliviana*, the earliest South American monkey. In: Plavcan, J.M., Kay, R.F., Jungers, W.L., van Schaik, C.P. (Eds.), Reconstructing Behavior in the Primate Fossil Record. Plenum, New York, pp. 339–370.

Kay, R.F., Gonzales, L.A., Salenbien, W., et al., 2019. Parvico materdei gen. et sp. nov.: a new platyrrhine from the Early Miocene of the Amazon Basin. Peru. J. Hum. Evol. 134, 102628. https://doi.org/10.1016/j.jhevol.2019.05.016..

Kay, R.F., Fleagle, J.G., Mitchell, T., Colbert, M., Bown, T.M., Powers, D.W., 2008. The anatomy of *Dolichocebus gaimanensis*, a stem platyrrhine monkey from Argentina. J. Hum. Evol. 54, 323–382.

Kay, R.F., Hunt, K.D., Beeker, C.D., Conrad, G.W., Johnson, C.C., Keller, J., 2011. Preliminary notes on a newly discovered skull of the extinct monkey Antillothrix from Hispaniola and the origin of the Greater Antillean monkeys. J. Hum. Evol. 60, 124–128.

Kay, R.F., Perry, J.G.M., Malinzak, M., et al., 2012. The paleobiology of Santacrucian primates. In: Vizcaino, S.F., Kay, R.F., Bargo, M.S. (Eds.), Early Miocene Paleobiology in Patagonia: High Latitude Paleocommunities of the Santa Cruz Formation. University Press, Cambridge.

Krupp, A., Cartelle, C., Fleagle, J.G., 2012. Size and external morphology of the brains of the large fossil platyrrhines Protopithecus and Caipora. Am. J. Phys. Anthropol. Supp. 54, 186.

Luchterhand, K., Kay, R.F., Madden, R.H., 1986. Mohanamico hershkovitzi, gen, et so, nov. un primate du Miocene moyen d'Amerique du Sud. C. R. Acad. Sci. (Paris), ser 3 (303), 1753–1758.

MacPhee, R.D.E., Fleagle, J.G., 1991. Postcranial remains of *Xenothrix mcgregori* (Primates, Xenotrichidae) and other Late Quaternary mammals from Long Mile Cave, Jamaica. Bull. Am. Mus. Nat. Hist. 206, 287–319.

MacPhee, R.D.E., Iturralde-Vinent, M.A., 1995. Earliest monkey from Greater Antilles. J. Hum. Evol. 28, 197–200.

MacPhee, R.D.E., Horovitz, I., 2002. Extinct Quaternary platyrrhines of the Greater Antilles and Brazil. In: Hartwig, W.C. (Ed.), The Primate Fossil Record. Cambridge University Press, Cambridge, pp. 189–200.

MacPhee, R.D.E., Horovitz, I., 2004. New craniodental remains of the Quaternary Jamaican monkey *Xenothrix mcgregori* (Xenotrichini, Callicebinae, Pitheciidae), with a reconsideration of the Aotus hypothesis. Am. Mus. Nov. 3434, 1–55.

MacPhee, R.D.E., Meldrum, D.J., 2006. Postcranial remains of the extinct monkeys of the Greater Antilles, with evidence for semiterrestriality in Paralouatta. Am. Mus. Nov. 3516, 1–65.

MacPhee, R.D.E., Horovitz, I., Arredondo, O., Vasquez, O.J., 1995. A new genus for the extinct Hispaniolan monkey Saimiri bernensis (Rimoli, 1977), with notes on its systematic position. Am. Mus. Nov. 3134, 1–21.

Marivaux, L., Salas-Gismondi, R., Tejada, J., et al., 2012. A platyrrhine talus from the Early Miocene of Peru (Amazonian Madre de Dios Sub-Andean Zone). J. Hum. Evol. 63, 696–703.

Marivaux, L., Adnet, S., Altamirano, A.J., et al., 2016. Neotropics provide insights into the emergence of New World monkeys: New dental evidence from the late Oligocene of Peruvian Amazonia. J. Hum. Evol. 97, 159–175.

Marivaux, L., Aguirre-Díaz, W., Benites-Palomino, A., et al., 2020. New record of Neosaimiri (Cebidae, Platyrrhini) from the late Middle Miocene of Peruvian Amazonia. J. Hum. Evol. 146, 102835.

Meldrum, D.J., Lemelin, P., 1991. The axial skeleton of Cebupithecia sarmientoi, a fossil platyrrhine from the middle Miocene of La Venta, Colombia. Am. J. Primatol. 25, 69–89.

Meldrum, D.J., Kay, R.F., 1997. Nuciruptor rubricae, a new pitheciin seed predator from the Miocene of Colombia. Am. J. Phys. Anthropol. 102, 407–427.

Ni, X., Flynn, J.J., Wyss, A.R., 2010. The bony labyrinth of the early platyrrhine primate Chilecebus. J. Hum. Evol. 59, 595–607.

Ni, X., Flynn, J.J., Wyss, A.R., Zhang, C., 2019. Cranial endocast of a stem platyrrhine primate and ancestral brain conditions in anthropoids. Sci. Adv. 5, eaav7913.

Perkins, M.E., Fleagle, J.G., Heitzler, M., Nash, B., Bown, T.M., 2012. Tephrochrology of the Miocene Santa Cruz and Pinturas Formations. In: Vizcaino, S.F., Kay, R.F., Bargo, S.M. (Eds.), Early Miocene Paleobiology in Patagonia high-latitude paleocommunities of the Santa Cruz Formation. Cambridge University Press, Cambridge, pp. 23–40.

Rosenberger, A.L., 1979. Cranial anatomy and implications of Dolichocebus, a Late Oligocene ceboid primate. Nature. 279, 416–418.

Rosenberger, A.L., 1984. Fossil New World monkeys dispute the molecular clock. J. Hum. Evol. 13, 737–742.

Rosenberger, A.L., 2002. Platyrrhine paleontology and systematics: The paradigm shifts. In: Hartwig, W.C. (Ed.), The Primate Fossil Record. Cambridge University Press, Cambridge, pp. 151–159.

Rosenberger, A.L., 2010. Platyrrhines, PAUP, parallelism, and the long lineage hypothesis: A reply to Kay et al. (2008). J. Hum. Evol. 59, 214–217.

Rosenberger, A.L., 2019. Dolichocebus gaimanensis is not a stem platyrrhine. Folia Primatol. 90, 494–506.

Rosenberger, A.L., 2020. New World Monkeys: The Evolutionary Odyssey. Princeton University Press, Princeton, NJ.

Rosenberger, A.L., Hartwig, W.C., Takai, M., Setoguchi, T., Shigehara, N., 1991a. Dental variability in Saimiri and the taxonomic status of Neosaimiri fieldsi, an early squirrel monkey from Colombia, South America. Int. J. Primatol. 12, 291–301.

Rosenberger, A.L., Hartwig, W.C., Wolff, R.G., 1991b. Szalatavus attricuspis and early platyrrhine primates. Folia Primatol. 56, 225–233.

Rosenberger, A.L., Setoguchi, T., Hartwig, W.C., 1991c. Laventiana annectens: New fossil evidence for the origin of callitrichine New World monkeys. Proc. Natl. Acad. Sci. USA. 88, 2137–2140.

Sears, K.E., Finarelli, J.A., Flynn, J.J., Wyss, A.R., 2008. Estimating body mass in New World "monkeys" (Platyrrhini, Primates), with consideration of the Miocene platyrrhine, Chilecebus carrascoensis. Am. Mus. Nov. 3617, 1–29.

Rosenberger, A.L., Tejedor, M., Cooke, S., Pekar, S., 2009. Platyrrhine ecophylogenetics in space and time. In: Garber, P., Estrada, A., Bicca-Marques, J.C., Heymann, E.W., Strier, K.B. (Eds.), South American Primates: Comparative Perspectives in the Study of Behavior, Ecology, and Conservation. Springer, New York, pp. 69–116.

Rosenberger, A.L., Cooke, S., Rimoli, R., Ni, X., Cardosa, L., 2011. First skull of Antillothrix bernensis, an extinct relic monkey from the Dominican Republic. Proc. Biol. Sci. 278, 67–74.

Seiffert, E.R., Tejedor, M.F., Fleagle, J.G., Novo, N.M., Cornejo, F.M., Bond, M., de Vries, D., Campbell Jr., K.E., 2020. A parapithecid stem anthropoid of African origin in the Paleogene of South America. Science. 368, 194–197.

Setoguchi, T., Rosenberger, A.L., 1985. Miocene marmosets: First fossil evidence. Int. J. Primatol. 6, 615–625.

Takai, M., Anaya, F., 1996. New specimens of the oldest fossil platyrrhine, Branisella boliviana, from Salla, Bolivia. Am. J. Phys. Anthropol. 99, 301–317.

Takai, M., Anaya, F., Shigehara, N., Setoguchi, T., 2000. New fossil material of the earliest New World monkey, Branisella boliviana, and the problems of platyrrhine origins. Am. J. Phys. Anthropol. 111, 263–281.

Takai, M., Anaya, F., Suzuki, H., Shigehara, N., Setoguchi, T., 2001. A new platyrrhine from the Middle Miocene of La Venta, Colombia, and the phyletic position of Callicebinae. Anthropol. Sci. 109, 289–307.

Tallman, M., Cooke, S.B., 2016. New endemic platyrrhine humerus from Haiti and the evolution of the Greater Antillean platyrrhines. J. Hum. Evol. 91, 144–166.

Tauber, A., 1991. Homunculus patagonicus Ameghino 1891 (Primates, Ceboidea), Mioceno temprano, de la costa Atlantica austral, Prov. de Santa Cruz, Republica Argentina. Acad. Nac. Ciencias Misc. 82, 1–32.

Tejedor, M.F., 2005. New specimens of Soriacebus adrianae Fleagle, 1990, with comments on pitheciin primates from the Miocene of Patagonia. Ameghiniana. 41, 249–251.

Tejedor, M.F., Rosenberger, A.L., 2008. A neotype for Homunculus patagonicus Amegnino, 1981, and a new interpretation of the taxon. Paleoanthro. 67, 82.

Tejedor, M., Tauber, A., Rosenberger, A.L., Swisher, C.C., Palacios, M.E., 2006. New primate genus from the Miocene of Argentina. Proc. Natl. Acad. Sci. USA. 103, 5437–5441.

Tejedor, M.F., Rosenberger, A.L., Cartelle, C., 2008. Nueva especie de Alouatta (Primates, Atelinae) del Pleistoceno Tardio de Bahia, Brasil. Ameghiniana. 45, 247–251.

Tejedor, M.F., Novo, N.M., In Press The Bryn Gwyn monkey, Dolichocebus gaimanensis, and its pivotal role in a phylogenetic controversy.

Wheeler, B.C., 2010. Community ecology of the Middle Miocene primates of La Venta, Colombia: The relationship between ecological diversity, divergence time, and phylogenetic richness. Primates. 51, 131–138.

Woods, R., Turvey, S.T., Brace, S., MacPhee, R.D.E., Barnes, I., 2018. Ancient DNA of the extinct Jamaican monkey Xenothrix reveals extreme insular change within a morphologically conservative radiation. Proc. Natl. Acad. Sci. USA. 115, 12769–12774.

Wyss, A.R., Flynn, J.J., Norell, M.A., et al., 1993. South America's earliest rodent and the recognition of a new interval of mammalian evolution. Nature. 365, 434–437.

Further Reading

Fleagle, J.G., Rosenberger, A.L., 1990. The Platyrrhine Fossil Record. Academic Press, New York.

Hartwig, W.C., Meldrum, D.J., 2002. Miocene platyrrhines of the northern Neotropics. In: Hartwig, W.C. (Ed.), The Primate Fossil Record. Cambridge University Press, Cambridge, pp. 151–159.

15

Primitive Catarrhines and Fossil Apes

Miocene Epoch

The Miocene was a relatively long epoch that began approximately 23 million years ago and ended about 5 million years ago. In the Early Miocene, world temperatures warmed appreciably from the cooler Oligocene, and there were minor fluctuations of warming and cooling periods throughout much of the epoch (see Fig. 10.4), as well as increasing aridity in Africa. Several major geophysical events took place during this epoch that affected both global climate and the biogeography of mammals throughout the Old World. The Tethys Sea contracted and was cut off from the Indian Ocean by the emergence of the Arabian Peninsula. However, southern Europe was a changing collection of islands and inland seas, the largest of which was the Paratethys. On at least one occasion in the Late Miocene, the Mediterranean remnant of the Tethys dried up completely. Further east, India continued to crash into Asia, leading to the rise of the Himalayas and the onset of the Monsoon weather pattern.

In East Africa, the Miocene was characterized by considerable volcanic activity in conjunction with the developing rift system. It is here, in the Early Miocene sediments of Kenya and Uganda, that we witness an impressive array of catarrhine primates. The cercopithecoid monkeys of the Early Miocene are not very diverse, and fossil evidence for their major radiation appears only in the later part of this epoch and in the succeeding Pliocene (see Chapter 16). In contrast, the Miocene deposits of Africa and Eurasia hold an extraordinary abundance and diversity of fossil apes (Fig. 15.1; Table 15.1).

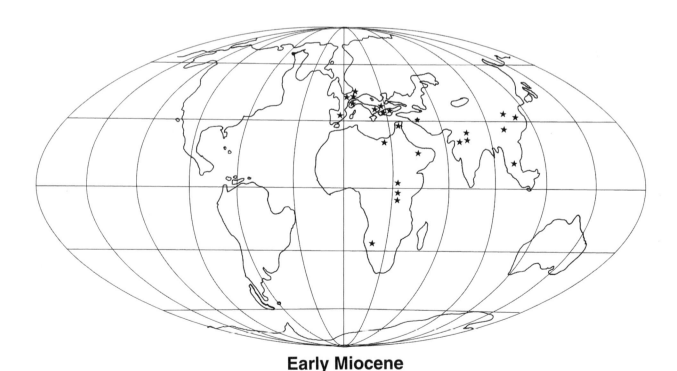

Early Miocene

FIGURE 15.1   Map of the Early Miocene world showing Miocene fossil ape locations. Map lines delineate study areas and do not necessarily depict accepted national boundaries.

Primate Adaptation and Evolution
https://doi.org/10.1016/B978-0-12-815809-8.00015-1

TABLE 15.1   Early and Middle Miocene catarrhines from Africa and Arabia

Species	Estimated mass (g)
Superfamily PROCONSULOIDEA **Family PROCONSULIDAE** **Subfamily PROCONSULINAE**	
Proconsul **(Early Miocene, Africa)**	
P. africanus	35,000
P. major	60,000–90,000
P. gitongai	60,000–90,000
P. meswae	?45,000
Ekembo **(Early Miocene, Africa)**	
E. heseloni	15,000
E. nyanzae	35,000
Subfamily AFROPITHECINAE	
Afropithecus **(Early Miocene, Africa)**	
A. turkanensis	?45,000
Morotopithecus **(Early Miocene, Africa)**	
M. bishopi	50,000
Heliopithecus **(Early Miocene, Arabia)**	
H. leakeyi	25,000
Nacholapithecus **(Middle Miocene, Africa)**	
N. kerioi	16,000
Equatorius **(Middle Miocene, Africa)**	
E. africanus	27,000
Otavipithecus **(Middle Miocene, Africa)**	
O. namibiensis	17,000
Subfamily NYANZAPITHECINAE	
Nyanzapithecus **(Early to Middle Miocene, Africa)**	
N. vancouveringorum	9500
N. pickfordi	9500
N. harrisoni	?7500
N. alesi	11,000
Rangwapithecus **(Early Miocene, Africa)**	
R. gordoni	?17,000
Mabokopithecus **(Middle Miocene, Africa)**	
M. clarki	9500
Turkanapithecus **(Early Miocene, Africa)**	
T. kalakolensis	10,000
Xenopithecus **(Early Miocene, Africa)**	
X. koruensis	?13,000
Rukwapithecus **(Late Oligocene, Africa)**	
R. fleaglei	12,000

TABLE 15.1   Cont'd

Species	Estimated mass (g)
Family DENDROPITHECIDAE	
Dendropithecus **(Early Miocene, Africa)**	
D. macinnesi	7000
D. ugandensis	?6000
Simiolus **(Early to Middle Miocene, Africa)**	
S. enjiessi	5000
S. leakeyorum	3750
S. cheptumoae	5000
S. andrewsi	6000
Micropithecus **(Early Miocene, Africa)**	
M. clarki	3750
M. chamtwaraensis	?5000
Family INCERTAE SEDIS	
Limnopithecus **(Early Miocene, Africa)**	
L. legetet	5500
L. evansi	5500
Kalepithecus **(Early Miocene, Africa)**	
K. songhoensis	5500
Kogolepithecus **(Early Miocene, Africa)**	
K. morotoensis	9000
Lomorupithecus **(Early Miocene, Africa)**	
L. harrisoni	4300
Iriripithecus **(Early Mionece, Africa)**	
I. alekileki	9500
Karamojapithecus **(Early Miocene, Africa)**	
K. akisimia	5000
Samburupithecus **(Late Miocene, Africa)**	
S. kiptalami	?100,000
Superfamily HOMINOIDEA **Family HOMINIDAE**	
Kenyapithecus **(Middle Miocene, Africa, and Eurasia)**	
K. wickeri	30,000
Nakalipithecus **(Late Miocene, Africa)**	
N. nakayami	100,000–140,000
Chororapithecus **(Late Miocene, Africa)**	
C. abyssinius	100,000–130,000
Superfamily INCERTAE SEDIS	
Saadanius **(Middle Oligocene, Arabia)**	
S. hijazensis	15,000–20,000
Kamoyapithecus **(Late Oligocene, Africa)**	
K. hamiltoni	35,000

Middle-Late Oligocene Primitive Catarrhines From Arabia and Africa

Saadanius hijazensis is a medium-sized catarrhine from the Middle Oligocene (29–28 Ma) of western Saudi Arabia (Fig. 15.2; Zalmout et al., 2010). The single specimen, a partial cranium preserving much of the face and upper dentition, is more advanced than *Aegyptopithecus* from the Fayum of Egypt, but less derived than the Early Miocene apes from East Africa discussed later in this chapter or the earliest monkeys described in the next chapter. In contrast to all of the Fayum anthropoids and pliopithecoids (discussed below), *Saadanius* has a complete tubular ectotympanic and is very close to the divergence of Old World monkeys and hominoids.

Kamoyapithecus is a poorly known genus from the latest Oligocene locality of Losodok Northern Kenya (Leakey et al., 1995, 2011). The single species *K. hamiltoni* resembles propliopithecids in having relatively broad upper molars with a restricted trigon but also shows similarities to *Afropithecus*, *Proconsul*, and other Early Miocene apes in its canine shape and large canine size (Hammond et al., 2019). It is not clear whether it precedes the monkey-ape divergence or is a primitive hominoid.

Rukwapithecus, from the late Oligocene of Tanzania, is more clearly a part of the Proconsulidae and is very similar to nyanzapithecines from the Miocene of Kenya (Stevens et al., 2013). There is one species, *R. fleaglei*.

Early and Middle Miocene Apes From Africa

In the latest Oligocene and Early to Middle Miocene sediments of Kenya and Uganda (25 million to 15 million years ago), we find evidence of an extensive radiation of primitive, ape-like catarrhines that are placed here in a single, probably paraphyletic, superfamily: the proconsuloids (Table 15.1). While cranial and postcranial remains are available for only a few of the many genera and species, all of these taxa seem to be more similar to extant catarrhines than are *Aegyptopithecus* and *Propliopithecus* from the Early Oligocene, and some show features that link them exclusively with living apes. However, many of the taxa are poorly known, and it is unclear whether they are stem catarrhines or stem hominoids (e.g., Harrison, 2010b, 2013). Thus, although they are often described as Miocene apes, the term ape is used very broadly to mean noncercopithecoid catarrhine. This is because cercopithecoids have a distinctive dental and skeletal anatomy, and the early anthropoids, as well as living hominoids, lack these derived anatomical features, making it difficult to distinguish basal catarrhines from stem hominoids. These Miocene apes ranged in size from the small, capuchin-size (3.5 kg) *Micropithecus clarki* to the female gorilla-size (50 kg) *Afropithecus* and *Proconsul major*. Fossil apes have been found in association with a variety of paleoenvironments, ranging from

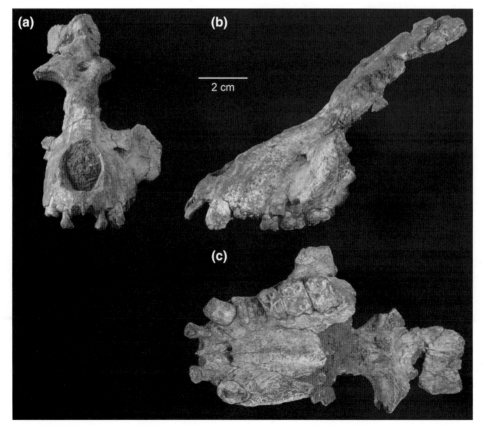

FIGURE 15.2   *Saadanius hijazensis* from the Middle Oligocene of Arabia. *(Courtesy of I. Zalmout.)*

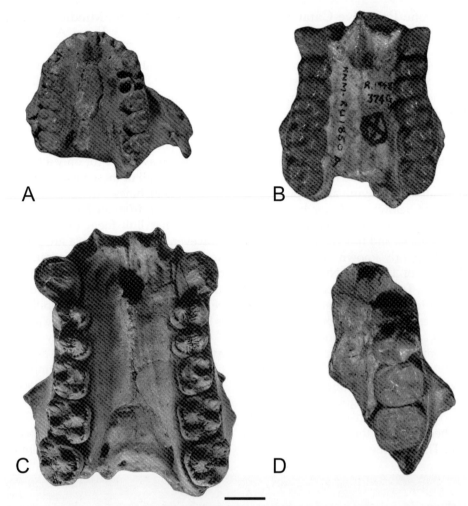

FIGURE 15.3   Upper dentitions of fossil apes from the Early Miocene of East Africa: (A) *Micropithecus clarki*; (B) *Dendropithecus macinnesi*; (C) *Rangwapithecus gordoni*; (D) *Ekembo heseloni*. Notice the large lingual cingulum on the upper molar of all except *Micropithecus* and the very long *Rangwapithecus* molars. Scale = 1 cm. *(Courtesy of R. L. Ciochon and P. Andrews.)*

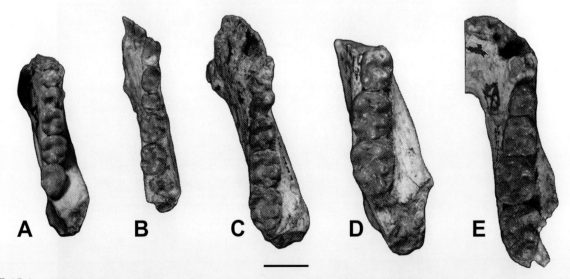

FIGURE 15.4   Lower dentitions of fossil apes from the Early Miocene of East Africa: (A) *Kalepithecus songhorensis*; (B) *Limnopithecus legetet*; (C) *Dendropithecus macinnesi*; (D) *Rangwapithecus gordoni*; (E) *Ekembo heseloni*. Notice the high cusps and well-developed crests on the *Rangwapithecus* molars. Scale = 1 cm. *(Courtesy of R. L. Ciochon and P. Andrews.)*

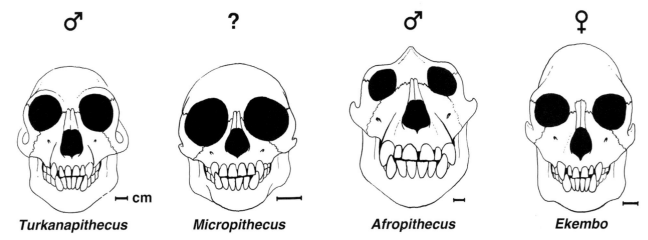

FIGURE 15.5   Reconstructed faces of four Early Miocene fossil apes from East Africa.

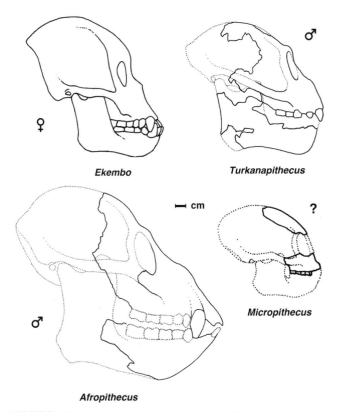

FIGURE 15.6   Reconstructed skulls of four Early Miocene fossil apes from East Africa.

tropical rainforests to open woodlands (Andrews, 1992), and seem to have spanned a range of ecological niches comparable to those occupied today by both Old World monkeys and apes.

The proconsuloids from East Africa are clearly derived from Oligocene propliopithecids but have several features that distinguish them from those more primitive catarrhines (Figs. 15.3 and 15.4). Like the propliopithecids, they have a dental formula of 2.1.2.3, with broad

upper central incisors and smaller upper laterals. The lower incisors of most species are taller and narrower than those of living apes. All species have relatively large, sexually dimorphic canines that shear against the lower anterior premolar. The upper premolars are relatively broad and bicuspid; the posterior lower premolar is a broad semimolariform tooth.

The upper molars distinguish proconsuloids from the earlier propliopithecids and are characterized by their quadrate shape with a relatively larger hypocone; many have a pronounced, often beaded, lingual cingulum and various conules. The lower molars have a broad talonid basin surrounded by five prism-like cusps, including a large hypoconulid. The major dental differences among the many genera and species are in overall size, the relative proportions of the anterior dentition, the development of cingulae, and the development of shearing crests on the molars, features that are often related to dietary differences.

Parts of the facial skeleton and other cranial bones are known for many of the Miocene apes of East Africa (Figs. 15.5 and 15.6; Rae, 1997). The shape of the face varies considerably, being short in some (*Micropithecus*), moderately long and broad in others (*Turkanapithecus*), and long and narrow in still others (*Afropithecus*). In most species, the nasal opening has been described as tall and relatively narrow, as in cercopithecoid monkeys, rather than broad and rounded, as in living apes. Orbit size suggests diurnal habits for all species, but the orbits are relatively larger in the smaller species. The only known auditory regions, those for *Ekembo heseloni* and *Nyanzapithecus alesi*, are identical to that region in living catarrhines (and *Saadanius*) in having a tubular tympanic extending laterally from the side of the bulla. A few specimens preserve enough of the neurocranium or frontal bones to enable some assessment of brain morphology. Relative brain size in *E. heseloni* and also in

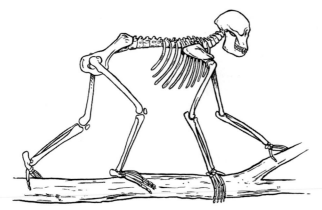

FIGURE 15.7   Juvenile skeleton of *Ekembo heseloni*.

nyanzapithecines (e.g., *Nyanzapithecus, Turkanapithecus*) seems to have been similar to that of living Old World monkeys or perhaps a little larger, but in *Afropithecus*, the brain seems to have been relatively small.

Hundreds of isolated skeletal elements have been attributed to various species of these primitive apes. Several limb elements are assigned to *Turkanapithecus, Morotopithecus, Dendropithecus*, and *Simiolus*, and relatively complete skeletons are available for *Equatorius, Nacholapithecus*, and several individuals of *E. heseloni* (Fig. 15.7). In overall limb proportions and many skeletal details, these primitive apes resemble living platyrrhines, and they lack many specialized skeletal features of the elbow or wrist that characterize either Old World monkeys or living hominoids (e.g., Fleagle, 1983). They have a more primitive, in some ways more behaviorally versatile, locomotor skeleton. However, the interspecific skeletal differences indicate considerable locomotor diversity among the different taxa (Rose, 1993).

The systematics of the Early and Middle Miocene apes from East Africa are poorly resolved. There are several reasons for this: comparable body parts are lacking for many species, many taxa are known primarily from teeth, and new taxa are being described or old ones rearranged each year (e.g., Pickford et al., 2010). The systematic scheme in Table 15.1 largely follows reviews by Harrison (2010b, 2013), but differs in the rank of many groups. Despite the numerous species listed, Early and Middle Miocene fossil catarrhines come almost totally from a limited part of East Africa and undoubtedly underestimate the true diversity of fossil apes from the Miocene of Africa (Fig. 15.8). In this chapter, we will concentrate on the best-known genera and species.

Proconsulids

The larger fossil apes from the Miocene of (mostly) Eastern Africa are usually placed in a distinct family, the Proconsulidae, which in turn can be divided into three groups: proconsulines, afropithecines, and nyanzapithecines, based primarily on aspects of their dentition.

Proconsul was the first Miocene ape described from Africa (Hopwood, 1933), and along with its sister genus *Ekembo*, it remains among the best known (Walker and Teaford, 1989; Walker and Shipman, 2007; McNulty et al., 2015). There are six species currently assigned to *Proconsul* and *Ekembo*. *P. africanus* (~20 kg), from the locality of Koru in Western Kenya, was the first described. *Ekembo heseloni*, a similar-sized species from Rusinga Island (formerly included in *P. africanus*), is known from many individuals and has become the comparative standard for proconsulids (Walker et al., 1993). *Ekembo nyanzae* (20–30 kg) is a larger species from Rusinga Island and nearby Mfwangano. *Proconsul major* (50 kg) is a large species from several sites in Kenya and Uganda that seems to show similarities to *P. africanus* rather than the Rusinga species (MacLatchy and Rossie, 2005; McNulty et al., 2015). *P. gitongai* is another large species from younger deposits in the Tugen Hills of Kenya (Pickford and Kunimatsu, 2005). *P. meswae* is from the locality of Meswa bridge in Western Kenya (Harrison and Andrews, 2009). Some authorities recognize a very different allocation of specimens to taxa and place *P. major* and *P. gitongai* into a separate genus, *Ugandapithecus*, along with other new species (Pickford et al., 2009). However, the more conservative scheme is followed here (Harrison, 2002, 2010b; McNulty et al., 2015). In many aspects of their dentition, these species differ mainly in size (Ruff et al., 1989). All have sexually dimorphic canines and a molar morphology indicating a predominantly frugivorous diet (Ungar and Kay, 1995; Kay and Ungar, 1996).

Many cranial parts are known for the smallest *Ekembo* species, *E. heseloni* (Fig. 15.6). It has a pronounced snout with prominent canine jugae and a relatively robust zygomatic bone. The brain is similar in size to that of a large monkey. As already noted, the auditory region is identical to that of extant apes and cercopithecoid monkeys. The external surface of the brain has a primitive sulcal morphology similar to that seen in gibbons and cercopithecoids, but it lacks many features seen in the brain of living great apes (Falk, 1983).

There is a nearly complete juvenile skeleton known for *E. heseloni* (Fig. 15.7). The limb proportions are monkey-like, with an intermembral index of 89. Compared to living catarrhines, it has short limbs for its estimated body size and a mixture of ape-like and more primitive, monkey-like features throughout the skeleton. It approaches the morphology of living apes in such features as the shape of the distal part of the humerus, the robustness of the fibula, the conformation of the tarsal bones, and the absence of a tail (Ward et al., 1991; Nakatsukasa et al., 2004; Russo, 2016). However, *E. heseloni* lacks many characteristic features of living apes,

FIGURE 15.8 Reconstruction of a fossil ape community from Rusinga Island, Kenya, approximately 18 million years ago; *upper left, Ekembo heseloni; upper right, Dendropithecus macinnesi; center, Limnopithecus legetet; lower, Ekembo nyanzae.*

such as a reduced ulnar styloid process, a short ulnar olecranon, and long curved digits. At the same time, *Ekembo* has none of the detailed skeletal features, such as a narrow elbow region, that characterize cercopithecoid monkeys. The skeleton indicates that *E. heseloni* was an arboreal quadruped and climber but lacked the suspensory abilities of living apes (Rose, 1993).

Ekembo nyanzae resembles *E. heseloni* in many general aspects of its skeleton and lacks any clear similarities to living hominoids in its trunk and hindlimbs (Ward et al., 1993). However, the olecranon process in *E. nyanzae* extends posteriorly rather than proximally, suggesting more terrestrial locomotion than in *E. heseloni*.

The canine sexual dimorphism of all *Ekembo* and *Proconsul* species suggests that there was competition for mates and that they did not live in monogamous social groups. However, in view of the diverse types of social groups found among extant apes, we cannot confidently speculate about many details of the social behavior of these Miocene species.

Afropithecus turkanensis is a large fossil ape from the Early to Middle Miocene of Kenya (Leakey and Leakey, 1986a; Leakey and Walker, 1997). *Afropithecus* has a long, narrow snout with parallel tooth rows, small orbits, and a broad interorbital area. The dentition is characterized by robust, procumbent incisors; short, round, tusk-like canines; and extremely broad upper premolars. A maxilla from Saudi Arabia classified as "*Heliopithecus*" is probably the same genus (Andrews and Martin, 1987b). The few postcranial elements associated with *Afropithecus* from Kenya are similar to those of *Ekembo* and suggest arboreal and quadrupedal habits (Rose, 1993).

There are very diverse views about the affinities of *Afropithecus*. The long snout, straight facial profile, and small frontal bone have suggested to some (Leakey et al., 1991) that *Afropithecus* is more closely related to primitive catarrhines such as *Aegyptopithecus*. However, *Afropithecus* is almost certainly a part of the proconsuloid radiation (Andrews, 1992).

Morotopithecus bishopi (Gebo et al., 1997; MacLatchy, 2004) is from the Early Miocene site of Moroto in Uganda. The Moroto site has been dated at 20.6 Ma, but it has been argued that faunal correlations suggest a younger age (Pickford et al., 2003). The type specimen is a large palate (Pilbeam, 1969) that was originally assigned to *Proconsul major* but which is strikingly similar to that of *Afropithecus* and also to a cranium attributed to *Ugandapithecus*. In contrast to the primitive quadrupedal nature of the limb elements of most Early Miocene hominoids from Africa, several skeletal elements from Moroto are much more similar to those of extant hominoids and indicate that *Morotopithecus* was to some degree suspensory. These include a lumbar vertebra with transverse processes that arise from the neural arch, as in living apes (Sanders and Bodenbender, 1994), and a rounded glenoid fossa of the scapula (Gebo et al., 1997). For many years, the striking

similarities among known dental remains of *Morotopithecus*, *Afropithecus*, and *Ugandapithecus*, combined with the lack of overlap in the known postcranial remains, made the distinctions among these large fossil apes unresolved (Pickford, 2002; Patel and Grossman, 2006; Harrison, 2010b). However, recent finds at Moroto help to clarify diagnostic differences among these taxa in the anatomy and proportions of the lower dentition. (MacLatchy et al., 2019).

Equatorius africanus is a medium-sized ape with many skeletal elements from several Middle Miocene sites in western and central Kenya (McCrossin et al., 1998; Ward et al., 1999). Like *Afropithecus* and *Morotopithecus*, the upper molars of *Equatorius* lack a pronounced lingual cingulum. While the postcranial skeleton of *Equatorius* is similar to that of most other proconsulids in lacking the derived features of extant hominoids, *Equatorius* show numerous features indicating more terrestrial locomotion than in other proconsulids.

Nacholapithecus kerioi, from the Middle Miocene of the Samburu Hills in Kenya, is very similar to *Equatorius*. Like *Afropithecus*, *Morotopithecus*, and *Equatorius*, *Nacholapithecus* has short, robust canines and upper molars with a reduced or absent lingual cingulum. The subnasal region has been suggested to be intermediate in morphology compared to earlier proconsulids like *Ekembo* and that of great apes (Nakatsukasa and Kunimatsu, 2009). The postcranial skeleton (Ishida et al., 2004; Nakatsukasa and Kunimatsu, 2009) is similar to that of *Ekembo*, suggesting quadrupedal locomotion with some vertical climbing (Pina et al., 2021). *Nacholapithecus* definitely lacks a tail (Nakatsukasa et al., 2003, 2004).

Otavipithecus namibiensis, from the Middle Miocene (approximately 13 million years ago) of Namibia, is the only Miocene ape currently known from southern Africa (Conroy et al., 1992). It is a medium-sized ape with an estimated body weight of 14 to 20 kg. The type specimen is a right mandible with four teeth. It shares a number of dental and mandibular similarities with *Kenyapithecus* and *Sivapithecus*. There is also a middle phalanx, which suggests that *Otavipithecus* was an arboreal quadruped (Conroy et al., 1993), a piece of frontal bone, and various other elements. *Otavipithecus* is too poorly known to determine its phylogenetic affinities with any certainty (e.g., Conroy, 1994; Harrison, 2010b). Although it was originally suggested that *Otavipithecus* showed similarities to Middle Miocene apes from Europe, subsequent studies indicate it is more probably related to the Early Miocene taxa from Eastern Africa, such as *Afropithecus* (e.g., Singleton, 1998).

One of the most distinctive of the Early Miocene apes is *Rangwapithecus gordoni*. This medium-sized (15 kg) species has relatively long, narrow molar teeth with numerous shearing crests that indicate a more folivorous diet than that of other Early Miocene apes (Kay, 1977). It also has a very deep mandible. *Rangwapithecus gordoni* seems to be found primarily in rainforest environments, and wrist bones attributed to this species are derived compared to *Ekembo*, suggesting that it perhaps practiced a unique form of quadrupedalism that could have included knuckle-walking or multiple positional behaviors (Wuthrich et al., 2019).

Nyanzapithecus is a small fossil ape known mostly from dental remains and one beautifully preserved infant cranium (Harrison, 1987; Nengo et al., 2017). There are four species: *N. vancouveringorum*, from the Early Miocene of Rusinga Island; *N. harrisoni*, from the Samburu Hills in Central Kenya (Kunimatsu, 1997); *N. pickfordi*, from the Middle Miocene of Maboko Island; and *N. alesi*, from the Middle Miocene of the Turkana Basin. *Nyanzapithecus* shows a number of similarities to *Rangwapithecus* and *Turkanapithecus*. Like *Rangwapithecus*, *Nyanzapithecus* is characterized by long upper premolars with similar-sized buccal and lingual cusps, long upper molars, and lower molars with deep lingual notches. It was a folivorous primate. Cranially, *Nyanzapithecus* shares features with *Turkanapithecus* that are reminiscent of pliopithecoids and gibbons, such as a relatively short snout, short cheek bones, a broad interorbital distance, and projecting orbital rims. More interesting, however, is that *Nyanzapithecus* shares these distinctive cranial and dental features with the European ape *Oreopithecus*. Although these similarities would seem to indicate an African origin for *Oreopithecus* (discussed later in this chapter), such an interpretation is complicated by the fact that a humeral fragment from Maboko, assigned to *Nyanzapithecus*, shows none of the modern skeletal features found in *Oreopithecus* and extant hominoids. These conflicting lines of evidence suggest that either the craniodental similarities between *Oreopithecus* and *Nyanzapithecus* or the postcranial similarities between *Oreopithecus* and living apes are due to convergence (Benefit and McCrossin, 1997; Nengo et al., 2017).

Mabokopithecus clarki is another *Oreopithecus*-like species from the Middle Miocene of Maboko Island, Kenya. It is similar to *Nyanzapithecus* but is known from only a few remains (see Benefit et al., 1998).

Turkanapithecus kalakolensis (Figs. 15.5 and 15.6) is a medium-sized fossil ape from Moruorot and Kalodirr, the same site in northern Kenya that yielded *Afropithecus* (Leakey and Leakey, 1986b; Rossie and Cote, 2022). Similar to *Rangwapithecus* and *Nyanzapithecus*, *Turkanapithecus* has relatively long molars with many extra cusps as well as relatively long and large anterior upper premolars. The mandible is relatively shallow with a broad ascending ramus. The cranium shows a broad, square snout; a broad interorbital region; large, rimmed orbits; and flaring zygomatic arches. Like other proconsulids, *Turkanapithecus* was primarily an arboreal quadruped, but some aspects of the limbs, including a reduced olecranon process, suggest that it was more suspensory than *Ekembo* (Rose, 1993).

Xenopithecus koruensis is based on a single maxillary specimen preserving two upper molars (Madden, 1980).

The specimen has generally been attributed to *Ekembo africanus*, but Harrison (2010b) has suggested that it may well be a distinct taxon related to *Turkanapithecus* and other nyanzapithecines.

Samburupithecus kiptalami is a large Late Miocene ape from Northern Kenya that is known from a single maxillary fragment (Ishida and Pickford, 1997). It differs from most other Miocene genera but resembles modern African apes and nyanzapithecines in having molar teeth that are mesiodistally elongated. *Samburupithecus* was originally argued to be a member of the clade including gorillas, chimpanzees, and humans, but more recent assessments have suggested that it is more similar to proconsuloids, particularly nyanzapithecines (Harrison, 2010b; Begun, 2015; Pugh, 2020, 2022).

Three small apes from Kenya and Uganda are known from moderate amounts of both cranial and postcranial remains, which suggest that they were probably more suspensory than other Miocene apes of Africa, and perhaps also more primitive. They are frequently placed in a separate clade, here recognized as a separate family, the Dendropithecidae. However, it is unclear how much of their difference from the proconsulids is a function of size and the relative completeness of the remains. For example, there is no evidence regarding the presence or absence of a tail in any of the Dendropithecidae. It is entirely possible that the Dendropithecidae are a paraphyletic assemblage of both stem catarrhine and/or stem hominoid taxa (Rossie and Hill, 2018).

Dendropithecus macinnesi is known from numerous jaws and teeth and much of a skeleton. It has tall, narrow incisors and broad molars with numerous crests, suggesting a frugivorous–folivorous diet. It was a medium-sized (9 kg) animal with long, slender limbs similar to those of the Neotropical spider monkey (*Ateles*). It was probably mainly quadrupedal but was the most suspensory of the earlier Miocene apes (Wuthrich et al., 2019). Although there is striking canine dimorphism, both sexes have relatively long, sharp canines, suggesting that the species was possibly monogamous.

Simiolus is a small ape (7 kg) from the locality of Kalodirr in northern Kenya (*S. enjiessi*), the Ngorora Formation of the Tugen Hills (*S. minutus*), and the Middle Miocene locality of Maboko Island (*S. leakeyorum*; see Gitau and Benefit, 1995; Rossie and Hill, 2018). *Simiolus* differs from other small Early Miocene apes in its very narrow upper canines, triangular P3, and long upper molars. It has a mosaic of characteristics found in various other genera but shows the greatest similarities to *Dendropithecus* and *Rangwapithecus*. The limbs of *Simiolus* are very similar to those of *Dendropithecus*, indicating that it was one of the more suspensory dendropithecids (Rose et al., 1992; Rossie et al., 2012).

Micropithecus clarki is the smallest Miocene ape from Africa, with an estimated body weight of 3 to 4 kg.

Micropithecus clarki is from the Early Miocene of Napak and Koru, and a new species *M. chamtwaraensis* is from near Koru (Pickford et al., 2021). Another species, *M. leakeyorum*, is from the Middle Miocene of Maboko Island (Harrison, 1989), but its attribution to *Micropithecus* has been questioned by several researchers who place it in *Simiolus* instead (Gitau and Benefit, 1995; Pickford et al., 2010; see Table 15.1). *Micropithecus* has distinctive dental proportions compared to the other Early Miocene apes. The dentition is characterized by relatively large incisors and canines and relatively small cheek teeth. There is a reduced cingulum on the upper molars. The face of *M. clarki* (Figs. 15.5 and 15.6) has a very short snout, a broad nasal opening, and large orbits, giving it a very gibbon-like appearance (Fleagle, 1975; Rae, 1993). A frontal bone attributed to this species has a smooth cranial surface and lacks any brow ridges.

Micropithecus is dentally very similar to another small "ape," **Dionysopithecus**, from the Miocene of China and other parts of Asia (Fleagle, 1975; Bernor et al., 1988). *Dionysopithecus* has since been identified as an early pliopithecoid (Harrison and Gu, 1999), suggesting a possible link between *Micropithecus* and pliopithecoids.

In addition to the Dendropithecidae, there are several other small apes whose relationships are unclear, in most cases because they are more poorly known than other Miocene taxa (see also Pickford et al., 2010).

The two species of **Limnopithecus**, *L. legetet* and *L. evansi*, were each about the size of a living gibbon (4–5 kg). They had a frugivorous diet. The few skeletal elements of these species indicate that they were arboreal quadrupeds.

Kalepithecus songhorensis is another small hominoid that was formerly included in *Dendropithecus*, but it differs from that taxon in numerous dental and gnathic features, including broad lower premolars, a deep mandible, and a broad palate. It was probably frugivorous.

Lomorupithecus harrisoni is a small ape from Napak in Uganda that has been identified by Rossie and MacLatchy (2006) as a possible pliopithecoid, the first from Africa. It is known from part of a face, a few upper teeth, and a juvenile mandible. Pliopithecoids are only known from a broad distribution in the Miocene of Eurasia (e.g., Begun, 2002). However, it has long been thought that pliopithecoids probably originated in Africa (e.g., Harrison and Gu, 1999), and as noted above and discussed below, *Dionysopithecus*, one of the earliest pliopithecoids, is very similar to *Micropithecus*, also from Napak, so the identification of a pliopithecoid among the small apes of East Africa is quite reasonable.

In recent years, Pickford and colleagues (e.g., 2003, 2010, 2021) have described many new Miocene ape taxa from sites in Uganda and have reassigned many specimens from older collections based on new information. In many cases, these revisions eliminate many of the differences in the primate faunas previously reported for distinct regions of East Africa. However, most of the reassessments are based on isolated teeth. Further study and more complete specimens are needed to clarify their role in our evolutionary understanding of this group.

Later Miocene Hominoids

Most of the African fossil apes discussed so far are generally thought to be stem hominoids, most likely outside the clade made up of extant hylobatids, great apes, and humans. However, there are several fossil apes from the Middle and Late Miocene of Eastern Africa that are widely believed to lie within the modern hominoid clade, most probably as stem great apes.

Kenyapithecus wickeri is from the Middle Miocene of Kenya at Fort Ternan. In the past, the genus *Kenyapithecus* also included the material from numerous sites in western and central Kenya now included in *Equatorius* and *Nacholapithecus*. Removal of this material to these other genera means that the single African species of *Kenyapithecus* is known only from teeth, a few mandibular and maxillary fragments, and the distal part of a humerus. The humerus indicates greater development of terrestrial adaptations than the Early Miocene species, with a prominent medial trochlear lip and a posteriorly directed medial epicondyle (McCrossin and Benefit, 1994). A second species of *Kenyapithecus*, *K. kizili*, has been described from the Middle Miocene site of Pasalar in Turkey (Kelley et al., 2008). Thus, *Kenyapithecus* is the only taxon of fossil ape found in both Africa and Eurasia, and provides an important datum for understanding the biogeography of primate dispersals between the two continents in the Miocene (e.g., Begun et al., 2012).

Nakalipithecus nakayamai is a female gorilla-sized ape from the Late Miocene of the Samburu hills known from a single mandible and about a dozen isolated teeth (Kunimatsu et al., 2007). The variation in size of the teeth suggests sexual size dimorphism. *Nakalipithecus* is extremely similar to *Ouranopithecus macedoniensis* from slightly younger deposits in Greece (Kunimatsu et al., 2007; Pugh, 2022), and is therefore critical to debates over whether the clade giving rise to African great apes and humans originated in Africa or in Eurasia.

Chororapithecus abyssinicus is a fossil ape from the Late Miocene Chorora Formation of Ethiopia and has been described as a direct lineal ancestor of the extant African gorilla (Suwa et al., 2007). This species is known from a small collection of isolated teeth that are similar in size to those of the extant gorilla and show well-developed shearing crests of the molars, suggesting a folivorous diet. Interestingly, in comparisons of dental metrics between the fossil teeth and those of extant apes, *Chororapithecus* seems to show no indication of similarities to gorillas, and other authors have questioned the likelihood that *Chororapithecus* is ancestral to any specific extant ape (e.g., Harrison, 2010a,b).

The fossil apes from the Miocene of East Africa exhibit a diversity in size comparable to that of the living Old World monkeys and most living apes (Fig. 15.9). Although their dental morphology indicates that most species were predominantly frugivorous, there were some genera with more folivorous tendencies, such as *Rangwapithecus*, and probably also *Limnopithecus* and *Dendropithecus* (Ungar and Kay, 1995). The skeletal anatomy of the Early Miocene apes is well known for only a few species, but the many isolated skeletal elements indicate that this radiation included arboreal quadrupeds (*Ekembo heseloni*), suspensory species (*Dendropithecus, Simiolus*, perhaps *Morotopithecus*), and more terrestrial species (*Ekembo nyanzae, Equatorius, Kenyapithecus*). There is no evidence of either the fast running or leaping abilities of living cercopithecoids or the brachiating habits of the lesser apes; rather, the skeletal anatomy suggests less specialized but more versatile locomotor abilities, such as those of the living spider monkeys or chimpanzees.

There is also evidence of diversity in the habitat preferences of different East African species (Pickford, 1983; Andrews, 1992). On the basis of the associated mammals, gastropods, and sediments, it seems that *Limnopithecus, Micropithecus*, and *Proconsul major* were more common in rainforest environments, whereas *Dendropithecus, Ekembo*

heseloni, and *E. nyanzae* were more common in woodland localities. In the overall breadth of their ecological adaptations, the Early Miocene apes seem to have filled more of the locomotor and dietary niches found among extant catarrhines than did the Early Oligocene propliopithecids from Egypt. Compared to the Egyptian fauna, the Early Miocene catarrhines were larger and had more terrestrial, suspensory, and folivorous species. Differences in size dimorphism and also differences between species in the type of sexual canine dimorphism suggest a diversity of social structures.

Phyletic Relationships of Proconsuloids and Other African Miocene Apes

The proconsuloids were much more similar to extant catarrhines in their dentition, cranium, and skeleton than were the Early Oligocene propliopithecids or the contemporaneous pliopithecoids of Eurasia. The one species that is well known, *Ekembo heseloni*, has a tubular tympanic, a larger braincase, and more modern skeletal anatomy than *Aegyptopithecus*. Although cranial and skeletal remains are rare for most other species, there is little evidence from the available remains to indicate that any of the other genera were less advanced. All seem to be more modern catarrhines compared to the propliopithecids.

While all of the proconsuloids seem more advanced than propliopithecids in some features, their relationship to the radiation of extant apes is more difficult to ascertain. Only a few demonstrate derived features that would clearly identify them as part of a lineage uniquely related to apes rather than as primitive catarrhines that precede the monkey–ape divergence.

In most dental and skeletal features, these Early Miocene apes are primitive compared to all living hominoids (Fig. 15.10) but nevertheless show various mosaics of primitive and derived features linking them with extant hominoids. In the same way that *Aegyptopithecus* is a stem catarrhine, the proconsuloids are best regarded as stem hominoids. Many seem to show facial features linking them with extant great apes (Rae, 1997, 1999, 2004; Rossie and Seiffert, 2006) but exhibit only a few ape-like postcranial features. *Ekembo heseloni*, for example, resembles living hominoids in having an incipiently spool-shaped articulation on the distal end of the humerus and probably in lacking a tail. At the same time, it retains an articulation between the ulna and carpal bones and "monkey-like" lumbar vertebrae. Thus, *Ekembo* possesses some of the derived features that characterize the ape lineage but lacks others. Likewise, *Afropithecus* has relatively large premolars that link this genus with extant hominoids but lacks derived features of the skeleton. *Morotopithecus* has hominoid features in its premolars, lumbar vertebrae, and glenoid articulation, but has

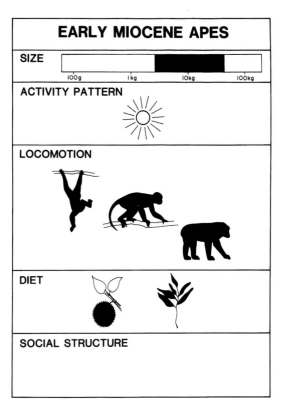

FIGURE 15.9 Adaptive diversity of Early Miocene apes.

FEATURES LINKING SOME EARLY MIOCENE APES WITH LIVING HOMINOIDS

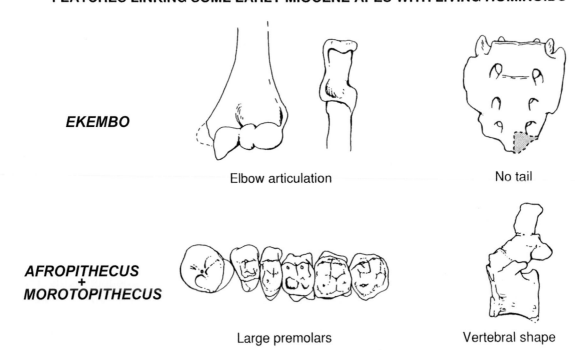

FIGURE 15.10   Features linking Early Miocene apes with living hominoids.

a more primitive femur. *Otavipithecus* from southern Africa is too poorly known to allow an accurate assessment of its affinities.

It is often assumed that the common anatomical specializations of extant apes in the postcranium characterized their last common ancestor and that the radiation of living apes came from a type of hominoid more advanced than any of the proconsuloids, even though several taxa may be in the common ancestry of later apes (e.g., Begun et al., 1997). Nevertheless, there has undoubtedly been considerable parallel evolution in the history of modern hominoid musculoskeletal anatomy, so the actual pathway of modern ape limb specializations is far from resolved (e.g., Larson, 1998; Fleagle and Lieberman, 2013).

Although these Miocene catarrhines are more primitive than living apes, they are also quite different from any cercopithecoid monkeys. The morphology of both these Early Miocene apes and the earlier Oligocene catarrhines, such as *Aegyptopithecus*, demonstrate that Old World monkeys are a very specialized group of higher primates. In their dentition, skull, and some aspects of their skeleton, the living apes have retained many more features from early catarrhines than have cercopithecoids. It is the retention of these primitive catarrhine features that often makes the distinction between stem catarrhines and stem hominoids difficult to ascertain in fragmentary fossil taxa.

While the fossil apes from the Early and Middle Miocene of East Africa are often treated as a single radiation and a single taxonomic group, the proconsuloids, this is undoubtedly a gradistic classification. Different taxa show different mosaics of primitive and derived features of the dentition, cranium, and limb skeleton (e.g., Begun et al., 1997), and most are known only from teeth and jaws, and perhaps a few other bits of anatomy. It is also possible that several more derived lineages of higher primates can be traced back to one or more of these Miocene apes. The lineage leading to *Oreopithecus*, for example, may have originated via *Nyanzapithecus* from a genus such as *Rangwapithecus*. It has also been suggested that the lineage leading to great apes and humans can be traced back uniquely to *Morotopithecus*, to the exclusion of other proconsuloids (MacLatchy et al., 2000; Young and MacLatchy, 2004). The position of the proconsuloid radiation relative to the origin of the gibbon lineage is more difficult to determine from current evidence. Most authors feel that all proconsuloids precede the origin of gibbons. However, if the great ape and human lineage can be traced to a genus from the Miocene of East Africa, as many believe, the more primitive gibbon lineage must have originated somewhere within this radiation as well.

Although various fossils from Eastern Asia have from time to time been linked to proconsuloids from Africa, more recently, these have all been identified as pliopithecoids (e.g., Harrison and Gu, 1999; Begun, 2002). Nevertheless, as discussed below, the origin of the Eurasian pliopithecoids is itself unknown, and most authorities suspect that they are somehow derived from somewhere in the radiation of proconsuloids. This view is supported by the identification of a possible pliopithecoid from the Early Miocene of Uganda (Rossie and MacLatchy, 2006; but see Harrison, 2010b).

Eurasian Pliopithecoids and Apes

Although fossil anthropoid primates were first discovered in Europe and Asia over 180 years ago, for most of the twentieth century, European fossil apes remained very poorly known compared to the abundant African fossils. However, the fossil record of ape evolution in both Europe and Asia has increased dramatically in recent years, including some relatively complete specimens, especially from Spain, Germany, and Hungary. Thus, many genera and species are currently recognized, but there are many differences of opinion as to the proper systematics (e.g., Alba, 2012; Begun et al., 2012; Begun, 2015; Pugh, 2022). The earliest fossil catarrhines in Eurasia date from the Early Miocene to early Middle Miocene, about 19 to 17 million years ago, but most are somewhat younger (e.g., Andrews et al., 1996; Harrison et al., 2020).

The noncercopithecoid catarrhines from the Miocene of Europe are generally recognized as belonging to two very distinct radiations that appear to have coexisted in Eurasia for over 10 million years. The Pliopithecoidea are primitive catarrhines related to the propliopithecids from the Oligocene of Egypt or possibly to some of the proconsuloids from sub-Saharan Africa. The fossil apes from Eurasia are all placed in Hominoidea, although there is considerable debate as to their relationship to extant hominoids.

Pliopithecoids

Pliopithecoids (Table 15.2) were initially described from France by Lartet in 1837, making them among the first primates ever identified in the fossil record (Begun, 2002; Fleagle and Hartwig, 1997). Until fairly recently, they were generally considered to be fossil gibbons, but now they are recognized as stem catarrhines preceding the monkey–ape divergence (Fleagle, 1984; Begun, 2002). In general, they lack two of the distinctive anatomical features – an external ear tube and the absence of an entepicondylar foramen – that characterize all crown catarrhines, hominoids, and Old World monkeys.

The teeth of pliopithecoids (Fig. 15.11) are quite primitive compared to those of other Eurasian apes. The anterior dentition has broad upper central incisors; smaller upper laterals; and tall, narrow lower incisors. The canines are very sexually dimorphic – long and

dagger-like in some individuals and short in others. The lower anterior premolar is similarly variable – narrow and sectorial in some individuals (presumably males) and broad in others (presumably females). The upper molars are quite broad and have a large lingual cingulum. The lower cheek teeth, including the posterior premolar, usually have a long and narrow occlusal surface and a prominent buccal cingulum. Many, but not all, have a characteristic "pliopithecine triangle" on the anterior buccal aspect of the talonid basin. Analysis of both shearing crests and microwear on the molars indicates that this radiation included both frugivorous and folivorous species (Ungar and Kay, 1995).

Parts of the skull and much of the skeleton from several individuals of *Epipliopithecus vindobonensis* are known from a fissure-fill at Devinska Nova (formerly Neudorf an der Marche), Slovakia (Fig. 15.11). The lower jaw is shallow with a broad ascending ramus, similar to that in extant gibbons. The skull is similar to that of a living gibbon in overall appearance but is more primitive in many details. The face has a short, narrow snout. The interorbital region is very broad, and the orbits are large and circular. The zygomatic region is relatively gracile. The frontal bone is high and rounded, suggesting a relatively large brain. Posteriorly, the temporal lines converge to form a sagittal crest in some individuals. The structure of the ear region is intermediate between that of *Aegyptopithecus* and that found in modern catarrhines (and proconsuloids). The tympanic bone forms a ring at the lateral surface of the bulla but does not form a complete tube as in living catarrhines. The inferior half of the tube is not ossified, suggesting a more primitive morphological condition than that found in either Old World monkeys or apes.

The skeleton of *Epipliopithecus* (Fig. 15.12) is much like that of a large living platyrrhine such as *Ateles* or *Lagothrix*. The intermembral index is 94. Both the forelimb and the hindlimb show adaptations in the joint surfaces that are characteristic of suspensory primates. Like the Fayum propliopithecids, *Epipliopithecus* lacks the distinguishing skeletal features of either living apes or living cercopithecoids, and it has many primitive skeletal features, such as an entepicondylar foramen on the humerus, a long ulnar styloid process, and a prehallux bone in the foot. Although the relatively small size of the sacral bodies would seem to suggest that *Epipliopithecus* was tailless, Ankel (1965) has demonstrated that the sacral canal has proportions indicating that it had a small tail (see also Russo, 2016). *Epipliopithecus* was an arboreal quadruped with suspensory abilities like those of the larger platyrrhines.

There are at least 8 genera and nearly 20 species of pliopithecoids from Eurasia, ranging from Early Miocene to Late Miocene in age (Harrison et al., 2020). Pliopithecoids are currently placed into four distinct

TABLE 15.2　Superfamily Pliopithecoidea

Species	Estimated mass (g)
Family DIONYSOPITHECIDAE	
Dionysopithecus (Early Miocene, Asia)	
D. shuangouensis	5500
D. orientalis	5750
Platydontopithecus (Early Miocene, Asia)	
P. jianghuaiensis	14,500
Family PLIOPITHECIDAE	
Pliopithecus (Middle to Late Miocene, Europe, Asia)	
P. antiquus	6500
P. pivetaui	7500
P. platyodon	10,000
P. zhanxiangi	13,500
P. canmatensis	5500
P. bii	11,000
Epipliopithecus (Middle Miocene, Europe)	
E. vindobonensis	8000
Family CROUZELIIDAE	
Subfamily CROUZELIINAE	
Plesiopliopithecus (=*Crouzelia*) (Middle Miocene, Europe)	
P. lockeri	5500
P. auscitanensis	5000
P. rhodanica	–
Barberapithecus (Late Miocene, Europe)	
B. huezleri	5000
Fanchangia (Early Miocene, Asia)	
F. jini	13,000
Subfamily ANAPITHECINAE	
Anapithecus (Middle Miocene, EUROPE)	
A. hernyaki	13,500
cf. A. priensis	15,000
Egarapithecus (Late Miocene, Spain)	
E. narcisoi	9500
Laccopithecus (Late Miocene, Asia)	
L. robustus	12,000
Family KRISHNAPITHECIDAE	
Krishnapithecus (Late Miocene, Asia)	
K. krishnaii	16,000
Family INCERTAE SEDIS	
Pliobates (Middle/Late Miocene, Europe)	
P. cataloniae	4500

groups based on details of dental morphology. Here they are placed in a single superfamily, Pliopithecoidea, with four families: Dionysopithecidae, Pliopithecidae, Krishnapithecidae, and Crouzelidae (Harrison et al., 2020; Table 15.2).

Dionysopithecus shuangouensis from the Early Miocene of China is the oldest and most primitive pliopithecoid. Like most pliopithecoids, *Dionysopithecus* has a pliopithecine triangle on the lower molars. As noted earlier, the upper molars are strikingly similar to the same teeth in *Micropithecus* from the Early Miocene of Uganda, and the lower premolars resemble fossils from Fort Ternan in Kenya. Other fossils from Thailand and Pakistan have also been referred to this genus (Harrison and Gu, 1999). *Platydontopithecus jianghuaiesis* is a larger dionysopithecine, also from the Early Miocene of China. The recently described *Fanchangia jini*, also from the Early Miocene of China, likely represents a primitive crouzelid and suggests that pliopithecoids may have initially diversified in Asia before migrating to Europe (Harrison et al., 2020).

Pliopithecus antiquus, originally named in 1849, is from numerous localities in France dating to about 15 Ma. Several other species of *Pliopithecus* are known from other parts of Europe, including a large sample of *P. platyodon* from the site of Goriach in Austria, *P. canmatensis* from Catalonia, Spain, and there are several species described from the Middle Miocene of China. As discussed above, *Epipliopithecus vindobonensis* from the Middle Miocene of Slovakia is the best-known pliopithecoid. *Krishnapithecus* is a little-known pliopithecoid from the Indian Siwaliks ~9 Ma with a mix of derived and primitive molar features. It is regarded as closely related to either the dionysopithecids or pliopithecids (Sankhyan et al., 2017).

Plesiopliopithecus is a small crouzelid pliopithecoid with numerous species from sites in Europe. *Anapithecus* is a large (13–15 kg) crouzelid best known from many cranial and postcranial specimens recovered at the site of Rudabanya in Hungary. The cranium is reported to be similar to that of *Epipliopithecus*. Detailed studies of dental development indicate that *Anapithecus* erupted its molar teeth much more rapidly than contemporary hominoids, a characteristic of many folivorous primates (Nargolwalla et al., 2005). *Egarapithecus* and *Barberapithecus* are young crouzelids from western Spain. *Laccopithecus robustus* (Fig. 15.13), from the latest Miocene site of Lufeng, China, is a large (12 kg), very late-surviving crouzelid. It has sexually dimorphic canines and anterior premolars. Like *Epipliopithecus*, *Laccopithecus* has large orbits and a short snout, but the zygomatic region is more robust. The auditory region is not preserved.

Pliobates cataloniae is an enigmatic catarrhine dating to ~11 to 12 Ma in Spain. Known from a skull and partial skeleton, it was originally described as a small-bodied ape close to the origins of gibbons due to a number of derived features in the postcranium. The humerus of *Pliobates* lacks an entepicondylar foramen as well as a capitular tail, and there are similarities to living hominoids in the wrist bones as well (Alba et al., 2015). However, *Pliobates* otherwise retains numerous primitive

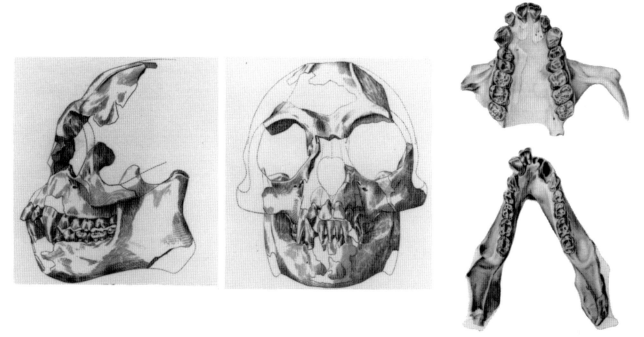

FIGURE 15.11   Cranial and dental remains of *Epipliopithecus vindobonensis*, from the Middle Miocene of Slovakia. Notice the gibbon-like face. *(From Zapfe, H., 1960. Die Primatenfunde aus der Miozanen Spaltenfullung von Neudorf an der march [Devinxka nova ves], Tschechoslowakev. Mit Anhang: Er Primatenfund aus dem Miozan von klein Hadersdorf in Niederoesterreich. Schweiz. Pal. Abh. 78, 4–293.).*

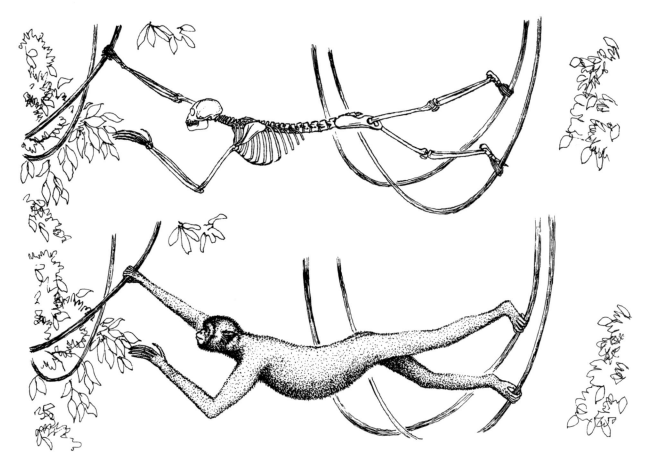

FIGURE 15.12   The skeleton of *Epipliopithecus* and a reconstruction of its locomotor habits.

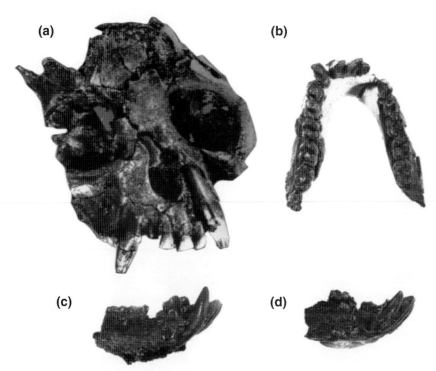

FIGURE 15.13   Cranial and dental remains of *Laccopithecus robustus*, from the Latest Miocene of Lufeng, China: (A) cranium; (B) mandible; (C) and (D) lateral views of male and female mandibles, showing sexual dimorphism. *(Courtesy of P. Yuerong.)*

features, more primitive than those seen in most known fossil apes, and shares striking similarities to pliopithecoids, including very broad molars with extensive cingulae, a cranium with a short snout and projecting orbital rims, and a relatively long ulnar styloid process. Most significantly, *Pliobates* exhibits an incompletely ossified tympanic (i.e., lacks an ear tube), a condition otherwise seen only in pliopithecoids among known catarrhine taxa, and recently discovered lower molars display a distinctive pliopithecine triangle (Bouchet et al., 2024). Thus, most authorities view *Pliobates* as a derived, late-occurring pliopithecoid that has converged on some crown hominoid forelimb features (Benefit, 2015; Nengo et al., 2017; Gilbert et al., 2020a,b; Bouchet et al., 2024).

Pliopithecoids were an extremely successful radiation of primitive catarrhines that ranged throughout much of Eurasia during much of the Miocene Epoch, from about 19 Ma until 8 Ma. Although they arrived in Eurasia before the earliest hominoids and were contemporary with hominoids throughout most of the Epoch, pliopithecoids were much more widespread and speciose, and species of the two radiations are rarely found at the same localities. It seems most likely that pliopithecoids originated in Africa sometime in the Late Oligocene or Early Miocene, and there are a few suggestions of pliopithecoid similarities in some of the many Early Miocene catarrhines from that continent (Rossie and MacLatchy, 2006). Much remains to be understood about the evolution and biogeography of this very successful group of catarrhine primates.

Hominidae

Slightly later than the first appearance of pliopithecoids in Asia, we find the first record of apes in Eurasia. The first Eurasian fossil apes are more similar to extant apes than the proconsuloids from Africa but share similarities with some of the poorly known later Miocene taxa from Africa such as *Kenyapithecus*. Unlike the proconsuloids, they are more confidently placed in the Hominoidea, but their relationships to the living great ape and human lineages are debated. In this chapter, most European ape fossils have been placed into the subfamily Dryopithecinae. The systematics and nomenclature of European Miocene taxa are often in flux, and different authorities often refer the same specimens to different genera and species (Begun et al., 2012; Alba, 2012). In addition, the dating of individual localities is often debated (but see Casanovas-Vilas et al., 2011; Figs. 15.14 and 15.15).

Griphopithecus is the oldest European fossil ape. Fossils attributed to this taxon range in age from a partial tooth from the end of the Early Miocene at Englesweis to the type specimens of *G. suessi* from early Late Miocene sites at Devinska Nova Ves in Slovakia (Casanovas-Vilar et al., 2011; Fig. 15.15). A nearly complete ulna and part of a humerus from a nearby site are often assigned to *Griphopithecus*. However, this genus is best known from larger samples at the Middle Miocene sites of Candir and Pasalar in Turkey (e.g.,

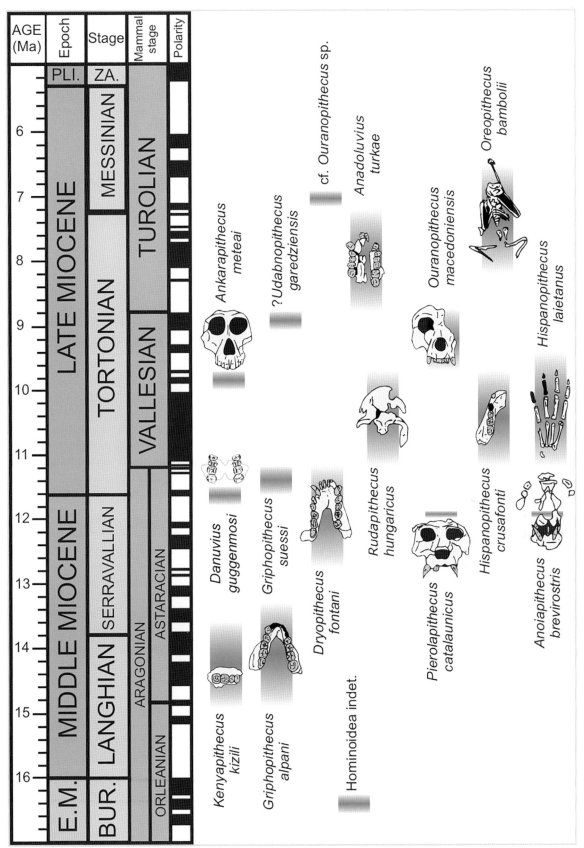

FIGURE 15.14 Chronology of European fossil apes. *(Courtesy of D. Alba and I. Casanovas-Vilar.)*

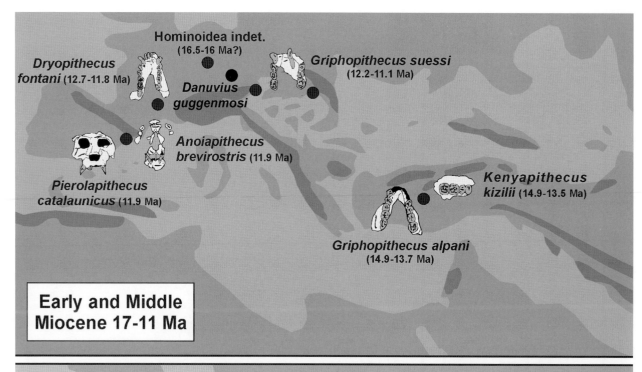

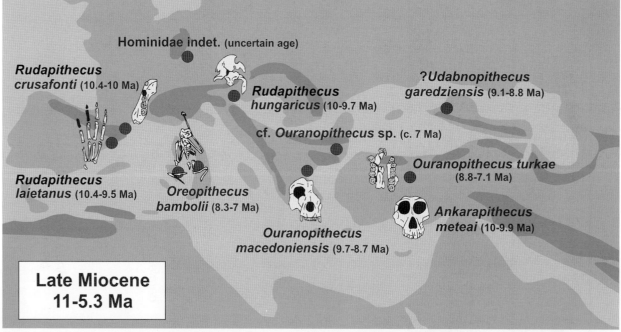

FIGURE 15.15   Biogeography of Early, Middle, and Late Miocene apes from Europe. Note that *Rudapithecus laietanus* and *Rudapithecus crusafonti* are placed in the genus *Hispanopithecus* by other authors, and *Ouranopithecus turkae* has recently been referred to a new genus, *Anadoluvius* (see also Table 15.3). Map lines delineate study areas and do not necessarily depict accepted national boundaries. *(Courtesy of I. Casanovas-Vilar.)*

Martin and Andrews, 1993; Begun et al., 2012). *Griphopithecus* has thicker tooth enamel than does *Dryopithecus*, and the attributed ulna has a moderately long olecranon. These and other dental features indicate that *Griphopithecus* is more primitive than the other large apes from Eurasia and similar to *Kenyapithecus*. It is debated whether it is a stem hominoid or a primitive great ape. Moreover, the wide range of ages and the

limited material from the European sites raise questions about the allocations of all of the fossils placed in this genus to a single taxon.

Dryopithecus is a Middle Miocene fossil ape that was first described nearly 150 years ago on the basis of a jaw (Fig. 15.16) and a humerus from Saint Gaudens in southern France. The number of species attributed to this genus has fluctuated wildly in the past century.

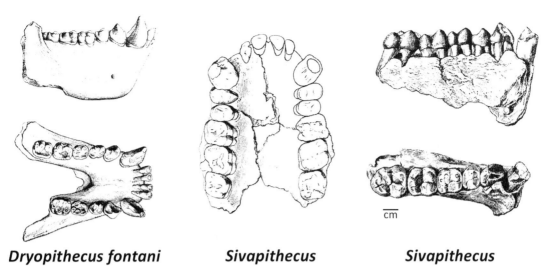

Dryopithecus fontani   **Sivapithecus**   **Sivapithecus**

FIGURE 15.16   Dental remains of two large Middle and Late Miocene fossil apes from Eurasia, *Dryopithecus* and *Sivapithecus*.

Several decades ago, when many fossil primates were known almost exclusively from jaws and teeth, virtually all large fossil apes were placed in this genus (Simons and Pilbeam, 1965). However, as fossil apes became better known, it has become clear that this was a gross oversimplification, and now *Dryopithecus* is just one of many genera of European fossil apes. The type species, *Dryopithecus fontani*, is known best from lower jaws and a humerus from Saint Gaudens, and a partial face and a femur from the site of Can Mata in Spain. A jaw from St. Stefan in southern Austria is sometimes placed in a separate species, *D. carinthiacus* (Begun et al., 2006, 2009).

The molars of *Dryopithecus* have more bulbous cusps than pliopithecoids and smaller cingulae and cingulids, and the mandibles are relatively robust. *Dryopithecus* differs from many other Eurasian apes in having thin, rather than thick, enamel on the cheek teeth; gracile canines; and a relatively gracile mandible (Figs. 15.16 and 15.17). The broad, rounded cusps on the cheek teeth indicate a predominantly frugivorous diet. The humeral shaft is relatively long and slender.

Pierolapithecus catalaunicus is a large Middle Miocene fossil ape from Eastern Spain that is known from a face and numerous parts of the skeleton, including ribs, hands, and vertebrae (Moyà-Solà et al., 2004; Fig. 15.17). *Pierolapithecus* appears to have had a moderately long face, but one that is superoinferiorly shorter than extant great apes, and differs from *Dryopithecus* in having molars with very thick enamel (Alba et al., 2010b; Pugh et al., 2023). It also lacks a frontal sinus. *Pierolapithecus* shares many derived features of the postcranium with great apes that are not found in any of the proconsuloids from Africa. The ribs suggest a broader chest than that of most monkeys, and there was no articulation between the styloid process of the ulna and the triquetrum at the wrist. However, the transverse processes on the lumbar vertebrae are intermediate between the condition found in monkeys and that of apes, and unlike extant great apes, *Pierolapithecus* retains an *os centrale* in the wrist and has relatively shorter digits than suspensory apes. All this suggests that *Pierolapithecus* was less suspensory than extant great apes, but there is considerable debate over the interpretation of the fingers of this fossil ape (Almécija et al., 2009; Deane and Begun, 2008, 2010; Alba et al., 2010a). *Pierolapithecus* is generally considered a stem hominid, although some have suggested it may be a basal member of the lineage leading to the orangutan, as discussed below.

Anoiapithecus brevirostris is another large, thick-enameled ape from the same area as *Pierolapithecus* in Eastern Spain (Moyà-Solà et al., 2009; Fig. 15.17). In contrast to the latter, however, *Anoiapithecus* seems to have a very orthognathic facial profile and a frontal sinus. The relationships between these two taxa and their place in hominoid phylogeny are the subject of ongoing debate. *Anoiapithecus* is usually regarded as a stem hominid.

Hispanopithecus is a Late Miocene fossil ape from Eastern Spain. Fossils attributed to this taxon were first described over 50 years ago but until recently were commonly regarded as belonging to the genus *Dryopithecus* (Moyà-Solà et al. 2009). The type species *H. laietanus* is known from many parts of a skeleton (Moyà-Solà and Kohler, 1995, 1996; Fig. 15.17). The cranium resembles that of modern hominoids, or specifically great apes, in many cranial features, including development of the brow ridges and a prominent glabella. The skeletal elements indicate a postcranial anatomy that is more similar to that of living hominoids than that of any of the proconsuloids or pliopithecoids, with a reduced olecranon process of the ulna, deep humeral trochlea, long hands, and short lumbar vertebrae with the transverse processes arising from

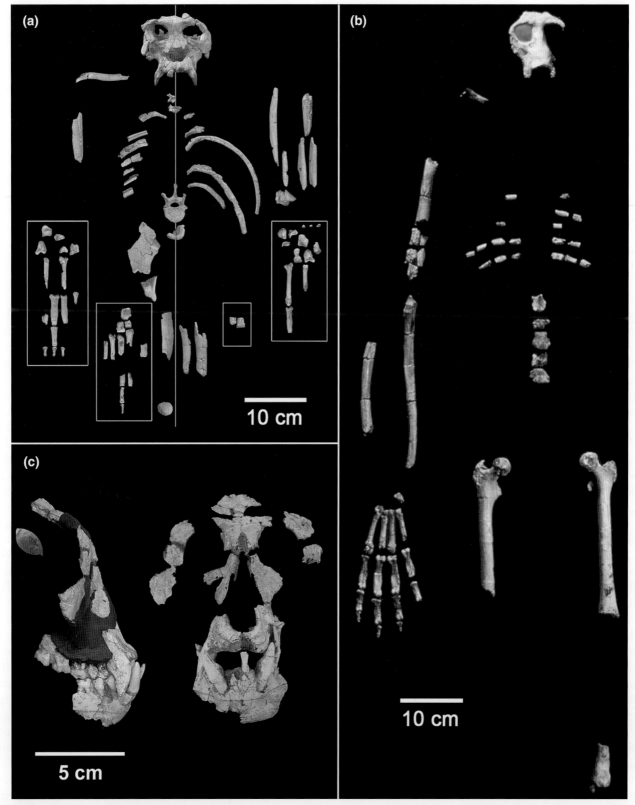

FIGURE 15.17 Fossil apes from the Middle and Late Miocene of the Valles-Penedes Basin in Spain: (A) *Pierolapithecus catalnicus*; (B) *Hispanopithecus laietanus*; (C) *Anoiapithecus brevirostirus*. *(Courtesy of D. Alba and S. Moyá-Solá.)*

the pedicle. Relative limb proportions of *H. laietanus* appear to be more like those of orangutans than African apes, with very long forelimbs and relatively short femora, suggesting that this species was suspensory.

Rudapithecus hungaricus is a fossil ape from the Late Miocene site of Rudabanya in Hungary (Fig. 15.15). Until recently, this species was commonly placed in the genus *Dryopithecus* and often considered synonymous with *D. brancoi* (e.g., Begun, 2002, 2007). *Rudapithecus* is known from many parts of the skeleton, including much of the cranium and numerous limb elements. A reconstruction of the cranium indicates that it is similar to that of extant great apes in overall shape, but the endocranium is more similar to that of gibbons (Gunz et al., 2020). Like *Dryopithecus* and *Hispanopithecus*, the teeth of *Rudapithecus* have thin enamel. The digits were long, with curved phalanges. The wrist shows a mixture of primitive catarrhine features, such as an unfused *os centrale*, and derived similarities to hominids (Kivell and Begun, 2009). The pelvis shows some features suggesting orthograde postures (Ward et al., 2019).

Danuvius guggenmosi is a recently discovered ape from ~11 to 12 Ma deposits in Germany (Bohme et al., 2019). It is known from partial jaws, isolated teeth, and a number of postcranial elements indicating it was a suspensory animal, similar to other dryopithecine great apes (see Williams et al., 2020)

Oreopithecus, from the Late Miocene of Italy, has been an enigma to paleontologists since its initial discovery in the latter part of the 19th century. The single species, **O. bambolii** (Table 15.3), is known from several sites in northern Italy, particularly from coal mines. The species is known from numerous remains, including cranial and skeletal elements (Figs. 15.18 and 15.19), but the most complete remains are crushed, making interpretation of their morphology sometimes difficult and controversial (Alba et al., 2024).

Oreopithecus has a dental formula of 2.1.2.3, like all catarrhines, but many aspects of its dentition are quite unique, hence the longstanding difficulties in determining its phyletic position among catarrhines. The upper central incisor is relatively large and round, and the lateral is smaller and peg-like; the lower incisors are narrow and spatulate. The canines are quite dimorphic, with presumed males having tall upper and lower canines and the females having very small canines. In the males, the upper canine shears against the anterior surface of the anterior lower premolar; in females, the lower premolars are more semimolariform. Upper premolars are characterized by two relatively tall cusps of similar size. The upper molars are long and narrow with a well-formed trigon, a large hypocone, and a lingual cingulum as in many other Miocene catarrhines. The paraconule is particularly well developed. The lower molars not only have the characteristic basic cusps found in all noncercopithecoid catarrhines but also have

an additional sixth cusp, the centroconid. The well-developed shearing crests clearly indicate a folivorous diet. Overall, the dentition is very similar to that found in the African Middle Miocene hominoid *Nyanzapithecus*.

The skull has a relatively short snout, a small brain, and a pronounced sagittal crest in some individuals. The auditory region indicates the presence of a tubular ectotympanic as in extant catarrhines and all known proconsuloids. The inner ear of *Oreopithecus* is more primitive than in any living ape (Urciuoli et al., 2020).

The skeleton of *Oreopithecus* has many indications of suspensory locomotor habits, including a relatively short trunk; a broad thorax; relatively long forelimbs; short hindlimbs; long, slender manual digits; and evidence of extensive mobility in virtually all joints. The elbow region is identical to that of extant great apes. The foot has long, curved phalanges and an abducted hallux. There have been suggestions that *Oreopithecus* was bipedal and that its hand proportions were suitable for manipulating tools. The more reasonable interpretation is that *Oreopithecus* was a suspensory animal, most similar to the orangutan in its adaptations.

Since its initial discovery, *Oreopithecus* has been identified by various authorities as being closely related to parapithecids, cercopithecoids, hominids, hominins, or as an ancient higher primate lineage not closely related to any modern group of anthropoids. In its limb structure, *Oreopithecus* shows more similarities to extant hominoids than does any other fossil ape, but with a unique pelvic anatomy (Hammond et al., 2020). It is certainly some type of specialized hominoid. The origin of the lineage leading to *Oreopithecus* is not clear. Dental and cranial evidence suggests that it may have arisen among the proconsuloids in East Africa, particularly the nyanzapithecines (e.g., Harrison, 1986; Benefit and McCrossin, 1997; Nengo et al., 2017; Rossie and Cote, 2022). However, skeletal remains of *Dryopithecus* have led some authors to suggest that *Oreopithecus* is closely related to *Dryopithecus* instead (e.g., Harrison and Rook, 1997; Begun, 2002). Thus, *Oreopithecus* is either regarded as a uniquely derived stem hominoid converging on great ape postcranial adaptations or as a strange great ape retaining many primitive features found among proconsuloids in its cranium and dentition.

Ouranopithecus macedoniensis is a large ape from several Late Miocene (9–10 million years ago) sites in northern Greece (Fig. 15.15) and is known from many jaws and a nearly complete face (DeBonis and Koufos, 1993; Figs. 15.20 and 15.21). A second species from the Late Miocene of Turkey has recently been transferred to a separate genus, **Anadoluvius** (Gulec et al., 2007; Sevim-Erol et al., 2023). The associated fauna of *O. macedoniensis* suggests a mosaic environment of woodland and savannah habitats (e.g., DeBonis and Koufos, 1994). *Ouranopithecus* was a large ape with an estimated body

TABLE 15.3   Middle and Late Miocene Apes from Eurasia

Species	Estimated mass (g)
Superfamily HOMINOIDEA	
Family INCERTAE SEDIS	
Oreopithecus **(Late Miocene, Europe)**	
O. bambolii	27,500
Buronius **(Late Miocene, Europe)**	
B. manfredschmidi	10,000
Family HOMINIDAE	
Subfamily INCERTAE SEDIS	
Kenyapithecus **(Middle Miocene, Africa, Eurasia)**	
K. kizili	35,000
Griphopithecus **(Middle Miocene, Eurasia)**	
G. alpani	28,000
G. darwini	-
G. seussi	-
Lufengpithecus **(Late Miocene, Asia)**	
L. hudiensis	40,000
Subfamily DRYOPITHECINAE	
Dryopithecus **(Middle to Late Miocene, Europe)**	
D. fontani	43,600
?Udabnopithecus **(Late Miocene, Eurasia)**	
U. garedziensis	-
Pierolapithecus **(Middle Miocene, Europe)**	
P. catalaunicus	32,500
Anoiapithecus **(Middle Miocene, Europe)**	
A. brevirostris	31,000
Hispanopithecus **(Late Miocene, Europe)**	
H. crusafonti	35,000
H. laietanus	33,000
Rudapithecus **(Late Miocene, Europe)**	
R. hungaricus	27,000
Ouranopithecus **(Late Miocene, Eurasia)**	
O. macedoniensis	105,000
Anadoluvius **(Late Miocene, Eurasia)**	
A. turkae	70,000
Graecopithecus **(Late Miocene, Europe)**	
G. freybergi	60,000
Danuvius **(Late Miocene, Europe)**	
D. guggenmosi	25,000
Subfamily PONGINAE	
Ankarapithecus **(Late Miocene, Turkey)**	
A. meteai	80,000
Sivapithecus **(Late Miocene, Europe, Asia)**	
S. indicus	30,000
S. sivalensis	35,000

TABLE 15.3   Cont'd

Species	Estimated mass (g)
S. parvada	60,000
Gigantopithecus **(Plio-Pleistocene, Asia)**	
G. blacki	250,000
Indopithecus **(Late Miocene, Asia)**	
I. giganteus	165,000
Khoratpithecus **(Middle Late Miocene, Asia)**	
K. chiangmuanensis	65,000
K. piriyai	70,000
K. magnus	90,000
K. ayeyarwadyensis	65,000
Lufengpithecus **(Late Miocene, Asia)**	
L. lufengensis	40,000
Family cf. HYLOBATIDAE	
Yuamoupithecus **(Late Miocene, Asia)**	
Y. xiaoyuan	7000
Kapi **(Middle Miocene, Asia)**	
K. ramnagarensis	6000

size of over 70 kg. The dentition of *Ouranopithecus* is characterized by relatively small canines and short, broad anterior lower premolars in both sexes (Fig. 15.20). The molar enamel is extraordinarily thick. Studies of both cusp morphology and microwear suggest a diet of hard or gritty objects, such as nuts or tubers (Ungar and Kay, 1995; Gulec et al., 2007). The dentition of *Ouranopithecus* was very sexually dimorphic in dental size (Koufos et al., 2016). The face of *Ouranopithecus* has a broad nasal opening, a large incisive foramen, pronounced brow ridges, and a broad interorbital region with a prominent glabella (Fig. 15.21). The cranial morphology of *Ouranopithecus* has been variously interpreted as showing derived similarities to many different clades of living apes, including orangutans. Despite these differing views, most authorities agree that the facial morphology of this genus (particularly the brow ridges) links it with the African ape and human clade. *Ouranopithecus* is a critical taxon for understanding the relationship between the Miocene apes of Europe and the origin of African great apes and humans (Kunimatsu et al., 2007; Begun et al., 2012), a debate that has been heightened by the discovery of a fossil ape with very similar teeth, *Nakalipithecus*, from similarly aged or slightly older deposits in Kenya.

Graecopithecus freybergi is a slightly younger taxon, also from Greece (Fig. 15.15), known from a single poorly

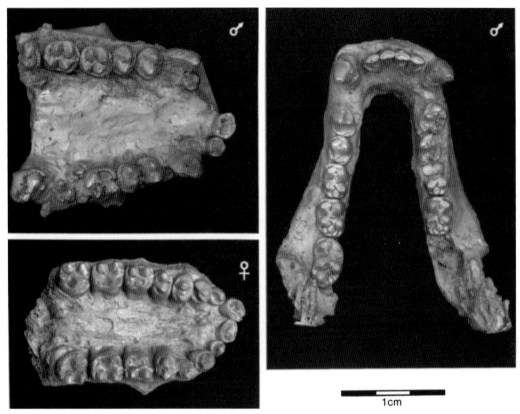

FIGURE 15.18   Upper dentition (*left*) and lower dentition (*right*) of *Oreopithecus bambolii*, from the Pliocene of Europe. *(Courtesy of E. Delson.)*

FIGURE 15.19   Skeleton of *Oreopithecus* and a reconstruction of its locomotor habits.

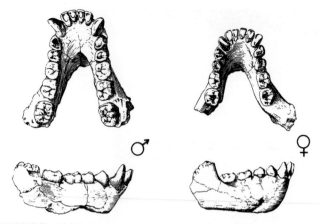

FIGURE 15.20   Male and female lower jaws of a Late Miocene fossil ape from Greece, *Ouranopithecus macedoniensis*.

preserved mandible that some workers place in *Ouranopithecus* (Andrews et al., 1996) but others (Begun et al., 2012) maintain as a distinct taxon. Most recently, largely on the basis of premolar root morphology, some authors have suggested *Graecopithecus* may be a hominin (Fuss et al., 2017).

Phylogenetic Relationships of Western Eurasian Apes

In contrast to the fossil "apes" from the Early and Middle Miocene of Africa, and the pliopithecoids, which have few, if any, features indicating that they are within the radiation of extant apes, the large fossil apes from Europe (and Turkey) are hominoids, or true apes. The one possible exception is *Griphopithecus*, which is known almost exclusively from dental remains and one attributed ulna. Rather, debates regarding the phylogenetic positions of *Dryopithecus, Pierolapithecus, Hispanopithecus, Rudapithecus, Oreopithecus,* and *Ouranopithecus* concern their relationships to the great apes of Africa and Asia. All the taxa for which there are relevant skeletal elements available have the characteristic elbow morphology of great apes; relatively long, curved digits; and a vertebral morphology more derived than that of Old World monkeys. Nevertheless, all show combinations of features that differ from the pattern found in any living ape, showing that there has been a greater diversity of ape locomotion in the past than what we see among the few apes that are alive today (e.g., Ward, 1997). Moreover, interpreting the phylogenetic significance of cranial features found in living and fossil apes is far more difficult than understanding the limbs.

According to Begun and colleagues (2012), *Griphopithecus* is a stem hominoid, and *Oreopithecus* and *Kenyapithecus* are stem great apes (hominids). *Ouranopithecus* is more closely related to African apes and humans than to the orangutan

and its relatives (i.e., a stem hominine, discussed further below), but the other European apes (*Dryopithecus, Pierolopithecus, Hispanopithecus,* and *Rudapithecus*) are the sister group to the extant African ape and human clade. The significant biogeographical implication of this relationship is that the lineage leading to African apes and humans originated in Europe rather than within any known group of hominoids from Africa. Others have noted that one or more of the European apes share features uniquely with the orangutan lineage and suggest that the Eurasian dryopithecine apes are best viewed as broadly ancestral to the orangutan, having no unique relationship with African apes and humans (Alba, 2012). The most recent comprehensive phylogenetic analysis of these taxa presents yet another alternative: *Oreopithecus* is a stem hominoid, *Griphopithecus* and *Kenyapithecus* are stem hominids, the European dryopithecines represent successive radiations of more derived stem hominids (i.e., broadly ancestral to both the living African and Asian great apes), and *Ouranopithecus* and *Nakalipithecus* are more closely related to the African great ape and human clade (Pugh, 2022). As is usually the case, all scenarios involve some degree of homoplasy in the evolution of traditional hominoid features.

Asian Apes

Ankarapithecus meteai, a large fossil ape from Turkey (Figs. 15.15 and 15.21) that is known from much of a cranium and several limb elements (Alpagut et al., 1996; Kappelman et al., 2003), has generally been considered a close relative of *Sivapithecus* (Andrews and Tekkaya, 1980). However, recent discoveries have demonstrated significant differences in the skull morphology of the two. The skull of *Ankarapithecus* (Fig. 15.21) is more flexed than that of *Sivapithecus*, and the face is relatively straighter with a broader interorbital dimension and rounder orbits, and it seems likely that *Ankarapithecus* retains the more primitive facial morphology from which the specializations characteristic of *Sivapithecus* and *Pongo* were derived (Alpagut et al., 1996). Thus, authors have debated whether *Ankarapaithecus* is best viewed as a stem hominid preceding the divergence of pongines from the great ape and human clade or whether it should be considered a stem pongine (Begun, 2015).

Sivapithecus is from the Siwalik Hills of northern India, Pakistan, and Nepal. Fossils come from sites ranging in age from approximately 13 million years ago to 8 million years ago. Although the systematics are not easy to resolve, at least three sexually dimorphic species are currently recognized: *S. indicus, S. sivalensis,* and *S. parvada*, in increasing order of size (Kelley, 2002, 2005). The dentition of *Sivapithecus* is characterized by cheek teeth with thick enamel and low cusp relief, the

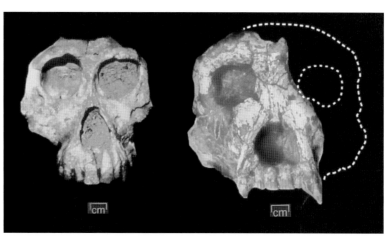

FIGURE 15.21   Facial skeleton of *Ankarapithecus meteai* (*left*) and *Ouranopithecus macedoniensis* (*right*). *(Courtesy of J. Kappelman and L. deBonis.)*

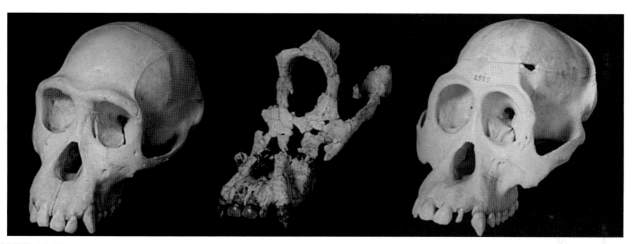

FIGURE 15.22   Cranial remains of *Sivapithecus* (*center*) and crania of *Pan* (*left*) and of *Pongo* (*right*). *(Courtesy of D. Pilbeam.)*

common absence of any cingulum on the molars, very broad lower premolars, robust canines, and a thick mandible (Fig. 15.16). The upper central incisors are much broader than the laterals. On the basis of the available material, *Sivapithecus* shows relatively little canine dimorphism compared with that found among living apes and monkeys. The combination of thick-enameled molars and low cusp relief is characteristic of living primates that eat seeds and nuts. It has been suggested that *Sivapithecus* had a diet of hard nuts, bark, or fruits with hard pits. The skull of *Sivapithecus* shows a striking resemblance to the living orangutan in such features as a narrow snout with a very large procumbent premaxilla; a small incisive foramen; broad zygomatic arches; a tall, narrow nasal aperture; and high orbits (Fig. 15.22).

There are various postcranial remains of *Sivapithecus*: humeri, a radius, part of a femur, and various hand and foot bones (Rose, 1997; Madar et al., 2002; Ward, 2007;

Begun, 2015). The distal articulation of the humeri of *Sivapithecus* resembles that of extant hominoids, particularly gorillas. The humeral shaft does not; it is bowed laterally rather than being straight like the humeral shafts of extant apes (see Richmond and Whalen, 2001; Alba et al., 2011). The distal part of the femur is also similar to that of extant hominoids. Manual and pedal phalanges are similar to those of extant apes, and *Sivapithecus* had a large grasping hallux, unlike *Pongo*. Although most features of the postcranium of *Sivapithecus* are similar to those of extant great apes, they indicate that *Sivapithecus* lacked the extreme adaptations for suspensory locomotion found in *Pongo* (e.g., Morgan et al., 2014). Rather, it was probably a more quadrupedal ape. In view of the size range, there was probably considerable locomotor diversity within the genus, but current evidence is too scanty to determine the exact nature of such differences. Despite considerable debate and difference of opinion (see Kelley, 2002; Begun, 2015), most

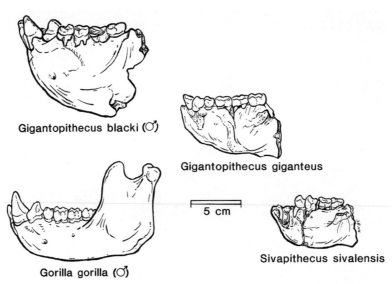

FIGURE 15.23   Lower jaws of *Gigantopithecus* and *Sivapithecus* compared with that of a male mountain gorilla, the largest living primate.

authorities identify *Sivapithecus* as a fossil related to the extant orangutan.

A close relative of *Sivapithecus* is **Gigantopithecus**, the largest-known fossil primate (Figs. 15.23 and 15.24). The two species of *Gigantopithecus* were almost certainly derived from a large Asian species of *Sivapithecus*. The earlier, smaller *G. giganteus* (= *G. bilaspurensis*) from the Latest Miocene of India and Pakistan, is often placed in its own genus, **Indopithecus**. The larger *G. blacki* is from Pleistocene caves in China and Vietnam. The smaller species probably weighed as much as a living gorilla (150 kg), and the Pleistocene species has an even larger estimated mass, perhaps as much as 300 kg (based primarily on the large mandible and dentition).

These extraordinary primates are known only from lower jaws and isolated teeth. They were initially discovered in Chinese drugstores, where the teeth were being sold as medicine (von Koenigswald, 1983). The lower incisors are very small and vertical. The canines are thick but relatively short. The lower anterior premolar is relatively broad, as in *Homo sapiens*, rather than elongated. Like those of *Sivapithecus*, the teeth of *Gigantopithecus* have thick enamel and low, flat cusps. In *G. blacki*, there are often accessory cusps. In both species, the mandible is very thick and extremely deep compared to the jaws of living apes (Fig. 15.23). The dental proportions, cheek tooth morphology, and robust mandibles indicate that *Gigantopithecus* ate some type of hard, fibrous material. One worker has suggested that they ate bamboo, like the living panda, but phytoliths recovered from the surface of fossil teeth suggest a more diverse diet including various fruits, leaves, and tubers (Ciochon et al., 1990; Zhang and Harrison, 2017). Their enormous size would seem to have precluded anything except a predominantly folivorous diet and terrestrial locomotion.

Lufengpithecus is a large ape from the Latest Miocene of southern China, known from more than 1000 dental remains and several skulls (Kelley and Gao, 2012). There are three recognized species from different sites (Kelley, 2002; but see Harrison et al., 2002; Harrison, 2006). The dentition of *Lufengpithecus* is characterized by tall, narrow incisors, slender canines, and narrow molars without extremely thick enamel. The teeth of *Lufengpithecus* are extremely sexually dimorphic (Kelley and Plavcan, 1998). The cranium, known from crushed cranial remains and two well-preserved juvenile specimens, shows distinct differences from orangutans or *Sivapithecus* and is more similar to European hominoids such as *Rudapithecus*. The phalanges of *Lufengpithecus* are long and curved, suggesting some suspensory behavior. The phyletic position of *Lufengpithecus* is subject to some debate, and it is possible that multiple genera are represented within the current *Lufengpithecus* grouping. Although some authorities feel that *Lufengpithecus* is likely related to *Pongo*, most feel that at least some *Lufengpithecus* species represent more basal hominoids, similar to European taxa (e.g., Kelley, 2002; Harrison, 2006; Kelley and Gao, 2012; Pugh, 2022).

Khoratpithecus is a fossil ape from middle Late Miocene deposits in Thailand and Myanmar that has been described as closely related to the orangutan. There are four species (Jaeger et al., 2011; Chaimanee et al., 2022). *Khoratpithecus* is known from numerous isolated teeth, two mandibles, and a maxilla (Chaimanee et al., 2004, 2019). The teeth are similar to those of orangutans and to *Lufengpithecus* and were initially attributed to that taxon (Chaimanee et al., 2003; Harrison, 2006). However, the mandibles of *Khoratpithecus* are distinctive and similar to *Pongo* in lacking a scar for the attachment of the anterior digastric muscle.

Yuanmoupithecus xiaoyuan is a smaller fossil ape from Late Miocene deposits in the Yuanmou Basin of China (Pan, 2006; Harrison et al., 2008). It is known

FIGURE 15.24 A reconstruction of *Gigantopithecus blacki* from the Pleistocene of China.

from teeth and a partial lower face (Ji et al., 2022). *Yuanmoupithecus* was originally compared with proconsuloids from Africa, including *Dendropithecus* and *Micropithecus*. However, following the recovery of additional material, it seems most likely that *Yuanmoupithecus* is a fossil gibbon. **Kapi ramnagarensis**, known from a single lower molar in the Lower Siwaliks, is even more similar to modern gibbons in some dental features than *Yuanmoupithecus* (Gilbert et al., 2020a). At ~12 to 13 Ma, it represents the earliest possible evidence of gibbons in Asia.

In addition to the genera and species of Asian fossil apes described above, there are also numerous fossils (mostly teeth) from Pleistocene deposits in many parts of Southeast Asia that have been attributed to the extant genera of gibbons and to the orangutan, *Pongo*. Two recently extinct gibbon genera are also often recognized: *Bunopithecus*, known from a single partial jaw from the Pleistocene of China, and *Junzi*, known from a partial skull found in a Chinese tomb dating to ~2300 years ago. These recent gibbon fossils are found in areas currently or historically occupied by lesser apes. However, Pleistocene orangutan fossils have been recovered not only on the islands of Borneo and Sumatra where they

live today, but also in many parts of mainland Southeast Asia and the island of Java, where there are no living orangutans. Some of the fossil orangutans were much larger than the living species.

Phyletic Relationships of Asian Apes

With the exception of *Yuanmoupithecus* and *Kapi*, which may be fossil gibbons, the fossil apes from Asia discussed above are generally thought to be part of a single clade, of which the extant orangutan, *Pongo*, is the only living representative. However, all of the fossil taxa differ from the extant genus in various distinct ways that make efforts to reconstruct phylogenetic relationships among them difficult. In addition, the fossil genera are each known from different body parts with different degrees of preservation. *Khoratpithecus* shows similarities to *Pongo* in its dental, maxillary and mandibular structure, especially in the lack of an anterior digastric attachment, and most authorities regard it as a sister taxon to *Pongo*. *Sivapithecus* shows striking similarities to the orangutan in many aspects of its dental and cranial anatomy (Kelley, 2002). However, although many parts of its skeleton are clearly those of a great ape, some postcranial remains suggest more quadrupedal locomotor habits for some species of *Sivapithecus*, leading a few authors to reconsider whether *Sivapithecus* was, indeed, related to living orangutans (Pilbeam et al., 1990; Pilbeam and Young, 2001; Andrews and Pilbeam, 1996). In total, it seems that the orangutan clade contained a greater diversity of apes in the past and that some locomotor features shared by living great apes undoubtedly reflect parallel evolution (Larson, 1992, 1998; Ward, 2007; Alba et al., 2011). *Gigantopithecus* is almost certainly a derived member of this radiation, based on recent proteomic data demonstrating close ties to extant *Pongo* (Welker et al., 2019). Furthermore, its dental and mandibular similarities to *Sivapithecus* add to the overall evidence that *Sivapithecus* is a close relative of *Pongo* as well (Kelley, 2002; Harrison, 2006; Alba et al., 2011; Alba, 2012; Begun et al., 2012; Begun, 2015; Pugh, 2022). Like *Sivapithecus*, *Ankarapithecus* also shows some orangutan-like features in the lower face (Andrews and Tekkaya, 1980), but a more complete skull shows that it is more primitive than *Sivapithecus* in other aspects of its cranial anatomy and its limb bones suggest a more quadrupedal rather than suspensory locomotion (Alpagut et al., 1996; Kappelman et al., 2003). The most difficult Asian ape to place is *Lufengpithecus*. For many years, *Lufengpithecus* teeth were considered those of an early hominin, and there are still some who place it in the human lineage. The teeth of *Lufengpithecus* are similar to those of orangutans, and the few postcranial elements seem to show suspensory

features. Although *Lufengpithecus* is known from partial skulls, one is squashed flat and the others are juveniles. The preserved juvenile crania are distinct enough to possibly represent different genera. Most authorities feel that *Lufengpithecus* species have a more primitive cranial morphology than other Asian hominoids and place them either as the sister group to *Ankarapithecus* and the other large fossil apes from Asia, as stem hominids, or possibly both (e.g., Harrison, 2006; Kelley and Gao, 2012; Begun, 2015; Pugh, 2022) (Fig. 15.25).

The Evolution of Living Hominoids

In the preceding sections of this chapter, we have reviewed the non-cercopithecoid catarrhines and fossil apes from the latest Oligocene, Miocene, and Plio-Pleistocene epochs. Like the Early Oligocene anthropoids from Egypt, the Miocene genera and species can be ordered on the basis of a suite of dental, cranial (Fig. 15.25), and postcranial features into some sort of phylogeny, though not without considerable uncertainty. It is quite evident that the radiation of hominoids during the Miocene was much more extensive and produced many more lineages than many earlier paleontologists imagined. A corollary of this increasingly complex picture of ape evolution during the Miocene is that the identification of unique lineages leading to particular extant genera is far more difficult than was previously thought. Extant apes are just disparate twigs in a

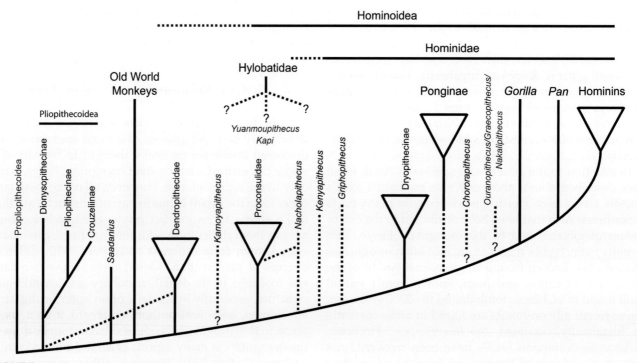

FIGURE 15.25   Phyletic radiation of fossil apes.

complex phylogeny (Fig. 15.25). Attempts to find the ancestry of unique lineages leading to gibbons, to the great apes, or to hominins have repeatedly been complicated by the discovery of more complete fossils with unsuspected primitive features or mosaic combinations of derived features, and by more careful consideration of comparative anatomy. Until very recently, the great temporal expanse of the Miocene, the diversity of Miocene environments and faunas, and the morphological diversity of the fossil apes from that epoch were all unknown and largely unsuspected. The extraordinary diversity of fossil apes from Africa, and especially Eurasia, that have come to light in just the past few decades remind us how little we really know about ape evolution and how many fossil apes are yet to be uncovered. To put our current understanding of ape evolution into perspective and to contrast it with earlier views, we now examine the fossil evidence specifically for what it tells about the evolution of extant hominoids, consider the alternative ways in which they could have evolved from the diverse radiations of Miocene apes, and compare these results with predictions about ape and human evolution derived from biomolecular studies.

Evolution of Gibbons

In each of the successive radiations of Oligocene and Miocene hominoids, there have been small apes that, at one time or another, have been identified as fossil gibbons. *Propliopithecus, Pliopithecus, Epipliopithecus, Dendropithecus, Micropithecus,* and *Dionysopithecus* all show various features (such as small size, short snouts, or large orbits) that cause them to resemble living lesser apes. As already discussed, however, most of these supposed fossil gibbons were extremely primitive in many detailed aspects of their cranial and skeletal anatomy, more so than we would expect in an ancestral gibbon based on the comparative anatomy of extant higher primates. For example, although *Propliopithecus* and *Epipliopithecus* were similar to living gibbons in their size and (in some species) had simple, superficially gibbon-like lower molars, they lacked such features as the tubular ectotympanic bone found in all living apes and Old World monkeys, and they retained primitive features in their limb bones that are lacking in the limbs of all living catarrhines. For other genera, such as *Micropithecus, Dendropithecus,* and *Laccopithecus*, there are suggestive cranial remains (although critical information about the ear region, and especially the limb skeleton, is not available), but they show other features that link them with other more primitive taxa. Only the poorly known *Kapi* and better known *Yuanmoupithecus* from the Middle and Late Miocene of India and China, respectively, seem to be viable candidates for real fossil gibbons prior to the Pleistocene.

All of the small apes from Africa and Eurasia were probably to some extent ecological vicars of the living lesser apes, but there is no real indication of a gibbon lineage prior to the last ~12 to 13 million years (Gilbert et al., 2020a). However, molecular estimates of the divergence date of gibbons from other hominoids suggest an origin of the lineage leading to the lesser apes around 20 million years ago (e.g., Perelman et al., 2011; Kuderna et al., 2023). This time range includes that of virtually all of the primitive catarrhines from the Early Miocene of Africa, as well as the fossil apes from Europe that seem to mark the appearance of the great ape and human clade, which must have been coincidental with the origin of the gibbon lineage. The absence of any evidence of a gibbon lineage throughout much of the Early to Middle Miocene is a major gap in our understanding of primate evolution that raises many unanswered questions. Are the similarities between gibbons and other hominoids the result of extensive homoplasy, and gibbons are derived from one of the primitive catarrhines from the Miocene of Africa or Eurasia? Is the small size of extant gibbons a recent phenomenon, such that some of the primitive "great apes" of Europe are really early hylobatids? Or was hylobatid evolution a small, regional phenomenon that took place outside the parts of the world with a fossil record?

Evolution of the Orangutan

The one living ape whose evolutionary history is now generally considered to be well established, in at least a broad sense, is the orangutan. Indeed, if anything, there seems to be an overabundance of fossil orangutan relatives from the Miocene of Asia and possibly Europe. The geographic and temporal gap between the Late Miocene fossils and the living great apes of Borneo and Sumatra is partly bridged by hundreds of fossil teeth from the Pleistocene of China and Java, but it seems clear that the lineage leading to the orangutan contained a greater diversity of species in the past. Molecular studies have indicated dates of about 16 million years ago for the orangutan divergence from the African ape and human clade (Steiper and Young, 2008; Perelman et al., 2011), a date which incorporates all of the Eurasian fossil apes that have been attributed by some authors to that lineage (e.g., Alba, 2012), all concordant with the fossil data (Andrews, 1986; Miyamoto et al., 1988; Pilbeam, 1996). The recent proteomic evidence derived from *Gigantopithecus* appears to confirm that at least some of the Miocene Asian hominoids are relatives of the living orangutan. Thus, all evidence indicates that extant *Pongo* is just one very specialized surviving twig in a very bushy tree of large Eurasian apes from the past 15 million years or so.

Evolution of African Apes

The evolutionary history of gorillas and chimpanzees is one of the most notable gaps in our understanding of ape and human evolution. A few fossil teeth less than 500 thousand years old from Kenya have been identified as those of a chimpanzee (McBrearty and Jablonski, 2005). *Samburupithecus* from the Late Miocene of Kenya (Ishida and Pickford, 1997) and *Chororopithecus* from the Late Miocene of Ethiopia (Suwa et al., 2007) have been identified as possible gorilla relatives, but neither shows any features besides size to support such a relationship (e.g., Ward and Duren, 2002; Harrison et al., 2010b; Begun, 2015; Pugh, 2022). All of the fossil apes from the Early and Middle Miocene of Africa seem too primitive to be uniquely ancestral to any living African ape. Molecular estimates indicate that the evolutionary divergence of the lineages leading to the living African great apes and to humans was probably sometime in the latest part of the Miocene, between 5 and 11 million years ago (Perelman et al., 2011; Kuderna et al., 2023). African fossil apes are rare from this time period, and as discussed in a later chapter, the hominid fossils from younger part of this time range are generally considered to be on the human lineage. However, some authors have suggested that some late Miocene hominins may actually be more closely related to African apes (Harrison, 2010a; Wood and Harrison, 2011). Nevertheless, for the present, the evolutionary history of chimpanzees and gorillas remains virtually undocumented. This absence of fossil apes probably reflects, in part, our lack of any substantial fossil record from western and central Africa where these apes live today (Almécija et al., 2021; Urciuoli and Alba, 2023).

The Biogeography of Ape Evolution

One of the most interesting, poorly understood, and intensely debated aspects of ape evolution is the biogeography of the clades that gave rise to the extant apes, including humans. All evidence indicates that catarrhine primates originated in Africa (including Arabia) and that the divergence of Old World monkeys and apes took place on that continent, most probably in the Middle or Late Oligocene. The proconsuloids almost certainly include the earliest stem hominoids, so it is likely that the common ancestors of extant hominoids lived in Africa. However, there are several alternative possibilities for reconstructing how the current geographical distribution positions of extant apes – gibbons and orangutans in Asia, African apes, and ancestral humans in Africa – came about (Stewart and Disotell, 1998; Begun et al., 2012; Gilbert et al., 2020b; Pugh, 2022).

The simplest scenario, conceptually at least, is that Africa is the source of all hominoid lineages and that hylobatids, pongines, gorillas, chimpanzees, and humans evolved successively from an African stem hominoid stock, with the first two subsequently migrating to Asia and the other remaining in Africa. The strengths of this scenario are that it seems likely that the hominoid clade originated in Africa, as there are numerous stem hominoids from the Miocene of East Africa and the fact that gorillas, chimpanzees, and the earliest fossil evidence of the human lineage are restricted to Africa. The difficulty is that, although stem hominoids are abundant in the fossil record of the African Miocene, the only fossil evidence of crown hominoids are human ancestors from the latest Miocene, Pliocene, and Pleistocene; a few large apes from the Late Miocene of Kenya and Ethiopia that are known only from teeth; and a few teeth from the Pleistocene of Kenya that have been identified as a chimpanzee (McBrearty and Jablonski, 2005). The most suggestive of the Miocene taxa is *Nakalipithecus* from the Late Miocene of Kenya, which is a possible member of the African ape and human clade that resembles the European *Ouranopithecus* but seems slightly older (Kunimatsu et al., 2007). There are no convincing hylobatids, pongines, or very convincing ancestors of either gorillas or chimpanzees from the Miocene fossil record of Africa.

An alternative view is that sometime in the Early Miocene, one or more stem hominoids dispersed to Eurasia, where they subsequently gave rise to hylobatids, pongines, and stem hominines, then a basal member of the African ape and human clade dispersed back to Africa and gave rise to gorillas, chimpanzees, and hominoids. As first noted by Stewart and Disotell (1998), the two scenarios are equally parsimonious when only the extant taxa are concerned, as each involves a minimum of two dispersals. However, when the fossil record of apes is considered, the latter alternative view becomes more likely, as there is a large array of fossil hominoids from the Middle and Late Miocene of Eurasia that not only differ from the Miocene apes of Africa in that they seem to be part of the crown group but which also show various combinations of derived features that place them in intermediate positions within the hominoid radiation (e.g., Begun et al., 2012; Begun, 2015; Pugh, 2022). Again, *Nakalipithecus* from the Late Miocene of Kenya complicates this scenario, as the close similarities between *Nakalipithecus* and *Ouranopithecus* from western Asia make it difficult to know whether *Ouranopithecus* represents a late Miocene dispersal out of Africa or whether *Nakalipithecus* represents a late Miocene dispersal into Africa.

These are simplified scenarios, and the actual phylogenetic and biogeographical history of ape evolution was likely much more complex, as there is evidence of mammalian dispersal back and forth between Africa and Eurasia throughout the Miocene. Moreover, our current appreciation of the diversity of fossil apes in the Miocene of both regions is undoubtedly a small sample of the actual diversity, as new taxa are being recovered from Africa, Europe, and Asia at a rapid pace. Nevertheless, most of the currently known fossil record supports the view that much of the radiation of crown hominoids took place in Eurasia.

References

Alba, D.M., 2012. Fossil apes from the Valles-Penedes Basin. Evol. Anthropol. 21, 225–269.

Alba, D.M., Almécija, S., Moyà-Solà, S., 2010a. Locomotor inferences in *Pierolapithecus* and *Hispanopithecus*: Reply to Deane and Begun. J. Hum. Evol. 59, 143–149.

Alba, D.M., Fortuny, J., Moyà-Solà, S., 2010b. Enamel thickness in Middle Miocene great apes *Anoiapithecus, Pierolapithecus* and *Dryopithecus*. Proc. R. Soc. B 277, 2237–2245.

Alba, D.M., Urciuoli, A., Hammond, A.S., Almecija, S., Rook, L., Zanolli, C., 2024. Miocene ape evolution: where does *Oreopithecus* fit in? Bolletino della Societa Paleontologica Italiana 63. https://ddd. uab.cat/pub/artpub/2024/289001/Alba_et_al_2024_BSPI_ONLINE. pdf.

Almécija, S., Alba, D.M., Moyà-Solà, S., 2009. *Pierolapithecus* and the functional morphology of Miocene ape hand phalanges: Paleobiological and evolutionary implications. J. Hum. Evol. 57, 284–297.

Alba, D.M., Moyà-Solà, S., Almécija, S., 2011. A partial hominoid humerus from the Middle Miocene of Castell de Barberà (Vallès-Penedès Basin, Catalonia, Spain). Am. J. Phys. Anthropol. 144, 365–381.

Alba, D.M., Almécija, S., DeMiguel, D., Fortuny, J., Pérez de los Rios, M., Pina, M., Robles, J.M., Moyà-Solà, S., 2015. Miocene small-bodied ape from Eurasia sheds light on hominoid evolution. Science 350, aab2625. https://doi.org/10.1126/science.aab2625

Almécija, S., Hammond, A.S., Thompson, N.E., Pugh, K.D., Moya-Sola, S., Alba, D.M., 2021. Fossil apes and human evolution. Science 372, 6542. https://doi.org/10.1126/science.abb4363

Alpagut, B., Andrews, P., Fortelius, M., et al., 1996. A new specimen of *Ankarapithecus meteai* from the Sinap Formation of central Anatolia. Nature 382, 349–351.

Andrews, P., 1986. Fossil evidence on human origins and dispersal. Cold Spring Harbor Symposia on Quantitative Biology 51, 419–428.

Andrews, P., 1992. Evolution and environment in the Hominoidea. Nature 360, 641–646.

Andrews, P., Tekkaya, I., 1980. A revision of the Turkish Miocene hominoid *Sivapithecus meteai*. Paleontol 23, 85.

Andrews, P., Martin, L., 1987b. The phyletic position of the Ad Dabtiyah hominoid. Bull. Br. Mus. Nat. Hist. (Geol.) 41, 383–393.

Andrews, P., Pilbeam, D.R., 1996. The nature of the evidence. Nature 379, 123–124.

Andrews, P., Harrison, T., Delson, E., Bernor, R., Martin, L., 1996. Distribution and biochronology of European and Southwest Asian Miocene Catarrhines. In: Bernor, R., Fahlbusch, V. (Eds.), Evolution of Neogene Continental Biotypes in Central Europe and the Eastern Mediterranean. Columbia University Press, New York.

Ankel, F., 1965. Der Canalis Sacralis als Indikator fur die Lange der Caudelregion der Primaten. Folia Primatol 3, 263–276.

Begun, D.R., 2002. European hominoids. In: Hartwig, W.C. (Ed.), The Primate Fossil Record. Cambridge University Press, Cambridge, pp. 339–368.

Begun, D.R., 2007. How to identify (as opposed to define) a homoplasy: Examples from fossil and living great apes. J. Hum. Evol. 52, 559–572.

Begun, D.R., 2009. Dryopithecins, Darwin, de Bonis, and the European origin of the African apes and human clade. Geodiversitas 31, 789–816.

Begun, D.R., 2015. Fossil record of Miocene hominoids. In: Henke, W., Tattersall, I. (Eds.), Handbook of Paleoanthropology. Springer Verlag, Berlin Heidelberg, pp. 1261–1332.

Begun, D.R., Ward, C.V., Rose, M.D., 1997. Events in hominoid evolution. In: Begun, D.R., Ward, C.V., Rose, M.D. (Eds.), Function, phylogeny and fossils: Miocene hominoid evolution and adaptation. Plenum Press, New York, pp. 389–415.

Begun, D.R., Nargolwalla, M.C., Hutchinson, M.P., 2006. Primate diversity in the Pannonian Basin: in situ evolution, dispersals, or both. Beit. Paläont 30, 43–56.

Begun, D.R., Nargolwalla, M.C., Kordos, L., 2012. European Miocene hominids and the origin of the African ape and human clade. Evol. Anthropol. 21, 10–23.

Benefit, B.R., McCrossin, M.L., 1997. New fossil evidence on the relationships of *Nyanzapithecus and Oreopithecus*. Am. J. Phys. Anthropol Supp. 24, 74.

Benefit, B.R., McCrossin, M.L., 2015. A window into ape evolution. Science 350, 515–516.

Benefit, B.R., Gitau, S.N., McCrossin, M.L., Palmer, A.K., 1998. A mandible of *Mabokopithecus clarki* sheds new light on oreopithecid evolution. Am. J. Phys. Anthropol Supp. 26, 109.

Bernor, R., Flynn, L.J., Harrison, T., Hussain, S.T., 1988. *Dionysopithecus* from southern Pakistan and the biochronology and biogeography of early Eurasian catarrhines. J. Hum. Evol. 17, 339–358.

Bohme, M.A., Spassov, N., Fuss, J., Troscher, A., Deane, A.S., Prieto, J., Kirscher, U., Lechner, T., Begun, D.R., 2019. A new Miocene ape and locomotion in the ancestor of great apes and humans. Nature 575, 489–493.

Bouchet, F., Zanolli, C., Urciuoli, A., et al., 2024. The Miocene primate Pliobates is a pliopithecoid. Nat. Commun. 15, 2822. https://doi.org/10.1038/s41467-024-47034-9.

Casanovas-Vilar, I., Alba, D.M., Garcés, M., Robles, J.M., Moyà-Solà, S., 2011. Updated chronology for the Miocene hominoid radiation in Western Eurasia. Proc. Natl. Acad. Sci. USA 108, 5554–5559.

Chaimanee, Y., Jolly, D., Benammi, M., Tafforeau, P., Duzer, D., Moussa, I., Jaeger, J.-J., 2003. A Middle Miocene hominoid from Thailand and orangutan origins. Nature 422, 61–65.

Chaimanee, Y., Suteethorn, V., Jintasakul, P., Vidthayanon, C., Marandat, B., Jaeger, J.-J., 2004. A new orang-utan relative from the Late Miocene of Thailand. Nature 427, 439–441.

Chaimanee, Y., Lazzari, V., Chaivanich, K., Jaeger, J.-J., 2019. FIrst maxilla of a late Miocene hominid from Thailand and the evolution of pongine derived characters. J. Hum. Evol. 134, 102636. https://doi.org/10.1016/j.jhevol.2019.06.007.

Chaimanee, Y., Lazzari, V., Yamee, C., Suraprasit, K., Rugbumrung, M., Chaivanich, K., Jaeger, J.-J., 2022. New materials of Khoratpithecus, a late Miocene hominoid from Nakhon Ratchasima Province, Northeastern Thailand, confirm its pongine affinities. Palaeontographica, Abteilung A 323, 147-186.

Ciochon, R.L., Piperno, D.R., Thompson, R.R., 1990. Opal phytoliths found on the teeth of the extinct ape, *Gigantopithecus blacki*: implications for paleodietary studies. Proc. Natl. Acad. Sci. USA 87, 8120–8124.

Conroy, G.C., 1994. *Otavipithecus*: or how to build a better hominid – not. J. Hum. Evol. 27, 373–383.

Conroy, G.C., Senut, B., Gommery, D., Pickford, M., Mein, P., 1992. Brief communication: New primate remains from the Miocene of Namibia, southern Africa. Am. J. Phys. Anthropol. 99, 487–492.

Conroy, G.C., Pickford, M., Senut, B., Mein, P., 1993. Diamonds in the desert: the discovery of *Otavipithecus namibiensis*. Evol. Anthropol. 2, 46–52.

Deane, A.S., Begun, D.R., 2008. Broken fingers: retesting locomotor hypotheses for fossil hominoids using fragmentary proximal phalanges and high-resolution polynomial curve fitting (HR-PCF). J. Hum. Evol. 55, 691–701.

Deane, A.S., Begun, D.R., 2010. *Pierolapithecus* locomotor adaptations: A reply to Alba et al.'s comment on Deane and Begun (2008). J. Hum. Evol. 59, 150–154.

DeBonis, L., Koufos, G.D., 1993. The face and the mandible of *Ouranopithecus macedoniensis*: Description of new specimens and comparisons. J. Hum. Evol. 24, 469–491.

DeBonis, L., Koufos, G.D., 1994. Our ancestors' ancestor: *Ouranopithecus* is a Greek link in human ancestry. Evol. Anthropol. 3, 75–83.

Falk, D., 1983. A reconsideration of the endocast of *Proconsul africanus* (R. S. 51). In: Ciochon, R.L., Corruccini, R.S. (Eds.), New Interpretations of Ape and Human Ancestry. Plenum Press, New York, pp. 239–248.

Fleagle, J.G., 1975. A small gibbon-like hominid from the Miocene of Uganda. Folia Primatol 24, 1–15.

Fleagle, J.G., 1983. Locomotor adaptations of Oligocene and Miocene hominoids and their phyletic implications. In: Ciochon, R.L., Corruccini, R.S. (Eds.), New Interpretations of Ape and Human Ancestry. Plenum Press, New York, pp. 301–324.

Fleagle, J.G., 1984. Are there any fossil gibbons? In: Chivers, D.J., Preuschoft, H., Creel, N., Brockelman, W. (Eds.), The Lesser Apes: Evolutionary and Behavior Biology. Edinburgh University Press, Edinburgh, pp. 431–477.

Fleagle, J.G., Hartwig, W.C., 1997. Paleoprimatology. In: Spencer, F. (Ed.), History of Physical Anthropology: An Encyclopedia. Garland Press, New York, pp. 796–811.

Fleagle, J.G., Lieberman, D.E., 2013. The evolution of primate locomotion: Many transformations, many ways. In: Dial, K., Shubin, N., Brainerd, E. (Eds.), Great Transformations: Major Events in the History of Vertebrate Life. University of California Press, Berkeley.

Fuss, J., Spassov, N., Begun, D.R., Bohme, M., 2017. Potential hominin affinities of *Graecopithecus* from the Late Miocene of Europe. PLoS One 12, e0177127. https://doi.org/10.1371/journal.pone.0177127

Gebo, D.L., MacLatchy, L., Kitto, R., Denio, A., Kingston, J., Pilbeam, D.R., 1997. A new hominoid genus from the Uganda Early Miocene. Science 276, 401–404.

Gilbert, C.C., Ortiz, A., Patel, B.A., Singh, N.P., Campisano, C.J., Fleagle, J.G., Pugh, K.D., Patnaik, R., 2020a. New Middle Miocene ape (Primates: Hylobatidae) from Ramnagar, India fills major gaps in the hominoid fossil record. Proceedings of the Royal Society B 287, 20201655. https://doi.org/10.1098/rspb.2020.1655

Gilbert, C.C., Pugh, K.D., Fleagle, J.G., 2020b. Chapter 17: Dispersal of Miocene hominoids (and pliopithecoids) from Africa to Eurasia in light of changing tectonics and climate. In: Prasad, G.V.R., Patnaik, R. (Eds.), Biological Consequences of Plate Tectonics: New Perspectives on Post-Gondwana and Break-up – A Tribute to Ashok Sahni. Springer, Cham, pp. 393–412.

Gitau, S.N., Benefit, B.R., 1995. New evidence concerning the facial morphology of *Simiolus leakeyorum* from Maboko Island. Am. J. Phys. Anthropol. Supp 20, 99.

Gulec, E.S., Sevim, A., Pehlevan, C., Kaya, F., 2007. A new great ape from the late Miocene of Turkey. Anthropol. Sci 115, 153–158.

Gunz, P., Kozakowski, S., Neubauer, S., Le Cabec, A., Kullmer, O., Benazzi, S., Hublin, J.-J., Begun, D.R., 2020. Skull reconstruction of the late Miocene ape *Rudapithecus hungaricus* from Rudabanya, Hungary. J. Hum. Evol 138, 102687.

Hammond, A.S., Foecke, K.K., Kelley, J., 2019. Hominoid anterior teeth from the late Oligocene site of Losodok, Kenya. J. Hum. Evol. 128, 59–75.

Hammond, A.S., Rook, L., Anaya, A.D., Almecija, S., 2020. Insights into the lower torso in late Miocene hominoid *Oreopithecus bambolii*. Proc. Natl. Acad. Sci. USA 116, 278–284.

Harrison, T., 1986. New fossil anthropoids from the Middle Miocene of East Africa and their bearing on the origin of the Oreopithecidae. Am. J. Phys. Anthropol. 71, 265–284.

Harrison, T., 1987. The phylogenetic relationships of the early catarrhine primates: A review of the current evidence. J. Hum. Evol. 16, 41–80.

Harrison, T., 1989. A new species of *Micropithecus* from the Middle Miocene of Kenya. J. Hum. Evol. 18, 537–557.

Harrison, T., 2002. Late Oligocene to Middle Miocene catarrhines from Afro-Arabia. In: Hartwig, W.C. (Ed.), The Primate Fossil Record. Cambridge University Press, Cambridge, pp. 311–338.

Harrison, T., 2006. Taxonomy, phylogenetic relationships, and biogeography of Miocene hominoids from Yunnan, China. In: Decong, Y. (Ed.), Collected Works for the 40th Anniversary of Yuanmou Man Discovery and the International Conference on Palaeoanthropological Studies. Yunnan Science and Technology Press, Kunming, pp. 233–249.

Harrison, T., 2010a. Apes among the tangled branches of human origins. Science 327, 532–534.

Harrison, T., 2010b. Dendropithecoidea, Proconsuloidea, and Hominoidea. In: Werdelin, L., Sanders, W.J. (Eds.), Cenozoic Mammals of Africa. University of California Press, Berkeley, pp. 429–469.

Harrison, T., 2013. Catarrhine origins. In: Begun, D.R. (Ed.), A Companion to Paleoanthropology. Wiley-Blackwell, Hoboken, pp. 376–396.

Harrison, T., Andrews, P., 2009. The anatomy and systematic position of the Early Miocene proconsulid from Meswa Bridge Kenya. J. Hum. Evol. 56, 479–496.

Harrison, T., Gu, Y., 1999. Taxonomy and phylogenetic relationships of Early Miocene catarrhines from Sihong, China. J. Hum. Evol. 37, 225–277.

Harrison, T., Rook, L., 1997. Enigmatic anthropoid or misunderstood ape? The phylogenetic status of *Oreopithecus bambolii* reconsidered. In: Begun, D.R., Ward, C.V., Rose, M.D. (Eds.), Function, Phylogeny, and Fossils: Miocene Hominoid Evolution and Adaptation. Plenum Press, New York, pp. 327–362.

Harrison, T., van der Made, J., Ribot, F., 2002. A new Middle Miocene Pliopithecid from Sant Quirze, northern Spain. J. Hum. Evol. 42, 371–377.

Harrison, T., Ji, X., Zheng, L., 2008. Renewed investigations at the Late Miocene hominoid locality of Leilao, Yunnan, China. Am. J. Phys. Anthropol Supp. 16, 113.

Harrison, T., Zhang, Y., Wei, G., Sun, C., Wang, Y., Liu, J., Tong, H., Huang, B., Xu, F., 2020. A new genus of pliopithecoid from the late Early Miocene of China and its implications for understanding the paleozoogeography of the Pliopithecoidea. J. Hum. Evol. 145, 102838. https://doi.org/10.1016/j.hevol.2020.102838

Hopwood, A.T., 1933. Miocene primates from Kenya. J. Linn. Soc. London, Zool. 38, 437–464.

Ishida, H., Pickford, M., 1997. A new Late Miocene hominoid from Kenya: *Samburupithecus kiptalami* gen. et sp. nov. C. R. Acad. Sci. (Paris). Earth Planetary Sci. 326, 823–829.

Ishida, H., Kunimatsu, Y., Takano, T., Nakano, Y., Nakatsukasa, M., 2004. *Nacholapithecus* skeleton from the Middle Miocene of Kenya. J. Hum. Evol. 46, 1–35.

Jaeger, J.-J., Soe, A.N., Chavasseau, O., 2011. First hominoid from the Late Miocene of the Irrawaddy Formation (Myanmar). PLoS One 6, e17065.

Ji, X., Harrison, T., Zhang, Y., Wu, Y., Zhang, C., Hu, J., Wu, D., Hou, Y., Li, S., Want, G., Wang, Z., 2022. The earliest hylobatid from the Late Miocene of China. J. Hum. Evol. 171103251.

Kappelman, J., Richmond, B.G., Seiffert, E.R., Maga, A.M., Ryan, T.M., 2003. Hominoidea (Primates). In: Fortelius, M., Kappleman, J., Sen, S., Bernor, R. (Eds.), Geology and Paleontology of the Miocene Sinap Formation. Columbia University Press, New York, pp. 90–124.

Kay, R.F., 1977. Diets of Early Miocene hominoids. Nature 268, 628–630.

Kay, R.F., Ungar, P.S., 1996. Dental evidence for diet in some Miocene catarrhines with comments on the effects of phylogeny on the interpretation of adaptation. In: Begun, D.R., Ward, C.V., Rose, M.D. (Eds.), Miocene Hominoid Fossils: Functional and Phylogenetic Implications. Plenum Press, New York, pp. 131–151.

Kelley, J., 2002. The hominoid radiation in Asia. In: Hartwig, W.C. (Ed.), The Primate Fossil Record. Cambridge University Press, Cambridge, pp. 369–384.

Kelley, J., 2005. Twenty-five years contemplating Sivapithecus taxonomy. In: Lieberman, D.E., Smith, R.J., Kelley, J. (Eds.), Interpreting the Past: Essays on Human, Primate, and Mammal Evolution: In Honor of David Pilbeam. Brill Academic Publishers, Boston, pp. 123–143.

Kelley, J., Gao, F., 2012. Juvenile hominoid cranium from the Late Miocene of southern China and hominoid diversity in Asia. Proc. Natl. Acad. Sci. USA 109, 6882–6885.

Kelley, J., Plavcan, J.M., 1998. A simulation test of hominoid species number at Lufeng, China: implications for the use of the coefficient of variation in paleotaxonomy. J. Hum. Evol. 35, 577–596.

Kelley, J., Andrews, P., Alpagut, B., 2008. A new hominoid species from the Middle Miocene site of Paşalar, Turkey. J. Hum. Evol. 54, 455–479.

Kivell, T., Begun, D.R., 2009. New primate carpal bones from Rudabanya (Late Miocene Hungary): Taxonomic and functional implications. J. Hum. Evol. 124, 697–709.

Koufos, G.D., de Bonis, L., Kugiumtzis, D., 2016. New material of the hominoid Ouranopithecus macedoniensis from the Late Miocene of the Axios Valley (Macedonia, Greece) with some remarks on its sexual dimorphism. Folia Primatol 87, 94–122.

Kuderna, L.F.K., Gao, H., Janiak, M.C., et al., 2023. A global catalog of whole-genome diversity from 233 primate species. Science 380, 906–913..

Kunimatsu, Y., 1997. New species of Nyanzapithecus from Nachola, northern Kenya. Anthropol. Sci. 105, 117–141.

Kunimatsu, Y., Nakatsukasa, M., Sawada, Y., et al., 2007. A new Late Miocene great ape from Kenya and its implications for the origins of African great apes and humans. Proc. Natl. Acad. Sci. USA 104, 19220–19225.

Larson, S.G., 1992. Parallel evolution of the ape forelimb. Evol. Anthropol. 6, 87–99.

Larson, S.G., 1998. Parallel evolution in the hominoid trunk and forelimb. Evol. Anthropol. 6, 87–99.

Leakey, R.E., Leakey, M.G., 1986a. A new Miocene hominoid from Kenya. Nature 324, 143–146.

Leakey, R.E., Leakey, M.G., 1986b. A second new Miocene hominoid from Kenya. Nature 324, 146–148.

Leakey, M.G., Walker, A., 1997. Afropithecus. In: Begun, D.R., Ward, C.V., Rose, M.D. (Eds.), Function, Phylogeny, and Fossils: Miocene Hominoid Evolution and Adaptation. Plenum Press, New York, pp. 225–240.

Leakey, M.G., Leakey, R.E., Richtsmeier, J.T., Simons, E.L., Walker, A.C., 1991. Similarities in Aegyptopithecus and Afropithecus facial morphology. Folia Primatol 56, 65–71.

Leakey, M.G., Ungar, P.S., Walker, A., 1995. A new genus of large primate from the Late Oligocene of Lothidok, Turkana District, Kenya. J. Hum. Evol. 28, 519–531.

Leakey, M.G., Grossman, A., Guiterrez, M., Fleagle, J.G., 2011. Faunal change in the Turkana Basin during the Late Oligocene and Miocene. Evol. Anthropol. 20, 238–253.

MacLatchy, L., 2004. The oldest ape. Evol. Anthropol. 13, 90–103.

MacLatchy, L., Gebo, D., Kityo, R., Pilbeam, D., 2000. Postcranial functional morphology of Morotopithecus bishopi, with implications for the evolution of modern ape locomotion. J. Hum. Evol. 39, 159–183.

MacLatchy, L., Rossie, J.B., 2005. The Napak hominoid: still Proconsul major. In: Lieberman, D.E., Smith, R.J., Kelley, J. (Eds.), Interpreting the Past: Essays on Human, Primate, and Mammal Evolution in Honor of David Pilbeam. Brill Academic Publishers, Boston, pp. 15–28.

MacLatchy, L., Rossie, J., Houssaye, A., Olejniczak, A.J., Smith, T.M., 2019. New hominoid fossils from Moroto II, Uganda and their bearing on the taxonomic and adaptive status of Morotopithecus bishopi. J. Hum. Evol. 132, 227–246.

Madar, S.I., Rose, M.D., Kelley, J., MacLatchy, L., Pilbeam, D.R., 2002. New Sivapithecus postcranial specimens from the Siwaliks of Pakistan. J. Hum. Evol. 42, 705–752.

Madden, R.H., 1980. New Proconsul (Xenopithecus) from the Miocene of Kenya. Primates 21, 241–252.

Martin, L., Andrews, P., 1993. Renaissance of Europe's ape. Nature 365, 494.

McBrearty, S., Jablonski, N.G., 2005. First fossil chimpanzee. Nature 437, 105–108.

McCrossin, M.L., Benefit, B.R., 1994. Maboko Island and the evolutionary history of Old World monkeys and apes. In: Corruccini, R.S., Ciochon, R.L. (Eds.), Integrative Paths to the Past: Paleoanthropology Advantages in Honor of F. Clark Howell. Prentice Hall, Englewood Cliffs, pp. 95–122.

McCrossin, M.L., Benefit, B.R., Gitau, S.N., 1998. Functional and phylogenetic analysis of the distal radius of Kenyapithecus, with comments on the origin of the African great ape and human clade. Am. J. Phys. Anthropol. Supp 26, 158.

McNulty, K.P., Begun, D.R., Kelley, J., Manthi, F.K., Mbua, E.N., 2015. A systematic revision of Proconsul with the description of a new genus of early Miocene hominoid. J. Hum. Evol. 84, 42–61.

Miyamoto, M., Koop, B., Slightom, J.L., Goodman, M., Tennant, M.R., 1988. Molecular systematics of higher primates: genealogical relations and classification. Proc. Natl. Acad. Sci. USA 85, 7627–7631.

Morgan, M.E., Lewton, K.L., Kelley, J., Otarola-Castillo, E., Barry, J.C., Flynn, L.J., Pilbeam, D., 2014. A partial hominoid innominate from the Miocene of Pakistan: description and preliminary analyses. Proc. Natl Acad. Sci. USA 112, 82–87.

Moyà-Solà, S., Köhler, M., 1995. New partial cranium of Dryopithecus Lartet, 1863 (Hominoidea, Primates) from the upper Miocene of Can Llobateres, Barcelona, Spain. J. Hum. Evol. 29, 101–139.

Moyà-Solà, S., Köhler, M., 1996. A Dryopithecus skeleton and the origins of great-ape locomotion. Nature 379, 156–159.

Moyà-Solà, S., Köhler, M., Alba, D.M., Casanovas-Vilar, I., Galindo, J., 2004. Pierolapithecus catalaunicus, a new Middle Miocene great ape from Spain. Science 306, 1339–1344.

Moyà-Solà, S., Alba, D.M., Almécija, S., et al., 2009. A unique Middle Miocene European hominoid and the origins of the great ape and human clade. Proc. Natl. Acad. Sci. USA 106, 9601–9606.

Moyà-Solà, S., Köhler, M., Alba, D.M., et al., 2009. First partial face and upper dentition of the Middle Miocene hominoid Dryopithecus fontani from Abocador de Can Mata (Vallès-Penedès Basin, Catalonia, NE Spain): taxonomic and phylogenetic implications. Am. J. Phys. Anthropol. 139, 126–145.

Nakatsukasa, M., Kunimatsu, Y., 2009. Nacholapithecus and its importance for understanding hominoid evolution. Evol. Anthropol. 18, 103–119.

Nakatsukasa, M., Tsujikawa, H., Shimizu, D., Talako, T., Kunimatsu, Y., Nakano, Y., Ishida, H., 2003. Definitive evidence for tail loss in Nacholapithecus, an East African Miocene hominoid. J. Hum. Evol. 45, 179–186.

Nakatsukasa, M., Ward, C.V., Walker, A., Teaford, M.F., Kunimatsu, Y., Ogihara, N., 2004. Tail loss in *Proconsul hesloni*. J. Hum. Evol. 46, 777–784.

Nargolwalla, M.C., Begun, D.R., Dean, M.C., Reid, D.J., Kordos, L., 2005. Dental development and life history in *Anapithecus hernyaki*. J. Hum. Evol. 49, 99–121.

Nengo, I., Tafforeau, P., Gilbert, C.C., Fleagle, J.G., Miller, E., Feibel, C., Fox, D., Feinberg, J., Pugh, K.D., Berruyer, C., Engle, Z., Spoor, F., 2017. New infant cranium from the African Miocene sheds light on ape evolution. Nature 548, 169–174.

Pan, Y., 2006. Mammalian fauna associated with *Lufengpithecus hudienensis*. In: Qi, G., Wei, D. (Eds.), *Lufengpithecus hudienensis* Site. Science Press, Beijing, pp. 113–148.

Patel, B., Grossman, A., 2006. Dental metric comparisons of *Morotopithecus* and *Afropithecus*: Implications for the validity of the genus *Morotopithecus*. J. Hum. Evol. 51, 506–512.

Perelman, P., Johnson, W.E., Roos, C., et al., 2011. A molecular phylogeny of living primates. PLoS Genet 7, e1001342.

Pickford, M., 1983. Sequence and environment of the Lower and Middle Miocene hominoids of western Kenya. In: Ciochon, R.L., Corruccini, R.S. (Eds.), New Interpretations of Ape and Miocene Hominoid Ancestry. Plenum Press, New York, pp. 421–440.

Pickford, M., 2002. New reconstruction of the Moroto hominoid snout and a reassessment of its affinities to *Afropithecus turkanensis*. Hum. Evol. 17, 1–19.

Pickford, M., Kunimatsu, Y., 2005. Anthropoids from the Middle Miocene (ca 14.5 Ma) of Kipsaraman, Tugen Hills, Kenya. Anthropol. Sci. 113, 189–224.

Pickford, M., Senut, B., Gommery, D., Musiime, E., 2003. New catarrhine fossils from Moroto II, Early Middle Miocene (ca 17.5 Ma) Uganda. Comptes Rendus Palevol 2, 649–662.

Pickford, M., Senut, B., Gommery, D., Misiime, E., 2009. Distinctiveness of *Ugandapithecus* from *Proconsul*. Estudios Geologicos 65, 183–241.

Pickford, M., Musalizi, S., Senut, B., Gommery, D., Musiime, E., 2010. Small apes from the Early Miocene of Napak, Uganda. Geo-Pal Uganda 2010, 1–111.

Pickford, M., Senut, B., Gommery, D., Musalizi, S., Ssebuyungo, C., 2021. Revision of the smaller-bodied anthropoids from Napak, early Miocene, Uganda: 2011–2020 collections. Munchner Geowissenschatich Abhandlungen, Reihe A, Geologie und Palaontologie 51, 1–135.

Pilbeam, D.R., 1969. Tertiary Pongidae of east Africa: Evolutionary relationships and taxonomy. Bull. Peabody Mus. Nat. Hist 31, 1–185.

Pilbeam, D.R., 1996. Genetic and morphological records of the Hominoidea and hominid origins: a synthesis. Mol. Phylogenet. Evol. 5, 155–168.

Pilbeam, D.R., Young, N.M., 2001. *Sivapithecus* and hominoid evolution: some brief comments. In: Koufos, G.D., Andrews, P. (Eds.), Hominoid Evolution and Climate Change in Europe Volume 2: Phylogeny of the Neogene Hominoid Primates of Eurasia. Cambridge University Press, Cambridge, pp. 349–364.

Pilbeam, D.R., Rose, M.D., Barry, J.C., Shah, S.M.I., 1990. New *Sivapithecus* humeri from Pakistan and the relationship of *Sivapithecus* and *Pongo*. Nature 348, 237–239.

Pina, M., Kikuchi, Y., Nakatsukasa, M., Nakano, Y., Kunimatsu, Y., Ogihara, N., Shimizu, D., Takano, T., Tsujikawa, H., Ishida, H., 2021. New femoral remains of *Nacholapithecus kerioi*: Implications for intraspecific variation and Miocene hominoid evolution. J. Hum. Evol. 155, 102982.

Pugh, K.D., 2020. Phylogenetic relationships of Middle-Late Miocene hominoids: implications for human evolution. PhD. Dissertation. City University of New York, New York.

Pugh, K.D., 2022. Phylogenetic analysis of Middle-Late Miocene apes. J. Hum. Evol 165, 103140.

PPugh, K.D., Catalano, S.A., Perez de los Rios, M., et al., 2023. The reconstructed cranium of Pierolapithecus and the evolution of the great ape face. Proc. Natl. Acad. Sci. USA 120, e2218778120. https://doi.org/10.1073/pnas.2218778120.

Rae, T.C., 1993. Phylogenetic Analysis of Proconsulid Facial Morphology. PhD. Dissertation. State University of New York at Stony Brook.

Rae, T.C., 1997. The face of Miocene hominoid. In: Begun, D.R., Ward, C.V., Rose, M.D. (Eds.), Miocene Hominoid Fossils: Functional and Phylogenetic Implications. Plenum Press, New York, pp. 59–77.

Rae, T.C., 1999. Mosaic evolution in the origin of the Hominoidea. Folia Primatol 70, 125–135.

Rae, T.C., 2004. Miocene hominoid craniofacial morphology and the emergence of great apes. Ann. Anat. 186, 417–421.

Richmond, B.G., Whalen, M., 2001. Forelimb function, bone curvature and phylogeny of *Sivapithecus*. In: Koufos, G.D., Andrews, P. (Eds.), Hominoid Evolution and Climate Change in Europe Volume 2: Phylogeny of the Neogene Hominoid Primates of Eurasia. Cambridge University Press, Cambridge, pp. 326–348.

Rose, M.D., 1993. Locomotor anatomy of Miocene hominids. In: Gebo, D.L. (Ed.), Postcranial Adaptation in Nonhuman Primates. NIU Press, DeKalb, pp. 252–272.

Rose, M.D., 1997. Functional and phylogenetic features of the forelimb in Miocene hominoids. In: Begun, D.R., Ward, C.V., Rose, M.D. (Eds.), Function, Phylogeny, and Fossils: Miocene Hominoid Evolution and Adaptations. Plenum Press, New York, pp. 79–100.

Rose, M.D., Leakey, M.G., Leakey, R.E., Walker, A., 1992. Postcranial specimens of *Simiolus enjiessi* and other small apes from the Early Miocene of Lake Turkana, Kenya. J. Hum. Evol. 22, 171–237.

Rossie, J.B., Cote, S.M., 2022. Additional hominoid fossils from the early Miocene of the Lothidok Formation, Kenya. Am. J. Biol. Anthropol. 179, 261–275.

Rossie, J.B., Hill, A., 2018. A new species of *Simiolus* from the middle Miocene of the Tugen Hills, Kenya. J. Hum. Evol. 125, 50–58.

Rossie, J.B., MacLatchy, L., 2006. A new pliopithecid genus from the Early Miocene of Uganda. J. Hum. Evol. 50, 568–586.

Rossie, J.B., Seiffert, E.R., 2006. Continental paleobiogeography as phylogenetic evidence. In: Lehman, S.M., Fleagle, J.G. (Eds.), Primate Biogeography, Plenum/Kluwer, New York, 469–522.

Rossie, J.B., Gutierrez, M., Goble, E., 2012. Fossil forelimbs of *Simiolus* from Morourot, Kenya. Am. J. Phys. Anthropol Suppl. 54, 252.

Ruff, C.B., Walker, A., Teaford, M.F., 1989. Body mass, sexual dimorphism and femoral proportions of *Proconsul* from Rusinga and Mfangano Islands, Kenya. J. Hum. Evol. 18, 515–536.

Russo, G.A., 2016. Comparative sacral morphology and the reconstructed tail lengths of five extinct primates: *Proconsul heseloni*, *Epipliopithecus vindobonensis*, *Archaeolemur edwardsi*, *Megaladapis grandidieri*, and *Palaeopropithecus kelyus*. J. Hum. Evol. 90, 135–162.

Sanders, W.J., Bodenbender, B.E., 1994. Morphometric analysis of lumbar vertebra UMP 67–28: Implications for spinal function and phylogeny of the Miocene Moroto hominoid. J. Hum. Evol. 26, 202–238.

Sankhyan, A.R., Kelley, J., Harrison, T., 2017. A highly derived pliopithecoid from the Late Miocene of Haritalyangar, India. J. Hum. Evol. 105, 1–12.

Sevim-Erol, A., Begun, D.R., Yavuz, A., et al., 2023. A new ape from Turkiye and the radiation of late Miocene hominines. Commun Biol. 6, 842. https://doi.org/10.1038/s42003-023-05210-5.

Simons, E.L., Pilbeam, D.R., 1965. Preliminary revision of the Dryopithecinae (Pongidae, Anthropoidea). Folia Primatol 3, 81–152.

Singleton, M., 1998. The phylogenetic affinities of *Otavipithecus namibiensis*: A parsimony analysis. Am. J. Phys. Anthropol. Supp 26, 202.

Steiper, M.E., Young, N.M., 2008. Timing primate evolution: Lessons from the discordance between molecular and paleontological estimates. Evol. Anthropol. 17, 179–188.

Stevens, N.J., Seiffert, E.R., O'Conner, P.M., Roberts, E.M., Scmitz, M.D., Krause, C., Gorscak, E., Ngasala, S., Hieronymus, T.L.,

Temu, J., 2013. Paleontological evidence for an Oligocene divergence between Old World monkeys and apes. Nature 497, 611–614.

Stewart, C.-B., Disotell, T.R., 1998. Primate evolution – in and out of Africa. Curr. Biol. 8, R582–R587.

Suwa, G., Kono, R.T., Katoh, S., Asfaw, B., Beyene, Y., 2007. A new species of great ape from the Late Miocene epoch in Ethiopia. Nature 448, 921–924.

Ungar, P.S., Kay, R.F., 1995. The dietary adaptations of European Miocene catarrhines. Proc. Natl. Acad. Sci. USA 92, 5479–5481.

Urciuoli, A., Alba, D.M., 2023. Systematics of Miocene apes: State of the art of a neverending controversy. J. Hum Evol. 175, 103309.

Urciuoli, A., Zanolli, C., Beaudet, A., Dumoncel, J., Santos, F., Moya-Sola, S., Alba, D.M., 2020. The evolution of the vestibular apparatus in apes and humans. eLife 9. e51261. https://doi.org/10.7554/eLife.51261

von Koenigswald, G.H.R., 1983. The significance of hitherto undescribed Miocene hominoids from the Siwaliks of Pakistan in the Senchenberg Museum, Frankfurt. In: Ciochon, R.L., Corruccini, R.S. (Eds.), New Interpretations of Ape and Human Ancestry. Plenum Press, New York, pp. 517–526.

Walker, A., Teaford, M.F., 1989. The hunt for Proconsul. Sci. Am. 260, 76–82.

Walker, A., Shipman, P., 2007. The Ape in the Tree. Harvard University Press, Cambridge.

Walker, A., Teaford, M.F., Martin, L., Andrews, P., 1993. A new species of Proconsul from the Early Miocene of Rusinga/Mfangano Islands, Kenya. J. Hum. Evol. 25, 43–56.

Ward, C.V., 1997. Functional anatomy and phyletic implications of the hominoid trunk and hindlimb. In: Begun, D.R., Ward, C.V., Rose, M.D. (Eds.), Function, Phylogeny and Fossils: Miocene Hominoid Evolution and Adaptations. Plenum Publishing, New York, pp. 101–130.

Ward, C.V., 2007. Postcranial and locomotor adaptations in hominoids. In: Henke, W., Tattersall, I. (Eds.), Handbook of Paleoanthropology. Springer Verlag, Heidelberg, pp. 889–920.

Ward, S.C., Duren, D.L., 2002. Middle and Late Miocene African hominoids. In: Hartwig, W.C. (Ed.), The Primate Fossil Record. Cambridge University Press, Cambridge, pp. 385–397.

Ward, C.V., Walker, A., Teaford, M.F., 1991. Proconsul did not have a tail. J. Hum. Evol. 21, 215–220.

Ward, C.V., Walker, A., Teaford, M.F., Odhiambo, I., 1993. Partial skeleton of Proconsul nyanzae from Mfangano Island, Kenya. Am. J. Phys. Anthropol. 90, 77–111.

Ward, C.V., Brown, B., Hill, A., Kelley, J., Downs, W., 1999. Equatorius: A new hominoid genus from the Middle Miocene of Kenya. Science 285, 1382–1386.

Ward, C.V., Hammond, A.S., Plavcan, J.M., Begun, D.R., 2019. A late Miocene hominid partial pelvis from Hungary. J. Hum. Evol. 136, 102645.

Welker, F., Ramos-Madrigal, J., Kuhlwilm, M., Llao, W., Gutenbrunner, P., de Manuel, M., Samodova, D., Mackle, M., Allentoft, M.E., Bacon, A.-M., Collins, M.J., Cox, J., Lalueza-Fox, C., Olsen, J.V., Dementer, F., Wang, W., Marques-Bonet, T., Cappellini, E., 2019. Enamel proteome shows that Gigantopithecus was an early diverging pongine. Nature 576, 262–265.

Williams, S.A., Prang, T.C., Meyer, M.R., Russo, G.A., Shapiro, L.J., 2020. Reevaluating bipedalism in Danuvius. Nature 586, E1–E3.

Wood, B.A., Harrison, T., 2011. The evolutionary context of the first hominins. Nature 470, 347–352.

Wuthrich, C., MacLatchy, L.M., Nengo, I.O., 2019. Wrist morphology reveals substantial locomotor diversity among early catarrhines: An analysis of capitates from the early Miocene of Tinderet (Kenya). Scientific Reports 9, 3728. https://doi.org/10.1038/s41598-019-39800-3

Young, N.M., MacLatchy, L., 2004. The phylogenetic position of Morotopithecus. J. Hum. Evol. 46, 163-184.

Zalmout, I.S., Sanders, W.J., MacLatchy, L., et al., 2010. New Oligocene primate from Saudi Arabia and the divergence of apes and Old World monkeys. Nature 466, 360–364.

Zhang, Y., Harrison, T., 2017. Gigantopithecus blacki: A giant ape from the Pleistocene of Asia revisited. Am. J. Phys. Anthropol. 162, 153–177.

Further Reading

Andrews, P., 2015. An Ape's View of Human Evolution. Cambridge University Press, Cambridge. pp. 328.

Andrews, P., 2020. Last common ancestor of apes and humans: Morphology and Environment. Folia Primatol 91, 122–148.

Begun, D.R., Ward, C.V., Rose, M.D., 1997. Function, phylogeny and fossils: Miocene hominoid evolution and adaptation. Plenum Press, New York.

Begun, D.R., 2013. The Miocene hominoid radiations. In: Begun, D.R. (Ed.), A Companion to Paleoanthropology. Wiley-Blackwell, Hoboken, pp. 398–416.

Begun, D.R., 2015. The Real Planet of the Apes: A New Story of Human Origins. Princeton University Press.

Harrison, T., 2005. The zoogeographic and phylogenetic relationships of early catarrhine primates in Asia. Anthropol. Sci. 113, 43–51.

Henke, W., Tattersall, I., 2015. Handbook of Paleontology. Springer Verlag, Heidelberg.

Zapfe, H., 1960. Die Primatenfunde aus der Miozanen Spaltenfullung von Neudorf an der march (Devinxka nova ves), Tschechoslowakev. Mit Anhang: Er Primatenfund aus dem Miozan von klein Hadersdorf in Niederoesterreich. Schweiz. Pal. Abh 78, 4–293.

CHAPTER

16

Fossil Old World Monkeys

Old World monkeys are the modern success story of catarrhine evolution. Although they first appear in the fossil record at approximately the same time as apes, the end of the Oligocene, they are quite rare throughout most of the Miocene epoch, and the major radiation of the group appears to have taken place later. From the Late Miocene to the Recent, Old World monkeys have an extensive fossil record from Africa, Europe, and Asia, including many parts of the world where they are absent or rare today (Fig. 16.1). Many new fossils from the time of the earliest Old World monkeys and from the period of their great diversity in the Plio-Pleistocene of Africa

and Asia have made Old World monkey evolution one of the most exciting areas in primate evolution.

The Earliest Old World Monkeys

The first record of cercopithecoid monkeys is *Nsungwepithecus gunnelli* from the late Oligocene of Tanzania (Stevens et al., 2013). It is a large primate, known from only one tooth, but it provides the earliest evidence of the divergence between cercopithecoids and hominoids. *Alophe metios*, known from a handful of

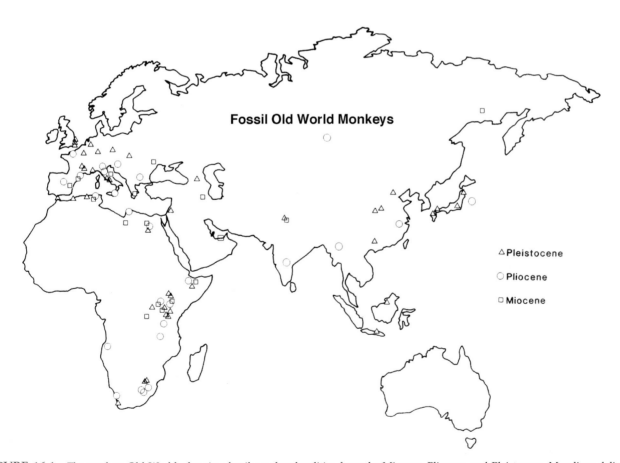

FIGURE 16.1   The modern Old World, showing fossil monkey localities from the Miocene, Pliocene, and Pleistocene. Map lines delineate study areas and do not necessarily depict accepted national boundaries.

Primate Adaptation and Evolution
https://doi.org/10.1016/B978-0-12-815809-8.00016-3

jaws in the Early Miocene of Kenya, is similar to *Nsungwepithecus* but is much smaller (Rasmussen et al., 2019). In some ways, *Alophe* is even more primitive. Together, *Nsungwepithecus* and *Alophe* display incomplete bilophodonty and appear to document the beginnings of the cercopithecoid radiation. They are best regarded as very early stem taxa.

From Early and Middle Miocene deposits in northern and eastern Africa, there are numerous primitive Old World monkeys that cannot be placed conveniently in either Colobinae or Cercopithecinae. Rather, they form a separate family, the victoriapithecids, which preceded the divergence of colobines and cercopithecines (Benefit and McCrossin, 2002; Miller et al., 2009).

Victoriapithecids are known primarily from jaws and teeth; a skull and limb elements are known for only one taxon, *Victoriapithecus macinnessi*. In victoriapithecids, as in all later cercopithecoids, the molar cusps are aligned in two rows, an anterior (mesial) and a posterior (distal) row, but their teeth are more primitive than those of later Old World monkeys in several respects. In some, there is no ridge or loph connecting the paired cusps in the lower molars. The upper molars frequently have a crista obliqua linking the metacone with the protocone, and the lower molars often have a small hypoconulid (Fig. 16.2). All of these dental features are present in primitive catarrhines and in apes, but are absent in extant Old World monkeys. As in the propliopithecids, the last lower premolar has an expanded buccal face and the lower molars have a relatively large base and a constricted occlusal surface. The trigonid is relatively short and the crown height is relatively low, as in colobines, but the molar cusps are relatively low, as in cercopithecines. They had sexually dimorphic canines, although only *Victoriapithecus* and *Noropithecus* are known from relatively large samples. There is considerable variation in the morphology of the mandible in different species. Overall, the dentition of these primitive Old World monkeys is intermediate between that of the early catarrhines from the Oligocene and later Old World monkeys.

At some Early and Middle Miocene localities, early monkeys are as common as fossil apes; at other localities in the same time period and geographical area, monkeys are absent. It has been suggested that early monkeys are more abundant in drier, more open habitats (Pickford and Senut, 1988), but the Wadi Moghara locality in Egypt was probably very wet. The fauna includes many water-adapted species, including a duck (Miller, 1998).

Victoriapithecids are currently placed in four genera (Table 16.1; Fig. 16.3), but this is likely an underestimate (Miller et al., 2009).

Prohylobates is known from Early Miocene deposits at the site of Wadi Moghara in northern Egypt. The type

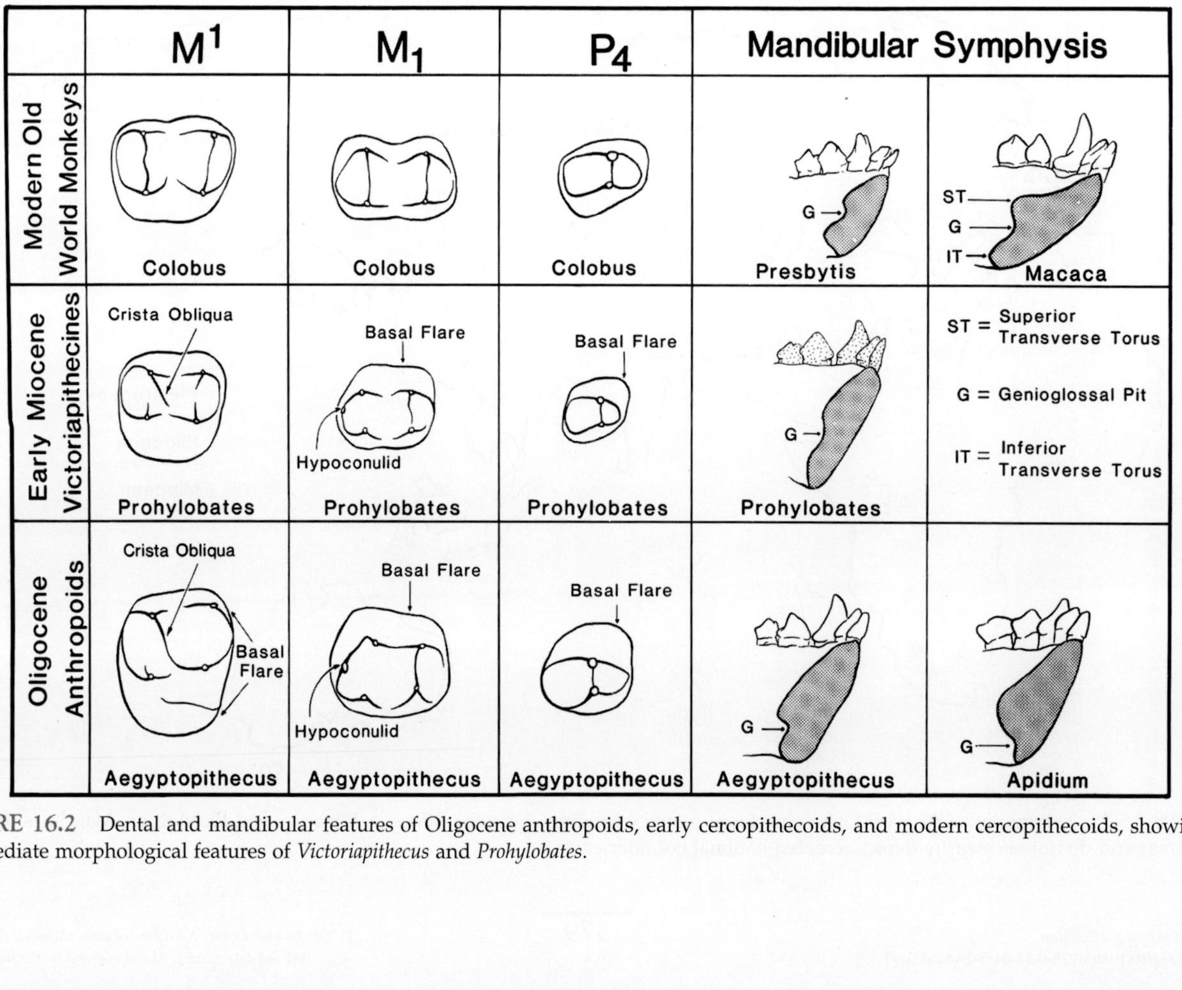

FIGURE 16.2   Dental and mandibular features of Oligocene anthropoids, early cercopithecoids, and modern cercopithecoids, showing the intermediate morphological features of *Victoriapithecus* and *Prohylobates*.

TABLE 16.1   Superfamily CERCOPITHECOIDEA
Family VICTORIAPITHECIDAE

Species	Estimated average mass (g)
Victoriapithecus (Middle Miocene, East Africa)	
V. macinnesi	7000
Prohylobates (Early Miocene, North Africa)	
P. tandyi	5500
Genus indet. (Early Miocene, North Africa)	
Gen. indet. *mogharensis*	8000
Genus et sp. indet. (Middle Miocene, East Africa)	
Gen. et sp. indet. Kipsaramon	8500
Zaltanpithecus (Middle Miocene, North Africa)	
Z. simonsi	20,000
Noropithecus (Early to Middle Miocene, East Africa)	
N. bulukensis	8500
N. fleaglei	9000
Family INCERTAE SEDIS	
Nsungwepithecus (Late Oligocene, East Africa)	
N. gunnelli	15,000
Alophe (Early Miocene, East Africa)	
A. metios	8500

species *P. tandyi* was based on a mandible with worn teeth. *Prohylobates tandyi* differs from other victoriapithecids in that the posterior cusps on the first lower molar are not connected by a lophid. It also has a very small third molar. There is a second monkey species from Wadi Moghara that has a very deep mandible. It was originally described as *Dryopithecus mogharensis* and later included in *P. tandyi*, but it is clearly a separate species. However, because there is no occlusal morphology preserved on either of the specimens, it is not clear to what genus it should be assigned.

Zaltanpithecus simonsi is a large monkey from the Early Miocene of Libya that was originally placed in the genus *Prohylobates* (Delson, 1979). It is known from a single mandible with two lower molars. The teeth of *Z. simonsi* are roughly 50% larger than those of *P. tandyi*, and the roots are widely splayed (Benefit, 2008).

Victoriapithecus macinnesi is known primarily from the Middle Miocene site of Maboko Island in Lake Victoria in Kenya, but fossil monkeys from other Miocene sites in Kenya and Uganda have also been assigned to this species (Miller et al., 2009; Pickford et al., 2019). There are hundreds of dental remains assigned to this species. *Victoriapithecus* has a small upper fourth premolar. The lower molars are completely bilophodont and increase in size posteriorly. The canines are sexually dimorphic. The cranial anatomy of *Victoriapithecus*

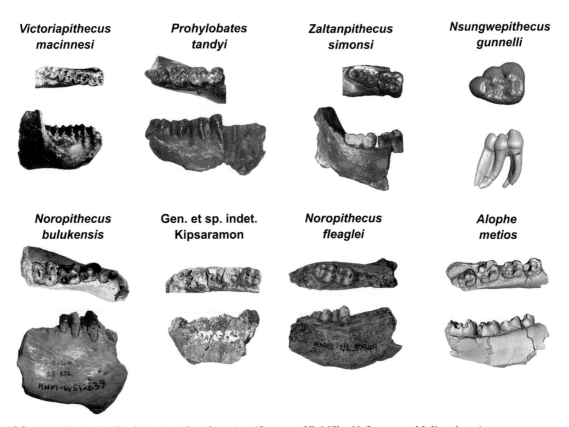

FIGURE 16.3   Mandibular fossils of victoriapithecid species. *(Courtesy of E. Miller, N. Stevens, and J. Kappelman.)*

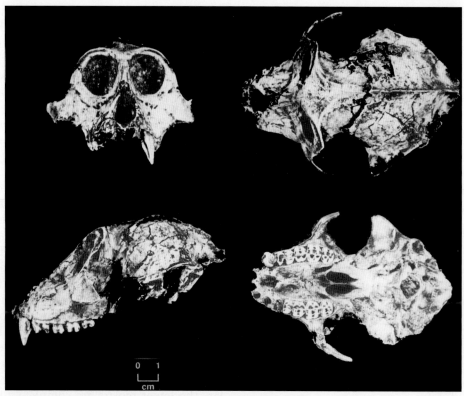

FIGURE 16.4   The skull of *Victoriapithecus. (Courtesy of B. Benefit and M. McCrossin.)*

(Fig. 16.4) shows a narrow interorbital pillar, a deep maxilla, tall orbits, and a small frontal trigon. *Victoriapithecus* had a relatively small brain compared with modern cercopithecoids, and the cranium is less flexed (Gonzales, 2015). Overall, the facial morphology of this early monkey resembles that of early catarrhines, such as *Aegyptopithecus* and living cercopithecines, and is strikingly different from the short face of colobines and gibbons (Benefit and McCrossin, 1993, 1997).

Postcranial remains of *Victoriapithecus* show a narrow articulation in the distal end of the humerus with a posteriorly directed medial epicondyle and a prominent medial trochlear lip. The ulnar notch is narrow and deep. Overall, the postcranial anatomy of *Victoriapithecus macinnesi* is most similar to that of living cercopithecoids, such as the vervet monkey or *Macaca fascicularis*, which are adept on either arboreal or terrestrial substrates, but in some features, *Victoriapithecus* resembles platyrrhines and stem catarrhines (Harrison, 1989; Blue et al., 2006). At the Maboko Island site, *Victoriapithecus* is most common in deposits that indicate riparian woodland and wooded grassland, rather than arid grasslands (Blue et al., 2006).

The genus **Noropithecus** (Miller et al., 2009) includes victoriapithecids from several localities in northern Kenya and southern Ethiopia that were formerly placed in either *Prohylobates* or *Victoriapithecus* (e.g., Leakey, 1985; Fleagle et al., 1997; Pickford and Kunimatsu, 2005).

Noropithecus differs from other victoriapithecids in having molar teeth with greater basal flare owing to closely approximated cusp tips. The mandible is also distinctive, with a deeper mandibular corpus and a more vertical mandibular symphysis (Locke et al., 2020). There are two species provisionally attributed to the species. The type species *Noropithecus bulukensis* is from the Early Miocene site of Buluk on the eastern side of Lake Turkana (Leakey, 1985). *Noropithecus fleaglei*, from the site of Nabwal, north of Buluk near the Kenya–Ethiopia border, is based on a broad and relatively deep mandible preserving two lower molars with very broad bases and cusps that nearly touch at their tips (Miller et al., 2009; Locke et al., 2020). Finally, there is a large and somewhat variable sample of victoriapithecid specimens from the Middle Miocene site of Kipsaramon in the Tugen Hills of northern Kenya that displays dental similarities with *Victoriapithecus* and mandibular similarities with *Noropithecus* (Miller et al., 2009; Gilbert et al., 2010; Locke et al., 2020). It is unclear what genus these specimens belong to, but a humerus from the area is similar to that of *Victoriapithecus*, and its features indicate terrestrial quadrupedal behavior (Gilbert et al., 2010).

Although victoriapithecids are definitely more similar to extant cercopithecids than to any other group of primates in their dental, cranial, and postcranial anatomy, they are distinctly more primitive than all later Old World monkeys in many aspects of their anatomy

(Benefit and McCrossin, 1997, 2002; Blue et al., 2006). The first Old World monkeys are intermediate links between early catarrhines and modern cercopithecoids, and provide clear evidence of the way in which characteristic features of both subfamilies of extant cercopithecoids evolved. The retention of a trigon on the upper molars and a small hypoconulid on the lower molars in these genera confirms what dental anatomists have known for years – that the bilophodont teeth of Old World monkeys are derived from an ancestor with more ape-like teeth. Likewise, the cranial anatomy of *Victoriapithecus* shows many similarities to that of primitive catarrhines and of some other fossil monkeys (Benefit and McCrossin, 1997). The limb skeleton suggests that Old World monkeys were probably partly terrestrial from their origin (see also Strasser, 1988; Gebo, 1989; Arenson et al., 2020).

Although we now know much about the morphology and diversity of this group, the identity of their closest catarrhine relatives is less clear. Although victoriapithecids (+ cercopithecids) are almost certainly the sister taxon to the more common and diverse proconsuloids, and both are close to the split between the major groups of catarrhines (Old World monkeys and hominoids), there are few shared similarities between the two groups of Early and Middle Miocene catarrhines. Rather, the cranial morphology of *Victoriapithecus* is more similar to that of the Oligocene stem catarrhines such as *Aegyptopithecus* (or *Saadanius*), while the limb skeleton is similar to that of later cercopithecines. It seems most likely that the common ancestor of monkeys and apes lies somewhere in the Middle-Late Oligocene fossil record of Africa or Arabia.

Fossil Cercopithecids

Although primitive monkeys such as victoriapithecids are the only cercopithecoids found in the Early and Middle Miocene of Africa, by the Late Miocene, there are numerous fossil taxa that can be clearly referred to the extant subfamilies of Old World monkeys – cercopithecines and colobines. In the latest Miocene and continuing through the Pliocene and Pleistocene, fossil monkeys are relatively abundant in deposits throughout Africa and Eurasia. Although there are differences in the relative abundance of individual clades and adaptive groups from what we see today, this radiation of monkeys was, for the most part, the same one that dominates living higher primate communities. Because many of the most striking features that distinguish the living subfamilies of Old World monkeys are soft tissues, such as the sacculate stomachs of colobines or the cheek pouches of cercopithecines, there are potential hazards in assigning fossil monkeys to one family or another solely on the basis of dental and cranial remains. Nevertheless, extant

cercopithecines can be distinguished from colobines by several dental and cranial features, including molars with long trigonids, higher crowns and relatively lower molar cusps, and skulls with longer snouts and narrower interorbital dimensions (Fig. 6.3). Although these same dental and cranial features seem to sort fossil members of the two subfamilies, many of the postcranial differences that characterize the living taxa do not so readily distinguish fossil cercopithecids, except for the tendency of cercopithecines to have longer thumbs and shorter digits than colobines.

Fossil Cercopithecines

Fossil cercopithecines, like those living today, can be readily divided into three major groups: macaques; mangabeys, mandrills, baboons, and geladas; and guenons (Table 16.2).

Macaques

The genus *Macaca* has the widest distribution of any living nonhuman primate, extending from North Africa and Gibraltar in the west to Japan and the Philippines in the east. Fossil macaques were even more widespread, especially in Europe and North Africa (Table 16.2). Although they are quite abundant and widespread, most fossil macaques are strikingly similar to the extant genus, indicating that *Macaca* has retained a very conservative morphology over the last ~7 million years or more (Delson and Rosenberger, 1984; Alba et al., 2014; Roos et al., 2019).

There are numerous, mostly dental, remains of macaques from latest Miocene or earliest Pliocene localities in Algeria, Libya, Egypt, Spain, Italy, and possibly China (Kohler et al., 2000; Alba et al., 2014; Roos et al., 2019). In the later Pliocene, macaques were widespread throughout much of North Africa and Europe (including Spain, France, Germany, Italy, The Netherlands, and Yugoslavia), and during the Middle Pleistocene, their range extended into Great Britain, southern Russia, and the Middle East. Most of these fossil populations cannot be distinguished in dental features from the living Barbary macaque *M. sylvanus* of Gibraltar and North Africa (Delson, 1980). The most distinctive fossil macaque, the Pleistocene "dwarf macaque," **Macaca majori**, from the island of Sardinia, was about 15% smaller in dental dimensions than living *M. sylvanus* (Delson et al., 2000).

Although two macaque-sized teeth of questionable provenience are suggested to be from the early Pliocene of China (Alba et al., 2014), macaques were not common in Asia until between 3.0 and 2.0 million years ago, probably in association with a faunal change at 2.5 million

TABLE 16.2   Fossil Cercopithecines
Family CERCOPITHECIDAE
Subfamily CERCOPITHECINAE

Species	Estimated average mass (g)
***Macaca* (Latest Miocene to Recent, North Africa, Europe, Asia)**	
M. sylvanus	11,250
M. prisca	11,000
M. majori	7750
M. libyca	11,500
M. anderssoni	13,750
M. palaeindica	13,000
M. jiangchuanensis	12,100
***Procyncocephalus* (Pliocene, Asia)**	
P. wimani	22,000
P. subhimalayensis	30,000
***Paradolichopithecus* (Pliocene, Europe, ?Asia)**	
P. arvernensis	25,000
P. sushkini	32,500
?P. gansuensis	28,000
***Papio* (Plio-Pleistocene to Recent, Africa)**	
P. robinsoni	23,500
?P. izodi	17,500
P. angusticeps	19,000
***Soromandrillus* (Plio-Pleistocene, Africa)**	
S. quadratirostris	30,500
***Dinopithecus* (Pleistocene, South Africa)**	
D. ingens	40,000
***Procercocebus* (Plio-Pleistocene, South Africa)**	
P. antiquus	16,500
***Pliopapio* (Miocene-Pliocene, East Africa)**	
P. alemui	10,000
***Parapapio* (Late Miocene to Early Pleistocene, Africa)**	
P. broomi	18,500
P. jonesi	15,000
P. whitei	22,500
P. lothagamensis	8750
P. ado	14,000
***Gorgopithecus* (Pleistocene, Africa)**	
G. major	33,500
***Theropithecus* (Plio-Pleistocene, Africa, Eurasia)**	
T. oswaldi ecki	21,000
T. oswaldi darti	28,500
T. oswaldi oswaldi	35,000

TABLE 16.2   Fossil Cercopithecines—Cont'd
Family CERCOPITHECIDAE
Subfamily CERCOPITHECINAE

Species	Estimated average mass (g)
T. oswaldi delsoni	47,500
T. oswaldi leakeyi	52,000
T. brumpti	29,000
?T. baringensis	26,000
***Cercopithecus* (Pliocene to Recent, Africa)**	
unnamed species	–
***Chlorocebus* (Pleistocene to Recent, Africa)**	
C. ngedere	5000
***Nanopithecus* (Pliocene, East Africa)**	
N. browni	1275

years ago (Barry, 1987). Macaques were relatively common throughout the Pleistocene of China and Southeast Asia. A larger species, **Macaca anderssoni** or **M. robusta**, is found in Early and Middle Pleistocene localities in northern China (including Zhoukoudian), and there are many younger, smaller remains from Middle to Late Pleistocene and Recent deposits in more southern, subtropical areas that are generally identified only as *Macaca* sp. There are also abundant remains of macaques from many Late Pleistocene and Recent deposits in other parts of Asia, including Japan, Korea, Vietnam, Java, Sumatra, and Borneo. Many of these are similar to species currently living in the same region. For example, the fossil macaques from Vietnam have been attributed to three of the five species currently living in that country. A more precise biochronology of fossil macaques offers the possibility of testing theories regarding the colonization of Asia by successive radiations of different species groups during the Pleistocene (Jablonski and Pan, 1988; Roos et al., 2019).

In addition to fossil representatives of the living *Macaca*, there are two genera of very large monkeys from the Late Pliocene and Pleistocene of Asia and Europe. **Procynocephalus** is a Late Pliocene Asian genus with one species from northern India and one from southern China. The appearance of *P. wimani* in the Late Pliocene of China is one of the earliest records of a cercopithecine in East Asia. It has a macaque-like dentition and skull, and its baboon-like skeleton suggests that its locomotor habits resembled those of the more terrestrial macaques, such as *M. nemestrina*. **Paradolichopithecus** is a similar baboon-like macaque from the Pliocene of Europe and western Asia. Both *Procynocephalus* and *Paradolichopithecus* are known from numerous localities, and some authors question whether they should be placed in the same genus (Jablonski, 2002; Takai et al., 2008; Kostopoulos et al., 2018).

Mangabeys, Mandrills, Baboons, and Geladas

Macaques are the only cercopithecines known to colonize Europe and Asia successfully. The other members of the subfamily are known almost totally from Africa, where they remain abundant today. As discussed in Chapter 6, the extant members of this group contain two distinct clades, one consisting of baboons, geladas, the kipunji, and *Lophocebus* mangabeys, and the other containing *Cercocebus* mangabeys, mandrills, and drills. Nearly all have fossil records from the Pliocene and Pleistocene of Africa.

Parapapio, a primitive papionin from the Late Miocene to Early Pleistocene of eastern and southern Africa, is the most primitive member of the baboon/gelada/mangabey/mandrill group and is probably near the ancestry of all living genera. *Parapapio* is intermediate in size between mangabeys and the largest savannah baboons. It is known from numerous species in both eastern and southern Africa that likely do not form a natural group. The earliest and most primitive species are from East Africa. The postcranial skeleton is best known for a specimen attributed to *Parapapio jonesi* from Hadar in Ethiopia (Frost and Delson, 2002). These bones indicate an arboreal monkey similar to the living *Lophocebus*. Other postcrania likely attributable to *Parapapio* in South Africa suggest more terrestrial species as well (Elton, 2001).

Pliopapio, from the Late Miocene and Early Pliocene of the Afar region of Ethiopia, is similar in size to a large macaque. Recently attributed postcrania from Ethiopia suggest that it was probably semiterrestrial (Frost et al., 2020a). It may be a basal African papionin or an early member of the *Papio/Rungwecebus/Lophocebus/Theropithecus* clade (Frost, 2001; Frost et al., 2009; Pugh and Gilbert, 2018).

Dinopithecus is a very large (35–50 kg), sexually dimorphic monkey known mainly from the Swartkrans cave deposits (Early Pleistocene) of South Africa. There are no skeletal remains assigned to the genus, and its exact phylogenetic position relative to living baboons, geladas, mangabeys, and mandrills is unclear (Gilbert, 2013; Pugh and Gilbert, 2018).

Although the extant savannah baboons of the genus *Papio* are widespread and abundant in Africa, the fossil record of this genus is relatively sparse. The oldest specimens attributed to *Papio* are slightly over 2 million years old from South Africa, and dental and cranial remains attributed to *Papio* have been described from the Plio-Pleistocene of both eastern and southern Africa. There are three often recognized fossil species from South Africa. *Papio robinsoni*, *P. izodi*, and *P. angusticeps* are all known from several South African cave deposits, but there is some disagreement about how distinct these are from one another and from extant species of *Papio* (Gilbert et al., 2018).

There are several other large monkeys from the past 3 million years in many parts of Africa whose likely relationships to extant baboons, mangabeys, geladas, and mandrills have become better established in recent years. *Gorgopithecus* is a large (30–40 kg) monkey from southern and eastern African Pleistocene deposits. *Gorgopithecus* seems to have little sexual dimorphism in the size of the cheek teeth, and it is probably a relative of *Papio* baboons and *Lophocebus* mangabeys (Gilbert, 2013; Gilbert et al., 2016; Pugh and Gilbert, 2018).

Theropithecus, the gelada from the Ethiopian highlands, is the only living representative of a group of large papionins that had an extensive radiation during the Pliocene and Pleistocene. Fossil *Theropithecus* remains have been recovered from many parts of Africa and also from India. Fossil species of *Theropithecus* resemble the living gelada in their complex cheek teeth, presumably related to a dietary specialization on grass blades, seeds, and tubers. *Theropithecus* is the only predominantly folivorous cercopithecine. Unlike folivorous colobines, however, the living *Theropithecus* exploits this dietary niche on the ground by specializing on grass. All *Theropithecus* have long forelimbs and short phalanges, indicating terrestrial quadrupedalism. The extinct species seem to have the same digital proportions as extant geladas, with relatively long thumbs compared to the size of the index finger, and were probably manual foragers. The extinct species are all much larger than the living gelada and show many more extreme dental, cranial, and skeletal specializations.

Fossil remains of *Theropithecus* are first known from the Early Pliocene (Frost et al., 2014, 2020b; Reda et al., 2024) and they extend throughout most of the Plio-Pleistocene. *Theropithecus* is commonly divided into two or three distinct lineages often identified as subgenera that have been separated since the beginning of the Pliocene. There is also disagreement over how many species should be recognized in each lineage (Jablonski, 1993; Getahun et al., 2023).

Theropithecus oswaldi, with numerous temporal and geographical subspecies, steadily increased in size over almost 4 million years of evolution in the African fossil record (Frost et al., 2014, 2017, 2022a). By the Middle-Late Pleistocene, it was an enormous monkey, with males probably weighing as much as 75+ kg. It was extremely abundant in many East African, North African, and South African Pliocene and Pleistocene sites (Fig. 16.5). Compared with *T. brumpti*, it has greatly reduced, laterally compressed incisors and canines; large molars; a short, deep face; and very long limbs. *Theropithecus o. darti* and *T. o. ecki* from the Pliocene are early members of this lineage (Eck, 1993; Frost et al.,

2022a; Getahun et al., 2023). From an ecological model based on the living gelada, Dunbar (1992, 1993) has suggested that these large lowland terrestrial grazers must have been dependent on standing water and especially vulnerable to local extinction through climatic fluctuation, a prediction that accords well with their sporadic distribution in the fossil record, especially in South Africa (Foley, 1993; Lee and Foley, 1993).

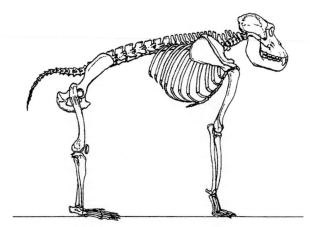

FIGURE 16.5   Skeleton of *Theropithecus oswaldi*.

The living gelada *Theropithecus (Theropithecus) gelada* has no fossil record. It is thought by some to be the most primitive lineage and is linked to the *T. oswaldi* lineage by others.

Theropithecus (Omopithecus) brumpti is an early species from the Late Pliocene of East Africa with extraordinary development of the zygomatic arches that must have given its face an extremely imposing appearance (Fig. 16.6). It is generally recognized as a separate subgenus or genus from other *Theropithecus*. Its molars have the greatest development of shearing crests of any known cercopithecine, suggesting even more folivorous habits than the extant gelada (Benefit and McCrossin, 1990). This species has been recovered from deposits indicating forested environments, but recent analyses suggest it was a terrestrial animal. The Early Pliocene ***T. baringensis*** is recognized by some authorities as a primitive member of this lineage (e.g., Jablonski, 1993, 2002).

One of the most surprising revelations from the fossil records is that during the Pleistocene, *Theropithecus* ranged as far as India. ***Theropithecus o. delsoni*** is a fossil gelada from the Pleistocene of northern India that seems related to the African *T. oswaldi* lineage. There is also suggestive evidence of a younger species of *Theropithecus* from cave deposits in southern India and fossil

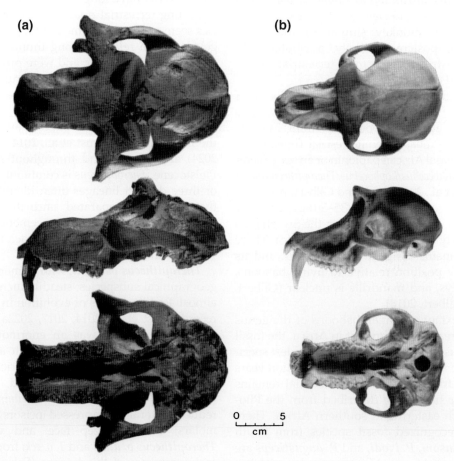

FIGURE 16.6   Skulls of (A) *Theropithecus brumpti* and (B) *Theropithecus gelada*. *(Courtesy of G. Eck.)*

Theropithecus from the Early Pleistocene of Israel and Spain as well (Gibert et al., 1995; Belmaker, 2010).

The extant mangabey genus *Lophocebus* is known from numerous fossils from East Africa, including East Turkana and Kanam East, and possibly from the Omo and Olduvai.

Fossil monkeys belonging to the *Cercocebus/Mandrillus* group are much rarer than those related to baboons and geladas. ***Procercocebus antiquus*** (Gilbert, 2007) from the South African site of Taung seems to be a member of that clade (Fig. 16.7), and there are some fragmentary specimens from South Africa that may represent *Cercocebus* mangabeys. However, virtually all of the fossils from East Africa previously described as *Cercocebus* are actually *Lophocebus*. The genus *Mandrillus*, currently restricted to West and Central Africa, has no fossil record; however, the large papionin ***"Papio"quadratirostris*** (Iwamoto, 1982) from the Early Pleistocene of eastern Africa and also Angola, placed by some authorities in *Papio* and others in *Theropithecus*, is more likely related to the living mandrills and drills and has been placed in its own genus, ***Soromandrillus*** (Gilbert, 2013).

Guenons

Despite their abundance in sub-Saharan Africa today, guenons are very rare in the fossil record. The earliest recognized guenon is known from an isolated molar found in the Late Miocene of the Arabian Peninsula, suggesting that guenons once ranged outside of mainland Africa (Gilbert et al., 2014). In the Plio-Pleistocene, there are guenon-like teeth from several localities in Kenya and Ethiopia, but most of the material is fragmentary and has not been assigned to any particular species (Leakey, 1988; Jablonski et al., 2008b). *Chlorocebus ngedere* from Laetoli, Tanzania is the earliest known vervet monkey (Arenson et al., 2022), and specimens similar to the living *Chlorocebus* and *Erythrocebus* have been described from the Middle Pleistocene of Ethiopia as well (Frost and Alemseged, 2007; Taylor et al., 2023). The new genus *Nanopithecus* is recognized from small teeth resembling those of the extant talapoin monkey at the Pliocene site of Kanapoi in Kenya (Plavcan et al., 2019). The dearth of fossil guenons in Plio-Pleistocene faunas of eastern and southern Africa containing an abundance of large baboons and colobines probably means that the radiation of this group took place in other parts of the continent.

Fossil Colobines

In contrast to the fossil cercopithecines, which are often relatively similar to extant genera, many fossil colobines from Miocene, Pliocene, and Pleistocene deposits are quite different from any living taxa. These extinct colobines had both a broader geographic range and more diverse ecological adaptations than the extant species (Table 16.3).

African Colobines

Although Africa has a relatively low diversity of colobines today, it has an abundant record of fossil colobines,

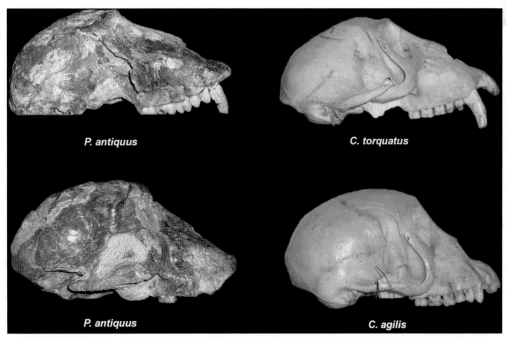

FIGURE 16.7   Crania of *Procercocebus antiquus*, an early member of the *Cercocebus-Mandrillus* lineage, compared with crania of extant *Cercocebus* mangabeys. *(Courtesy of C. Gilbert.)*

TABLE 16.3   Fossil Colobines
Family CERCOPITHECIDAE
Subfamily COLOBINAE

Species	Estimated average mass (g)
Cercopithecoides (Late Miocene to Pleistocene, Africa)	
?C. bruneti	12,000
?C. kerioensis	10,000
C. williamsi	19,500
C. coronatus (= kimeui)	38,000
C. meavae	16,000
C. alemayehui	16,000
C. haasgati	15,000
Sawecolobus (Late Miocene, Africa)	
S. lukeinoensis	7000
Kuseracolobus (Late Miocene to Pliocene, Africa)	
K. aramisi	15,000
K. hafu	30,000
Microcolobus (Late Miocene, East Africa)	
M. tugenensis	5000
Libypithecus (Late Miocene, North Africa)	
L. markgrafi	10,000
Colobus (Late Miocene to Recent, Africa)	
Many unnamed species	–
C. flandrini	21,000
C. freedmani	10,000
Paracolobus (Late Miocene-Plio-Pleistocene, Africa)	
P. chemeroni	32,500
P. mutiwa	40,000
P. enkorikae	13,500
Rhinocolobus (Plio-Pleistocene, Africa)	
R. turkanensis	25,000
Mesopithecus (Late Miocene to Pliocene, Eurasia)	
M. pentelicus	11,000
M. monspessulanus	8500
?M. sivalensis	8000
Dolichopithecus (Pliocene, Europe)	
D. ruscinensis	20,500
D. balcanicus	24,000
Parapresbytis (Plio-Pleistocene, Asia)	
P. eohanumani	28,500
Kanagawapithecus (Late Pliocene, Asia)	
K. leptopostorbitalis	30,000

TABLE 16.3   Fossil Colobines—Cont'd
Family CERCOPITHECIDAE
Subfamily COLOBINAE

Species	Estimated average mass (g)
Myanmarcolobus (Late Miocene/Early Pliocene, Asia)	
M. yawensis	11,000
Semnopithecus (Late Pliocene to Recent, Asia)	
S. gwebinensis	16,000
Presbytis (Pleistocene to Recent, Asia)	6250
Trachypithecus (Pleistocene to Recent, Asia)	6750
Rhinopithecus (Early Pleistocene to Recent, Asia)	
R. lantainensis	24,000

beginning in the Late Miocene and extending into the Pliocene and Pleistocene. During this time, there was an extensive radiation of African leaf-eating monkeys, many of which were unlike anything living today.

The oldest recognized colobine monkeys are found in ~12.5 million-year-old deposits from the Ngorora Formation of the Tugen Hills, Kenya (Rossie et al. 2013). Two isolated teeth, one premolar and one molar, demonstrate derived features such as sharp, complete lophids and cusps that are positioned on the margins of the occlusal surface, suggesting affinities with modern colobines rather than cercopithecines or victoriapithecids. However, the teeth also retain primitive characteristics relative to later colobines, particularly in the relatively low height of the cusps compared to the overall height of the crown. The weak development of folivorous adaptations in these early colobines suggests that they relied on a more omnivorous combination of young leaves, seeds, and fruit compared to modern colobines (Rossie et al., 2013).

Microcolobus from the Late Miocene of northern Kenya is the best-known early colobine. The type species is *M. tugenensis* from the Tugen Hills (Benefit and Pickford, 1986), and there is a second unnamed species from the site of Nakali (Nakatsukasa et al., 2010). Both are approximately 10 million years old. *Microcolobus* was a small monkey of 4 to 5 kg and is known from dental, mandibular, and postcranial remains. Similar to the Ngorora specimens and *Mesopithecus* (discussed below), it differs from later colobines in having slightly lower molar cusps and more crushing surfaces on the lower premolars. It is also unusual among colobines in the shape of the mandibular symphysis. In view of its small body size and less-developed shearing crests, it has been suggested that it was probably less folivorous than many later colobines. Postcranial bones of *Microcolobus* differ considerably from those of *Victoriapithecus* and indicate that it was an arboreal monkey. However, unlike extant colobines, it does not seem to have possessed a reduced thumb. Thus, *Microcolobus* is more primitive than all later colobines and probably preceded the modern radiations in Africa and Asia.

Libypithecus markgrafi (Fig. 16.8), from the Late Miocene of Wadi Natrun in Egypt, was a medium-sized, Late Miocene colobine. The species is known from a relatively complete skull and an isolated molar. The skull has a long snout compared to most extant colobines and well-developed sagittal and nuchal crests. In many aspects of its cranial morphology, *Libypithecus* resembles *Victoriapithecus* and probably retains the primitive cercopithecoid condition (Benefit and McCrossin, 1997). Its relationship to later colobines is less clear. Some authors have suggested that it is closely allied with the European *Mesopithecus*; others have argued that it shows similarities to *Colobus* from sub-Saharan Africa. Because *Libypithecus* is known only from a skull, there is no suitable material for a direct comparison with *Microcolobus*.

Kuseracolobus is a relatively large, sexually dimorphic colobine from Late Miocene and Pliocene deposits in the Afar region of Ethiopia. There are at least two species: the smaller *K. aramisi* (Frost, 2001), in which males were approximately 18kg and females 12kg, and the larger *K. hafu*, with estimated male body sizes of 30 to 40kg (Hlusko, 2006). Postcranial elements attributed to *Kuseracolobus* indicate that it was an arboreal monkey with habits similar to those of extant *Colobus* (Frost et al., 2009).

Paracolobus is a collection of medium to large (30+ kg) colobines from the Late Miocene and Pliocene of eastern Africa with several named species. *Paracolobus chemeroni* has a moderately long cranium, broad muzzle, and wide face as well as a deep and slender mandible (Fig. 16.9; Jablonski and Frost, 2010). Dentally, it is similar to living colobines, suggesting a largely folivorous diet. *P. chemeroni* has an intermembral index of 92, similar to that of the living proboscis monkey and red colobus. The skeleton indicates that *Paracolobus chemeroni* was probably an arboreal quadruped, although features found in the upper limb suggest some terrestriality in its locomotor repertoire as well (Birchette, 1982; Frost, 2001; Frost et al., 2022b). The late Miocene *P. enkorikae* is dentally similar to *P. chemeroni*, but much smaller (~13–14kg) and probably more arboreal (Hlusko, 2007).

?Paracolobus mutiwa is similar in dental size to *P. chemeroni* but very different in its cranial and postcranial anatomy. It almost certainly represents a distinct genus. The face is very long for a colobine, with maxillary ridges and fossae more reminiscent of papionin monkeys (Frost et al., 2022b). In addition, the mandible is very deep with a notably expanded gonial region (Harris et al., 1988; Jablonski et al., 2008a). The skeleton displays numerous features indicating a highly terrestrial monkey (Anderson, 2021).

Rhinocolobus turkanensis is another large monkey from the later Pliocene and Early Pleistocene of eastern Africa. It was smaller than *Paracolobus* and probably weighed about 20 to 25kg. As the name indicates, *Rhinocolobus* has a pronounced snout on its relatively deep face (Fig. 16.9). Its dentition indicates a folivorous diet, and the few skeletal remains suggest that it was an arboreal monkey. It was common in woodland and gallery forest environments.

Sawecolobus lukeinoensis is a recently described taxon from the Latest Miocene of Kenya (Gommery et al., 2022). It is known from a partial skull and a number of jaws and teeth. Overall, it shares many features with the better-known *Cercopithecoides*, particularly the earliest *Cercopithecoides* taxa (i.e., *C. bruneti* and *C. kerioensis*; see below), but the mandible of *Sawecolobus* was deeper and less robust compared to most *Cercopithecoides* species. The interorbital pillar was likely narrower than *Cercopithecoides* as well. Postcrania attributed to *S. lukeinoensis* suggest it was probably semiterrestrial.

Cercopithecoides is a widespread and very successful monkey with numerous species from the Miocene through Pleistocene of both eastern and southern Africa. In recent years, a number of new species have been placed in the genus *Cercopithecoides*, but some of these

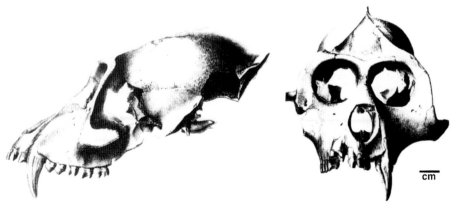

FIGURE 16.8   Skull of *Libypithecus markgrafi*, a Pliocene colobine from Egypt.

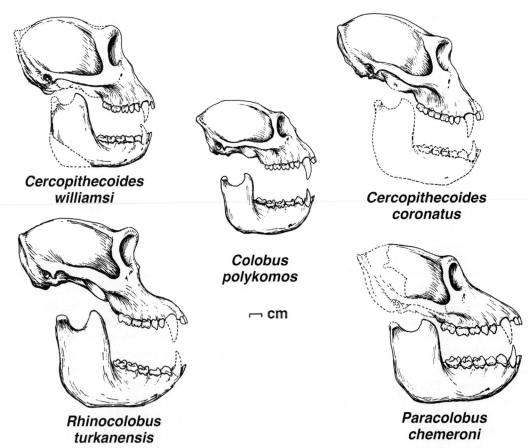

FIGURE 16.9 Skulls of Plio-Pleistocene colobines from East Africa and the extant *Colobus polykomos*. Note the greater size of the fossil monkeys.

taxa almost certainly belong in a different genus based on notable differences in the mandible, among other features (e.g., ?*C. bruneti*, ?*C. kerioensis*). Among the remaining taxa, *Cercopithecoides* typically displays relatively broad molars and a short-snouted skull associated with a relatively shallow and transversely thick mandible (Fig. 16.9). In the largest species, *C. coronatus* (= *kimeui*), the broad molars have an inflated, baboon-like appearance and are heavily worn on all of the individuals, suggesting a softer but perhaps gritty diet compared to that of most extant colobines.

The most striking adaptations of *Cercopithecoides* are in its limbs, which resemble a terrestrial cercopithecine more than any extant colobine. Significantly, a partial skeleton of *C. williamsi* from Koobi Fora demonstrates that the thumb was greatly reduced in this species, most similar to the condition seen in modern African colobines and suggesting a phyletic link with this group (Frost et al., 2015). *Cercopithecoides* was presumably a terrestrial forager and was particularly common in grassland environments.

From Pliocene and Pleistocene deposits in Algeria, Libya, Sudan, Ethiopia, Kenya, and Tanzania, there are numerous fossils that are loosely assigned to the genus *Colobus*. Although some have been attributed to the living species *C. guereza* and others to a fossil taxon (*C. freedmani*), most are known only from isolated teeth or single jaws, and both their habits and their affinities with later forms are indeterminate at present. They probably include specimens attributable to both *Colobus* and *Piliocolobus* as well as other undescribed taxa (Jablonski and Frost, 2010). One isolated astragalus from Kenya suggests the modern *Colobus* and *Piliocolobus* clades appeared as early as the Late Miocene (Gilbert et al., 2010).

Eurasian Colobines

The oldest fossil colobine from Eurasia is *Mesopithecus* (Fig. 16.10). This langur-sized monkey is known from many localities in the Late Miocene through Pliocene of southern and central Europe. The genus ranged as far west as England and as far east as China (Jablonski et al., 2020). There are two species: *M. pentelicus* (about 11 kg) and a younger, smaller species, *M. monspessulanus* (8 kg). *Mesopithecus* resembles living colobines in most dental and cranial features, including relatively small incisors, high-crowned cheek teeth, a deep mandible, a

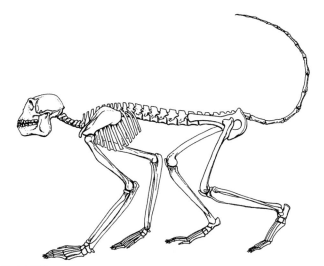

FIGURE 16.10   A reconstructed skeleton of *Mesopithecus*.

short face with large orbits, a narrow nasal opening, and a broad interorbital distance. It was probably a relatively folivorous monkey.

The limb skeleton of *Mesopithecus* resembles that of living colobines in numerous features. However, in the older species *M. pentelicus*, the limbs are more robust than those of most living colobines, and the digits are relatively shorter, suggesting that it was partly terrestrial, like the Hanuman langur of India. Furthermore, the thumb of *M. pentelicus* appears unreduced, which is likely a primitive retention (Frost et al., 2015). The localities that have yielded remains of *M. pentelicus* seem to be characterized by woodland savannah environments (Delson, 1975). The later species *M. monspessulanus* is more like living colobines in its limb skeleton and also has been found in more wooded environments. Both species are sexually dimorphic and presumably lived in polygynous social groups.

Dolichopithecus is a larger Pliocene colobine that is definitively known only from Europe, but several new colobines from northeastern Asia are debated as *Dolichopithecus* species or close relatives of the genus (see below and Delson, 1994). Dentally, *Dolichopithecus* is similar to *Mesopithecus*, but it has a longer snout and a larger overall size (20–25 kg). It also is sexually dimorphic in tooth and skull size.

In its skeleton, *Dolichopithecus* has more extensive adaptations for terrestrial quadrupedalism than any other colobine. Its limb proportions and many of its joint articulations are baboon-like, and it has short, stout phalanges. The genus seems to have been associated with humid forests and probably foraged on the forest floor, a habitus that would have separated it ecologically from the sympatric, more arboreal *Mesopithecus*.

It is not clear whether *Mesopithecus* and *Dolichopithecus* represent late-occurring stem colobines or whether they are more closely related to either the living colobines of Africa or Asia. There are few diagnostic features to link them unequivocally with either extant group, and the unreduced thumb in *Mesopithecus*, in particular, may suggest that a stem position is more likely (Frost et al., 2015, 2022b). Indeed, one recent phylogenetic analysis placed both *Dolichopithecus* and *Mesopithecus* as late-occurring stem colobine taxa (Arenson, 2024). However, their Eurasian distribution has been interpreted by some authors as being suggestive of affinities with Asian langurs (Simons, 1970; Jablonski, 1998).

Asian Colobines

The fossil record of Asian colobines is relatively poor, considering the diversity and abundance of leaf-eating monkeys on that continent today (Jablonski, 1993). ***Parapresbytis eohanuman*** is a large (30 kg) colobine from Pliocene deposits in northeast Asia near Lake Baikal. It is known from cranial, dental, and postcranial specimens. The relationship of *Parapresbytis* to other colobines is debated. Delson (1994) has suggested that it is not distinct from *Dolichopithecus*, and Jablonski (2002) suggested that *Parapresbytis* shows no particular similarity to any single taxon. Arenson (2024) has recently hypothesized a relationship to the odd-nosed colobines. It shows an unusual mosaic of features not currently found in any single extant or fossil colobine taxon. The limb elements indicate that it was not terrestrial like *Dolichopithecus*, and despite its large size, it was a predominantly arboreal monkey (Egi et al., 2007).

Kanagawapithecus is another large colobine known from a partial cranium in the late Pliocene of Japan. Similar to *Parapresbytis*, it has been argued to be a species of *Dolichopithecus* in the past (e.g., Delson, 1994). However, *Kanagawapithecus* has a shorter snout and much more gracile orbital region compared to known *Dolichopithecus* specimens (Iwamoto et al., 2005; Nishimura et al., 2012). It also exhibits a maxillary sinus, another feature not typically found in *Dolichopithecus* (Nishimura et al., 2012). *Kanagawapithecus* may also be related to extant odd-nosed colobines (Arenson, 2024).

Myanmarcolobus yawensis is a recently discovered colobine from the late Miocene or early Pliocene of Myanmar. It is represented by a single mandibular molar row with distinctively shaped molars (Takai et al., 2015).

Other early Asian colobines include ?***Mesopithecus sivalensis*** from Late Miocene deposits about 6 million years old in the Siwaliks of northern India and Pakistan, and possibly *"Semnopithecus" palaeindicus* represented by an astragalus from the Plio-Pleistocene of the Siwaliks (Barry, 1987; Szalay and Delson, 1979; Jablonski, 2002; Harrison and Delson, 2007; Takai et al., 2015; Khan et al., 2020). Fossils attributable to extant taxa of Asian colobines are known from a variety of Pliocene, Pleistocene, and Recent deposits in China, Burma, Vietnam, Myanmar,

Java, Sumatra, and Borneo. The recently described *Semnopithecus gwebinensis* is known from the late Pliocene of Myanmar (Takai et al., 2016), *Pygathrix* and several species of *Rhinopithecus* are known from Pleistocene localities in China (Jablonski, 2002), and several species of *Trachypithecus* and *Presbytis* have been described from the fossil records of Borneo, Sumatra, and Java. Understanding the radiation of Asian colobines in light of the geomorphological and climatic changes of that region during the past 6 million years is a major challenge for paleoanthropology.

Fossil Record of Cercopithecoids

The fossil record of Old World monkeys (Fig. 16.11) is quite different from that of apes, the other major catarrhine group. For apes, we have abundant remains in the Early and Middle Miocene until about 10 million years ago and very little from the latest Miocene to Recent. In contrast, Old World monkeys have a modest fossil record of early victoriapithecids from the Early and Middle Miocene and increasing numbers of fossil monkeys in the Latest Miocene through Pleistocene (Fig. 16.11). For apes, there are far more extinct genera and species than there are living taxa, and the many extinct species suggest diverse adaptive and phyletic radiations in the past that have left few, if any, descendants. In contrast, living monkeys outnumber the extinct taxa, and many of the fossil monkeys seem to be part of the present-day radiation.

Several authors have argued that the temporal pattern of change in the relative abundance of monkeys and apes during the last 20 million years indicates an ecological replacement of early apes by Old World monkeys (Fig. 16.12). It is equally likely, however, that this apparent change in the primate fauna reflects climatic changes during the Miocene of Africa and Europe, rather than simply competition between monkeys and apes in a stable environment (see Harrison, 2010; also Kelley, 1997).

The earliest fossil monkeys, like the earliest fossil apes, provide evidence of intermediate stages in catarrhine evolution. The victoriapithecids and *Libypithecus* demonstrate that both colobines and cercopithecines preserve a mosaic of both primitive features from the earliest monkeys as well as derived features unique to their respective subfamilies. The fossil record also suggests that the arboreal nature of most living colobines has not characterized all members of that subfamily. Both *Dolichopithecus* and *Cercopithecoides* were very terrestrial colobines. However, the arboreally adapted postcrania of the primitive *Microcolobus* suggest that the terrestrially adapted fossil colobines may be secondarily derived for the subfamily.

A particularly striking feature of the cercopithecoid fossil record is the size difference between extinct and living monkeys. Many extinct colobines and cercopithecines were larger than related living genera. Like the extant

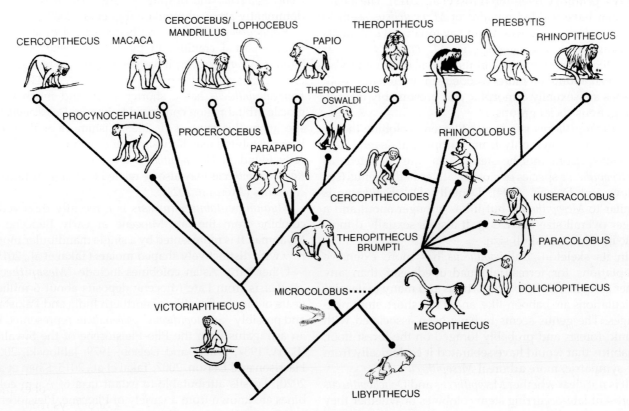

FIGURE 16.11   Generalized cladogram of extant and fossil Old World monkeys.

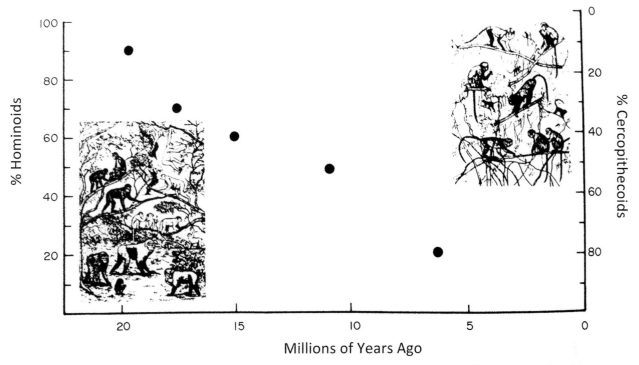

FIGURE 16.12 Relative species diversity of hominoids and cercopithecoids during the past 20 million years in Africa. The diversity of monkeys has increased as the diversity of hominoids has decreased. *(After Andrews, 1981).*

Malagasy fauna, the living cercopithecoids are the smaller genera from the Pliocene and Early Pleistocene. This Pleistocene extinction of relatively large species is a common phenomenon around the world that cannot clearly or exclusively be attributed to either climatic changes or hominin hunting (Martin and Klein, 1984).

Despite the presence of numerous well-preserved fossils, the phylogenetic relationships between many of the extinct fossil taxa and the extant radiations of Old World monkeys are still poorly understood. Although almost all fossil monkeys, except the victoriapithecids, can be allocated to either Colobinae or Cercopithecinae with little question, there have been very few attempts to determine the relationships among the fossil and extinct genera (but see Gilbert, 2013; Pugh and Gilbert, 2018; Arenson, 2024). The broad strokes of papionin phylogenetic history have become much better understood in recent years, and a number of fossil papionin monkeys can be assigned to the modern clades with some confidence (e.g., *Procercocebus, Soromandrillus, Gorgopithecus, Paradolichopithecus*; see Nishimura et al., 2007; Gilbert, 2007, 2013; Gilbert et al., 2016; O'Shea et al., 2016; Pugh and Gilbert, 2018). Among colobines, the significantly reduced thumb seen in *Cercopithecoides williamsi*, along with subtle cranial features, suggests that *Cercopithecoides* is a close relative of the living African colobines (Frost et al., 2015; Arenson, 2024). *Mesopithecus* and *Dolichopithecus* have been described previously as possible sister taxa, but it is unclear how they are related to each other or the extant and fossil African and Asian

radiations, or how the many fossil colobines from Africa are related to the extant African genera or preceded the divergence of the modern forms. In the recent analyses by Arenson (2024), most fossil colobines in Africa seem to be stem taxa (with the exception of *Cercopithecoides*) and many of these stem taxa retain a mosaic of primitive facial features, such as a relatively longer snout and narrower interorbital breadth (e.g., *Libypithecus, Rhinocolobus*). Depending on the phylogenetic position of *Mesopithecus*, it is possible that thumb reduction occurred independently in the living African and Asian colobine lineages (Frost et al., 2015).

The fossil record of Old World monkeys expands our understanding of the biogeography of the group and is full of both surprises and questions. It shows that both colobines (*Mesopithecus* and *Dolichopithecus*) and cercopithecines (*Macaca* and *Paradolichopithecus*) ranged over much of Europe during the last 5 million years, and that *Theropithecus* was once found as far south as South Africa and as far east as India. Similarly, relatives of the living *Cercocebus* mangabeys, mandrills, and drills once ranged over a much greater portion of Africa than they do today, and guenons apparently ranged into the Arabian peninsula at one time. Nevertheless, while fossil apes flourished in Eurasia in the Middle Miocene, monkeys only appear in Europe during the latest Miocene, and in Asia even later. Moreover, in contrast to the abundance of extinct monkeys from Africa, many fossil monkeys from Asia can be assigned to living genera.

References

Alba, D.M., Delson, E., Carnevale, G., Colombero, S., Delfino, M., Giuntelli, P., Pavia, M., Pavia, G., 2014. First joint record of Mesopithecus and cf. Macaca in the Miocene of Europe. J. Hum. Evol. 67, 1–18.

Andrews, P., 1981. Species diversity and diet in monkeys and apes during the Miocene. In: Stringer, C.B. (Ed.), Aspects of Human Evolution. Taylor and Francis, London, pp. 25–61.

Anderson, M., 2021. An assessment of the postcranial skeleton of the *Paracolobus mutiva* (Primates: Colobinae) specimen KNM-WT 16827 from Lomekwi, West Turkana, Kenya. J. Hum. Evol. 156, 103012.

Arenson, J.L., 2024. Evolution of the extant and fossil Colobinae (Primates, Cercopithecidae). Ph.D. Dissertation. The City University of New York, New York.

Arenson, J.L., Sargis, E.J., Hart, J.A., Hart, T.B., Detwiler, K.M., Gilbert, C.C., 2020. Skeletal morphology of the lesula (*Cercopithecus lomamiensis*) and the evolution of guenon locomotor behavior. Am. J. Phys. Anthropol. 172, 3–24. https://doi.org/10.1002/ajpa.24025

Arenson, J.L., Harrison, T., Sargis, E.J., Taboada, H.G., Gilbert, C.C., 2022. A new species of fossil guenon (Cercopithecini, Cercopithecidae) from the Early Pleistocene Lower Ngaloba Beds, Laetoli, Tanzania. J. Hum. Evol. 163, 103136. https://doi.org/10.1016/j.jhevol.2021.103136

Barry, J.C., 1987. The history and chronology of Siwalik cercopithecids. Hum. Evol. 2, 47–58.

Belmaker, M., 2010. The presence of a large cercopithecine (cf. *Theropithecus* sp.) in the 'Ubeidiya formation (Early Pleistocene, Israel). J. Hum Evol. 58, 79–89.

Benefit, B.R., 2008. The biostratigraphy and palaeontology of fossil cercopithecoids from eastern Libya. Geology of East Libia. vol. 3, 247–266.

Benefit, B.R., McCrossin, M.L., 1990. Diet, species diversity and distribution of African fossil baboons. Kroeber Anthropological Society Papers, 71–72, 77–93.

Benefit, B.R., McCrossin, M.L., 1993. Facial anatomy of Victoriapithecus and its relevance to the ancestral cranial morphology of Old World monkeys and apes. Am. J. Phys. Anthropol. 92, 329–370.

Benefit, B.R., McCrossin, M.L., 1997. Earliest known Old World monkey skull. Nature 388, 368–371.

Benefit, B.R., McCrossin, M.L., 2002. The Victoriapithecidae. In: Hartwig, W.C. (Ed.), The Primate Fossil Record. Cambridge University Press, Cambridge, pp. 241–253.

Benefit, B.R., Pickford, M., 1986. Miocene fossil cercopithecoids from Kenya. Am. J. Phys. Anthropol. 69, 441–464.

Birchette, M.G., 1982. The postcranial skeleton of *Paracolobus chemeroni*. Ph.D. dissertation. Harvard University.

Blue, K., McCrossin, M., Benefit, B.R., 2006. Terrestriality in a Middle Miocene context: Victoriapithecus from Maboko, Kenya. In: Ishida, H., Tuttle, R., Pickford, M., Ogihara, N., Nakatsukasa, M. (Eds.), Human Origins and Environmental Background. Springer, USA, pp. 45–58.

Delson, E., 1975. Evolutionary history of the Cercopithecidae. In: Szalay, F.S. (Ed.),. In: Approaches to Primate Paleobiology, Contributions to Primatology, vol. 5. Karger, Basel, pp. 167–217.

Delson, E., 1979. Prohylobates (Primates) from the Early Miocene of Libya: A new species and its implications for cercopithecid origins. Geobios 12, 725–733.

Delson, E., 1980. Fossil macaques, phyletic relationships and a scenario of deployment, Lindburg, D.G. (Ed.), The Macaques: Studies in Ecology, Behaviour and Evolution. Van Nostrand, New York, pp. 10–29.

Delson, E., 1994. Evolutionary history of the colobine monkeys in paleoenvironmental perspective, Davies, G., Oates, J.F. (Eds.), Their Ecology, Behaviour, and Evolution. Cambridge University Press, Cambridge, pp. 11–43.

Delson, E., Rosenberger, A.L., 1984. Are there any anthropoid primate "living fossils"? In: Eldredge, N., Stanley, S. (Eds.), Living Fossils. Springer-Verlag, New York, pp. 50–61.

Delson, E., Terranova, C.J., Jungers, W.L., Sargis, E.J., Jablonski, N.G., Dechow, P.C., 2000. Body mass in Cercopithecidae (Primates, Mammalia): estimation and scaling in extinct and extant taxa. Anthropological Papers of the American Museum of Natural History 8, 1–159.

Dunbar, R.I.M., 1992. Behavioural ecology of the extinct papionines. J. Hum. Evol. 22, 407–422.

Dunbar, R.I.M., 1993. Socioecology of the extinct theropiths: a modeling approach. In: Jablonski, N.G. (Ed.), Theropithecus: The Rise and Fall of a Primate Genus. Cambridge University Press, Cambridge, pp. 465–486.

Eck, G.G., 1993. *Theropithecus darti* from the Hadar Formation, Ethiopia. In: Jablonski, N.G. (Ed.), Theropithecus: The Rise and Fall of a Primate Genus. Cambridge University Press, Cambridge, pp. 15–83.

Egi, N., Naktsukasa, M., Kalmykov, N.P., Maschenko, E.N., Takai, M., 2007. Distal humerus and ulna of Parapresbytis (Colobinae) from the Pliocene of Russia and Mongolia: Phylogenetic and ecological implications based on elbow morphology. Anthropol. Sci. 115, 107–117.

Elton, S., 2001. Locomotor and habitat classifications of cercopithecoid postcranial material from Sterkfontein Member 4, Bolt's Farm, and Swartkrans Members 1 and 2, South Africa. Palaeontologia Africana 37, 115–126.

Fleagle, J.G., Bown, T.M., Harris, J.M., Watkins, R.W., Leakey, M.G., 1997. Fossil monkeys from northern Kenya. Am. J. Phys. Anthropol. S24, 111.

Foley, R.A., 1993. African terrestrial primates: the comparative evolutionary biology of Theropithecus and the Hominidae. In: Jablonski, N.G. (Ed.), Theropithecus: The Rise and Fall of a Primate Genus. Cambridge University Press, Cambridge, pp. 245–270.

Frost, S.R., 2001. Fossil Cercopithecidae of the Afar Depression, Ethiopia: Species systematics and comparison to the Turkana Basin. Ph.D. Dissertation. The City University of New York, New York.

Frost, S.R., Alemseged, Z., 2007. Middle Pleistocene fossil Cercopithecidae from Asbole, Afar Region, Ethiopia. J. Hum. Evol. 53, 227–259.

Frost, S.S., Delson, E., 2002. Fossil Cercopithecidae from the Hadar Formation and surrounding areas of the Afar Depression, Ethiopia. J. Hum. Evol. 43, 687–748.

Frost, S.R., Haile-Selassie, Y., Hlusko, L., 2009. Cercopithecidae. In: Haile-Selassie, Y., Woldegabriel, G. (Eds.), *Ardipithecus kadabba*: Late Miocene evidence from the Middle Awash, Ethiopia. University of California Press, Berkeley, pp. 135–158.

Frost, S.R., Jablonski, N.G., Haile-Selassie, Y., 2014. Early Pliocene Cercopithecidae from Woranso-Mille (Central Afar, Ethiopia) and the origins of the *Theropithecus oswaldi* lineage. J. Hum. Evol. 76, 39–53.

Frost, S.R., Gilbert, C.C., Pugh, K.D., Guthrie, E.H., Delson, E., 2015. The hand of *Cercopithecoides williamsi* (Mammalia, Primates): Earliest evidence for thumb reduction among colobine monkeys. PLoS One. 10, e0125030.

Frost, S.R., Saanane, C., Starkovich, B.M., Schwartz, H., Schrenk, F., Harvati, K., 2017. New cranium of the large cercopithecid primate *Theropithecus oswaldi* leakeyi (Hopwood, 1934) from the paleoanthropological site of Makuyuni, Tanzania. J. Hum. Evol. 109, 46–56.

Frost, S.R., Simpson, S.W., Levin, N.E., Quade, J., Rogers, M.J., Semaw, S., 2020a. Fossil Cercopithecidae from the Early Pliocene Sagantole Formation at Gona, Ethiopia. J. Hum. Evol. 144, 102789.

Frost, S.R., Ward, C.V., Manthi, F.K., Plavcan, J.M., 2020b. Cercopithecid fossils from Kanapoi, West Turkana, Kenya (2007–2015). J. Hum. Evol. 140, 102642.

Frost, S.R., White, F., Reda, H., Gilbert, C.C., 2022a. Biochronology of South African hominin-bearing sites: A reassessment using cercopithecid primates. Proc. Natl. Acad. Sci. USA. 119 e2210627119: https://doi.org/10.1073/pnas.2210627119

Frost, S.R., Gilbert, C.C., Nakatsukasa, M.N., 2022b. The colobine fossil record. In: Matsuda, I. (Ed.), The Colobines: Natural History. Behaviour and Ecological Diversity. Cambridge University Press, Cambridge, pp. 13–31.

Gebo, D.L., 1989. Locomotor and phylogenetic considerations in anthropoid evolution. J. Hum. Evol. 18, 201–233.

Getahun, D.A., Delson, E., Seyoum, C.M., 2023. A review of Theropithecus oswaldi with the proposal of a new subspecies. J. Hum. Evol. 180, 103373. https://doi.org/10.1016/j.jhevol.2023.103373.

Gilbert, C.C., 2007. Craniomandibular morphology supporting the diphyletic origin of mangabeys and a new genus of the Cercocebus/Mandrillus clade, Procercocebus. J. Hum. Evol. 53, 69–102.

Gilbert, C.C., 2013. Cladistic analysis of extant and fossil African papionins using craniodental data. J. Hum. Evol. 64, 399–433.

Gibert, J., Ribot, F., Gilbert, L., Leakey, M., Arribas, A., Martinez, B., 1995. Presence of the cercopithecid genus Theropithecus in Cueva Victoria (Murcia, Spain). J. Hum. Evol. 28, 487–493.

Gilbert, C.C., Goble, E.D., Hill, A., 2010. Miocene Cercopithecoidea from the Tugen Hills, Kenya. J. Hum. Evol. 59, 465–483.

Gilbert, C.C., Bibi, F., Hill, A., Beech, M.J., 2014. Early guenon from the late Miocene Baynunah Formation, Abu Dhabi, with implications for cercopithecoids biogeography and evolution. Proceedings of the National Academy of Sciences, USA 111, 10119–10124.

Gilbert, C.C., Frost, S.R., Delson, E., 2016. Reassessment of the Olduvai Bed I cercopithecoids: A new biochronological and biogeographical link to the South African fossil record. J. Hum. Evol. 92, 50–59.

Gilbert, C.C., Frost, S.R., Pugh, K.D., Anderson, M., Delson, E., 2018. Evolution of the modern baboon (Papio hamadryas): A reassessment of the African Plio-Pleistocene record. J. Hum. Evol. 122, 38–69.

Gommery, D., Senut, B., Pickford, M., Nishimura, T.D., Kipkech, J., 2022. The Late Miocene monkeys from Aragai (Lukeino Formation, Tugen Hills, Kenya). Geodiversitas 44, 471–504.

Harris, J.M., Brown, F.H., Leakey, M.G., 1988. Stratigraphy and paleontology of Pliocene and Pleistocene localities west of Lake Turkana, Kenya. Nat. Hist. Mus. Los Angeles Cty. Contrib. Sci. 399, 1–128.

Harrison, T., 1989. New postcranial remains of Victoriapithecus from the Middle Miocene of Kenya. J. Hum. Evol. 18, 3–54.

Harrison, T., 2010. New estimates of hominoid taxonomic diversity in Africa during the Neogene and its implications for understanding catarrhine community structure. J. Vert. Paleontol 30 (Suppl. 3), 102A.

Harrison, T., Delson, E., 2007. Mesopithecus sivalensis from the Late Miocene of the Siwaliks [Abstract]. Am. J. Phys. Anthropol. 132 (S44), 126.

Hlusko, L.J., 2006. A new large Pliocene species (Mammalia: Primates) from Asa Issie, Ethiopia. Geobios. 39, 57–69.

Hlusko, L.J., 2007. A new Late Miocene species of Paracolobus and other Cercopithecoidea (Mammalia: Primates) fossils from Lemudong'o, Kenya. Kirtlandia 56, 72–85.

Iwamoto, M., 1982. A fossil baboon skull from the lower Omo Basin, southwestern Ethiopia. Primates 23, 533–541.

Iwamoto, M., Hasegawa, Y., Koizumi, A., 2005. A Pliocene colobine from the Nakatsu group, Kanagawa, Japan. Anthropol. Sci. 113, 123–127.

Jablonski, N.G., 1993. The phylogeny of Theropithecus. In: Jablonski, N.G. (Ed.), Theropithecus: The Rise and Fall of a Primate Genus. Cambridge University Press, Cambridge, pp. 209–224.

Jablonski, N.G., 1998. The evolution of the doucs and snub-nosed monkeys and the question of the phyletic unity of the odd-nosed colobines. In: Jablonski, N.G. (Ed.), The Natural History of the Doucs and Snub-Nosed Monkeys. World Scientific, Singapore, pp. 13–52.

Jablonski, N.G., 2002. Fossil Old World monkeys: the Late Neogene radiation. In: Hartwig, W.C. (Ed.), The Primate Fossil Record. Cambridge University Press, Cambridge, pp. 255–299.

Jablonski, N.G., Frost, S., 2010. Cercopithecoidea. In: Werdelin, L., Sanders, W.J. (Eds.), Cenozoic Mammals of Africa. University of California Press, Berkeley, pp. 393–428.

Jablonski, N.G., Leakey, M.G., 2008. Systematic paleontology of the small colobines. In: Jablonski, N.G., Leakey, M.G. (Eds.), Koobi Fora Research Project, The Fossil Monkeys, vol 6. California Academy of Sciences, San Francisco, pp. 12–30.

Jablonski, N.G., Pan, Y.R., 1988. The evolution and palaeobiogeography of monkeys in China. In: Aigner, J.S., Jablonski, N.G., Taylor, G., Walker, D., Pinxian, W. (Eds.), The Palaeoenvironment of East Asia from the Mid-Tertiary, Oceanography, Palaeozoology and Palaeoanthropology, vol. II. University of Hong Kong Press, Hong Kong, pp. 849–867.

Jablonski, N.G., Leakey, M.G., Ward, C.V., Antón, M., 2008a. Systematic paleontology of the large colobines. In: Jablonski, N.G., Leakey, M.G. (Eds.), Koobi Fora Research Project. The Fossil Monkeys, vol. 6. California Academy of Sciences, San Francisco, pp. 31–102.

Jablonski, N.G., Leakey, M.G., Anton, M., 2008b. Systematic paleontology of the cercopithecines. In: Jablonski, N.G., Leakey, M.G. (Eds.), Koobi Fora Research Project, The Fossil Monkeys, vol 6. California Academy of Sciences, San Francisco, pp. 103–300.

Jablonski, N.G., Ji, X., Kelley, J., Flynn, L.J., Deng, C., Su, D.F., 2020. Mesopithecus pentelicus from Zhaotong, China, the easternmost representative of a widespread Miocene cercopithecoid species. J. Hum. Evol. 146, 102851.

Kelley, J., 1997. Paleobiological and phylogenetic significance of life history in Miocene hominoids. In: Begun, D., Ward, C.V., Rose, M.D. (Eds.), Function, Phylogeny, and Fossils: Miocene Hominoid Evolution and Adaptations. Plenum Press, New York, pp. 173–208.

Khan, M.A., Kelley, J., Flynn, L., Babar, M.A., Jablonski, N.G., 2020. New fossils of Mesopithecus from Hasnot, Pakistan. J. Hum. Evol. 145, 102818.

Kohler, M., Moya-Sola, S., Alba, D.M., 2000. Macaca (Primates, Cercopithecidae) from the Late Miocene of Spain. J. Hum. Evol. 38, 447–452.

Kostopoulos, D.S., Guy, F., Kynigopoulou, Z., Koufos, G.D., Valentin, X., Merceron, G., 2018. A 2 Ma old baboon-like monkey from Northern Greece and new evidence to support the Paradolichopithecus-Procynocephalus synonymy (Primates: Cercopithecidae). J. Hum. Evol. 121, 178–192.

Leakey, M., 1985. Early Miocene cercopithecids from Buluk, Northern Kenya. Folia Primatol 44, 1–14.

Leakey, M.G., 1988. Fossil evidence for the evolution of the guenons. In: Gautier-Hion, A., Bourliere, F., Gautier, J.-P. (Eds.), A Primate Radiation: Evolutionary Biology of the African Guenons. Cambridge University Press, Cambridge, pp. 7–12.

Lee, P.C., Foley, R.A., 1993. Ecological energetics and extinction of giant baboons. In: Jablonski, N.G. (Ed.), Theropithecus: The Rise and Fall of a Primate Genus. Cambridge University Press, Cambridge, pp. 487–498.

Locke, E.M., Benefit, B.R., Kimock, C.M., Miller, E.R., Nengo, I., 2020. New dentognathic fossils of Noropithecus bulukensis (Primates, Victoriapithecidae) from the late Early Miocene of Buluk, Kenya. J. Hum. Evol. 148, 102886.

Martin, P.S., Klein, R.G., 1984. Quaternary Extinctions: A Prehistoric Revolution. University of Arizona Press, Tucson.

Miller, E.R., 1998. Faunal correlation of Wadi Moghara, Egypt – implications for the age of Prohylobates tandyi. J. Hum. Evol. 36, 519–533.

Miller, E.R., Benefit, B.R., McCrossin, M.L., et al., 2009. Systematics of Early and Middle Miocene Old World monkeys. J. Hum. Evol. 57, 195–334.

Nakatsukasa, M., Mbua, E., Yoshihiro, S., et al., 2010. Earliest colobine skeletons from Nakali, Kenya. Am. J. Phys. Anthropol. 143, 365–382.

Nishimura, T.D., Takai, M., Maschenko, E.N., 2007. The maxillary sinus of *Paradolichopithecus sushkini* (late Pliocene, southern Tajikistan) and its phyletic implications. J. Hum. Evol. 52, 637–646.

Nishimura, T.D., Takai, M., Senut, B., Taru, H., Maschenko, E.N., Prieur, A., 2012. Reassessment of *Dolichopithecus* (*Kanagawapithecus*) *leptopostorbitalis*, a colobine monkey from the Late Pliocene of Japan. J. Hum. Evol. 62, 548–561.

O'Shea, N., Delson, E., Pugh, K.D., Gilbert, C.C., 2016. Phylogenetic analysis of Paradolichopithecus: fossil baboon or macaque? Am. J. Phys. Anthropol Suppl. 62, 244.

Pickford, M., Senut, B., 1988. Habitat and locomotion in Miocene cercopithecoids. In: Gautier-Hion, A., Bourliere, F., Gautier, J.-P. (Eds.), A Primate Radiation: Evolutionary Biology of the African Guenons. Cambridge University Press, Cambridge, pp. 35–53.

Pickford, M.H.L., Kunimatsu, Y., 2005. Catarrhines from the Middle Miocene (ca. 14.5 Ma) of Kipsaraman, Tugen Hills, Kenya. Anthropol. Sci. 113, 189–224.

Piickford, M., Senut, B., Musalizi, S., Gommery, D., Ssbuyungo, C., 2019. Early Miocene victoriapithecid monkey from Napak, Uganda. Geo-Pal Uganda 12, 1–17.

Plavcan, J.M., Ward, C.V., Kay, R.F., Manthi, F.K., 2019. A diminutive Pliocene guenon from Kanapoi, West Turkan, Kenya. J. Hum. Evol. 135, 102623.

Pugh, K.D., Gilbert, C.C., 2018. Phylogenetic relationships of living and fossil African papionins: Combined evidence from morphology and molecules. J. Hum. Evol. 123, 35–51.

Rasmussen, D.T., Friscia, A.R., Guitierrez, M., Kappelman, J., Miller, E.R., Muteti, S., Reynoso, D., Rossie, J.B., Spell, T.L., Tabor, N.J., Gierlowski-Kordesch, E., Jacobs, B.F., Kyongo, B., Macharwas, M., Muchemi, F., 2019. Primitive Old World monkey from the earliest Miocene of Kenya and the evolution of cercopithecoid bilophodonty. Proceedings of the National Academy of Sciences, USA 116, 6051–6056.

Reda, H.G., Frost, S.R., Simons, E.A., Quade, J., Simpson, S.W., 2024. Description and taxonomic assessment of fossil Cercopithecidae from the Pliocene Galili Formation (Ethiopia). *J. Hum. Evol.* 190, 103508. https://doi.org/10.1016/j.jhevol.2024.103508.

Roos, C., Kothe, M., Alba, D.M., Delson, E., Zinner, D., 2019. The radiation of macaques out of Africa: Evidence from mitogenome divergence times and the fossil record. J. Hum. Evol. 133, 114–132.

Rossie, J.B., Gilbert, C.C., Hill, A., 2013. Early cercopithecid monkeys from the Tugen Hills, Kenya. Proceedings of the National Academy of Sciences, USA 110, 5818–5822.

Simons, E.L., 1970. The deployment and history of Old World monkeys (Cercopithecidae, Primates). In: Napier, J.R., Napier, P.H. (Eds.), Old World Monkeys. Academic Press, New York, pp. 97–137.

Stevens, N.J., Seiffert, E.R., O'Conner, P.M., Roberts, E.M., Scmitz, M.D., Krause, C., Gorscak, E., Ngasala, S., Hieronymus, T.L., Temu, J., 2013. Paleontological evidence for an Oligocene divergence between Old World monkeys and apes. Nature 497, 611–614.

Strasser, E., 1988. Pedal evidence for the origin and diversification of cercopithecid clades. J. Hum. Evol. 17, 225–245.

Szalay, F.S., Delson, E., 1979. Evolutionary History of the Primates. Academic Press, New York, 580p.

Takai, M., Maschenko, E.N., Nishimura, T.D., Anezaki, T., Suzuki, T., 2008. Phylogenetic relationships and biogeographic history of *Paradolichopithecus sushkini* Trofimov 1977. A large-bodied cercopithecine monkey from the Pliocene of Eurasia. Quat. Int. 179, 108–119.

Takai, M., Htike, T., Thein, Z.-M.-M., Soe, A.N., Maung, M., Tsubamoto, T., Egi, N., Nishimura, T.D., Nishioka, Y., 2015. First discovery of colobine fossils from the Late Miocene/Early Pliocene in central Myanmar. J. Hum. Evol. 84, 1–15.

Takai, M., Nishioka, Y., Htike, T., Maung, M., Khaing, K., Thein, Z.-M.-M., Tsubamoto, T., Egi, N., 2016. Late Pliocene Semnopithecus fossils from central Myanmar: Rethinking of the evolutionary history of cercopithecid monkeys in Southeast Asia. Historical Biology 28, 172–188.

Taylor, C.E., Brasil, M.F., Monson, T.A., Yohler, R.M., Hlusko, L.J., 2023. Halibee fossil assemblages reveal later Pleistocene cercopithecins (Cercopithecidae: Primates) in the Middle Awash of Ethiopia. *Am. J. Biol. Anthropol.* 180, 6–47.

Further Reading

Benefit, B.R., McCrossin, M.L., 2002. The Victoriapithecidae. In: Hartwig, W.C. (Ed.), The Primate Fossil Record. Cambridge University Press, Cambridge, pp. 241–253.

Frost, S.R., Gilbert, C.C., Nakatsukasa, M.N., 2022. The colobine fossil record. In: Matsuda, I. (Ed.), The Colobines: Natural History, Behaviour and Ecological Diversity. Cambridge University Press, Cambridge.

Jablonski, N.G., 2002. Fossil Old World monkeys: the Late Neogene radiation. In: Hartwig, W.C. (Ed.), The Primate Fossil Record. Cambridge University Press, Cambridge, pp. 255–299.

Jablonski, N.G., Frost, S., 2010. Cercopithecoidea. In: Werdelin, L., Sanders, W.J. (Eds.), Cenozoic Mammals of Africa. University of California Press, Berkeley, pp. 393–394.

Szalay, F.S., Delson, E., 1979. Evolutionary History of the Primates. Academic Press, New York.

Whitehead, P.F., Jolly, C.J., 2000. Old World Monkeys. Cambridge University Press, Cambridge.

CHAPTER

17

Fossil Hominins, the Bipedal Primates

Pliocene Epoch

The short Pliocene epoch was a time of considerable faunal change in many parts of the world in association with major geographical and climatic events. The most significant tectonic event was the completion of the Panama land bridge between North and South America, which led to the exchange of faunas between those two previously separated continents. In the Old World, the Mediterranean Sea refilled at the beginning of the Pliocene after drying up in the Late Miocene. In general, sea levels were higher and temperatures were warmer

in the Early Pliocene than in the Late Miocene. In primate evolution, the Pliocene is characterized by two major events: the spread of cercopithecoid monkeys throughout many parts of the Old World (see Chapter 16) and the radiation of hominins in Africa (Fig. 17.1).

The separation of the lineages leading to the living African apes on the one hand and to humans on the other took place sometime in the Late Miocene, between 10 million and 5 million years ago. There are three genera from the Late Miocene and Early Pliocene that have been put forth as the earliest members of our lineage. They are known from different body parts, so there is

FIGURE 17.1   Map of sites yielding fossils of early hominins. Map lines delineate study areas and do not necessarily depict accepted national boundaries.

some debate about how they are related to one another and to more definitive, younger hominins such as *Australopithecus, Paranthropus,* and *Homo* (e.g., Wood and Harrison, 2011).

Sahelanthropus

Sahelanthropus tchadensis from the Late Miocene (6–7 Ma) of the western Djurab Desert of Chad in Central Africa (Fig. 17.1) is the oldest of the taxa identified as a basal hominin (Brunet et al., 2002). The type specimen is a nearly complete, but distorted, cranium (Fig. 17.2). The species is also known from other dental and gnathic remains. The cranium has an unusually small braincase set behind a large, relatively flat face. In both cranial and dental features, *Sahelanthropus* appears to group with later hominins rather than apes (Brunet et al., 2005; Franck et al., 2005; Zollikofer et al., 2005). There is no indication of a honing mechanism between the upper canine and lower first premolar as in apes. The orientations of the nuchal plane and the foramen magnum suggest that *Sahelanthropus* was bipedal, a hallmark of hominins. There are limited postcranial remains, including a partial femur and two ulnae. The ulnae attributed to *Sahelanthropus* suggest arboreal locomotor activities, and the partial femur is quite primitive in its neck angle and cross-sectional shape. There is a debate over whether *Sahelanthropus* habitually walked on two legs (Macchiarelli et al., 2020; Daver et al., 2022).

Orrorin

Orrorin tugenensis is a Late Miocene species from the Tugen Hills of Central Kenya that is dated to approximately 6 Ma (Senut et al., 2001). It is known from 13 dental, gnathic, and postcranial specimens drawn from at least five individuals (Fig. 17.3). The cheek teeth of *Orrorin* are relatively small, with thick enamel. The upper canine has a distinct mesial groove as in living and fossil apes, but not later hominins. The femur indicates bipedal habits and resembles the same bone in *Australopithecus,* with a long neck and relatively small head compared to the femur of our own genus, *Homo* (Richmond and Jungers, 2008; Almécija et al., 2013). An isolated finger bone is curved, and the humerus exhibits strong muscle markings for the brachioradialis, suggesting climbing activity.

Ardipithecus

There are two species of **Ardipithecus,** both from northeastern Ethiopia. *Ardipithecus kadabba,* a Late Miocene species dated at 5.2 to 5.8 Ma, is known only from a small collection of teeth and bits of the skeleton. The upper canine has several primitive features. It has a

FIGURE 17.2 The reconstructed cranium of "Toumai," *Sahelanthropus tchadensis. (Courtesy of M. Brunet.)*

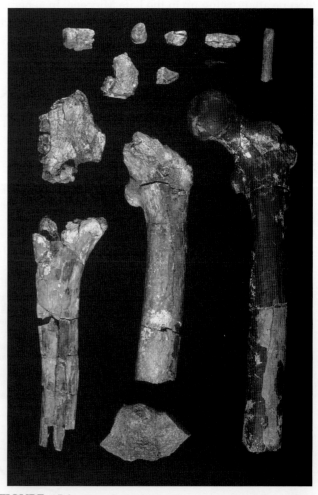

FIGURE 17.3 Dental, gnathic, and postcranial fossils of *Orrorin tugenensis. (Courtesy of M. Pickford.)*

mesial groove as in *Orrorin* and apes, and there is some indication of honing between the upper canine and lower anterior premolar, as found in apes but not any later hominins. A toe bone resembles apes in being somewhat curved, but in the orientation of the base, it resembles later bipedal hominins.

The younger species, *Ardipithecus ramidus*, is from deposits approximately 4.4 million years old. The type locality is Aramis near the Awash River in Ethiopia (White et al., 1994). It is known by over 100 specimens from more than 30 different individuals, including one relatively complete, greatly crushed and subsequently restored female (White et al., 2009). *Ardipithecus ramidus* has relatively smaller cheek teeth with thinner enamel and larger canines than later hominins, but there is no honing between the upper canine and the lower anterior premolar. This species was apparently monomorphic in both canine size and body size. The reconstructed body size is ~50kg. *Ar. ramidus* had a small brain and less robust face than later hominins but a shorter face than chimpanzees. The basicranium is

similar to *Australopithecus* in its proportions and possesses a relatively anterior foramen magnum (Suwa et al., 2009b; Kimbel et al., 2014). Evidence from the thin molar enamel, dental microwear, and stable isotopes suggests that *Ar. ramidus* consumed a diet of relatively soft food that was almost exclusively C_3 vegetation such as fruits and leaves (e.g., Suwa et al., 2009a; Grine, 2013).

The most unusual aspect of *Ar. ramidus* is its postcranial skeleton and reconstructed locomotor behavior (Fig. 17.4). This Pliocene species has been described as an arboreal quadruped with the ability to walk and run bipedally on the ground, but lacking the suspensory abilities of either extant apes or the Miocene apes of Eurasia. This interpretation is controversial, and other studies have debated whether *Ardipithecus* engaged in more forelimb dominated and suspensory behaviors than initially believed (e.g., Rolian et al., 2013; Prang et al., 2021; Chaney et al., 2021; Prang, 2022). The hands have been described as having relatively short digits and a longer thumb than extant apes, but the highly curved phalanges and overall hand proportions are more similar to

FIGURE 17.4   A reconstruction of the locomotor behavior of *Ardipithecus ramidus*. (*Drawing courtesy of Mauricio Anton.*)

living great apes than to later hominins (Prang et al., 2021). Likewise, the foot has African great ape-like proportions and a large, abducted big toe for arboreal grasping, but it is also capable of bipedal behavior (Prang, 2019; but see Chaney et al., 2021). The pelvis is a mosaic with a long distally projecting ischium, like that of a chimpanzee, which is well suited for quadrupedal walking, but a short, broad ilium for bipedal walking. Studies of the associated fauna, including isotopic analyses, indicate that *Ar. ramidus* inhabited forested environments, but other interpretations favor more open woodland habitats (White et al., 2009; Cerling et al., 2010, 2011b).

Phylogenetic relationships between *Ardipithecus* and fossil and living hominoids are very uncertain. In dental and cranial features, it seems to fit reasonably as the sister taxon to later *Australopithecus* (Strait and Grine, 2004; Strait et al., 2007; Mongle et al., 2019). However, the descriptions of its postcranial morphology as being more primitive than that of any fossil or living ape except *Ekembo* would suggest that hominins and living great apes are both derived relative to their last common ancestor (White et al., 2015). Alternatively, it is possible that numerous autapomorphic skeletal features evolved in *Ardipithecus* relative to its common ancestor with *Australopithecus*.

There is also debate about the relationships among *Sahelanthropus*, *Orrorin*, and *Ardipithecus*. Some have suggested that they may sample a single taxon (Cela-Conde and Ayala, 2003; White et al., 2009), but most authorities continue to recognize three distinct genera, perhaps with very different relationships to later hominins (Pickford, 2012). The absence of many complete overlapping body parts for the three limits the ability to resolve this issue at present (e.g., Simpson, 2013). They are all best considered to occupy unresolved sister taxa relationships to *Australopithecus*.

Australopithecus

Australopithecus (southern ape), from central, eastern, and southern parts of Africa (Reed et al., 2013), is the best-known and most widespread genus of early hominin (Table 17.1). As a genus, it is unquestionably a paraphyletic wastebasket taxon, as it is likely that the genera *Paranthropus* and *Homo* both originated somewhere within the species of *Australopithecus*.

The earliest species, *Au. anamensis* and *Au. afarensis*, have been found in sites over 4 million years old, and some specimens of this genus persist until the latest Pliocene, approximately 2 million years ago. *Australopithecus* species have big teeth and small brains compared to modern humans. They were relatively short in stature, but their size range is much greater than that of any living human population and more comparable to our entire species, with estimated body weights between 30 kg (the size of an Ituri pygmy) for the smallest

TABLE 17.1 Superfamily Hominoidea
Family HOMINIDAE
Tribe HOMININI

Species	Estimated average mass (kg)
Subtribe Australopithecina	
***Sahelanthropus* (Latest Miocene, Africa)**	
S. tchadensis	?50
***Orrorin* (Latest Miocene, Africa)**	
O. tugenensis	46
***Ardipithecus* (Latest Miocene to Early Pliocene, Africa)**	
Ar. kadabba	–
Ar. ramidus	50
***Australopithecus* (Pliocene to Early Pleistocene, Africa)**	
Au. anamensis	61
Au. afarensis	45
Au. africanus	39
Au. bahrelghazali	–
Au. garhi	–
Au. sediba	41
Au. prometheus	–
***Kenyanthropus* (Pliocene, Africa)**	
K. platyops	–
***Paranthropus* (Late Pliocene to Early Pleistocene, Africa)**	
P. aethiopicus	31?
P. boisei	45
P. robustus	42
Subtribe Hominina	
***Homo* (Late Pliocene to Recent, Cosmopolitan)**	
H. habilis	42
H. rudolfensis	–
H. erectus	62
H. heidelbergensis	70
H. neanderthalensis	68
H. floresiensis	36.5
H. luzonensis	–
H. naledi	55
H. sapiens	63.5

individuals and 85 kg (the size of a small college football player) for the largest. Most species were highly sexually dimorphic in body size.

Compared to living apes, all *Australopithecus* species have small incisors and canines relative to their body weight. The lower anterior premolar does not function as a sharpening blade for the upper canine. The molars of *Australopithecus* are large and are characterized by thick enamel and bulbous cusps, features shared with some Miocene apes. The mandible is thick and has a high ascending ramus.

Cranially, *Australopithecus* is more ape-like than human-like in proportions, with a large face and relatively small brain. Details of brain morphology in *Australopithecus* have been debated since the first discovery of the genus. Our understanding of early hominin brain evolution is hindered by problems in estimating the body size of the different species and by the lack of clear impressions on the internal surface of the cranium. In general, it appears that their brains were relatively larger than those of nonhuman primates but much smaller than those of later fossil hominins or living humans. In external morphology, their brains are generally ape-like, with few human features.

There are isolated skeletal elements attributed to most species of *Australopithecus* and relatively complete associated skeletons for *Au. afarensis* (Fig. 17.5), *Au. africanus*, and *Au. sediba* (Fig. 17.9). Like all later hominins, *Australopithecus* was bipedal. This is evident from many aspects of its skeleton, including the relatively long legs; the short, broad ilium; and the angulation of the knee joint. The bipedal habits of *Australopithecus*, inferred from skeletal morphology, are confirmed by a series of footprints preserved at Laetoli, Tanzania (Fig. 17.6). However, *Australopithecus* was not just like modern humans, and species of *Australopithecus* show similarities to living apes in many features of the skeleton, including the shoulder, the hands, the feet, and some details of the pelvis, femur, and tibia (e.g., Stern, 2000; Ward, 2013). The skeletal anatomy of *Australopithecus* is in many ways intermediate between that of living apes and humans, suggesting that these early hominins were terrestrial bipeds that retained some abilities for arboreal climbing (Figs. 17.7 and 17.8).

The diversity in locomotor abilities among *Australopithecus* species is difficult to determine because of a lack of skeletal material for some species and because of our weakness in creating models for interpreting a locomotor radiation of bipeds with only one extant analogue: ourselves. Species of *Australopithecus* are found associated with faunas suggesting a variety of habitats, including both open and wooded (Behrensmeyer and Reed, 2013).

The systematics and biogeographical history of *Australopithecus* species are complicated by the same factors that cause confusion in the systematics of most other groups of fossil primates: inadequate dating of some sites, fragmentary remains, and differing taxonomic philosophies (e.g., Lockwood, 2013). Seven species of *Australopithecus* are generally recognized, and more are

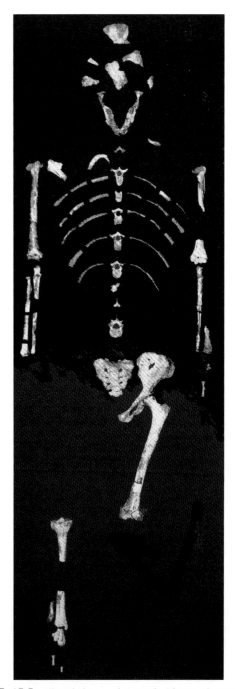

FIGURE 17.5 The skeleton of *Australopithecus afarensis*, "Lucy" (AL-288), from Hadar, Ethiopia. This is one of the most complete skeletons of an early hominin.

likely waiting to be uncovered. *Australopithecus anamensis*, *Au. afarensis*, *Au. deyiremeda*, and *Au. garhi* are from eastern Africa (Ethiopia, Kenya, and Tanzania); *Au. africanus* and *Au. sediba* are from South Africa; and *Au. bahrelghazali* is from Chad in central Africa. The large-toothed species of early hominins are placed in a separate genus, *Paranthropus*, described in a later section.

Australopithecus anamensis is the oldest species of *Australopithecus*, found at several sites in Northern Kenya and Ethiopia that date between 3.8 million and 4.2 million years ago. It is known primarily from dental remains, a nearly complete male cranium, and a few limb elements, including a humerus and a tibia (Leakey et al., 1995, 1998; Haile-Selassie et al., 2019). The cranium is long and low with great postorbital constriction, similar to earlier hominins such as *Sahelanthropus*, but with a broad upper face and zygomatic region more similar to *Au. afarensis* (Haile-Selassie et al., 2019). The face exhibits increased prognathism relative to the earliest hominins, again more similar to *Au. afarensis* than *Ardipithecus* and *Sahelanthropus*. The tibia suggests that it was bipedal. Whereas the canine roots were quite dimorphic, the crowns were relatively short in both sexes (Manthi et al., 2012). Faunal and geological information suggests that *Au. anamensis* is associated with a variety of habitats.

Australopithecus afarensis (Figs. 17.5–17.8) is the best-known species of *Australopithecus*. It has been recovered from Pliocene deposits in many parts of eastern Africa (Ethiopia, Tanzania, and possibly Kenya). This species has a long temporal span from 3.8 Ma to at least 2.9 Ma (Alemseged, 2013). Although it is best known from 3.4 to 3.0-million-year-old deposits at Hadar in Ethiopia, older remains (3.7 million years old) have been described from Laetoli in Tanzania and from several sites in Ethiopia (e.g., Haile-Selassie et al., 2010). Analysis of the dental remains of *Au. afarensis* and *Au. anamensis* suggests that they may represent a single continuous lineage (Kimbel et al., 2006); however, the male cranium assigned to *Au. anamensis* overlaps in time with some specimens assigned to *Au. afarensis*, perhaps refuting the hypothesis of a single anagenetic lineage (Haile-Selassie et al., 2019). Although there is some debate, analyses of individual limb elements suggest that *Au. afarensis* was extremely sexually dimorphic in body size, with the smallest individuals weighing no more than 30 kg and the largest probably twice as much.

In dental proportions, *Au. afarensis* has larger canines and incisors than later hominins. The molars are larger than those of living apes and have low cusps and thick enamel. The relatively large anterior dentition suggests that this species was frugivorous, and the thick enamel indicates that nuts, grains, or hard fruit pits may have been part of its diet. Although *Au. afarensis* was probably very sexually dimorphic in body size, it had little canine dimorphism compared to living great apes, but more than modern humans.

The cranial anatomy of *Au. afarensis*, known from several relatively complete specimens (Kimbel et al., 2004), is similar to that of living chimpanzees in many aspects. This species has a longer snout and shallower face than later hominins and an ape-like nuchal region. The brain is small. There is a sagittal crest both anteriorly, as in other australopithecines, and posteriorly, as in apes.

The skeletal anatomy of *Au. afarensis* has been particularly well documented (e.g., Johanson et al., 1982; Drapeau et al., 2005) but debated (e.g., Stern and Susman, 1983; Stern, 2000; Ward et al., 2012). The most famous fossil of this species, "Lucy" (AL-288) from Hadar, is known from 40% of a skeleton, including large portions of almost all long bones (Fig. 17.5). In limb proportions, Lucy is intermediate between living chimpanzees and humans. Based on an estimated body weight of 30 kg, she has relatively short hindlimbs but forelimbs similar in length to those of a small human. Compared to a pygmy chimpanzee of the same size, she has relatively short arms but similar-sized hindlimbs. There is also an infant skeleton of *Au. afarensis* that provides information about the ontogeny of early hominins (e.g., Alemseged et al., 2006; Green and Alemseged, 2012; Ward et al., 2017; DeSilva et al., 2018).

The forelimb remains of Lucy, although human-like in proportions, are more chimpanzee-like in some features. The curved phalanges, large pisiform bone, and cranially oriented shoulder joint all suggest some suspensory abilities for this early hominin, as do other chimpanzee-like features of the humerus and ulna (Susman et al., 1984; Stern, 2000; see also Green and Alemseged, 2012). The pelvis of *Au. afarensis*, like that of all later hominins, has a short, broad ilium and a relatively short ischium, resembling those of bipedal humans more than that of any living ape, but the iliac blade faces posteriorly as in nonhuman primates, rather than laterally as in humans. The distal part of the femur is strikingly human-like in its **valgus** (knock-kneed) angulation, yet it lacks the enlarged lateral lip of the patellar groove found in bipedal humans. The foot shows features indicative of bipedal walking and a human-like transverse

FIGURE 17.6 The 3.5 million-year-old footprints from Laetoli, Tanzania, presumably made by *Australopithecus afarensis*. (*Photograph by P. Jones and T. White.*)

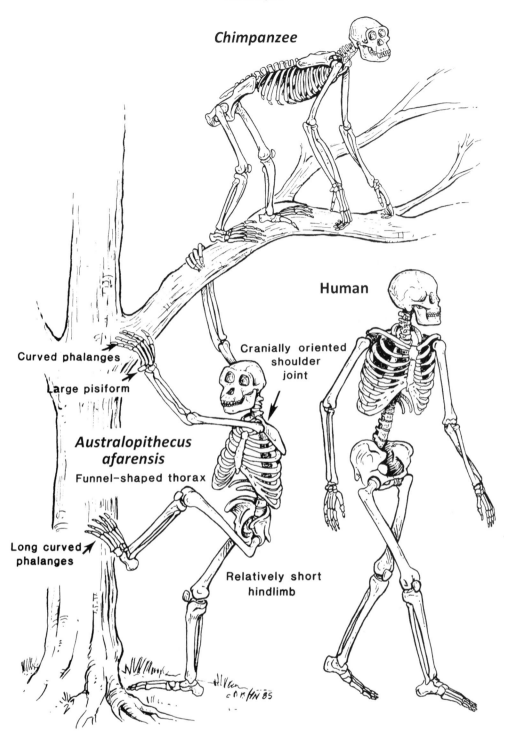

Chimpanzee

Human

Curved phalanges

Cranially oriented shoulder joint

Large pisiform

Australopithecus afarensis

Funnel-shaped thorax

Long curved phalanges

Relatively short hindlimb

FIGURE 17.7 The skeletons of *Australopithecus afarensis, Pan troglodytes,* and *Homo sapiens*. Note the ape-like features in *Au. afarensis* that suggest climbing behavior.

arch, but other details of the ankle and foot, such as the relatively long, curved pedal phalanges, are more chimpanzee-like and suggest grasping behavior. The extent to which the locomotion of *Au. afarensis* differed from that of modern humans and the significance of the primitive features of its skeleton have been the subject of extensive and ongoing debate (e.g., Stern, 2000; Ward,

2002, 2013; Ruff et al., 2016). Although *Au. afarensis* was clearly a biped, it seems likely that its locomotion differed significantly from that of modern humans; even the relatively human-like hindlimb elements are different in detail from those of all later hominins, suggesting some sort of arboreal abilities and a different gait from that of extant humans (Fig. 17.6).

FIGURE 17.8   A small *Australopithecus afarensis* group.

As with the dental and cranial remains of *Au. afarensis*, there is considerable variability in both size and morphological detail among the skeletal remains attributed to this species. However, there is no indication of behavioral or functional differences between the large and small individuals (Ward et al., 2012). The size differences in the limb elements attributed to this species are comparable to those of the most dimorphic living apes, gorillas and orangutans (Gordon, 2013; but see Reno et al., 2003, 2010). Overall, *Au. afarensis* seems very much like a missing link between the living African apes and later hominins in its dental, cranial, and skeletal morphology. Microwear studies suggest it differed from any living primate in several ways. The thick enamel and large mandible indicate that *Au. afarensis* (and *Au. anamensis*) was adapted for heavy chewing, but it is unclear what this tells us about the diet of *Au. afarensis*. It traveled bipedally on the ground but probably slept in and perhaps foraged in the trees. It seems likely that *Au. afarensis* occupied a variety of habitats (Behrensmeyer and Reed, 2013).

The social structure of *Au. afarensis* is difficult to reconstruct since the combination of little canine dimorphism with considerable body size dimorphism is unique among living primates (Plavcan and van Schaik,

1993) but most like the pattern found among modern humans. Several authors have inferred monogamous habits (Reno et al., 2003, 2010). Based on the canines, Lovejoy (1981) has suggested that the size dimorphism reflects different foraging patterns and anti-predator strategies for the two sexes. However, these are minority views. The majority of researchers have argued that *Au. afarensis* exhibited considerable size dimorphism and that this indicates a polygynous social structure (e.g., Gordon et al., 2008; Gordon, 2013). It is also debated whether early hominin canine reduction is related primarily to social structure (Lovejoy, 2009) or to some aspect of dietary adaptation, as seen in the Miocene apes which also exhibit reduced canines (e.g., *Ouranopithecus*, *Gigantopithecus*, etc.).

The recently described hominin *Australopithecus deyiremeda* is known from Woranso-Mille, Ethiopia from 3.3 to 3.5 Ma and was contemporaneous with *Au. afarensis* (Haile-Selassie et al., 2015). Although only represented by jaws, palates, and teeth, it differs from *Au. afarensis* in possessing a more anterior zygomatic root on the maxilla, a thicker mandible, and a smaller postcanine dentition. Moreover, pedal remains from the site of Woranso-Mille also indicate a taxon with a more primitive foot, like that

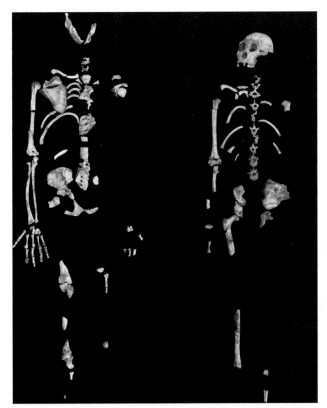

FIGURE 17.9   Fossils of *Australopithecus sediba* from South Africa. *(Courtesy of L. Berger.)*

of *Ardipithecus*, at this locality, either *A. deyiremeda* or something else (Haile-Selassie et al., 2012, 2015, 2016).

Kenyanthropus platyops is a taxon similar to *Au. afarensis* that was described by Leakey and colleagues (2001) from deposits west of Lake Turkana dated to between 3.2 and 3.5 Ma. It is known from a relatively complete, but crushed, cranium. As the name suggests, *K. platyops* is distinguished from other australopiths by its short face. It is almost certainly a close relative of *Au. afarensis* (Strait et al., 2007). One author has suggested that the distinguishing features are merely the result of the distorted nature of the fossil (White, 2003), but Spoor et al. (2016) have argued that it is significantly different and provides evidence of a diversity of forms among australopiths.

Australopithecus garhi is from the Middle Awash area of Ethiopia and has a date of 2.5 Ma (Asfaw et al., 1999). It is known from a maxilla with a complete dentition and some cranial fragments. Several parts of an upper limb and part of a femur were recovered at other sites in the region. The postcanine teeth of *Au. garhi* are very large, and the maxilla suggests a prognathic face. The unassociated limb bones, if belonging to one individual, would indicate a high brachial index, as in apes, but a low intermembral index, as in humans. Bones with cut marks suggesting butchery have also been found in the area, but it is unclear who made the tools, as *Homo* was also in Ethiopia at that time. In their original description, the

authors suggested that *A. garhi* was a "candidate ancestor for early *Homo*," but subsequent analyses (Strait and Grine, 2004) have failed to link *Au. garhi* with *Homo*. It has been suggested that some fossils from Pliocene deposits in the Omo region of southernmost Ethiopia may also be attributable to *Au. garhi* (White, 2002).

Australopithecus bahrelghazali is from deposits 3.0 million to 3.5 million years old in the central African country of Chad (Brunet et al., 1995, 1996). So far, it is known only from a lower jaw, an isolated upper premolar, and a maxilla. This species is unusual in having a more vertical mandibular symphysis than other species of *Australopithecus*. It extends the geographic range of early hominins 2500 km farther west than previously documented.

Originally described by Raymond Dart (1925) from the limeworks at Taung in the Cape Province of South Africa, *Australopithecus africanus* is best known from the caves at Sterkfontein and Makapansgat. Because the limestone caves are not amenable to radiometric dating, the absolute age of *Au. africanus* has been argued, and the results of numerous dating methods, including faunal associations, suggest that at Sterkfontein the fossils range somewhere between 2 million and 4 million years old (Herries et al., 2013; Granger et al., 2022; Frost et al., 2022).

Compared to *Au. afarensis*, *Au. africanus* has more similar-sized central and lateral upper incisors and larger cheek teeth. The relatively smaller anterior dentition resulted in a shorter snout in *Au. africanus*. The occipital region and the tympanic bones of the South African species are more like those of later hominins. There is considerable variability in the cranial and dental morphology of the many specimens assigned to *Au. africanus*, and there is no consensus regarding how this morphological diversity might reflect sexual dimorphism or multiple species (e.g., Clarke, 2013; Grine, 2013).

The skeleton of *Au. africanus* is similar to that of *Au. afarensis* in many features, including relatively large upper limb elements and relatively short legs. As in *Au. afarensis*, there is considerable size dimorphism in limb elements attributed to *Au. africanus* (McHenry, 1994). Behaviorally, *Au. africanus* was probably very similar to *Au. afarensis*. Dental microwear studies suggest that *Au. africanus* had a more variable diet than other taxa, and isotopic studies show a mixed diet of both C_3 and C_4 plants (Grine et al., 2013) (Fig. 17.13). Many aspects of its limb skeleton suggest that, like *Au. afarensis*, it was both a biped and an adept climber.

A virtually complete skeleton from Sterkfontein known as "Little Foot" greatly enhances our knowledge of this taxon. Some researchers consider it a distinct species, *A. prometheus* (but see Grine, 2019; Ward and Zipfel, 2020). The cranium of "Little Foot" possesses notable features such as a broad interorbital breadth, expansion of the braincase in the parietal region ("bossing"), a relatively long palate, and a sagittal crest in males (Clarke and Kuman, 2019). Postcranially and behaviorally, it exhibits many of the same features seen in other *Au. africanus* and *Au. afarensis* skeletons, including relatively long arms and an upward-facing shoulder joint suggesting retained climbing abilities (Heaton et al., 2019; Carlson et al., 2021).

The age of the "Little Foot" is controversial, with some arguing the skeleton is 3.67 million years old on the basis of new radiometric dating methods (e.g., Granger et al., 2015, 2022), while others argue that the skeleton is the same age as other Sterkfontein *Au. africanus* specimens (~2.8–2.0 Ma) based on paleomagnetism, faunal correlation, and other radiometric methods (Herries et al., 2013; Kramers and Dirks, 2017; Pickering et al., 2019; Frost et al., 2022).

Australopithecus sediba (Fig. 17.9) is a recently described species from the site of Malapa in South Africa dated to slightly younger than 2 million years ago (Dirks et al., 2010). The species is known so far from two partial skeletons (Berger et al., 2010; de Ruiter et al., 2013). The teeth of *Au. sediba* are near the lower end or smaller than the teeth of *Au. africanus* and within the range of species of *Homo*, with very little indication of sexual dimorphism. Body size is estimated at 30 kg and 37 kg for two individuals. The cranial remains have a mosaic of features characteristic of *Au. africanus* and early *Homo*, with a small cranial capacity of 420 mL, smaller than any species of *Homo* except *H. floresiensis*. Similarly, the skeletal remains resemble *Australopithecus* in some respects: relatively long upper limbs with a high brachial index, upward-facing shoulder joint, large joint surfaces, and a primitive calcaneus. They also show derived features characteristic of early *Homo*, including a long thumb, vertical iliac blades, and a modern talocrural joint. Overall, *Au. sediba* seems to be morphologically intermediate between fossils currently classified as *Australopithecus* and those attributed to early *Homo*, and some researchers differ on which genus is more appropriate (e.g., de Ruiter et al., 2013). Others suggest that *Au. sediba* is simply a close relative of *Au. africanus* (Kimbel and Rak, 2017).

Paranthropus

The robust australopiths (Fig. 17.10), here placed in the genus *Paranthropus*, are from Late Pliocene and Early Pleistocene deposits in eastern and southern Africa (e.g., Wood and Schroer, 2013). Compared to *Australopithecus*, *Paranthropus* species have larger molars and premolars combined with relatively smaller canines and incisors. They have much flatter, broader, dished faces, and later species have a more flexed cranial base. The skeletal anatomy of *Paranthropus* is less well known than that of *Australopithecus*, but *Paranthropus* was almost certainly

some type of primitive biped. Despite the large teeth, body size estimates based on limbs suggest body weights similar to those of *Australopithecus* and high levels of sexual dimorphism. There are three species currently recognized. All seem to have been most common in relatively wet habitats characterized by edaphic grasslands (Reed, 1997).

Paranthropus aethiopicus is the oldest and most primitive species of *Paranthropus*, from deposits between 2.7 million and 2.3 million years old in southern Ethiopia, northern Kenya, Tanzania, and Malawi. The best specimen, KNM-WT 17000, "the Black Skull" (Fig. 17.11), has a massive face combined with a relatively long snout, a primitive cranial base like *Au. afarensis*, and very large sagittal and nuchal crests.

Paranthropus robustus (Figs. 17.10 and 17.12) is from the cave sites of Swartkrans, Kromdraai, Gondolin, and Drimolen in South Africa. It is, estimated at between 2.3 million and just under 1 million years old and was contemporaneous with members of the genus *Homo*. *Paranthropus robustus* was probably similar in size to *Au. africanus* and was also sexually dimorphic in size.

Paranthropus robustus has smaller incisors and canines, larger cheek teeth with thicker enamel, and a thicker mandible than *Australopithecus* species (Fig. 17.10). These differences in dental morphology are associated with differences in both gross and microscopic tooth wear (Fig. 17.13). Individuals of *P. robustus* wore down their teeth flatter and used more crushing than shearing, and microwear studies show a greater complexity suggesting a diverse diet including hard food items. However, the isotopic studies indicate a diet comprising a mixture of C_3 and C_4 plants similar to that of *Au. africanus* (Ungar and Sponheimer, 2011; Grine et al., 2013). *Paranthropus robustus* has a shorter, broader face with deeper zygomatic arches and a larger temporal fossa than seen in the skull of *Au. africanus* (Fig. 17.10). The larger individuals (males?) have sagittal and nuchal crests. Like the molar differences, the cranial differences seem related to more powerful chewing in *P. robustus*. Lockwood and colleagues (2007) have argued that the development of cranial structures in *P. robustus* indicates a pattern of extended growth in males associated with a male reproductive structure involving extreme sexual dimorphism and the monopolization of multiple females by individual males, as with gorillas.

The skeletal differences between *P. robustus* and *Australopithecus* are difficult to assess because there are few well-associated complete limb elements. Susman (1988) suggests that *P. robustus* was more human-like in both hands and feet than *Au. afarensis*. The hand bones show evidence of manipulative abilities, suggesting that this species was capable of using and making tools (Susman, 1991, 1994). The foot bones indicate that it was bipedal and less arboreal than *Au. afarensis*.

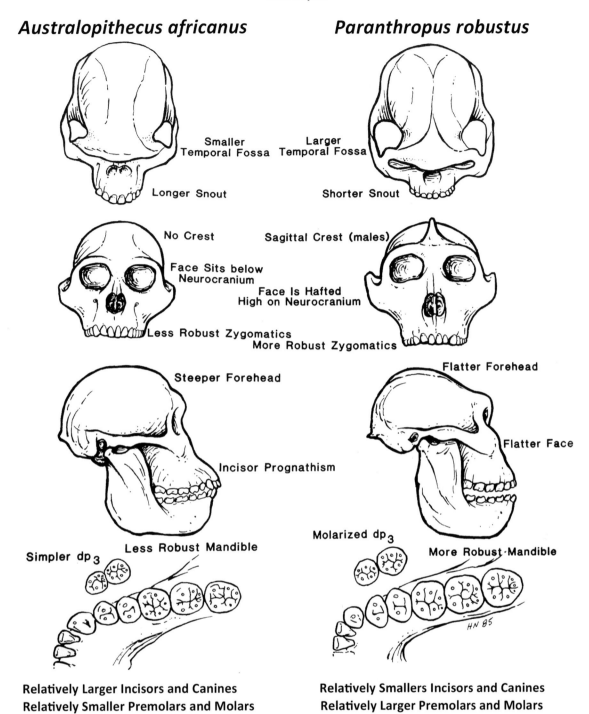

Australopithecus africanus

Smaller Temporal Fossa

Longer Snout

No Crest

Face Sits below Neurocranium

Less Robust Zygomatics

Steeper Forehead

Incisor Prognathism

Simpler dp$_3$

Less Robust Mandible

Relatively Larger Incisors and Canines
Relatively Smaller Premolars and Molars

Paranthropus robustus

Larger Temporal Fossa

Shorter Snout

Sagittal Crest (males)

Face Is Hafted High on Neurocranium

More Robust Zygomatics

Flatter Forehead

Flatter Face

Molarized dp$_3$

More Robust Mandible

Relatively Smallers Incisors and Canines
Relatively Larger Premolars and Molars

FIGURE 17.10 Cranial and dental features of *Australopithecus* and *Paranthropus*.

Habitats for *P. robustus* have been reconstructed as secondary grasslands associated with rivers and wetlands (Reed, 1997).

The dental and cranial features that characterize the South African *P. robustus* are more strongly developed in the "hyper-robust" ***Paranthropus boisei*** from East Africa. This species, known from deposits in Ethiopia, Kenya, and Tanzania between approximately 2.2 million and 1.2 million years old, was contemporaneous with members of our own genus, *Homo. Paranthropus boisei* was similar in size to *P. robustus*, with an estimated body weight of 30 to 50 kg. This species is sexually dimorphic in both size and cranial shape. Compared to *P. robustus*, *P. boisei* has smaller incisors and canines, absolutely larger cheek teeth, and a heavier mandible. The skull has an extremely broad, short face with a large temporal

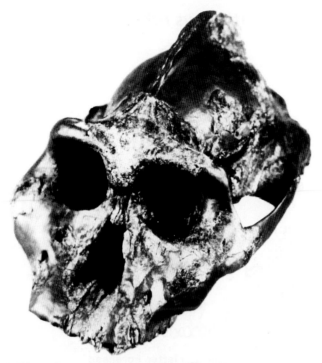

FIGURE 17.11   The cranium of *Paranthropus aethiopicus* from West Turkana. *(Photo by A. Walker, copyright National Museums of Kenya.)*

fossa between the flaring zygomatic arches and the relatively small brain. There are pronounced sagittal and nuchal crests in large males. Based on microwear analyses, the diet of *P. boisei* has been argued to range from soft fruits to tough foods, and recent studies of enamel isotopes indicate that *Paranthropus boisei* was unusual among Pliocene hominins in having a diet made up almost exclusively of plants using a C_4 pathway, such as grasses and sedges (Fig. 17.13). Thus, it seems likely that the large teeth and heavy masticatory apparatus were adaptations for processing large amounts of nutritionally poor food, rather than specifically hard items (Cerling et al., 2011a; Ungar and Sponheimer, 2011; Grine et al., 2013).

Although there are few limb bones that can definitely be attributed to *P. boisei*, several very large forelimb bones and an upper limb partial skeleton from eastern African sites are often assigned to the species. These bones suggest suspensory behaviors but perhaps less frequent than in *Australopithecus* (Green et al., 2018; Lague et al., 2019; Richmond et al., 2020). Tools are often found in association with *P. boisei* in eastern Africa, and hand bones attributed to *P. boisei* suggest a precision grip (Richmond et al., 2020), but the existence of other hominins (*Homo* sp.) from the same timespan precludes unequivocal determinations as to which or how many species made or used tools. Like *P. robustus*, *P. boisei* seems to have become extinct before 1 million years ago.

Early Hominin Adaptations and Hominin Origins

Dental and Cranial Adaptations

The early hominins from the Late Miocene through Early Pleistocene show considerable diversity in dental and cranial morphology. *Ardipithecus* and *Sahelanthropus* have been reported to have thin enamel on their cheek teeth; *Orrorin* has been reported to have thick enamel by some (Pickford, 2012) and thin by others (White et al., 2009). In general, *Australopithecus* species have thick enamel, and *Paranthropus* species have even thicker. Despite differences in enamel thickness and microwear patterns, all species that have been studied, except for *Paranthropus boisei*, have dental isotope values indicating a diet composed primarily of C_3 plants. Similarly, there are dramatic differences in cranial and mandibular morphology. There is certainly some evidence of dietary diversity among the many species (Fig. 17.13), often synchronic and sometimes sympatric, but what these differences may have been in terms of specific food items is very difficult to reconstruct (Ungar and Sponheimer, 2011; Grine et al., 2013; Sponheimer, 2013).

A striking feature of early hominins is that the height of the canine teeth and sexual dimorphism in canine height are reported to be low in all species for which it can be calculated, although there may be more dimorphism in the earliest (Late Miocene) species. Correspondingly, early hominins lack the sectorial morphology of the lower anterior premolar related to honing of the upper canine. This is an unusual feature for any primate and is presumably related to aspects of social behavior and/or diet. All evidence suggests that the timing of dental development and eruption in early hominins was more similar to that of extant apes than of modern humans, suggesting a relatively fast life history, although canine eruption may have been earlier in the fossil hominins than in extant apes. All early hominins had small brains, similar in size or slightly larger than those of extant great apes.

Postcranial Adaptations

The skeleton of modern humans is characterized by many features related to our abilities to walk and run bipedally. Early hominins display features of the pelvis, limb proportions, and feet, which indicate that they were capable of some type of bipedal progression (Harcourt-Smith, 2007; Ward, 2013). At the same time, they differ in various ways from modern humans and among themselves, so there must have been considerable locomotor diversity in the hominin lineage(s) between 6 and 2 million years ago. Unfortunately, our ability to detail and understand this likely diversity is greatly limited by

FIGURE 17.12 Reconstruction of a group of *Paranthropus robustus* from Swartkrans in southern Africa. Faunal evidence suggests a wet habitat (Reed, 1997). Dental studies suggest an herbivorous, gritty diet, and anatomical studies of the hands and feet indicate bipedal behavior and possible tool use. *(Courtesy of F. Grine.)*

both the lack of complete, associated skeletal material and the lack of more than a single extant primate biped that might be used to test functional hypotheses.

Modern humans are also unusual in the manipulative abilities of the hands with our relatively long, opposable thumb and relatively straight phalanges with a broad apical tuft on the distal phalanx. The hand bones of *Australopithecus* species show a range of mosaics of primitive ape-like features and derived human features, but they are largely intermediate in many features, including phalangeal curvature and features of the carpal bones, likely reflecting intermediate levels of

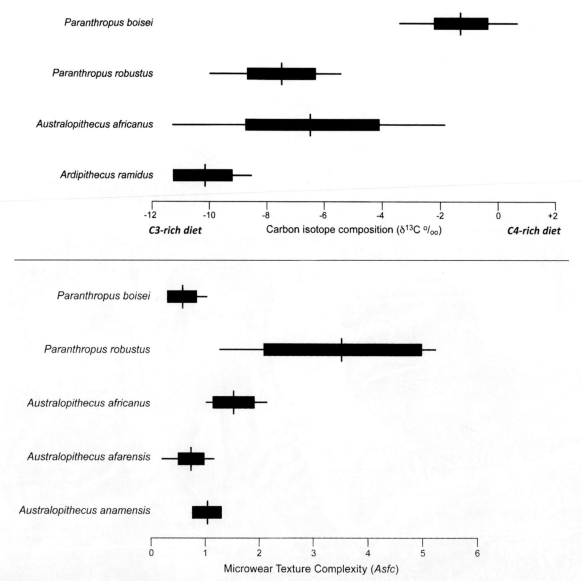

FIGURE 17.13   Comparative analyses of dental microwear and isotopic studies of dental enamel for early hominins. *(Courtesy of F. Grine.)*

arboreal behavior (e.g., Alba et al., 2003; Tocheri et al., 2008; Kivell et al., 2011). The nearly complete hand of *Au. sediba* is unusual in having a very long thumb in addition to features more similar to the hand of *Homo*.

Tool Use in Early Hominins

As many authors have emphasized, all early hominins may have made and used some type of perishable tools, since such tools are used by both chimpanzees and later hominins (Panger et al., 2002; Wynn et al., 2011). However, evidence of stone tools or tool use associated with hominins other than *Homo* is limited and controversial. There is some evidence that *Paranthropus* was a tool user (Fig. 17.12). Bone and stone tools are often found in deposits with species of *Paranthropus boisei* from eastern Africa and *P. robustus*

from South Africa, and it has been argued that *Paranthropus* has anatomical features of its hand consistent with tool use (Susman, 1991, 1994). Tool use attributable to *Australopithecus* is rarer and more controversial. However, bones with cut marks similar to those made by stone tools have been reported in deposits containing *Australopithecus garhi* (Asfaw et al., 1999) and *Au. afarensis* (McPherron et al., 2010), and the extensive stone tools at Gona (Semaw, 2000) are several hundred thousand years earlier than any fossils attributed to *Homo*. Most recently, primitive stone tools were discovered dated to ~3.3 Ma at Lomekwi in Kenya (Harmand et al., 2015), at least a half million years before the appearance of *Homo* or *Paranthropus* in the fossil record. These early stone tools and cut marks are suggestive of *Australopithecus* as a tool-maker, but definitive association remains elusive.

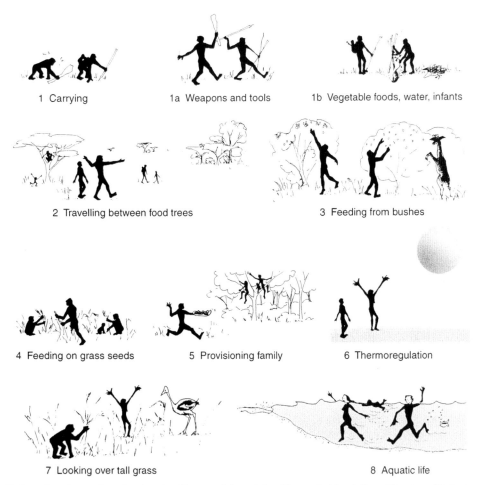

1 Carrying   1a Weapons and tools   1b Vegetable foods, water, infants

2 Travelling between food trees   3 Feeding from bushes

4 Feeding on grass seeds   5 Provisioning family   6 Thermoregulation

7 Looking over tall grass   8 Aquatic life

FIGURE 17.14   Various theories on the adaptive significance of the origin of hominin bipedalism. *(Courtesy of J. Sept and L. Betti-Nash.)*

Evolution of Bipedalism

The evidence from *Sahelanthropus*, *Orrorin*, and *Ardipithecus* suggests that bipedalism of some sort was present in the human lineage by the beginning of the Pliocene. Like all other members of our order, each species of early hominin was a primate with a particular suite of adaptations that enabled it to make a living in its particular time and place. What, then, is the behavioral and ecological context in which our ancestors evolved this unique primate adaptation (Fig. 17.14)?

Attempts to "explain" the origin of hominin bipedalism have come from many different perspectives and addressed different levels of explanation. Some authors have asked what kind of locomotor behavior preceded and facilitated the evolution of bipedalism in early hominins. Fleagle and colleagues (1981; also Prost, 1980) showed that vertical climbing in nonhuman primates involves similar patterns of hindlimb excursions and muscle recruitment as human bipedalism, and that this behavior is, in some sense, preadaptive for bipedal walking. In another study that emphasized the extended hindlimbs characteristic of human bipedal walking,

Thorpe and colleagues (2007) noted that orangutans employ extended hindlimb postures when walking on flexible branches and suggest that this arboreal behavior led to human bipedalism (but see Begun et al., 2007). Thus, these studies (also Stern, 1975) argue that the limb mechanics used by modern humans in terrestrial bipedalism have their basis, in a functional sense, in arboreal behaviors rather than being modified versions of terrestrial quadrupedalism.

Most efforts to understand the evolution of bipedalism have focused on the adaptive significance of the behavior. One of the most striking advantages of bipedal locomotion is that it frees the hands from locomotor activities to be useful for carrying things, and freeing of the hands has long been considered the major selective advantage of bipedalism (Hewes, 1961). Because stone tools do not normally seem to be associated with *Australopithecus*, some authors have suggested that the earliest hominins must have been under selection to free their hands for manipulating some type of perishable wooden tools, or perhaps spears or clubs.

Other authors have looked to primate-like foraging adaptations associated with bipedalism. Zihlman and

Tanner (1978) see considerable similarity between australoiths and living chimpanzees in many aspects of their behavior. They suggest that in modern hunter–gatherer communities, it is the gathering of plants by females that provides most of the food for subsistence, just as plants provide most of the food for living apes. In their view, we should pay more attention to the function of characteristic hominin features in a gathering rather than a hunting context. In such a context, bipedalism would enable females to carry hominin infants more easily (although it is unlikely that these infants would have been as helpless as our own). Free hands would also enable gatherers to carry extra water, thus extending foraging range into dry areas, and to carry surplus food in a way that living apes cannot. Zihlman and Tanner suggest that, if the earliest hominins used tools, those tools were probably digging sticks or natural containers for transporting food and water, not clubs or spears.

Another common ecological argument for the origin of bipedalism is that it is related to life in open habitats and a need to see over tall grass. Although it is true that many primates living in open habitats, such as patas monkeys and vervet monkeys, frequently stand to look for conspecifics or look out for predators, none seem to have adopted bipedalism as a normal pattern of movement. Although the fossil sites that have yielded most early hominins suggest wooded rather than open habitats (Reed, 1997), it is clear that open habitats were never far away (Cerling et al., 2011b).

Rodman and McHenry (1980) associate bipedalism with foraging in an open woodland habitat. However, in their view, the earliest hominins may well have fed in trees but adopted bipedal postures for traveling between trees. They suggest that bipedal walking would be a more efficient method of locomotion than quadrupedal travel and cite studies showing that modern human bipedalism is indeed a very efficient pattern of movement. However, as numerous studies have shown (Taylor and Rowntree, 1973; Steudel, 1994; Sockol et al., 2007), bipedal and quadrupedal chimpanzees show no differences in cost of locomotion, so bipedalism per se is not inherently a cheap way of moving. Because the earliest hominins lacked many of the distinctive anatomical features associated with the striding bipedalism of modern humans, an energetic argument for the origin of this pattern of locomotion seems unrealistic, even though small changes in locomotor morphology can lower the cost of bipedal locomotion below that of quadrupedal apes (Pontzer et al., 2009).

There are several theories suggesting that bipedalism arose as a postural, rather than a locomotor, adaptation for foraging, either on the ground or in the trees. Several authors have suggested that the evolution of hominin bipedalism began as a feeding posture which enabled early hominins to feed on tall bushes or small trees, and

Hunt (e.g., 1994, 1996) has argued from comparisons with living chimpanzees that early hominins began their bipedal behavior by standing during feeding in an arboreal setting. Like the open-habitat hypothesis, these suggestions fail to explain why the other primates that regularly use bipedal postures have never adopted bipedal locomotion.

Other theories address this problem of hominin uniqueness directly. Jolly (1970) has argued that bipedalism in an herbivorous primate is advantageous only when the animal is feeding on small, evenly distributed objects such as nuts, grains, or small seeds. In this type of feeding situation, an individual's foraging efficiency is linked directly to the speed with which it can pluck and ingest food items. A squatting or partly bipedal animal with both hands free for foraging is best adapted to this type of diet. Furthermore, small, hard objects such as seeds may have been the type of food that the *Australopithecus* dentition, with its broad, flat, thick-enameled molars, was designed to chew. Finally, Jolly argues that many foods of this nature are common in open habitats, in which bipedal locomotion seems most appropriate. Jolly's model of hominin origins, the "seed-eater" model, is based largely on the habits of the gelada, an open-savannah small-object-feeding monkey.

In a set of highly provocative papers on hominin origins, Lovejoy (1981, 2009) suggests that the major adaptive feature of the hominin lineage from its origin lies in our reproductive system. In his view, bipedalism, canine reduction, and reduced male–male competition form a unique adaptive suite that is dramatically different from that of African apes. Living great apes, he argues, are characterized by male–male competition that is reflected in their canine dimorphism. The human lineage, he suggests, has been characterized by an absence of overt male–male competition, as indicated by the near absence of canine dimorphism in *Ardipithecus*. Rather, in Lovejoy's view, hominin social behavior has long been characterized by a monogamous mating system in which males provisioned their mates and offspring, even though this took place in the context of a large multi-male, multi-female social system and later hominins in the genus *Australopithecus* appear to be strongly dimorphic in body size. According to Lovejoy, bipedalism, by freeing the hands, allowed the males to bring extra food back to the less widely foraging members of their family, a behavior unknown in other primates.

Peter Wheeler (e.g., 1993) has advanced a quite different, physiological theory for the origin of bipedalism. In his view, bipedalism would have provided an early hominin considerable thermoregulatory advantages in limiting direct exposure to sun during the middle of the day and facilitating convective heat loss in a stressful, open equatorial habitat.

Jablonski and Chaplin (1993) have argued that hominin bipedalism arose as an extension of bipedal displays of threat and gestures of appeasement found among extant African apes. In their view, it is this aspect of bipedalism that links human bipedal locomotion with the bipedal behavior of our closest relatives, and thus would have provided the selective advantage for this behavior in the earliest hominin bipeds prior to any energetic or physiological benefits that later hominins may have experienced.

Finally, there are several theories that view aquatic behavior of some sort as a major selective factor in the evolution of hominin bipedal locomotion. The aquatic ape theory has been championed most vigorously by Morgan (1999), based on earlier work by Hardy (1960), who noted physiological similarities between humans and aquatic mammals. More recently, Wrangham and colleagues (Wrangham, 2005; Wrangham et al., 2009) have argued that human bipedalism may have evolved as an adaptation for foraging in wetlands habitats such as rivers and lakes.

Although no current theory seems satisfactory on all counts, most attempt to explain the evolution of a bipedal, herbivorous primate with small canines based on our understanding of the socioecology and physiology of all primates, such as feeding, travel, and reproduction, rather than only on modern human behavior. Moreover, many incorporate an impressive breadth and sophistication of information, including naturalistic field observations and experimental studies of locomotor energetics. All are necessarily speculative, but they represent a changing perspective on human origins. Today, we see hominins as one of many peculiar radiations in primate history, the evolution of which should be explicable in terms of adaptation to ecological surroundings. This is in significant contrast to the more traditional views, in which nonhuman primates were, at best, stepping stones or, at worst, failed experiments on the road to humanity. This new view of hominin origins has come in part from an increased appreciation that the earliest bipedal hominins were neither like living apes nor like living modern humans. Rather, they were both morphologically and adaptively distinct from both and no doubt showed considerable diversity among the different genera and species.

Although it is important to see early hominins in the context of hominoid evolution, it is equally important to realize that, in the same way that they were not little people, they also were not just bipedal chimpanzees but the beginning of a new radiation of very different hominoids. It is this uniqueness of bipedalism, with only one extant species showing a commitment to this behavior, that makes reconstructing hominin origins so difficult (e.g., Cartmill, 1990). Human bipedalism is morphologically and behaviorally different from the occasional facultative bipedal behaviors occasionally seen in other primates. The morphological and behavioral commitment to bipedalism that characterized early hominins suggests unique ecological and historical circumstances

as well. Identifying the ecological and phylogenetic conditions surrounding hominin origins is a major challenge for paleoanthropologists and will be for many years to come. Ultimately, those factors can be identified only through a more complete fossil record that better documents the actual transformations and ecological conditions during early hominin evolution.

Humans differ from living apes in numerous morphological and behavioral features, and there has been a tendency in the study of human evolution to see all human features, including bipedalism, large brains, manipulative hands, tool use, and language, as integrally related to a single adaptive complex extending back to the origin of the hominin lineage (Darwin, 1871). Such an approach was reasonable when there was only one known hominin, *Homo sapiens*, and no fossil record of more primitive taxa lacking this complete suite of features. But such completely integrated models provide no insight into the evolution of these features individually. The fossil record of early hominins provides direct evidence that the features characterizing living humans are not inseparably linked but, rather, evolved in mosaic fashion over many millions of years.

Phyletic Relationships of Early Hominins

There are several long-standing debates, and many new ones, concerning the phyletic relationships among australopith species and the genus *Homo*. One rapidly changing new issue is the phylogeny of the very earliest hominins. Until fairly recently, most authorities agreed that *Au. afarensis* was the most primitive hominin species, ancestral to all later forms. Although *Au. afarensis* is the best known of the earliest hominins, *Sahelanthropus*, *Orrorin*, *Ardipithecus*, and *Australopithecus anamensis* are older and more primitive in many features. Apart from the possibility that *Au. anamensis* and *Au. afarensis* are a single lineage, there is no consensus as to how these earliest hominins are related to one another, but there is increasing evidence of a bushy phylogeny at the base of our family tree (Kappelman and Fleagle, 1995; Haile-Selassie et al., 2012).

Although the robust australopithecine species have traditionally been considered a single clade, the discovery of the Black Skull (KNM-WT 17000) led many researchers to argue that the robust species were not a natural group, but rather they evolved their dental and cranial features in parallel (Grine, 1988). There is still a diversity of views on this topic (Wood and Schroer, 2013), but the most robust analyses indicate that all three robust species are a monophyletic natural group (Strait et al., 1997, 2007; Strait and Grine, 2004; Mongle et al., 2019).

Perhaps the most unsettled issue in early hominin evolution is the relationship of different australopithecine genera and species to our own genus, *Homo*, and to *Paranthropus*. *Australopithecus afarensis, Au. africanus*, and *Au. sediba* have been advanced by various authorities (e.g., Johanson and White, 1979; Grine, 1988, 1993; Berger et al., 2010) as the species closest to the ancestry of *Homo*. Phylogenetic analyses have repeatedly identified all or part of the robust clade as the sister group of *Homo*. These phylogenetic analyses suggest the existence of an ancestor common to the *Paranthropus* clade and the *Homo* clade that has yet to be identified in the fossil record, or that *Au. africanus* or a similar taxon, such as *Au. sediba*, is the most likely ancestor of the *Homo* and *Paranthropus* clades (Strait et al., 1997, 2007; Strait and Grine, 2004; Mongle et al., 2019). Obviously, our understanding of early hominin phylogeny will continue to be modified as new fossils are recovered and new taxa are identified through more detailed analyses. At present, there is perhaps no group of primate fossils more complex and poorly sorted out than those identified as early members of the genus *Homo*.

Pleistocene Epoch

The Pleistocene is the epoch of human evolution. There is currently some disagreement about the date that should be assigned to the beginning of the epoch. Since the middle of the last century, the beginning of the Pleistocene was recognized as from approximately 1.7 million years ago; however, more recently, it has been established as 2.6 Ma, but many authorities still use the younger date, which will be followed here. The Pleistocene was characterized geologically by repeated glaciations of the Northern Hemisphere. The initial onset of dramatic cooling seems to have begun around 2.5 million years ago and there is evidence of another extreme cooling after 1 million years ago. Some researchers have argued for major shifts in the flora and fauna in Africa at 2.5 million years ago and have attributed these to worldwide changes from relatively warm, wet climates to cooler, drier climates (Vrba et al., 1996). Others find more gradual trends in faunas and reconstructed environments until approximately 1.8 million years ago, at the beginning of the Pleistocene, when there was a major appearance of secondary grassland habitats (Cerling, 1992; deMenocal, 1995; Reed, 1997).

In hominin evolution, the Pleistocene is characterized by the radiation and geographical expansion of the genus *Homo*, which began in the Late Pliocene and continues to the present, and by the extinction of the robust australopithecines approximately 1 million years ago. The correlation between major events in human evolution and patterns of global climate change is a topic of considerable interest and controversy (e.g., deMenocal, 2004, 2011; Behrensmeyer, 2006; Vrba, 2007; Cerling et al., 2011b).

Early *Homo*

The fossil record of our own genus begins in the Late Pliocene, before 2 million years ago. There are many different views about the history and biogeography of human diversity in the Pleistocene and no consensus on how many species of our genus should be recognized at any time period before about 30,000 years ago. However, the number of species in our genus is definitely growing, with many more likely to be discovered or recognized. A relatively conservative assessment suggests at least nine species, three of which are found in the earliest Pleistocene deposits of Africa (Table 17.1).

Compared to *Australopithecus* and *Paranthropus*, *Homo* is characterized by smaller molars and premolars and a more slender mandible. Throughout the evolution of the genus, there has been a trend toward reduction in the size of the cheek teeth. The anterior teeth, canines and incisors, are larger than those of *Paranthropus*. The cranium of *Homo* (Fig. 17.15) is characterized by a relatively larger brain and a smaller face than that of *Australopithecus*. The genus *Homo* resembles species of *Paranthropus* and differs from *Australopithecus* in having a flexed skull base, related at least in part to a relatively larger brain and smaller face.

There is evidence of considerable diversity in limb proportions and femoral anatomy among the earliest species of the genus *Homo*. Our genus seems to be characterized by a narrower thorax, a narrower ilium, relatively longer legs, and a larger femoral head than *Australopithecus*. The foot of *Homo* has shorter digits than the feet of more primitive hominins (Figs. 17.6 and 17.7).

Our genus has long been characterized by the presence of culture. The first appearance of stone tools in Africa precedes the appearance of *Homo* in the fossil record. Many authorities, nevertheless, argue that *Homo* is most closely associated with the manufacture and use of stone tools. The best-known early stone tools are crude choppers and scrapers, collectively called the Oldowan industry because of the original discovery at Olduvai Gorge, Tanzania (Fig. 17.16). Wear on the cutting edges indicates that these tools were used in a variety of activities, including butchering of small animals, trimming of leather, and preparation of plan remains. Although the core choppers were probably used for some activities, it is the smaller, razor-sharp flakes that are more effective in food processing.

The earliest specimen confidently attributed to the genus *Homo* is a mandible from Ethiopia dated approximately 2.8 million years ago (Villmoare et al., 2015). This specimen has not been allocated to any particular species. However, by 1.8 million years ago, there was a considerable diversity of advanced hominins present in eastern and southern Africa (e.g., Leakey et al., 2012). Three species are commonly recognized: *Homo habilis*,

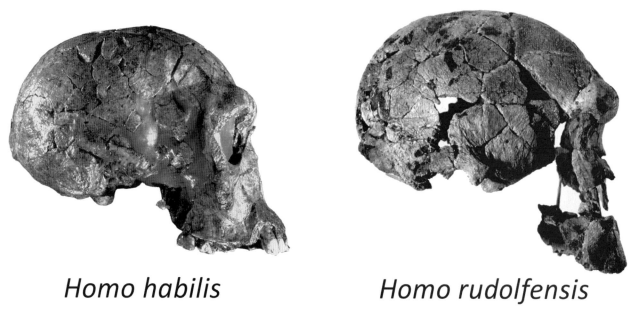

Homo habilis   Homo rudolfensis

FIGURE 17.15   Comparison of *Homo habilis* (KNM ER 1813) and *Homo rudolfensis* (KNM ER 1470). *(Courtesy of M. Leakey and the Turkana Basin Institute.)*

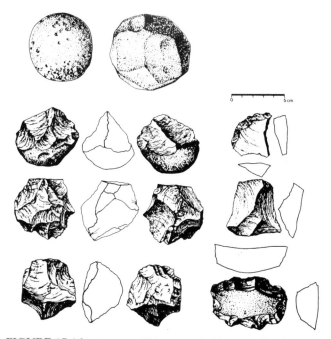

FIGURE 17.16   Primitive Oldowan tools. *(Courtesy of Kathy Schick.)*

Homo rudolfensis, and *Homo erectus* (or *Homo ergaster*), and probably an additional species from southern Africa (Fig. 17.18; Grine et al., 1993, 2009; Grine, 2005; Curnoe, 2010). Although the details of the systematics of *Homo* in the Early Pleistocene are far from resolved, it seems clear that the origin of this genus was very bushy, with considerable morphological and ecological diversity soon after its initial appearance, a common phenomenon in primate evolution.

Homo habilis first appeared around 2.3 million years ago in the latest Pliocene and earliest Pleistocene of eastern Africa, where it is best known from Olduvai Gorge in Tanzania and the Turkana Basin of northern Kenya and southern Ethiopia (Fig. 17.15). Compared to *Australopithecus*, *H. habilis* has narrower premolars and first molars, a narrower mandible, a more coronal orientation of the petrous part of the temporal bone, and delayed eruption of the canines. The average cranial capacity is larger than that of the more primitive *Australopithecus* and smaller than that of most *H. erectus*, although individual specimens overlap with both taxa. The hand bones are more robust than those of later hominins and suggest that this species retained some suspensory abilities. The foot (Fig. 17.17) is more advanced than that of *Australopithecus* and resembles the foot of extant humans in most features, suggesting a similar bipedal gait (Susman, 1988). The skeletal proportions of *H. habilis* are known largely from unassociated elements and fragmentary bones. It has been suggested that they were more similar to *Australopithecus* than to *Homo erectus* in having long forelimbs, but there is little evidence to support this.

The sites that have yielded *Homo habilis* (and *Paranthropus boisei*) regularly include Oldowan stone tools (Fig. 17.16). Indeed, this species was originally named on the assumption that it was a tool user. Some of the broken animal bones found in association with Oldowan tools and *H. habilis* fossil remains show cut marks that appear to have been made by the stone tools. Furthermore, the concentrations of stone tools and broken bones suggest to some workers that animal parts were transported to the role of meat eating and hunting in early hominins has been overemphasized in past archaeology. Unfortunately, plant foods leave few fossilized remains, so it is not possible to reconstruct the relative proportions of meat and plant material in the diet.

Homo rudolfensis is a species of early hominin that is roughly contemporary with *Homo habilis* (Wood, 1991). It is known mainly from the Turkana Basin of northern Kenya and possibly from Malawi as well (Schrenk et al., 1993). *Homo rudolfensis* differs from *H. habilis* in having a flatter, broader face and broader cheek teeth with more complex crowns and thicker enamel (Fig. 17.15). The holotype and best-known specimen is a cranium, KNM-ER 1470, but it is also known from a well-preserved mandible (Leakey et al., 2012). The limbs of *H. rudolfensis* are not well known, but the species seems to be characterized by longer hindlimbs and a larger femoral head than those of *H. habilis*. Some workers argue that *Homo habilis* and *Homo rudolfensis* are better considered as advanced australopiths because they are much more similar to that radiation than *Homo erectus* from the same time period (Fig. 17.19; Walker and Shipman, 1996; Wood and Collard, 1999a,b; Collard and Wood, 2007; Wood, 2009). However, both are almost certainly near the ancestry of *Homo erectus* and other later hominins, and most authorities keep them in our genus.

Homo erectus

In addition to the "more primitive" species of *Homo*, such as *H. habilis* and *H. rudolfensis*, there was a more advanced hominin in Africa at the beginning of the Pleistocene. This species is ***Homo erectus***, which has long been known from the Middle Pleistocene of Asia (Rightmire, 1990). The African version of *Homo erectus* is considered by some authorities as a separate species, *H. ergaster*, with distinctive cranial proportions (e.g., Wood, 1992). However, analyses comparing African and Asian samples have not been able to find consistent differences between them. This has led most authorities to attribute both the African and the Asian fossils to *Homo erectus* (e.g., Antón, 2003; Baab, 2010). Equally difficult is the proper species designation for the ever-growing collection of Early Pleistocene hominins from the site of Dmanisi in Georgia (e.g., Gabunia et al., 2001; Rightmire and Lordkipanidze, 2009, 2010; Lordkipanidze et al., 2007). Originally attributed to *Homo erectus*, they were subsequently placed in a separate species, *H. georgicus*. More recent analyses have suggested that there may be more than one taxon present at Dmanisi. For the present, however, *H. georgicus* is best considered a population of a single, geographically and temporally diverse species, *Homo erectus*.

It is not surprising that the systematics of "*Homo erectus*" is the subject of so much debate. This "species"

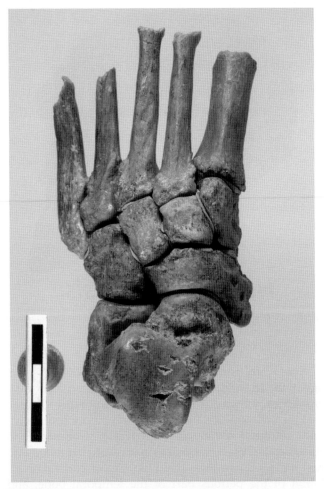

FIGURE 17.17   The foot of *Homo habilis* from Olduvai Gorge. *(Courtesy of R. Susman.)*

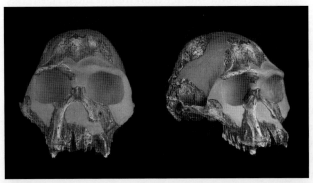

FIGURE 17.18   *Homo* sp. from South Africa. *(Courtesy of F. Grine.)*

sites by hominins (Isaac, 1983); others see both the cut marks and the accumulations as possible results of geological processes (Binford, 1981). Isaac (1983) has argued that these concentrations indicate the emergence of some type of home base or "central-place foraging" and possibly food sharing among early hominins. Whether the animal parts are the result of hunting or of scavenging activities cannot be determined. It seems likely that the

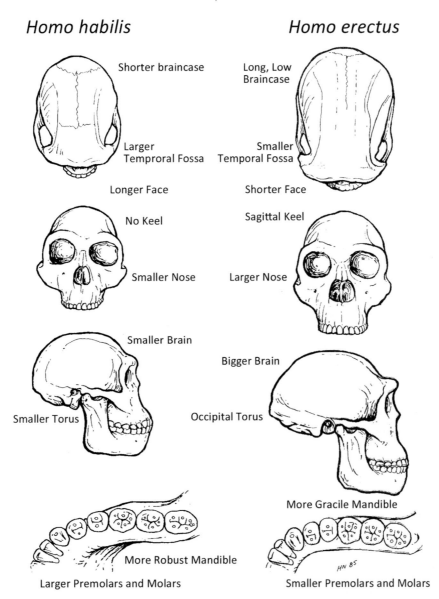

Homo habilis   Homo erectus

Shorter braincase   Long, Low Braincase

Larger Temproral Fossa   Smaller Temporal Fossa

Longer Face   Shorter Face

No Keel   Sagittal Keel

Smaller Nose   Larger Nose

Smaller Brain   Bigger Brain

Smaller Torus   Occipital Torus

More Gracile Mandible

More Robust Mandible

Larger Premolars and Molars   Smaller Premolars and Molars

FIGURE 17.19   Cranial and dental characteristics of *Homo habilis* and *Homo erectus*.

is characterized by many factors that should lead to morphological diversity (e.g., Antón, 2003, 2013; Antón et al., 2007). The geographic distribution of *H. erectus* exceeds that of any other primate except *Homo sapiens* from Africa through western Asia to eastern and southeastern Asia, including populations isolated on the island of Java. The temporal span of fossils attributed to this taxon covers nearly 2 million years. Attempts to divide the complex pattern of morphological diversity into any simple scheme of two or three species have never yielded satisfactory results. Thus, placing everything in a single widespread and variable taxon is, perhaps, unsatisfactory, in that it does not acknowledge the morphological diversity that is present, but it is, nevertheless, a workable allocation (e.g., Tattersall, 2007).

Compared to *Australopithecus* or *H. habilis*, *H. erectus* has smaller cheek teeth and a more slender mandible, in keeping with the general trend of masticatory reduction within the genus (Fig. 17.19). Brain size is larger than in earlier hominins, with considerable variation among specimens. The cranium of *H. erectus* is characterized by very thick bones (less so in some early African specimens); a long, low vault with sagittal keeling; projecting brow ridges; and a prominent occipital torus. The face of this species was relatively broad and had a large nasal opening. *Homo erectus* seems to have had a mean body size larger than that of australopithecines and less sexual dimorphism.

The most complete skeleton of *H. erectus* is from the west side of Lake Turkana in northern Kenya from deposits dated at approximately 1.6 million years ago (Brown et al., 1985; Walker and Leakey, 1994; Fig 17.20).

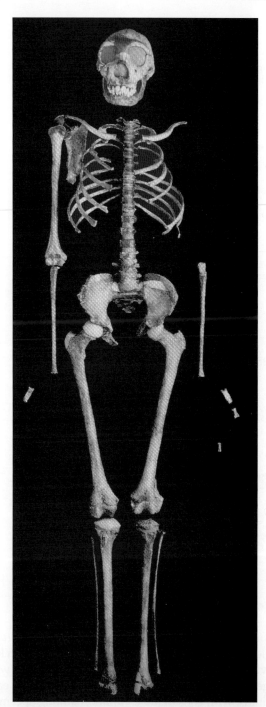

FIGURE 17.20   The 1.6-million-year-old skeleton of a *Homo erectus* boy from West Turkana, Kenya. *(Photograph by D. Brill, National Geographic Society.)*

The limb proportions are similar to those of *H. sapiens*, but most of the limb bones are more robust. The chest is more conical, as in apes. The femoral neck is long, as in *Australopithecus*, but the femoral head is large, as in modern humans. It had a relatively narrow trunk and slender limbs, associated with an equatorial climate. Populations of *Homo erectus* in more northern latitudes seem to have had shorter limbs and broader trunks (Ruff, 1993).

Adaptations of *Homo erectus*

In Africa, *H. erectus* remains are associated with Acheulean hand axes; in Asia, including Dmanisi, the species is found with more primitive chopping tools, similar to earlier Oldowan artifacts. *Homo erectus* is the first fossil primate with a substantial archaeological record. The species developed a wide range of stone implements for different purposes, many of which are still manufactured and used today by modern humans. Archaeological sites attributed to *H. erectus* are widespread and diverse. Some seem to have been camps, others were sites of animal kills, and others were butchering sites. Some of the later sites show evidence of simple structures. The variation in size of the camps suggests a social organization of individual families that sometimes camped (and presumably foraged) alone and at other times joined with other families, a social structure similar to that of living hunter–gatherers.

Homo erectus hunted and scavenged, and successfully preyed on a variety of medium-sized and large mammals, including elephants, antelopes, horses, and deer. At later sites, the archaeological evidence indicates that they exploited virtually all available animals in the area. Like both their primate forebears and living hunter–gatherers, *H. erectus* probably relied on plant parts of some sort for most of their diet. There are remains of berries at Zhoukoudian and other sites. As with other hominins, this part of the diet of *H. erectus* is very difficult to reconstruct, and our view of their subsistence behavior is certainly distorted by an overemphasis on hunting because of abundant animal bones.

Because of the many ways in which *Homo erectus* resembles modern humans and differs from earlier hominins, several authors have suggested that this species was probably the first hominin to exhibit two major adaptations that distinguish humans from all other animals: long-distance running and cooking. Bramble and Lieberman (2004) have argued that the ecological success of *Homo erectus* was in part due to their abilities as long-distance runners. Running involved different musculoskeletal adaptations from walking, and many of these first appear in *Homo erectus* (Fig. 17.21). In the view of Lieberman and colleagues (2009), these adaptations would have enabled *Homo erectus* to be an effective predator with a unique niche in the Early Pleistocene of Africa.

Wrangham and colleagues (Wrangham, 2009; Wrangham and Carmody, 2010) have argued that *Homo erectus* was also the first hominin to regularly cook its food. Cooking of food dramatically increases the energy that an animal can extract from a given amount of either meat or vegetables, thereby greatly reducing the amount of time an individual needs to spend in procuring food. It also reduces the effort and time involved in preparing food through chewing. Many of the morphological features that distinguish *Homo erectus* from earlier homi-nins, such as reduced breadth of the abdomen and reduction in the size of the teeth, accord with what would be expected to result from an ability to cook food (Fig. 17.22). A major problem with the view that *Homo erectus* cooked its food is the difficulty in identifying solid evidence for controlled fire in the archaeological record before the Middle Pleistocene. There is clear archaeological evidence for human control of fire at the site of Gesher Benot Ya'aqov in Israel, dated to almost 800,000 years ago, and at Wonderwerk Cave in South Africa approximately 1 million years ago as well (Berna et al., 2012). Earlier documentation of fire is more debatable, but evidence of controlled fire has been presented for several sites in South Africa, including Swartkrans, dated to nearly 2 million years ago (Brain, 1993).

Homo floresiensis, Homo luzonensis, and Homo naledi

Homo floresiensis is known from a collection of at least four individuals, preserving most of the skeleton, from the Liang Bua cave on the Asian island of Flores (Fig. 17.23). The hominin fossils are dated to between 60,000 and 100,000 years ago (Sutikna et al., 2016), but archaeological excavations have recovered stone tools on Flores dated to more than a million years ago. *Homo floresiensis*, often dubbed "The Hobbit," was a small hominin, slightly over 1 m in height, with estimated body sizes of 25 to 35 kg. In size, it was similar to Lucy, the well-known fossil of *Au. afarensis*, and smaller in both stature and brain size than any modern humans, including various pygmy populations. *Homo floresiensis* was originally described as a dwarf taxon derived from *Homo erectus*, which lived in Southeast Asia for much of the last 1.8 million years. Others have argued that the Liang Bua fossils are just modern humans suffering from some growth disorder or congenital deformities such as microcephaly or dwarfism, although these arguments have largely been disproven.

Even more recently, *Homo luzonensis* has been described as another small-bodied hominin species

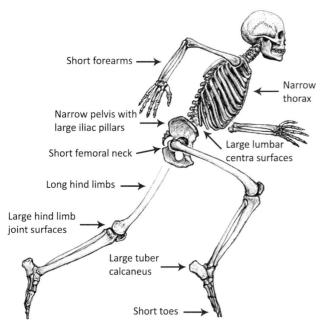

FIGURE 17.21   Skeletal features indicating adaptation for running. *(Adapted from Bramble, D.M., Lieberman, D.E., 2004. Endurance running and the evolution of Homo. Nature 432, 345–352.)*

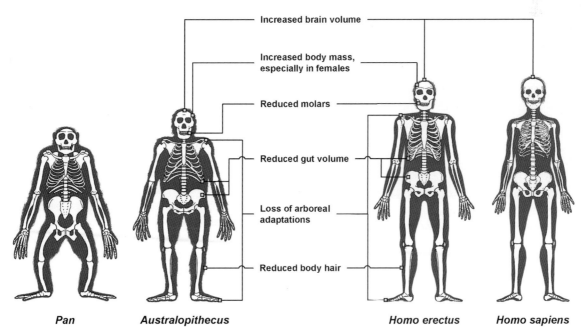

FIGURE 17.22   Comparison of the skeletons of *Pan*, *Australopithecus*, *Homo erectus*, and *Homo sapiens* showing features of *Homo* that could be related to the use of cooked food. *Drawing by L. Betti-Nash. (Courtesy of R. Wrangham.)*

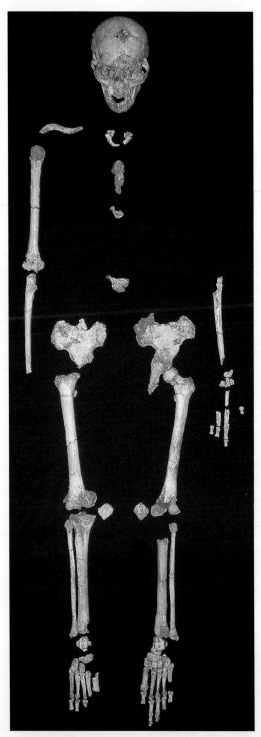

FIGURE 17.23 The "Hobbit," *Homo floresiensis* from Flores in southeast Asia. *(Courtesy of William Jungers.)*

possessing distinctly shaped premolars and the retention of two to three premolar roots. Despite the young age of the known fossils, in aspects of dental, cranial, and particularly postcranial morphology, *H. floresiensis* and *H. luzonensis* are most similar to fossils of *Australopithecus* or early *Homo* from East Africa and Dmanisi, rather than to either modern humans or Asian *Homo erectus* (Aiello, 2010; Jungers, 2013; Detroit et al., 2019).

The newly discovered diversity of Pleistocene hominin taxa is not limited to Asia. In the South African Rising Star Cave system, dated between 335 and 235 ka, a large number of fossil bones have recently been described as belonging to a new species, **Homo naledi** (Berger et al., 2015). *Homo naledi* combines features seen in early *Homo* such as a relatively small brain (513 cc), small teeth, a relatively high and rounded cranium, large brow ridges, curved phalanges, and a primitive shoulder joint, with those seen in later *Homo* species such as a more modern looking hand and foot. It also shares a more flexed occipital and occipital torus with *H. erectus*.

Like the Denisova fossil (see below), *H. floresiensis*, *H. luzonensis*, and *H. naledi* suggest that there is much about the diversity and biogeography of the genus *Homo* during the past 2 million years that we do not understand.

Homo heidelbergensis

In deposits dating from much of the Middle Pleistocene, approximately 700,000 to 130,000 years ago, we find archaic hominins from Africa and Europe that are more modern than *Homo erectus* but which lack the diagnostic features of either Neandertals or modern humans. These are generally placed in a separate taxon, **Homo heidelbergensis** (Rightmire, 1998, 2007; Stringer, 2012). Fossils attributed to this species are known from southern Africa (Broken Hill or Kabwe, and Berg Aukas), eastern Africa (Bodo), Greece (Petralona), Germany (Heidelberg), and possibly China (Dali). Less complete remains probably attributable to this taxon are known from many other sites in Europe and Africa. These include remains of a young individual from the TD6 level at the Gran Dolina site in Atapuerca, Spain that has been described as **Homo antecessor** (Bermudez de Castro et al., 1997). The cranial similarities between the African and some European fossils are striking. However, many researchers have argued that the European specimens should be separated from others because the boundary between some fossils attributed to *H. heidelbergensis* and those considered as early Neandertals is difficult to identify. In that scheme, African fossils intermediate between *Homo erectus* and *Homo sapiens* would be placed in the taxon *Homo rhodesiensis*, the type specimen of which is the cranium from Kabwe, Zambia. Here, we are using *H. heidelbergensis* in the broader sense.

Compared to *Homo erectus*, *H. heidelbergensis* has smaller teeth and a mandible with a broader ramus and a deeper corpus. The cranium (Fig. 17.24) is characterized by a

dating to a similar time period as *H. floresiensis*, between 80,000 and 50,000 years ago (Detroit et al., 2019). Although only known from dental remains and a few isolated postcranial bones, *H. luzonensis* is slightly smaller than *H. floresiensis* in overall dental size but shares primitive postcranial features such as curved fingers and toes. It differs from other *Homo* species in

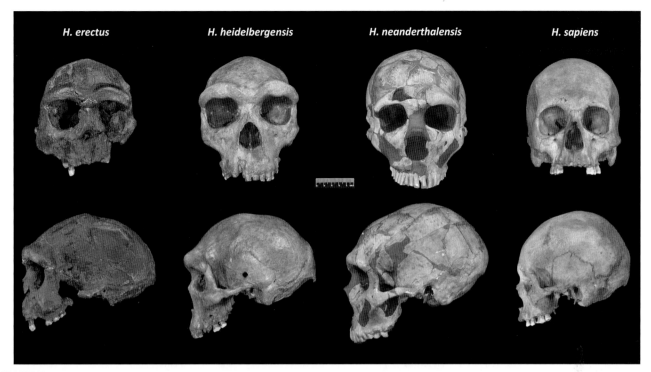

FIGURE 17.24 Facial and lateral views of the crania of *Homo erectus*, *H. heidelbergensis*, *H. neanderthalensis*, and *H. sapiens*, showing distinctive features. All pictures from The Natural History Museum London. *(Courtesy of C. Stringer.)*

greater cranial capacity, a divided brow, and an upright nasal aperture, among other features (Rightmire, 1996). Many of the better-known crania of *H. heidelbergensis* are enormous. This was a very large species of hominin, with estimated body sizes of roughly 100kg, based on orbit dimensions (Kappelman, 1996) and postcranial remains (Grine et al., 1995; Ruff et al., 1997).

Postcranial elements of *H. heidelbergensis* are rare. There are several elements from Europe; from southern Africa, there is a pelvis from Kabwe and a femur and parts of an ulna and radius, all possibly attributed to this taxon (Grine et al., 1995). The femur is massive.

Fossils attributed to *H. heidelbergensis* are associated with Acheulian hand axes in both Africa and Europe. Like earlier *Homo erectus*, these archaic humans were hunters and gatherers or scavengers (Klein, 2009).

Homo neanderthalensis

Neandertals, now commonly placed in a distinct species, **Homo neanderthalensis**, are a distinct group of Middle to Early Upper Pleistocene hominins from Europe and the Middle East. Neandertals seem to have originated among European populations of *Homo heidelbergensis*, and Middle Pleistocene hominins such as those from Atapuerca, Spain (Arsuaga et al., 1993), are often considered transitional between the two species (e.g., Stringer, 2012). The latest Neandertals are contemporaneous with modern humans in Europe approximately 35,000 years ago. Neandertals are characterized by molars with a large pulp cavity, a low mandibular ramus that is separated from the last molar by a gap, a prognathic face with a large and unusually formed nasal region, large brow ridges that arch over both orbits, and a large brain in a long and low, but rounded cranium with a prominent bun posteriorly (Figs. 17.24 and 17.25). Their limbs are quite robust, with shortened distal elements (Fig. 17.26).

Neandertals were hunter–gatherers that used a wide range of Mousterian flake stone tools as well as wooden implements. They show considerable diversity in both prey selection and tool use in different geographical areas and temporally within regions. There is evidence that they used free-standing shelters as well as caves. They used fire, although cooking hearths are relatively rare. They are the first hominins known to actively bury their dead.

After a quarter-millennium of considerable ecological and demographic success, Neandertals disappeared approximately 30,000 years ago. However, it seems that their genes live on in low frequencies among some populations of modern humans (Green et al., 2010; Currat and Excoffier, 2011). The nature of the ecological and demographic relationships between Neandertals and modern humans and what led to the disappearance of Neandertals as a distinct group of hominins are among the most hotly debated issues in human evolution, as their disappearance roughly coincides with the initial appearance in Europe of modern humans belonging to our own species, *Homo sapiens*.

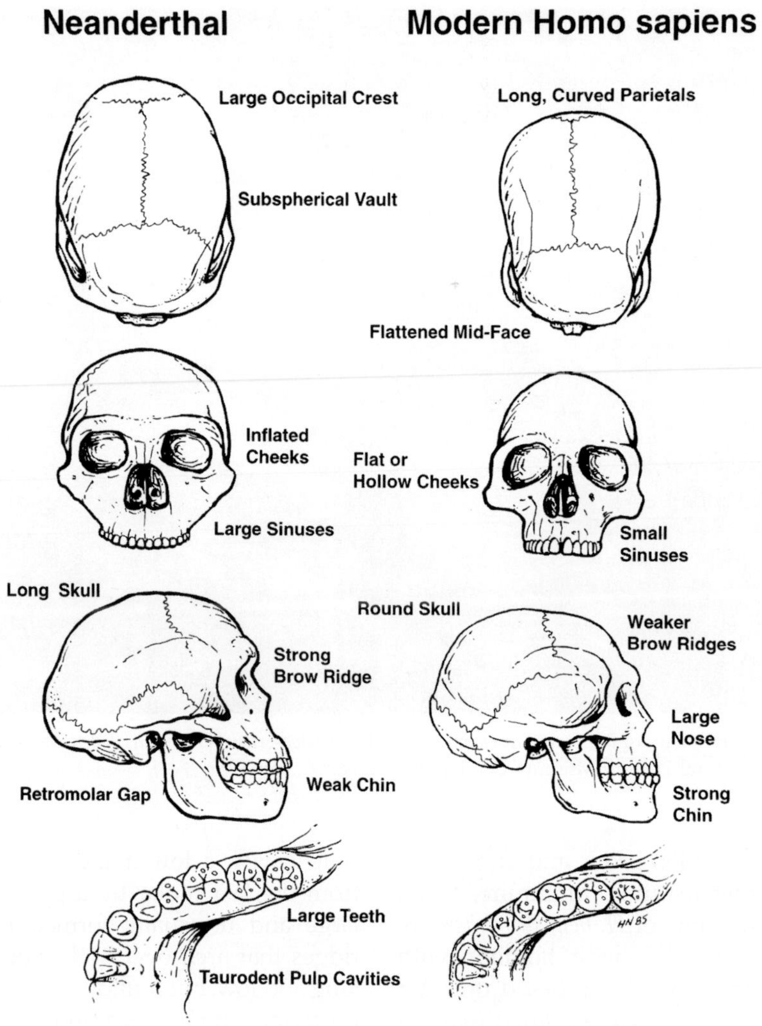

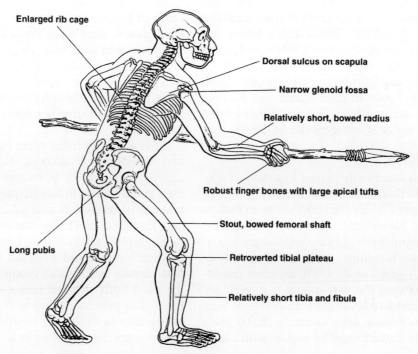

Neanderthal

Modern Homo sapiens

Large Occipital Crest

Long, Curved Parietals

Subspherical Vault

Flattened Mid-Face

Inflated Cheeks

Flat or Hollow Cheeks

Large Sinuses

Small Sinuses

Long Skull

Round Skull

Strong Brow Ridge

Weaker Brow Ridges

Retromolar Gap

Large Nose

Weak Chin

Strong Chin

Large Teeth

Taurodent Pulp Cavities

HN 85

FIGURE 17.25   Comparative dental and cranial features of *Homo neanderthalensis* and *Homo sapiens*.

Enlarged rib cage

Dorsal sulcus on scapula

Narrow glenoid fossa

Relatively short, bowed radius

Robust finger bones with large apical tufts

Stout, bowed femoral shaft

Long pubis

Retroverted tibial plateau

Relatively short tibia and fibula

FIGURE 17.26   The skeleton of *Homo neanderthalensis*, with distinctive features indicated.

Denisovans

In 2010, the finger bone and a molar tooth of a fossil hominin were recovered from a cave roughly 40,000 years old in Denisova, Russia (Reich et al., 2010). Since the cave was filled with tools commonly found associated with Neandertals, it was assumed that the bone belonged to a late population of Neandertals. However, analysis of ancient DNA from the bone indicated that it had a genetic composition that is distinct from that of either modern humans or any of the Neandertals that had previously been sequenced. Thus, it is evident there was another population of hominins in Eurasia contemporary with Neandertals and modern humans, known as Denisovans. A recently described skull called **Homo longi** from Harbin, China may, in fact, be that of a Denisovan (Ji et al., 2021). It possesses facial features seen in *H. heidelbergensis* and Neandertals such as a large face with large nasal aperture, massive and thick double-arched brow ridges, a low-sloping forehead, large molars, and a large brain. Subsequent and ongoing analyses of the ancient DNA of Neandertals, Denisovans, and modern humans indicate a very complicated history that is far from understood at present (Meyer et al., 2012; Stringer and Barnes, 2015; Teixeira et al., 2021).

Homo sapiens

Fossils attributable to **H. sapiens** because of their similarity to living populations of our species first appear in Africa around 200,000 to 300,000 years ago. The earliest are from the Kibish Formation of southern Ethiopia (Leakey et al., 1969; Stringer and McKie, 1996; McDougall et al., 2005; Fleagle et al., 2008; Vidal et al., 2022; Fig. 17.27), Jebel Irhoud cave in Morocco (Hublin et al., 2017),

and from the site of Herto in the Middle Awash region of Ethiopia (White et al., 2003). The exact timing, the location, and the pattern of morphological transformations involved in the origin of our species are difficult to reconstruct, as there are many different Late Pleistocene fossils from many parts of Africa that show various mosaics of archaic and modern features (Trinkaus, 2005; Gunz et al., 2009; Hublin et al., 2017). Our species is distinguished from earlier members of the genus *Homo* by our smaller teeth, vertical mandibular ramus, the development of a chin, a shortened face, reduced brow ridges, and high vertical forehead (Figs. 17.24 and 17.25). Compared to Neandertals, we have smaller brains. Our limbs are more gracile and have much thinner bony shafts (Churchill, 1996). The markedly less robust skeleton in modern *H. sapiens* in the Late Pleistocene seems to be related to the replacement of physical exertion with technological skill.

For the first 100,000 to 200,000 years of our existence as a distinct species, *Homo sapiens* appears to have been largely restricted to Africa, although some temporary expansion around the Mediterranean appears to have taken place between 180,000 and 200,000 as documented by modern-looking specimens in Israel and Greece (Hershkovitz et al., 2018; Harvati et al., 2019). Sometime between 100,000 and 50,000 years ago, however, modern humans left Africa and permanently expanded their range into Asia and Europe. It seems likely that this was associated with major advances in human cultural abilities, including more diverse tools and the development of art (e.g., Klein, 2009). It is not clear whether the final expansion of *Homo sapiens* out of Africa took place in one or several waves, or what route, or routes, the dispersal followed (e.g., Lahr and Foley, 1994). The fossil record of

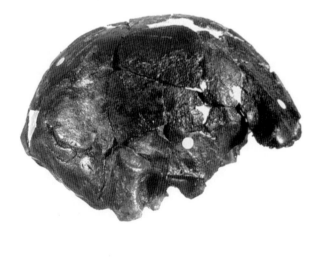

Omo 1   Omo 2

FIGURE 17.27   Cranial remains of fossils from the Kibish Formation of southern Ethiopia attributed to early *Homo sapiens*. *(Photos courtesy of M. Day.)*

Homo sapiens from Africa between 100,000 and 40,000 years ago is relatively poor, but a cranium from Hofmeyr in southern Africa (Fig. 17.28) dated to 36,000 years ago is strikingly similar to the earliest *Homo sapiens* in Europe and may well sample a part of the dispersing population (Grine et al., 2007). Recent genetic analyses have also identified southern Africa as the likely source of non-African modern humans.

Although the earliest *H. sapiens* were probably similar to earlier hominins in being hunters and gatherers, many populations specialized in fish, shellfish, small mammals, and undoubtedly many other foods before developing the sophisticated habits of agriculture and animal husbandry in the very recent past. The technological skill of our species has enabled us to exploit virtually all available habitats on earth, from the more traditional woodlands and savannahs to tropical forests, oceanic islands, and the Arctic. From as early as about 50,000 years ago, *Homo sapiens* have been distinguished from more archaic hominins in the range of cultural and artistic abilities (e.g., Klein, 1992). *Homo sapiens* considerably extended the geographic range of *H. erectus* and earlier hominins by successfully colonizing areas that had previously been beyond the range of catarrhines, such as North America and South America, islands such as Madagascar that had never been colonized by higher primates, and Australia, which had never seen a primate. We are still expanding our range. Our own species, which has been studied more thoroughly than any other primate species, demonstrates a complex history of geographic differentiation and subsequent recombination.

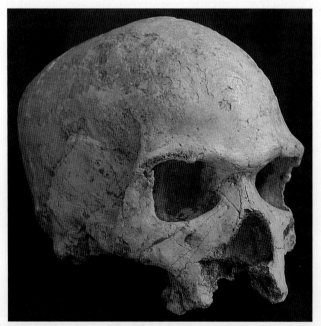

FIGURE 17.28 A cranium from Hofmeyr in southern Africa is very similar to that of the modern humans that dispersed to Eurasia approximately 50,000 years ago. *(Courtesy of F. Grine.)*

Homo sapiens as an Adaptive Radiation

In some respects, it seems inappropriate to discuss a single species as an adaptive radiation. Yet, from a primate perspective, our species is remarkable in both the geographic spread and the diversity of ecological conditions we have occupied over the past 100,000 years. This has been accomplished with relatively few morphological differences, such as skin color and various blood polymorphisms, in association with different environments and perhaps other factors, such as random genetic drift and geographic isolation (e.g., Harcourt, 2012). Like macaques, humans are an example of what one scientist has called "specialized generalists" (Rose, 1983), or perhaps we are best considered a weed species (Richard et al., 1989) that has a morphological and especially a behavioral repertoire (culture) and is able to occupy a remarkable diversity of ecologies.

The evolutionary specializations that have permitted modern humans to exploit such a wide range of environments in many different ways are those that characterize the genus: a large brain that enhances learning and memory, our uniquely proportioned hand with its very mobile thumb, and our uniquely shaped vocal tract, which, in conjunction with our brain, permits a wide range of linguistic communication.

Human Phylogeny

Perhaps no aspect of primate evolution has seen more surprises in the last decade than the paleontological and genetic record of the evolution of the genus *Homo*. It is generally accepted that there were at least three, and maybe more, species of *Homo* living in Africa approximately 2 million years ago. However, there is no consensus on how these species are related to one another. Of these Early Pleistocene hominins, it has long been assumed that *Homo erectus* is the species that gave rise to later hominin species and that this was the hominin that first left Africa. However, the fossils from Dmanisi and the primitive morphology of *Homo floresiensis* and *Homo luzonensis* suggest that there may have been a more primitive taxon in Asia in the Early Pleistocene and that *Homo erectus* may have originated outside Africa.

The geographically widespread hominins from the Middle Pleistocene are also poorly understood. In the past, they have been variously considered as late *Homo erectus*, "early (archaic) *Homo sapiens*," or several, perhaps many, distinct species. Now, they are often placed in a single species, *Homo heidelbergensis* (e.g., Rightmire, 1998, 2007; Stringer, 2012). In this scheme, *H. heidelbergensis* is something of a primitive, wastebasket taxon that likely includes the ancestors of *Homo sapiens* and *Homo neanderthalensis* (Fig. 17.29, *right*).

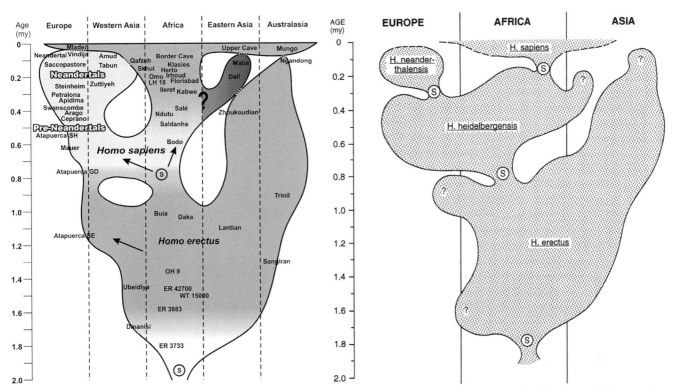

FIGURE 17.29 Alternative views of speciation events in human evolution since the Middle Pleistocene. On the left, Bräuer (2012) recognizes only two species of the genus *Homo* in the last million years. The figure on the right, from Rightmire (e.g., 1998), recognizes *Homo heidelbergensis* as a taxon that is ancestral to both *Homo neanderthalensis* and *Homo sapiens*. Other authors recognize even more taxa. Map lines delineate study areas and do not necessarily depict accepted national boundaries. *(Courtesy of the authors.)*

Alternatively, some authors recognize two taxa: *H. heidelbergensis* for the European forms ancestral to *H. neanderthalensis* and *H. rhodesiensis* in Africa for forms ancestral to *Homo sapiens* as discussed above (Fig. 17.30). However there seem to be gradual (but mosaic) transitions between time-successive populations within both geographical areas: *H. heidelbergensis* and *H. neanderthalensis* in Europe (Hublin, 2009), and *H. rhodesiensis* and *H. sapiens* in Africa (Bräuer, 2008, 2012; Mbua and Bräuer, 2012; Hublin, 2009; Hublin et al., 2017). Accordingly, some authors recognize only a single species of *Homo*, *Homo sapiens*, for all populations (apart from *H. floresiensis*, *H. luzonensis*, *H. naledi*, and the Denisovans) in addition to *Homo erectus* (Fig. 17.29, *left*; Bräuer, 2008, 2012). In other areas, such as the Levant or Asia, it seems that very different populations or species of hominins persisted synchronously or alternately for long periods of time. Moreover, recent genetic studies suggest that the phylogenetic relationships among Middle Pleistocene hominins and later, more well-defined groups, such as Neandertals and anatomically modern *Homo sapiens*, are complex and involved gene exchange between different lineages. The geographical and phylogenetic history of humans in the Middle Pleistocene is a long way from being resolved (Figs. 17.29–17.31).

One of the most longstanding and well-publicized debates in later human evolution is the relationship

between *Homo sapiens* and Neandertals, here recognized as a distinct species, *H. neanderthalensis*. Information from a wide range of disciplines, including paleontology, genetics, archeology, and geochronology, has been brought to bear on this debate. Indeed, the more we learn, the more complex the likely scenarios turn out to be. Morphological studies have consistently supported the view that Neandertals as a group can be readily distinguished from all *Homo sapiens* (e.g., Lahr, 1996; Harvati, 2007). Genetic studies regularly point to a genetic origin for our species between 100,000 and 200,000 years ago, and early studies of ancient DNA from Neandertals indicated that the Neandertal lineage had been distinct from modern humans for over 500,000 years (e.g., Krings et al., 1997; Ward and Stringer, 1997). However, recent studies, based on a more complete Neandertal genome, indicate that there was definitely some interbreeding between the modern human and Neandertal lineages (Green et al., 2010). While the discovery of interbreeding between modern humans and Neandertals was at first surprising, it is important to place it in the broader context of primate evolution. The more we learn about speciation and the lineage-splitting process among our living primate relatives, the more it becomes clear that hybridization between well-defined species and even genera is a relatively common

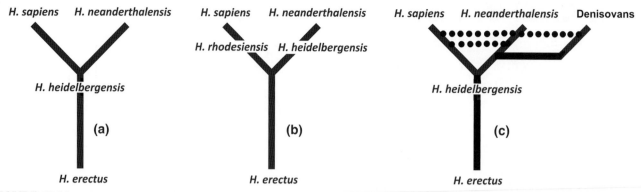

FIGURE 17.30   Alternative views of the position of *Homo heidelbergensis* in human evolution. (a) *H. heidelbergensis* is broadly ancestral to both *H. sapiens* and *H. neanderthalensis*. (b) *H. heidelbergensis* in Eurasia is specifically ancestral to *H. neanderthalensis* while another "archaic" species, *H. rhodesiensis*, is ancestral to *H. sapiens* in Africa. (c) *H. heidelbergensis* is broadly ancestral to *H. sapiens*, *H. neanderthalensis*, and the Denisovans. Dotted lines illustrate limited gene flow between *H. sapiens*, *H. neanderthalensis*, and Denisovans.

occurrence (Tung and Barreiro, 2017). Thus, we should probably not be surprised that the speciation process among hominins involved hybridization as well.

Hominin Evolution in Perspective

By any criteria, the fossil record of human evolution (Fig. 17.31) is better documented than that of any other group of taxa in the order Primates. Since our lineage diverged from that of our nearest extant relatives, the chimpanzees and bonobos, sometime in the past 10 million years, we have littered the fossil record with thousands of fossils, including many skulls, and numerous skeletons, stretching back to the latest Miocene in Africa and from later time periods on five other continents. These have been placed in seven genera and twenty or more species. In a primate perspective, human evolution is filled with debates and unresolved issues precisely because we have such a rich fossil record and can ask detailed questions that would never be raised about the evolutionary history of a species of fossil galago or pitheciine monkey. In addition, galagos and pitheciine monkeys don't write and read textbooks!

Thus, despite many unresolved questions about the details of where to draw the boundary between *Australopithecus* and *Homo* or how many species should be recognized between *Homo heidelbergensis* and *Homo sapiens*, there are many aspects of the broader picture on human evolution that are evident from this collection of fossils. One pattern is that human evolution was not a linear sequence of anagenetic species stretching back over seven million years. Rather, at virtually all time periods older than about thirty thousand years ago for which we have a fossil record, there was a diversity of species exhibiting a diversity of morphologies indicating different adaptations in diet, and for the past couple of million years, in culture, as exhibited by archaeological remains. The fact that there is but a single extant species today emphasizes the fact that hominins have been

characterized by both numerous radiations and numerous extinctions of not only individual species, but long-lived, successful lineages such as the robust australopiths or Neandertals.

We also have a relatively good idea of the timing and sequence of the initial appearance of the features that distinguish *Homo sapiens* from our ape relatives. Thus, our nondimorphic canines and some type of bipedalism extend back to the earliest recognizable members of our lineage, in the latest Miocene and early Pliocene, although both underwent additional modifications through time. Recognizable stone tools and increased brain size came later in the Pliocene and early Pleistocene, again, with ongoing changes in both during the past two million years. The most significant feature of modern humans that has thus far proved impossible to clearly date from the fossil record is language; cooking and the use of fire are also difficult to clearly pin down.

In terms of biogeography, all current zoological and paleontological evidence indicates that our lineage originated in Africa and was restricted to that continent until roughly two million years ago. The genus *Homo* is probably also of African origin, although the presence of primitive members of our genus in the Republic of Georgia and on the Islands of Flores and the Philippines raises the possibility of very early dispersals. Finally, both paleontological and genetic evidence indicates that our own species *Homo sapiens* originated in Africa and subsequently spread to other continents, first to Eurasia and Australia, much later to North and South America, and most recently to Antarctica and the moon for short visits. Genetic studies indicate that there were other taxa and other dispersals that have not previously been identified by archaeologists and paleontologists, and also suggest that fossil species interbred.

As with all areas of zoology and paleontology, the study of human evolution is growing at an incredible pace as new discoveries and new techniques continue to provide surprises. Thus, in light of the rich fossil record that we already have available and the extraordinary

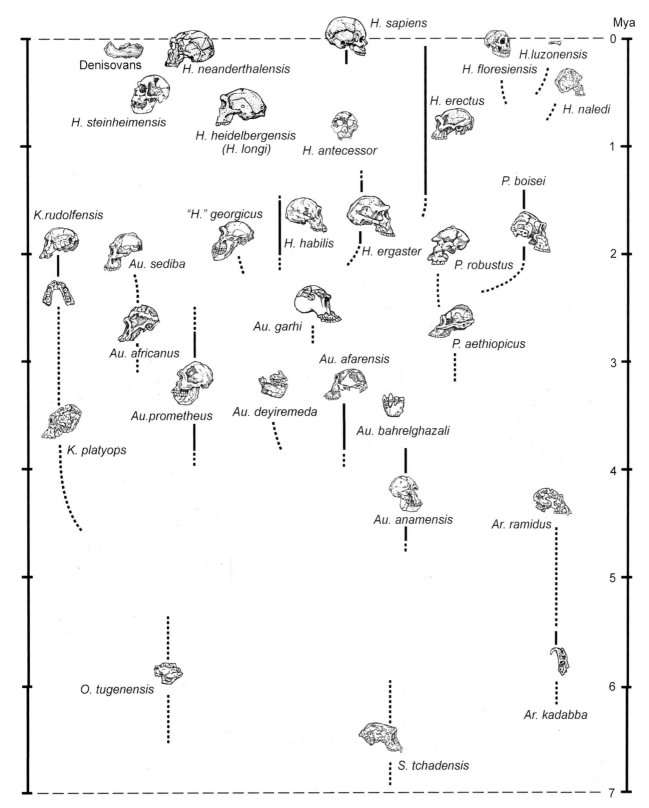

FIGURE 17.31 Summary of the temporal range of hominin species and likely relationships. *(Modified from Tattersall, I., 2012. Masters of the Planet: The Search for Human Origins. Palgrave Macmillan, New York. Courtesy of I. Tattersall.)*

wealth of genetic information available for humans and other primates, our understanding of human evolution will continue to expand, and current interpretations will be replaced by newer ones.

References

Aiello, L.C., 2010. Five years of *Homo floresiensis*. Am. J. Phys. Anthropol. 142, 167–179.

Alba, D.M., Moyà-Solà, S., Kohler, M., 2003. Morphological affinities of the *Australopithecus afarensis* hand on the basis of manual proportions and relative thumb length. J. Hum. Evol. 44, 225–254.

Alemseged, Z., 2013. *Australopithecus* in Ethiopia. In: Reed, K.E., Fleagle, J.G., Leakey, R.E. (Eds.), The Paleobiology of *Australopithecus*. Springer, New York.

Alemseged, Z., Spoor, F., Kimbel, W.H., Bobe, R., Geraads, D., Reed, D., Wynn, J.G., 2006. A juvenile early hominin skeleton from Dikika, Ethiopia. Nature. 443, 296–301.

Almécija, S., Tallman, M., Alba, D., Pina, M., Moya-Sola, S., Jungers, W.L., 2013. The femur of *Orrorin tugenensis* exhibits morphometric affinities with both Miocene apes and later hominins. Nature Comm. 4, 2888.

Antón, S.C., 2003. A natural history of H. *erectus*. Yrbk. Phys. Anthropol. 46, 126–170.

Antón, S.C., 2013. *Homo erectus* and related taxa. In: Begun, D.R. (Ed.), A Companion to Paleoanthropology. Wiley-Blackwell, Hoboken, pp. 497–516.

Antón, S.C., Spoor, F., Fellmann, C.D., Swisher III, C.C., 2007. Defining *Homo erectus*: size considered. In: Henke, R., Tattersall, Handbook of Paleoanthropology, vol. 3, Chapter 11. Springer-Verlag, Berlin, pp. 1655–1693.

Arsuaga, J.L., Lorenzo, C., Carretero, J.-M., et al., 1993. Three new human skulls from the Sima de los Huesos Middle Pleistocene site in Sierra de Atapuerca, Spain. Nature. 362, 534–537.

Asfaw, B., White, T.D., Lovejoy, C.O., Latimer, B.M., Simpson, S., Suwa, G., 1999. *Australopithecus garhi*: A new species of early hominid from Ethiopia. Science. 284, 629–635.

Baab, K., 2010. Cranial shape in Asian *Homo erectus*: geographic, anagenetic, and size-related variation. In: Norton, C.J., Braun, D.R. (Eds.), Asian Paleoanthropology: From Africa to China and Beyond. Springer, New York, pp. 57–79.

Begun, D.R., Richmond, B.G., Strait, D.S., 2007. Comment on "Origin of human bipedalism as an adaptation for locomotion on flexible branches". Science. 318, 1066d.

Behrensmeyer, A.K., 2006. Climate change and human evolution. Science. 311, 476–478.

Behrensmeyer, A.K., Reed, K.E., 2013. Reconstructing the paleoecology of *Australopithecus*: Site taphonomy, paleoenvironments, and faunas. In: Reed, K.E., Fleagle, J.G., Leakey, R.E. (Eds.), The Paleobiology of *Australopithecus*. Springer, New York, pp. 41–62.

Berger, L.R., de Ruiter, D.J., Churchill, S.E., et al., 2010. *Australopithecus sediba*: A new species of *Homo*-like australopith from South Africa. Science. 328, 195–204.

Berger, L.R., Hawks, J., de Ruiter, D.J., et al., 2015. *Homo naledi*, a new species of the genus *Homo* from the Dinaledi Chamber, South Africa. eLife 2015 (4), e09560.

Bermúdez de Castro, J.M., Arsuaga, J.L., Carbonell, E., Rosas, A., Martínez, I., Mosqueraet, M., 1997. A hominid from the Lower Pleistocene of Atapuerca, Spain: Possible ancestor to Neandertals and modern humans. Science. 276, 1392–1395.

Berna, F., Goldberg, P., Horwitz, L.K., Brink, J., Holt, S., Bamford, M., Chazan, M., 2012. Microstratigraphic evidence of in situ fire in the Acheulean strata of Wonderwerk Cave, Northern Cape province, South Africa. Proc. Natl. Acad. Sci. USA 109, E1215–E1220.

Binford, L.R., 1981. Bones: Ancient Men and Modern Myths. Academic Press, New York.

Brain, C.K., 1993. Swartkrans: A cave's chronicle of early man. Transvaal Mus. Monograph. 8, 1–270.

Bramble, D.M., Lieberman, D.E., 2004. Endurance running and the evolution of *Homo*. Nature. 432, 345–352.

Bräuer, G., 2008. The origin of modern anatomy: by speciation or intraspecific evolution? Evol. Anthropol. 17, 22–23.

Bräuer, G., 2012. Middle Pleistocene diversity in Africa and the origin of modern humans. In: Hublin, J.-J., McPherron, S.P. (Eds.), Modern Human Origins: A North African Perspective. Springer, New York, pp. 221–240.

Brown, F., Harris, J.R., Leakey, R.F.F., Walker, A., 1985. Early *Homo erectus* skeleton from west Lake Turkana, Kenya. Nature. 316, 788–792.

Brunet, M., Beauvilain, A., Coppens, Y., Heintz, E., Moutaye, A.H.E., Pilbeam, D., 1995. The first australopithecine 2,500 kilometres west of the Rift Valley (Chad). Nature. 378, 373–374.

Brunet, M., Beauvilain, A., Coppens, Y., Heintz, E., Moutaye, A.H.E., Pilbeam, D., 1996. *Australopithecus bahrelghazali*, une novelle espece d'Hominide ancien de la region de Koro Toro (Tchad). C. R. Acad. Sci. 322, 907–913.

Brunet, M., Guy, F., Pilbeam, D., et al., 2002. A new hominid from the Upper Miocene of Chad, Central Africa. Nature. 418, 145–151.

Brunet, M., Guy, F., Pilbeam, D., et al., 2005. New material of the earliest hominid from the Upper Miocene of Chad. Nature. 434, 752–755.

Carlson, K.J., Green, D.J., Jashashvili, T., Pickering, T.R., Heaton, J.L., Beaudet, A., Stratford, D., Crompton, R., Kuman, K., Bruxelles, L., Clarke, R.J., 2021. The pectoral girdle of StW 573 ("Little Foot") and its implications for shoulder evolution in the Hominina. J. Hum. Evol. 158, 102983.

Cartmill, M., 1990. Human uniqueness and theoretical content in paleoanthropology. Int. J. Primatol. 11, 173–192.

Cela-Conde, C.J., Ayala, F.J., 2003. Genera of the human lineage. Proc. Natl. Acad. Sci. USA 100, 7684–7689.

Cerling, T.E., 1992. Development of grasslands and savannas in East Africa during the Neogene. Paleogeogr. Paleoclimatol. Paleoecol. (Global Planetary Change Section) 97, 241–247.

Cerling, T.E., Levin, N.E., Quade, J., et al., 2010. Comment on the paleoenvironment of *Ardipithecus ramidus*. Science. 328, 1105-d.

Cerling, T.E., Mbua, E., Kirera, F.M., et al., 2011a. Diet of *Paranthropus boisei* in the early Pleistocene of East Africa. Proc. Natl. Acad. Sci. USA 108, 9337–9341.

Cerling, T.E., Wynn, J.G., Andanje, S.A., et al., 2011b. Woody cover and hominin environments in the past 6 million years. Nature. 476, 51–56.

Chaney, M.E., Ruiz, C.A., Meindl, R.S., Lovejoy, C.O., 2021. The foot of the human-chimpanzee last common ancestor was not African ape-like: A response to Prang (2019). J. Hum. Evol. https://doi.org/10.1016/j.jhevol.2020.102940

Churchill, S.E., 1996. Particulate versus integrated evolution of the upper body in Late Pleistocene humans: a test of two models. Am. J. Phys. Anthropol. 100, 559–583.

Clarke, R., 2013. *Australopithecus* from Sterkfontein Caves, South Africa. In: Reed, K.E., Fleagle, J.G., Leakey, R.E. (Eds.), The Paleobiology of *Australopithecus*. Springer, New York.

Clarke, R., Kuman, K., 2019. The skull of StW 573, a 3.67 Ma Australopithecus Prometheus skeleton from Sterkfontein Caves, South Africa. J. Hum. Evol. 134, 102634.

Collard, M., Wood, B.A., 2007. Defining the genus Homo. In: Henke, W. Tattersall, I. (Eds.),. In: Handbook of Paleoanthropology, vol. 3. Springer-Verlag, Berlin, pp. 1575–1610.

Curnoe, D., 2010. A review of early *Homo* in southern Africa focusing on cranial, mandibular and dental remains, with the description of a new species (*Homo gautengensis* sp. nov.). Homo 61, 151–177.

Currat, M., Excoffier, L., 2011. Strong reproductive isolation between humans and Neanderthals inferred from observed patterns of introgression. *Proc. Natl. Acad. Sci.* USA 108, 15129–15134..

Dart, R.A., 1925. *Australopithecus africanus*: the man-ape of South Africa. Nature 115, 195–199.

Darwin, C., 1871. The Descent of Man and Selection in Relation to Sex. Murray, London.

Daver, G., Guy, F., Makcaye, H.T., Likius, A., Boisserie, J.-R., Moussa, A., Pallas, L., Vignaud, P., Clarisse, N.D., 2022. Postcranial evidence of late Miocene hominin bipedalism in Chad. Nature. 609, 94–100.

deMenocal, P.B., 1995. Plio-Pleistocene African climate. Science. 270, 53–58.

deMenocal, P.B., 2004. African climate change and faunal evolution during the Plio-Pleistocene. Earth Planet. Sci. Lett. 220, 3–24.

deMenocal, P.B., 2011. Climate and human evolution. Science. 311, 540–541.

DeSilva, J.M., Gill, C.M., Prang, T.C., Bredella, M.A., Alemseged, Z., 2018. A nearly complete foot from Dikika, Ethiopia and its implications for the ontogeny and function of *Australopithecus afarensis*. Sci. Adv. 4 eaar7723

de Ruiter, D.J., Churchill, S.E., Berger, L.R., 2013. *Australopithecus sediba* from Malapa, South Africa. In: Reed, K.E., Fleagle, J.G., Leakey, R.E. (Eds.), The Paleobiology of *Australopithecus*. Springer, New York, pp. 147–162.

Detroit, F., Mijares, A.S., Corny, J., Daver, G., Zanolli, C., Dizon, E., Robles, E., Grun, R., Piper, P.J., 2019. A new species of *Homo* from the Late Pleistocene of the Philippines. Nature. 568, 181–186.

Dirks, P.H.G.M., Kibii, J.M., Kuhn, B.F., et al., 2010. Geological setting and age of *Australopithecus sediba* from southern Africa. Science. 328, 205–208.

Drapeau, M.S., Ward, C.V., Kimbel, W.H., Johanson, D.C., Rak, Y., 2005. Associated cranial and forelimb remains attributed to *Australopithecus afarensis* from Hadar, Ethiopia. J. Hum. Evol. 48, 593–642.

Fleagle, J.G., Stern Jr., J.T., Jungers, W.L., Susman, R.L., Vangor, A.K., Wells, J.P., 1981. Climbing: A biomechanical link with brachiation and with bipedalism. Symp. Zool. Soc. Lond. 48, 359–375.

Fleagle, J.G., Assefa, Z., Brown, F.H., Shea, J.J., 2008. Paleoanthropology of the Kibish Formation, southern Ethiopia: Introduction. J. Hum. Evol. 55, 360–365.

Franck, G., Lieberman, D.E., Pilbeam, D.R., et al., 2005. Morphological affinities of the *Sahelanthropus tchadensis* (Late Miocene hominid from Chad) cranium. Proc. Natl. Acad. Sci. USA 52, 18836–18841.

Frost, S.R., White, F., Reda, H., Gilbert, C.C., 2022. Biochronology of South African hominin-bearing sites: a reassessment using cercopithecid primates. Proc. Natl. Acad. Sci. USA 119, e2210627119. https://doi.org/10.1073/pnas.2210627119

Gabunia, L., Antón, S.C., Lordkipanidze, D., Vekua, A., Justus, A., Swisher III, C.C., 2001. Dmanisi and dispersal. Evol. Anthropol. 10, 158–170.

Gordon, A.D., 2013. Sexual size dimorphism in *Australopithecus*: Current understanding and new directions. In: Reed, K.E., Fleagle, J.G., Leakey, R.E. (Eds.), The Paleobiology of *Australopithecus*. Springer, New York, pp. 195–212.

Gordon, A.D., Green, D.J., Richmond, B.G., 2008. Strong postcranial size dimorphism in *Australopithecus afarensis*: Results from two new multivariate resampling methods for multivariate data sets with missing data. Am. J. Phys. Anthropol. 135, 311–328.

Granger, D.E., Gibbon, R.J., Kuman, K., Clarke, R.J., Bruxelles, L., Caffee, M.W., 2015. New cosmogenic burial ages for Sterkfontein Member 2 *Australopithecus* and Member 5 Oldowan. Nature. 522, 85–88.

Granger, D.E., Stratford, D., Bruxelles, L., Gibbon, R.J., Clarke, R.J., Kuman, K., 2022. 2563 Cosmogenic nuclide dating of *Australopithecus* at Sterkfontein, South Africa. Proc. Natl. Acad. Sci. USA 119 (27) e21235216119.

Green, D.J., Alemseged, Z., 2012. *Australopithecus afarensis* scapular ontogeny, function, and the role of climbing in human evolution. Science. 338, 514–517.

Green, R.E., Krause, J., Briggs, A.W., et al., 2010. A draft sequence of the Neandertal genome. Science. 328, 710–722.

Green, D.J., Chirchir, H., Mbua, E., Harris, J.W.K., Braun, D.R., Griffin, N.L., Richmond, B.G., 2018. Scapular anatomy of *Paranthropus boisei* from Ileret, Kenya. J. Hum. Evol. 125, 181–192.

Grine, F.E., 1988. Evolutionary History of the Robust Australopithecines. Aldine, Hawthorne, New York, pp. 73–104.

Grine, F.E., 1993. Australopithecine taxonomy and phylogeny: Historical background and recent interpretation. In: Ciochon, R.L., Fleagle, J.G. (Eds.), The Human Evolution Sourcebook. Prentice-Hall, Englewood Cliffs, New Jersey, pp. 196–208.

Grine, F.E., 2005. Early *Homo* at Swartkrans, South Africa: A review of the evidence and an evaluation of recently proposed morphs. S. Afr. J. Sci. 101, 43–52.

Grine, F.E., 2013. The alpha taxonomy of *Australopithecus africanus*. In: Reed, K.E., Fleagle, J.G., Leakey, R.E. (Eds.), The Paleobiology of Australopithecus. Springer, New York, pp. 73–104.

Grine, F.E., 2019. The alpha taxonomy of *Australopithecus* at Sterkfontein: The postcranial evidence. C. R. Palevol. 18, 335–352.

Grine, F.E., Demes, B., Jungers, W.L., Cole III, T.M., 1993. Taxonomic affinity of the early *Homo* cranium from Swartkrans, South Africa. Am. J. Phys. Anthropol. 92, 411–426.

Grine, F.E., Jungers, W.L., Tobias, P.V., Pearson, O.M., 1995. Fossil *Homo* femur from Berg Aukus, Northern Namibia. Am. J. Phys. Anthropol. 97, 151–185.

Grine, F.E., Bailey, R.M., Harvati, K., et al., 2007. Late Pleistocene human skull from Hofmeyr, South Africa, and modern human origins. Science. 315, 226–229.

Grine, F.E., Smith, H.F., Heesy, C.P., Smith, E.J., 2009. Phenetic affinities of Plio-Pleistocene *Homo* fossils from South Africa: Molar cusp proportions. In: Grine, F.E., Fleagle, J.G., Leakey, R.E. (Eds.), The First Humans: Origin and Early Evolution of the Genus *Homo*. Springer, New York, pp. 49–62.

Grine, F.E., Ungar, P.S., Teaford, M.F., El-Zaatari, S., 2013. Molar microwear, diet and adaptation in a purported hominin species lineage from the Pliocene of East Africa. In: Reed, K.E., Fleagle, J.G., Leakey, R.E. (Eds.), The Paleobiology of *Australopithecus*. Springer, New York.

Gunz, P., Bookstein, F.L., Mitteroecker, P., Stadlmayr, A., Seidler, H., Weber, G.W., 2009. Early modern human diversity suggests subdivided population structure and a complex out-of-Africa scenario. Proc. Natl. Acad. Sci. USA 106, 6094–6098.

Haile-Selassie, Y., Saylor, B.Z., Deino, A., Alene, M., Latimer, B.M., 2010. New hominid fossils from Woranso-Mille (Central Afar, Ethiopia) and taxonomy of early *Australopithecus*. Am. J. Phys. Anthropol. 141, 406–417.

Haile-Selassie, Y., Saylor, B.Z., Deino, A., Levin, N.E., Alene, M., Latimer, B.M., 2012. A new hominin foot from Ethiopia shows multiple Pliocene bipedal adaptations. Nature. 483, 565–569.

Haile-Selassie, Y., Gibert, L., Melillo, S.M., Ryan, T.M., Alene, M., Deino, A., Levin, N.E., Scott, G., Saylor, B.Z., 2015. New species from Ethiopia further expands Middle Pliocene hominin diversity. Nature. 521, 483–488.

Haile-Selassie, Y., Melillo, S.M., Su, D.F., 2016. The Pliocene hominin diversity conundrum: Do more fossils mean less clarity? Proc. Natl. Acad. Sci. 113, 6364–6371.

Haile-Selassie, Y., Melillo, S.M., Vazzana, A., Benazzi, S., Ryan, T.M., 2019. A 3.8-million-year-old hominin cranium from Woranso-Mille, Ethiopia. Nature. 573, 214–219.

Harcourt, A.H., 2012. Human Biogeography. University of California Press, Berkeley.

Harcourt-Smith, W.E.H., 2007. The origins of bipedal locomotion. In: Henke, W. Tattersall, I. (Eds.),. In: Handbook of Paleoanthropology, vol. 3. Springer-Verlag, Berlin, pp. 1483–1518.

Hardy, A., 1960. Was man more aquatic in the past? New Sci. 7, 642–645.

Harmand, S., Lewis, J.E., Feibel, C.S., Lepre, C.J., Prat, S., Lenoble, A., Boes, X., Quinn, R.L., Brenet, M., Arroyo, A., Taylor, N., Clement, S., Daver, G., Brugal, J.-P., Leakey, L., Morlock, R.A., Wright, J.D., Lokorodi, S., Kirwa, C., Kent, D.V., Roche, H., 2015. 3.3-million-year-old stone tools from Lomekwi 3, West Turkana, Kenya. Nature. 521, 310–315.

Harvati, K., 2007. Neanderthals and their contemporaries. In: Henke, W. Tattersall, I. (Eds.),. In: Handbook of Paleoanthropology, vol. 3. Springer-Verlag, Berlin, pp. 1717–1748.

Harvati, K., Roding, C., Bosman, A.M., et al., 2019. Apidima Cave fossils provide earliest evidence of *Homo sapiens* in Eurasia. Nature. 571, 500–504.

Heaton, J.L., Pickering, T.R., Carlson, K.J., Crompton, R.H., Jashashvili, T., Beaudet, A., Bruxelles, L., Kuman, K., Heile, A.J., Stratford, D., Clarke, R.J., 2019. The long limb bones of the StW 573 *Australopithecus* skeleton from Sterkfontein Member 2: Descriptions and proportions. J. Hum. Evol. 133, 167–197.

Herries, A.I.R., Pickering, R., Adams, J.W., et al., 2013. A multidisciplinary perspective on the age of *Australopithecus* in southern Africa. In: Reed, K.E., Fleagle, J.G., Leakey, R.E. (Eds.), The Paleobiology of *Australopithecus*. Springer, New York, pp. 21–40.

Hershkovitz, I., Weber, G.W., Quam, R., et al., 2018. Earliest modern humans out of Africa. Science. 359, 456–459.

Hewes, G.W., 1961. Food transport and the origin of hominid bipedalism. Am. Anthropol. 69, 63–67.

Hublin, J.-J., 2009. The origin of Neandertals. Proc. Natl. Acad. Sci. USA 106, 16022–16027.

Hublin, J.-J., Ben-Ncer, A., Bailey, S.E., Freidline, S.E., Neubauer, S., Skinner, M.M., Bergmann, I., Le Cabec, A., Benazzi, S., Harvati, K., Gunz, P., 2017. New fossils from Jebel Irhoud, Morocco and the Pan-African origin of *Homo sapiens*. Nature. 546, 289–292.

Hunt, K.D., 1994. The evolution of human bipedality: Ecology and functional morphology. J. Hum. Evol. 26, 183–202.

Hunt, K.D., 1996. The postural feeding hypothesis: An ecological model for the evolution of bipedalism. S. Afr. J. Sci. 9, 77–90.

Isaac, G., 1983. Aspects of human evolution. In: Bendall, D.S. (Ed.), Evolution from Molecules to Men. Cambridge University Press, Cambridge, pp. 509–543.

Jablonski, N.G., Chaplin, G., 1993. Origin of habitual terrestrial bipedalism in the ancestor of the Hominidae. J. Hum. Evol. 24, 259–280.

Ji, Q., Wu, W., Ji, Y., Li, Q., Ni, X., 2021. Late Middle Pleistocene Harbin cranium represents a new *Homo* species. The Innovation 2, 100132.

Johanson, D.C., White, T.D., 1979. A systematic assessment of early African hominids. Science. 203, 321–330.

Johanson, D.C., Taieb, M., Coppens, Y., 1982. Pliocene hominids from the Hadar Formation, Ethiopia (1973–1977): Stratigraphic, chronologic, and paleoenvironmental contexts, with notes on hominid morphology and systematics. Am. J. Phys. Anthropol. 57, 373–402.

Jolly, C.F., 1970. The seed-eaters: A new model of hominid differentiation based on a baboon analogy. Man. 5, 5–28.

Jungers, W.L., 2013. Homo floresiensis. In: Begun, D.R. (Ed.), Companion to Paleoanthropology. Wiley-Blackwell, Malden, Massachusetts.

Kappelman, J., 1996. The evolution of body mass and relative brain size in fossil hominids. J. Hum. Evol. 30, 243–276.

Kappelman, J., Fleagle, J.G., 1995. Age of early hominids. Nature. 376, 558–559.

Kimbel, W.H., Rak, Y., 2017. *Australopithecus sediba* and the emergence of *Homo*: Questionable evidence from the cranium of the juvenile holotype MH 1. J. Hum. Evol. 107, 94–106.

Kimbel, W.H., Rak, Y., Johanson, D.C., 2004. Oxford The Skull of *Australopithecus afarensis*. Oxford University Press.

Kimbel, W.H., Lockwood, C.A., Ward, C.V., Leakey, M.G., Rak, Y., Johanson, D.C., 2006. Was *Australopithecus anamensis* ancestral to A.

afarensis? A case of anagenesis in the hominin fossil record. J. Hum. Evol. 51, 134–152.

Kimbel, W.H., Suwa, G., Asfaw, B., Rak, Y., White, T.D., 2014. *Ardipithecus ramidus* and the evolution of the human cranial base. Proc. Natl. Acad. Sci. USA 111, 948–953.

Kivell, T.L., Kibii, J.M., Churchill, S.E., Schmid, P., Berger, L.R., 2011. *Australopithecus sediba* hand demonstrates mosaic evolution of locomotor and manipulative abilities. Science. 333, 1411–1417.

Klein, R.G., 1992. The archeology of modern human origins. Evol. Anthropol. 1, 5–14.

Klein, R.G., 2009. The Human Career: Human Biological and Cultural Origins. University of Chicago Press, Chicago.

Kramers, J.D., Dirks, P.H.G.M., 2017. The age of fossil StW573 ("Little Foot"): An alternative interpretation of ^{26}Al/^{10}Be burial data. S. Afr. J. Sci. 113 https://doi.org/10.17159/sajs.2017/20160085

Krings, M., Stune, A., Schmitz, R.W., Krainitzkijlt, H., Stoneking, M., Pääbo, S., 1997. Neandertal DNA sequences and the origin of modern humans. Cell. 90, 14–30.

Lahr, M.M., 1996. The Evolution of Modern Human Diversity: A Study in Cranial Variation. Cambridge University Press, Cambridge.

Lahr, M.M., Foley, R., 1994. Multiple dispersals and modern human origins. Evol. Anthropol. 3, 48–60.

Leakey, R.E.F., Butzer, K., Day, M.H., 1969. Early *Homo sapiens* remains from the Omo River region of south-west Ethiopia. Nature. 222, 1132–1138.

Leakey, M.G., Feibel, C.S., McDougall, I., Walker, A., 1995. New four-million-year-old hominid species from Kanapoi and Allia Bay, Kenya. Nature. 376, 565–571.

Leakey, M.G., Feibel, C.S., McDougall, I., Ward, C., Walker, A., 1998. New specimens and confirmation of an early age for *Australopithecus anamensis*. Nature. 393, 62–66.

Leakey, M.G., Spoor, F., Brown, F.H., et al., 2001. New hominin genus from eastern Africa shows diverse Middle Pliocene lineages. Nature. 410, 433–440.

Leakey, M.G., Spoor, F., Dean, M.C., et al., 2012. New fossils from Koobi Fora in northern Kenya confirm taxonomic diversity in early *Homo*. Nature. 488, 201–204.

Lague, M.R., Chirchir, H., Green, D.J., Mbua, E., Harris, J.W.K., Braun, D.R., Griffin, N.L., Richmond, B.G., 2019. Cross-sectional properties of the humeral diaphysis of *Paranthropus boisei*: Implications for upper limb function. J. Hum. Evol. 126, 51–70.

Lieberman, D.E., Bramble, D.M., Raichlen, D.A., Shea, J.J., 2009. Brains, brawn, and the evolution of human endurance running capabilities. In: Grine, F.E., Fleagle, J.G., Leakey, R.E. (Eds.), The First Humans: Origin and Early Evolution of the Genus Homo. Springer, New York, pp. 77–92.

Lockwood, C.A., 2013. Whence *Australopithecus africanus*? Comparing the skulls of South African and East African *Australopithecus*. In: Reed, K.E., Fleagle, J.G., Leakey, R.E. (Eds.), The Paleobiology of Early *Australopithecus*. Springer, New York, pp. 175–182.

Lockwood, C.A., Menter, C.G., Moggi-Cecchi, J., Keyser, A.W., 2007. Extended male growth in a fossil hominin species. Science. 318, 1443–1446.

Lordkipanidze, D., Jashashvili, T., Vekua, A., et al., 2007. Postcranial evidence from early *Homo* from Dmanisi, Georgia. Nature. 449, 305–310.

Lovejoy, C.O., 1981. The origin of man. Science. 211, 341–350.

Lovejoy, C.O., 2009. Reexamining human origins in light of *Ardipithecus ramidus*. Science. 326, 74e1–74e8.

Macchiarelli, R., Bergeret-Medina, A., Marchi, D., Wood, B., 2020. Nature and relationships of *Sahelanthropus tchadensis*. J. Hum. Evol. 149, 102898.

Manthi, F.K., Plavcan, J.M., Ward, C.V., 2012. New hominin fossils from Kanapoi, Kenya, and the mosaic evolution of canine teeth in early hominins. S. Afr. J. Sci. 108, 724.

Mbua, E., Bräuer, G., 2012. Patterns of Middle Pleistocene hominin evolution in Africa and the emergence of modern humans. In:

Reynolds, S.C., Gallagher, A. (Eds.), African Genesis: Perspectives on Hominin Evolution. Cambridge University Press, Cambridge, pp. 394–422.

McDougall, I., Brown, F.H., Fleagle, J.G., 2005. Stratigraphic placement and age of modern humans from Kibish, Ethiopia. Nature. 433, 733–736.

McHenry, H.M., 1994. Behavioral ecological implications of early hominid body size. J. Hum. Evol. 27, 77–87.

McPherron, S.P., Alemseged, Z., Marean, C.W., et al., 2010. Evidence for stone-tool-assisted consumption of animal tissues before 3.39 million years ago at Dikika, Ethiopia. Nature. 466, 857–860.

Meyer, M., Kircher, M., Gansauge, M.T., et al., 2012. A high-coverage genome sequence from an archaic Denisovan individual. Science. 338, 222–226.

Mongle, C.S., Strait, D.S., Grine, F.E., 2019. Expanded character sampling underscores phylogenetic stability of *Ardipithecus ramidus* as a basal hominin. J. Hum. Evol. 131, 28–39.

Morgan, E., 1999. The Aquatic Ape Hypothesis: The Most Critical Theory of Human Evolution. Souvenir Press, London.

Panger, M.A., Brooks, A.S., Richmond, B.G., Wood, B., 2002. Older than the Oldowan? Rethinking the emergence of hominin tool use. Evol. Anthropol. 11, 235–245.

Pickering, R., Herries, A.I.R., Woodhead, J.D., Hellstrom, J.C., Green, H.E., Paul, B., Ritzman, T., Strait, D.S., Schoville, B.J., Hancox, P.J., 2019. U-Pb-dated flowstones restrict South African early hominin record to dry climate phases. Nature. 565, 226–229.

Pickford, M., 2012. *Orrorin* and the African ape/hominid dichotomy. In: Reynolds, S.C., Gallagher, A. (Eds.), African Genesis: Perspectives on Hominin Evolution. Cambridge University Press, Cambridge, pp. 99–119.

Plavcan, J.M., van Schaik, C., 1993. Intrasexual competition and canine dimorphism in anthropoid primates. Am. J. Phys. Anthropol. 87, 461–477.

Pontzer, H., Raichlen, D.A., Sockol, M.D., 2009. The metabolic cost of walking in humans, chimpanzees, and early hominins. J. Hum. Evol. 56, 43–54.

Prang, T.C., 2019. The African ape-like foot of *Ardipithecus ramidus* and its implications for the origin of bipedalism. eLife 8, e44433. https://doi.org/10.7554/eLife.44433

Prang, T.C., 2022. New analyses of the Ardipithecus ramidus foot provide additional evidence of its African ape-like affinities: A reply to Chaney et al. (2021). J. Hum. Evol. 164, 103135.

Prang, T.C., Ramirez, K., Grabowski, M., Williams, S.A., 2021. *Ardipithecus* hand provides evidence that humans and chimpanzees evolved from an ancestor with suspensor adaptations. Sci. Adv. 7. https://doi.org/10.1126/sciadv.abf2474

Prost, J.H., 1980. Origin of bipedality. Yrbk. Phys. Anthropol. 19, 59–68.

Reed, K.E., 1997. Early hominid evolution and ecological change through the African Plio-Pleistocene. J. Hum. Evol. 32, 289–322.

Reed, K.E., Fleagle, J.G., Leakey, R.E., 2013. The Paleobiology of *Australopithecus*. Springer, New York.

Reich, D., Green, R.E., et al., 2010. Genetic history of an archaic hominin group from Denisova Cave in Siberia. Nature. 468, 1053–1060.

Reno, P.L., Meindl, R.S., McCollum, M.A., Lovejoy, C.O., 2003. Sexual dimorphism in *Australopithecus afarensis* was similar to that of modern humans. Proc. Natl. Acad. Sci. USA 100, 9404–9409.

Reno, P.L., McCollum, M.A., Meindl, R.S., Lovejoy, C.O., 2010. An enlarged postcranial sample confirms *Australopithecus afarensis* dimorphism was similar to modern humans. Phil. Trans. R. Soc. B. 365, 3355–3363.

Richard, A.F., Goldstein, S.J., Dewar, R.E., 1989. Weed macaques: The evolutionary implications of macaque feeding ecology. Int. J. Primatol. 10, 569–594.

Richmond, B.G., Jungers, W.L., 2008. *Orrorin tugenensis* femoral morphology and the evolution of hominin bipedalism. Science. 319, 1662–1665.

Richmond, B.G., Green, D.J., Lague, M.R., Chirchir, H., Behrensmeyer, A.K., Bobe, R., Bamford, M.K., Griffin, N.L., Gunz, P., Mbua, E., Merritte, S.R., Pobiner, B., Kiura, P., Kibunjia, M., Harris, J.W.K., Braun, D.R., 2020. The upper limb of *Paranthropus boisei* from Ileret, Kenya. J. Hum. Evol. 141, 102727.

Rightmire, G.P., 1990. Cambridge The Evolution of *Homo erectus*. Cambridge University Press.

Rightmire, G.P., 1996. The human cranium from Bodo, Ethiopia: Evidence for speciation in the Middle Pleistocene? J. Hum. Evol. 31, 21–39.

Rightmire, G.P., 1998. Human evolution in the Middle Pleistocene: The role of *Homo heidelbergensis*. Evol. Anthropol. 6, 218–227.

Rightmire, G.P., 2007. Later Middle Pleistocene *Homo*. In: Henke, W., Tattersall, I. (Eds.), Handbook of Paleoanthropology, vol. 3. Springer-Verlag, Berlin, PP. 1695–1715.

Rightmire, G.P., Lordkipanidze, D., 2009. Comparisons of Early Pleistocene skulls from East Africa and the Georgian Caucasus: evidence bearing on the origin and systematics of genus *Homo*. In: Grine, F.E., Fleagle, J.G., Leakey, R.E. (Eds.), The First Humans: Origin and Early Evolution of the Genus *Homo*. Springer, New York, pp. 39–48.

Rightmire, G.P., Lordkipanidze, D., 2010. Fossil skulls from Dmanisi: A paleodeme representing earliest *Homo* in Eurasia. In: Fleagle, J.G., Shea, J.J., Grine, F.E., Baden, A.L., Leakey, R.E. (Eds.), Out of Africa I: The First Hominin Colonization of Eurasia. Springer, New York, pp. 225–243.

Rodman, P.S., McHenry, H.M., 1980. Bioenergetics and the origin of hominid bipedalism. Am. J. Phys. Anthropol. 52, 103–106.

Rolian, C., Dunsworth, H.M., McNulty, K., Lemelin, P., Jungers, W.L., 2013. More than the sum of its parts? Multivariate analysis of locomotor behavior in *Ardipithecus ramidus*. Am. J. Phys. Anthropol. 150 (S56), 235.

Rose, M.D., 1983. Miocene hominoid postcranial morphology: Monkey-like, ape-like, neither, or both?. In: Ciochon, R.L., Corrucini, R. (Eds.), New Interpretations of Ape and Human Ancestry. Plenum Press, New York, pp. 405–420.

Ruff, C.B., 1993. Climatic adaptation and hominid evolution: The thermoregulatory imperative. Evol. Anthropol. 2, 53–60.

Ruff, C.B., Trinkaus, E., Holliday, T.W., 1997. Body mass and encephalization in Pleistocene *Homo*. Nature. 387, 173–176.

Ruff, C.B., Burgess, M.L., Ketcham, R.A., Kappelman, J., 2016. Limb bone structural proportions and locomotor behavior in A.L. 288-1 ("Lucy"). PLoS One. 11, e0166095.

Schrenk, F., Bromage, T.G., Betzler, C.G., Ring, U., Juwayeyl, Y.M., 1993. Oldest *Homo* and Pliocene biogeography of the Malawi Rift. Nature. 365, 833–836.

Semaw, S., 2000. The world's oldest stone artefacts from Gona, Ethiopia: their implications for understanding stone technology and patterns of human evolution between 2.6–1.5 million years ago. J. Archaeol. Sci. 27, 1197–1214.

Senut, B., Pickford, M., Gommery, D., Mein, P., Cheboi, K., Coppens, Y., 2001. First hominoid from the Miocene (Lukeino Formation, Kenya). C. R. Acad. Sci. 332, 137–144.

Simpson, S.W., 2013. Before *Australopithecus*: the earliest hominins. In: Begun, D.R. (Ed.), A Companion to Paleoanthropology. Wiley-Blackwell, West Sussex, UK, pp. 417–433.

Sockol, M.D., Raichlen, D.A., Pontzer, H., 2007. Chimpanzee locomotor energetics and the origin of human bipedalism. Proc. Natl. Acad. Sci. USA 104, 12265–12269.

Sponheimer, M., 2013. Some ruminations of australopith diets. In: Reed, K.E., Fleagle, J.G., Leakey, R.E. (Eds.), The Paleobiology of *Australopithecus*. Springer, New York, pp. 225–234.

Spoor, F., Leakey, M.G., O'Higgins, P., 2016. Middle Pliocene hominin diversity: *Australopithecus deyiremeda* and *Kenyanthropus platyops*. Phil. Trans. R. Soc. B. 371, 20150231.

Stern J.T., Jr., 1975. Before bipedality. Yrbk. Phys. Anthropol. 19, 59–68.

Stern J.T., Jr., 2000. Climbing to the top: A personal memoir of *Australopithecus afarensis*. Evol. Anthropol. 9, 113–133.

Stern J.T., Jr., Susman, R.L., 1983. The locomotor anatomy of *Australopithecus afarensis*. Am. J. Phys. Anthropol. 60, 279–317.

Steudel, K.L., 1994. Locomotor energetics and hominid evolution. Evol. Anthropol. 3, 42–48.

Strait, D.S., Grine, F.E., 2004. Inferring hominoid and early hominid phylogeny using craniodental data: the role of fossil taxa. J. Hum. Evol. 47, 399–452.

Strait, D.S., Grine, F.E., Moniz, M.A., 1997. A reappraisal of early hominid phylogeny. J. Hum. Evol. 32, 17–82.

Strait, D., Grine, F.E., Fleagle, J.G., 2007. Analyzing hominid phylogeny. In: Henke, W., Tattersall, I. (Eds.), Handbook of Paleoanthropology, vol. 3. Springer-Verlag, Berlin.1781–1806.

Stringer, C., 2012. The status of *Homo heidelbergensis* (Schoetensack 1908). Evol. Anthropol. 21, 101–107.

Stringer, C., Barnes, I., 2015. Deciphering the Denisovans. Proc. Natl. Acad. Sci. USA 112, 15542–15543.

Stringer, C., McKie, R., 1996. African Exodus. Henry Holt, New York.

Susman, R.L., 1988. Hand of *Paranthropus robustus* from Member 1, Swartkrans: fossil evidence for tool behavior. Science. 240, 781–784.

Susman, R.L., 1991. Who made the Oldowan tools? Fossil Evidence for tool behavior in Plio-Pleistocene hominids. J. Anthropol. Res. 47, 129–149.

Susman, R.L., 1994. Fossil evidence for early hominid tool use. Science. 265, 1570–1573.

Susman, R.L., Stern J.T., Jr., Jungers, W.L., 1984. Arboreality and bipedality in the Hadar hominids. Folia Primatol. 43, 113–156.

Sutikna, T., Tocheri, M.W., Morwood, M.J., et al., 2016. Revised stratigraphy and chronology for *Homo floresiensis* at Liang Bua in Indonesia. Nature. 532, 366–369.

Suwa, G., Kono, R.T., Simpson, S.W., Asfaw, B., Lovejoy, C.O., White, T.D., 2009a. Paleobiological implications of the *Ardipithecus ramidus* dentition. Science. 326, 94–99.

Suwa, G., Asfaw, B., Kono, R.T., Kubo, D., Lovejoy, C.O., White, T.D., 2009b. The *Ardipithecus ramidus* skull and its implications for hominid origins. Science. 326, 57–63.

Tattersall, I., 2007. *Homo ergaster* and its contemporaries. In: Henke, W. Tattersall, I. (Eds.),. In: Handbook of Paleoanthropology, vol. 3. Springer-Verlag, Berlin, pp. 1633–1653.

Taylor, C.R., Rowntree, V.J., 1973. Running on two or on four legs: Which consumes more energy? Science. 179, 186–187.

Teixeira, J.C., Jacobs, G.S., Stringer, C., et al., 2021. Widespread Denisovan ancestry in Island Southeast Asia but no evidence o substantial super-archaic hominin admixture. *Nat. Ecol. Evol.* 5, 616–624.

Thorpe, S.K.S., Holder, R.L., Crompton, R.H., 2007. Origin of human bipedalism as an adaptation for locomotion on flexible branches. Science. 316, 1328–1331.

Tocheri, M.W., Orr, C.M., Jacofsky, M.C., Marzke, M.W., 2008. The evolutionary history of the hominin hand since the last common ancestor of *Pan* and *Homo*. J. Anat. 212, 544–562.

Trinkaus, E., 2005. Early modern humans. Annu. Rev. Anthropol. 34, 207–230.

Tung, J., Barreiro, L.B., 2017. The contribution of admixture to primate evolution. Curr. Opin. Genet. Dev. 47, 61–68.

Ungar, P.S., Sponheimer, M., 2011. The diets of early hominins. Science 334, 190–193.

Vidal, C.M., Lane, C.S., Asrat, A., Barfod, D.N., Mark, D.F., Tomlinson, E.L., Tadesse, A.Z., Yirgu, G., Deino, A., Hutchison, W., Mounier, A., Oppenheimer, C., 2022. Age of the oldest known *Homo sapiens* from eastern Africa. Nature. 601, 579–583.

Villmoore, B., Kimbel, W.H., Seyoum, C., Campisano, C.J., Dimaggio, E.N., Rowan, J., Braun, D.R., Arrowsmith, J.R., Reed, K.E., 2015. Early *Homo* at 2.8 Ma from Ledi-Geraru, Afar, Ethiopia. Science. 347, 1352–1355.

Vrba, E.S., 2007. Role of environmental stimuli in hominid origins. In: Henke, W. Tattersall, I. (Eds.),. In: Handbook of Paleoanthropology, vol. 3. Springer-Verlag, Berlin, pp. 1441–1481.

Vrba, E.S., Denton, G.H., Partridge, T.C., Burckle, L.H., 1996. Paleoclimate and Evolution with Emphasis on Human Origins. Yale University Press, New Haven.

Walker, A.C., Leakey, R.E.F., 1994. The Nariokotome *Homo erectus* Skeleton. Harvard University Press, Cambridge.

Walker, A.C., Shipman, P., 1996. The Wisdom of the Bones. A. A. Knopf, New York.

Ward, C.V., 2002. Interpreting the posture and locomotion of *Australopithecus afarensis*: Where do we stand? Yrbk. Phys. Anthropol. 45, 185–215.

Ward, C.V., 2013. Locomotion and limb use in *Australopithecus*. In: Reed, K.E., Fleagle, J.G., Leakey, R.E. (Eds.), The Paleobiology of *Australopithecus*. Springer, New York.

Ward, R., Stringer, C., 1997. A molecular handle on the Neanderthals. Nature. 388, 225–226.

Ward, C.V., Zipfel, B., 2020. Summary and synthesis. In: Zipfel, B., Richmond, B.G., Ward, C.V. (Eds.), Hominin Postcranial remains from Sterkfontein, South Africa, 1936–1995. Oxford University Press, Oxford, pp. 335–337.

Ward, C.V., Kimbel, W.H., Harmon, E.H., Johanson, D.C., 2012. New postcranial fossils of *Australopithecus afarensis* from Hadar, Ethiopia (1990–2007). J. Hum. Evol. 63, 1–51.

Ward, C.V., Nalley, T.K., Spoor, F., Tafforeau, P., Alemseged, Z., 2017. Thoracic vertebral count and thoracolumbar transition in *Australopithecus afarensis*. Proc. Natl. Acad. Sci. USA 114, 6000–6004.

Wheeler, P.E., 1993. The influence of stature and body form on hominid energy and water budgets: a comparison of *Australopithecus* and early *Homo* physiques. J. Hum. Evol. 24, 13–28.

White, T.D., 2002. Earliest hominids. In: Hartwig, W.C. (Ed.), The Primate Fossil Record. Cambridge University Press, Cambridge, pp. 407–417.

White, T.D., 2003. Early hominids – diversity or distortion? Science. 299, 1994–1997.

White, T.D., Suwa, G., Asfaw, B., 1994. *Australopithecus ramidus*, a new species of early hominid from Aramis, Ethiopia. Nature. 371, 306–333.

White, T.D., Asfaw, B., DeGusta, D., et al., 2003. Pleistocene *Homo sapiens* from Middle Awash, Ethiopia. Nature. 423, 742–747.

White, T.D., Asfaw, B., Beyene, Y., Haile-Selassie, Y., Lovejoy, C.O., Suwa, G., WoldeGabriel, G., 2009. *Ardipithecus ramidus* and the paleobiology of early hominids. Science. 326, 75–86.

White, T.D., Lovejoy, C.O., Asfaw, B., Carlson, J.P., Suwa, G., 2015. Neither chimpanzee nor human, Ardipithecus reveals the surprising ancestry of both. Proc. Natl. Acad. Sci. 112, 4877–4884.

Wood, B., 1991. Koobi Fora Research Project. Hominid Cranial Remains, vol. 4. Oxford University Press, Oxford.

Wood, B., 1992. Origin and evolution of the genus *Homo*. Nature 355, 783–790.

Wood, B.A., 2009. Where does the genus *Homo* begin, and how would we know? In: Grine, F.E., Fleagle, J.G., Leakey, R.E. (Eds.), The First Humans: Origin and Early Evolution of the Genus *Homo*. Springer, New York, pp. 17–28.

Wood, B.A., Collard, M., 1999a. The human genus. Science. 284, 65–71.

Wood, B.A., Collard, M., 1999b. The changing face of genus *Homo*. Evol. Anthropol. 8, 195–207.

Wood, B., Harrison, T., 2011. The evolutionary context of the first hominins. Nature. 470, 347–352.

Wood, B., Schroer, K., 2013. Paranthropus. In: Begun, D.R. (Ed.), A Companion to Paleoanthropology. Wiley-Blackwell, Hoboken, pp. 457–478.

Wrangham, R.W., 2005. The delta hypothesis: Hominoid ecology and hominin origins. In: Lieberman, D.E., Smith, R.J., Kelley, J. (Eds.), Interpreting the Past: Essays on Human, Primate and Mammal

Evolution in Honor of David Pilbeam. Brill Academic, Boston, pp. 231–242.

Wrangham, R., 2009. Catching Fire: How Cooking Made Us Human. Basic Books, New York.

Wrangham, R., Carmody, R., 2010. Human adaptation to the control of fire. Evol. Anthropol. 19, 187–199.

Wrangham, R., Cheney, D., Seyfarth, R., Sarmiento, E., 2009. Shallow-water habitats as sources of fallback foods for hominins. Am. J. Phys. Anthropol. 140, 630–642.

Wynn, T., Hernandez-Aguilar, R.A., Marchant, L.F., McGrew, W.C., 2011. "An ape's view of the Oldowan" revisited. Evol. Anthropol. 20, 181–197.

Zihlman, A., Tanner, N.M., 1978. Gathering and the hominid adaptation. In: Tiger, L., Fowler, H. (Eds.), Female Hierarchies. Beresford, Chicago, pp. 163–194.

Zollikofer, C.P.E., Ponce de León, M.S., Lieberman, D.E., et al., 2005. Virtual cranial reconstruction of *Sahelanthropus tchadensis*. Nature. 434, 755–759.

Further reading

Cartmill, M., Smith, F.H., 2022. The Human Lineage. Wiley-Blackwell, Hoboken.

Conroy, G.C., Pontzer, H., 2012. Reconstructing Human Origins: A Modern Synthesis. W. W. Norton, New York.

Harvati-Papatheodorou, K., 2013. Neanderthals. In: Begun, D.R. (Ed.), A Companion to Paleoanthropology. Wiley-Blackwell, Hoboken, pp. 538–556.

Henke, W., Tattersall, I., 2015.. Handbook of Paleoanthropology, vols. Springer Verlag, Heidelberg. 1–3.

Hublin, J.-J., 2013. The Middle Pleistocene record: On the ancestry of Neandertals, modern humans and others. In: Begun, D.R. (Ed.), A Companion to Paleoanthropology. Wiley-Blackwell, Hoboken, pp. 517–537.

Lieberman, D.E., 2011. The Evolution of the Human Head. Cambridge, Belknap.

MacLatchy, L.M., DeSilva, J., Sanders, W.J., Wood, B., 2010. Hominini. In: Werdelin, L., Sanders, W.J. (Eds.), Cenozoic Mammals of Africa. University of California Press, Berkeley, pp. 471–540.

Stringer, C.B., 2012. Lone Survivors. Macmillan, London.

Tattersall, I., 2022. Understanding Human Evolution. Cambridge University Press, Cambridge.

Tattersall, I., 2012. Masters of the Planet: The Search for Human Origins. *Palgrave Macmillan*, New York.

Wood, B., 2019. Human Evolution: A Very Short Introduction. Oxford University Press, Oxford.

Wood, B., Leakey, M., 2011. The Omo-Turkana Basin fossil hominins and their contribution to our understanding of human evolution in Africa. Evol. Anthropol. 20, 254–265.

Zipfel, B., Richmond, B.G., Ward, C.V., 2020. Hominin Postcranial remains from Sterkfontein, South Africa, 1936–1995. Oxford University Press, Oxford.

18

Patterns in Primate Evolution

Throughout the preceding chapters, we have examined primates – their phyletic relationships and ecological and behavioral adaptations – in more or less chronological and taxonomic order, family by family. Now that we have outlined and described the primate radiations, we are in a position to look for general trends. How can we characterize primate evolution as a whole? Are there repetitive patterns in the evolution of this order? With a good account of primate history and phylogeny at hand, we can also begin to examine theoretical questions about evolutionary processes. How do the various theories of evolutionary mechanisms, of speciation and of species extinction, fit the primate evidence? The fossil record of primates is particularly appropriate for such investigations since it is as complete as that of any group of mammals and it has certainly been more thoroughly studied.

Primate Adaptive Radiations

The primate fossil record documents an extraordinary diversity of extinct forms, not just isolated species and genera, but major radiations of families. There are over 80 genera and 500 species of living primates. More than twice as many fossil species have been discovered and described, and many more remain to be uncovered. The vast majority of primate taxa that have ever lived are now extinct.

Unfortunately, comparing the number of living species with the total number of extinct species cannot give us a good indication about how the current diversity of primates compares with that in the past. Living primates sample a single slice in time over a broad geographic area, whereas our knowledge of the fossil record is derived from samples of very restricted geographic areas and relatively long periods of time. For example, there is virtually no paleontological record from the areas in which primate diversity is greatest today – the Amazon Basin, the Congo Basin, and Southeast Asia. Nevertheless, we can still compare the taxonomic diversity of primates in the past with that of today, as

can we compare patterns of morphological and reconstructed ecological diversity of selected primate faunas from the past with what we find among primate faunas today. In this way, we can at least speculate about how the types of adaptations exploited by the primate faunas of the past were similar to or different from those characterizing the living radiations.

Body Size Changes

As we discussed in Chapter 9, body size is closely correlated with many aspects of a species' ecology, including diet and locomotion, and is also an easy parameter by which to compare species. Comparing the size distributions of living and fossil primate faunas gives us some indication of the adaptive diversity across groups and the extent to which they seem to occupy similar ecological niches in a very broad sense. However, there are many difficulties in comparing changes in primate body size over time. These include problems (including statistical error bars) with estimating body size in fossils often known from fragmentary remains, poor and biased sampling of diversity in the past, and disagreements over the numbers of species alive today. As a result, there is considerable debate over patterns of body size evolution in primates and in efforts to reconstruct the size of the common ancestor of all primates (e.g., Soligo, 2006; Soligo and Martin, 2006, 2007; Silcox et al., 2007). Nevertheless, there have been changes in the size distributions in primates through time (Fig. 18.1).

Several patterns are evident. First, there has been considerable size diversity among primates throughout their evolution. This is certainly associated with a considerable ecological diversity, as discussed in earlier chapters. Like many other groups of mammals, primates as an order seem to have increased in size during the past 56 million years. An overall size increase for primates during their evolution is reflected in two different aspects of their size distributions. First, very tiny primates were relatively more common in the Eocene, including species that are much smaller than any extant

Primate Adaptation and Evolution
https://doi.org/10.1016/B978-0-12-815809-8.00018-7

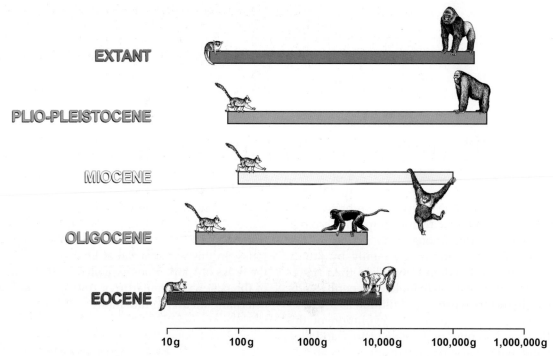

FIGURE 18.1   Body size ranges of primates through time.

primates (Gebo et al., 2000, 2012), but tiny primates are less common in most Oligocene to Recent faunas (Fleagle, 1978; Covert, 1986). Second, Miocene through Recent primates include very large species that are unknown from earlier periods (Fleagle and Kay, 1985). Both of these size changes suggest that the adaptive space occupied by primates has changed through time and that primates of the past showed different ecological adaptations from those found among living primate species. In this regard, the size changes corroborate the indications from specific morphological features, such as teeth and limbs.

Dietary Diversity

Although there is evidence of ecological diversity (especially in diet) throughout primate evolution, the expression of these adaptations among different groups has probably varied considerably. A frugivorous plesiadapid was probably not much like a frugivorous adapoid in many of its nondental adaptations, just as a frugivorous gibbon is very different in many aspects of its biology and foraging strategy from a frugivorous macaque. The detailed differences in foraging strategy that have been documented for extant primates in earlier chapters of this book are, of course, far beyond the realistic scope of paleontological studies, but we can see some general trends in the adaptive diversity of primate radiations

through time. On the basis of dental morphology, it seems most likely that the plesiadapiforms were predominantly insectivorous and frugivorous, and probably included some gum specialists. There are few species that show indications of extensive adaptation for folivory. In contrast, the fossil prosimians from the Eocene of North America and Europe include many folivorous species (especially among the adapoids), as well as others adapted for insectivorous (especially the omomyoids) and frugivorous diets (see Covert, 1986; Strait, 1997). We know relatively little about the early radiation of modern groups of strepsirrhines, except that the ancestors of galagos and lorises (and daubentoniids) probably appeared in the Eocene of Africa and were relatively diverse by middle Miocene times, and that the diversity of Malagasy primates has been dramatically reduced in just the last few thousand years. In contrast, the totally faunivorous extant tarsiers have limited diversity today and seem to be very similar to fossil tarsiers from the Eocene of East Asia. However, tarsier-like primates had a much larger geographical diversity in the past.

Old World anthropoids have undergone dramatic changes in dietary diversity during the past 35 million years. The Late Eocene and Early Oligocene higher primates from the rich Fayum deposits of Egypt were predominantly frugivorous anthropoids (Kirk and Simons, 2001). There are no anthropoid species that show dental adaptations to folivory comparable to those of many modern leaf eaters. Rather, this niche seems to have been occupied by adapoid primates (e.g., Seiffert et al., 2009). In the early part of the Miocene epoch, there are a few species with dentitions suggesting folivory; however, it seems that most of the species were likely frugivorous (Kay and Ungar, 1997). The contemporaneous early Miocene lorises and galagos were presumably frugivores, insectivores, and gum eaters, much like their modern relatives. By the latter part of the Miocene, there were more higher primates with teeth suggesting folivorous habits, and the proportions of frugivores to folivores were comparable to that among living Old World anthropoids. Moreover, there are indications that the fossil apes developed increasing specializations for shearing of leafy vegetation throughout the Miocene, perhaps as a result of competition with cercopithecoids (Kay and Ungar, 1997).

Less is known about the evolution of dietary diversity among fossil platyrrhines. Although the record for the Oligocene and earliest Miocene is relatively poor, the late middle Miocene fauna from La Venta in Colombia shows a diversity comparable to that of modern platyrrhines, with a predominance of frugivores, a few specialized seed-eaters and insectivores, and few folivores (Fleagle et al., 1997; Wheeler, 2010; Cooke, 2011). The increasingly well-documented fauna of fossil platyrrhines from the Caribbean seems to have been predominantly frugivorous (Cooke et al., 2011).

Locomotor Diversity

Temporal changes in primate locomotor habits are very difficult to document in the paleontological record because of the rarity of fossil skeletons. Still, there seem to be some general patterns. The increasingly well-documented skeletal remains of plesiadapiforms indicate largely arboreal habits (Bloch et al., 2007; Kirk et al., 2008; Chester et al., 2015, 2017), but the widespread presence of long, curved claws suggests that their arboreal behavior was qualitatively different from that of Eocene to Recent primates. They were probably not leapers but rather largely quadrupedal climbers and clingers. In contrast, most Eocene fossil prosimians are similar to extant prosimians in their general skeletal anatomy. Most seem to have been arboreal quadrupeds and quadrupedal leapers similar to cheirogaleids or *Lepilemur* (Covert, 1995). Many of the diagnostic features of modern primates that first appear in the early Eocene seem to reflect an adaptation to arboreal leaping (Dagosto, 2007). Apart from fossil tarsiers from China, there are few indications among the Eocene prosimians of the specialized vertical clinging behaviors that characterize living indriids or tarsiers (Fleagle and Anapol, 1992; Gebo, 2011; Fleagle and Lieberman, 2015). There is also no evidence among Eocene taxa of either terrestrial

species or large species with extreme suspensory abilities, such as those found in more recent strepsirrhines such as *Palaeopropithecus*.

Old World anthropoids show locomotor change over the past 30 million years. Late Eocene and Oligocene higher primates were arboreal, platyrrhine-like species, and all of the skeletal remains indicate leaping and quadrupedal habits. There is no evidence of terrestrial species, suspensory species, or specialized clingers. However, the first appearance of hominoids and cercopithecoids in the early Miocene is associated with evidence of more terrestrial and more suspensory species. The few remains from the last half of the Miocene indicate the presence of essentially modern locomotor adaptations in the fossil monkeys, but the evolution of ape skeletal adaptations is more difficult to interpret (e.g., Larson, 1998; Young, 2003). Miocene fossil apes are generally more primitive than modern apes, and many seem to be more quadrupedal without extreme adaptations for suspension, as found in extant gibbons or orangutans (Fleagle and Lieberman, 2015). However, the hominoids of the Miocene were a much more diverse group than their living relatives, and the fossil apes from Europe during the later parts of the Miocene show more adaptations for suspensory behavior, in mosaic combinations, than earlier apes from Africa. In fact, the latest Miocene taxon, *Oreopithecus*, is the fossil ape that is most similar to living hominoids in its postcranial anatomy and likely locomotor behavior, but it is also the most divergent in its dental anatomy. The skeletal specializations of gibbons that are associated with brachiation are unknown prior to the Pleistocene, and modern human bipedal walking and running seem to have evolved slowly, and perhaps in many different ways, during the past 7 million years.

The patterns in body size, diet, and locomotion described in the previous paragraphs are interrelated. It is not surprising that the first appearance of relatively large higher primates in the early Miocene is associated with evidence of more folivorous and terrestrial habits since folivory and terrestriality are functionally linked with relatively large size.

Adaptation and Phylogeny

Aspects of size and adaptation are clearly associated with particular taxonomic groups. The major taxonomic groups of living and fossil primates often have characteristic adaptive features that permit them to exploit a unique array of resources. Thus, if extant primates are clustered on the basis of features of basic ecology such as body size, diet, and locomotion, the result is a plot with closely related taxa grouping together (Fig. 18.2; Fleagle and Reed, 1996, 1999). In most cases, changes in

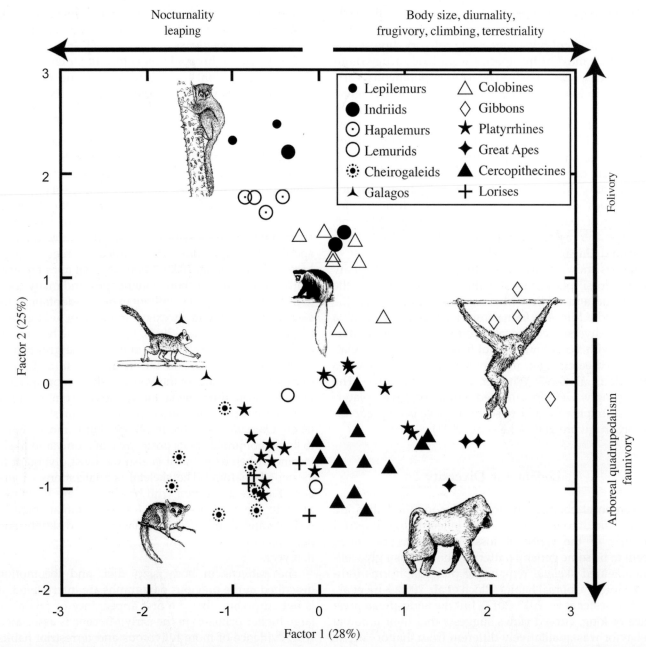

FIGURE 18.2 When extant primates are grouped according to ecological variables, phylogenetic groups tend to occupy specific ecological niches. *(Redrawn from Fleagle, J.G., Reed, K.E., 1999. Phylogenetic and temporal perspectives on primate ecology. In: Fleagle, J.G., Janson, C.H., Reed, K.E. [Eds.], Primate Communities. Cambridge University Press, Cambridge, pp. 92–115.)*

the "primate" adaptive zone through time are linked with the appearance or disappearance of particular groups of primates (Fig. 18.3). For example, the increase in folivory and terrestriality among higher primates since the Miocene is largely due to the radiation of Old World monkeys.

Even though different phylogenetic groups tend to show distinct and different combinations of morphological adaptations in a broad sense, there have nevertheless been some striking examples of parallel or convergent

adaptations in primate evolution between members of very different phylogenetic radiations. Bilophodont molar teeth have evolved independently in the strepsirrhine indriid *Archaeolemur* and in Old World monkeys, and the indriid *Hadropithecus* shows striking similarities to the hominine genus *Paranthropus* in both cranial morphology and isotopic signatures associated with diet (Fig. 18.4; e.g., Dumont et al., 2011; Godfrey et al., 2011).

Locomotor behavior shows many examples of parallel or convergent adaptation in different phylogenetic (and

Taxonomic Composition

Body Size Distribution

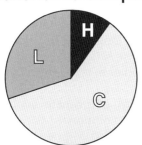

Present
Kibale Forest

ℍ = Hominoids
ℂ = Cercopithecoids
𝕃 = Lorisoids

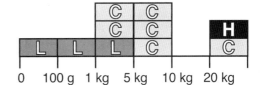

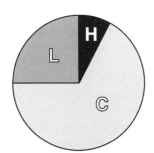

Pliocene   3.0 mya
Omo, Shungura B

ℍ = Hominoids
ℂ = Cercopithecoids
𝕃 = Lorisoids

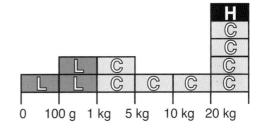

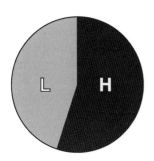

Early Miocene   20 mya
Songhor, Kenya

ℍ = Hominoids
𝕃 = Lorisoids

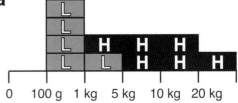

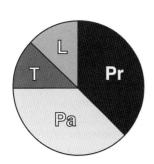

Early Oligocene   33 mya
Fayum, Quarries I,M

ℙ𝕒 = Parapithecoids
ℙ𝕣 = Propliopithecoids
𝕃 = Lorisoids
𝕋 = Tarsioids

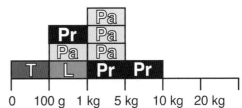

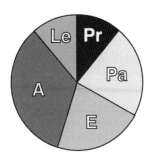

Late Eocene   36 mya
Fayum, L-41

ℙ𝕒 = Parapithecoids
ℙ𝕣 = Propliopithecoids
𝔼 = Early Anthropoids
𝕃𝕖 = Lemuriformes
𝔸 = Adapoids

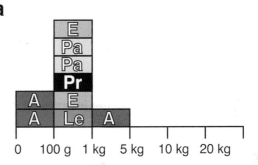

FIGURE 18.3   Changes in body size in African primate faunas are the result of the change in the taxonomic groups present.

Hadropithecus   Paranthropus

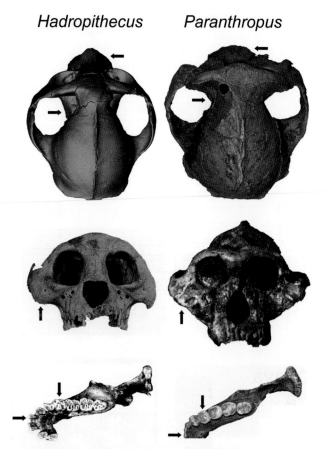

FIGURE 18.4 Similar cranial and dental adaptations between a subfossil Malagasy lemur, *Hadropithecus*, and the robust early hominin *Paranthropus* associated with similar dietary habits eating C4 plants such as grasses or underground storage organs such as tubers. *(Courtesy of L. Godfrey and T. Ryan.)*

geographic) radiations. Leapers, arboreal quadrupeds, and suspensory species have evolved independently in Old World primates (lorisoids, tarsiers, and catarrhines), Malagasy lemurs, and New World platyrrhines (Fig. 18.5; Fleagle and Lieberman, 2015). However, the morphological details of the adaptations of leapers or quadrupeds in different radiations are usually different. In addition, similar locomotor adaptations are not necessarily linked to parallel adaptations in other aspects of behavior. Thus, suspensory apes and platyrrhines are largely frugivorous, whereas the suspensory sloth lemurs are very folivorous.

Patterns in Primate Phylogeny

In the previous chapters, we considered the evolution of the major groups of both living and fossil primates one at a time, with particular consideration of their adaptive diversity. It is also interesting to compare patterns of diversity in the evolutionary history of these different primate groups during the past 66 million years. Theoretically, there are many different evolutionary patterns that we might expect to find. One pattern would be a series of distinct, long-lived old lineages. Alternatively, we might find a series of evolutionary radiations succeeding one another in time, or maybe one slowly replacing another.

Not surprisingly, the record of primate evolution shows evidence of all of these evolutionary patterns in various groups at various times. At a gross level, the major pattern of primate evolution seems to be one of succeeding adaptive radiations occupying the "primate adaptive zone": a radiation of plesiadapiforms in the Paleocene, followed by a radiation of prosimians at the beginning of the Eocene, and finally the radiation of anthropoids beginning in the later Eocene and early Oligocene. However, this sequence is a summary of our knowledge of all primates rather than an account of the faunal succession that took place in any one locality. Only from Eurasia do we actually have a fossil record in which all three radiations succeed one another. In other continental areas, one or more of these major radiations is absent from the fossil record, either because the animals were never there or because we have not uncovered the fossils of some particular period. Higher primates other than humans have never successfully invaded North America as far as we know; only higher primates are known from South America, and on Madagascar, strepsirrhines were the only primates present until the recent arrival of humans. It is possible, indeed likely, that some of the global patterns that we see in the primate fossil record reflect our available sample of fossil primates more than the actual timing or biogeography of the evolution of particular taxa. Regional changes through time have been much more diverse.

If we examine evolutionary radiations of subfamilies and families of primates within restricted geographic areas, we obtain a less distorted but more parochial view of evolutionary changes in primate history. Again, we find a diversity of evolutionary patterns. Among the platyrrhine monkeys of South America, we see evidence of many distinctive lineages going back to the later Miocene, none of which seem ever to have been very diverse (e.g., Rosenberger, 1984, 1992). The history of higher primate evolution in Africa seems to have been very different, with a succession of different anthropoids in the Eocene (many small species of debated affinities), Oligocene (parapithecids and propliopithecids), the early Miocene (proconsulids and a few cercopithecoids), and the late Miocene to Recent (hominins and cercopithecoid monkeys) (Fig. 18.3), with considerable differences among the continents of Africa, Europe, and Asia associated with geological and climatic changes.

An analysis at the generic level shows similar heterogeneity. There are a few primate genera that seem to

Eurasia & Africa

Madagascar

New World

Leapers   Quadrupeds   Hangers

FIGURE 18.5   Evolution of similar locomotor adaptations in primates of different continental areas. *(Images by S. Nash.)*

have persisted for millions of years with very little change: *Tarsius* (Beard et al., 1994; Rossie et al., 2006), *Saimiri, Aotus* (Setoguchi and Rosenberger, 1987), and *Macaca* (Delson and Rosenberger, 1984). Many other lineages have undergone dramatic morphological changes in a relatively short time, the most notable being *Homo*.

Much of our perception of the patterns in primate evolution is determined by the density of the fossil record. When only a few fossils are known from widely separated time periods, the simplest scenario is a single transformational lineage through time. However, as more fossils are recovered, phyletic trees become more complex, defying any simple linear sequence. In general, it seems that the initial appearance of most taxa is characterized by a rapid proliferation of forms, yielding an evolutionary tree with a bushy base. This has been clearly demonstrated with increasing discoveries of omomyoid prosimians, proconsuloid apes, early hominins, and early humans of the genus *Homo*.

Primate Evolution at the Species Level

One of the most poorly understood issues in evolutionary biology is the same one that preoccupied Darwin: the origin of species. In Darwin's day, the major issue was over the mechanism leading to evolutionary change and the appearance of new species. Darwin (1859) resolved this issue with his "discovery" and description of natural selection. In recent decades, there has been debate over the *tempo* of evolutionary change – i.e., whether new species appear by gradual modification of earlier types or through rapid changes in form (e.g., Eldredge and Gould, 1972; Gould and Eldredge, 1993; Carroll, 1997); currently there are major issues over how species are identified among both living and extinct primates (Chapter 1).

There are several periods in primate evolution for which the fossil record is sufficiently well sampled, allowing questions about the pattern of evolution to be fruitfully examined. Two of the best examples come from western North America, where Gingerich, Bown, Rose and colleagues have carefully documented the evolutionary history of fossil mammals through a long, continuous series of late Paleocene and early Eocene sediments in northern Wyoming. Overall, there is evidence for many different patterns of morphological change, from gradual change through time to abrupt morphological shifts. For the instances in which a new species appears abruptly, it is impossible to determine whether this abrupt appearance is the result of rapid, discontinuous change from another local form, immigration from another area, or an absence of linking forms because of missing fossils. This is one of the major difficulties in testing theories of evolutionary change with

fossil evidence. A record of gradual change is positive evidence for gradual evolution, but a record of discontinuous change can be interpreted as the result of several very different phenomena.

Bown and Rose (e.g., 1987) have produced extraordinarily detailed documentation of evolutionary change within lineages of early Eocene omomyoid and adapoid prosimians in northern Wyoming (Figs. 18.6 and 18.7). Charting gradual change in many aspects of dental morphology, including reduction in size and loss of teeth, changes in the size and shape of cusps, and changes in the number and size of tooth roots, shows that, within a lineage, the transition from one paleospecies to another is characterized by changing frequencies of the diagnostic features within intermediate populations. However, the morphological differences that characterize the end products rarely change at the same rate. Thus, species-specific differences in morphology are a dynamic phenomenon and are easiest to characterize in the absence

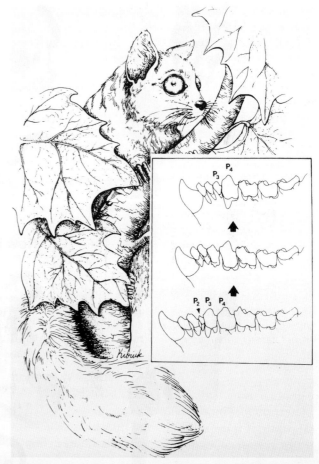

FIGURE 18.6 Changes through time in the lower dentition of a lineage of fossil prosimians from the early Eocene of Wyoming. Note the loss of P$_2$ and the gradual change in the size and shape of P$_3$ and P$_4$. *(Modified from Bown, B., Rose, K.D., 1987. Patterns of dental evolution in early Eocene anaptomorphine primates (Omomyidae) from the Bighorn Basin, Wyoming. Paleontological Society Memoir no. 23. J. Paleontol. 61, 1–62.)*

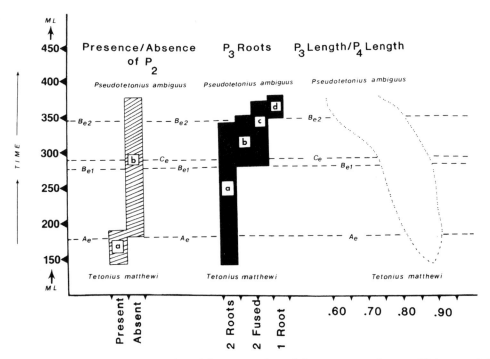

FIGURE 18.7   Changes in the size of P_3/P_4, the number of the roots on P_3, and the presence or absence of P_3 in an evolving lineage of early Eocene omomyids. The morphological changes take place at different rates and at different times in the stratigraphic sequence. Because of the mosaic nature of morphological change in this lineage, identification of the boundary between the older species, *Tetonius matthewi*, and the younger species, *Pseudotetonius ambiguous*, must be based on an arbitrary choice of criteria. *(Modified from Modified from Bown, B., Rose, K.D., 1987. Patterns of dental evolution in early Eocene anaptomorphine primates (Omomyidae) from the Bighorn Basin, Wyoming. Paleontological Society Memoir no. 23. J. Paleontol. 61, 1–62.)*

of intermediate forms. For a very good fossil record, the identification of species boundaries becomes arbitrary, depending on which morphological feature is being used. *Tetonius matthewi*, for example, gradually evolves into *Pseudotetonius ambiguus* in the early Eocene deposits of northern Wyoming. However, the diagnostic features of the latter appear at different levels in the stratigraphic section and at different rates. Loss of P_2 occurs relatively low in the geological section, reduction of the roots of P_3 takes place in a series of steps, and changes in the size of P_3/P_4 take place gradually. Depending on which criterion or combination of criteria is used to define *Pseudotetonius ambiguus*, one can place the taxonomic transition at almost any time within the evolutionary lineage. There is no abrupt appearance of a new species; rather, there is a mosaic appearance of morphological changes that can be used, in retrospect, to define a different species. It seems likely that this is a common pattern of evolutionary change, and the distinctiveness of many species in the fossil record is, to a large degree, an artifact of the incompleteness of the record. In contrast, taxa that are well sampled in the fossil record often pose problems in identifying species boundaries.

Among fossil monkeys, the genus *Theropithecus* is extremely abundant in the fossil record of eastern and southern Africa, and the "species" *Theropithecus oswaldi*

is well sampled over more than three million years. Although the endpoints of this evolutionary sequence are very different, there is no clear break in the temporal sequence of changing morphologies (Jablonski, 1993; Jablonski and Frost, 2010; Frost et al., 2022).

The fossil record of human evolution contains many examples of gradual morphological change that makes the identification of individual species difficult. In East Africa, the pattern of morphological change in the teeth of *Australopithecus anamensis* and *Australopithecus afarensis* has been argued to be consistent with the two "species" being part of a continuously evolving lineage (Kimbel et al., 2006). Similarly, fossils attributed to *Homo heidelbergensis* in Europe seem to grade continuously into *Homo neanderthalensis* (Hublin, 2009; Brauer, 2012; Stringer, 2012). In contrast, other parts of the hominin fossil record seem to show abrupt change and/or the rapid appearance of new species, often associated with climatic change. This seems to be the case in the Middle East over the past 100,000 years. Overall, hominin evolution seems to show examples of many types of evolutionary change (e.g., Kimbel, 1995).

Until recently, species of both extant and fossil primates were identified on the basis of morphological features, often cranial and dental features, although for extant taxa, pelage coloration often played a role. In the beginning of the twenty-first century, this is no longer the case. Many extant species are based on genetic information that is not available for most fossils. It is unclear how this will impact our view of primate diversity through time, but it is certainly a topic that deserves more attention.

Mosaic Evolution

Perhaps the clearest pattern evident in the fossil record is that the evolution of primate groups has not taken place through large, abrupt morphological transformations, but rather through a pattern of mosaic evolution (e.g., McHenry, 1975; Cartmill, 2012). The characteristic anatomical features of living species, genera, families, and subfamilies have more commonly appeared, one by one, over long expanses of time rather than in sudden bursts of many coordinated changes. This is an important fact to remember when trying to explain the origin of major groups in terms of the adaptive characteristics of the living members. The large anatomical differences we see between many living taxa today are clearly the result of the extinction of intermediate forms. A major contribution of the fossil record is in providing us with intermediate forms that document the

sequence through which modern differences appeared and the timing of the appearances of the successive novelties (Figs. 18.8 and 18.9). For example, all living catarrhines are distinguished from platyrrhines by a suite of characters, including reduction of the premolar count from three to two in each quadrant, a connection of the frontal and sphenoid bone on the side wall of the skull, and a bony tube extending laterally from the ectotympanic ring (Fig. 10.8). However, the early Oligocene *Aegyptopithecus zeuxis* shows an intermediate condition between catarrhines and platyrrhines, having two premolars, the catarrhine skull wall, but no ear tube, and demonstrates both the sequence in which modern catarrhine features appeared and also provides a minimum age for the first appearance of the premolar loss and appearance of the bony pattern on the skull wall. *Aegyptopithecus* is indeed a missing link as the name *zeuxis* suggests.

Thus, humans (Fig. 18.9) and all other organisms are made up of a mosaic of features that have been acquired at various times and for various purposes during their evolutionary past. For example, our fingernails, which we share with all primates, first appeared approximately 56 million years ago with the earliest record of primates in the early Eocene. The posterior wall of our orbit, which we share with all anthropoids, first appeared

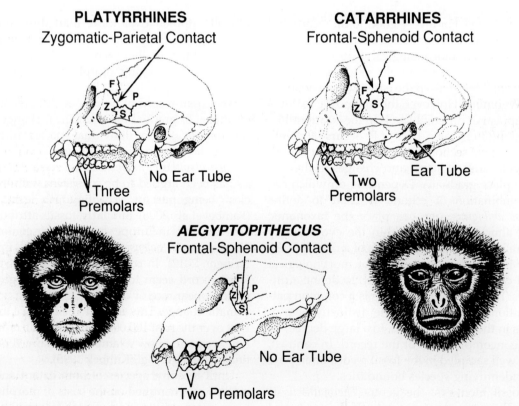

FIGURE 18.8   *Aegyptopithecus zeuxis,* from the early Oligocene of Egypt, displays an intermediate mixture of cranial features when compared with living platyrrhines and catarrhines.

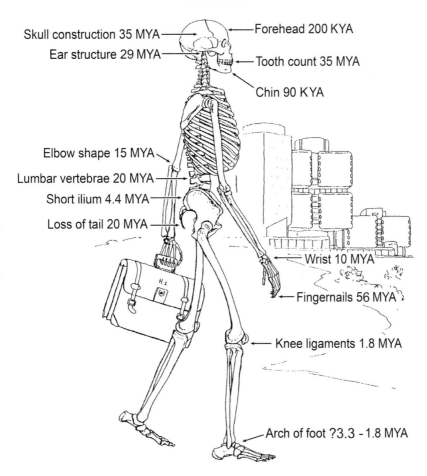

Skull construction 35 MYA

Forehead 200 KYA

Ear structure 29 MYA

Tooth count 35 MYA

Chin 90 KYA

Elbow shape 15 MYA

Lumbar vertebrae 20 MYA

Short ilium 4.4 MYA

Loss of tail 20 MYA

Wrist 10 MYA

Fingernails 56 MYA

Knee ligaments 1.8 MYA

Arch of foot ?3.3 - 1.8 MYA

FIGURE 18.9 The timing of the appearance of distinctive anatomical features of the human skeleton that have evolved during the past 56 million years.

approximately 35 million years ago, whereas our tailless condition, which we share with apes, is a much later development and probably dates to around 20 million years ago. Although living humans can be distinguished from chimpanzees by a whole suite of anatomical differences, we know that these were not evolved in a single change but slowly over millions of years. On the basis of the current record, it seems that our shortened ilium was the first distinctive feature to appear over 4 million years ago. The characteristic arch of our foot and our unusual knee cartilages appeared later. Features such as our prominent chin and high forehead are much more recent, dating to the last one or two hundred thousand years. This pattern of mosaic evolution can only be elucidated by a well-dated fossil record.

Primate Extinctions

Over the past 56 million years, many new primate species have appeared, and slightly fewer have become extinct. In many cases, a species disappeared by evolving into an animal with a different morphology that we recognize as a new species; this phenomenon is called **pseudoextinction**. In other cases, species and lineages disappeared, leaving no descendants. Three primary reasons are commonly given to account for the major extinctions in the primate fossil record: climatic changes, competition with other primates or other mammals, and predation.

Climatic Changes

Living primates are, for the most part, tropical animals, with only a few hardy genera and species found in temperate climates. Most primates are arboreal animals; only some baboons and humans have successfully abandoned forested areas for a life in open habitats largely devoid of trees. Climate clearly plays a role in the distribution and diversity of living primates. In general, primate species are more common at lower latitudes and in areas with increased rainfall (see Chapter 8). Thus, climatic change has frequently been put forth to explain the disappearance of a primate group (Fig. 18.10). Climate has played

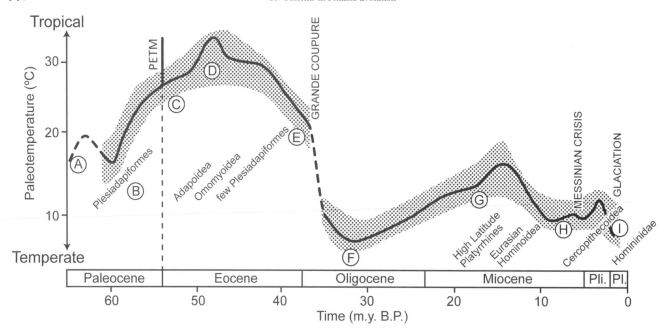

FIGURE 18.10 Temperatures during the Cenozoic, with major events in the primate fossil record of the northern hemisphere marked: A, appearance of *Purgatorius*; B, radiation of plesiadapiforms; C, appearance of modern primates, adapoids, and omomyoids following the Late Paleocene Thermal Excursion (PETM); D, height of primate diversity in Europe and North America associated with the Eocene Thermal Maximum; E, decline and extinction of primates in Europe and North America associated with global cooling in the early Oligocene; F, nadir of primate diversity in northern continents; G, appearance of hominoids in Europe and Asia as well as fossil platyrrhines in southernmost South America; H, disappearance of northern hominoids, extensive ecological radiation of cercopithecoids, and emergence of hominids; I, evolution and dispersal of humans. *(Modified from Gingerich, P.D., 1986. Plesiadapis and the delineation of the order Primates. In: Wood, B., Martin, L., Andrews, P. (Eds.), Major Topics in Primate and Human Evolution. Cambridge University Press, Cambridge, pp. 32–46.)*

a major role in the extinction of many primate groups from northern continents, including the adapoids and omomyoids of Europe at the end of the Eocene and the European hominoids at the end of the Miocene (e.g., Gingerich, 1986; Merceron et al., 2010). In addition, increased aridity has been suggested as a contributor to the disappearance of the proconsuloids in the middle Miocene of Africa (Pickford, 1983). On the other hand, climate change has also played a major role in facilitating primate dispersals, such as the spread of early prosimians during the Paleocene-Eocene Thermal Maximum (Fig. 12.25). Thus, climate has been a major factor in both the appearance and the extinction of species as well as major clades (e.g., Gunnell, 1997; Springer et al., 2012).

Competition

Since climatic changes evident in the fossil record are invariably associated with faunal change resulting from the appearance or spread of new groups of primates and other mammals, we must attempt to distinguish the effects of climate per se from the effects of ecological competition with other species. The extinction of the plesiadapiforms, for example, coincides roughly with the radiation of both rodents and early prosimians, and the decline of proconsuloids in the middle and late Miocene of Africa has been associated with an increased abundance of cercopithecoid monkeys (e.g., Andrews, 1981). Unfortunately, it is very difficult to determine

whether a new group contributes to the extinction of earlier groups through direct competition for resources or merely fills the gap left by their extinction. For example, the replacement of apes as the dominant catarrhines by monkeys in the later Miocene of Africa may be the result of their differential ability to thrive in a drier climate rather than due to direct competition (Harrison, 2010). Careful analysis is required to sort out the alternative possibilities (see Maas et al., 1988). Although there is some evidence for ecological competition among extant primates (e.g., Ganzhorn, 1988, 1989), it has been suggested that in most cases, the fossil record may not provide adequate temporal resolution to permit the identification of ecological competition (Sepkoski, 1996; but see Gingerich, 1996). In the case of the extinction of plesiadapiforms, it appears that competition with rodents may well have contributed to the demise of the group, whereas the radiation of the prosimians occurred after the decline or disappearance of most plesiadapiforms and so seems a less likely cause of extinction but rather a case of opportunistic replacement.

Predation

One cause of primate extinctions that seems well documented, in at least one case, is human predation. The disappearance of the large diurnal lemurs from the fauna of Madagascar clearly postdates the appearance of humans on the island and is most likely due in large part to human hunting and habitat destruction (Dewar, 1984, 1997; Godfrey et al., 2010). The lemur extinction on Madagascar was a relatively recent event, dating to the last 5000 years, but it is also not unlikely that hominin predation was also responsible for the extinction of most of the large monkeys from East Africa during the early and middle Pleistocene, for the disappearance of the orangutan from mainland Asia, and for the very recent extinction of New World monkeys in the Caribbean.

Limiting Primate Extinctions

We often have difficulty in reconstructing the events that led to the extinctions of particular species, genera, and/or lineages of primates in the past. In contrast, the factors that will probably lead to the extinction of many primate species alive today are relatively easy to identify: habitat destruction, hunting, and live capture for the pet or research markets. Of these, habitat destruction due to our expanding human population is the greatest and most immediate threat. More than 90% of all primate species live in the tropical forests of Asia, Africa, and South and Central America, and their fate is intimately linked with the future of these forests. There are endangered primates in virtually every part of the tropical world, but the greatest threats are in Madagascar and eastern Brazil where faunas of unique primates with limited geographic ranges are under extreme pressure from rapidly growing local human populations that rely on these forests and land to survive (Fig. 18.11). What is more, increasing globalization means that commercial interests in these areas are also on the rise. Foreign interests in agriculture (e.g., palm oil); timber exploitation, including the illegal logging of precious hardwoods; and strip mining for everything from precious gemstones to the minerals that help power our cellular phones and hybrid cars, further threaten these already precariously balanced ecosystems. To this end, Madagascar's lemurs are among the most endangered mammals alive today and risk extinction within less than two generations unless current patterns of human exploitation change.

FIGURE 18.11 Four of the 25 most endangered primates in 2012. Clockwise from *upper left*: the Lion-tailed macaque (*Macaca silenus*) from Asia; the Ruffed lemur (*Varecia variegata*) from Madagascar; the Roloway guenon (*Cercopithecus diana roloway*) from Africa; and the Yellow-tailed woolly monkey (*Lagothrix flavicauda*) from South America.

With an expanding global human population and the developmental demands of many parts of the tropical world, it is inevitable that many primate populations and species will disappear in the coming decades (e.g., Jernvall and Wright, 1998; Tecot and Wright, 2010). The goal of conservationists is to limit the losses as much as possible. This involves (1) protecting areas for particularly endangered and vulnerable species; (2) creating large national parks and reserves in areas of high primate diversity or abundance; (3) maintaining parks and reserves that already exist and enforcing protective legislature in them; (4) creating public awareness of the need for primate conservation and the importance of primates both as a national heritage and as a resource determining ways in which people and other primates can coexist in multiple-use areas (Mittermeier and Cheney, 1986). Only such a broad approach will ensure that future generations are able to appreciate the living remnants of 56 million years of primate adaptation and evolution.

References

Andrews, P., 1981. Species diversity and diet in monkeys and apes during the Miocene. In: Stringer, C.B. (Ed.), Aspects of Human Evolution. Taylor and Francis, London, pp. 25–61.

Beard, K.C., Tao, Q., Dawson, M.R., Wang, B., ChuanKuei, L. (Eds.), 1994. A diverse new primate fauna from the middle Eocene fissure-fillings in southeastern China Nature. 368, 604–609.

Bloch, J.I., Silcox, M.T., Boyer, D.M., Sargis, E.J., 2007. New Paleocene skeletons and the relationship of plesiadapiforms to crown-clade primates. Proc. Nat. Acad. Sci. USA. 104, 1159–1164.

Bown, B., Rose, K.D., 1987. Patterns of dental evolution in early Eocene anaptomorphine primates (Omomyidae) from the Bighorn Basin, Wyoming. Paleontological Society Memoir no. 23. J. Paleontol. 61, 1–62.

Brauer, G., 2012. Middle Pleistocene diversity in Africa and the origin of modern humans. In: Hublin, J.-J., McPherron, S.P. (Eds.), Modern Human Origins: A North African Perspective. Springer, New York, pp. 221–240.

Carroll, R.I., 1997. Patterns and Processes of Vertebrate Evolution. Cambridge Paleobiology Series. Cambridge University Press, Cambridge.

Cartmill, M., 2012. Primate origins, human origins, and the end of higher taxa. Evol. Anthropol. 21, 208–220.

Chester, S.G., Bloch, J.I., Boyer, D.M., Clemens, W.A., 2015. Oldest known euarchontan tarsals and affinities of Paleocene Purgatorius to Primates. P. Natl. Acad. Sci. USA. 112, 1487–1492.

Chester, S.G.B., Williamson, T.E., Bloch, J.I., Silcox, M.T., 2017. Oldest skeleton of a plesiadapiform provides additional evidence for an exclusively arboreal radiation of stem primates in the Palaeocene. R. Soc. Open Sci. 4, 170329. https://doi.org/10.1098/rsos.170329

Cooke, S., 2011. Paleodiet of extinct platyrrhines with emphasis on the Caribbean forms: Three-dimensional geometric morphometrics of mandibular second molars. Anat. Rec. 294, 2073–2091.

Cooke, S., Rosenberger, A.L., Turvey, S., 2011. An extinct monkey from Haiti and the origins of the Greater Antillean primates. Proc. Natl. Acad. Sci. USA 108, 2699–2704.

Covert, H.H., 1986. Biology of Cenozoic primates. In: Swindler, D.W., Erwin, J. (Eds.), Comparative Primate Biology, vol. 1, Systematics. Evolution, and Anatomy. Alan R. Liss, New York, pp. 335–359.

Covert, H.H., 1995. Locomotor adaptations of Eocene primates: Adaptive diversity among the earliest prosimians. In: Alternman, L., Doyle, G.A., Izard, M.K. (Eds.), Creatures of the Dark: The Nocturnal Prosimians. Plenum Press, New York, pp. 495–509.

Dagosto, M., 2007. The postcranial morphotype of primates. In: Ravosa, M.J., Dagosto, M. (Eds.), Primate Origins. Kluwer, New York, pp. 489–534.

Delson, E., Rosenberger, A.L., 1984. Are there any anthropoid primate living fossils? In: Eldredge, N., Stanley, S.M. (Eds.), Living Fossils. Springer-Verlag, New York, pp. 50–61.

Dewar, R.E., 1984. Extinctions in Madagascar: The loss of the subfossil fauna. In: Martin, P.S., Klein, R.G. (Eds.), Quaternary Extinctions: A Prehistoric Revolution. University of Arizona Press, Tucson, pp. 574–593.

Dewar, R.E., 1997. Were people responsible for the extinction of Madagascar's subfossils, and how will we ever know? In: Goodman, S.M., Patterson, B.D. (Eds.), Natural Change and Human Impact in Madagascar. Smithsonian Institution Press, Washington, D.C., pp. 364–380.

Dumont, E.R., Ryan, T.M., Godfrey, L.R., 2011. The Hadropithecus conundrum reconsidered, with implications for interpreting diet in fossil hominins. Proc. R. Soc. B 278, 3654–3661.

Eldredge, N., Gould, S.J., 1972. Punctuated equilibria: An alternative to phyletic gradualism. In: Schopf, T.J.M. (Ed.), Models in Paleobiology. Freeman Cooper, San Francisco, pp. 82–115.

Fleagle, J.G., 1978. Size distributions of living and fossil primate faunas. Paleobiol. 4, 67–76.

Fleagle, J.G., Anapol, F.C., 1992. The indriid ischium and the hominid hip. J. Hum. Evol. 22, 285–305.

Fleagle, J.G., Kay, R.F., 1985. The paleobiology of catarrhines. In: Delson, E. (Ed.), Ancestors: The Hard Evidence. Alan R. Liss, New York, pp. 23–36.

Fleagle, J.G., Kay, R.F., Anthony, M.R.L., 1997. Fossil New World monkeys. In: Kay, R.F., Madar, S.I., Cifelli, R.L., Flynn, J.J. (Eds.), Vertebrate Paleontology in the Neotropics. The Miocene Fauna of La Venta, Columbia. Smithsonian Institution Press, Washington, D.C., pp. 473–495.

Fleagle, J.G., Reed, K.E., 1996. Comparing primate communities: A multivariate approach. J. Hum. Evol. 30, 489–510.

Fleagle, J.G., Reed, K.E., 1999. Phylogenetic and temporal perspectives on primate ecology. In: Fleagle, J.G., Janson, C.H., Reed, K.E. (Eds.), Primate Communities. Cambridge University Press, Cambridge, pp. 92–115.

Feagle, J.G., Lieberman, D.E., 2015. Major transformations in the evolution of primate locomotion. In: Dial, K., Shubin, N., Brainerd, E. (Eds.), Great Transformations in Vertebrate Evolution. University of Chicago Press, Chicago, pp. 257–278.

Frost, S.R., White, F., Reda, H., Gilbert, C.C., 2022. Biochronology of South African hominin-bearing sites: A reassessment using cercopithecid primates. Proc. Natl. Acad. Sci. USA. 119, e2210627119.

Ganzhorn, J.U., 1988. Food partitioning among Malagasy primates. Oecologia. 75, 436–450.

Ganzhorn, J.U., 1989. Primate species separation in relation to secondary plant chemicals. Hum. Evol. 4, 125–132.

Gebo, D.L., 2011. Vertical clinging and leaping revisited: vertical support use as the ancestral condition of strepsirrhine primates. Am. J. Phys. Anthropol. 146, 323–335.

Gebo, D.L., Dagosto, M., Beard, K.C., Qi, T., 2000. The smallest primates. J. Hum. Evol. 38, 585–594.

Gebo, D.L., Dagosto, M., Ni, X., Beard, K.C., 2012. Species diversity and postcranial anatomy of Eocene primates from Shanghuang, China. Evol. Anthropol. 21, 224–238.

Gingerich, P.D., 1986. Plesiadapis and the delineation of the order Primates. In: Wood, B., Martin, L., Andrews, P. (Eds.), Major Topics in Primate and Human Evolution. Cambridge University Press, Cambridge, pp. 32–46.

Gingerich, P.D., 1996. Rates of evolution in divergent species lineages as a test of character displacement in the fossil record: tooth size in Paleocene Plesiadapis (Mammalia: Proprimates). Paleovertebrata 25, 193–204.

Godfrey, L.R., Crowley, B.E., Dumont, E.R., 2011. Thinking outside the box: A lemur's take on hominin craniodental evolution. Proc. Natl. Acad. Sci. USA. 108, E742.

Godfrey, L.R., Jungers, W.L., Burney, D.A., 2010. Subfossil lemurs of Madagascar. In: Werdelin, L., Sanders, W.J. (Eds.), Cenozoic Mammals of Africa. University of California Press, Berkley, pp. 351–367.

Gould, S.J., Eldredge, N., 1993. Punctuated equilibrium comes of age. Nature 366, 223–227.

Gunnell, G.F., 1997. Wasatchian-Bridgerian (Eocene paleoecology of the western interior of North America: changing paleoenvironments and taxonomic composition of omomyid (Tarsiiformes) primates. J. Hum. Evol. 32, 105–132.

Harrison, T., 2010. New estimates of hominoid diversity in Africa during the Neogene and its implications for understanding catarrhine community structure. Abstracts of the 70th Annual Meetings of the Society of Vertebrate Paleontology, 102a.

Hublin, J.-J., 2009. The origin of Neandertals. Proc. Natl. Acad. Sci. USA 106, 16022–16027.

Jablonski, N.G., 1993. The phylogeny of Theropithecus. In: Jablonski, N.G. (Ed.), Theropithecus: The Rise and Fall of a Primate Genus. Cambridge University Press, Cambridge, pp. 209–224.

Jablonski, N.G., Frost, S., 2010. Cercopithecoidea. In: Werdelin, L., Sanders, W.J. (Eds.), Cenozoic Mammals of Africa. University of California Press, Berkeley, pp. 393–428.

Jernvall, J., Wright, P.C., 1998. Diversity components of impending primate extinctions. Proc. Natl. Acad. Sci. USA 95, 11279–11283.

Kay, R.F., Ungar, P.S., 1997. Dental evidence for diet in some Miocene catarrhines with comments on the effects of phylogeny on the interpretation of adaptation. In: Begun, D.R., Ward, C.V., Rose, K.D. (Eds.), Function, Phylogeny, and Fossils: Miocene Hominoid Evolution and Adaptations. Plenum Press, New York, pp. 131–151.

Kimbel, W.H., 1995. Hominoid speciation and Pliocene climatic change. In: Vrba, E.S., Denton, G.H., Partridge, T.C., Burckle, L.H. (Eds.), Paleoclimate and Evolution, With Emphasis on Human Evolution. Yale University Press, New Haven, pp. 425–437.

Kimbel, W.H., Lockwood, C.A., Ward, C.V., Leakey, M.G., Rak, Y., Johanson, D.C., 2006. Was Australopithecus anamensis ancestral to A. afarensis? A case of anagenesis in the hominin fossil record. J. Hum. Evol. 51, 134–152.

Kirk, E.C., Simons, E.L., 2001. Diets of fossil primates from the Fayum Depression of Egypt: A quantitative analysis of molar shearing. J. Hum. Evol. 40, 203–229.

Kirk, E.C., Lemelin, P., Hamrick, M.W., Boyer, D.M., Bloch, J.I., 2008. Intrinsic hand proportions of euarchontans and other mammals: implications for the locomotor behavior of plesiadapiforms. J. Hum. Evol. 55, 278–299.

Larson, S.G., 1998. Parallel evolution in the hominoid trunk and forelimb. Evol. Anthropol. 6, 87–99.

Maas, M.C., Krause, D.W., Strait, S.G., 1988. Decline and extinction of Plesiadapiformes (Mammalia: Primates) in North America: displacement or replacement? Paleobiol. 14, 410–431.

McHenry, H.M., 1975. Fossils and the mosaic nature of human evolution. Science 190, 425–431.

Merceron, G., Kaiser, T.M., Kostopoulos, D.S., Schule, E., 2010. Ruminant diets and the Miocene extinction of European great apes. Proc. R. Soc. B 277, 3105–3112.

Mittermeier, R.A., Cheney, D.L., 1986. Conservation of primates and their habitats. In: Smuts, B.B., Cheney, D.L., Seyfarth, R.M., Wrangham, R.W., Struhsaker, T.T. (Eds.), Primate Societies. University of Chicago Press, Chicago, pp. 477–490.

Pickford, M., 1983. Sequence and environments of the lower and middle Miocene hominoids of western Kenya. In: Ciochon, R.L., Corruccini, R.S. (Eds.), New Interpretation of Ape and Human Ancestry. Plenum Press, New York, pp. 421–439.

Rosenberger, A.L., 1984. Fossil New World monkeys dispute the molecular clock. J. Hum. Evol. 13, 737–742.

Rosenberger, A.L., 1992. Evolution of feeding niches in New World monkeys. Am. J. Phys. Anthropol. 88, 525–562.

Rossie, J.B., Ni, X., Beard, K.C., 2006. Cranial remains of an Eocene tarsier. Proc. Natl. Acad. Sci. USA 103, 4381–4385.

Seiffert, E.R., Perry, J.G.M., Simons, E.L., Boyer, D.M., 2009. Convergent evolution of anthropoid-like adaptations in Eocene adapiform primates. Nature 461, 1118–1121.

Sepkoski, J.J., 1996. Competition in macroevolution: the double wedge revisited. In: Jablonski, N.G., Erwin, D.H., Lipps, J.H. (Eds.), Evolutionary Paleobiology. University of Chicago Press, Chicago.

Setoguchi, T., Rosenberger, A.L., 1987. A fossil owl monkey from La Venta, Colombia. Nature 326, 692–694.

Silcox, M.T., Boyer, D.M., Bloch, J.I., Sargis, E.J., 2007. Revisiting the adaptive origins of primates (again). J. Hum. Evol. 53, 321–324.

Soligo, C., 2006. Correlates of body mass evolution in primates. Am. J. Phys. Anthropol. 130, 283–293.

Soligo, C., Martin, R.D., 2006. Adaptive origins of primates revisited. J. Hum. Evol. 50, 414–430.

Soligo, C., Martin, R.D., 2007. The first primates: a reply to Silcox et al. (2007). J. Hum. Evol. 53, 325–328.

Springer, M.S., Meredith, R.W., Gatesy, J., Emerling, C.A., Park, J., et al., 2012. Macroevolutionary dynamics and historical biogeography of primate diversification inferred from a species supermatrix. PLoS One 7, e49521.

Strait, S.G., 1997. Tooth use and the physical properties of food. Evol. Anthropol. 5, 199–211.

Stringer, C., 2012. The status of Homo heidelbergensis (Schoetensack 1908). Evol. Anthropol. 21, 101–107.

Tecot, S., Wright, P.C., 2010. Primate conservation efforts Yearbook of Science and Technology. McGraw Hill, New York. 310–315.

Wheeler, B.C., 2010. Community ecology of the Middle Miocene primates of La Venta, Colombia: The relationship between ecological diversity, divergence time, and phylogenetic richness. Primates 51, 131–138.

Young, N.M., 2003. A reassessment of living hominoid postcranial variability: Implications for ape evolution. J. Hum. Evol. 45, 441–464.

Glossary

abduction: movement of a limb or part of a limb away from the midline of the body

absolute dating: determination of the age, in years, of a fossil or fossil site, usually on the basis of the amount of change in radioactive elements in rocks

adaptation: evolutionary process whereby a population changes (usually through selection) to better survive in its given environment; or, a specific new characteristic that enables survival

adaptive radiation: a group of closely related organisms that have evolved morphological and behavioral features enabling them to exploit different ecological niches

adduction: movement of a limb or part of a limb toward the midline of the body

age-graded group: a social group with several adult females and several adult males who differ in social and reproductive status according to age. Thus, with time, an age-graded group can change from a one-male reproductive system to a multimale system

Allen's Rule: an ecogeographic principle that states that warm-blooded species from colder climates tend to have shorter limbs than those living in warmer climates

allometry: the relationship between the size and shape of an organism, or, more broadly, the relationship between an organism's size and various aspects of its biology, such as morphology, ecology, or behavior; also, the study of such relationships

alloparenting (aunting): assistance in care of infants and juveniles by individuals other than parents

allopatry: the absence of overlap in the geographical range of two species or populations

anagenesis: the pattern in which a lineage undergoes gradual evolutionary change through time

arboreal: living in trees

arboreal quadrupedalism: mode of locomotion in which an animal moves along horizontal branches with a regular gait pattern involving all four limbs

articulation: a joint between two or more bones

basal metabolism: the energy requirements of an animal at rest

Bergmann's Rule: an ecogeographic principle that states that within a broadly distributed genus, species tend to be larger at higher latitudes (colder environments)

bilateral symmetry: a type of developmental shape in which right and left sides of the organism are mirror images of one another

bilophodonty: a condition of the molar teeth in which the mesial and distal pairs of cusps form two ridges or lophs

biogeography: the study of the distribution of species, organisms, and ecosystems in geographic space and through geological time

biomass: the sum of the weights of the organisms in a particular area

bipedalism: mode of locomotion using only the hindlimbs, usually alternately rather than together

brachiation: arboreal locomotion in which the animal progresses below branches by using only the forelimbs

buccal: the cheek side of a tooth

bunodont: (teeth) have low, rounded cusps

canopy: a layer of forest foliage that is laterally continuous and usually distinct vertically from other layers. A tropical forest often has one or more distinct canopies

cathemeral: active intermittently throughout the 24-hour day rather than active only during the day (diurnal) or only during the night (nocturnal)

clade: a group composed of all the species descended from a single common ancestor; a monophyletic group

cladogenesis: division of an ancestral species into two separate descendent species

cladogram: branching tree diagram used to represent phyletic relationships

conspecific: belonging to the same species

convergent evolution: the independent evolution of similar morphological features from different ancestral conditions. The wings of bats and birds are an example of convergent evolution

core area: the part of a group's home range that is used most intensively

cranial capacity: the volume of the brain, usually determined by measuring the volume of the inside of the neurocranium

crepuscular: active primarily during the hours around dawn and dusk

cryptic: hidden; not normally visible

day range: the distance a group of animals travels during a single day

deciduous dentition: the milk teeth or first set of teeth in the mammalian jaw. The deciduous dentition is replaced by the permanent dentition

dental eruption sequence: the order in which the different teeth erupt or come into use

dental formula: a notation of the number of incisors, canines, premolars, and molars in the upper and lower dentition of a species. In humans, the adult dental formula is 2.1.2.3/2.1.2.3

derived feature: a specialized morphological (or behavioral) characteristic that departs from the condition found in the ancestors of a species or group of species

diagnostic: distinguishing or characteristic, as the diagnostic features of a group of organisms

diurnal: active primarily during daylight hours

dorsal: toward the back side of the body; the opposite of ventral

dizygotic twins (fraternal twins): twins that develop from two fertilized eggs or zygotes. This contrasts with monozygotic (or identical) twins, which develop from a single fertilized egg or zygote

ecological niche: the complex of features (such as diet, forest type preference, canopy preference, activity pattern) that characterize the position a species occupies in the ecosystem

ecology: the study of the relationship between an organism and all aspects of its environment; or all aspects of the environment of an organism which affect its way of life

emergent trees: the trees in a tropical forest which extend above the relatively continuous canopy

Primate Adaptation and Evolution
https://doi.org/10.1016/B978-0-12-815809-8.00020-5

endocast: an impression of the inside of the cranium, often preserving features of the surface of the brain

evolution: modification by descent, or genetic change in a population through time

extant: living, as opposed to extinct

extension: a movement in which the angle of a limb joint increases

exudate: a substance, such as gum, sap, or resin, which flows from the vascular system of a plant

faunal correlation: determination of the relative ages of different geological strata by comparing the fossils within the strata and assigning similar ages to strata with similar fossils; a method of relative dating

faunivore: an animal that eats primarily other animals (includes insectivores and carnivores)

fission–fusion dynamics: a ranging pattern found among several primate species (though to varying degrees) in which group composition and cohesion varies during the course of hours, days, and even weeks; the variation in grouping usually depends on the types and distribution of food resources

fitness: an individual's reproductive success, or its relative success in passing on its genes to the next generation compared with other individuals

flexion: a movement in which the angle of a limb joint decreases; the opposite of extension

folivore: an animal that feeds primarily on leaves

foraging strategy: the behavioral adaptations of a species related to its acquisition of food items

Forster effect: a biogeographic phenomenon in which species diversity becomes greater near the equator than at higher latitudes

founder effect: changes in an allele frequency that are the result of a small initial population

frugivore: an animal that feeds primarily on fruit

gallery forest: a forest along a river or stream

genetic drift: change in allele frequencies in a population due to chance rather than selection

genotype: the genetic makeup of an organism

Gloger's Rule: a biogeographic pattern in which species living in warm wet habitats tend to have darker pelage than those in dry habitats

gracile: relatively slender or delicately built

grade: a level or stage of organization, or a group of organisms sharing a suite of features (either primitive or derived) that distinguishes them from more advanced or more primitive animals but does not necessarily define a clade

gradistic classification: a classification in which organisms are grouped according to grade or level of organization rather than according to ancestry or phylogeny

graminivore: an animal that eats primarily grains; often also used to describe an animal that eats seeds

gregarious: living in regular social groups; contrasted to solitary living

grooming: the cleaning of the body surface by licking, biting, picking with fingers or claws, or other kinds of manipulation

growth allometry: the relationship between size and shape during the growth (or ontogeny) of an organism

gummivore: an animal that eats exudates – gum, saps, or resins

holophyletic group: a taxonomic group of organisms which has a single common ancestor and which includes all descendants of that ancestor

home range: the area of land that is regularly used by a group of animals for a year or longer

homologous (homology): having the same developmental and evolutionary origin. The bones in the hands of primates and the wings of bats are homologous

homoplasy: morphological similarity in two species that is not the result of common ancestry; it includes convergent evolution and **parallel evolution**

infanticide: the killing of infants

insectivore: an animal that eats primarily insects (and other invertebrates); also used to refer to the mammalian order Insectivora (which includes shrews, moles, and hedgehogs)

insertion: the attachment of a muscle or ligament farthest from the trunk or center of the body

intermembral index: a measure of the relative length of the forelimbs and hindlimbs of an animal: (humerus plus radius length)/(femur plus tibia length) *100

interspecific allometry: the relationship between size and shape among a range of different species; for example, a comparison between mouse and elephant

ischial callosity: a fatty sitting pad on the ischium of all Old World monkeys and gibbons

Kay's threshold: the body weight (approximately 500 g) that is roughly the upper size limit of predominantly insectivorous primates and the lower size limit of predominantly folivorous primates

keystone resources: critical elements in a species' ecology that limit its population size and distribution

knuckle-walking: a type of quadrupedal walking, used by chimpanzees and gorillas, in which the upper body is supported by the dorsal surface of the middle phalanges of the hands

kyphosis: dorsally convex curvature of the back

life history parameter: a characteristic of the growth and development of an organism such as the length of gestation, timing of sexual maturity, length of reproductive period, or lifespan

locomotion: movement from one place to another

lordosis: ventrally convex curvature of the back

mandible: jawbone housing the lower dentition

mandibular symphysis: the joint between the right and left halves of the mandible. In human and other higher primates, this joint is fused

monogamy: a mating system consisting of an adult male, adult female and their offspring

monophyletic group: a taxonomic group of organisms which has a single common ancestor

morphology: the shape of anatomical structures

multilevel society: a type of social organization in which smaller groups (usually one male–multifemale) are subsets of larger social bands

multimale-multifemale group: a social organization describing a group of animals in which several adult males and several adult females are reproductively active

nasolacrimal duct: a tube within the nasal cavity that drains tears from the eye into the nasal cavity

natural selection: nonrandom differential preservation of genotypes from one generation to the next which leads to changes in the genetic structure of a population

nectivore: an animal that eats nectar

neoteny: the retention of the features of a juvenile animal of one species in the adult form of a different species

neotropics: the tropical regions of North America, Central America, and South America

nocturnal: active primarily during the night

noyau: a type of social organization in which adult individuals are more or less solitary and have separate home ranges; ranges of individuals do not overlap with those of other individuals of the same sex, but they do overlap with ranges of individuals of the opposite sex

olfaction: the sense of smell

one-male group (or one-male unit): a social group containing several reproductively active females but only one reproductively active male

ontogeny: the development of an organism from conception to adulthood

Organ of Jacobson: an organ for chemical reception found in the anterior part of the roof of the mouth of many vertebrates; also known as the vomeronasal organ

paleomagnetism: study of the magnetism of rocks that were formed in earlier time periods. More broadly, the study of changes in the earth's magnetic fields during geological time

palmar: pertaining to the palm side of the hand

parallel evolution: independent evolution of similar (and homologous) morphological features in separate lineages

paraphyletic classification: a classification in which a taxonomic group contains some, but not all, of the members of a clade

phyletic classification: a classification in which taxonomic groups correspond to monophyletic groups

phyletic gradualism: a model of evolution in which change takes place slowly in small steps, in contrast to the punctuated equilibrium model

phylogeny: the evolutionary or genealogical relationships among a group of organisms

plantar: pertaining to the sole of the foot

polyandry: a type of mating system in which there are two or more reproductively active males and a single reproductively active female

polygyny: any type of mating system in which one male mates with more than one female

prehensile: capable of grasping; for example, the prehensile tail of some platyrrhine monkeys

primary rain forest: rain forest characterized by the later stages of the vegetational succession cycle

primitive feature: a behavioral or morphological feature that is characteristic of a species and its ancestors

procumbent: inclined forward, protruding, as in the procumbent incisors of some primates

prognathism: prominence of the snout

pronation: rotation of the forearm so that the palm faces dorsally or downward; the reverse movement from supination

punctuated equilibrium: a model of evolution in which change takes place primarily by abrupt genetic shifts, in contrast to phyletic gradualism

quadrumanous: four-handed; as in quadrumanous climbing, in which many suspensory primates use their feet in the same manner that they use their hands

ramus: the vertical part of the mandible, often called the ascending ramus

Rapaport's Rule: a biogeographical pattern in which species found closer to the equator have smaller geographical ranges than those found farther from the equator

relative dating: a determination of whether a fossil or fossil site is younger or older than other fossils or sites, usually through study of the stratigraphic position or evolutionary relationships of the fauna; contrasts with absolute dating

reproductive strategy: an organism's complex of behavioral and physiological features concerned with reproduction. The reproductive strategy of oysters, for example, is characterized by the production of large numbers of offspring and no parental care, whereas the reproductive strategy of humans is often characterized by production of relatively few offspring and investment of large amounts of parental care by both parents

reproductive success: the contribution of an individual to the gene pool of the next generation

reversal: refers to an evolutionary change in an organism that resembles the condition found in an earlier ancestor

sagittal crest: a bony ridge on the top of the neurocranium formed by the attachment of the temporalis muscles

saltation: leaping; either a type of locomotion, or a description of rapid evolutionary change characterized by a lack of intermediate forms, i.e., "leaping" from one distinct species to another

savannah: a type of vegetation zone characterized by grasslands with scattered trees

schizodactyly: grasping between the second and third digits of the hand rather than between the pollex (thumb) and second digit

secondary compounds: poisons produced by plants which exist in leaves, flowers, etc., and deter animals from eating them

secondary rain forest: rain forest characterized by immature stages of the succession cycle, commonly found on the edges of forests, along rivers, and around tree falls

sexual dichromatism: the condition in which males and females of a species differ in color

sexual dimorphism: any condition in which males and females of a species differ in some aspect of their nonreproductive anatomy such as body size, canine tooth size, or snout length

single-species hypothesis: the theory that there has never been more than one hominin lineage at any time because all hominins are characterized by culture and thus all occupy the same ecological niche

social organization: the demographic composition of a group – numbers of males and females

social structure: the types of interactions among individuals in a social group

socioecology: the scientific study of how social structure and organization are influenced by an organism's environment

speciation: appearance of new species

species–area relationship: a biogeographical rule that larger geographical areas contain more species

species selection: a type of natural selection which hypothetically operates at the species level so that the pattern of evolution is shaped by the differential survival or extinction of species

subfossil: recently extinct, often from historical time periods. Some strepsirrhines from Madagascar, for example, have become extinct in the past thousand years

supination: rotation of the forearm such that the palmar surface faces anteriorly or upward; the reverse movement from pronation

suspensory behavior: locomotor and postural habits characterized by hanging or suspension of the body below or among branches rather than walking, running, or sitting on top of branches

suture: a joint between two bones in which the bones interdigitate and are separated by fibrous tissue. The joints between most of the bones of the skull are sutures

sympatry: overlap in the geographical range of two species or populations

systematics: the science of classifying organisms and the study of their genealogical relationships

tapetum lucidum: a reflective layer in the eye that reduces an animal's visual acuity, but enhances its ability to see at night; among primates, this feature is present only in strepsirrhines

taphonomy: study of the processes that affect the remains of organisms from the death of the organism through its fossilization

taxonomy: the science of describing, naming, and classifying organisms

terrestrial: on the ground

terrestrial quadrupedalism: four-limbed locomotion on the ground

territory: part of a home range that is exclusive to a group of animals and is actively defended from other groups of the same species

tooth comb: a formation of the lower incisors (and sometimes canines) into a comb-like structure for grooming

tympanic bone: the bone that forms the bony ring for the eardrum

type specimen: a single designated individual of an organism which serves as the basis for the original name and description of the species

understory: the part of a forest that lies below the canopy layers

valgus: an angulation of the femur such that the knees are closer together than the hip joints; "knock-kneed"

ventral: toward the belly side of an animal; the opposite of dorsal

vertical clinging and leaping: a type of locomotion and posture in which animals cling to vertical supports and move by leaping between these vertical supports

woodland: a vegetation type characterized by discontinuous stands of relatively short trees separated by grassland

Appendix 1

Classification of the Order Primates

Genera in Bold Contain Extant Members

ORDER Primates
SEMIORDER Strepsirrhini
SUBORDER Strepsirrhini
INFRAORDER Lemuriformes
SUPERFAMILY Lemuroidea
FAMILY Lemuridae
    Lemur
    Hapalemur
    Prolemur
    Eulemur
    Varecia
    Pachylemur

FAMILY Megaladapidae
    Megaladapis

FAMILY Indriidae
SUBFAMILY Indriinae
    Avahi
    Propithecus
    Indri

SUBFAMILY Archaeolemurinae
    Archaeolemur
    Hadropithecus

SUBFAMILY Palaeopropithecinae
    Mesopropithecus
    Babakotia
    Palaeopropithecus
    Archaeoindris

FAMILY Cheirogaleidae
    Microcebus
    Cheirogaleus
    Mirza
    Allocebus
    Phaner

FAMILY Lepilemuridae
    Lepilemur

SUPERFAMILY Daubentonioidea
FAMILY Daubentoniidae
    Daubentonia
    Plesiopithecus
    Propotto

INFRAORDER Lorisiformes
SUPERFAMILY Lorisoidea
FAMILY Lorisidae
    Arctocebus
    Perodicticus
    Mioeuoticus
    Loris
    Nycticebus
    Nycticeboides
    Microloris

FAMILY Galagidae
    Galago
    Otolemur
    Galagoides
    Sciurocheirus
    Euoticus
    Paragalago
    Komba
    Progalago
    Wadilemur
    Saharagalago
    Namaloris
    Laetolia

FAMILY *Incertae sedis*
    Karanisia

SUPERFAMILY Incertae sedis
FAMILY Azibiidae
    Azibius
    Algeripithecus

FAMILY Djebelemuridae
    Djebelemur
    "Anchomomys"
    Omanodon
    Shizarodon

Primate Adaptation and Evolution
https://doi.org/10.1016/B978-0-12-815809-8.00019-9

SUBORDER Adapiformes
SUPERFAMILY Adapoidea
FAMILY Notharctidae
  Cantius
  Copelemur
  Notharctus
  Pelycodus
  Smilodectes
  Hesperolemur

FAMILY Cercamoniidae
  Donrussellia
  Panobius
  Protoadapis
  Barnesia
  Periconodon
  Buxella
  Agerinia
  Anchomomys
  Mazateronodon
  Nivesia
  Pronycticebus

FAMILY Adapidae
  Adapis
  Cryptadapis
  Microadapis
  Leptadapis
  Palaeolemur
  Magnadapis
  Paradapis

FAMILY Caenopithecidae
  Caenopithecus
  Europolemur
  Godinotia
  Darwinius
  Mahgarita
  Mescalerolemur
  Afradapis
  Aframonius
  Masradapis
  Namadapis
  Notnamaia
  Adapoides

FAMILY Asiadapidae
  Asiadapis
  Marcgodinotius

FAMILY Sivaladapidae
SUBFAMILY Sivaladapinae
  Sivaladapis
  Indraloris
  Siamoadapis
  Sinoadapis
  Ramadapis

SUBFAMILY Hoanghoniinae
  Hoanghonius
  Rencunius
  Paukkaungia
  Kyitchaungia
  Laomaki

SUBFAMILY Wailekiinae
  Wailekia
  Guangxilemur
  Yunnanadapis

SUBFAMILY Incertae Sedis
  Lushius

FAMILY Incertae sedis
  Bugtilemur
  Muangthanhinius
  Sulaimanius
  Ekgmowechashala

SEMIORDER Haplorhini
SUBORDER Anthropoidea
INFRAORDER Platyrrhini
SUPERFAMILY Pithecioidea
FAMILY Pitheciidae
SUBFAMILY Pitheciinae
  Pithecia
  Chiropotes
  Cacajao
  Proteropithecia
  Nuciruptor
  Cebupithecia

SUBFAMILY Callicebinae
  Callicebus
  Plecturocebus
  Cheracebus
  Miocallicebus

SUBFAMILY Xenotrichinae
  Xenothrix
  Antillothrix
  Insulacebus

SUPERFAMILY Ceboidea
FAMILY Cebidae
SUBFAMILY Cebinae
  Cebus
  Sapajus
  Acrecebus
  Saimiri
  Neosaimiri
  Laventiana
  Chilecebus
  Panamacebus

SUBFAMILY Aotinae
Aotus
Tremacebus

SUBFAMILY Callitrichinae
Callithrix
Mico
Cebuella
Callibella
Callimico
Saguinus
Leontopithecus
Leontocebus
Micodon
Patasola
Lagonimico

FAMILY Atelidae
SUBFAMILY Atelinae
Ateles
Caipora
Protopithecus
Solimoea
Lagothrix
Brachyteles

SUBFAMILY Alouattinae
Alouatta
Paralouatta
Stirtonia
Cartelles

SUPERFAMILY Incertae sedis
FAMILY Homunculidae
Dolichocebus
Soriacebus
Carlocebus
Mazzonicebus
Homunculus
Killikaike
Canaanimico

FAMILY Incertae sedis
Branisella
Szalatavus
Mohanimico
Parvimico
Perupithecus

INFRAORDER Catarrhini
SUPERFAMILY Cercopithecoidea
FAMILY Cercopithecidae
SUBFAMILY Cercopithecinae
Macaca
Procynocephalus
Paradolichopithecus
Cercocebus

Mandrillus
Procercocebus
Soromandrillus
Lophocebus
Rungwecebus
Papio
Parapapio
Pliopapio
Dinopithecus
Gorgopithecus
Theropithecus
Erythrocebus
Chlorocebus
Nanopithecus
Allochrocebus
Miopithecus
Allenopithecus
Cercopithecus

SUBFAMILY Colobinae
Colobus
Piliocolobus
Procolobus
Microcolobus
Libypithecus
Kuseracolobus
Cercopithecoides
Sawecolobus
Paracolobus
Rhinocolobus
Mesopithecus
Dolichopithecus
Semnopithecus
Trachypithecus
Presbytis
Nasalis
Simias
Pygathrix
Rhinopithecus
Parapresbytis
Kanagawapithecus
Myanmarcolobus

FAMILY Victoriapithecidae
Victoriapithecus
Prohylobates
Zaltanpithecus
Noropithecus

FAMILY Incertae sedis
Nsungwepithecus
Alophe

SUPERFAMILY Hominoidea
FAMILY Hylobatidae
Hylobates

Hoolock
Nomascus
Symphalangus
Yuanmoupithecus
Kapi

FAMILY Hominidae
SUBFAMILY Ponginae
Pongo
Ankarapithecus
Sivapithecus
Gigantopithecus
Indopithecus
Khoratpithecus
Lufengpithecus

SUBFAMILY Homininae
Pan
Gorilla
Homo
Australopithecus
Paranthropus
Kenyanthropus
Ardipithecus
Orrorin
Sahelanthropus

SUBFAMILY Dryopithecinae
Dryopithecus
Pierolapithecus
Anoiapithecus
Hispanopithecus
Rudapithecus
Ouranopithecus
Graecopithecus
Danuvius
Anadoluvius

SUBFAMILY Incertae sedis
Kenyapithecus
Griphopithecus
Chororapithecus
Lufengpithecus
Nakalipithecus

SUPERFAMILY Proconsuloidea
FAMILY Proconsulidae
SUBFAMILY Proconsulinae
Proconsul
Ekembo

SUBFAMILY Afropithecinae
Afropithecus
Heliopithecus
Morotopithecus
Nacholapithecus
Equatorius
Otavipithecus

SUBFAMILY Nyanzapithecinae
Nyanzapithecus
Rangwapithecus
Mabokopithecus
Turkanapithecus
Xenopithecus
Rukwapithecus

FAMILY Dendropithecidae
Dendropithecus
Simiolus
Micropithecus

FAMILY Incertae sedis
Limnopithecus
Kalepithecus
Kogolepithecus
Lomorupithecus
Iriripithecus
Karamojapithecus
Samburupithecus
Oreopithecus

SUPERFAMILY Propliopithecoidae
FAMILY Propliopithecidae
Propliopithecus
Moeripithecus
Aegyptopithecus

FAMILY Oligopithecidae
Oligopithecus
Catopithecus
Talahpithecus

SUPERFAMILY Pliopithecoidea
FAMILY Dionysopithecidae
SUBFAMILY Dionysopithecinae
Dionysopithecus
Platydontopithecus

FAMILY Pliopithecidae
SUBFAMILY Pliopithecinae
Pliopithecus
Epipliopithecus

SUBFAMILY Crouzeliinae
Plesiopliopithecus (=Crouzelia)
Anapithecus
Barberapithecus
Fanchangia

SUBFAMILY Anapithecinae
Anapithecus
Egarapithecus
Laccopithecus

SUBFAMILY Krishnapithecinae
Krishnapithecus

FAMILY Incertae sedis
Pliobates

SUPERFAMILY Incertae sedis
Saadanius
Kamoyapithecus

INFRAORDER Parapithecoidea
SUPERFAMILY Parapithecoidea
FAMILY Parapithecidae
Abuqatrania
Qatrania
Ucayalipithecus
Apidium
Parapithecus
Lokonepithecus

FAMILY Incertae sedis
Biretia
Arsinoea

SUPERFAMILY Proteopithecoidea
FAMILY Proteopithecidae
Proteopithecus
Serapia

INFRAORDER Incertae sedis
FAMILY Eosimiidae
Eosimias
Phenacopithecus
Bahinia
?Anthrasimias

FAMILY Amphipithecidae
Amphipithecus
Pondaungia
Ganlea
Myanmarpithecus
Siamopithecus
Bugtipithecus
Krabia

FAMILY Incertae sedis
Phileosimias
Amamria
Aseanpithecus

SUBORDER Tarsiiformes
INFRAORDER Tarsiiformes
SUPERFAMILY Tarsioidea
FAMILY Tarsiidae
Tarsius
Cephalopachus
Carlito
Hesperotarsius
Xanthorhysis
Afrotarsius

Afrasia
Oligotarsius

SUBORDER Omomyiformes
SUPERFAMILY Omomyoidea
FAMILY Omomyidae
SUBFAMILY Anaptomorphinae
Teilhardina
Archicebus
Baataromomys
Bownomomys
Anaptomorphus
Gazinius
Tetonius
Pseudotetonius
Absarokius
Tatmanius
Strigorhysis
Aycrossia
Trogolemur
Walshina
Sphacorhysis
Anemorhysis
Arapahovius
Chlororhysis
Artimonius

SUBFAMILY Omomyinae
Omomys
Chumashius
Steinius
Uintanius
Jemezius
Macrotarsius
Hemiacodon
Yaquius
Ekwiiyemakius
Gunnelltarsius
Brontomomys
Ourayia
Wyomomys
Ageitodendron
Utahia
Stockia
Asiomomys
Chipetaia
Washakius
Tarka
Tarkadectes
Tarkops
Shoshonius
Dyseolemur
Loveina

FAMILY Microchoeridae
Nannopithex
Pseudoloris
Necrolemur
Microchoerus
Vectipithex
Melaneremia
Indusomys
Quercyloris

FAMILY Incertae sedis
Rooneyia
Kohatius
Vastanomys

SEMIORDER Incertae sedis
Altiatlasius
Altanius
Nosmips

Appendix 2

Timescale of Primate Evolution

Age (Ma)	Period		Epoch	Age	Major Events in Primate Evolution
	QUATERNARY		HOLOCENE		
			PLEISTOCENE	CALABRIAN	
				GELASIAN	
			PLIOCENE	PIACENZIAN	
5				ZANCLEAN	
	NEOGENE	TERTIARY	MIOCENE	MESSINIAN	**First Hominins** *(6-7 Ma)*
10				TORTONIAN	
				SERRAVALLIAN	
15				LANGHIAN	
				BURDIGALIAN	
20				AQUITANIAN	
25	PALEOGENE		OLIGOCENE	CHATTIAN	**First OWM and Apes** *(about 25 Ma)*
30				RUPELIAN	**First ?Platyrrhines** *(about 32 Ma)*
35			EOCENE	PRIABONIAN	**First Catarrhines** *(about 35 Ma)*
40				BARTONIAN	
45				LUTETIAN	**First Anthropoids** *(about 45 Ma)*
50				YPRESIAN	
55					**First Prosimians** *(about 56 Ma)*
			PALEOCENE	THANETIAN	
60				SELANDIAN	
				DANIAN	
65					**First ?Plesiadapiforms** *(about 66 Ma)*

Primate Adaptation and Evolution
https://doi.org/10.1016/B978-0-12-815809-8.00021-7

Index

Note: Page numbers with '*f*' denote figures and '*t*' tables. **Bold** names are genera with living members.